Protein Folding Handbook

Edited by J. Buchner and T. Kiefhaber

Further Titles of Interest

K. H. Nierhaus, D. N. Wilson (eds.)

Protein Biosynthesis and Ribosome Structure

ISBN 3-527-30638-2

R. J. Mayer, A. J. Ciechanover, M. Rechsteiner (eds.)

Protein Degradation

ISBN 3-527-30837-7 (Vol. 1)
ISBN 3-527-31130-0 (Vol. 2)

G. Cesareni, M. Gimona, M. Sudol, M. Yaffe (eds.)

Modular Protein Domains

ISBN 3-527-30813-X

S. Brakmann, A. Schwienhorst (eds.)

Evolutionary Methods in Biotechnology

ISBN 3-527-30799-0

Protein Folding Handbook

Edited by Johannes Buchner and Thomas Kiefhaber

WILEY-VCH Verlag GmbH & Co. KGaA

Editors

Prof. Dr. Johannes Buchner
Institut für Organische Chemie und
Biochemie
Technische Universität München
Lichtenbergstrasse 4
85747 Garching
Germany
johannes.buchner@ch.tum.de

Prof. Dr. Thomas Kiefhaber
Biozentrum der Universität Basel
Division of Biophysical Chemistry
Klingelbergstrasse 70
4056 Basel
Switzerland
t.kiefhaber@unibas.ch

Cover
Artwork by Prof. Erich Gohl, Regensburg

■ This book was carefully produced. Nevertheless, authors, editors and publisher do not warrant the information contained therein to be free of errors. Readers are advised to keep in mind that statements, data, illustrations, procedural details or other items may inadvertently be inaccurate.

Library of Congress Card No.: applied for
A catalogue record for this book is available from the British Library.
Bibliographic information published by Die Deutsche Bibliothek
Die Deutsche Bibliothek lists this publication in the Deutsche Nationalbibliografie; detailed bibliographic data is available in the Internet at http://dnb.ddb.de

© 2005 WILEY-VCH Verlag GmbH & Co. KGaA, Weinheim
All rights reserved (including those of translation in other languages). No part of this book may be reproduced in any form – by photoprinting, microfilm, or any other means – nor transmitted or translated into machine language without written permission from the publishers. Registered names, trademarks, etc. used in this book, even when not specifically marked as such, are not to be considered unprotected by law.

Printed in the Federal Republic of Germany.
Printed on acid-free paper.

Typesetting   Asco Typesetters, Hong Kong
Printing   betz-druck gmbh, Darmstadt
Bookbinding   Litges & Dopf Buchbinderei GmbH, Heppenheim

ISBN-13 978-3-527-30784-5
ISBN-10 3-527-30784-2

Contents

Part I, Volume 1

    Preface *LVIII*

    Contributors of Part I *LX*

I/1   **Principles of Protein Stability and Design** *1*

1   **Early Days of Studying the Mechanism of Protein Folding** *3*
    Robert L. Baldwin
1.1   Introduction *3*
1.2   Two-state Folding *4*
1.3   Levinthal's Paradox *5*
1.4   The Domain as a Unit of Folding *6*
1.5   Detection of Folding Intermediates and Initial Work on the Kinetic Mechanism of Folding *7*
1.6   Two Unfolded Forms of RNase A and Explanation by Proline Isomerization *9*
1.7   Covalent Intermediates in the Coupled Processes of Disulfide Bond Formation and Folding *11*
1.8   Early Stages of Folding Detected by Antibodies and by Hydrogen Exchange *12*
1.9   Molten Globule Folding Intermediates *14*
1.10   Structures of Peptide Models for Folding Intermediates *15*
    Acknowledgments *16*
    References *16*

2   **Spectroscopic Techniques to Study Protein Folding and Stability** *22*
    Franz Schmid
2.1   Introduction *22*
2.2   Absorbance *23*
2.2.1   Absorbance of Proteins *23*
2.2.2   Practical Considerations for the Measurement of Protein Absorbance *27*

2.2.3	Data Interpretation 29
2.3	Fluorescence 29
2.3.1	The Fluorescence of Proteins 30
2.3.2	Energy Transfer and Fluorescence Quenching in a Protein: Barnase 31
2.3.3	Protein Unfolding Monitored by Fluorescence 33
2.3.4	Environmental Effects on Tyrosine and Tryptophan Emission 36
2.3.5	Practical Considerations 37
2.4	Circular Dichroism 38
2.4.1	CD Spectra of Native and Unfolded Proteins 38
2.4.2	Measurement of Circular Dichroism 41
2.4.3	Evaluation of CD Data 42
	References 43

3	**Denaturation of Proteins by Urea and Guanidine Hydrochloride** 45
	C. Nick Pace, Gerald R. Grimsley, and J. Martin Scholtz
3.1	Historical Perspective 45
3.2	How Urea Denatures Proteins 45
3.3	Linear Extrapolation Method 48
3.4	$\Delta G(H_2O)$ 50
3.5	m-Values 55
3.6	Concluding Remarks 58
3.7	Experimental Protocols 59
3.7.1	How to Choose the Best Denaturant for your Study 59
3.7.2	How to Prepare Denaturant Solutions 59
3.7.3	How to Determine Solvent Denaturation Curves 60
3.7.3.1	Determining a Urea or GdmCl Denaturation Curve 62
3.7.3.2	How to Analyze Urea or GdmCl Denaturant Curves 63
3.7.4	Determining Differences in Stability 64
	Acknowledgments 65
	References 65

4	**Thermal Unfolding of Proteins Studied by Calorimetry** 70
	George I. Makhatadze
4.1	Introduction 70
4.2	Two-state Unfolding 71
4.3	Cold Denaturation 76
4.4	Mechanisms of Thermostabilization 77
4.5	Thermodynamic Dissection of Forces Contributing to Protein Stability 79
4.5.1	Heat Capacity Changes, ΔC_p 81
4.5.2	Enthalpy of Unfolding, ΔH 81
4.5.3	Entropy of Unfolding, ΔS 83
4.6	Multistate Transitions 84
4.6.1	Two-state Dimeric Model 85

4.6.2	Two-state Multimeric Model *86*	
4.6.3	Three-state Dimeric Model *86*	
4.6.4	Two-state Model with Ligand Binding *88*	
4.6.5	Four-state (Two-domain Protein) Model *90*	
4.7	Experimental Protocols *92*	
4.7.1	How to Prepare for DSC Experiments *92*	
4.7.2	How to Choose Appropriate Conditions *94*	
4.7.3	Critical Factors in Running DSC Experiments *94*	
	References *95*	
5	**Pressure–Temperature Phase Diagrams of Proteins** *99*	
	Wolfgang Doster and Josef Friedrich	
5.1	Introduction *99*	
5.2	Basic Aspects of Phase Diagrams of Proteins and Early Experiments *100*	
5.3	Thermodynamics of Pressure–Temperature Phase Diagrams *103*	
5.4	Measuring Phase Stability Boundaries with Optical Techniques *110*	
5.4.1	Fluorescence Experiments with Cytochrome c *110*	
5.4.2	Results *112*	
5.5	What Do We Learn from the Stability Diagram? *116*	
5.5.1	Thermodynamics *116*	
5.5.2	Determination of the Equilibrium Constant of Denaturation *117*	
5.5.3	Microscopic Aspects *120*	
5.5.4	Structural Features of the Pressure-denatured State *122*	
5.6	Conclusions and Outlook *123*	
	Acknowledgment *124*	
	References *124*	
6	**Weak Interactions in Protein Folding: Hydrophobic Free Energy, van der Waals Interactions, Peptide Hydrogen Bonds, and Peptide Solvation** *127*	
	Robert L. Baldwin	
6.1	Introduction *127*	
6.2	Hydrophobic Free Energy, Burial of Nonpolar Surface and van der Waals Interactions *128*	
6.2.1	History *128*	
6.2.2	Liquid–Liquid Transfer Model *128*	
6.2.3	Relation between Hydrophobic Free Energy and Molecular Surface Area *130*	
6.2.4	Quasi-experimental Estimates of the Work of Making a Cavity in Water or in Liquid Alkane *131*	
6.2.5	Molecular Dynamics Simulations of the Work of Making Cavities in Water *133*	
6.2.6	Dependence of Transfer Free Energy on the Volume of the Solute *134*	
6.2.7	Molecular Nature of Hydrophobic Free Energy *136*	

6.2.8	Simulation of Hydrophobic Clusters	137
6.2.9	ΔC_p and the Temperature-dependent Thermodynamics of Hydrophobic Free Energy	137
6.2.10	Modeling Formation of the Hydrophobic Core from Solvation Free Energy and van der Waals Interactions between Nonpolar Residues	142
6.2.11	Evidence Supporting a Role for van der Waals Interactions in Forming the Hydrophobic Core	144
6.3	Peptide Solvation and the Peptide Hydrogen Bond	145
6.3.1	History	145
6.3.2	Solvation Free Energies of Amides	147
6.3.3	Test of the Hydrogen-Bond Inventory	149
6.3.4	The Born Equation	150
6.3.5	Prediction of Solvation Free Energies of Polar Molecules by an Electrostatic Algorithm	150
6.3.6	Prediction of the Solvation Free Energies of Peptide Groups in Different Backbone Conformations	151
6.3.7	Predicted Desolvation Penalty for Burial of a Peptide H-bond	153
6.3.8	Gas–Liquid Transfer Model	154
	Acknowledgments	156
	References	156
7	**Electrostatics of Proteins: Principles, Models and Applications**	**163**
	Sonja Braun-Sand and Arieh Warshel	
7.1	Introduction	163
7.2	Historical Perspectives	163
7.3	Electrostatic Models: From Microscopic to Macroscopic Models	166
7.3.1	All-Atom Models	166
7.3.2	Dipolar Lattice Models and the PDLD Approach	168
7.3.3	The PDLD/S-LRA Model	170
7.3.4	Continuum (Poisson-Boltzmann) and Related Approaches	171
7.3.5	Effective Dielectric Constant for Charge–Charge Interactions and the GB Model	172
7.4	The Meaning and Use of the Protein Dielectric Constant	173
7.5	Validation Studies	176
7.6	Systems Studied	178
7.6.1	Solvation Energies of Small Molecules	178
7.6.2	Calculation of pK_a Values of Ionizable Residues	179
7.6.3	Redox and Electron Transport Processes	180
7.6.4	Ligand Binding	181
7.6.5	Enzyme Catalysis	182
7.6.6	Ion Pairs	183
7.6.7	Protein–Protein Interactions	184
7.6.8	Ion Channels	185
7.6.9	Helix Macrodipoles versus Localized Molecular Dipoles	185
7.6.10	Folding and Stability	186
7.7	Concluding Remarks	189

Acknowledgments 190
References 190

8 Protein Conformational Transitions as Seen from the Solvent: Magnetic Relaxation Dispersion Studies of Water, Co-solvent, and Denaturant Interactions with Nonnative Proteins 201

Bertil Halle, Vladimir P. Denisov, Kristofer Modig, and Monika Davidovic

8.1 The Role of the Solvent in Protein Folding and Stability 201
8.2 Information Content of Magnetic Relaxation Dispersion 202
8.3 Thermal Perturbations 205
8.3.1 Heat Denaturation 205
8.3.2 Cold Denaturation 209
8.4 Electrostatic Perturbations 213
8.5 Solvent Perturbations 218
8.5.1 Denaturation Induced by Urea 219
8.5.2 Denaturation Induced by Guanidinium Chloride 225
8.5.3 Conformational Transitions Induced by Co-solvents 228
8.6 Outlook 233
8.7 Experimental Protocols and Data Analysis 233
8.7.1 Experimental Methodology 233
8.7.1.1 Multiple-field MRD 234
8.7.1.2 Field-cycling MRD 234
8.7.1.3 Choice of Nuclear Isotope 235
8.7.2 Data Analysis 236
8.7.2.1 Exchange Averaging 236
8.7.2.2 Spectral Density Function 237
8.7.2.3 Residence Time 239
8.7.2.4 ^{19}F Relaxation 240
8.7.2.5 Coexisting Protein Species 241
8.7.2.6 Preferential Solvation 241
References 242

9 Stability and Design of α-Helices 247

Andrew J. Doig, Neil Errington, and Teuku M. Iqbalsyah

9.1 Introduction 247
9.2 Structure of the α-Helix 247
9.2.1 Capping Motifs 248
9.2.2 Metal Binding 250
9.2.3 The 3_{10}-Helix 251
9.2.4 The π-Helix 251
9.3 Design of Peptide Helices 252
9.3.1 Host–Guest Studies 253
9.3.2 Helix Lengths 253
9.3.3 The Helix Dipole 253
9.3.4 Acetylation and Amidation 254
9.3.5 Side Chain Spacings 255

9.3.6	Solubility 256
9.3.7	Concentration Determination 257
9.3.8	Design of Peptides to Measure Helix Parameters 257
9.3.9	Helix Templates 259
9.3.10	Design of 3_{10}-Helices 259
9.3.11	Design of π-helices 261
9.4	Helix Coil Theory 261
9.4.1	Zimm-Bragg Model 261
9.4.2	Lifson-Roig Model 262
9.4.3	The Unfolded State and Polyproline II Helix 265
9.4.4	Single Sequence Approximation 265
9.4.5	N- and C-Caps 266
9.4.6	Capping Boxes 266
9.4.7	Side-chain Interactions 266
9.4.8	N1, N2, and N3 Preferences 267
9.4.9	Helix Dipole 267
9.4.10	3_{10}- and π-Helices 268
9.4.11	AGADIR 268
9.4.12	Lomize-Mosberg Model 269
9.4.13	Extension of the Zimm-Bragg Model 270
9.4.14	Availability of Helix/Coil Programs 270
9.5	Forces Affecting α-Helix Stability 270
9.5.1	Helix Interior 270
9.5.2	Caps 273
9.5.3	Phosphorylation 276
9.5.4	Noncovalent Side-chain Interactions 276
9.5.5	Covalent Side-chain interactions 277
9.5.6	Capping Motifs 277
9.5.7	Ionic Strength 279
9.5.8	Temperature 279
9.5.9	Trifluoroethanol 279
9.5.10	pK_a Values 280
9.5.11	Relevance to Proteins 281
9.6	Experimental Protocols and Strategies 281
9.6.1	Solid Phase Peptide Synthesis (SPPS) Based on the Fmoc Strategy 281
9.6.1.1	Equipment and Reagents 281
9.6.1.2	Fmoc Deprotection and Coupling 283
9.6.1.3	Kaiser Test 284
9.6.1.4	Acetylation and Cleavage 285
9.6.1.5	Peptide Precipitation 286
9.6.2	Peptide Purification 286
9.6.2.1	Equipment and Reagents 286
9.6.2.2	Method 286
9.6.3	Circular Dichroism 287
9.6.4	Acquisition of Spectra 288

9.6.4.1	Instrumental Considerations	288
9.6.5	Data Manipulation and Analysis	289
9.6.5.1	Protocol for CD Measurement of Helix Content	291
9.6.6	Aggregation Test for Helical Peptides	291
9.6.6.1	Equipment and Reagents	291
9.6.6.2	Method	292
9.6.7	Vibrational Circular Dichroism	292
9.6.8	NMR Spectroscopy	292
9.6.8.1	Nuclear Overhauser Effect	293
9.6.8.2	Amide Proton Exchange Rates	294
9.6.8.3	^{13}C NMR	294
9.6.9	Fourier Transform Infrared Spectroscopy	295
9.6.9.1	Secondary Structure	295
9.6.10	Raman Spectroscopy and Raman Optical Activity	296
9.6.11	pH Titrations	298
9.6.11.1	Equipment and Reagents	298
9.6.11.2	Method	298
	Acknowledgments	299
	References	299

10	**Design and Stability of Peptide β-Sheets**	**314**
	Mark S. Searle	
10.1	Introduction	314
10.2	β-Hairpins Derived from Native Protein Sequences	315
10.3	Role of β-Turns in Nucleating β-Hairpin Folding	316
10.4	Intrinsic ϕ, ψ Propensities of Amino Acids	319
10.5	Side-chain Interactions and β-Hairpin Stability	321
10.5.1	Aromatic Clusters Stabilize β-Hairpins	322
10.5.2	Salt Bridges Enhance Hairpin Stability	325
10.6	Cooperative Interactions in β-Sheet Peptides: Kinetic Barriers to Folding	330
10.7	Quantitative Analysis of Peptide Folding	331
10.8	Thermodynamics of β-Hairpin Folding	332
10.9	Multistranded Antiparallel β-Sheet Peptides	334
10.10	Concluding Remarks: Weak Interactions and Stabilization of Peptide β-Sheets	339
	References	340

11	**Predicting Free Energy Changes of Mutations in Proteins**	**343**
	Raphael Guerois, Joaquim Mendes, and Luis Serrano	
11.1	Physical Forces that Determine Protein Conformational Stability	343
11.1.1	Protein Conformational Stability [1]	343
11.1.2	Structures of the N and D States [2–6]	344
11.1.3	Studies Aimed at Understanding the Physical Forces that Determine Protein Conformational Stability [1, 2, 8, 19–26]	346
11.1.4	Forces Determining Conformational Stability [1, 2, 8, 19–27]	346

11.1.5	Intramolecular Interactions	347
11.1.5.1	van der Waals Interactions	347
11.1.5.2	Electrostatic Interactions	347
11.1.5.3	Conformational Strain	349
11.1.6	Solvation	350
11.1.7	Intramolecular Interactions and Solvation Taken Together	350
11.1.8	Entropy	351
11.1.9	Cavity Formation	352
11.1.10	Summary	353
11.2	Methods for the Prediction of the Effect of Point Mutations on in vitro Protein Stability	353
11.2.1	General Considerations on Protein Plasticity upon Mutation	353
11.2.2	Predictive Strategies	355
11.2.3	Methods	356
11.2.3.1	From Sequence and Multiple Sequence Alignment Analysis	356
11.2.3.2	Statistical Analysis of the Structure Databases	356
11.2.3.3	Helix/Coil Transition Model	357
11.2.3.4	Physicochemical Method Based on Protein Engineering Experiments	359
11.2.3.5	Methods Based only on the Basic Principles of Physics and Thermodynamics	364
11.3	Mutation Effects on in vivo Stability	366
11.3.1	The N-terminal Rule	366
11.3.2	The C-terminal Rule	367
11.3.3	PEST Signals	368
11.4	Mutation Effects on Aggregation	368
	References 369	

I/2 Dynamics and Mechanisms of Protein Folding Reactions 377

12.1	**Kinetic Mechanisms in Protein Folding**	**379**
	Annett Bachmann and Thomas Kiefhaber	
12.1.1	Introduction	379
12.1.2	Analysis of Protein Folding Reactions using Simple Kinetic Models	379
12.1.2.1	General Treatment of Kinetic Data	380
12.1.2.2	Two-state Protein Folding	380
12.1.2.3	Complex Folding Kinetics	384
12.1.2.3.1	Heterogeneity in the Unfolded State	384
12.1.2.3.2	Folding through Intermediates	388
12.1.2.3.3	Rapid Pre-equilibria	391
12.1.2.3.4	Folding through an On-pathway High-energy Intermediate	393
12.1.3	A Case Study: the Mechanism of Lysozyme Folding	394
12.1.3.1	Lysozyme Folding at pH 5.2 and Low Salt Concentrations	394
12.1.3.2	Lysozyme Folding at pH 9.2 or at High Salt Concentrations	398
12.1.4	Non-exponential Kinetics	401

12.1.5	Conclusions and Outlook	*401*
12.1.6	Protocols – Analytical Solutions of Three-state Protein Folding Models *402*	
12.1.6.1	Triangular Mechanism *402*	
12.1.6.2	On-pathway Intermediate *403*	
12.1.6.3	Off-pathway Mechanism *404*	
12.1.6.4	Folding Through an On-pathway High-Energy Intermediate *404*	
	Acknowledgments *406*	
	References *406*	
12.2	**Characterization of Protein Folding Barriers with Rate Equilibrium Free Energy Relationships** *411*	
	Thomas Kiefhaber, Ignacio E. Sánchez, and Annett Bachmann	
12.2.1	Introduction *411*	
12.2.2	Rate Equilibrium Free Energy Relationships *411*	
12.2.2.1	Linear Rate Equilibrium Free Energy Relationships in Protein Folding *414*	
12.2.2.2	Properties of Protein Folding Transition States Derived from Linear REFERs *418*	
12.2.3	Nonlinear Rate Equilibrium Free Energy Relationships in Protein Folding *420*	
12.2.3.1	Self-Interaction and Cross-Interaction Parameters *420*	
12.2.3.2	Hammond and Anti-Hammond Behavior *424*	
12.2.3.3	Sequential and Parallel Transition States *425*	
12.2.3.4	Ground State Effects *428*	
12.2.4	Experimental Results on the Shape of Free Energy Barriers in Protein Folding *432*	
12.2.4.1	Broadness of Free Energy Barriers *432*	
12.2.4.2	Parallel Pathways *437*	
12.2.5	Folding in the Absence of Enthalpy Barriers *438*	
12.2.6	Conclusions and Outlook *438*	
	Acknowledgments *439*	
	References *439*	
13	**A Guide to Measuring and Interpreting ϕ-values** *445*	
	Nicholas R. Guydosh and Alan R. Fersht	
13.1	Introduction *445*	
13.2	Basic Concept of ϕ-Value Analysis *445*	
13.3	Further Interpretation of ϕ *448*	
13.4	Techniques *450*	
13.5	Conclusions *452*	
	References *452*	
14	**Fast Relaxation Methods** *454*	
	Martin Gruebele	
14.1	Introduction *454*	

14.2	Techniques	455
14.2.1	Fast Pressure-Jump Experiments	455
14.2.2	Fast Resistive Heating Experiments	456
14.2.3	Fast Laser-induced Relaxation Experiments	457
14.2.3.1	Laser Photolysis	457
14.2.3.2	Electrochemical Jumps	458
14.2.3.3	Laser-induced pH Jumps	458
14.2.3.4	Covalent Bond Dissociation	459
14.2.3.5	Chromophore Excitation	460
14.2.3.6	Laser Temperature Jumps	460
14.2.4	Multichannel Detection Techniques for Relaxation Studies	461
14.2.4.1	Small Angle X-ray Scattering or Light Scattering	462
14.2.4.2	Direct Absorption Techniques	463
14.2.4.3	Circular Dichroism and Optical Rotatory Dispersion	464
14.2.4.4	Raman and Resonance Raman Scattering	464
14.2.4.5	Intrinsic Fluorescence	465
14.2.4.6	Extrinsic Fluorescence	465
14.3	Protein Folding by Relaxation	466
14.3.1	Transition State Theory, Energy Landscapes, and Fast Folding	466
14.3.2	Viscosity Dependence of Folding Motions	470
14.3.3	Resolving Burst Phases	471
14.3.4	Fast Folding and Unfolded Proteins	472
14.3.5	Experiment and Simulation	472
14.4	Summary	474
14.5	Experimental Protocols	475
14.5.1	Design Criteria for Laser Temperature Jumps	475
14.5.2	Design Criteria for Fast Single-Shot Detection Systems	476
14.5.3	Designing Proteins for Fast Relaxation Experiments	477
14.5.4	Linear Kinetic, Nonlinear Kinetic, and Generalized Kinetic Analysis of Fast Relaxation	477
14.5.4.1	The Reaction $D \rightleftharpoons F$ in the Presence of a Barrier	477
14.5.4.2	The Reaction $2A \rightleftharpoons A_2$ in the Presence of a Barrier	478
14.5.4.3	The Reaction $D \rightleftharpoons F$ at Short Times or over Low Barriers	479
14.5.5	Relaxation Data Analysis by Linear Decomposition	480
14.5.5.1	Singular Value Decomposition (SVD)	480
14.5.5.2	χ-Analysis	481
	Acknowledgments	481
	References	482
15	**Early Events in Protein Folding Explored by Rapid Mixing Methods**	491
	Heinrich Roder, Kosuke Maki, Ramil F. Latypov, Hong Cheng, and M. C. Ramachandra Shastry	
15.1	Importance of Kinetics for Understanding Protein Folding	491
15.2	Burst-phase Signals in Stopped-flow Experiments	492
15.3	Turbulent Mixing	494

15.4	Detection Methods *495*	
15.4.1	Tryptophan Fluorescence *495*	
15.4.2	ANS Fluorescence *498*	
15.4.3	FRET *499*	
15.4.4	Continuous-flow Absorbance *501*	
15.4.5	Other Detection Methods used in Ultrafast Folding Studies *502*	
15.5	A Quenched-Flow Method for H-D Exchange Labeling Studies on the Microsecond Time Scale *502*	
15.6	Evidence for Accumulation of Early Folding Intermediates in Small Proteins *505*	
15.6.1	B1 Domain of Protein G *505*	
15.6.2	Ubiquitin *508*	
15.6.3	Cytochrome *c* *512*	
15.7	Significance of Early Folding Events *515*	
15.7.1	Barrier-limited Folding vs. Chain Diffusion *515*	
15.7.2	Chain Compaction: Random Collapse vs. Specific Folding *516*	
15.7.3	Kinetic Role of Early Folding Intermediates *517*	
15.7.4	Broader Implications *520*	
	Appendix *521*	
A1	Design and Calibration of Rapid Mixing Instruments *521*	
A1.1	Stopped-flow Equipment *521*	
A1.2	Continuous-flow Instrumentation *524*	
	Acknowledgments *528*	
	References *528*	
16	**Kinetic Protein Folding Studies using NMR Spectroscopy** *536*	
	Markus Zeeb and Jochen Balbach	
16.1	Introduction *536*	
16.2	Following Slow Protein Folding Reactions in Real Time *538*	
16.3	Two-dimensional Real-time NMR Spectroscopy *545*	
16.4	Dynamic and Spin Relaxation NMR for Quantifying Microsecond-to-Millisecond Folding Rates *550*	
16.5	Conclusions and Future Directions *555*	
16.6	Experimental Protocols *556*	
16.6.1	How to Record and Analyze 1D Real-time NMR Spectra *556*	
16.6.1.1	Acquisition *556*	
16.6.1.2	Processing *557*	
16.6.1.3	Analysis *557*	
16.6.1.4	Analysis of 1D Real-time Diffusion Experiments *558*	
16.6.2	How to Extract Folding Rates from 1D Spectra by Line Shape Analysis *559*	
16.6.2.1	Acquisition *560*	
16.6.2.2	Processing *560*	
16.6.2.3	Analysis *561*	
16.6.3	How to Extract Folding Rates from 2D Real-time NMR Spectra *562*	

16.6.3.1	Acquisition 563
16.6.3.2	Processing 563
16.6.3.3	Analysis 563
16.6.4	How to Analyze Heteronuclear NMR Relaxation and Exchange Data 565
16.6.4.1	Acquisition 566
16.6.4.2	Processing 567
16.6.4.3	Analysis 567
	Acknowledgments 569
	References 569

Part I, Volume 2

17	Fluorescence Resonance Energy Transfer (FRET) and Single Molecule Fluorescence Detection Studies of the Mechanism of Protein Folding and Unfolding 573
	Elisha Haas
	Abbreviations 573
17.1	Introduction 573
17.2	What are the Main Aspects of the Protein Folding Problem that can be Addressed by Methods Based on FRET Measurements? 574
17.2.1	The Three Protein Folding Problems 574
17.2.1.1	The Chain Entropy Problem 574
17.2.1.2	The Function Problem: Conformational Fluctuations 575
17.3	Theoretical Background 576
17.3.1	Nonradiative Excitation Energy Transfer 576
17.3.2	What is FRET? The Singlet–Singlet Excitation Transfer 577
17.3.3	Rate of Nonradiative Excitation Energy Transfer within a Donor–Acceptor Pair 578
17.3.4	The Orientation Factor 583
17.3.5	How to Determine and Control the Value of R_o? 584
17.3.6	Index of Refraction n 584
17.3.7	The Donor Quantum Yield Φ_D^o 586
17.3.8	The Spectral Overlap Integral J 586
17.4	Determination of Intramolecular Distances in Protein Molecules using FRET Measurements 586
17.4.1	Single Distance between Donor and Acceptor 587
17.4.1.1	Method 1: Steady State Determination of Decrease of Donor Emission 587
17.4.1.2	Method 2: Acceptor Excitation Spectroscopy 588
17.4.2	Time-resolved Methods 588
17.4.3	Determination of E from Donor Fluorescence Decay Rates 589
17.4.4	Determination of Acceptor Fluorescence Lifetime 589
17.4.5	Determination of Intramolecular Distance Distributions 590

17.4.6	Evaluation of the Effect of Fast Conformational Fluctuations and Determination of Intramolecular Diffusion Coefficients	592
17.5	Experimental Challenges in the Implementation of FRET Folding Experiments	594
17.5.1	Optimized Design and Preparation of Labeled Protein Samples for FRET Folding Experiments	594
17.5.2	Strategies for Site-specific Double Labeling of Proteins	595
17.5.3	Preparation of Double-labeled Mutants Using Engineered Cysteine Residues (strategy 4)	596
17.5.4	Possible Pitfalls Associated with the Preparation of Labeled Protein Samples for FRET Folding Experiments	599
17.6	Experimental Aspects of Folding Studies by Distance Determination Based on FRET Measurements	600
17.6.1	Steady State Determination of Transfer Efficiency	600
17.6.1.1	Donor Emission	600
17.6.1.2	Acceptor Excitation Spectroscopy	601
17.6.2	Time-resolved Measurements	601
17.7	Data Analysis	603
17.7.1	Rigorous Error Analysis	606
17.7.2	Elimination of Systematic Errors	606
17.8	Applications of trFRET for Characterization of Unfolded and Partially Folded Conformations of Globular Proteins under Equilibrium Conditions	607
17.8.1	Bovine Pancreatic Trypsin Inhibitor	607
17.8.2	The Loop Hypothesis	608
17.8.3	RNase A	609
17.8.4	Staphylococcal Nuclease	611
17.9	Unfolding Transition via Continuum of Native-like Forms	611
17.10	The Third Folding Problem: Domain Motions and Conformational Fluctuations of Enzyme Molecules	611
17.11	Single Molecule FRET-detected Folding Experiments	613
17.12	Principles of Applications of Single Molecule FRET Spectroscopy in Folding Studies	615
17.12.1	Design and Analysis of Single Molecule FRET Experiments	615
17.12.1.1	How is Single Molecule FRET Efficiency Determined?	615
17.12.1.2	The Challenge of Extending the Length of the Time Trajectories	617
17.12.2	Distance and Time Resolution of the Single Molecule FRET Folding Experiments	618
17.13	Folding Kinetics	619
17.13.1	Steady State and trFRET-detected Folding Kinetics Experiments	619
17.13.2	Steady State Detection	619
17.13.3	Time-resolved FRET Detection of Rapid Folding Kinetics: the "Double Kinetics" Experiment	621
17.13.4	Multiple Probes Analysis of the Folding Transition	622
17.14	Concluding Remarks	625

Acknowledgments   626
References   627

18   Application of Hydrogen Exchange Kinetics to Studies of Protein Folding   *634*

Kaare Teilum, Birthe B. Kragelund, and Flemming M. Poulsen

18.1   Introduction   634
18.2   The Hydrogen Exchange Reaction   638
18.2.1   Calculating the Intrinsic Hydrogen Exchange Rate Constant, k_{int} 638
18.3   Protein Dynamics by Hydrogen Exchange in Native and Denaturing Conditions   641
18.3.1   Mechanisms of Exchange   642
18.3.2   Local Opening and Closing Rates from Hydrogen Exchange Kinetics   642
18.3.2.1   The General Amide Exchange Rate Expression – the Linderstrøm-Lang Equation   643
18.3.2.2   Limits to the General Rate Expression – EX1 and EX2   644
18.3.2.3   The Range between the EX1 and EX2 Limits   646
18.3.2.4   Identification of Exchange Limit   646
18.3.2.5   Global Opening and Closing Rates and Protein Folding   647
18.3.3   The "Native State Hydrogen Exchange" Strategy   648
18.3.3.1   Localization of Partially Unfolded States, PUFs   650
18.4   Hydrogen Exchange as a Structural Probe in Kinetic Folding Experiments   651
18.4.1   Protein Folding/Hydrogen Exchange Competition   652
18.4.2   Hydrogen Exchange Pulse Labeling   656
18.4.3   Protection Factors in Folding Intermediates   657
18.4.4   Kinetic Intermediate Structures Characterized by Hydrogen Exchange   659
18.5   Experimental Protocols   661
18.5.1   How to Determine Hydrogen Exchange Kinetics at Equilibrium   661
18.5.1.1   Equilibrium Hydrogen Exchange Experiments   661
18.5.1.2   Determination of Segmental Opening and Closing Rates, k_{op} and k_{cl} 662
18.5.1.3   Determination of ΔG_{fluc}, m, and $\Delta G°_{unf}$ 662
18.5.2   Planning a Hydrogen Exchange Folding Experiment   662
18.5.2.1   Determine a Combination of t_{pulse} and pH_{pulse} 662
18.5.2.2   Setup Quench Flow Apparatus   662
18.5.2.3   Prepare Deuterated Protein and Chemicals   663
18.5.2.4   Prepare Buffers and Unfolded Protein   663
18.5.2.5   Check pH in the Mixing Steps   664
18.5.2.6   Sample Mixing and Preparation   664
18.5.3   Data Analysis   664
Acknowledgments   665
References   665

19	**Studying Protein Folding and Aggregation by Laser Light Scattering**	**673**
	Klaus Gast and Andreas J. Modler	
19.1	Introduction 673	
19.2	Basic Principles of Laser Light Scattering 674	
19.2.1	Light Scattering by Macromolecular Solutions 674	
19.2.2	Molecular Parameters Obtained from Static Light Scattering (SLS) 676	
19.2.3	Molecular Parameters Obtained from Dynamic Light Scattering (DLS) 678	
19.2.4	Advantages of Combined SLS and DLS Experiments 680	
19.3	Laser Light Scattering of Proteins in Different Conformational States – Equilibrium Folding/Unfolding Transitions 680	
19.3.1	General Considerations, Hydrodynamic Dimensions in the Natively Folded State 680	
19.3.2	Changes in the Hydrodynamic Dimensions during Heat-induced Unfolding 682	
19.3.3	Changes in the Hydrodynamic Dimensions upon Cold Denaturation 683	
19.3.4	Denaturant-induced Changes of the Hydrodynamic Dimensions 684	
19.3.5	Acid-induced Changes of the Hydrodynamic Dimensions 685	
19.3.6	Dimensions in Partially Folded States – Molten Globules and Fluoroalcohol-induced States 686	
19.3.7	Comparison of the Dimensions of Proteins in Different Conformational States 687	
19.3.8	Scaling Laws for the Native and Highly Unfolded States, Hydrodynamic Modeling 687	
19.4	Studying Folding Kinetics by Laser Light Scattering 689	
19.4.1	General Considerations, Attainable Time Regions 689	
19.4.2	Hydrodynamic Dimensions of the Kinetic Molten Globule of Bovine α-Lactalbumin 690	
19.4.3	RNase A is Only Weakly Collapsed During the Burst Phase of Folding 691	
19.5	Misfolding and Aggregation Studied by Laser Light Scattering 692	
19.5.1	Overview: Some Typical Light Scattering Studies of Protein Aggregation 692	
19.5.2	Studying Misfolding and Amyloid Formation by Laser Light Scattering 693	
19.5.2.1	Overview: Initial States, Critical Oligomers, Protofibrils, Fibrils 693	
19.5.2.2	Aggregation Kinetics of $A\beta$ Peptides 694	
19.5.2.3	Kinetics of Oligomer and Fibril Formation of PGK and Recombinant Hamster Prion Protein 695	
19.5.2.4	Mechanisms of Misfolding and Misassembly, Some General Remarks 698	
19.6	Experimental Protocols 698	
19.6.1	Laser Light Scattering Instrumentation 698	

19.6.1.1	Basic Experimental Set-up, General Requirements	698
19.6.1.2	Supplementary Measurements and Useful Options	700
19.6.1.3	Commercially Available Light Scattering Instrumentation	701
19.6.2	Experimental Protocols for the Determination of Molecular Mass and Stokes Radius of a Protein in a Particular Conformational State	701
	Protocol 1 702	
	Protocol 2 704	
	Acknowledgments 704	
	References 704	

20	**Conformational Properties of Unfolded Proteins**	**710**
	Patrick J. Fleming and George D. Rose	
20.1	Introduction	710
20.1.1	Unfolded vs. Denatured Proteins	710
20.2	Early History	711
20.3	The Random Coil	712
20.3.1	The Random Coil – Theory	713
20.3.1.1	The Random Coil Model Prompts Three Questions	716
20.3.1.2	The Folding Funnel	716
20.3.1.3	Transition State Theory	717
20.3.1.4	Other Examples	717
20.3.1.5	Implicit Assumptions from the Random Coil Model	718
20.3.2	The Random Coil – Experiment	718
20.3.2.1	Intrinsic Viscosity	719
20.3.2.2	SAXS and SANS	720
20.4	Questions about the Random Coil Model	721
20.4.1	Questions from Theory	722
20.4.1.1	The Flory Isolated-pair Hypothesis	722
20.4.1.2	Structure vs. Energy Duality	724
20.4.1.3	The "Rediscovery" of Polyproline II Conformation	724
20.4.1.4	P_{II} in Unfolded Peptides and Proteins	726
20.4.2	Questions from Experiment	727
20.4.2.1	Residual Structure in Denatured Proteins and Peptides	727
20.4.3	The Reconciliation Problem	728
20.4.4	Organization in the Unfolded State – the Entropic Conjecture	728
20.4.4.1	Steric Restrictions beyond the Dipeptide	729
20.5	Future Directions	730
	Acknowledgments 731	
	References 731	

21	**Conformation and Dynamics of Nonnative States of Proteins studied by NMR Spectroscopy**	**737**
	Julia Wirmer, Christian Schlörb, and Harald Schwalbe	
21.1	Introduction	737
21.1.1	Structural Diversity of Polypeptide Chains	737

21.1.2	Intrinsically Unstructured and Natively Unfolded Proteins	739
21.2	Prerequisites: NMR Resonance Assignment	740
21.3	NMR Parameters	744
21.3.1	Chemical shifts δ	745
21.3.1.1	Conformational Dependence of Chemical Shifts	745
21.3.1.2	Interpretation of Chemical Shifts in the Presence of Conformational Averaging	746
21.3.2	J Coupling Constants	748
21.3.2.1	Conformational Dependence of J Coupling Constants	748
21.3.2.2	Interpretation of J Coupling Constants in the Presence of Conformational Averaging	750
21.3.3	Relaxation: Homonuclear NOEs	750
21.3.3.1	Distance Dependence of Homonuclear NOEs	750
21.3.3.2	Interpretation of Homonuclear NOEs in the Presence of Conformational Averaging	754
21.3.4	Heteronuclear Relaxation (^{15}N R_1, R_2, hetNOE)	757
21.3.4.1	Correlation Time Dependence of Heteronuclear Relaxation Parameters	757
21.3.4.2	Dependence on Internal Motions of Heteronuclear Relaxation Parameters	759
21.3.5	Residual Dipolar Couplings	760
21.3.5.1	Conformational Dependence of Residual Dipolar Couplings	760
21.3.5.2	Interpretation of Residual Dipolar Couplings in the Presence of Conformational Averaging	763
21.3.6	Diffusion	765
21.3.7	Paramagnetic Spin Labels	766
21.3.8	H/D Exchange	767
21.3.9	Photo-CIDNP	767
21.4	Model for the Random Coil State of a Protein	768
21.5	Nonnative States of Proteins: Examples from Lysozyme, α-Lactalbumin, and Ubiquitin	771
21.5.1	Backbone Conformation	772
21.5.1.1	Interpretation of Chemical Shifts	772
21.5.1.2	Interpretation of NOEs	774
21.5.1.3	Interpretation of J Coupling Constants	780
21.5.2	Side-chain Conformation	784
21.5.2.1	Interpretation of J Coupling Constants	784
21.5.3	Backbone Dynamics	786
21.5.3.1	Interpretation of ^{15}N Relaxation Rates	786
21.6	Summary and Outlook	793
	Acknowledgments	794
	References	794
22	**Dynamics of Unfolded Polypeptide Chains**	**809**
	Beat Fierz and Thomas Kiefhaber	
22.1	Introduction	809

22.2	Equilibrium Properties of Chain Molecules	809
22.2.1	The Freely Jointed Chain	810
22.2.2	Chain Stiffness	810
22.2.3	Polypeptide Chains	811
22.2.4	Excluded Volume Effects	812
22.3	Theory of Polymer Dynamics	813
22.3.1	The Langevin Equation	813
22.3.2	Rouse Model and Zimm Model	814
22.3.3	Dynamics of Loop Closure and the Szabo-Schulten-Schulten Theory	815
22.4	Experimental Studies on the Dynamics in Unfolded Polypeptide Chains	816
22.4.1	Experimental Systems for the Study of Intrachain Diffusion	816
22.4.1.1	Early Experimental Studies	816
22.4.1.2	Triplet Transfer and Triplet Quenching Studies	821
22.4.1.3	Fluorescence Quenching	825
22.4.2	Experimental Results on Dynamic Properties of Unfolded Polypeptide Chains	825
22.4.2.1	Kinetics of Intrachain Diffusion	826
22.4.2.2	Effect of Loop Size on the Dynamics in Flexible Polypeptide Chains	826
22.4.2.3	Effect of Amino Acid Sequence on Chain Dynamics	829
22.4.2.4	Effect of the Solvent on Intrachain Diffusion	831
22.4.2.5	Effect of Solvent Viscosity on Intrachain Diffusion	833
22.4.2.6	End-to-end Diffusion vs. Intrachain Diffusion	834
22.4.2.7	Chain Diffusion in Natural Protein Sequences	834
22.5	Implications for Protein Folding Kinetics	837
22.5.1	Rate of Contact Formation during the Earliest Steps in Protein Folding	837
22.5.2	The Speed Limit of Protein Folding vs. the Pre-exponential Factor	839
22.5.3	Contributions of Chain Dynamics to Rate- and Equilibrium Constants for Protein Folding Reactions	840
22.6	Conclusions and Outlook	844
22.7	Experimental Protocols and Instrumentation	844
22.7.1	Properties of the Electron Transfer Probes and Treatment of the Transfer Kinetics	845
22.7.2	Test for Diffusion-controlled Reactions	847
22.7.2.1	Determination of Bimolecular Quenching or Transfer Rate Constants	847
22.7.2.2	Testing the Viscosity Dependence	848
22.7.2.3	Determination of Activation Energy	848
22.7.3	Instrumentation	849
	Acknowledgments	849
	References	849

23 Equilibrium and Kinetically Observed Molten Globule States 856
Kosuke Maki, Kiyoto Kamagata, and Kunihiro Kuwajima

- 23.1 Introduction 856
- 23.2 Equilibrium Molten Globule State 858
- 23.2.1 Structural Characteristics of the Molten Globule State 858
- 23.2.2 Typical Examples of the Equilibrium Molten Globule State 859
- 23.2.3 Thermodynamic Properties of the Molten Globule State 860
- 23.3 The Kinetically Observed Molten Globule State 862
- 23.3.1 Observation and Identification of the Molten Globule State in Kinetic Refolding 862
- 23.3.2 Kinetics of Formation of the Early Folding Intermediates 863
- 23.3.3 Late Folding Intermediates and Structural Diversity 864
- 23.3.4 Evidence for the On-pathway Folding Intermediate 865
- 23.4 Two-stage Hierarchical Folding Funnel 866
- 23.5 Unification of the Folding Mechanism between Non-two-state and Two-state Proteins 867
- 23.5.1 Statistical Analysis of the Folding Data of Non-two-state and Two-state Proteins 868
- 23.5.2 A Unified Mechanism of Protein Folding: Hierarchy 870
- 23.5.3 Hidden Folding Intermediates in Two-state Proteins 871
- 23.6 Practical Aspects of the Experimental Study of Molten Globules 872
- 23.6.1 Observation of the Equilibrium Molten Globule State 872
- 23.6.1.1 Two-state Unfolding Transition 872
- 23.6.1.2 Multi-state (Three-state) Unfolding Transition 874
- 23.6.2 Burst-phase Intermediate Accumulated during the Dead Time of Refolding Kinetics 876
- 23.6.3 Testing the Identity of the Molten Globule State with the Burst-Phase Intermediate 877
- References 879

24 Alcohol- and Salt-induced Partially Folded Intermediates 884
Daizo Hamada and Yuji Goto

- 24.1 Introduction 884
- 24.2 Alcohol-induced Intermediates of Proteins and Peptides 886
- 24.2.1 Formation of Secondary Structures by Alcohols 888
- 24.2.2 Alcohol-induced Denaturation of Proteins 888
- 24.2.3 Formation of Compact Molten Globule States 889
- 24.2.4 Example: β-Lactoglobulin 890
- 24.3 Mechanism of Alcohol-induced Conformational Change 893
- 24.4 Effects of Alcohols on Folding Kinetics 896
- 24.5 Salt-induced Formation of the Intermediate States 899
- 24.5.1 Acid-denatured Proteins 899
- 24.5.2 Acid-induced Unfolding and Refolding Transitions 900
- 24.6 Mechanism of Salt-induced Conformational Change 904
- 24.7 Generality of the Salt Effects 906

24.8	Conclusion 907	
	References 908	
25	**Prolyl Isomerization in Protein Folding** 916	
	Franz Schmid	
25.1	Introduction 916	
25.2	Prolyl Peptide Bonds 917	
25.3	Prolyl Isomerizations as Rate-determining Steps of Protein Folding 918	
25.3.1	The Discovery of Fast and Slow Refolding Species 918	
25.3.2	Detection of Proline-limited Folding Processes 919	
25.3.3	Proline-limited Folding Reactions 921	
25.3.4	Interrelation between Prolyl Isomerization and Conformational Folding 923	
25.4	Examples of Proline-limited Folding Reactions 924	
25.4.1	Ribonuclease A 924	
25.4.2	Ribonuclease T1 926	
25.4.3	The Structure of a Folding Intermediate with an Incorrect Prolyl Isomer 928	
25.5	Native-state Prolyl Isomerizations 929	
25.6	Nonprolyl Isomerizations in Protein Folding 930	
25.7	Catalysis of Protein Folding by Prolyl Isomerases 932	
25.7.1	Prolyl Isomerases as Tools for Identifying Proline-limited Folding Steps 932	
25.7.2	Specificity of Prolyl Isomerases 933	
25.7.3	The Trigger Factor 934	
25.7.4	Catalysis of Prolyl Isomerization During de novo Protein Folding 935	
25.8	Concluding Remarks 936	
25.9	Experimental Protocols 936	
25.9.1	Slow Refolding Assays ("Double Jumps") to Measure Prolyl Isomerizations in an Unfolded Protein 936	
25.9.1.1	Guidelines for the Design of Double Jump Experiments 937	
25.9.1.2	Formation of U_S Species after Unfolding of RNase A 938	
25.9.2	Slow Unfolding Assays for Detecting and Measuring Prolyl Isomerizations in Refolding 938	
25.9.2.1	Practical Considerations 939	
25.9.2.2	Kinetics of the Formation of Fully Folded IIHY-G3P* Molecules 939	
	References 939	
26	**Folding and Disulfide Formation** 946	
	Margherita Ruoppolo, Piero Pucci, and Gennaro Marino	
26.1	Chemistry of the Disulfide Bond 946	
26.2	Trapping Protein Disulfides 947	
26.3	Mass Spectrometric Analysis of Folding Intermediates 948	
26.4	Mechanism(s) of Oxidative Folding so Far – Early and Late Folding Steps 949	

26.5	Emerging Concepts from Mass Spectrometric Studies	950
26.5.1	Three-fingered Toxins	951
26.5.2	RNase A	953
26.5.3	Antibody Fragments	955
26.5.4	Human Nerve Growth Factor	956
26.6	Unanswered Questions	956
26.7	Concluding Remarks	957
26.8	Experimental Protocols	957
26.8.1	How to Prepare Folding Solutions	957
26.8.2	How to Carry Out Folding Reactions	958
26.8.3	How to Choose the Best Mass Spectrometric Equipment for Your Study	959
26.8.4	How to Perform Electrospray (ES)MS Analysis	959
26.8.5	How to Perform Matrix-assisted Laser Desorption Ionization (MALDI) MS Analysis	960
	References	961
27	**Concurrent Association and Folding of Small Oligomeric Proteins**	**965**
	Hans Rudolf Bosshard	
27.1	Introduction	965
27.2	Experimental Methods Used to Follow the Folding of Oligomeric Proteins	966
27.2.1	Equilibrium Methods	966
27.2.2	Kinetic Methods	968
27.3	Dimeric Proteins	969
27.3.1	Two-state Folding of Dimeric Proteins	970
27.3.1.1	Examples of Dimeric Proteins Obeying Two-state Folding	971
27.3.2	Folding of Dimeric Proteins through Intermediate States	978
27.4	Trimeric and Tetrameric Proteins	983
27.5	Concluding Remarks	986
	Appendix – Concurrent Association and Folding of Small Oligomeric Proteins	987
A1	Equilibrium Constants for Two-state Folding	988
A1.1	Homooligomeric Protein	988
A1.2	Heterooligomeric Protein	989
A2	Calculation of Thermodynamic Parameters from Equilibrium Constants	990
A2.1	Basic Thermodynamic Relationships	990
A2.2	Linear Extrapolation of Denaturant Unfolding Curves of Two-state Reaction	990
A2.3	Calculation of the van't Hoff Enthalpy Change from Thermal Unfolding Data	990
A2.4	Calculation of the van't Hoff Enthalpy Change from the Concentration-dependence of T_m	991
A2.5	Extrapolation of Thermodynamic Parameters to Different Temperatures: Gibbs-Helmholtz Equation	991

A3	Kinetics of Reversible Two-state Folding and Unfolding: Integrated Rate Equations *992*
A3.1	Two-state Folding of Dimeric Protein *992*
A3.2	Two-state Unfolding of Dimeric Protein *992*
A3.3	Reversible Two-state Folding and Unfolding *993*
A3.3.1	Homodimeric protein *993*
A3.3.2	Heterodimeric protein *993*
A4	Kinetics of Reversible Two-state Folding: Relaxation after Disturbance of a Pre-existing Equilibrium (Method of Bernasconi) *994*
	Acknowledgments *995*
	References *995*

28	**Folding of Membrane Proteins** *998*
	Lukas K. Tamm and Heedeok Hong
28.1	Introduction *998*
28.2	Thermodyamics of Residue Partitioning into Lipid Bilayers *1000*
28.3	Stability of β-Barrel Proteins *1001*
28.4	Stability of Helical Membrane Proteins *1009*
28.5	Helix and Other Lateral Interactions in Membrane Proteins *1010*
28.6	The Membrane Interface as an Important Contributor to Membrane Protein Folding *1012*
28.7	Membrane Toxins as Models for Helical Membrane Protein Insertion *1013*
28.8	Mechanisms of β-Barrel Membrane Protein Folding *1015*
28.9	Experimental Protocols *1016*
28.9.1	SDS Gel Shift Assay for Heat-modifiable Membrane Proteins *1016*
28.9.1.1	Reversible Folding and Unfolding Protocol Using OmpA as an Example *1016*
28.9.2	Tryptophan Fluorescence and Time-resolved Distance Determination by Tryptophan Fluorescence Quenching *1018*
28.9.2.1	TDFQ Protocol for Monitoring the Translocation of Tryptophans across Membranes *1019*
28.9.3	Circular Dichroism Spectroscopy *1020*
28.9.4	Fourier Transform Infrared Spectroscopy *1022*
28.9.4.1	Protocol for Obtaining Conformation and Orientation of Membrane Proteins and Peptides by Polarized ATR-FTIR Spectroscopy *1023*
	Acknowledgments *1025*
	References *1025*

29	**Protein Folding Catalysis by Pro-domains** *1032*
	Philip N. Bryan
29.1	Introduction *1032*
29.2	Bimolecular Folding Mechanisms *1033*
29.3	Structures of Reactants and Products *1033*
29.3.1	Structure of Free SBT *1033*

29.3.2	Structure of SBT/Pro-domain Complex	*1036*
29.3.3	Structure of Free ALP	*1037*
29.3.4	Structure of the ALP/Pro-domain Complex	*1037*
29.4	Stability of the Mature Protease	*1039*
29.4.1	Stability of ALP	*1039*
29.4.2	Stability of Subtilisin	*1040*
29.5	Analysis of Pro-domain Binding to the Folded Protease	*1042*
29.6	Analysis of Folding Steps	*1043*
29.7	Why are Pro-domains Required for Folding?	*1046*
29.8	What is the Origin of High Cooperativity?	*1047*
29.9	How Does the Pro-domain Accelerate Folding?	*1048*
29.10	Are High Kinetic Stability and Facile Folding Mutually Exclusive? *1049*	
29.11	Experimental Protocols for Studying SBT Folding	*1049*
29.11.1	Fermentation and Purification of Active Subtilisin	*1049*
29.11.2	Fermentation and Purification of Facile-folding Ala221 Subtilisin from *E. coli* *1050*	
29.11.3	Mutagenesis and Protein Expression of Pro-domain Mutants	*1051*
29.11.4	Purification of Pro-domain	*1052*
29.11.5	Kinetics of Pro-domain Binding to Native SBT	*1052*
29.11.6	Kinetic Analysis of Pro-domain Facilitated Subtilisin Folding	*1052*
29.11.6.1	Single Mixing	*1052*
29.11.6.2	Double Jump: Renaturation–Denaturation	*1053*
29.11.6.3	Double Jump: Denaturation–Renaturation	*1053*
29.11.6.4	Triple Jump: Denaturation–Renaturation–Denaturation	*1054*
	References *1054*	
30	**The Thermodynamics and Kinetics of Collagen Folding**	*1059*
	Hans Peter Bächinger and Jürgen Engel	
30.1	Introduction *1059*	
30.1.1	The Collagen Family *1059*	
30.1.2	Biosynthesis of Collagens *1060*	
30.1.3	The Triple Helical Domain in Collagens and Other Proteins	*1061*
30.1.4	N- and C-Propeptide, Telopeptides, Flanking Coiled-Coil Domains *1061*	
30.1.5	Why is the Folding of the Triple Helix of Interest?	*1061*
30.2	Thermodynamics of Collagen Folding *1062*	
30.2.1	Stability of the Triple Helix *1062*	
30.2.2	The Role of Posttranslational Modifications	*1063*
30.2.3	Energies Involved in the Stability of the Triple Helix	*1063*
30.2.4	Model Peptides Forming the Collagen Triple Helix	*1066*
30.2.4.1	Type of Peptides *1066*	
30.2.4.2	The All-or-none Transition of Short Model Peptides	*1066*
30.2.4.3	Thermodynamic Parameters for Different Model Systems	*1069*
30.2.4.4	Contribution of Different Tripeptide Units to Stability	*1075*

30.2.4.5	Crystal and NMR Structures of Triple Helices 1076
30.2.4.6	Conformation of the Randomly Coiled Chains 1077
30.2.4.7	Model Studies with Isomers of Hydroxyproline and Fluoroproline 1078
30.2.4.8	Cis ⇌ trans Equilibria of Peptide Bonds 1079
30.2.4.9	Interpretations of Stabilities on a Molecular Level 1080
30.3	Kinetics of Triple Helix Formation 1081
30.3.1	Properties of Collagen Triple Helices that Influence Kinetics 1081
30.3.2	Folding of Triple Helices from Single Chains 1082
30.3.2.1	Early Work 1082
30.3.2.2	Concentration Dependence of the Folding of $(PPG)_{10}$ and $(POG)_{10}$ 1082
30.3.2.3	Model Mechanism of the Folding Kinetics 1085
30.3.2.4	Rate Constants of Nucleation and Propagation 1087
30.3.2.5	Host–guest Peptides and an Alternative Kinetics Model 1088
30.3.3	Triple Helix Formation from Linked Chains 1089
30.3.3.1	The Short N-terminal Triple Helix of Collagen III in Fragment Col1–3 1089
30.3.3.2	Folding of the Central Long Triple Helix of Collagen III 1090
30.3.3.3	The Zipper Model 1092
30.3.4	Designed Collagen Models with Chains Connected by a Disulfide Knot or by Trimerizing Domains 1097
30.3.4.1	Disulfide-linked Model Peptides 1097
30.3.4.2	Model Peptides Linked by a Foldon Domain 1098
30.3.4.3	Collagen Triple Helix Formation can be Nucleated at either End 1098
30.3.4.4	Hysteresis of Triple Helix Formation 1099
30.3.5	Influence of *cis–trans* Isomerase and Chaperones 1100
30.3.6	Mutations in Collagen Triple Helices Affect Proper Folding 1101
	References 1101

31	**Unfolding Induced by Mechanical Force** **1111**
	Jane Clarke and Phil M. Williams
31.1	Introduction 1111
31.2	Experimental Basics 1112
31.2.1	Instrumentation 1112
31.2.2	Sample Preparation 1113
31.2.3	Collecting Data 1114
31.2.4	Anatomy of a Force Trace 1115
31.2.5	Detecting Intermediates in a Force Trace 1115
31.2.6	Analyzing the Force Trace 1116
31.3	Analysis of Force Data 1117
31.3.1	Basic Theory behind Dynamic Force Spectroscopy 1117
31.3.2	The Ramp of Force Experiment 1119
31.3.3	The Golden Equation of DFS 1121
31.3.4	Nonlinear Loading 1122

31.3.4.1	The Worm-line Chain (WLC)	*1123*
31.3.5	Experiments under Constant Force	*1124*
31.3.6	Effect of Tandem Repeats on Kinetics	*1125*
31.3.7	Determining the Modal Force	*1126*
31.3.8	Comparing Behavior	*1127*
31.3.9	Fitting the Data	*1127*
31.4	Use of Complementary Techniques	*1129*
31.4.1	Protein Engineering	*1130*
31.4.1.1	Choosing Mutants	*1130*
31.4.1.2	Determining $\Delta\Delta G_{D-N}$	*1131*
31.4.1.3	Determining $\Delta\Delta G_{TS-N}$	*1131*
31.4.1.4	Interpreting the Φ-values	*1132*
31.4.2	Computer Simulation	*1133*
31.5	Titin I27: A Case Study	*1134*
31.5.1	The Protein System	*1134*
31.5.2	The Unfolding Intermediate	*1135*
31.5.3	The Transition State	*1136*
31.5.4	The Relationship Between the Native and Transition States	*1137*
31.5.5	The Energy Landscape under Force	*1139*
31.6	Conclusions – the Future	*1139*
	References	*1139*
32	**Molecular Dynamics Simulations to Study Protein Folding and Unfolding** *1143*	
	Amedeo Caflisch and Emanuele Paci	
32.1	Introduction	*1143*
32.2	Molecular Dynamics Simulations of Peptides and Proteins	*1144*
32.2.1	Folding of Structured Peptides	*1144*
32.2.1.1	Reversible Folding and Free Energy Surfaces	*1144*
32.2.1.2	Non-Arrhenius Temperature Dependence of the Folding Rate	*1147*
32.2.1.3	Denatured State and Levinthal Paradox	*1148*
32.2.1.4	Folding Events of Trp-cage	*1149*
32.2.2	Unfolding Simulations of Proteins	*1150*
32.2.2.1	High-temperature Simulations	*1150*
32.2.2.2	Biased Unfolding	*1150*
32.2.2.3	Forced Unfolding	*1151*
32.2.3	Determination of the Transition State Ensemble	*1153*
32.3	MD Techniques and Protocols	*1155*
32.3.1	Techniques to Improve Sampling	*1155*
32.3.1.1	Replica Exchange Molecular Dynamics	*1155*
32.3.1.2	Methods Based on Path Sampling	*1157*
32.3.2	MD with Restraints	*1157*
32.3.3	Distributed Computing Approach	*1158*
32.3.4	Implicit Solvent Models versus Explicit Water	*1160*
32.4	Conclusion	*1162*
	References	*1162*

33	**Molecular Dynamics Simulations of Proteins and Peptides: Problems, Achievements, and Perspectives** *1170*
	Paul Tavan, Heiko Carstens, and Gerald Mathias
33.1	Introduction *1170*
33.2	Basic Physics of Protein Structure and Dynamics *1171*
33.2.1	Protein Electrostatics *1172*
33.2.2	Relaxation Times and Spatial Scales *1172*
33.2.3	Solvent Environment *1173*
33.2.4	Water *1174*
33.2.5	Polarizability of the Peptide Groups and of Other Protein Components *1175*
33.3	State of the Art *1177*
33.3.1	Control of Thermodynamic Conditions *1177*
33.3.2	Long-range Electrostatics *1177*
33.3.3	Polarizability *1179*
33.3.4	Higher Multipole Moments of the Molecular Components *1180*
33.3.5	MM Models of Water *1181*
33.3.6	Complexity of Protein–Solvent Systems and Consequences for MM-MD *1182*
33.3.7	What about Successes of MD Methods? *1182*
33.3.8	Accessible Time Scales and Accuracy Issues *1184*
33.3.9	Continuum Solvent Models *1185*
33.3.10	Are there Further Problems beyond Electrostatics and Structure Prediction? *1187*
33.4	Conformational Dynamics of a Light-switchable Model Peptide *1187*
33.4.1	Computational Methods *1188*
33.4.2	Results and Discussion *1190*
	Summary *1194*
	Acknowledgments *1194*
	References *1194*

Part II, Volume 1

Contributors of Part II *LVIII*

1	**Paradigm Changes from "Unboiling an Egg" to "Synthesizing a Rabbit"** *3*
	Rainer Jaenicke
1.1	Protein Structure, Stability, and Self-organization *3*
1.2	Autonomous and Assisted Folding and Association *6*
1.3	Native, Intermediate, and Denatured States *11*
1.4	Folding and Merging of Domains – Association of Subunits *13*
1.5	Limits of Reconstitution *19*
1.6	In Vitro Denaturation-Renaturation vs. Folding in Vivo *21*

1.7	Perspectives 24	
	Acknowledgements 26	
	References 26	
2	**Folding and Association of Multi-domain and Oligomeric Proteins** 32	
	Hauke Lilie and Robert Seckler	
2.1	Introduction 32	
2.2	Folding of Multi-domain Proteins 33	
2.2.1	Domain Architecture 33	
2.2.2	γ-Crystallin as a Model for a Two-domain Protein 35	
2.2.3	The Giant Protein Titin 39	
2.3	Folding and Association of Oligomeric Proteins 41	
2.3.1	Why Oligomers? 41	
2.3.2	Inter-subunit Interfaces 42	
2.3.3	Domain Swapping 44	
2.3.4	Stability of Oligomeric Proteins 45	
2.3.5	Methods Probing Folding/Association 47	
2.3.5.1	Chemical Cross-linking 47	
2.3.5.2	Analytical Gel Filtration Chromatography 47	
2.3.5.3	Scattering Methods 48	
2.3.5.4	Fluorescence Resonance Energy Transfer 48	
2.3.5.5	Hybrid Formation 48	
2.3.6	Kinetics of Folding and Association 49	
2.3.6.1	General Considerations 49	
2.3.6.2	Reconstitution Intermediates 50	
2.3.6.3	Rates of Association 52	
2.3.6.4	Homo- Versus Heterodimerization 52	
2.4	Renaturation versus Aggregation 54	
2.5	Case Studies on Protein Folding and Association 54	
2.5.1	Antibody Fragments 54	
2.5.2	Trimeric Tail Spike Protein of Bacteriophage P22 59	
2.6	Experimental Protocols 62	
	References 65	
3	**Studying Protein Folding in Vivo** 73	
	I. Marije Liscaljet, Bertrand Kleizen, and Ineke Braakman	
3.1	Introduction 73	
3.2	General Features in Folding Proteins Amenable to in Vivo Study 73	
3.2.1	Increasing Compactness 76	
3.2.2	Decreasing Accessibility to Different Reagents 76	
3.2.3	Changes in Conformation 77	
3.2.4	Assistance During Folding 78	
3.3	Location-specific Features in Protein Folding 79	
3.3.1	Translocation and Signal Peptide Cleavage 79	
3.3.2	Glycosylation 80	

3.3.3	Disulfide Bond Formation in the ER	81
3.3.4	Degradation	82
3.3.5	Transport from ER to Golgi and Plasma Membrane	83
3.4	How to Manipulate Protein Folding	84
3.4.1	Pharmacological Intervention (Low-molecular-weight Reagents)	84
3.4.1.1	Reducing and Oxidizing Agents	84
3.4.1.2	Calcium Depletion	84
3.4.1.3	ATP Depletion	85
3.4.1.4	Cross-linking	85
3.4.1.5	Glycosylation Inhibitors	85
3.4.2	Genetic Modifications (High-molecular-weight Manipulations)	86
3.4.2.1	Substrate Protein Mutants	86
3.4.2.2	Changing the Concentration or Activity of Folding Enzymes and Chaperones	87
3.5	Experimental Protocols	88
3.5.1	Protein-labeling Protocols	88
3.5.1.1	Basic Protocol Pulse Chase: Adherent Cells	88
3.5.1.2	Pulse Chase in Suspension Cells	91
3.5.2	(Co)-immunoprecipitation and Accessory Protocols	93
3.5.2.1	Immunoprecipitation	93
3.5.2.2	Co-precipitation with Calnexin ([84]; adapted from Ou et al. [85])	94
3.5.2.3	Co-immunoprecipitation with Other Chaperones	95
3.5.2.4	Protease Resistance	95
3.5.2.5	Endo H Resistance	96
3.5.2.6	Cell Surface Expression Tested by Protease	96
3.5.3	SDS-PAGE [13]	97
	Acknowledgements	98
	References	98
4	**Characterization of ATPase Cycles of Molecular Chaperones by Fluorescence and Transient Kinetic Methods**	**105**
	Sandra Schlee and Jochen Reinstein	
4.1	Introduction	105
4.1.1	Characterization of ATPase Cycles of Energy-transducing Systems	105
4.1.2	The Use of Fluorescent Nucleotide Analogues	106
4.1.2.1	Fluorescent Modifications of Nucleotides	106
4.1.2.2	How to Find a Suitable Analogue for a Specific Protein	108
4.2	Characterization of ATPase Cycles of Molecular Chaperones	109
4.2.1	Biased View	109
4.2.2	The ATPase Cycle of DnaK	109
4.2.3	The ATPase Cycle of the Chaperone Hsp90	109
4.2.4	The ATPase Cycle of the Chaperone ClpB	111
4.2.4.1	ClpB, an Oligomeric ATPase With Two AAA Modules Per Protomer	111

4.2.4.2	Nucleotide-binding Properties of NBD1 and NBD2 *111*	
4.2.4.3	Cooperativity of ATP Hydrolysis and Interdomain Communication *114*	
4.3	Experimental Protocols *116*	
4.3.1	Synthesis of Fluorescent Nucleotide Analogues *116*	
4.3.1.1	Synthesis and Characterization of (P_β)MABA-ADP and (P_γ)MABA-ATP *116*	
4.3.1.2	Synthesis and Characterization of N8-MABA Nucleotides *119*	
4.3.1.3	Synthesis of MANT Nucleotides *120*	
4.3.2	Preparation of Nucleotides and Proteins *121*	
4.3.2.1	Assessment of Quality of Nucleotide Stock Solution *121*	
4.3.2.2	Determination of the Nucleotide Content of Proteins *122*	
4.3.2.3	Nucleotide Depletion Methods *123*	
4.3.3	Steady-state ATPase Assays *124*	
4.3.3.1	Coupled Enzymatic Assay *124*	
4.3.3.2	Assays Based on [α-^{32}P]-ATP and TLC *125*	
4.3.3.3	Assays Based on Released P_i *125*	
4.3.4	Single-turnover ATPase Assays *126*	
4.3.4.1	Manual Mixing Procedures *126*	
4.3.4.2	Quenched Flow *127*	
4.3.5	Nucleotide-binding Measurements *127*	
4.3.5.1	Isothermal Titration Calorimetry *127*	
4.3.5.2	Equilibrium Dialysis *129*	
4.3.5.3	Filter Binding *129*	
4.3.5.4	Equilibrium Fluorescence Titration *130*	
4.3.5.5	Competition Experiments *132*	
4.3.6	Analytical Solutions of Equilibrium Systems *133*	
4.3.6.1	Quadratic Equation *133*	
4.3.6.2	Cubic Equation *134*	
4.3.6.3	Iterative Solutions *138*	
4.3.7	Time-resolved Binding Measurements *141*	
4.3.7.1	Introduction *141*	
4.3.7.2	One-step Irreversible Process *142*	
4.3.7.3	One-step Reversible Process *143*	
4.3.7.4	Reversible Second Order Reduced to Pseudo-first Order *144*	
4.3.7.5	Two Simultaneous Irreversible Pathways – Partitioning *146*	
4.3.7.6	Two-step Consecutive (Sequential) Reaction *148*	
4.3.7.7	Two-step Binding Reactions *150*	
	References *152*	
5	**Analysis of Chaperone Function in Vitro** *162*	
	Johannes Buchner and Stefan Walter	
5.1	Introduction *162*	
5.2	Basic Functional Principles of Molecular Chaperones *164*	
5.2.1	Recognition of Nonnative Proteins *166*	

5.2.2	Induction of Conformational Changes in the Substrate	167
5.2.3	Energy Consumption and Regulation of Chaperone Function	169
5.3	Limits and Extensions of the Chaperone Concept	170
5.3.1	Co-chaperones	171
5.3.2	Specific Chaperones	171
5.4	Working with Molecular Chaperones	172
5.4.1	Natural versus Artificial Substrate Proteins	172
5.4.2	Stability of Chaperones	172
5.5	Assays to Assess and Characterize Chaperone Function	174
5.5.1	Generating Nonnative Conformations of Proteins	174
5.5.2	Aggregation Assays	174
5.5.3	Detection of Complexes Between Chaperone and Substrate	175
5.5.4	Refolding of Denatured Substrates	175
5.5.5	ATPase Activity and Effect of Substrate and Cofactors	176
5.6	Experimental Protocols	176
5.6.1	General Considerations	176
5.6.1.1	Analysis of Chaperone Stability	176
5.6.1.2	Generation of Nonnative Proteins	177
5.6.1.3	Model Substrates for Chaperone Assays	177
5.6.2	Suppression of Aggregation	179
5.6.3	Complex Formation between Chaperones and Polypeptide Substrates	183
5.6.4	Identification of Chaperone-binding Sites	184
5.6.5	Chaperone-mediated Refolding of Test Proteins	186
5.6.6	ATPase Activity	188
	Acknowledgments	188
	References	189
6	**Physical Methods for Studies of Fiber Formation and Structure**	**197**
	Thomas Scheibel and Louise Serpell	
6.1	Introduction	197
6.2	Overview: Protein Fibers Formed in Vivo	198
6.2.1	Amyloid Fibers	198
6.2.2	Silks	199
6.2.3	Collagens	199
6.2.4	Actin, Myosin, and Tropomyosin Filaments	200
6.2.5	Intermediate Filaments/Nuclear Lamina	202
6.2.6	Fibrinogen/Fibrin	203
6.2.7	Microtubules	203
6.2.8	Elastic Fibers	204
6.2.9	Flagella and Pili	204
6.2.10	Filamentary Structures in Rod-like Viruses	205
6.2.11	Protein Fibers Used by Viruses and Bacteriophages to Bind to Their Hosts	206
6.3	Overview: Fiber Structures	206

6.3.1	Study of the Structure of β-sheet-containing Proteins	207
6.3.1.1	Amyloid	207
6.3.1.2	Paired Helical Filaments	207
6.3.1.3	β-Silks	207
6.3.1.4	β-Sheet-containing Viral Fibers	208
6.3.2	α-Helix-containing Protein Fibers	209
6.3.2.1	Collagen	209
6.3.2.2	Tropomyosin	210
6.3.2.3	Intermediate Filaments	210
6.3.3	Protein Polymers Consisting of a Mixture of Secondary Structure 211	
6.3.3.1	Tubulin	211
6.3.3.2	Actin and Myosin Filaments	212
6.4	Methods to Study Fiber Assembly	213
6.4.1	Circular Dichroism Measurements for Monitoring Structural Changes Upon Fiber Assembly	213
6.4.1.1	Theory of CD	213
6.4.1.2	Experimental Guide to Measure CD Spectra and Structural Transition Kinetics	214
6.4.2	Intrinsic Fluorescence Measurements to Analyze Structural Changes	215
6.4.2.1	Theory of Protein Fluorescence	215
6.4.2.2	Experimental Guide to Measure Trp Fluorescence	216
6.4.3	Covalent Fluorescent Labeling to Determine Structural Changes of Proteins with Environmentally Sensitive Fluorophores	217
6.4.3.1	Theory on Environmental Sensitivity of Fluorophores	217
6.4.3.2	Experimental Guide to Labeling Proteins With Fluorophores	218
6.4.4	1-Anilino-8-Naphthalensulfonate (ANS) Binding to Investigate Fiber Assembly	219
6.4.4.1	Theory on Using ANS Fluorescence for Detecting Conformational Changes in Proteins	219
6.4.4.2	Experimental Guide to Using ANS for Monitoring Protein Fiber Assembly	220
6.4.5	Light Scattering to Monitor Particle Growth	220
6.4.5.1	Theory of Classical Light Scattering	221
6.4.5.2	Theory of Dynamic Light Scattering	221
6.4.5.3	Experimental Guide to Analyzing Fiber Assembly Using DLS	222
6.4.6	Field-flow Fractionation to Monitor Particle Growth	222
6.4.6.1	Theory of FFF	222
6.4.6.2	Experimental Guide to Using FFF for Monitoring Fiber Assembly	223
6.4.7	Fiber Growth-rate Analysis Using Surface Plasmon Resonance	223
6.4.7.1	Theory of SPR	223
6.4.7.2	Experimental Guide to Using SPR for Fiber-growth Analysis	224
6.4.8	Single-fiber Growth Imaging Using Atomic Force Microscopy	225

6.4.8.1	Theory of Atomic Force Microscopy	225
6.4.8.2	Experimental Guide for Using AFM to Investigate Fiber Growth	225
6.4.9	Dyes Specific for Detecting Amyloid Fibers	226
6.4.9.1	Theory on Congo Red and Thioflavin T Binding to Amyloid	226
6.4.9.2	Experimental Guide to Detecting Amyloid Fibers with CR and Thioflavin Binding	227
6.5	Methods to Study Fiber Morphology and Structure	228
6.5.1	Scanning Electron Microscopy for Examining the Low-resolution Morphology of a Fiber Specimen	228
6.5.1.1	Theory of SEM	228
6.5.1.2	Experimental Guide to Examining Fibers by SEM	229
6.5.2	Transmission Electron Microscopy for Examining Fiber Morphology and Structure	230
6.5.2.1	Theory of TEM	230
6.5.2.2	Experimental Guide to Examining Fiber Samples by TEM	231
6.5.3	Cryo-electron Microscopy for Examination of the Structure of Fibrous Proteins	232
6.5.3.1	Theory of Cryo-electron Microscopy	232
6.5.3.2	Experimental Guide to Preparing Proteins for Cryo-electron Microscopy	233
6.5.3.3	Structural Analysis from Electron Micrographs	233
6.5.4	Atomic Force Microscopy for Examining the Structure and Morphology of Fibrous Proteins	234
6.5.4.1	Experimental Guide for Using AFM to Monitor Fiber Morphology	234
6.5.5	Use of X-ray Diffraction for Examining the Structure of Fibrous Proteins	236
6.5.5.1	Theory of X-Ray Fiber Diffraction	236
6.5.5.2	Experimental Guide to X-Ray Fiber Diffraction	237
6.5.6	Fourier Transformed Infrared Spectroscopy	239
6.5.6.1	Theory of FTIR	239
6.5.6.2	Experimental Guide to Determining Protein Conformation by FTIR	240
6.6	Concluding Remarks	241
	Acknowledgements	242
	References	242
7	**Protein Unfolding in the Cell**	**254**
	Prakash Koodathingal, Neil E. Jaffe, and Andreas Matouschek	
7.1	Introduction	254
7.2	Protein Translocation Across Membranes	254
7.2.1	Compartmentalization and Unfolding	254
7.2.2	Mitochondria Actively Unfold Precursor Proteins	256
7.2.3	The Protein Import Machinery of Mitochondria	257
7.2.4	Specificity of Unfolding	259

7.2.5	Protein Import into Other Cellular Compartments	*259*
7.3	Protein Unfolding and Degradation by ATP-dependent Proteases	*260*
7.3.1	Structural Considerations of Unfoldases Associated With Degradation	*260*
7.3.2	Unfolding Is Required for Degradation by ATP-dependent Proteases	*261*
7.3.3	The Role of ATP and Models of Protein Unfolding	*262*
7.3.4	Proteins Are Unfolded Sequentially and Processively	*263*
7.3.5	The Influence of Substrate Structure on the Degradation Process	*264*
7.3.6	Unfolding by Pulling	*264*
7.3.7	Specificity of Degradation	*265*
7.4	Conclusions	*266*
7.5	Experimental Protocols	*266*
7.5.1	Size of Import Channels in the Outer and Inner Membranes of Mitochondria	*266*
7.5.2	Structure of Precursor Proteins During Import into Mitochondria	*266*
7.5.3	Import of Barnase Mutants	*267*
7.5.4	Protein Degradation by ATP-dependent Proteases	*267*
7.5.5	Use of Multi-domain Substrates	*268*
7.5.6	Studies Using Circular Permutants	*268*
	References	*269*
8	**Natively Disordered Proteins**	*275*
	Gary W. Daughdrill, Gary J. Pielak, Vladimir N. Uversky, Marc S. Cortese, and A. Keith Dunker	
8.1	Introduction	*275*
8.1.1	The Protein Structure-Function Paradigm	*275*
8.1.2	Natively Disordered Proteins	*277*
8.1.3	A New Protein Structure-Function Paradigm	*280*
8.2	Methods Used to Characterize Natively Disordered Proteins	*281*
8.2.1	NMR Spectroscopy	*281*
8.2.1.1	Chemical Shifts Measure the Presence of Transient Secondary Structure	*282*
8.2.1.2	Pulsed Field Gradient Methods to Measure Translational Diffusion	*284*
8.2.1.3	NMR Relaxation and Protein Flexibility	*284*
8.2.1.4	Using the Model-free Analysis of Relaxation Data to Estimate Internal Mobility and Rotational Correlation Time	*285*
8.2.1.5	Using Reduced Spectral Density Mapping to Assess the Amplitude and Frequencies of Intramolecular Motion	*286*
8.2.1.6	Characterization of the Dynamic Structures of Natively Disordered Proteins Using NMR	*287*
8.2.2	X-ray Crystallography	*288*
8.2.3	Small Angle X-ray Diffraction and Hydrodynamic Measurements	*293*

8.2.4	Circular Dichroism Spectropolarimetry	297
8.2.5	Infrared and Raman Spectroscopy	299
8.2.6	Fluorescence Methods	301
8.2.6.1	Intrinsic Fluorescence of Proteins	301
8.2.6.2	Dynamic Quenching of Fluorescence	302
8.2.6.3	Fluorescence Polarization and Anisotropy	303
8.2.6.4	Fluorescence Resonance Energy Transfer	303
8.2.6.5	ANS Fluorescence	305
8.2.7	Conformational Stability	308
8.2.7.1	Effect of Temperature on Proteins with Extended Disorder	309
8.2.7.2	Effect of pH on Proteins with Extended Disorder	309
8.2.8	Mass Spectrometry-based High-resolution Hydrogen-Deuterium Exchange	309
8.2.9	Protease Sensitivity	311
8.2.10	Prediction from Sequence	313
8.2.11	Advantage of Multiple Methods	314
8.3	Do Natively Disordered Proteins Exist Inside Cells?	315
8.3.1	Evolution of Ordered and Disordered Proteins Is Fundamentally Different	315
8.3.1.1	The Evolution of Natively Disordered Proteins	315
8.3.1.2	Adaptive Evolution and Protein Flexibility	317
8.3.1.3	Phylogeny Reconstruction and Protein Structure	318
8.3.2	Direct Measurement by NMR	320
8.4	Functional Repertoire	322
8.4.1	Molecular Recognition	322
8.4.1.1	The Coupling of Folding and Binding	322
8.4.1.2	Structural Plasticity for the Purpose of Functional Plasticity	323
8.4.1.3	Systems Where Disorder Increases Upon Binding	323
8.4.2	Assembly/Disassembly	325
8.4.3	Highly Entropic Chains	325
8.4.4	Protein Modification	327
8.5	Importance of Disorder for Protein Folding	328
8.6	Experimental Protocols	331
8.6.1	NMR Spectroscopy	331
8.6.1.1	General Requirements	331
8.6.1.2	Measuring Transient Secondary Structure in Secondary Chemical Shifts	332
8.6.1.3	Measuring the Translational Diffusion Coefficient Using Pulsed Field Gradient Diffusion Experiments	332
8.6.1.4	Relaxation Experiments	332
8.6.1.5	Relaxation Data Analysis Using Reduced Spectral Density Mapping	333
8.6.1.6	In-cell NMR	334
8.6.2	X-ray Crystallography	334
8.6.3	Circular Dichroism Spectropolarimetry	336

Acknowledgements *337*
References *337*

9   The Catalysis of Disulfide Bond Formation in Prokaryotes *358*
Jean-Francois Collet and James C. Bardwell

9.1   Introduction *358*
9.2   Disulfide Bond Formation in the *E. coli* Periplasm *358*
9.2.1   A Small Bond, a Big Effect *358*
9.2.2   Disulfide Bond Formation Is a Catalyzed Process *359*
9.2.3   DsbA, a Protein-folding Catalyst *359*
9.2.4   How is DsbA Re-oxidized? *361*
9.2.5   From Where Does the Oxidative Power of DsbB Originate? *361*
9.2.6   How Are Disulfide Bonds Transferred From DsbB to DsbA? *362*
9.2.7   How Can DsbB Generate Disulfide by Quinone Reduction? *364*
9.3   Disulfide Bond Isomerization *365*
9.3.1   The Protein Disulfide Isomerases DsbC and DsbG *365*
9.3.2   Dimerization of DsbC and DsbG Is Important for Isomerase and Chaperone Activity *366*
9.3.3   Dimerization Protects from DsbB Oxidation *367*
9.3.4   Import of Electrons from the Cytoplasm: DsbD *367*
9.3.5   Conclusions *369*
9.4   Experimental Protocols *369*
9.4.1   Oxidation-reduction of a Protein Sample *369*
9.4.2   Determination of the Free Thiol Content of a Protein *370*
9.4.3   Separation by HPLC *371*
9.4.4   Tryptophan Fluorescence *372*
9.4.5   Assay of Disulfide Oxidase Activity *372*
References *373*

10   Catalysis of Peptidyl-prolyl *cis/trans* Isomerization by Enzymes *377*
Gunter Fischer

10.1   Introduction *377*
10.2   Peptidyl-prolyl *cis/trans* Isomerization *379*
10.3   Monitoring Peptidyl-prolyl *cis/trans* Isomerase Activity *383*
10.4   Prototypical Peptidyl-prolyl *cis/trans* Isomerases *388*
10.4.1   General Considerations *388*
10.4.2   Prototypic Cyclophilins *390*
10.4.3   Prototypic FK506-binding Proteins *394*
10.4.4   Prototypic Parvulins *397*
10.5   Concluding Remarks *399*
10.6   Experimental Protocols *399*
10.6.1   PPIase Assays: Materials *399*
10.6.2   PPIase Assays: Equipment *400*
10.6.3   Assaying Procedure: Protease-coupled Spectrophotometric Assay *400*

10.6.4	Assaying Procedure: Protease-free Spectrophotometric Assay 401
	References 401

11 Secondary Amide Peptide Bond *cis/trans* Isomerization in Polypeptide Backbone Restructuring: Implications for Catalysis   415
Cordelia Schiene-Fischer and Christian Lücke

11.1	Introduction 415
11.2	Monitoring Secondary Amide Peptide Bond *cis/trans* Isomerization 416
11.3	Kinetics and Thermodynamics of Secondary Amide Peptide Bond *cis/trans* Isomerization 418
11.4	Principles of DnaK Catalysis 420
11.5	Concluding Remarks 423
11.6	Experimental Protocols 424
11.6.1	Stopped-flow Measurements of Peptide Bond *cis/trans* Isomerization 424
11.6.2	Two-dimensional ^{1}H-NMR Exchange Experiments 425
	References 426

12 Ribosome-associated Proteins Acting on Newly Synthesized Polypeptide Chains   429
Sabine Rospert, Matthias Gautschi, Magdalena Rakwalska, and Uta Raue

12.1	Introduction 429
12.2	Signal Recognition Particle, Nascent Polypeptide–associated Complex, and Trigger Factor 432
12.2.1	Signal Recognition Particle 432
12.2.2	An Interplay between Eukaryotic SRP and Nascent Polypeptide–associated Complex? 435
12.2.3	Interplay between Bacterial SRP and Trigger Factor? 435
12.2.4	Functional Redundancy: TF and the Bacterial Hsp70 Homologue DnaK 436
12.3	Chaperones Bound to the Eukaryotic Ribosome: Hsp70 and Hsp40 Systems 436
12.3.1	Sis1p and Ssa1p: an Hsp70/Hsp40 System Involved in Translation Initiation? 437
12.3.2	Ssb1/2p, an Hsp70 Homologue Distributed Between Ribosomes and Cytosol 438
12.3.3	Function of Ssb1/2p in Degradation and Protein Folding 439
12.3.4	Zuotin and Ssz1p: a Stable Chaperone Complex Bound to the Yeast Ribosome 440
12.3.5	A Functional Chaperone Triad Consisting of Ssb1/2p, Ssz1p, and Zuotin 440
12.3.6	Effects of Ribosome-bound Chaperones on the Yeast Prion [PSI^+] 442
12.4	Enzymes Acting on Nascent Polypeptide Chains 443

12.4.1	Methionine Aminopeptidases 443
12.4.2	N^α-acetyltransferases 444
12.5	A Complex Arrangement at the Yeast Ribosomal Tunnel Exit 445
12.6	Experimental Protocols 446
12.6.1	Purification of Ribosome-associated Protein Complexes from Yeast 446
12.6.2	Growth of Yeast and Preparation of Ribosome-associated Proteins by High-salt Treatment of Ribosomes 447
12.6.3	Purification of NAC and RAC 448
	References 449

Part II, Volume 2

13	**The Role of Trigger Factor in Folding of Newly Synthesized Proteins** 459
	Elke Deuerling, Thomas Rauch, Holger Patzelt, and Bernd Bukau
13.1	Introduction 459
13.2	In Vivo Function of Trigger Factor 459
13.2.1	Discovery 459
13.2.2	Trigger Factor Cooperates With the DnaK Chaperone in the Folding of Newly Synthesized Cytosolic Proteins 460
13.2.3	In Vivo Substrates of Trigger Factor and DnaK 461
13.2.4	Substrate Specificity of Trigger Factor 463
13.3	Structure–Function Analysis of Trigger Factor 465
13.3.1	Domain Structure and Conservation 465
13.3.2	Quaternary Structure 468
13.3.3	PPIase and Chaperone Activity of Trigger Factor 469
13.3.4	Importance of Ribosome Association 470
13.4	Models of the Trigger Factor Mechanism 471
13.5	Experimental Protocols 473
13.5.1	Trigger Factor Purification 473
13.5.2	GAPDH Trigger Factor Activity Assay 475
13.5.3	Modular Cell-free *E. coli* Transcription/Translation System 475
13.5.4	Isolation of Ribosomes and Add-back Experiments 483
13.5.5	Cross-linking Techniques 485
	References 485

14	**Cellular Functions of Hsp70 Chaperones** 490
	Elizabeth A. Craig and Peggy Huang
14.1	Introduction 490
14.2	"Soluble" Hsp70s/J-proteins Function in General Protein Folding 492
14.2.1	The Soluble Hsp70 of *E. coli*, DnaK 492
14.2.2	Soluble Hsp70s of Major Eukaryotic Cellular Compartments 493
14.2.2.1	Eukaryotic Cytosol 493
14.2.2.2	Matrix of Mitochondria 494
14.2.2.3	Lumen of the Endoplasmic Reticulum 494

14.3	"Tethered" Hsp70s/J-proteins: Roles in Protein Folding on the Ribosome and in Protein Translocation 495	
14.3.1	Membrane-tethered Hsp70/J-protein 495	
14.3.2	Ribosome-associated Hsp70/J-proteins 496	
14.4	Modulating of Protein Conformation by Hsp70s/J-proteins 498	
14.4.1	Assembly of Fe/S Centers 499	
14.4.2	Uncoating of Clathrin-coated Vesicles 500	
14.4.3	Regulation of the Heat Shock Response 501	
14.4.4	Regulation of Activity of DNA Replication-initiator Proteins 502	
14.5	Cases of a Single Hsp70 Functioning With Multiple J-Proteins 504	
14.6	Hsp70s/J-proteins – When an Hsp70 Maybe Isn't Really a Chaperone 504	
14.6.1	The Ribosome-associated "Hsp70" Ssz1 505	
14.6.2	Mitochondrial Hsp70 as the Regulatory Subunit of an Endonuclease 506	
14.7	Emerging Concepts and Unanswered Questions 507	
	References 507	
15	**Regulation of Hsp70 Chaperones by Co-chaperones** 516	
	Matthias P. Mayer and Bernd Bukau	
15.1	Introduction 516	
15.2	Hsp70 Proteins 517	
15.2.1	Structure and Conservation 517	
15.2.2	ATPase Cycle 519	
15.2.3	Structural Investigations 521	
15.2.4	Interactions With Substrates 522	
15.3	J-domain Protein Family 526	
15.3.1	Structure and Conservation 526	
15.3.2	Interaction With Hsp70s 530	
15.3.3	Interactions with Substrates 532	
15.4	Nucleotide Exchange Factors 534	
15.4.1	GrpE: Structure and Interaction with DnaK 534	
15.4.2	Nucleotide Exchange Reaction 535	
15.4.3	Bag Family: Structure and Interaction With Hsp70 536	
15.4.4	Relevance of Regulated Nucleotide Exchange for Hsp70s 538	
15.5	TPR Motifs Containing Co-chaperones of Hsp70 540	
15.5.1	Hip 541	
15.5.2	Hop 542	
15.5.3	Chip 543	
15.6	Concluding Remarks 544	
15.7	Experimental Protocols 544	
15.7.1	Hsp70s 544	
15.7.2	J-Domain Proteins 545	
15.7.3	GrpE 546	
15.7.4	Bag-1 547	

15.7.5	Hip	548
15.7.6	Hop	549
15.7.7	Chip	549
	References	550

16	**Protein Folding in the Endoplasmic Reticulum Via the Hsp70 Family**	**563**
	Ying Shen, Kyung Tae Chung, and Linda M. Hendershot	
16.1	Introduction	563
16.2	BiP Interactions with Unfolded Proteins	564
16.3	ER-localized DnaJ Homologues	567
16.4	ER-localized Nucleotide-exchange/releasing Factors	571
16.5	Organization and Relative Levels of Chaperones in the ER	572
16.6	Regulation of ER Chaperone Levels	573
16.7	Disposal of BiP-associated Proteins That Fail to Fold or Assemble	575
16.8	Other Roles of BiP in the ER	576
16.9	Concluding Comments	576
16.10	Experimental Protocols	577
16.10.1	Production of Recombinant ER Proteins	577
16.10.1.1	General Concerns	577
16.10.1.2	Bacterial Expression	578
16.10.1.3	Yeast Expression	580
16.10.1.4	Baculovirus	581
16.10.1.5	Mammalian Cells	583
16.10.2	Yeast Two-hybrid Screen for Identifying Interacting Partners of ER Proteins	586
16.10.3	Methods for Determining Subcellular Localization, Topology, and Orientation of Proteins	588
16.10.3.1	Sequence Predictions	588
16.10.3.2	Immunofluorescence Staining	589
16.10.3.3	Subcellular Fractionation	589
16.10.3.4	Determination of Topology	590
16.10.3.5	*N*-linked Glycosylation	592
16.10.4	Nucleotide Binding, Hydrolysis, and Exchange Assays	594
16.10.4.1	Nucleotide-binding Assays	594
16.10.4.2	ATP Hydrolysis Assays	596
16.10.4.3	Nucleotide Exchange Assays	597
16.10.5	Assays for Protein–Protein Interactions in Vitro/in Vivo	599
16.10.5.1	In Vitro GST Pull-down Assay	599
16.10.5.2	Co-immunoprecipitation	600
16.10.5.3	Chemical Cross-linking	600
16.10.5.4	Yeast Two-hybrid System	601
16.10.6	In Vivo Folding, Assembly, and Chaperone-binding Assays	601
16.10.6.1	Monitoring Oxidation of Intrachain Disulfide Bonds	601
16.10.6.2	Detection of Chaperone Binding	602

Acknowledgements 603
References 603

17　Quality Control In Glycoprotein Folding　617
　　E. Sergio Trombetta and Armando J. Parodi
17.1　Introduction　617
17.2　ER N-glycan Processing Reactions　617
17.3　The UDP-Glc:Glycoprotein Glucosyltransferase　619
17.4　Protein Folding in the ER　621
17.5　Unconventional Chaperones (Lectins) Are Present in the ER Lumen　621
17.6　In Vivo Glycoprotein-CNX/CRT Interaction　623
17.7　Effect of CNX/CRT Binding on Glycoprotein Folding and ER Retention　624
17.8　Glycoprotein-CNX/CRT Interaction Is Not Essential for Unicellular Organisms and Cells in Culture　627
17.9　Diversion of Misfolded Glycoproteins to Proteasomal Degradation　629
17.10　Unfolding Irreparably Misfolded Glycoproteins to Facilitate Proteasomal Degradation　632
17.11　Summary and Future Directions　633
17.12　Characterization of N-glycans from Glycoproteins　634
17.12.1　Characterization of N-glycans Present in Immunoprecipitated Samples　634
17.12.2　Analysis of Radio-labeled N-glycans　636
17.12.3　Extraction and Analysis of Protein-bound N-glycans　636
17.12.4　GII and GT Assays　637
17.12.4.1　Assay for GII　637
17.12.4.2　Assay for GT　638
17.12.5　Purification of GII and GT from Rat Liver　639
　　References　641

18　Procollagen Biosynthesis in Mammalian Cells　649
　　Mohammed Tasab and Neil J. Bulleid
18.1　Introduction　649
18.1.1　Variety and Complexity of Collagen Proteins　649
18.1.2　Fibrillar Procollagen　650
18.1.3　Expression of Fibrillar Collagens　650
18.2　The Procollagen Biosynthetic Process: An Overview　651
18.3　Disulfide Bonding in Procollagen Assembly　653
18.4　The Influence of Primary Amino Acid Sequence on Intracellular Procollagen Folding　654
18.4.1　Chain Recognition and Type-specific Assembly　654
18.4.2　Assembly of Multi-subunit Proteins　654
18.4.3　Coordination of Type-specific Procollagen Assembly and Chain Selection　655

18.4.4	Hypervariable Motifs: Components of a Recognition Mechanism That Distinguishes Between Procollagen Chains?	656
18.4.5	Modeling the C-propeptide	657
18.4.6	Chain Association	657
18.5	Posttranslational Modifications That Affect Procollagen Folding	658
18.5.1	Hydroxylation and Triple-helix Stability	658
18.6	Procollagen Chaperones	658
18.6.1	Prolyl 4-Hydroxylase	658
18.6.2	Protein Disulfide Isomerase	659
18.6.3	Hsp47	660
18.6.4	PPI and BiP	661
18.7	Analysis of Procollagen Folding	662
18.8	Experimental Part	663
18.8.1	Materials Required	663
18.8.2	Experimental Protocols	664
	References	668

19   Redox Regulation of Chaperones   677
Jörg H. Hoffmann and Ursula Jakob

19.1	Introduction	677
19.2	Disulfide Bonds as Redox-Switches	677
19.2.1	Functionality of Disulfide Bonds	677
19.2.2	Regulatory Disulfide Bonds as Functional Switches	679
19.2.3	Redox Regulation of Chaperone Activity	680
19.3	Prokaryotic Hsp33: A Chaperone Activated by Oxidation	680
19.3.1	Identification of a Redox-regulated Chaperone	680
19.3.2	Activation Mechanism of Hsp33	681
19.3.3	The Crystal Structure of Active Hsp33	682
19.3.4	The Active Hsp33-Dimer: An Efficient Chaperone Holdase	683
19.3.5	Hsp33 is Part of a Sophisticated Multi-chaperone Network	684
19.4	Eukaryotic Protein Disulfide Isomerase (PDI): Redox Shuffling in the ER	685
19.4.1	PDI, A Multifunctional Enzyme in Eukaryotes	685
19.4.2	PDI and Redox Regulation	687
19.5	Concluding Remarks and Outlook	688
19.6	Appendix – Experimental Protocols	688
19.6.1	How to Work With Redox-regulated Chaperones in Vitro	689
19.6.1.1	Preparation of the Reduced Protein Species	689
19.6.1.2	Preparation of the Oxidized Protein Species	690
19.6.1.3	In Vitro Thiol Trapping to Monitor the Redox State of Proteins	691
19.6.2	Thiol Coordinating Zinc Centers as Redox Switches	691
19.6.2.1	PAR-PMPS Assay to Quantify Zinc	691
19.6.2.2	Determination of Zinc-binding Constants	692
19.6.3	Functional Analysis of Redox-regulated Chaperones in Vitro/in Vivo	693
19.6.3.1	Chaperone Activity Assays	693

19.6.3.2	Manipulating and Analyzing Redox Conditions in Vivo	694
	Acknowledgements 694	
	References 694	

20	**The *E. coli* GroE Chaperone** 699	
	Steven G. Burston and Stefan Walter	
20.1	Introduction 699	
20.2	The Structure of GroEL 699	
20.3	The Structure of GroEL-ATP 700	
20.4	The Structure of GroES and its Interaction with GroEL 701	
20.5	The Interaction Between GroEL and Substrate Polypeptides 702	
20.6	GroEL is a Complex Allosteric Macromolecule 703	
20.7	The Reaction Cycle of the GroE Chaperone 705	
20.8	The Effect of GroE on Protein-folding Pathways 708	
20.9	Future Perspectives 710	
20.10	Experimental Protocols 710	
	Acknowledgments 719	
	References 719	

21	**Structure and Function of the Cytosolic Chaperonin CCT** 725	
	José M. Valpuesta, José L. Carrascosa, and Keith R. Willison	
21.1	Introduction 725	
21.2	Structure and Composition of CCT 726	
21.3	Regulation of CCT Expression 729	
21.4	Functional Cycle of CCT 730	
21.5	Folding Mechanism of CCT 731	
21.6	Substrates of CCT 735	
21.7	Co-chaperones of CCT 739	
21.8	Evolution of CCT 741	
21.9	Concluding Remarks 743	
21.10	Experimental Protocols 743	
21.10.1	Purification 743	
21.10.2	ATP Hydrolysis Measurements 744	
21.10.3	CCT Substrate-binding and Folding Assays 744	
21.10.4	Electron Microscopy and Image Processing 744	
	References 747	

22	**Structure and Function of GimC/Prefoldin** 756	
	Katja Siegers, Andreas Bracher, and Ulrich Hartl	
22.1	Introduction 756	
22.2	Evolutionary Distribution of GimC/Prefoldin 757	
22.3	Structure of the Archaeal GimC/Prefoldin 757	
22.4	Complexity of the Eukaryotic/Archaeal GimC/Prefoldin 759	
22.5	Functional Cooperation of GimC/Prefoldin With the Eukaryotic Chaperonin TRiC/CCT 761	

22.6	Experimental Protocols *764*	
22.6.1	Actin-folding Kinetics *764*	
22.6.2	Prevention of Aggregation (Light-scattering) Assay *765*	
22.6.3	Actin-binding Assay *765*	
	Acknowledgements *766*	
	References *766*	
23	**Hsp90: From Dispensable Heat Shock Protein to Global Player** *768*	
	Klaus Richter, Birgit Meinlschmidt, and Johannes Buchner	
23.1	Introduction *768*	
23.2	The Hsp90 Family in Vivo *768*	
23.2.1	Evolutionary Relationships within the Hsp90 Gene Family *768*	
23.2.2	In Vivo Functions of Hsp90 *769*	
23.2.3	Regulation of Hsp90 Expression and Posttranscriptional Activation *772*	
23.2.4	Chemical Inhibition of Hsp90 *773*	
23.2.5	Identification of Natural Hsp90 Substrates *774*	
23.3	In Vitro Investigation of the Chaperone Hsp90 *775*	
23.3.1	Hsp90: A Special Kind of ATPase *775*	
23.3.2	The ATPase Cycle of Hsp90 *780*	
23.3.3	Interaction of Hsp90 with Model Substrate Proteins *781*	
23.3.4	Investigating Hsp90 Substrate Interactions Using Native Substrates *783*	
23.4	Partner Proteins: Does Complexity Lead to Specificity? *784*	
23.4.1	Hop, p23, and PPIases: The Chaperone Cycle of Hsp90 *784*	
23.4.2	Hop/Sti1: Interactions Mediated by TPR Domains *787*	
23.4.3	p23/Sba1: Nucleotide-specific Interaction with Hsp90 *789*	
23.4.4	Large PPIases: Conferring Specificity to Substrate Localization? *790*	
23.4.5	Pp5: Facilitating Dephosphorylation *791*	
23.4.6	Cdc37: Building Complexes with Kinases *792*	
23.4.7	Tom70: Chaperoning Mitochondrial Import *793*	
23.4.8	CHIP and Sgt1: Multiple Connections to Protein Degradation *793*	
23.4.9	Aha1 and Hch1: Just Stimulating the ATPase? *794*	
23.4.10	Cns1, Sgt2, and Xap2: Is a TPR Enough to Become an Hsp90 Partner? *796*	
23.5	Outlook *796*	
23.6	Appendix – Experimental Protocols *797*	
23.6.1	Calculation of Phylogenetic Trees Based on Protein Sequences *797*	
23.6.2	Investigating the in Vivo Effect of Hsp90 Mutations in *S. cerevisiae* *797*	
23.6.3	Well-characterized Hsp90 Mutants *798*	
23.6.4	Investigating Activation of Heterologously Expressed Src Kinase in *S. cerevisiae* *800*	
23.6.5	Investigation of Heterologously Expressed Glucocorticoid Receptor in *S. cerevisiae* *800*	

23.6.6	Investigation of Chaperone Activity	801
23.6.7	Analysis of the ATPase Activity of Hsp90	802
23.6.8	Detecting Specific Influences on Hsp90 ATPase Activity	803
23.6.9	Investigation of the Quaternary Structure by SEC-HPLC	804
23.6.10	Investigation of Binding Events Using Changes of the Intrinsic Fluorescence	806
23.6.11	Investigation of Binding Events Using Isothermal Titration Calorimetry	807
23.6.12	Investigation of Protein-Protein Interactions Using Cross-linking	807
23.6.13	Investigation of Protein-Protein Interactions Using Surface Plasmon Resonance Spectroscopy	808
	Acknowledgements	810
	References	810

24 Small Heat Shock Proteins: Dynamic Players in the Folding Game 830
Franz Narberhaus and Martin Haslbeck

24.1	Introduction	830
24.2	α-Crystallins and the Small Heat Shock Protein Family: Diverse Yet Similar	830
24.3	Cellular Functions of α-Hsps	831
24.3.1	Chaperone Activity in Vitro	831
24.3.2	Chaperone Function in Vivo	835
24.3.3	Other Functions	836
24.4	The Oligomeric Structure of α-Hsps	837
24.5	Dynamic Structures as Key to Chaperone Activity	839
24.6	Experimental Protocols	840
24.6.1	Purification of sHsps	840
24.6.2	Chaperone Assays	843
24.6.3	Monitoring Dynamics of sHsps	846
	Acknowledgements	847
	References	848

25 Alpha-crystallin: Its Involvement in Suppression of Protein Aggregation and Protein Folding 858
Joseph Horwitz

25.1	Introduction	858
25.2	Distribution of Alpha-crystallin in the Various Tissues	858
25.3	Structure	859
25.4	Phosphorylation and Other Posttranslation Modification	860
25.5	Binding of Target Proteins to Alpha-crystallin	861
25.6	The Function of Alpha-crystallin	863
25.7	Experimental Protocols	863
25.7.1	Preparation of Alpha-crystallin	863
	Acknowledgements	870
	References	870

26	**Transmembrane Domains in Membrane Protein Folding, Oligomerization, and Function** *876*	

Anja Ridder and Dieter Langosch

26.1 Introduction *876*
26.1.1 Structure of Transmembrane Domains *876*
26.1.2 The Biosynthetic Route towards Folded and Oligomeric Integral Membrane Proteins *877*
26.1.3 Structure and Stability of TMSs *878*
26.1.3.1 Amino Acid Composition of TMSs and Flanking Regions *878*
26.1.3.2 Stability of Transmembrane Helices *879*
26.2 The Nature of Transmembrane Helix-Helix Interactions *880*
26.2.1 General Considerations *880*
26.2.1.1 Attractive Forces within Lipid Bilayers *880*
26.2.1.2 Forces between Transmembrane Helices *881*
26.2.1.3 Entropic Factors Influencing Transmembrane Helix–Helix Interactions *882*
26.2.2 Lessons from Sequence Analyses and High-resolution Structures *883*
26.2.3 Lessons from Bitopic Membrane Proteins *886*
26.2.3.1 Transmembrane Segments Forming Right-handed Pairs *886*
26.2.3.2 Transmembrane Segments Forming Left-handed Assemblies *889*
26.2.4 Selection of Self-interacting TMSs from Combinatorial Libraries *892*
26.2.5 Role of Lipids in Packing/Assembly of Membrane Proteins *893*
26.3 Conformational Flexibility of Transmembrane Segments *895*
26.4 Experimental Protocols *897*
26.4.1 Biochemical and Biophysical Techniques *897*
26.4.1.1 Visualization of Oligomeric States by Electrophoretic Techniques *898*
26.4.1.2 Hydrodynamic Methods *899*
26.4.1.3 Fluorescence Resonance Transfer *900*
26.4.2 Genetic Assays *901*
26.4.2.1 The ToxR System *901*
26.4.2.2 Other Genetic Assays *902*
26.4.3 Identification of TMS-TMS Interfaces by Mutational Analysis *903*
References *904*

Part II, Volume 3

27	**SecB** *919*	

Arnold J. M. Driessen, Janny de Wit, and Nico Nouwen

27.1 Introduction *919*
27.2 Selective Binding of Preproteins by SecB *920*
27.3 SecA-SecB Interaction *925*
27.4 Preprotein Transfer from SecB to SecA *928*
27.5 Concluding Remarks *929*
27.6 Experimental Protocols *930*
27.6.1 How to Analyze SecB-Preprotein Interactions *930*

27.6.2	How to Analyze SecB-SecA Interaction *931*
	Acknowledgements *932*
	References *933*

28 Protein Folding in the Periplasm and Outer Membrane of *E. coli* *938*
Michael Ehrmann

28.1	Introduction *938*
28.2	Individual Cellular Factors *940*
28.2.1	The Proline Isomerases FkpA, PpiA, SurA, and PpiD *941*
28.2.1.1	FkpA *942*
28.2.1.2	PpiA *942*
28.2.1.3	SurA *943*
28.2.1.4	PpiD *943*
28.2.2	Skp *944*
28.2.3	Proteases and Protease/Chaperone Machines *945*
28.2.3.1	The HtrA Family of Serine Proteases *946*
28.2.3.2	*E. coli* HtrAs *946*
28.2.3.3	DegP and DegQ *946*
28.2.3.4	DegS *947*
28.2.3.5	The Structure of HtrA *947*
28.2.3.6	Other Proteases *948*
28.3	Organization of Folding Factors into Pathways and Networks *950*
28.3.1	Synthetic Lethality and Extragenic High-copy Suppressors *950*
28.3.2	Reconstituted in Vitro Systems *951*
28.4	Regulation *951*
28.4.1	The Sigma E Pathway *951*
28.4.2	The Cpx Pathway *952*
28.4.3	The Bae Pathway *953*
28.5	Future Perspectives *953*
28.6	Experimental Protocols *954*
28.6.1	Pulse Chase Immunoprecipitation *954*
	Acknowledgements *957*
	References *957*

29 Formation of Adhesive Pili by the Chaperone-Usher Pathway *965*
Michael Vetsch and Rudi Glockshuber

29.1	Basic Properties of Bacterial, Adhesive Surface Organelles *965*
29.2	Structure and Function of Pilus Chaperones *970*
29.3	Structure and Folding of Pilus Subunits *971*
29.4	Structure and Function of Pilus Ushers *973*
29.5	Conclusions and Outlook *976*
29.6	Experimental Protocols *977*
29.6.1	Test for the Presence of Type 1 Piliated *E. coli* Cells *977*
29.6.2	Functional Expression of Pilus Subunits in the *E. coli* Periplasm *977*
29.6.3	Purification of Pilus Subunits from the *E. coli* Periplasm *978*

29.6.4	Preparation of Ushers 979
	Acknowledgements 979
	References 980

30	**Unfolding of Proteins During Import into Mitochondria** 987
	Walter Neupert, Michael Brunner, and Kai Hell
30.1	Introduction 987
30.2	Translocation Machineries and Pathways of the Mitochondrial Protein Import System 988
30.2.1	Import of Proteins Destined for the Mitochondrial Matrix 990
30.3	Import into Mitochondria Requires Protein Unfolding 993
30.4	Mechanisms of Unfolding by the Mitochondrial Import Motor 995
30.4.1	Targeted Brownian Ratchet 995
30.4.2	Power-stroke Model 995
30.5	Studies to Discriminate between the Models 996
30.5.1	Studies on the Unfolding of Preproteins 996
30.5.1.1	Comparison of the Import of Folded and Unfolded Proteins 996
30.5.1.2	Import of Preproteins With Different Presequence Lengths 999
30.5.1.3	Import of Titin Domains 1000
30.5.1.4	Unfolding by the Mitochondrial Membrane Potential $\Delta\Psi$ 1000
30.5.2	Mechanistic Studies of the Import Motor 1000
30.5.2.1	Brownian Movement of the Polypeptide Within the Import Channel 1000
30.5.2.2	Recruitment of mtHsp70 by Tim44 1001
30.5.2.3	Import Without Recruitment of mtHsp70 by Tim44 1002
30.5.2.4	MtHsp70 Function in the Import Motor 1003
30.6	Discussion and Perspectives 1004
30.7	Experimental Protocols 1006
30.7.1	Protein Import Into Mitochondria in Vitro 1006
30.7.2	Stabilization of the DHFR Domain by Methotrexate 1008
30.7.3	Import of Precursor Proteins Unfolded With Urea 1009
30.7.4	Kinetic Analysis of the Unfolding Reaction by Trapping of Intermediates 1009
	References 1011

31	**The Chaperone System of Mitochondria** 1020
	Wolfgang Voos and Nikolaus Pfanner
31.1	Introduction 1020
31.2	Membrane Translocation and the Hsp70 Import Motor 1020
31.3	Folding of Newly Imported Proteins Catalyzed by the Hsp70 and Hsp60 Systems 1026
31.4	Mitochondrial Protein Synthesis and the Assembly Problem 1030
31.5	Aggregation versus Degradation: Chaperone Functions Under Stress Conditions 1033
31.6	Experimental Protocols 1034

31.6.1	Chaperone Functions Characterized With Yeast Mutants	1034
31.6.2	Interaction of Imported Proteins With Matrix Chaperones	1036
31.6.3	Folding of Imported Model Proteins	1037
31.6.4	Assaying Mitochondrial Degradation of Imported Proteins	1038
31.6.5	Aggregation of Proteins in the Mitochondrial Matrix	1038
	References	1039

32 Chaperone Systems in Chloroplasts 1047
Thomas Becker, Jürgen Soll, and Enrico Schleiff

32.1	Introduction	1047
32.2	Chaperone Systems within Chloroplasts	1048
32.2.1	The Hsp70 System of Chloroplasts	1048
32.2.1.1	The Chloroplast Hsp70s	1049
32.2.1.2	The Co-chaperones of Chloroplastic Hsp70s	1051
32.2.2	The Chaperonins	1052
32.2.3	The HSP100/Clp Protein Family in Chloroplasts	1056
32.2.4	The Small Heat Shock Proteins	1058
32.2.5	Hsp90 Proteins of Chloroplasts	1061
32.2.6	Chaperone-like Proteins	1062
32.2.6.1	The Protein Disulfide Isomerase (PDI)	1062
32.2.6.2	The Peptidyl-prolyl *cis* Isomerase (PPIase)	1063
32.3	The Functional Chaperone Pathways in Chloroplasts	1065
32.3.1	Chaperones Involved in Protein Translocation	1065
32.3.2	Protein Transport Inside of Plastids	1070
32.3.3	Protein Folding and Complex Assembly Within Chloroplasts	1071
32.3.4	Chloroplast Chaperones Involved in Proteolysis	1072
32.3.5	Protein Storage Within Plastids	1073
32.3.6	Protein Protection and Repair	1074
32.4	Experimental Protocols	1075
32.4.1	Characterization of Cpn60 Binding to the Large Subunit of Rubisco via Native PAGE (adopted from Ref. [6])	1075
32.4.2	Purification of Chloroplast Cpn60 From Young Pea Plants (adopted from Ref. [203])	1076
32.4.3	Purification of Chloroplast Hsp21 From Pea (*Pisum sativum*) (adopted from [90])	1077
32.4.4	Light-scattering Assays for Determination of the Chaperone Activity Using Citrate Synthase as Substrate (adopted from [196])	1078
32.4.5	The Use Of *Bis*-ANS to Assess Surface Exposure of Hydrophobic Domains of Hsp17 of *Synechocystis* (adopted from [202])	1079
32.4.6	Determination of Hsp17 Binding to Lipids (adopted from Refs. [204, 205])	1079
	References	1081

33 An Overview of Protein Misfolding Diseases 1093
Christopher M. Dobson

33.1	Introduction	1093

33.2	Protein Misfolding and Its Consequences for Disease 1094
33.3	The Structure and Mechanism of Amyloid Formation 1097
33.4	A Generic Description of Amyloid Formation 1101
33.5	The Fundamental Origins of Amyloid Disease 1104
33.6	Approaches to Therapeutic Intervention in Amyloid Disease 1106
33.7	Concluding Remarks 1108
	Acknowledgements 1108
	References 1109

34 Biochemistry and Structural Biology of Mammalian Prion Disease 1114
Rudi Glockshuber

34.1	Introduction 1114
34.1.1	Prions and the "Protein-Only" Hypothesis 1114
34.1.2	Models of PrP^{Sc} Propagation 1115
34.2	Properties of PrP^{C} and PrP^{Sc} 1117
34.3	Three-dimensional Structure and Folding of Recombinant PrP 1120
34.3.1	Expression of the Recombinant Prion Protein for Structural and Biophysical Studies 1120
34.3.2	Three-dimensional Structures of Recombinant Prion Proteins from Different Species and Their Implications for the Species Barrier of Prion Transmission 1120
34.3.2.1	Solution Structure of Murine PrP 1120
34.3.2.2	Comparison of Mammalian Prion Protein Structures and the Species Barrier of Prion Transmission 1124
34.3.3	Biophysical Characterization of the Recombinant Prion Protein 1125
34.3.3.1	Folding and Stability of Recombinant PrP 1125
34.3.3.2	Role of the Disulfide Bond in PrP 1127
34.3.3.3	Influence of Point Mutations Linked With Inherited TSEs on the Stability of Recombinant PrP 1129
34.4	Generation of Infectious Prions in Vitro: Principal Difficulties in Proving the Protein-Only Hypothesis 1131
34.5	Understanding the Strain Phenomenon in the Context of the Protein-Only Hypothesis: Are Prions Crystals? 1132
34.6	Conclusions and Outlook 1135
34.7	Experimental Protocols 1136
34.7.1	Protocol 1 [53, 55] 1136
34.7.2	Protocol 2 [54] 1137
	References 1138

35 Insights into the Nature of Yeast Prions 1144
Lev Z. Osherovich and Jonathan S. Weissman

35.1	Introduction 1144
35.2	Prions as Heritable Amyloidoses 1145
35.3	Prion Strains and Species Barriers: Universal Features of Amyloid-based Prion Elements 1149

35.4	Prediction and Identification of Novel Prion Elements	*1151*
35.5	Requirements for Prion Inheritance beyond Amyloid-mediated Growth	*1154*
35.6	Chaperones and Prion Replication	*1157*
35.7	The Structure of Prion Particles	*1158*
35.8	Prion-like Structures as Protein Interaction Modules	*1159*
35.9	Experimental Protocols	*1160*
35.9.1	Generation of Sup35 Amyloid Fibers in Vitro	*1160*
35.9.2	Thioflavin T–based Amyloid Seeding Efficacy Assay (Adapted from Chien et al. 2003)	*1161*
35.9.3	AFM-based Single-fiber Growth Assay	*1162*
35.9.4	Prion Infection Protocol (Adapted from Tanaka et al. 2004)	*1164*
35.9.5	Preparation of Lyticase	*1165*
35.9.6	Protocol for Counting Heritable Prion Units (Adapted from Cox et al. 2003)	*1166*
	Acknowledgements *1167*	
	References *1168*	
36	**Polyglutamine Aggregates as a Model for Protein-misfolding Diseases** *1175*	
	Soojin Kim, James F. Morley, Anat Ben-Zvi, and Richard I. Morimoto	
36.1	Introduction *1175*	
36.2	Polyglutamine Diseases *1175*	
36.2.1	Genetics *1175*	
36.2.2	Polyglutamine Diseases Involve a Toxic Gain of Function	*1176*
36.3	Polyglutamine Aggregates *1176*	
36.3.1	Presence of the Expanded Polyglutamine Is Sufficient to Induce Aggregation in Vivo	*1176*
36.3.2	Length of the Polyglutamine Dictates the Rate of Aggregate Formation	*1177*
36.3.3	Polyglutamine Aggregates Exhibit Features Characteristic of Amyloids	*1179*
36.3.4	Characterization of Protein Aggregates in Vivo Using Dynamic Imaging Methods	*1180*
36.4	A Role for Oligomeric Intermediates in Toxicity	*1181*
36.5	Consequences of Misfolded Proteins and Aggregates on Protein Homeostasis	*1181*
36.6	Modulators of Polyglutamine Aggregation and Toxicity	*1184*
36.6.1	Protein Context *1184*	
36.6.2	Molecular Chaperones *1185*	
36.6.3	Proteasomes *1188*	
36.6.4	The Protein-folding "Buffer" and Aging	*1188*
36.6.5	Summary *1189*	
36.7	Experimental Protocols *1190*	
36.7.1	FRAP Analysis *1190*	
	References *1192*	

37	**Protein Folding and Aggregation in the Expanded Polyglutamine Repeat Diseases** *1200*	
	Ronald Wetzel	
37.1	Introduction *1200*	
37.2	Key Features of the Polyglutamine Diseases *1201*	
37.2.1	The Variety of Expanded PolyGln Diseases *1201*	
37.2.2	Clinical Features *1201*	
37.2.2.1	Repeat Expansions and Repeat Length *1202*	
37.2.3	The Role of PolyGln and PolyGln Aggregates *1203*	
37.3	PolyGln Peptides in Studies of the Molecular Basis of Expanded Polyglutamine Diseases *1205*	
37.3.1	Conformational Studies *1205*	
37.3.2	Preliminary in Vitro Aggregation Studies *1206*	
37.3.3	In Vivo Aggregation Studies *1206*	
37.4	Analyzing Polyglutamine Behavior With Synthetic Peptides: Practical Aspects *1207*	
37.4.1	Disaggregation of Synthetic Polyglutamine Peptides *1209*	
37.4.2	Growing and Manipulating Aggregates *1210*	
37.4.2.1	Polyglutamine Aggregation by Freeze Concentration *1210*	
37.4.2.2	Preparing Small Aggregates *1211*	
37.5	In vitro Studies of PolyGln Aggregation *1212*	
37.5.1	The Universe of Protein Aggregation Mechanisms *1212*	
37.5.2	Basic Studies on Spontaneous Aggregation *1213*	
37.5.3	Nucleation Kinetics of PolyGln *1215*	
37.5.4	Elongation Kinetics *1218*	
37.5.4.1	Microtiter Plate Assay for Elongation Kinetics *1219*	
37.5.4.2	Repeat-length and Aggregate-size Dependence of Elongation Rates *1220*	
37.6	The Structure of PolyGln Aggregates *1221*	
37.6.1	Electron Microscopy Analysis *1222*	
37.6.2	Analysis with Amyloid Dyes Thioflavin T and Congo Red *1222*	
37.6.3	Circular Dichroism Analysis *1224*	
37.6.4	Presence of a Generic Amyloid Epitope in PolyGln Aggregates *1225*	
37.6.5	Proline Mutagenesis to Dissect the Polyglutamine Fold Within the Aggregate *1225*	
37.7	Polyglutamine Aggregates and Cytotoxicity *1227*	
37.7.1	Direct Cytotoxicity of PolyGln Aggregates *1228*	
37.7.1.1	Delivery of Aggregates into Cells and Cellular Compartments *1229*	
37.7.1.2	Cell Killing by Nuclear-targeted PolyGln Aggregates *1229*	
37.7.2	Visualization of Functional, Recruitment-positive Aggregation Foci *1230*	
37.8	Inhibitors of polyGln Aggregation *1231*	
37.8.1	Designed Peptide Inhibitors *1231*	
37.8.2	Screening for Inhibitors of PolyGln Elongation *1231*	
37.9	Concluding Remarks *1232*	
37.10	Experimental Protocols *1233*	

37.10.1	Disaggregation of Synthetic PolyGln Peptides	*1233*
37.10.2	Determining the Concentration of Low-molecular-weight PolyGln Peptides by HPLC	*1235*
	Acknowledgements	*1237*
	References	*1238*

38 **Production of Recombinant Proteins for Therapy, Diagnostics, and Industrial Research by in Vitro Folding** *1245*
Christian Lange and Rainer Rudolph

38.1	Introduction	*1245*
38.1.1	The Inclusion Body Problem	*1245*
38.1.2	Cost and Scale Limitations in Industrial Protein Folding	*1248*
38.2	Treatment of Inclusion Bodies	*1250*
38.2.1	Isolation of Inclusion Bodies	*1250*
38.2.2	Solubilization of Inclusion Bodies	*1250*
38.3	Refolding in Solution	*1252*
38.3.1	Protein Design Considerations	*1252*
38.3.2	Oxidative Refolding With Disulfide Bond Formation	*1253*
38.3.3	Transfer of the Unfolded Proteins Into Refolding Buffer	*1255*
38.3.4	Refolding Additives	*1257*
38.3.5	Cofactors in Protein Folding	*1260*
38.3.6	Chaperones and Folding-helper Proteins	*1261*
38.3.7	An Artificial Chaperone System	*1261*
38.3.8	Pressure-induced Folding	*1262*
38.3.9	Temperature-leap Techniques	*1263*
38.3.10	Recycling of Aggregates	*1264*
38.4	Alternative Refolding Techniques	*1264*
38.4.1	Matrix-assisted Refolding	*1264*
38.4.2	Folding by Gel Filtration	*1266*
38.4.3	Direct Refolding of Inclusion Body Material	*1267*
38.5	Conclusions	*1268*
38.6	Experimental Protocols	*1268*
38.6.1	Protocol 1: Isolation of Inclusion Bodies	*1268*
38.6.2	Protocol 2: Solubilization of Inclusion Bodies	*1269*
38.6.3	Protocol 3: Refolding of Proteins	*1270*
	Acknowledgements	*1271*
	References	*1271*

39 **Engineering Proteins for Stability and Efficient Folding** *1281*
Bernhard Schimmele and Andreas Plückthun

39.1	Introduction	*1281*
39.2	Kinetic and Thermodynamic Aspects of Natural Proteins	*1281*
39.2.1	The Stability of Natural Proteins	*1281*
39.2.2	Different Kinds of "Stability"	*1282*
39.2.2.1	Thermodynamic Stability	*1283*

39.2.2.2	Kinetic Stability	*1285*
39.2.2.3	Folding Efficiency	*1287*
39.3	The Engineering Approach	*1288*
39.3.1	Consensus Strategies	*1288*
39.3.1.1	Principles	*1288*
39.3.1.2	Examples	*1291*
39.3.2	Structure-based Engineering	*1292*
39.3.2.1	Entropic Stabilization	*1294*
39.3.2.2	Hydrophobic Core Packing	*1296*
39.3.2.3	Charge Interactions	*1297*
39.3.2.4	Hydrogen Bonding	*1298*
39.3.2.5	Disallowed Phi-Psi Angles	*1298*
39.3.2.6	Local Secondary Structure Propensities	*1299*
39.3.2.7	Exposed Hydrophobic Side Chains	*1299*
39.3.2.8	Inter-domain Interactions	*1300*
39.3.3	Case Study: Combining Consensus Design and Rational Engineering to Yield Antibodies with Favorable Biophysical Properties	*1300*
39.4	The Selection and Evolution Approach	*1305*
39.4.1	Principles	*1305*
39.4.2	Screening and Selection Technologies Available for Improving Biophysical Properties	*1311*
39.4.2.1	In Vitro Display Technologies	*1313*
39.4.2.2	Partial in Vitro Display Technologies	*1314*
39.4.2.3	In Vivo Selection Technologies	*1315*
39.4.3	Selection for Enhanced Biophysical Properties	*1316*
39.4.3.1	Selection for Solubility	*1316*
39.4.3.2	Selection for Protein Display Rates	*1317*
39.4.3.3	Selection on the Basis of Cellular Quality Control	*1318*
39.4.4	Selection for Increased Stability	*1319*
39.4.4.1	General Strategies	*1319*
39.4.4.2	Protein Destabilization	*1319*
39.4.4.3	Selections Based on Elevated Temperature	*1321*
39.4.4.4	Selections Based on Destabilizing Agents	*1322*
39.4.4.5	Selection for Proteolytic Stability	*1323*
39.5	Conclusions and Perspectives	*1324*
	Acknowledgements	*1326*
	References	*1326*

Index *1334*

Preface

During protein folding, the linear information encoded in the amino acid sequence of a polypeptide chain is transformed into a defined three-dimensional structure. Although it was established decades ago that protein folding is a spontaneous reaction, a complete understanding of this fundamental process is still lacking. Since its humble beginnings in the fields of biophysical chemistry and biochemistry, the analysis of protein folding and stability has spread into a wide range of disciplines including physics, cell biology, medicine and biotechnology. The recent discovery of the sophisticated cellular machinery of molecular chaperones and folding catalysts which assists protein structure formation in vivo and the analysis of protein folding diseases have added additional twists to the study of protein folding. Consequently, the field has undergone a significant transformation over the last decade.

The increasing interest in protein folding considerably improved our knowledge of the physical principles underlying protein stability and folding and stimulated the development of powerful new techniques for their analysis both in vitro and in vivo. Aspects that had been studied previously in splendid isolation such as basic principles and protein folding diseases are now growing together, experimentally and conceptually.

Over the past years, a number of excellent books on protein folding were published, starting with the famous 'red bible of protein folding' (Jaenicke, 1980). The *Protein Folding Handbook* takes a different approach as it combines in-depth reviews with detailed experimental protocols. Thus, a comprehensive, multi-faceted view of the entire field of protein folding from basic physical principles to molecular chaperones, protein folding diseases and the biotechnology of protein folding – is presented. The *Protein Folding Handbook* intends to provide those who become interested in a specific aspect of protein folding with both the scientific background and the experimental approaches. Ideally, the protocols of the methods sections will allow and stimulate even the novice to boldly enter new experimental territory.

We are grateful to the authors who embarked with us on this endeavour. It was a pleasure to see how, chapter by chapter, the current state of the art in this field emerged and the common themes became apparent.

Finally, we would like to thank Dr. Frank Weinreich from Wiley-VCH for his never-waning efforts to transform our ideas into a book, Prof. Erich Gohl for the cover artwork, Brigitte Heiz-Wyss and Susanne Hilber for excellent assistance and Annett Bachmann and Stefan Walter for help with creating the index.

München and Basel
October 2004

Johannes Buchner
Thomas Kiefhaber

Reference

Jaenicke, R. (1980) Protein Folding. Elsevier, Amsterdam, New York

Contributors of Part I

Hans-Peter Bächinger
Shriners Hospital for Children &
Department of Biochemistry and Molecular
Biology
Oregon Health & Science University
3101 SW Sam Jackson Park Road
Portland, OR 97239-3009
USA

Annett Bachmann
Division of Biophysical Chemistry
Biozentrum, University of Basel
Klingelbergstrasse 70
4056 Basel
Switzerland

Jochen Balbach
Biochemie III
Universität Bayreuth
Universitätsstr. 30
95440 Bayreuth
Germany

Robert L. Baldwin
Department of Biochemistry
Stanford University School of Medicine
279 Campus Drive West
Stanford, CA 94305
USA

Hans-Rudolf Bosshard
Department of Biochemistry
University of Zurich
Winterthurerstrasse 190
8057 Zurich
Switzerland

Sonja Braun-Sand
Department of Chemistry
University of Southern California
3620 S. McClintock Ave.
Los Angeles, CA 90089-1062
USA

Philip N. Bryan
Center for Advanced Research in
Biotechnology
University of Maryland Biotechnology
Institute
9600 Gudelsky Drive
Rockville, MD 20850
USA

Amedeo Caflisch
Department of Biochemistry
University of Zürich
Winterthurerstrasse 190
8057 Zürich
Switzerland

Heiko Carstens
Department für Physik
Ludwig-Maximilians-Universität München
Oettingenstrasse 67
80538 München
Germany

Hong Cheng
Basic Science Division
Fox Chase Cancer Center
333 Cottman Avenue
Philadelphia, PA 19111
USA

Jane Clarke
University Chemical Laboratories
MRC Centre for Protein Engineering
Lensfield Road
Cambridge CB2 1EW
United Kingdom

Monika Davidovic
Department of Biophysical Chemistry
Lund University
22100 Lund
Sweden

Vladimir P. Denisov
Department of Biophysical Chemistry
Lund University
22100 Lund
Sweden

Andrew J. Doig
Department of Biomolecular Sciences
UMIST
PO Box 88
Manchester M60 1QD
United Kingdom

Wolfgang Doster
Department of Physics E13
Technische Universität München
85748 Garching
Germany

Jürgen Engel
Division of Biophysical Chemistry
Biozentrum, University of Basel
Klingelbergstrasse 70
4056 Basel
Switzerland

Neil Errington
Department of Biomolecular Sciences
UMIST
PO Box 88
Manchester M60 1QD
United Kingdom

Alan R. Fersht
Department of Chemistry
University of Cambridge
Lensfield Road
Cambridge CB2 1EW
United Kingdom

Beat Fierz
Division of Biophysical Chemistry
Biozentrum, University of Basel
Klingelbergstrasse 70
4056 Basel
Switzerland

Patrick J. Fleming
Jenkins Department of Biophysics
Johns Hopkins University
3400 N. Charles Street
Baltimore, MD 21218
USA

Josef Friedrich
Department of Physics E14
Technische Universität München
85350 Freising
Germany

Klaus Gast
Institut für Biochemie und Biologie
Universität Potsdam
Karl-Liebknecht-Str. 24–25
14476 Golm
Germany

Yuji Goto
Institute for Protein Research
Osaka University
3-2 Yamadaoka, Suita
Osaka 565-0871
Japan

Gerald R. Grimsley
Dept. of Medical Biochemistry & Genetics
Texas A&M University
College Station, TX 77843-1113
USA

Martin Gruebele
School of Chemistry
University of Illinois, Urbana-Champaign
600 S Mathews Avenue Box 5–6
Urbana, Illinois 61801
USA

Raphael Guerois
DBJC/SBFM
CEA Saclay
91191 Gif-sur-Yvette
France

Nicholas R. Guydosh
Department of Biological Sciences
Stanford University
Stanford, CA 94305-5020
USA

Elisha Haas
Faculty of Life Sciences4
Bar Ilan University
Ramat Gan 52900
Israel

Bertil Halle
Department of Biophysical Chemistry
Lund University
22100 Lund
Sweden

Daizo Hamada
Osaka Medical Center for Maternal and Child Health
840 Murodo, Izumi
Osaka 594-1011
Japan

Heedeok Hong
Department of Molecular Physiology and
Biological Physics
University of Virginia
Charlottesville, VA 22908
USA

Teuku M. Iqbalsyah
Department of Biomolecular Sciences
UMIST
PO Box 88
Manchester M60 1QD
United Kingdom

Kiyoto Kamagata
Department of Physics, Graduate School of
Science
University of Tokyo
7-3-1 Hongo, Bunkyo-ku
Tokyo 113-0033
Japan

Thomas Kiefhaber
Division of Biophysical Chemistry
Biozentrum, University of Basel
Klingelbergstrasse 70
4056 Basel
Switzerland

Birthe B. Kragelund
Institute of Molecular Biology
University of Copenhagen
Øster Farimagsgade 2 A
1353 Copenhagen
Denmark

Kunihiro Kuwajima
Department of Physics, Graduate School of
Science
University of Tokyo
7-3-1 Hongo, Bunkyo-ku
Tokyo 113-0033
Japan

Ramil F. Latypov
Amgen Inc.
Thousand Oaks, CA 91320
USA

George I. Makhatadze
Department of Biochemistry and Molecular
Biology
Penn State University College of Medicine
500 University Drive
Hershey, PA 17033-0850
USA

Kosuke Maki
Department of Physics, Graduate School of
Science
University of Tokyo
7-3-1 Hongo, Bunkyo-ku
Tokyo 113-0033
Japan

Gennaro Marino
Facoltà die Scienze Biotecnologiche
Università degli Studi di Napoli "Federico II"
Via S. Pansini, 5
80131 Napoli
Italy

Gerald Mathias
Department für Physik
Ludwig-Maximilians-Universität München
Oettingenstr. 67
80538 München
Germany

Joaquim Mendes
EMBL Heidelberg
Meyerhofstrasse 1
69117 Heidelberg
Germany

Kristofer Modig
Institute of Molecular Biology
University of Copenhagen
Øster Farimagsgade 2A
1353 Copenhagen
Denmark

Andreas J. Modler
Institut für Biologie und Biochemie
Universität Potsdam
Karl-Liebknecht-Str. 24-25
14476 Golm
Germany

C. Nick Pace
Dept. of Medical Biochemistry & Genetics
Texas A&M University
College Station, TX 77843-1113
USA

Emanuele Paci
Department of Biochemistry
University of Zürich
Winterthurerstrasse 190
8057 Zürich
Switzerland

Flemming M. Poulsen
Institute of Molecular Biology
Department of Protein Chemistry
University of Copenhagen
Øster Farimagsgade 2 A
1353 Copenhagen
Denmark

Piero Pucci
Dipartimento di Chimica Organica e Biochimica
Università degli Studi di Napoli "Federico II"
Via S. Pansini, 5
80131 Napoli
Italy

Heinrich Roder
Basic Science Division
Fox Chase Cancer Center
333 Cottman Avenue
Philadelphia, PA 19111
USA

George D. Rose
Jenkins Department of Biophysics
Johns Hopkins University
3400 N. Charles Street
Baltimore, MD 21218
USA

Margherita Ruoppolo
Dipartimento di Biochimica e Biotecnologie Mediche
Università degli Studi di Napoli "Federico II"
Via S. Pansini, 5
80131 Napoli
Italy

Ignacio E. Sánchez
Division of Biophysical Chemistry
Biozentrum, University of Basel
Klingelbergstrasse 70
4056 Basel
Switzerland

Christian Schlörb
Institute for Organic Chemistry and Chemical Biology
Center for Biomolecular Magnetic Resonance
Johann Wolfgang Goethe-University
Marie-Curie-Strasse 11
60439 Frankfurt
Germany

Franz Schmid
Biochemisches Laboratorium
Universität Bayreuth
95440 Bayreuth
Germany

J. Martin Scholtz
Dept. of Medical Biochemistry & Genetics
Texas A&M University
College Station, TX 77843-1113
USA

Harald Schwalbe
Institute for Organic Chemistry and Chemical Biology
Center for Biomolecular Magnetic Resonance
Johann Wolfgang Goethe-University
Marie-Curie-Strasse 11
60439 Frankfurt
Germany

Mark S. Searle
School of Chemistry
University of Nottingham
University Park
Nottingham NG7 2RD
United Kingdom

Luis Serrano
EMBL Heidelberg
Meyerhofstrasse 1
69117 Heidelberg
Germany

M. C. Ramachandra Shastry
Basic Science Division
Fox Chase Cancer Center
333 Cottman Avenue
Philadelphia, PA 19111
USA

Lukas K. Tamm
Department of Molecular Physiology and Biological Physics
University of Virginia
Charlottesville, VA 22908
USA

Paul Tavan
Department of Physik
Ludwig-Maximilians-Universität München
Oettingenstrasse 67
80538 München
Germany

Kaare Teilum
Institute of Molecular Biology
University of Copenhagen
Øster Farimagsgade 2 A
1353 Copenhagen
Denmark

Arieh Warshel
Department of Chemistry
University of Southern California
3620 S. McClintock Avenue
Los Angeles, CA 90089-1062
USA

Phil M. Williams
Laboratory of Biophysics and Surface Analysis
University of Nottingham School of Pharmacy
Nottingham, NG7 2RD
United Kingdom

Julia Wirmer
Institute for Organic Chemistry and Chemical Biology
Center for Biomolecular Magnetic Resonance
Johann Wolfgang Goethe-University
Marie-Curie-Strasse 11
60439 Frankfurt
Germany

Markus Zeeb
Biochemie III
Universität Bayreuth
Universitätsstr. 30
95440 Bayreuth
Germany

Part I
1 Principles of Protein Stability and Design

1
Early Days of Studying the Mechanism of Protein Folding

Robert L. Baldwin

1.1
Introduction

Modern work on the mechanism of protein folding began with Chris Anfinsen. He recognized the folding problem, and he asked: How does the amino acid sequence of a protein determine its three-dimensional structure? From this basic question came various research problems, including (1) What is the mechanism of the folding process? (2) How can the three-dimensional structure be predicted from the amino acid sequence? and (3) What is the relation between the folding process in vivo and in vitro? Only the early history of the first problem will be considered here.

The basic facts needed to state the folding problem were already in place before Anfinsen's work. He knew this, well before 1973 when he received the Nobel Prize, and he was somewhat embarrassed about it. In his Nobel address [1], he says in the opening paragraph "Many others, including Anson and Mirsky in the '30s and Lumry and Eyring in the '50s, had observed and discussed the reversibility of the denaturation of proteins." Anfinsen's statement of the folding problem may be dated to 1961 [2], when his laboratory found that the amino acid sequence of ribonuclease A (RNase A) contains the information needed to make the correct four disulfide bonds of the native protein. There are eight –SH groups in the unfolded RNase A chain which could make 105 different S–S bonds. Although in 1961 the reversibility of protein denaturation was recognized by protein chemists, the knowledge that protein denaturation equals protein unfolding had been gained only a few years earlier, in a series of papers from Walter Kauzmann's laboratory, beginning with Ref. [3]. The first protein structure, that of sperm whale myoglobin (2 Å resolution), determined by John Kendrew and his coworkers [4], became available only in 1960. The myoglobin structure confirmed the proposal that proteins possess 3D structures held together by weak, noncovalent bonds, and consequently they might unfold in denaturing conditions. Wu suggested this explanation of protein denaturation as early as 1931 [5].

The other thread in Anfinsen's proposal was, of course, the recognition that protein folding is part of the coding problem. The basic dogma of molecular biology,

"DNA makes RNA makes protein," was already in place in 1958, and Anfinsen knew that the newly synthesized product of RNA translation is an inactive, unfolded polypeptide chain. How does it fold up to become active? In the 1960s and 1970s there was speculation about a code for folding, and some workers even proposed a three-letter code (i.e., three amino acid residues) [6]. Anfinsen proposed a thermodynamic hypothesis [1, 7]: the newly synthesized polypeptide chain folds up under the driving force of a free energy gradient and the protein reaches its thermodynamically most stable conformation. For protein chemists familiar with reversible denaturation, this appeared obvious: what else should drive a reversible chemical reaction besides the free energy difference between reactant and product? But for molecular biologists interested in knowing how an unfolded, newly synthesized protein is able to fold, Anfinsen's hypothesis represented a considerable leap of faith. In fact, biology does introduce subtle complexities, and more is said below about the thermodynamic hypothesis (see Section 1.3). Michael Levitt and Arieh Warshel made a farseeing proposal in 1975: they argued that an unfolded protein folds under the influence of a molecular force field, and someday the folding process will be simulated with the use of a force field [8].

Starting from Anfinsen's insights, this review examines what happened in the 1960s, 1970s, and 1980s to lay the groundwork for the modern study of how proteins fold in vitro. My review ends when the literature balloons out at the end of the 1980s with the study of new problems, such as: (1) experimental study of the transition state, (2) whether hydrophobic collapse precedes secondary structure, (3) the nature of the conformational reactions that allow hydrogen exchange in proteins, (4) using molecular dynamics to simulate the folding process, (5) the speed limit for folding, (6) helix and β-strand propensities, and (7) the mechanism of forming amyloid fibrils. This review is not comprehensive. The aim is to follow the threads that led to the prevalent view at the end of the 1980s. A few references are added after the 1980s in order to complete the picture for topics studied earlier.

1.2
Two-state Folding

An important achievement of early work on the mechanism of protein folding was the recognition that small proteins commonly show two-state equilibrium denaturation reactions and there are no observable intermediates. After the initial observation by Harrington and Schellman in 1956 that RNase A undergoes reversible thermal denaturation [9], at least five laboratories then examined the nature of the unfolding transition curve (see Ref. [10] for references). In the 1950s and 1960s RNase A was widely studied because of its small size (124 residues), purity and availability. The shape of its unfolding transition curve puzzled almost everyone, because the van't Hoff plot of ln K versus $(1/T)$ (K = equilibrium constant for unfolding) is unmistakably curved, which gives a small but observable asymmetry to the unfolding curve. The slope of the van't Hoff plot, $\Delta H/R$ (ΔH = enthalpy of unfolding, R = gas constant) should be constant if ΔH is constant. The typical explanation then was that stable intermediates are present during protein unfolding and they explain the shape of the unfolding curve. In 1965 John Brandts [10] recognized that the correct explanation for the peculiar shape of the unfolding curve lies in the unusual thermodynamics of protein folding. Brandts argued that the thermodynamics of unfolding is dominated by hydrophobic free energy, as proposed by Walter Kauzmann [11] in 1959. Then there should be a large positive value of ΔC_p for unfolding, which explains the curvature of the van't Hoff plot [10]. ΔC_p is the difference between the heat capacities of the native and denatured forms of a protein, and a large value for ΔC_p causes a strong dependence of ΔH on temperature.

Brandts' proposal that thermal denaturation of RNase A is a two-state reaction without intermediates [10] was strongly supported, also in 1965, by Ginsburg and Carroll [12], who introduced the superposition test for intermediates and found no populated intermediates in the unfolding reaction of RNase A. In the superposition test, two or more normalized unfolding curves are superimposed and tested for coincidence. They are monitored by at least two probes that report on fundamentally different molecular properties, such as specific viscosity that reports on molecular volume and optical rotation that reports on secondary structure.

In 1966 Lumry, Biltonen and Brandts [13] introduced the calorimetric ratio test for intermediates. In this test, the ratio of the calorimetric and van't Hoff values of ΔH should equal 1 if there are no populated intermediates. In the 1970s, after the development of differential scanning calorimetry by Peter Privalov [14], the calorimetric ratio test became widely used as a criterion for two-state folding.

Charles Tanford and his laboratory undertook a wide-ranging study of whether the denaturation reactions of small proteins are truly two-state reactions. In 1968 Tanford summarized the results in a long and widely quoted review [15]. He recognized that conditions must be found in which denatured proteins are completely unfolded, without residual structure, before one can confidently determine if denaturation is a two-state reaction. His laboratory found that 6 M GdmCl (guanidinium chloride) is a denaturant that eliminates residual structure in water-soluble proteins, whereas thermally denatured proteins retain significant residual structure [15]. (Whether or not the residual structure is related to the structure of the native protein was left for future study.) A key finding was that the reversible denaturation reactions of several small proteins are indeed two-state reactions when 6 M GdmCl is the denaturant [15]. This work opened the way to later study of two basic questions: (1) Are there intermediates in protein folding reactions and, if so, how can they be detected? and (2) How can the energetics of protein folding be measured experimentally? In discussing these problems, I use the terms "unfolded" and "denatured" interchangeably, and a "denatured" protein typically has some residual structure that depends on solvent conditions and temperature.

1.3
Levinthal's Paradox

In 1968 Cyrus Levinthal released a bombshell that became known as Levinthal's paradox [16, 17]. He had begun work on prediction of the 3D structures of proteins from their amino acid sequences. He observed that, if protein folding is truly a two-state reaction without intermediates, then the time needed to fold can be estimated from the time needed to search randomly all possible backbone conformations. Levinthal estimated that the time needed for folding by a random search is far longer than the life of the universe. A plausible conclusion from his calculation is that there must be folding intermediates and pathways [16, 17]. When Levinthal's calculation was repeated in 1992, with the addition of a small free energy bias as the driving force for folding, the time needed to search all conformations by a random search process was reduced to a few seconds [18]. Note, however, that a free energy bias in favor of the native structure is likely to produce intermediates in the folding process, although not necessarily ones that are populated.

In 1969 Levinthal pointed out [17] that, if it has a choice, a protein folds to the structure dictated by the fastest folding pathway and not to the most stable structure, in contrast to Anfinsen's thermodynamic hypothesis. His proposal was confirmed experimentally in 1996 for a protein from the serpin family, whose members form two different stable structures. The folding pathway of the serpin, plasminogen activator inhibitor 1, was found to be under kinetic control [19]. In 1998 Agard and coworkers [20] found that the stability of the folded structure of α-lytic protease appears to be under kinetic control; i.e., the denatured protein is not only kinetically unable to refold but also thermodynamically more stable than the native form [20]. This surprising deduction does not contradict Anfinsen's thermodynamic hypothesis. The enzymatically active form of this protein is formed after complete folding of a much longer polypeptide whose long Pro sequence is cleaved off after folding is complete.

1.4
The Domain as a Unit of Folding

Knowledge that polypeptide chains are synthesized starting from the N-terminus led to speculation that folding begins from the N-terminus of the chain. In 1970, Taniuchi [21] observed that the correct four S–S bonds of RNase A are not formed and the protein remains unfolded if the four C-terminal residues are deleted from the 124-residue polypeptide chain. Consequently, almost the entire polypeptide chain (maybe all of it) is required for stable folding (but see, however, Section 1.10). In 1969 Goldberg [22] found that intracistronic complementation occurs in β-galactosidase via the presence of at least two independently folding units ("globules") in each of the four identical polypeptide chains. In 1973 Wetlaufer [23] found contiguous folded regions in the X-ray structures of several proteins, in some cases apparently connected by flexible linkers. These three observations taken together gave rise to the concept that the domain ($\sim$100 amino acids) is the unit of stable folding.

Wetlaufer [23] pointed out that a contiguous folded region of the polypeptide chain is likely to arise from a structural folding nucleus. He noted that a structural nucleus might also serve as a kinetic nucleus for the folding process, with the consequence that successive folding events occur rapidly after the nucleus is formed, so that folding intermediates are never populated. His suggestion was often used in the early 1970s to explain why folding intermediates could not be found. In 1981 Lesk and Rose [24] pointed out that each protein domain can typically be divided into two subdomains, each of which is also folded from a contiguous segment of polypeptide chain – although the subdomains are not separated by flexible linkers. Their observation favors a hierarchic mechanism of folding [24].

In 1974 Goldberg and coworkers [25] proposed domain swapping as the explanation for the concentration-dependent formation of large aggregates of refolding tryptophanase, formed at a critical urea concentration. Their work has been taken as a model for understanding the formation of inclusion bodies and the need for chaperones to improve the yield in many folding reactions. Also in the 1970s Jaenicke and coworkers [26] began a systematic investigation of folding coupled to subunit association in the concentration-dependent folding reactions of oligomeric proteins. These subjects are discussed elsewhere in this book.

1.5
Detection of Folding Intermediates and Initial Work on the Kinetic Mechanism of Folding

Demonstration of two-state equilibrium denaturation by Brandts [10] and by Tanford [15] made clear the difficulty of detecting any folding intermediates that might exist. In 1971, reports of complex kinetics of unfolding/refolding by Ikai and Tanford for cytochrome c (cyt c) [27] and by Tsong, Baldwin, and Elson for RNase A [28] raised hope that fast-reaction methods would succeed in detecting and characterizing kinetic folding intermediates. Complexity in the refolding kinetics of staphylococcal nuclease (SNase) had already been reported in 1970 [29] from Anfinsen's laboratory (see also Ref. [30]).

Ikai and Tanford used a stopped-flow apparatus to measure the folding kinetics of cyt c [27, 31] and hen lysozyme [32]. These studies laid the groundwork for systematic investigation of folding reactions that can be represented by a simple sequential model, $U \leftrightarrow I_1 \leftrightarrow I_2 \leftrightarrow \bullet\bullet\bullet N$, with one unfolded form U, one native form, N, and intervening intermediates. Off-pathway intermediates are described by a branched pathway. A mathematical framework for such studies is given in Ref. [33]. Ikai and Tanford concluded initially that they had evidence for incorrectly folded, off-pathway intermediates in the refolding of cyt c [27, 31], but they assumed there was only one unfolded form. Concurrent studies of RNase A refolding showed that two or more unfolded forms are present, which suggested that the same is true of other denatured proteins (see Section 1.6). Ikai and Tanford made a further important contribution by developing tests for stopped-flow artifacts that result from mixing concentrated GdmCl with buffer; this problem was a serious issue at the time. In 1973, when Ikai finished his PhD work, Charles Tanford left the field of folding mechanisms to take up the new study of membrane proteins.

Earlier, in 1968/69, Fritz Pohl began using a slow temperature-jump ("T-jump") method, capable of observing kinetic changes only in the time range of seconds and longer, to study the kinetics of protein folding. He reported apparent two-state kinetics for the unfolding/refolding reactions of chymotrypsin [34], trypsin [35], chymotrypsinogen-A [36], and RNase A [36]. He found that the entire kinetic progress curve for unfolding or refolding could be represented by a single exponential time course and the kinetic amplitude agreed with the value expected from the equilibrium unfolding curve. He interpreted his results as showing that folding is highly cooperative, which strengthened the general view that folding intermediates would be very difficult to detect, perhaps impossible. Ikai and Tanford undertook their stopped-flow study of the unfolding/refolding kinetics of cyt c, despite Pohl's evidence that the kinetic folding reactions of small proteins are two-state, because Wayne Fish in Tanford's laboratory had evidence suggesting an equilibrium folding intermediate (A. Ikai, personal communication, 2003).

Tsong, Baldwin, and Elson [28, 37] undertook their fast T-jump study (dead time, 10 µs) of the unfolding kinetics of RNase A because they believed that fast kinetics might reveal intermediates in apparent two-state unfolding reactions. Pörschke and Eigen had observed [38] that short RNA helices show unfolding intermediates in the msec time range even though both the major folding and unfolding reactions follow a single exponential time course in the seconds time range. The unfolding kinetics of RNase A could be fitted to a similar type of nucleation-dependent folding model [28, 37, 40]. The model predicts kinetics in which the relative amplitude of the fast unfolding phase increases rapidly with increasing temperature, in agreement with experiment [28, 37]. A fast T-jump study of the unfolding kinetics of chymotrypsinogen-A [40] gave results like those of RNase A, suggesting that the nucleation model might be generally applicable. For chymotrypsinogen, as for RNase A, there is a fast (milliseconds) phase in unfolding, in addition to the slow unfolding reaction (seconds) studied earlier by Pohl [36]. The fast phases in the unfolding of RNase A [28, 37] and of chymotrypsinogen [40] were missed by Pohl [36] because they account for only a few percent of the total kinetic amplitude in his conditions, and because he measured only the slow unfolding reactions. The relative amplitude of the fast unfolding phase approaches 100% at temperatures near the upper end of the thermal unfolding transition of RNase A (see Section 1.6), but of course the total amplitude becomes small.

The T-jump and stopped-flow unfolding studies of RNase A were monitored by tyrosine absorbance [28, 37] and RNase A has six tyrosine residues. Consequently, the fast and slow unfolding reactions of RNase A might be detecting unfolding reactions in different parts of the molecule that occur in different time ranges. This hypothesis was tested and ruled out by studying a chemically reacted derivative of RNase A containing a single, partly buried, dinitrophenyl (DNP) group [41]. The unfolding kinetics monitored by the single DNP group are biphasic, exactly like the kinetics observed by the six tyrosyl groups. Later work by Paul Hagerman (see below) showed that the fast phase in the unfolding of RNase A arises from an unfolding intermediate that is a minor species in the ensemble of unfolded forms.

Thus, in 1973 the stage was set for kinetic studies of the mechanism of protein folding/unfolding. However, three basic questions required answers before folding mechanisms could be analyzed. (1) Are the observed folding intermediates on-pathway? (2) Are the intermediates partly folded forms or are they completely unfolded, while the complex unfolding/refolding kinetics result from the interconversion of different denatured forms? (3) How can the structures be determined of folding intermediates whose lifetimes are as short as milliseconds?

1.6 Two Unfolded Forms of RNase A and Explanation by Proline Isomerization

In 1973 Garel and Baldwin found that unfolded RNase A contains two different major denatured forms, a fast-folding species (U_F, ~20%) and a slow-folding species (U_S, ~80%) [42] that refolds 50 times more slowly than U_F in some conditions. The two different denatured forms were discovered when refolding was monitored with a probe specific for the enzymatically active protein, namely binding of the specific inhibitor 2′-CMP. Refolding was studied initially after a pH jump (pH 2.0 → pH 5.8) at high temperatures (to obtain complete unfolding at pH 2.0) [42] and later after dilution from 6 M GdmCl [43]. When the unfolding transition is complete in the initial conditions, either at low pH and high temperature [42] or in 6 M GdmCl [43], both the fast and slow kinetic phases of refolding yield native RNase A as product. In 6 M GdmCl, the denatured protein should be completely unfolded and therefore the fast-folding and slow-folding forms correspond to two different denatured species.

In 1976 Hagerman and Baldwin [44] studied the kinetic mechanism of RNase A unfolding by using a stopped-flow apparatus and pH jumps to analyze the unfolding/refolding kinetics as a function of temperature throughout the thermal unfolding zone at pH 3.0. Their analysis is based on a four-species mechanism, $U_S \leftrightarrow U_F \leftrightarrow I \leftrightarrow N$, in which U_S and U_F are the slow-folding and fast-folding forms of the denatured protein discussed above and I is a new unfolding intermediate observed above T_m. Because I is completely unfolded, as judged either by tyrosine absorbance or by enthalpy content [44], I may be labeled instead as U_3 and the unfolding/refolding mechanism written as $U_S \leftrightarrow U_F \leftrightarrow U_3 \leftrightarrow N$. The proportions of $U_S:U_F:U_3$ in denatured RNase A were predicted in 1976 to be 0.78:0.20:0.02 [44]. Although U_3 was studied initially only as an unfolding intermediate [44], U_3 should be the immediate precursor of N in refolding experiments according to the sequential unfolding/refolding mechanism. This prediction was confirmed next with experiments using a sequential mixing apparatus [45]. U_3 and U_F are populated transiently by unfolding N with a first mixing step and then U_3 and U_F are allowed to refold after a second mixing step. U_3 forms N much more rapidly than U_F does [45], and in 1994 U_3 was detected in the equilibrium population of denatured RNase A species [46].

The 1976 analysis predicts correctly the equilibrium curve for thermal denaturation from the kinetic data [44], and also shows that refolding of U_F and U_S to N must occur by the sequential mechanism $U_S \leftrightarrow U_F \leftrightarrow N$ and not by the split mechanism $U_S \leftrightarrow N \leftrightarrow U_F$. The latter conclusion follows from the behavior of the kinetic amplitudes when the relative rates of the fast and slow kinetic refolding reactions are varied by changing the temperature [44]. The issue of a sequential versus a split mechanism was tested in a different manner by Brandts and coworkers [47], who were aware of Hagerman's work (see their discussion). They introduced the interrupted unfolding (or "double-jump") experiment in which the species present at each time of unfolding are assayed by refolding measurements.

Brandts and coworkers [47] proposed a proline isomerization model as an explanation for the two different forms of RNase A. Their model includes two separate hypotheses. The first is that the fast-folding and slow-folding forms of a denatured protein are produced by slow cis–trans isomerization of proline peptide bonds after unfolding occurs. The second is that the fast-folding and slow-folding denatured species account entirely for the complex unfolding/refolding kinetics and no structural folding intermediates are populated. Thus, their unfolding/refolding mechanism for a protein with only one proline residue is $N \leftrightarrow U_F \leftrightarrow U_S$. The fast-folding species U_F has the same prolyl isomer (cis or trans) as N and the slow-folding species U_S contains the other prolyl isomer.

Nuclear magnetic resonance (NMR) studies of proline-containing peptides show that the cis:trans ratio of a prolyl peptide bond commonly lies between 30:70 and 10:90, depending on neighboring residues. Because the proline ring sometimes clashes sterically with neighboring side chains, proline peptide bonds (X-Pro) are quite different from ordinary peptide bonds, for which the % cis is only $\sim$0.1–1%. RNase A contains two cis proline residues, as well as as two trans proline residues, and therefore denatured RNase A should have an unusually high fraction of $[U_S]$ (as observed), because the two cis residues isomerize to trans after unfolding and produce U_S species. A later NMR study of the trans $\rightarrow$ cis isomerization rate of Gly-Pro in water [48] places it in the same time range as the $U_S \leftrightarrow U_F$ reaction of RNase A.

The proline isomerization model of Brandts and coworkers was very persuasive but it proved difficult to test, particularly because one of its two hypotheses turned out to be wrong, namely that no structural intermediates are populated during the kinetics of unfolding or refolding. In 1978 Schmid and Baldwin [49] found that the $U_F \leftrightarrow U_S$ reaction in unfolded RNase A is acid-catalyzed, although very strong acid, >5 M $HClO_4$, is required for catalysis. Very strong acid is needed for the cis $\rightarrow$ trans isomerization of both prolyl peptide bonds [50] and ordinary peptide bonds, and the slow rate of the $U_F \leftrightarrow U_S$ reaction of RNase A implies that the critical bonds are prolyl peptide bonds rather than ordinary peptide bonds. The high activation enthalpy ($\sim$85 kJ mol^{-1}) expected for isomerization of prolyl peptide bonds was found for the $U_F \leftrightarrow U_S$ reaction of denatured RNase A, both in 3.3 M $HClO_4$ and in 5 M GdmCl [49]. The acid catalysis results were widely accepted as evidence that the $U_F \leftrightarrow U_S$ reaction of RNase A is proline isomerization, and the later discovery of prolyl isomerases (see below) ended any doubts. In further work, the role of proline isomerization in protein folding kinetics has been thoroughly analyzed, especially for RNase T1 [51, 52], by combining mutagenesis of specific proline residues with sequential mixing experiments and with accurate measurement and analysis of kinetic amplitudes and relaxation times.

The $U_F \leftrightarrow U_S$ reaction of unfolded RNase A is quite slow (~ 1000 s) at 0 °C and it is straightforward to ask whether partial folding precedes proline isomerization at 0 °C. In 1979, Cook, Schmid, and Baldwin [53] tested this issue. They found that that partial folding does precede proline isomerization and they obtained two quite surprising results. (1) The major partly folded form (I_N) has properties closely resembling those of native RNase A (thus, I_N refers to a native-like intermediate). I_N even has RNase A catalytic activity [54]! (2) Not only does proline isomerization occur within the folded structure of I_N, but also the isomerization rate is speeded up in I_N by as much as 40-fold, compared with the rate of the $U_S \leftrightarrow U_F$ reaction in denatured RNase A [53]. These experiments gave the first clear indication that partly folded, noncovalent intermediates are sometimes populated during the kinetic process of protein folding.

Because U_S species fold slowly in physiological conditions, which is likely to make them susceptible to proteolytic cleavage, prolyl isomerases seemed needed to speed up proline isomerization in vivo. Gunter Fischer and coworkers found the first prolyl isomerase and in 1985 they showed that it catalyzes the $U_S \leftrightarrow U_F$ reaction of RNase A [55]. At least three classes of prolyl isomerases are known today. The role of prolyl isomerases in folding in vivo is discussed elsewhere in this book (see Chapter 25).

1.7
Covalent Intermediates in the Coupled Processes of Disulfide Bond Formation and Folding

Disulfide bonds stabilize the folded structures of proteins that contain S–S bonds and, when they are reduced, the protein typically unfolds. This observation was the starting point of Anfinsen's work [2] when he showed that the folding process directs the formation of the four unique S–S bonds of native RNase A. Tom Creighton had the basic insight that S–S intermediates can be covalently trapped, purified by chromatography, and structurally characterized. Because formation of S–S bonds is linked to the folding process, these S–S intermediates should also be folding intermediates. Beginning in 1974, Creighton reported the isolation and general properties of both the one-disulfide [56] and two-disulfide [57] intermediates of the small protein BPTI (bovine pancreatic trypsin inhibitor, 58 residues, three S–S bonds). He later measured the equilibrium constants for forming each of the three S–S bonds in BPTI [58, 59]. For example, he found that the effective concentration of the two –SH groups that form the S–S bond between cysteine residues 5 and 55 is $\sim 10^7$ M [58]. This value is three orders of magnitude higher than the effective concentration of the two adjacent –SH groups in dithiothreitol [58], and it illustrates the rigid alignment of these two –SH groups by the folded structure of BPTI. Later, after the development of two-dimensional NMR made it possible to determine the structures of protein species that are difficult to crystallize, Creighton and coworkers determined the structures of various S–S intermediates of BPTI (see, for example, Ref. [60]). Stabilization of the BPTI structure by a given

S–S bond depends on the effective concentration of the two –SH groups before the bond is formed, and the increase in effective concentration as successive S–S bonds are formed illustrates strikingly how the cooperativity of protein folding operates [58, 59]. The same principle has been used at an early stage in the folding of BPTI to examine the interplay between S–S bond and reverse turn formation [61]. The pathway of disulfide bond formation in BPTI is complex and has been the subject of considerable discussion [62, 63].

Early work [2] from Anfinsen's laboratory showed that nonnative S–S bonds are often formed during the kinetic process in which unfolded RNase A folds and eventually makes the correct S–S bonds. Because the S–S bond is covalent, some mechanism is needed to break nonnative S–S bonds during folding and allow formation of new S–S bonds. Anfinsen and coworkers reported in 1963 that enzymes such as protein disulfide isomerase [7] have a major role in ensuring that correct S–S bonds are formed. The role of disulfide isomerases in folding in vivo is discussed elsewhere in this book (see Chapter 26).

1.8
Early Stages of Folding Detected by Antibodies and by Hydrogen Exchange

Initial studies of the folding of peptide fragments, and also of proteins made to unfold by removing a stabilizing linkage or cofactor, gave the following generalization: the tertiary structures of proteins are easily unfolded and little residual structure remains afterwards. In 1956 Harrington and Schellman [9] found that breaking the four disulfide bonds of RNase A causes general unfolding of the tertiary structure and also destroys the helical structure, which should be local structure that could in principle survive loss of the tertiary structure. In 1968 Epand and Scheraga [64] tested by circular dichroism (CD) whether peptides from helix-containing segments of myoglobin still form helices in aqueous solution. They studied two long peptides and found they have very low helix contents; they did not pursue the problem. In 1969 Taniuchi and Anfinsen [65] made a similar experiment with SNase. They cleaved the polypeptide chain between residues 126 and 127, which causes the protein to unfold. Both fragments 1–126 and 127–149 were found to lack detectable native-like structure by various physical methods, including circular dichroism. For SNase as for RNase A, circular dichroism indicates that the native protein has some helical structure and at that time (1968) the X-ray structure of myoglobin was known [4], which gives the detailed structures of its eight helices. Thus, the overall conclusion from these experiments was that the helical secondary structures of the three proteins are stable only when the tertiary structures are present. In 1971 Lewis, Momany and Scheraga [66] proposed a hierarchic mechanism of folding in which β-turns play a directing role at early stages of folding by increasing the effective concentrations of locally formed structures, such as helices, that later interact in the native structure.

Anfinsen considered it likely that proteins fold by a hierarchic mechanism [1], and he developed an antibody method for sensitively detecting any native-like structure still present in a denatured protein [67]. His method is simple in principle. He and his coworkers took fragments 1–126 and 99–149 of SNase, which were devoid of detectable structure by physical criteria, and bound them covalently to individual sepharose columns. Antisera were prepared against both native SNase and the two polypeptide fragments, and polyclonal antibodies were purified by immunoabsorption against each homologous antigen. The antibodies developed against native SNase were tested for their ability to cross-react with the two denatured fragments. The results indicate that a weak cross-reaction occurs and a denatured fragment reacts with antibodies made against native SNase as if the denatured fragment exists in the "native format" a small fraction ($\sim 0.02\%$) of the time. These experiments are the forerunner of modern ones in which monoclonal antibodies are made against short peptides, and some of the monoclonal antibodies cross-react significantly with the native protein from which the peptide is derived. In 1975 Anfinsen and coworkers made the converse experiment [68]. They found that antibodies directed against denatured fragments of SNase are able to cross-react with native SNase. They conclude that antibodies made against denatured fragments detect unfolded SNase in equilibrium with native SNase, even though only a tiny fraction of the native protein, less than 0.01%, is found to be unfolded.

In 1979 Schmid and Baldwin [69] developed a competition method for detecting H-bonded secondary structure formed at an early stage in the refolding of denatured RNase A. Native proteins were known at that time, from Linderstrøm-Lang's development of the hydrogen exchange method [70], to contain large numbers of highly protected peptide NH protons. Shortly afterwards, in 1982, NMR hydrogen exchange experiments by Wagner and Wüthrich [71] demonstrated that, as expected, the highly protected peptide NH protons of BPTI are ones involved in H-bonded secondary structure. In this period the exchange rates of freely exchanging, unprotected peptide NH protons were already known from earlier studies of dipeptides by Englander and coworkers [72]. The peptide NH exchange rates are base-catalyzed and the rates become faster above pH 7 than the measured folding rates of the two major U_S species of denatured RNase A.

Thus, the principle of the competition experiment is straightforward [69]. Refolding of denatured, ^{3}H-labeled RNase A is performed at pH values where the rate of exchange-out of the ^{3}H label from denatured RNase A is either faster or slower, depending on pH, than the observed rate of refolding to form native RNase A. (The exchange rates of peptide NH protons in denatured RNase A can be computed from the peptide data [72].) When exchange-out is slower than folding, the folding process traps many ^{3}H-labeled protons. When exchange-out is faster than the formation of native RNase A, the observed folding rate can be used to predict the number of ^{3}H-labeled protons that should be trapped by folding. However, the observed number of ^{3}H-labeled protons trapped by folding is always much larger than the number predicted in this way. Control experiments, made at the same pH values but in the presence of modest GdmCl concentrations added to destabilize folding intermediates, show no trapped ^{3}H label. The first conclusion is that one or more folding intermediates are formed rapidly and they give protection against exchange-out of ^{3}H label. The second conclusion is that some form of early structure, probably H-bonded secondary structure, is stable before the tertiary structure is formed. In 1980, a more convenient and informative pulse-labeling version of the competition experiment was tested [73] and found superior to the competition method. Methods of resolving and assigning the proton spectra of native proteins were being developed rapidly in the early 1980s and it was evident that the secondary structures of early folding intermediates would be determined by this approach, probably within a few years. A stopped-flow apparatus could be used to trap protected peptide NH protons by means of ^{2}H–^{1}H exchange, and 2D ^{1}H-NMR could then be used to determine the structural locations of the protected N^{1}H protons.

1.9
Molten Globule Folding Intermediates

In 1981, Oleg Ptitsyn and coworkers released a bombshell [74] that was comparable in its impact to Levinthal's paradox. They proposed that the folding intermediates everyone had been searching for were sitting under our noses in plain sight, in the form of partly folded structures formed when certain native proteins are exposed to mildly destabilizing conditions. A few proteins were known to form these curious, partly folded, structures, particularly at acid pH. Ptitsyn and coworkers proposed that the partly folded forms, or "acid forms," were structurally related to authentic folding intermediates. The acid forms were supposed to differ from true folding intermediates essentially only by protonation reactions resulting from pH titration to acid pH. The acid forms were found to have surprising properties which suggested that their secondary structures are stable and native-like and their conformations are compact, even though the acid forms lack fixed tertiary structures (for reviews, see Refs [75, 76]). Until then, most workers had taken it for granted that folding intermediates should simply be "partial replicas" of native proteins. They should contain some unfolded segments plus some other folded segments whose tertiary and secondary structures are native-like. Later work has verified essential features of Ptitsyn's proposal, although argument about the details continues.

Ptitsyn's background was in polymer physics, and he was accustomed to analyzing problems involving the conformations of polymers. In 1973, he gave his forecast of a plausible model for the kinetic process of protein folding [77], which resembles the 1971 hierarchic mechanism of Scheraga and coworkers [66] but includes also later stages in the folding process. Ptitsyn later used the term "framework model," coined in a 1982 review by Kim and Baldwin [78], to describe his model. In making his 1981 proposal [74], Ptitsyn was impressed by the resemblance between the physical properties of acid forms and the properties he had hypothesized for early folding intermediates. He must also have been impressed by Kuwajima's results for the acid form of α-LA, which revealed some striking properties of acid forms (see below).

The name "molten globule" was given to these acid forms by Ohgushi and Wada [79] in 1983: "globule" meaning compact, "molten" meaning no fixed tertiary structure. Ohgushi and Wada were studying two acid forms of cyt c, which are converted from one form to the other by varying the salt concentration [79, 80].

In 1981 the best studied of these partly folded acid forms was bovine α-lactalbumin (α-LA), which was chosen by Ptitsyn and coworkers [74] for their initial study of an acid form. In 1976 and earlier, Kuwajima and coworkers had analyzed the pH-dependent interconversion between the acid form (or "A-state") and native α-LA [81]. They found a very unusual equilibrium folding intermediate in GdmCl-induced denaturation at neutral pH [81], which by continuity – as the pH is varied – is the same species as the acid form of α-LA. The explanation for this unusual folding intermediate is that native α-LA is a calcium metalloprotein, a property discovered only in 1980 [82]. The 1976 [81] and earlier studies of α-LA were made with the apoprotein in the absence of Ca^{2+}, and the apoprotein is much less stable than the holoprotein. When GdmCl-induced unfolding of the more stable holoprotein is studied in the presence of Ca^{2+}, no stable folding intermediate is observed [76]. The near-UV and far-UV CD spectra of the 1976 folding intermediate [81] reveal some basic properties of molten globule intermediates. The α-LA folding intermediate has no fixed tertiary structure, as judged by its near-UV CD spectrum, but its secondary structure resembles that of native α-LA, as judged by its far-UV CD spectrum. In 1990, when 2D ^{1}H-NMR was used together with ^{2}H–^{1}H exchange to determine the locations and stability of the helices in a few acid forms, the helices were found at the same locations as in the native structures. Particularly clear results were found for the helices in the acid forms of cyt c [83] and apomyoglobin [84].

1.10
Structures of Peptide Models for Folding Intermediates

By 1979 a paradox was evident concerning the stability of helices in folding intermediates. The experiments of Epand and Scheraga and of Taniuchi and Anfinsen indicated that the helices of myoglobin [64] and SNase [65] were unstable when the intact polypeptide chains of these proteins were cut into smaller fragments. On the other hand, the ^{3}H-labeling experiments of Schmid and Baldwin [69] indicated that an early folding intermediate – probably a H-bonded intermediate – of RNase A is stable before the tertiary structure is formed. In 1971 Brown and Klee [85] had found partial helix formation by CD at 0 °C in the "C-peptide" of RNase A. C-peptide is formed by cyanogen bromide cleavage at Met13 and contains residues 1–13, while residues 3–12 form a helix in native RNase A.

In 1978, Blum, Smallcombe and Baldwin [86] used the four His residues of RNase A as probes for structure in an NMR study in real time of the kinetics of RNase A folding at pH 2, 10 °C. These are conditions in which the folding rate is sufficiently slow to take 1D NMR spectra during folding. The carbon-bound protons of the imidazole side chains could be resolved by 1D NMR in that period, provided the peptide N^1H protons are first exchanged for ^{2}H. By chemical shift, His12 appears to be part of a rapidly formed, folded structure at 10 °C, although it is unfolded at 45 °C, pH 2 [86]. This result suggests that the N-terminal helix of RNase

A is partly folded at low temperatures, in agreement with the C-peptide study of Brown and Klee [85]. An ensuing study by Bierzynski et al. [87] confirmed that temperature-dependent helix formation does occur. Interestingly, C-peptide helix formation was found to be strongly pH dependent with apparent pK values, indicating that the ionized forms of His12 and either Glu9 or Glu2 are needed for helix formation [87]. Many peptides later, two specific side-chain interactions were found to contribute substantially to C-peptide helix stability [88]: an amino–aromatic interaction between Phe8 and His12$^+$, and a salt bridge between Glu2$^-$ and Arg10$^+$. Both interactions could be seen in the 1970 X-ray structure of RNase S [89], although the Phe 8•••His 12$^+$ interaction was not recognized as such in 1970.

In 1989, Marqusee et al. [90] found that alanine-based peptides form stable helices in water without the help of any specific interactions. Because alanine has only a -CH$_3$ side chain, this result indicates that the helix backbone itself is stable in water, although the helix has only marginal stability. Side chains longer than that of Ala detract from, rather than increasing, helix stability. In 1985 Dyson and coworkers [91] found that even reverse turns can be detected in short peptides by ^{1}H-NMR, and 10 years later Serrano and coworkers [92] observed formation of a stable β-hairpin in water. Thus, it is possible for all classes of secondary structure to be present, and to aid in directing the folding process, at very early stages of the folding process.

In 1988 a landmark experiment by Oas and Kim [93] showed that peptides are able to model more advanced folding intermediates. Oas and Kim took two peptides from BPTI: P$_\alpha$, a 16-residue peptide from residues 43–58 which includes the C-terminal helix, and P$_\beta$, a 14-residue peptide from residues 20–33 which includes part of the central β-hairpin. Separately, each peptide appears structureless by physical criteria. When the two peptides are joined by forming the 30–51 S–S bond, both helical and β structures appear. NMR characterization indicates that the 3D structure of the peptide complex resembles that of BPTI [93]. The central question about folding intermediates in that period was [78]: are they formed according to the framework model (secondary structure forms before tertiary structure) or the subdomain model (secondary and tertiary structures form simultaneously in local subdomains)? Before the 1988 experiment of Oas and Kim [93], all evidence seemed to support the framework model. After their experiment, the subdomain model became the focus of much further work.

Acknowledgments

I thank Atsushi Ikai for information and Thomas Kiefhaber for discussion.

References

1 ANFINSEN, C. B. (1973). Principles that govern the folding of protein chains. Science **181**, 223–230.

2 ANFINSEN, C. B., HABER, E., SELA, M., & WHITE, JR., F. H. (1961). The kinetics of formation of native

ribonuclease, during oxidation of the reduced polypeptide chain. *Proc. NatL Acad. Sci. USA* **47**, 1309–1314.
3. SIMPSON, R. B. & KAUZMANN, W. (1953). The kinetics of protein denaturation. I. The behavior of the optical rotation of ovalbumin in urea solutions. *J. Am. Chem. Soc.* **75**, 5139–5152.
4. KENDREW, J. C., DICKERSON, R. E., STRANDBERG, B. E. et al. (1960). Structure of myoglobin. A three-dimensional Fourier synthesis at 2 Å resolution. *Nature (London)* **185**, 422–427.
5. WU, H. (1931). Studies on denaturation of proteins. XIII. A theory of denaturation. *Chin. J. Physiol.* **5**, 321–344.
6. KABAT, E. A. & WU, T. T. (1972). Construction of a three-dimensional model of the polypeptide backbone of the variable region of kappa immunoglobulin light chains. *Proc. Natl Acad. Sci. USA* **69**, 960–964.
7. EPSTEIN, C. J., GOLDBERGER, R. F., & ANFINSEN, C. B. (1963). The genetic control of tertiary protein structure: studies with model systems. *Cold Spring Harbor Symp. Quant. Biol.* **28**, 439–444.
8. LEVITT, M. & WARSHEL, A. (1975). Computer simulation of protein folding. *Nature (London)* **253**, 694–698.
9. HARRINGTON, W. F. & SCHELLMAN, J. A. (1956). Evidence for the instability of hydrogen-bonded peptide structures in water based on studies of ribonuclease and oxidized ribonuclease. *C.R. Trav. Lab. Carlsberg, Sér. Chim.* **30**, 21–43.
10. BRANDTS, J. F. (1965). The nature of the complexities in the ribonuclease conformational transition and the implications regarding clathrating. *J. Am. Chem. Soc.* **87**, 2759–2760.
11. KAUZMANN, W. (1959). Factors in interpretation of protein denaturation. *Adv. Protein Chem.* **14**, 1–63.
12. GINSBURG, A. & CARROLL, W. R. (1965). Some specific ion effects on the conformation and thermal stability of ribonuclease. *Biochemistry* **4**, 2159–2174.
13. LUMRY, R., BILTONEN, R., & BRANDTS, J. F. (1966). Validity of the "two-state" hypothesis for conformational transitions of proteins. *Biopolymers* **4**, 917–944.
14. PRIVALOV, P. L. (1979). Stability of proteins. Small globular proteins. *Adv. Protein Chem.* **33**, 167–241.
15. TANFORD, C. (1968). Protein denaturation. A. Characterization of the denatured state. B. The transition from native to denatured state. *Adv. Protein Chem.* **23**, 121–282.
16. LEVINTHAL, C. (1968). Are there pathways for protein folding? *J. Chim. Phys.* **65**, 44–45.
17. LEVINTHAL, C. (1969). How to fold graciously. In *Mössbauer Spectroscopy in Biological Systems* (FRAUENFELDER, H., GUNSALUS, I. C., TSIBRIS, J. C. M., DEBRUNNER, P. G., & MÜNCK, E., eds), pp. 22–24, University of Illinois Press, Urbana.
18. ZWANZIG, R., SZABO, A., & BAGCHI, B. (1992). Levinthal's paradox. *Proc. Natl Acad. Sci. USA* **89**, 20–22.
19. WANG, Z., MOTTONEN, J., & GOLDSMITH, E. J. (1996). Kinetically controlled folding of the serpin plasminogen activator inhibitor 1. *Biochemistry* **35**, 16443–16448.
20. SOHL, J. L., JASWAL, S. S., & AGARD, D. A. (1998). Unfolded conformations of α-lytic protease are more stable than its native state. *Nature (London)* **395**, 817–819.
21. TANIUCHI, H. (1970). Formation of randomly paired disulfide bonds in des-(121–124)-ribonuclease after reduction and reoxidation. *J. Biol. Chem.* **245**, 5459–5468.
22. GOLDBERG, M. E. (1969). Tertiary structure of *Escherichia coli* β-D-galactosidase. *J. Mol. Biol.* **46**, 441–446.
23. WETLAUFER, D. B. (1973). Nucleation, rapid folding and globular intrachain regions in proteins. *Proc. Natl Acad. Sci. USA* **70**, 697–701.
24. LESK, A. M. & ROSE, G. D. (1981). Folding units in globular proteins. *Proc. Natl Acad. Sci. USA* **78**, 4304–4308.
25. LONDON, J., SKRZYNIA, C., & GOLDBERG, M. E. (1974). Renaturation

of *Escherichia coli* tryptophanase after exposure to 8 M urea: evidence for the existence of nucleation centers. *Eur. J. Biochem.* **47**, 409–415.

26 JAENICKE, R. & RUDOLPH, R. (1980). Folding and association of oligomeric enzymes. In *Protein Folding* (JAENICKE, R., ed.), pp. 525–546, Elsevier/North Holland, Amsterdam.

27 IKAI, A. & TANFORD, C. (1971). Kinetic evidence for incorrectly folded intermediate states in the refolding of denatured proteins. *Nature (London)* **230**, 100–102.

28 TSONG, T. Y., BALDWIN, R. L., & ELSON, E. L. (1971). The sequential unfolding of ribonuclease A: detection of a fast initial phase in the kinetics of unfolding. *Proc. Natl Acad. Sci. USA* **68**, 2712–2715.

29 SCHECHTER, A. N., CHEN, R. F., & ANFINSEN, C. B. (1970). Kinetics of folding of staphylococcal nuclease. *Science* **167**, 886–887.

30 EPSTEIN, H. F., SCHECHTER, A. N., CHEN, R. F., & ANFINSEN, C. B. (1971). Folding of staphylococcal nuclease: kinetic studies of vtwo processes in acid denaturation. *J. Mol. Biol.* **60**, 499–508.

31 IKAI, A., FISH, W. W., & TANFORD, C. (1973). Kinetics of unfolding and refolding of proteins. II. Results for cytochrome *c*. *J. Mol. Biol.* **73**, 165–184.

32 TANFORD, C., AUNE, K. C., & IKAI, A. (1973). Kinetics of unfolding and refolding of proteins. III. Results for lysozyme. *J. Mol. Biol.* **73**, 185–197.

33 IKAI, A. & TANFORD, C. (1973). Kinetics of unfolding and refolding of proteins. I. Mathematical analysis. *J. Mol. Biol.* **73**, 145–163.

34 POHL, F. M. (1968). Einfache Temperatursprung-Methode im Sekunden-bis Stundenbereich und die reversible Denaturierung von Chymotrypsin. *Eur. J. Biochem.* **4**, 373–377.

35 POHL, F. M. (1968). Kinetics of reversible denaturation of trypsin in water and water-ethanol mixtures. *Eur. J. Biochem.* **7**, 146–152.

36 POHL, F. M. (1969). On the kinetics of structural transition I of some pancreatic proteins. *FEBS Lett.* **3**, 60–64.

37 TSONG, T. Y., BALDWIN, R. L., & ELSON, E. L. (1972). Properties of the unfolding and refolding reactions of ribonuclease A. *Proc. Natl Acad. Sci. USA* **69**, 1809–1812.

38 PÖRSCHKE, D. & EIGEN, M. (1971). Cooperative non-enzymic base recognition. III. Kinetics of the helix-coil transition of the oligoribouridylic-oligoriboadenylic acid system and of oligoriboadenylic acid alone at acid pH. *J. Mol. Biol.* **62**, 361–381.

39 TSONG, T. Y., BALDWIN, R. L., & McPHIE, P. (1972). A sequential model of nucleation-dependent protein folding: kinetic studies of ribonuclease A. *J. Mol. Biol.* **63**, 453–475.

40 TSONG, T. Y. & BALDWIN, R. L. (1972). Kinetic evidence for intermediate states in the unfolding of chymotrypsinogen A. *J. Mol. Biol.* **69**, 145–148.

41 TSONG, T. Y. & BALDWIN, R. L. (1972). Kinetic evidence for intermediate states in the unfolding of ribonuclease A. II. Kinetics of exposure to solvent of a specific dinitrophenyl group. *J. Mol. Biol.* **69**, 149–153.

42 GAREL, J.-R. & BALDWIN, R. L. (1973). Both the fast and slow refolding reactions of ribonuclease A yield native enzyme. *Proc. Natl Acad. Sci. USA* **70**, 3347–3351.

43 GAREL, J.-R., NALL, B. T., & BALDWIN, R. L. (1976). Guanidine-unfolded state of ribonuclease A contains both fast- and slow-refolding species. *Proc. Natl Acad. Sci. USA* **73**, 1853–1857.

44 HAGERMAN, P. J. & BALDWIN, R. L. (1976). A quantitative treatment of the kinetics of the folding transition of ribonuclease A. *Biochemistry*, **15**, 1462–1473.

45 HAGERMAN, P. J., SCHMID, F. X., & BALDWIN, R. L. (1979). Refolding behavior of a kinetic intermediate observed in the low pH unfolding of ribonuclease A. *Biochemistry* **18**, 293–297.

46 HOURY, W. A., ROTHWARF, D. M., & SCHERAGA, H. A. (1994). A very fast phase in the refolding of disulfide-intact ribonuclease A: implications for the refolding and unfolding pathways. *Biochemistry* **33**, 2516–2530.

47 BRANDTS, J. F., HALVORSON, H. R., & BRENNAN, M. (1975). Consideration of the possibility that the slow step in protein denaturation reactions is due to cis-trans isomerism of proline residues. *Biochemistry* **14**, 4953–4963.

48 CHENG, H. N. & BOVEY, F. A. (1977). Cis-trans equilibrium and kinetic studies of acetyl-L-proline and glycyl-L-proline. *Biopolymers* **16**, 1465–1472.

49 SCHMID, F. X. & BALDWIN, R. L. (1978). Acid catlysis of the formation of the slow-folding species of RNase A: evidence that the reaction is proline isomerization. *Proc. Natl Acad. Sci. USA* **75**, 4764–4768.

50 STEINBERG, I. Z., HARRINGTON, W. F., BERGER, A., SELA, M., & KATCHALSKI, E. (1960). The configurational changes of poly-L-proline in solution. *J. Am. Chem. Soc.* **82**, 5263–5279.

51 KIEFHABER, T. & SCHMID, F. X. (1992). Kinetic coupling between protein folding and prolyl isomerization. II. Folding of ribonuclease A and ribonuclease T_1. *J. Mol. Biol.* **224**, 231–240.

52 MAYR, L. M., ODEFEY, C., SCHUTKOWSKI, MIKE, & SCHMID, F. X. (1996). Kinetic analysis of the unfolding and refolding reactions of ribonuclease T1 by a stopped-flow double-mixing technique. *Biochemistry* **35**, 5550–5561.

53 COOK, K. H., SCHMID, F. X., & BALDWIN, R. L. (1979). Role of proline isomerization in folding of ribonuclease A at low temperatures. *Proc. Natl Acad. Sci. USA* **76**, 6157–6161.

54 SCHMID, F. X. & BLASCHEK, H. (1981). A native-like intermediate on the ribonuclease A folding pathway. 2. Comparison of its properties to native ribonuclease A. *Eur. J. Biochem.* **114**, 111–117.

55 FISCHER, G. & BANG, H. (1985). The refolding of urea-denatured ribonuclease A is catalyzed by peptidyl-prolyl cis-trans isomerase. *Biochim. Biophys. Acta* **828**, 39–42.

56 CREIGHTON, T. E. (1974). The single-disulfide intermediates in the refolding of reduced pancreatic trypsin inhibitor. *J. Mol. Biol.* **87**, 603–624.

57 CREIGHTON, T. E. (1974). The two-disulfide intermediates and the folding pathway of reduced pancreatic trypsin inhibitor. *J. Mol. Biol.* **95**, 167–199.

58 CREIGHTON, T. E. (1983). An empirical approach to protein conformation stability and flexibility. *Biopolymers* **22**, 49–58.

59 CREIGHTON, T. E. (1988). Disulfide bonds and protein stability. *BioEssays* **8**, 57–64.

60 VAN MIERLO, C. P. M., DARBY, N. J., NEUHAUS, D., & CREIGHTON, T. E. (1991). Two-dimensional ^{1}H nuclear magnetic resonance study of the (5–55) single-disulfide folding intermediate of pancreatic trypsin inhibitor. *J. Mol. Biol.* **222**, 373–390.

61 ZDANOWSKI, K. & DADLEZ, M. (1999). Stability of the residual structure in unfolded BPTI in different conditions of temperature and solvent composition measured by disulfide kinetics and double mutant cycle analysis. *J. Mol. Biol.* **287**, 433–445.

62 CREIGHTON, T. E. (1992). The disulfide folding pathway of BPTI. *Science* **256**, 111–112.

63 WEISSMAN, J. S. & KIM, P. S. (1992). Response. *Science* **256**, 112–114.

64 EPAND, R. M. & SCHERAGA, H. A. (1968). The influence of long-range interactions on the structure of myoglobin. *Biochemistry* **7**, 2864–2872.

65 TANIUCHI, H. & ANFINSEN, C. B. (1969). An experimental approach to the study of the folding of staphylococcal nuclease. *J. Biol. Chem.* **244**, 3864–3875.

66 LEWIS, P. N., MOMANY, F. A., & SCHERAGA, H. A. (1971). Folding of polypeptide chains in proteins: a proposed mechanism for folding. *Proc. Natl Acad. Sci. USA* **68**, 2293–2297.

67 SACHS, D. H., SCHECHTER, A. N., EASTLAKE, A., & ANFINSEN, C. B. (1972). An immunologic approach to

the conformational equilibria of polypeptides. *Proc. Natl Acad. Sci. USA* **69**, 3790–3794.

68 FURIE, B., SCHECHTER, A. N., SACHS, D. H., & ANFINSEN, C. B. (1975). An immunological approach to the conformational equilibrium of staphylococcal nuclease. *J. Mol. Biol.* **92**, 497–506.

69 SCHMID, F. X. & BALDWIN, R. L. (1979). Detection of an early intermediate in the folding of ribonuclease A by protection of amide protons against exchange. *J. Mol. Biol.* **135**, 199–215.

70 LINDERSTRØM-LANG, K. (1958). Deuterium exchange and protein structure. In *Symposium on Protein Structure* (NEUBERGER, A., ed.) Methuen, London.

71 WAGNER, G. A. & WÜTHRICH, K. (1982). Amide proton exchange and surface conformation of BPTI in solution. *J. Mol. Biol.* **160**, 343–361.

72 MOLDAY, R. S., ENGLANDER, S. W., & KALLEN, R. G. (1972). Primary structure effects on peptide group hydrogen exchange. *Biochemistry* **11**, 150–158.

73 KIM, P. S. & BALDWIN, R. L. (1980). Structural intermediates trapped during the folding of ribonuclease A by amide proton exchange. *Biochemistry* **19**, 6124–6129.

74 DOLGIKH, D. A., GILMANSHIN, R. I., BRAZHNIKOV, E. V. et al. (1981). α-Lalbumin: Compact state with fluctuating tertiary structure? *FEBS Lett.* **136**, 311–315.

75 PTITSYN, O. B. (1995). Molten globule and protein folding. *Adv. Protein Chem.* **47**, 83–229.

76 KUWAJIMA, K. (1989). The molten globule state as a clue for understanding the folding and cooperativity of globular-protein structure. *Proteins Struct. Funct. Genet.* **6**, 87–103.

77 PTITSYN, O. B. (1973). The stepwise mechanism of protein self-organization. *Dokl. Nauk SSSR* **210**, 1213–1215.

78 KIM, P. S. & BALDWIN, R. L. (1982). Specific intermediates in the folding reactions of small proteins and the mechanism of protein folding. *Annu. Rev. Biochem.* **51**, 459–489.

79 OHGUSHI, M. & WADA, A. (1983). 'Molten-globule state': a compact form of globular proteins with mobile side chains. *FEBS Lett.* **164**, 21–24.

80 KURODA, Y., KIDOKORO, S., & WADA, A. (1992). Thermodynamic characterization of cytochrome *c* at low pH. Observation of the molten globule state and of the cold denaturation process. *J. Mol. Biol.* **223**, 1139–1153.

81 KUWAJIMA, K., NITTA, K., YONEYAMA, M., & SUGAI, S. (1976). Three-state denaturation of α-lactalbumin by guanidine hydrochloride. *J. Mol. Biol.* **106**, 359–373.

82 HIRAOKA, Y., SEGAWA, T., KUWAJIMA, K., SUGAI, S., & MURAI, N. (1980). α-Lactalbumin: a calcium metallo-protein. *Biochem. Biophys. Res. Commun.* **95**, 1098–1104.

83 JENG, M.-F., ENGLANDER, S. W., ELÖVE, G., WAND, A. J., & RODER, H. (1990). Structural description of acid-denatured cytochrome *c* by hydrogen exchange and 2D NMR. *Biochemistry* **29**, 10433–10437.

84 HUGHSON, F. M., WRIGHT, P. E., & BALDWIN, R. L. (1990). Structural characterization of a partly folded apomyoglobin intermediate. *Science* **249**, 1544–1548.

85 BROWN, J. E. & KLEE, W. A. (1971). Helix-coil transition of the isolated amino terminus of ribonuclease. *Biochemistry* **10**, 470–476.

86 BLUM, A. D., SMALLCOMBE, S. H., & BALDWIN, R. L. (1978). Nuclear magnetic resonance evidence for a structural intermediate at an early stage in the refolding of ribonuclease A. *J. Mol. Biol.* **118**, 305–316.

87 BIERZYNSKI, A., KIM, P. S., & BALDWIN, R. L. (1982). A salt bridge stabilizes the helix formed by isolated C-peptide of RNase A. *Proc. Natl Acad. Sci. USA* **79**, 2470–2474.

88 SHOEMAKER, K. R., FAIRMAN, R., KIM, P. S., YORK, E. J., STEWART, J. M., & BALDWIN, R. L. (1987). The C-peptide helix from ribonuclease A considered

as an autonomous folding unit. *Cold Spring Harbor Symp. Quant. Biol.* **52**, 391–398.

89 WYCKOFF, H. W., TSERNOGLOU, D., HANSON, A. W., KNOX, J. R., LEE, B., & RICHARDS, F. M. (1970). The 3-dimensional structure of ribonuclease S: interpretation of an electron density map at a nominal resolution of 2 Å. *J. Biol. Chem.* **245**, 305–328.

90 MARQUSEE, S., ROBBINS, V. H., & BALDWIN, R. L. (1989). Unusually stable helix formation in short alanine-based peptides. *Proc. Natl Acad. Sci. USA* **86**, 5286–5290.

91 DYSON, H. J., CROSS, K. J., HOUGHTEN, R. A., WILSON, I. A., WRIGHT, P. E., & LERNER, R. A. (1985). The immunodominant site of a synthetic immunogen has a conformational preference in water for a type-II reverse turn. *Nature (London)* **318**, 480–483.

92 BLANCO, F. J., RIVAS, G., & SERRANO, L. (1995). A short linear peptide that folds into a stable β-hairpin in aqueous solution. *Nature Struct. Biol.* **1**, 584–590.

93 OAS, T. G. & KIM, P. S. (1988). A peptide model of a protein folding intermediate. *Nature (London)* **336**, 42–48.

2
Spectroscopic Techniques to Study Protein Folding and Stability

Franz Schmid

2.1
Introduction

Optical spectroscopy provides the standard techniques for measuring the conformational stabilities of proteins and for following the kinetics of unfolding and refolding reactions. Practically all unfolding and refolding reactions are accompanied by changes in absorbance, fluorescence, or optical activity (usually measured as the circular dichroism). These optical properties scale linearly with protein concentration. Spectra and spectral changes can be measured with very high accuracy in dilute protein solutions, and spectrophotometers are standard laboratory equipment. Therefore the methods of optical spectroscopy will remain the basic techniques for studying protein folding. Light absorbance occurs in about 10^{-15} s and the excited states show lifetimes of around 10^{-8} s. Optical spectroscopy is thus well suited for measuring the kinetics of ultrafast processes in folding (e.g., in temperature-jump or pressure-jump experiments).

Proteins absorb and emit light in the UV range of the spectrum. The absorbance originates from the peptide groups, from the aromatic amino acid side chains and from disulfide bonds. The fluorescence emission originates from the aromatic amino acids. Some proteins that carry covalently linked cofactors, such as the heme proteins, show absorbance in the visible range.

In absorption, light energy is used to promote electrons from the ground state to an excited state, and when the excited electrons revert back to the ground state they can lose their energy in the form of emitted light, which is called fluorescence. When a chromophore is part of an asymmetric structure, or when it is immobilized in an asymmetric environment, left-handed and right-handed circularly polarized light are absorbed to different extents, and we call this phenomenon circular dichroism (CD).

The energies of the ground and the activated state depend on the molecular environment and the mobility of the chromophores, and therefore the optical properties of folded and unfolded proteins are usually different. Spectroscopy is a nondestructive method, and, if necessary, the samples can be recovered after the experiments.

Protein Folding Handbook. Part I. Edited by J. Buchner and T. Kiefhaber
Copyright © 2005 WILEY-VCH Verlag GmbH & Co. KGaA, Weinheim
ISBN: 3-527-30784-2

This chapter provides an introduction into the spectral properties of amino acids and proteins as well as practical advice for the planning and the interpretation of spectroscopic measurements of protein folding and stability.

2.2 Absorbance

Light absorption can occur when an incoming photon provides the energy that is necessary to promote an electron from the ground state to an excited state. The energy (E) of light depends on its frequency v ($E = hv$) and is thus inversely correlated with its wavelength λ ($E = hc/\lambda$). Light in the 200-nm region has sufficient energy for exciting electrons that participate in small electronic systems such as double bonds. Aromatic molecules have lower lying excited states, and therefore they also absorb light in the region between 250 and 290 nm, which is called the "near UV" or "aromatic" region of a spectrum. The amino acids phenylalanine, tyrosine, and tryptophan absorb here, as well as the bases in nucleic acids. The aromatic region is used extensively to follow the conformational transitions of proteins and nucleic acids. Light in the visible region of the spectrum can only excite electrons that participate in extended delocalized π systems. Such systems do not occur in normal proteins, but can be introduced by site-specific labeling. The basic aspects of protein absorption are very well covered in Refs [1–6].

The absorbance (A) is related to the intensity of the light before (I_o) and after (I) passage through the protein solution:

$$A = -\log_{10}(I/I_o) \tag{1}$$

The absorbance depends linearly on the concentration of the absorbing molecules in a solution, and absorbance can be measured very easily and with very high accuracy. Therefore absorbance spectroscopy is the most common method for determining the concentration of biological macromolecules in solution. The relation between A and the molar concentration c is given by the Beer-Lambert relationship:

$$A = \varepsilon \times c \times l \tag{2}$$

where l is the pathlength in cm, and ε is the molar absorption coefficient, which can be determined experimentally or calculated for proteins by adding up the contributions of the constituent aromatic amino acids of a protein [7–10].

2.2.1 Absorbance of Proteins

Proteins absorb light in the UV range. As an example, the absorption spectrum of lysozyme is shown in Figure 2.1. The peptide bonds of the protein backbone ab-

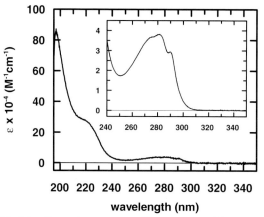

Fig. 2.1. Absorption spectrum of lysozyme. The spectrum was measured at a protein concentration of 0.65 μM in 10 mM K-phosphate buffer pH 7.0, 25 °C, in a 1-cm cell. The inset provides an expanded view of the spectrum in the near-UV region. It was measured with 5.5 μM lysozyme under otherwise identical conditions.

sorb strongly between 180 and 230 nm. Figure 2.1 shows the descending slope of this absorption. The shoulder near 220 nm also contains significant contributions from the aromatic residues of lysozyme. The inset of Figure 2.1 provides an expanded view of the near-UV region of the spectrum. The absorption here is caused by the aromatic side chains of tyrosine, tryptophan, and phenylalanine. Disulfide bonds display a very weak absorbance band around 250 nm. Figure 2.1 emphasizes that proteins absorb much more strongly in the far-UV region than in the near-UV (the "aromatic") region, which is commonly used to measure protein concentrations and to follow protein unfolding.

Spectra of the individual aromatic amino acids are shown in Figure 2.2; their molar absorptions at the respective maxima in the near UV are compared in Table 2.1. The contributions of phenylalanine, tyrosine, and tryptophan to the absorbance of a protein are strongly different. In the aromatic region the molar absorption of phenylalanine ($\lambda_{max} = 258$ nm, $\varepsilon_{258} = 200$ M^{-1} cm^{-1}) is smaller by an order of magnitude compared with those of tyrosine ($\lambda_{max} = 275$ nm, $\varepsilon_{275} = 1400$ M^{-1} cm^{-1}) and tryptophan ($\lambda_{max} = 280$ nm, $\varepsilon_{280} = 5600$ M^{-1} cm^{-1}), and it is virtually zero above 270 nm. The near-UV spectrum of a protein is therefore dominated by the contributions of tyrosine and tryptophan. The shape of the absorbance spectrum of a protein reflects the relative contents of the three aromatic residues in the molecule.

The absorbance maximum near 280 nm is used for determining the concentrations of protein solutions and the 280–295 nm range is used to follow protein unfolding reactions. In this wavelength range only tyrosine and tryptophan contribute to the observed absorbance and to the changes caused by conformational changes. With the distinct fine structure of their absorbance (Figure 2.2) the phenylalanine residues contribute "wiggles" in the 250–260 nm region, and they influence the

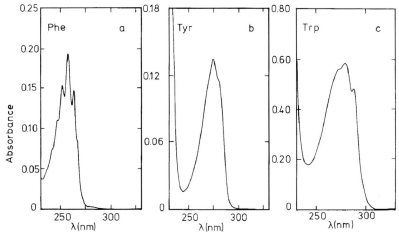

Fig. 2.2. Ultraviolet absorption spectra of the aromatic amino acids in a 1-cm cell in 0.01 M potassium phosphate buffer, pH 7.0 at 25 °C. (a) 1 mM phenylalanine, (b) 0.1 mM tyrosine, (c) 0.1 mM tryptophan. The figure is taken from Ref. [6].

ratio A_{260}/A_{280}, which is often used to identify contaminating nucleic acids in protein preparations.

The aromatic amino acids absorb only below 310 nm, and therefore pure protein solutions should not show absorbance at wavelengths greater than 310 nm. Proteins that are devoid of tryptophan should not absorb above 300 nm (cf. Figure 2.2). A sloping baseline in the 310–400 nm region usually originates from light scattering when large particles, such as aggregates are present in the protein solution. A plot of $\log_{10} A$ as a function of $\log_{10}$ wavelength (above 310 nm) reveals the contribution of light scattering, and the linear extrapolation to shorter wavelength can be used to correct the measured absorbance for the contribution of scattering [5].

Figure 2.3 reflects the strong influence of tryptophan residues on absorbance spectra. It compares the absorption spectra of the wild-type form (Figure 2.3a) and the Trp59Tyr variant (Figure 2.3b) of ribonuclease T1. The wild-type protein

Tab. 2.1. Absorbance and fluorescence properties of the aromatic amino acids[a].

Compound	Absorbance		Fluorescence		Sensitivity
	λ_{max} (nm)	ε_{max} (M^{-1} cm^{-1})	λ_{max} (nm)	ϕ_F[b]	$\varepsilon_{max} \times \phi_F$[b] ($M^{-1}$ cm^{-1})
Tryptophan	280	5600	355	0.13	730
Tyrosine	275	1400	304	0.14	200
Phenylalanine	258	200	282	0.02	4

[a] In water at neutral pH; data are from Ref. [26].
[b] ϕ_F, fluorescence quantum yield.

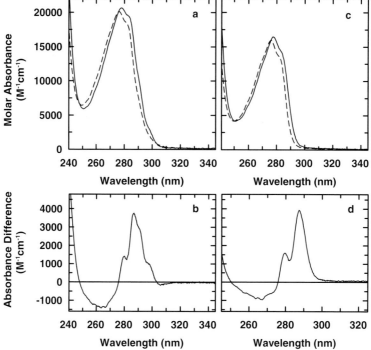

Fig. 2.3. Ultraviolet absorption spectra of a) the wild-type form, and c) the Trp59Tyr variant of ribonuclease T1. The spectra of the native proteins (in 0.1 M sodium acetate, pH 5.0) are shown by the continuous lines. The spectra of the unfolded proteins (in 6.0 M GdmCl in the same buffer) are shown by broken lines. The difference spectra between the native and unfolded forms are shown in b) and d). Spectra of 15 µM protein were measured at 25 °C in 1-cm cuvettes in a double-beam instrument with a bandwidth of 1 nm at 25 °C. The spectra of the native and unfolded proteins were recorded successively, stored, and subtracted. The figure is taken from Ref. [6].

contains one tryptophan and nine tyrosines, the Trp59Tyr variant contains 10 tyrosines and no tryptophan. The Trp59Tyr replacement reduces the absorbance in the 280–290 nm region significantly (Figure 2.3c), and the absorbance is essentially zero above 300 nm.

In folded proteins, most of the aromatic residues are buried in the hydrophobic core of the molecule, and upon unfolding they become exposed to the aqueous solvent. The spectral changes that accompany unfolding are dominated by a shift of the absorption spectra of the aromatic amino acids towards lower wavelengths (a blue shift). In addition, small changes in intensity and peak widths are observed as well. The blue shift upon unfolding is typically in the range of 3 nm. It leads to maxima in the difference spectra between native and unfolded proteins in the descending slope of the original spectrum, that is in the 285–288 nm region for tyrosine and around 291–294 nm for tryptophan [11].

Figure 2.3 shows also the spectra for the unfolded forms of the two variants of ribonuclease T1. The difference spectra in Figure 2.3b,d show that the maximal differences in absorbance are indeed observed in the descending slopes of the original spectra in the 285–295 nm region. The difference spectrum of the Trp59Tyr variant (Figure 2.3d) shows a prominent maximum at 287 nm. It is typical for proteins that contain tyrosine residues only. In the wild-type protein with its single tryptophan (Figure 2.3b) additional signals are observed between 290 and 300 nm. They originate from the absorbance bands of tryptophan, which are visible as shoulders in the protein spectrum (Figure 2.3a). Difference spectra between the native and the unfolded form, such as those for the variants of ribonuclease T1 in Figure 2.3b,d are observed for most proteins. Generally, these differences are small, but they can be determined with very high accuracy. The contributions of phenylalanine residues to difference spectra are very small. In some cases they are apparent as ripples in the spectral region around 260 nm (cf. Figure 2.3b,d). They are easily mistaken as instrumental noise.

Proteins are usually unfolded by increasing the temperature or by adding destabilizing agents such as urea or guanidinium chloride (GdmCl) (see Chapter 3). These changes not only unfold the protein, but they also influence the spectral properties of the protein solution. The refractive index of the solution decreases slightly with increasing temperature, the protein concentration decreases because the solution expands upon heating, and the ionization state of dissociable groups can change. All these effect are very small, and in combination they often lead to an almost temperature independent protein absorbance in the absence of unfolding.

Chemical denaturants have a significant influence on the absorbances of tyrosine and tryptophan at 287 nm and at 291 nm, respectively, where protein unfolding is usually monitored. The refractive index of the solvent increases as a function of the concentration of GdmCl or urea, which leads to a slight shift to higher wavelengths (a red shift) of the spectra and thus to an increase in absorbance at 287 and 291 nm. A_{291} of tryptophan increases by about 50% when 7 M GdmCl is added. Similar increases are also observed for tyrosine at 287 nm and when urea is used as the denaturant. The GdmCl and urea dependences of tyrosine and tryptophan absorbance are found in Ref. [6]. The spectrum of the unfolded form of a protein and its denaturant dependence can be modeled by mixing tyrosine and tryptophan in a ratio as in the protein under investigation and by measuring the absorbance of this mixture as a function of the denaturant concentration. Such data provide adequate base lines for denaturant-induced unfolding transitions.

The absorbance changes upon unfolding are measured in the steeply descending slopes of the absorbance spectra (cf. Figure 2.3) and consequently they depend on the instrumental settings (in particular on the bandwidth). The reference data must therefore be determined under the same experimental settings as the actual unfolding transition.

2.2.2
Practical Considerations for the Measurement of Protein Absorbance

The solvents used for spectroscopic measurements should be transparent in the wavelength range of the experiment. Water absorbs only below 170 nm, but several of the commonly used buffers absorb significantly already in the region below

Tab. 2.2. Absorbance of various salt and buffer solutions in the far-UV region[a].

Compound		No absorbance above (nm)	Absorbance of a 0.01 M solution in a 0.1-cm cell at:			
			210 nm	200 nm	190 nm	180 nm
NaClO$_4$		170	0	0	0	0
NaF, KF		170	0	0	0	0
Boric acid		180	0	0	0	0
NaCl		205	0	0.02	>0.5	>0.5
Na$_2$HPO$_4$		210	0	0.05	0.3	>0.5
NaH$_2$PO$_4$		195	0	0	0.01	0.15
Na-acetate		220	0.03	0.17	>0.5	>0.5
Glycine		220	0.03	0.1	>0.5	>0.5
Diethylamine		240	0.4	>0.5	>0.5	>0.5
NaOH	pH 12	230	≥0.5	>2	>2	>2
Boric acid, NaOH	pH 9.1	200	0	0	0.09	0.3
Tricine	pH 8.5	230	0.22	0.44	>0.5	>0.5
Tris	pH 8.0	220	0.02	0.13	0.24	>0.5
Hepes	pH 7.5	230	0.37	0.5	>0.5	>0.5
Pipes	pH 7.0	230	0.20	0.49	0.29	>0.5
Mops	pH 7.0	230	0.10	0.34	0.28	>0.5
Mes	pH 6.0	230	0.07	0.29	0.29	>0.5
Cacodylate	pH 6.0	210	0.01	0.20	0.22	>0.5

[a] Buffers were titrated with 1 M NaOH or 0.5 M H$_2$SO$_4$ to the indicated pH values. Data are from Ref. [6].

230 nm. It is therefore important to select appropriate buffers for measurements in this spectral region and to carefully match buffer concentrations in the sample and the reference cell. Absorbance values of commonly used salt and buffer solutions are found in Table 2.2.

Denaturants such as urea and GdmCl are used in very high concentrations (typically 1–10 M) in protein unfolding experiments. Therefore, the purity of these unfolding agents is of utmost importance. Even impurities in the 1 ppm range are present in the same range of concentration as the protein to be studied (often near 1 µM). Such impurities (such as cyanate in urea solutions) might lead to chemical modifications of the protein. In addition, they may absorb light and thus interfere with the measurement of protein difference spectra.

Solutions for spectroscopy should be filtered through 0.45-µm filters or centrifuged to remove dust particles. For measuring difference spectra such as in protein unfolding studies, a double-beam spectrophotometer and a set of matched quartz cells should be used. The equivalence of two cuvettes is easily tested by placing aliquots of the same solution (e.g., a solution of the protein of interest) into sample and reference cuvette. The difference in absorbance between both cells should ideally be zero (or <0.5% of the measured absorbance) in the wavelength range of interest.

2.2.3
Data Interpretation

Absorbance spectroscopy is an accurate method, but usually spectral changes such as those observed upon protein unfolding cannot be interpreted in molecular terms. The size and shape of difference spectra depends on the nature and the number of the aromatic amino acids as well as on the degree of burial of their side chains in the interior of the native protein. Changes can be assigned to particular aromatic residues only for proteins that carry only a single tyrosine or tryptophan residue, for example, or for mutants that differ from the reference protein only in a single aromatic residue (such as the ribonuclease T1 variants in Figure 2.3).

In the early days of protein spectroscopy there were efforts to determine by difference spectroscopy the number of aromatic residues that are buried in the native protein and become exposed upon unfolding. Donovan [3] presented estimates for the change in absorbance produced by the transfer of a residue from the protein interior to water. He suggested numbers of -700 M^{-1} cm^{-1} at 287 nm for tyrosine and -1600 M^{-1} cm^{-1} at 292 nm for tryptophan. These values were derived by assuming that the transfer from the interior of a protein to an aqueous environment is equivalent to a transfer from 120% ethylene glycol to water. In a protein the absorbance changes upon unfolding depend on the environment of the aromatic residues in the folded protein, which are different for the individual residues, and therefore the numbers given above are, at best, reasonable first approximations.

In summary, the molecular interpretation of absorption differences is often vague. The usefulness of absorption spectroscopy for studies of protein folding rather lies in the favourable properties of absorbance and the ease of the measurements. Absorbance changes linearly with concentration, and even small differences in absorbance can be measured with very high accuracy and reproducibility. Moreover, absorbance can be followed with a very high time resolution after rapid mixing or after perturbations such as a temperature or a pressure jump. Thus, absorbance spectroscopy remains to be a powerful technique for determining the stability of proteins (Chapter 3) and for following the kinetics of protein unfolding and refolding (Chapter 12.1).

2.3
Fluorescence

A molecule emits fluorescence when, after excitation, an electron returns from the first excited state (S_1) back to the ground state (S_0). At about 10^{-8} s, the lifetime of the excited state is very long compared with the time for excitation, which takes only about 10^{-15} s (on the human time scale this would be equivalent to the times of 1 year and 1 second). An excited molecule thus has ample time to relax into the lowest accessible vibrational and rotational states of S_1 before it returns to the ground state. As a consequence, the energy of the emitted light is always smaller than that of the absorbed light. In other words, the fluorescence emission of a chromophore always occurs at a wavelength that is higher than the wavelength of its absorption.

Electrons can revert to the ground state S_0 also by nonradiative processes (such as vibrational transitions), and in some instances transitions to the excited triplet state T_1 occur. The S_1 to T_1 transition is facilitated by magnetic perturbations, such as the collision of an activated molecule with paramagnetic molecules or with molecules with polarizable electrons. This leads to the phenomenon of fluorescence quenching. Good collisional quenchers are iodide ions, acrylamide, or oxygen. Quenching can also occur intramolecularly, in proteins by the side chains of Cys and His and by amides and carboxyl groups. All conformational transitions that change the exposure to quenchers in the solvent or within the protein will thus lead to fluorescence changes.

Another factor that leads to a decrease in emission is fluorescence resonance energy transfer (FRET), also called Förster transfer. The efficiency of this energy transfer between two chromophores depends, among other factors, on the extent of overlap between the fluorescence spectrum of the donor and the absorption spectrum of the acceptor, and, in particular, on the distance between donor and acceptor.

In combination, these radiationless routes of deactivation compete with light emission, and therefore the quantum yields (ϕ_F) of most chromophores are much smaller than 1. ϕ_F is equal to the ratio of the emitted to the absorbed photons. Its maximal value is thus 1. Good introductions into the principles of fluorescence of biological samples are found in Refs [2, 12, 13].

2.3.1
The Fluorescence of Proteins

The fluorescence of proteins originates from the phenylalanine, tyrosine, and tryptophan residues. Emission spectra for the three aromatic amino acids are shown in Figure 2.4, and their absorption and emission properties are summarized in Table 2.1. The excitation spectra correspond to the respective absorption spectra (Figure 2.2).

The fluorescence intensity of a particular chromophore depends on both its absorption at the excitation wavelength and its quantum yield at the wavelength of the emission. The product $\varepsilon_{max} \times \phi_F$ is thus a good parameter to characterize the fluorescence "sensitivity" of this chromophore. In proteins that contain all three aromatic amino acids, fluorescence is usually dominated by the contribution of the tryptophan residues because its sensitivity parameter is 730, which is much higher than the corresponding values which are 200 for tyrosine and only 4 for phenylalanine (Table 2.1). In proteins, phenylalanine fluorescence is practically not observable, because the other two aromatic residues absorb strongly around 280 nm, where phenylalanine emits, and because its emission is almost completely transferred to tyrosine and tryptophan by FRET.

The fluorescence of tyrosines is often undetectable in proteins that also contain

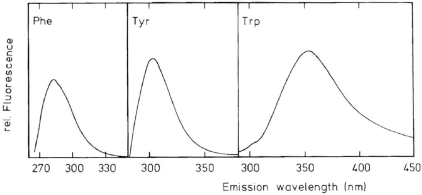

Fig. 2.4. Fluorescence spectra of the aromatic amino acids in 0.01 M potassium phosphate buffer, pH 7.0 at 25 °C. For the measurements, 100 μM phenylalanine, 6 μM tyrosine, and 1 μM tryptophan were used. Excitation was at 257 nm (Phe), 274 nm (Tyr), and 278 nm (Trp). The figure is taken from Ref. [6].

tryptophan, (i) because tryptophan shows a much stronger emission than tyrosine (Table 2.1), (ii) because in folded proteins tryptophan emission is often shifted to lower wavelengths and thus masks the contribution of tyrosine at 303 nm, and (iii) because FRET can occur from tyrosine to tryptophan in the compact native state of a protein [13, 14].

2.3.2
Energy Transfer and Fluorescence Quenching in a Protein: Barnase

The small bacterial ribonuclease barnase provides an instructive example for how strongly protein fluorescence is influenced by the local environment of the chromophores by quenching and by energy transfer, in this case transfer between individual tryptophan residues. Barnase contains three tryptophan residues at positions 35, 71, and 94 (Figure 2.5a). Trp94 is in close contact with His18, Trp71 is about 10 Å away from Tyr92, and Trp35 is about 25 Å away from the other two tryptophan residues. The fluorescence of wild-type barnase increases strongly between pH 6 and pH 9, following a titration curve with an apparent pK of 7.75. This was suggested to reflect the quenching of Trp94 emission by His18.

Fersht and coworkers [15] constructed single mutants of barnase, one in which His18 was mutated to a glycine and then three mutants in which one tryptophan residue at a time was mutated to another aromatic or aliphatic residue. They found that the fluorescence of barnase increased more than 2.5-fold when His17 was mutated to Gly (Figure 2.5b), which suggests that His18 does indeed act as an intramolecular quencher of Trp94 fluorescence. The observed fluorescence increase was, however, much higher than expected under the implicit assumption that all three tryptophan residues contribute equally to the fluorescence of the wild-type protein.

(a)

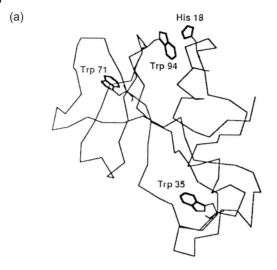

(b)

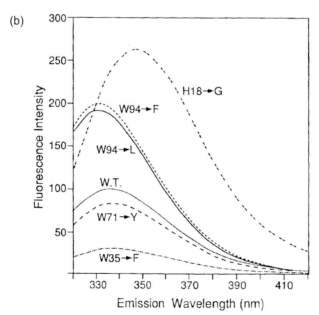

Fig. 2.5. a) The structure of barnase showing the locations of the three Trp residues and of His18. b) Fluorescence emission spectra of barnase and the variants with single mutations of His18 and of its three Trp residues. All spectra were recorded in 10 mM Bis/Tris buffer, pH 5.5 at the same protein concentration of 4 μM. Under these conditions His18 is protonated. The maximal emission of the wild-type protein (W.T.) is set as 100. The figure is taken from Ref. [15].

The subsequent mutations of the individual tryptophan residues revealed that their contributions to the overall fluorescence are indeed strongly different. The Trp35Phe mutation alone decreased the emission by 80% (Figure 2.5b), which indicates that Trp35 is the major contributor to the fluorescence of wild-type barnase, and that the emission of the other two tryptophan residues is largely abolished in the folded protein. Indeed, the replacement of Trp71 by a tyrosine decreased the emission only by 15% (Figure 2.5b). Trp71 is close to Trp94 (Figure 2.5a) and loses most of its emission because of energy transfer to Trp94. Trp94 in turn is efficiently quenched by the neighboring His18. The replacement of Trp94 by Phe or Leu prevents both energy transfer from Trp71 and the quenching of Trp94 by His18. As a consequence, the substitution of Trp94 led to a twofold increase in fluorescence (Figure 2.5b). It reflects the increase in the fluorescence of Trp71, which, in the Trp94Phe variant, no longer loses emission by energy transfer to Trp94.

In summary, this mutational analysis reveals how, in wild-type barnase, Trp71 and Trp94 lose most of their emission via quenching by His18. It demonstrates nicely that protein fluorescence is strongly influenced by energy transfer and by intramolecular quenching, which render fluorescence an excellent probe for protein folding.

2.3.3
Protein Unfolding Monitored by Fluorescence

The extended lifetime of the excited state is a key to understanding the changes that occur during a protein folding transition. As described above, a broad range of interactions or perturbations can influence the fluorophore while it is in the activated state. In particular, singulet-to-triplet inter system crossing can occur, or energy transfer to an acceptor. Fluorescence emission is consequently much more sensitive to changes in the environment of the chromophore than absorption.

The changes in fluorescence during the unfolding of a protein are often very large. In proteins that contain tryptophan both shifts in wavelength and changes in intensity are usually observed. Depending on the environment in the folded protein the tryptophan emission of a native protein can be greater or smaller than the emission of free tryptophan in aqueous solution, and fluorescence can increase or decrease upon unfolding, depending on the protein under investigation. The emission maximum of solvent-exposed tryptophan is near 350 nm, but in a hydrophobic environment, such as in the interior of a folded protein, tryptophan emission occurs at lower wavelengths (indole shows an emission maximum of 320 nm in hexane [16]). As an example the emission spectra of native and of unfolded ribonuclease T1 are shown in Figure 2.6. As mentioned, ribonuclease T1 contains nine tyrosine residues and only one tryptophan (Trp59), which is inaccessible to solvent in the native protein [17].

The fluorescence of the tryptophan residues can be investigated selectively by excitation at wavelengths higher than 295 nm. Tyrosine does not absorb above 295 nm because its absorbance spectrum is blue shifted (as compared with tryptophan, Figure 2.2) and is thus not excited at 295 nm. A comparison of the emission spec-

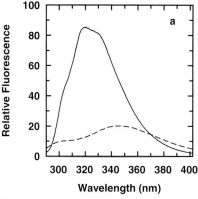
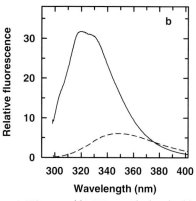

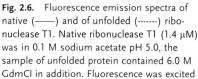

Fig. 2.6. Fluorescence emission spectra of native (——) and of unfolded (-------) ribonuclease T1. Native ribonuclease T1 (1.4 μM) was in 0.1 M sodium acetate pH 5.0, the sample of unfolded protein contained 6.0 M GdmCl in addition. Fluorescence was excited at a) 278 nm and b) 295 nm. The bandwidths were 3 nm for excitation and 5 nm for emission. Spectra were recorded at 25 °C in 1 × 1 cm cells in a Hitachi F-4010 fluorimeter. The figure is taken from Ref. [6].

tra observed after excitation at 280 nm and at 295 nm gives information about the contributions of the tryptophan and the tyrosine residues to the fluorescence of a protein. The data for ribonuclease T1 in Figure 2.6 show that the shapes of the fluorescence spectra observed after excitation at 278 nm and 295 nm are virtually identical. The measured emission originates almost completely from the single Trp59, which is inaccessible to solvent and hence displays a strongly blue-shifted emission maximum near 320 nm. Tyrosine emission is barely detectable in the spectrum of the native protein, because energy transfer to tryptophan occurs. Unfolding by GdmCl results in a strong decrease in tryptophan fluorescence and a concomitant red shift of the maximum to about 350 nm. The distances between the tyrosine residues and Trp59 increase, and energy transfer becomes less efficient. As a consequence, the tyrosine fluorescence near 303 nm becomes visible in the spectrum of the unfolded protein when excited at 278 nm (Figure 2.6a), but not when excited at 295 nm (Figure 2.6b). The examples in Figure 2.6 indicate that, unlike the absorbance changes, the fluorescence changes upon folding can be very large and the contribution of the tryptophan residues can be studied selectively by changing the excitation wavelength. These features, together with its very high sensitivity, make fluorescence measurements extremely useful for following conformational changes in proteins.

Multiple emission bands do not necessarily originate from different tryptophan residues of a folded protein. Figure 2.6 shows that even single tryptophan residues, such as Trp59 of ribonuclease T1 can give rise to several emission bands.

In contrast to ribonuclease T1, where we found a fivefold decrease of tryptophan fluorescence, the fluorescence of the FK 506 binding protein FKBP12 increases about 12-fold upon unfolding [18]. Figure 2.7 shows fluorescence spectra recorded

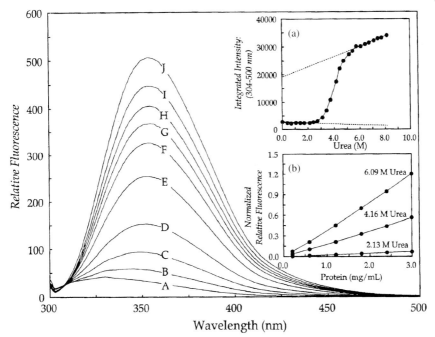

Fig. 2.7. Urea-induced denaturation of FKBP12 measured by intrinsic tryptophan fluorescence at 25 °C. A) Samples containing 50 μM FKBP12 in 75 mM phosphate/25 mM glycine buffer, 1 mM DTT, pH 7.2 and the following concentrations of urea. A)–J) 0.0, 3.1, 3.4, 3.7, 4.1, 4.4, 4.7, 5.1, 6.1, and 7.1 M. Excitation was at 294 nm to minimize inner filter effects. Inset a) Equilibrium unfolding curve based on the integrated emission intensities between 304 and 500 nm. The dashed lines represent linear fits to the pre- and post-transitional baselines. Inset b) A plot of the fluorescence intensity at 350 nm versus protein concentration to examine the linear dependence of the emission on protein concentration. The figure is taken from Ref. [18].

between 0 M and 7.1 M urea. Incidentally both ribonuclease T1 and FKBP12 have a single tryptophan at position 59. The emission of Trp59 of FKBP12 is barely detectable in the folded protein (Figure 2.7), but as soon as the protein enters the transition zone (around 3 M urea, cf. inset a in Figure 2.7) the fluorescence increases strongly and its maximum shifts from 330 nm to 350 nm.

In the experiments shown in Figure 2.7 the fluorescence of the tyrosine residues of FKBP12 was suppressed by exciting Trp59 selectively by light of 294 nm. At this wavelength tyrosine absorption is almost zero (cf. Figure 2.2). In addition, at 294 nm the absorbance of the samples is very low, the inner filter effect (cf. Section 2.3.5) becomes negligible, and the fluorescence emission remains linear over a wide range of protein concentrations (cf. inset b of Figure 2.7).

The fluorescence maximum of tyrosine remains near 303 nm, irrespective of its molecular environment. Therefore the unfolding of proteins that contain only tyrosine is usually accompanied by changes in the intensity, but not in the wavelength of emission. As an example, the fluorescence emission spectra of the Trp59Tyr

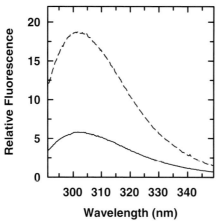

Fig. 2.8. Fluorescence emission spectra of 1.5 µM of the native (———) and of the unfolded (-------) Trp59Tyr variant of ribonuclease T1. This protein contains 10 Tyr residues. The native protein was in 0.1 M sodium acetate, pH 5.0; the unfolded sample contained 6.0 M GdmCl in addition. Fluorescence was excited at 278 nm, the spectra were recorded as in Figure 2.6. The figure is taken from Ref. [6].

variant of ribonuclease T1 in the native and unfolded states are shown in Figure 2.8. A decreased tyrosine fluorescence in the native state (as in Figure 2.8) is frequently observed. It is thought to originate from hydrogen bonding of the tyrosyl hydroxyl group and/or the proximity of quenchers, such as disulfide bonds in the folded state [19].

2.3.4
Environmental Effects on Tyrosine and Tryptophan Emission

The fluorescence of the exposed aromatic amino acids of a protein depends on the solvent conditions. Solvent additives that promote the $S_1 \rightarrow T_1$ transition (such as acrylamide or iodide) quench the fluorescence, and this effect can be used to measure the exposure of tryptophan side chains to the solvent [20].

The fluorescence generally decreases with increasing temperature. This decrease is substantial. To a first approximation, tyrosine and tryptophan emission decrease $\geq 1\%$ per degree increase in temperature [6]. Fluorescence is therefore generally not suited for following the heat-induced unfolding of a protein.

Denaturants such as GdmCl and urea also influence the fluorescence of tyrosine and tryptophan. The dependences on the concentration of GdmCl and urea of tryptophan and tyrosine emission are given in Ref. [6]. As in absorbance experiments (cf. Section 2.2.1) the emission of the unfolded protein can be represented well by an appropriate mixture of tyrosine and tryptophan. The dependence of the emission of this mixture on denaturant concentration is often very useful to determine the baseline equivalent to the unfolded protein for the quantitative analysis of unfolding transition curves.

Protein fluorescence can also be affected by salts. Charged denaturants, such as GdmCl, can thus act in a twofold manner: as a denaturant and as a salt. To identify the salt effect, it is advisable to titrate the protein in control experiments with the appropriate concentrations of NaCl or KCl. Salt effects of GdmCl are often strongest at low GdmCl concentration and thus they can severely affect the baseline of the native protein in equilibrium unfolding experiments.

2.3.5
Practical Considerations

Fluorescence is an extremely sensitive technique, so it is essential to avoid contamination of cuvettes and glassware with fluorescing substances. Deionized and quartz-distilled water should be used and plastic containers should be avoided because they may leach out fluorescing additives. Also, laboratory detergents usually contain strongly fluorescing substances. Fluorescing impurities are easily identified when running buffer controls.

In water, a Raman scattering peak is observed. It originates from the excitation of an O–H vibrational mode of the H_2O molecule, and thus it is present in all aqueous solvents. The Raman peak of water is weak and occurs at a constant frequency difference, not at a constant wavelength difference from the exciting light. The position of the Raman peak as a function of the excitation wavelength can be found in Ref. [6]. After excitation at 280 nm the Raman peak occurs at 310 nm, and therefore it overlaps with the emission spectra of the aromatic amino acids, in particular with tyrosine emission. The Raman peak of the solvent should always be measured separately and subtracted from the measured spectra.

As in absorption spectroscopy impurities and dust particles should be avoided as much as possible and transparent buffers and solvents should be used (cf. Section 2.2). A single dust particle moving slowly through the light beam can cause severe distortions of the fluorescence signal. Continuous stirring during the fluorescence measurement is strongly recommended. Stirring keeps dust particles in rapid motion and thereby minimizes their distorting effect on the signal. In addition it continually transports new protein molecules into the small volume of the cell that is illuminated by the excitation light beam and thereby averages-out photochemical decomposition reactions over the entire sample. Stirring also improves the thermal equilibration in the cuvette, which is important, regarding the strong temperature dependence of fluorescence.

Fluorescence increases with fluorophore concentration in a linear fashion only at very low concentration. Nonlinearity at higher concentration is caused by inner-filter effects. The exciting light is attenuated by the absorption of the protein and of the solvent. The extent of this attenuation depends on the absorbance of the sample at the wavelength of excitation, and therefore it can be varied by a shift in the excitation wavelength. Further attenuation originates from reabsorption of fluorescent light by protein and solvent molecules. This effect is usually small, because the absorption of proteins and solvents beyond 300 nm is almost zero. Inner-filter effects are negligible when the absorbance of the sample at the excitation wavelength is smaller than 0.1 absorbance units. Additional practical advice on how to measure protein fluorescence is found in Ref. [6].

2.4
Circular Dichroism

Circular dichroism (CD) is a measure for the optical activity of asymmetric molecules in solution and reflects the unequal absorption of left-handed and right-handed circularly polarized light. The CD signals of a particular chromophore are therefore observed in the same spectral region as its absorption bands. A good introduction to the basic principles of CD is provided by Refs [2] and [21].

Proteins show CD bands in two spectral regions. The CD in the far UV (170–250 nm) originates largely from the dichroic absorbance of the amide bonds, and therefore this region is usually termed the amide region. The aromatic side chains also absorb in the far UV and accordingly they also contribute to the CD in this region. These contributions may dominate the spectra of proteins that show a weak amide CD and/or a high content of aromatic residues. The CD bands in the near UV (250–300 nm) originate from the aromatic amino acids and from small contributions of disulfide bonds. Figure 2.9 shows the CD spectra of pancreatic ribonuclease for the native and the denaturant-unfolded forms.

The two spectral regions give different kinds of information about protein structure. The CD in the amide region reports on the backbone (i.e., the secondary) structure of a protein and is used to characterize the secondary structure and changes therein. In particular the α-helix displays a strong and characteristic CD spectrum in the far-UV region. The contributions of the other elements of secondary structure are less well defined. The content of the different secondary structures in a protein can be calculated from its CD spectrum in the amide regions. Methods for these calculations are found in Ref. [22].

The aromatic residues have planar chromophores and are intrinsically symmetric. When they are mobile, such as in short peptides or in unfolded proteins, their CD is almost zero. In the presence of ordered structure, such as in a folded protein, the environment of the aromatic side chains becomes asymmetric, and therefore they show CD bands in the near UV. The signs and the magnitudes of the aromatic CD bands cannot be calculated; they depend on the immediate structural and electronic environment of the immobilized chromophores. Therefore the individual peaks in the complex near-UV CD spectrum of proteins usually cannot be assigned to specific amino acid side chains. However, the near-UV CD spectrum represents a highly sensitive criterion for the native state of a protein. It can thus be used as a fingerprint of the correctly folded conformation.

2.4.1
CD Spectra of Native and Unfolded Proteins

Figure 2.10 shows representative spectra in the far-UV region for all-helical (a), for all-β (b), and for unstructured proteins (c). Helical proteins show rather strong CD

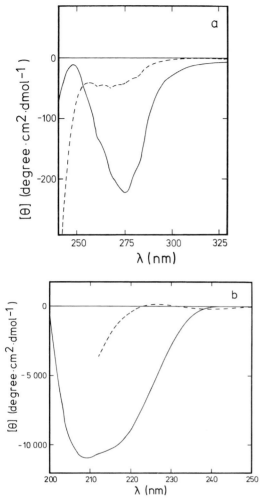

Fig. 2.9. CD spectra of pancreatic ribonuclease (from red deer) at 10 °C in the native (——) and in the unfolded state (-------). The native protein was in 0.1 M sodium cacodylate/HCl, pH 6.0, the unfolded protein was in the same buffer plus 6.0 M GdmCl. a) CD in the aromatic region. The protein concentration was 78 µM in a 1-cm cell. b) CD in the peptide region. 28 µM protein in a 0.1-cm cell. The figure is taken from Ref. [6].

bands with minima near 222 and 208 nm and a maximum near 195 nm. β proteins show weaker and more dissimilar spectra. The shape of the CD spectrum of a β protein depends, among other factors, on the length and orientation of the strands and on the twist of the sheet. Most spectra show, however, a positive ellipticity between 190 and 200 nm. In proteins with helices and sheets usually the helical contributions dominate the CD spectrum. Unordered proteins and peptides show weak CD signals above 210 nm, but a pronounced minimum between 195 and 200 nm.

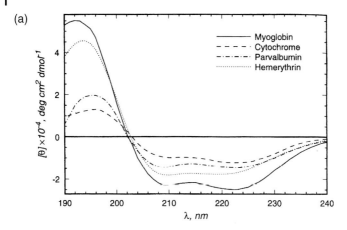
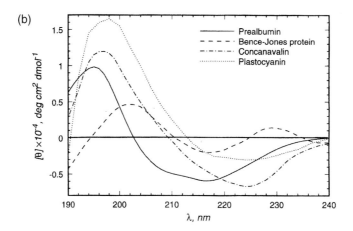
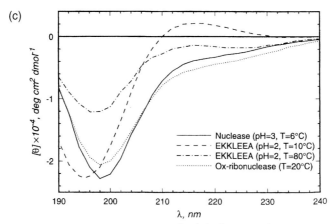

Fig. 2.10. Representative CD spectra of a) helical proteins, b) proteins with β-sheet structure, and c) unordered proteins and peptides. The figure is taken from Ref. [22].

The strong difference near 195 nm between the CD of folded and unfolded proteins should, in principle, provide an ideal probe for monitoring protein unfolding transitions. In practice, however, most unfolding reactions are monitored near 220 nm. Wavelengths below 200 nm are almost never used, because most buffers and virtually all denaturants absorb strongly below 200 nm (Table 2.2), and therefore the advantage of having a strong change in signal is more than offset by the very high background absorbance of the solvent.

The aromatic amino acid side chains absorb not only in the near-UV, but also in the far-UV region [23]. Accordingly, in native proteins they also contribute to the CD in the 180–230 nm region, which has traditionally been ascribed to only the amide bonds in the protein backbone. The aromatic contributions to the far-UV CD are often modest, and for helical proteins it is usually adequate to neglect them in the analyses, because the CD of helices is very strong (cf. Figure 2.10, top). For β proteins with their small and diverse CD signals (cf. Figure 2.10, middle), the aromatic bands can lead to problems and therefore the contents of β structure as calculated from the CD spectra can be seriously wrong. Carlsson and coworkers employed a mutational approach to analyze the contributions of the seven tryptophan residues to the amide CD spectrum of carbonic anhydrase [24]. They found strong contributions of these tryptophan residues to the far-UV spectrum of this protein.

2.4.2
Measurement of Circular Dichroism

The difference between the absorbances of right- and left-handed circularly polarized light of a protein sample is extremely small. In the far-UV region it lies in the range of 10^{-4}–10^{-6} absorbance units in samples with a total absorbance of about 1.0. This requires that less than 0.1% of the absorbance signal be measured accurately and reproducibly. Therefore highly sensitive instruments are necessary and careful sample preparation is important.

CD instruments use a high-frequency photoelectric modulator to generate alternatively the two circularly polarized light components. This leads to an alternating current contribution to the photomultiplier signal that is proportional to the CD of the sample. A detailed description of the principles of CD measurements is found in Refs [6] and [25].

Since the measured CD is so small, an extremely high stability of the CD signal is required. In general the instrument should be allowed to warm up for 30–60 min, and the stability of the baseline with and without the cuvette should be checked regularly.

When CD is measured in the far-UV region particular attention should be paid to a careful selection of solvents and buffers. The contributions of buffers and salts to the total absorbance of the sample should be as small as possible. This poses severe restrictions on the choice of solvents for CD measurements because many commonly used salts and buffers absorb strongly in the far-UV region. Absorbance values for several of them are given in Table 2.2. Common denaturants, such as urea and GdmCl also absorb strongly in the far UV. Therefore they cannot be used for CD measurements below 210 nm. Oxygen also absorbs in the far UV, but precautions such as purging the cell compartment with nitrogen is necessary only when measurements are performed below 190 nm.

Protein concentration, pathlength and solvent must be carefully selected to obtain good CD spectra and to avoid artifacts. A good signal-to-noise ratio is achieved when the total absorbance is around 1.0. The magnitude of the CD depends on the protein concentration and therefore its contribution to the total optical density of the sample should be as high as possible, and the absorbance of the solvent should be as small as possible. Therefore, for measurements in the far-UV region, short path lengths (0.2–0.01 cm) and increased protein concentrations should be used. Cells of 0.1 cm are usually sufficient for measurements down to about 200 nm. When preparing samples for CD measurements it should be remembered that protein absorbance is about tenfold higher in the far UV when compared with the aromatic region. Artifacts can arise easily from an excessive absorbance of the sample or from light scattering caused by protein aggregates in the sample. Whenever the optical density of the sample becomes too high, the instrument cannot discriminate properly between the right and the left circularly polarized light components, and the CD signal decreases in size. This phenomenon, called absorption flattening, is a serious problem below 200 nm. It is good practice to ascertain that the measured CD is linearly related with path length and protein concentration.

2.4.3
Evaluation of CD Data

CD is recorded either as the difference in absorbance of right- and left-handed circularly polarized light, $\Delta A = A_L - A_R$ or as ellipticity, Θ_{obs}, in degrees. Data in both formats can be converted to the molar values, i.e., to the differential molar circular dichroic extinction coefficient, $\Delta \varepsilon = \varepsilon_L - \varepsilon_R$ and to the molar ellipticity, $[\Theta]$. $\Delta \varepsilon$ and $[\Theta]$ are interrelated by Eq. (3).

$$[\Theta] = 3300 \times \Delta \varepsilon \tag{3}$$

It should be noted that the concentration standards are different for $[\Theta]$ and $\Delta \varepsilon$. $\Delta \varepsilon$ is the differential absorbance of a 1 mol L^{-1} solution in a 1-cm cell, whereas $[\Theta]$ is the rotation in degrees of a 1 dmol cm^{-3} solution and a path length of 1 cm.

CD of proteins in the amide region is usually given as mean residue ellipticity, $[\Theta]_{MRW}$, which is based on the concentration of the sum of the amino acids in the protein solution under investigation. The concentration of residues is obtained by multiplying the molar protein concentration with the number of amino acids. For data in the aromatic region, different units are used in the literature. Frequently, aromatic CD is given as $\Delta \varepsilon$, but $[\Theta]$, based on the protein concentration, and $[\Theta]_{MRW}$, based on the residue concentration are also found.

The molar ellipticity, $[\Theta]$, or the residue ellipticity, $[\Theta]_{MRW}$, are calculated from the measured Θ (in degrees) by using Eqs (4) or (5):

$$[\Theta] = \frac{\Theta \times 100 \times M_r}{c \times l} \quad (4)$$

$$[\Theta]_{\text{MRW}} = \frac{\Theta \times 100 \times M_r}{c \times l \times N_A} \quad (5)$$

where Θ is the measured ellipticity in degrees, c is the protein concentration in mg mL^{-1}, l is the pathlength in cm, and M_r is the protein molecular weight. N_A is the number of amino acids per protein. $[\Theta]$ and $[\Theta]_{\text{MRW}}$ have the units degrees × cm^2 × dmol^{-1}. The factor 100 in Eqs (4) and (5) originates from the conversion of the molar concentration to the dmol cm^{-3} concentration unit.

The relation between $[\Theta]$ and $\Delta\varepsilon$, given in Eq. (3) allows an immediate transformation of raw Θ data into ΔA values and vice versa by the relationship: $\Theta = 33(A_L - A_R)$. This implies that a measured ellipticity of 10 millidegrees is equivalent to a ΔA of only 0.0003 absorbance units. Practical aspects of how to measure and analyze CD spectra are discussed in detail in Ref. [6].

The analysis of CD spectra is often used to determine the secondary structure of folded proteins or of folding intermediates. References to computer programs and critical evaluations of the methods can be found in Ref. [22].

References

1 D. B. Wetlaufer, *Adv. Protein Chem.*, **1962**, *17*, 303–421.
2 C. R. Cantor, P. R. Schimmel, *Biophysical Chemistry*. 1980, San Francisco: W. H. Freeman.
3 J. W. Donovan, *Meth. Enzymol.*, **1973**, *27*, 497–525.
4 S. B. Brown, Ultraviolet and visible spectroscopy. In *An Introduction to Spectroscopy for Biochemists*, S. B. Brown, Editor. 1980, Academic Press: London. pp. 14–69.
5 W. Colón, *Meth. Enzymol.*, **1999**, *309*, 605–632.
6 F. X. Schmid, Optical spectroscopy to characterize protein conformation and conformational changes. In *Protein Structure: A practical approach*, T. E. Creighton, Editor. 1997, Oxford University Press: Oxford. pp. 261–298.
7 H. Edelhoch, *Biochemistry*, **1967**, *6*, 1948–1954.
8 S. C. Gill, P. H. von Hippel, *Anal. Biochem.*, **1989**, *182*, 319–326.
9 C. N. Pace, F. X. Schmid, How to determine the molar absorbance coefficient of a protein. In *Protein Structure: A practical approach*, T. E. Creighton, Editor. 1997, Oxford University Press: Oxford. pp. 253–260.
10 C. N. Pace, F. Vajdos, L. Fee, G. Grimsley, T. Gray, *Protein Sci.*, **1995**, *4*, 2411–2423.
11 S. Yanari, F. A. Bovey, *J. Biol. Chem.*, **1960**, *235*, 2818–2826.
12 G. R. Penzer, Molecular emission spectroscopy. In *An Introduction to Spectroscopy for Biochemists*, S. B. Brown, Editor. 1980, Academic Press: London. pp. 70–114.
13 J. R. Lakowicz, *Principles of Fluorescence Spectroscopy*. 1999, New York: Kluwer/Plenum Publishers.
14 P. Wu, L. Brand, *Anal. Biochem.*, **1994**, *218*, 1–13.
15 R. Loewenthal, J. Sancho, A. R. Fersht, *Biochemistry*, **1991**, *30*, 6775–6779.
16 F. W. J. Teale, *Biochem. J.*, **1960**, *76*, 381–387.

17 U. Heinemann, W. Saenger, *Nature*, **1982**, *299*, 27–31.
18 D. A. Egan, T. M. Logan, H. Liang, E. Matayoshi, S. W. Fesik, T. F. Holzman, *Biochemistry*, **1993**, *32*, 1920–1927.
19 R. W. Cowgill, *Biochim. Biophys. Acta*, **1967**, *140*, 37–43.
20 M. R. Eftink, C. A. Ghiron, *Anal. Biochem.*, **1981**, *114*, 199–227.
21 G. D. Fasman, ed. *Circular Dichroism and the Conformational Analysis of Biomolecules*. 1996, Plenum Press: New York.
22 S. Y. Venyaminov, J. T. Yang, Determination of protein secondary structure. In *Circular Dichroism and the Conformational Analysis of Biomolecules*, G. D. Fasman, Editor. 1996, Plenum Press: New York. pp. 69–107.
23 R. W. Woody, A. K. Dunker, Aromatic and cystine side chain circular dichroism in proteins. In *Circular Dichroism and the Conformational Analysis of Biomolecules*, G. D. Fasman, Editor. 1996, Plenum Press: New York. pp. 109–157.
24 P. O. Freskgard, L. G. Martensson, P. Jonasson, B. H. Jonsson, U. Carlsson, *Biochemistry*, **1994**, *33*, 14281–14288.
25 W. C. Johnson, Circular dichroism and its empirical application to biopolymers. In *Methods of Biochemical Analysis*, D. Glick, Editor. 1985, John Wiley: New York. pp. 61–163.
26 M. R. Eftink, Fluorescence techniques for studying protein structure. In *Methods of Biochemical Analysis*, C. H. Suelter, Editor. 1991, John Wiley: New York. pp. 127–205.

3
Denaturation of Proteins by Urea and Guanidine Hydrochloride

C. Nick Pace, Gerald R. Grimsley, and J. Martin Scholtz

3.1
Historical Perspective

The ability of urea to denature proteins has been known since 1900 [1]. Urea is also effective in denaturing more complex biological systems: "A dead frog placed in saturated urea solution becomes translucent and falls to pieces in a few hours" [2]. (It was not reported if frog denaturation is reversible!) The even greater effectiveness of guanidine hydrochloride (guanidinium chloride (GdmCl)) as a protein denaturant was first reported in 1938 by Greenstein [3]. In this chapter we discuss why urea and GdmCl have proven to be so useful to protein chemists and summarize what we have learned over the past 100 years.

Tanford [4] was the first to study quantitatively the unfolding of proteins by urea. The diagram in Figure 3.1 is from Tanford's paper; it shows that when a protein unfolds many nonpolar side chains and peptide groups that were buried in the folded protein are exposed to solvent in the unfolded state. Figure 3.1 also gives the measured values of the free energy of transfer, ΔG_{tr}, of a leucine side chain and a peptide group from water to the solvents shown. Note that ΔG_{tr} of both are negative in the presence of urea and GdmCl. This is true of almost all the constituent groups of a protein, and explains why urea and GdmCl denature proteins, and why GdmCl is a stronger denaturant than urea. It is also clear why both folded and unfolded proteins are generally more soluble in urea and GdmCl solutions than they are in water. Most of this chapter will focus on urea denaturation, however GdmCl denaturation will be discussed where appropriate.

3.2
How Urea Denatures Proteins

Urea is remarkably soluble, 10.5 M, in water at 25 °C. Experimental [5–7] and theoretical studies [8–11] have improved our understanding of urea solutions and how urea and water interact with peptide groups and nonpolar side chains. Based on neutron diffraction studies, Soper et al. [5] show that urea incorporates readily

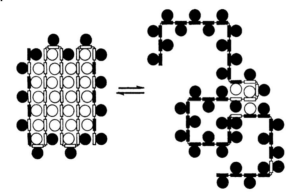

Fig. 3.1. Schematic from Tanford [4] illustrating the changes in accessibility of peptide groups (rectangles) and side chains (circles) on protein unfolding. The closed symbols represent groups that are accessible to solvent and the open symbols represent groups that are not accessible to solvent. See Wang and Bolen [88] for osmolyte, and Pace et al. [56] for denaturant, ΔG_{tr} values.

$\Delta G_{tr}(H_2O \rightarrow \text{Solvent})$ (cal/mol)

Solvent	Peptide group	Leu side chain
Urea (2 M)	-70	-110
GdnHCl (2 M)	-135	-210
Ethanol (40%)	+270	-375
Glycerol (20%)	+55	+16
Sucrose (2 M)	+52	+74
Sarcosine (2 M)	+89	+77
TMAO (2 M)	+177	+18

into water through hydrogen bonding and there is great similarity in the spatial distribution of water around water, urea around water, urea around urea, and water around urea. They estimate that each urea forms ≈5.7 hydrogen bonds. In addition, urea hydrogen bonds with itself, forming chains and clusters in water. (Using a similar approach, Mason et al. [12] show that the Gdm ion has no recognizable hydration shell and is one of the most weakly hydrated cations yet characterized. The Gdm ion can form only three undistorted hydrogen bonds with water.) Urea is nearly three times larger than water, and can form more hydrogen bonds. This may account for the favorable ΔG_{tr} for the transfer of peptide groups from water to urea solutions. This question is considered experimentally and theoretically in Zou et al. [7].

Why ΔG_{tr} values are favorable for the transfer of nonpolar side chains from water to urea solutions was more difficult to explain. Muller [13] suggested that urea molecules replace some of the water molecules in the hydration shell formed around nonpolar solutes, and this changes the hydrogen bonding and van der Waals interactions in such a way as to increase the solubility of the nonpolar solute. This idea is supported by the experimental results of Soper et al. [5], and by the theoretical calculations of Laidig and Daggett [14].

In the absence of denaturant, Timasheff states [15]: "The fact is that there is no rigid shell of water around a protein molecule, but rather there is a fluctuating cloud of water molecules that are thermodynamically affected more or less strongly by the protein molecule." Crystallographers are able to see some water molecules that are bound tightly enough to be observed, but the fact that no water molecules are observed near many polar groups shows that they do not see them all [16]. When crystal structures are determined in the presence of urea and GdmCl, some denaturant molecules are seen bound to the protein, mainly by hydrogen bonds [17, 18]. These denaturant molecules sometimes link adjacent polar groups on the protein and reduce the conformational flexibility, as shown by a decrease in the temperature factors.

There is also experimental evidence that denaturants (and water) reduce the flexibility of the denatured state by transiently linking parts of the protein molecule through hydrogen bonds to polar groups in the protein [19]. Halle's group has used magnetic relaxation dispersion techniques (MRD) to gain a better understanding of protein hydration at the molecular level (see Chapter 8). For example, Modig et al. [6] summarize:

... the urea ^{2}H and water ^{17}O MRD data support a picture of the denatured state where much of the polypeptide chain participates in clusters that are more compact and more ordered than a random coil but nevertheless are penetrated by large numbers of water and urea molecules. These solvent-penetrated clusters must be sufficiently compact to allow side-chains from different polypeptide segments to come into hydrophobic contact, while, at the same time, permitting solvent molecules to interact favorably with peptide groups and with charged and polar side chains. The exceptional hydrogen-bonding capacity and small size of water and urea molecules are likely essential attributes in this regard. In such clusters, many water and urea molecules will simultaneously interact with more than one polypeptide segment and their rotational motions will therefore be more strongly retarded than at the surface of the native protein.

We now have a good understanding of how urea denatures proteins because of careful studies and thoughtful analyses from the labs of Timasheff [20], Record [21], and Schellman [22]. If the excluded volume of the protein is taken into account, the hydrodynamic properties of urea denatured proteins suggest that they approach a randomly coiled conformation [23]. Consequently, the accessible surface area of the denatured state is considerably greater than that of the native state. (This is shown clearly in figure 9 in Goldenberg [23].) Studies by Timasheff [20] and Record [21] have shown that urea binds preferentially to proteins over water, so the concentration of urea near the protein is greater than the concentration in the bulk solvent. Thus, as the urea concentration increases, the denatured state will be favored because it has a much greater surface to interact with the urea (Figure 3.1). Both groups suggest that the dominant contribution is due to preferential binding of urea (or Gdm ions) to the newly exposed peptide groups [20, 21].

Schellman [22] has extended his previous studies to show clearly how osmolytes such as trimethylamine-N-oxide (TMAO) or denaturants such as urea act to stabilize or destabilize proteins. The two key factors are the excluded volumes of the co-solvents and their contact interactions with the protein. Because the molecules of interest are all larger than water, the excluded volume effect will always favor the native state. In contrast, contact interactions will generally favor the denatured state. For urea, the contact interaction term is greater than the excluded volume effect, so urea acts as a protein denaturant. For osmolytes, the reverse is true and osmolytes stabilize proteins. Schellman [24] was the first to show that because the direct interaction of urea molecules with proteins is so weak, denaturation should be represented by a solvent exchange mechanism rather than solely by the binding of the denaturant to the protein molecule. His $K_{av} - 1$ term is a measure of the contact interaction for the average site on a protein. For RNase T1, this term is 0.12 for TMAO, 0.22 for urea, and 0.40 for guanidinium ion. The larger value of the contact term reflects the greater preferential binding of the Gdm ion compared to urea, and this is what makes GdmCl a better denaturant than urea [22].

In summary, it is now clear that urea denatures proteins because there is a preferential interaction with urea compared with water, and the denatured state has more sites for interaction. Consequently, proteins unfold as the urea concentration is increased.

3.3
Linear Extrapolation Method

A typical urea denaturation curve is shown in Figure 3.2A. When a protein unfolds by a two-state mechanism, the equilibrium constant, K, can be calculated from the experimental data using:

$$K = [(y)_N - (y)]/[(y) - (y)_D] \tag{1}$$

where (y) is the observed value of the parameter used to follow unfolding, and $(y)_N$ and $(y)_D$ are the values (y) would have for the native state and the denatured state under the same conditions where (y) was measured. In the original analyses of urea denaturation curves [25, 26], log K was found to vary linearly as a function of log [urea] and the slope of the plot was denoted by n, and the midpoint of the curve by $(urea)_{1/2}$ (where log $K = 0$). These parameters could then be used to calculate the standard free energy of denaturation

$$\Delta G° = -RT \ln K \tag{2}$$

and its dependence on urea concentration using [4]:

$$d(\Delta G°)/d(urea) = RTn/(urea)_{1/2} \tag{3}$$

This equation was used by Alexander and Pace [25] to estimate the differences in

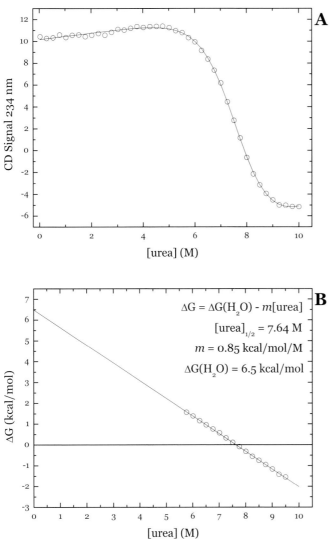

Fig. 3.2. A) Urea denaturation curve for RNase Sa determined at 25 °C, pH 5.0, 30 mM sodium acetate buffer as described by Pace et al. [78]. The solid line is based on an analysis of the data using Eq. (5). B) ΔG as a function of urea molarity. The ΔG values were calculated from the data in the transition region as described in the text. The $\Delta G(H_2O)$ and m-values can be determined by fitting the data in B to Eq. (4), or by analyzing the data in A with Eq. (5).

stability among three genetic variants of a protein for the first time, and was the forerunner of the linear extrapolation method.

The linear extrapolation method (LEM) was first used to analyze urea and GdnHCl denaturation curves by Greene and Pace [27]. When $\Delta G°$ is calculated as a function of urea concentration using data such as those shown in Figure 3.2A, $\Delta G°$ is found to vary linearly with urea concentration as shown in Figure 3.2B. Based on results such as these for several proteins, this equation was proposed for analyzing the data:

$$\Delta G = \Delta G(H_2O) - m[\text{urea}] \tag{4}$$

where $\Delta G(H_2O)$ is an estimate of the conformational stability of a protein that assumes that the linear dependence continues to 0 M denaturant, and m is a measure of the dependence of ΔG on urea concentration, i.e., the slope of the plot shown in Figure 3.2B. The same approach has recently been used for measuring the stability of RNA molecules [28].

In the early days, $(y)_N$ and $(y)_D$ were obtained by extrapolating the pre- and post-transition baselines into the transition region, and then using Eq. (1) to calculate K and then $\Delta G°$. However, Santoro and Bolen [29] had a better idea and proposed that nonlinear least squares be used to directly fit data such as those shown in Figure 3.2A. With their approach, six parameters are used to fit the data: a slope and an intercept for the pretransition and posttransition regions, which are generally linear, and $\Delta G(H_2O)$ and m for the transition region, leading to

$$y = \{(y_F + m_F[\text{urea}]) + (y_U + m_U[\text{urea}]) \cdot \exp - ((\Delta G(H_2O) - m \cdot [\text{urea}])/RT)\}$$
$$/(1 + \exp - (\Delta G(H_2O) - m \cdot [\text{urea}])/RT)) \tag{5}$$

where y_F and y_U are the intercepts and m_F and m_U the slopes of the pre- and post-transition baselines, and $\Delta G(H_2O)$ and m are defined by Eq. (4).

The linear extrapolation method is now widely used for estimating the conformational stability of proteins and for measuring the difference in stability between proteins differing slightly in structure. (The original how-to-do-it paper on solvent denaturation [30] has been cited over 1200 times.) In addition, the m-value has taken on a life of its own [31]. These topics will be discussed further below.

3.4
$\Delta G(H_2O)$

The conformational stability of RNase Sa at 25 °C and pH 5 is ~ 7.0 kcal mol^{-1}. This corresponds to one unfolded molecule for each 135 000 folded molecules ($K = 7.4 \times 10^{-6}$). To study the equilibrium between the folded and unfolded conformations by conventional techniques requires destabilizing the protein so that both conformations are present at measurable concentrations. With urea denaturation this is done by increasing the urea concentration. It is clear from Figure 3.2A that the unfolding equilibrium can be studied only near 7 M urea, and that a long extrapolation is needed to estimate $\Delta G°$ in the absence of urea, $\Delta G(H_2O)$. The same is true for thermal denaturation. The unfolding equilibrium can be studied only near 55 °C (Figure 3.3A). Thermal denaturation curves can be analyzed to ob-

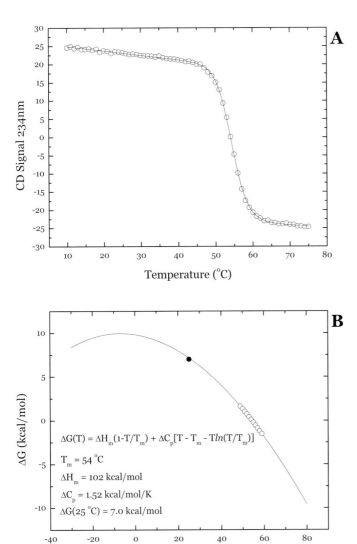

Fig. 3.3. A) Thermal denaturation curve for RNase Sa determined at pH 5.0 in 30 mM sodium acetate buffer as described [78]. The solid line is based on a nonlinear least squares analysis of the data similar to that described for Eq. (5), but for thermal denaturation curves [78]. B) ΔG as a function of temperature. The ΔG values were calculated from data in the transition region as described in the text, and the data were analyzed to determine T_m and ΔH_m as described [78]. The ΔC_p value is from Pace et al. [78]. The $\Delta G(25\,°C)$ value (filled circle) was calculated using Eq. (6) which is also shown in B.

tain the midpoint of the thermal denaturation curve, T_m, and the enthalpy change at T_m, ΔH_m. These two parameters plus the heat capacity change for folding, ΔC_p, can then be used in the Gibbs-Helmholtz equation [32]:

$$\Delta G(T) = \Delta H_m(1 - T/T_m) + \Delta C_p[T - T_m - T\ln(T/T_m)] \quad (6)$$

to calculate ΔG at any temperature T, $\Delta G(T)$. The temperature dependence of ΔG and the value of $\Delta G(25\,°C)$ for the denaturation of RNase Sa are shown in Figure 3.3B. Note that the value of $\Delta G(25\,°C)$ from the thermal denaturation experiment agrees with the $\Delta G(H_2O)$ value from the urea denaturation experiments within experimental error, which is about ± 0.3 kcal mol^{-1} (see Eftink and Ionescu [33] for an excellent discussion of the problems most frequently encountered in analyzing solvent and thermal denaturation curves).

Previously, we used RNase T1 to show that estimates of $\Delta G(T)$ based on DSC experiments were in good agreement with estimates of $\Delta G(H_2O)$ from urea denaturation [34, 35]. In a more recent study using RNase A, we compared $\Delta G(T)$ values from DSC determined in Makhatadze's laboratory with $\Delta G(H_2O)$ values determined using urea denaturation and the LEM [36] in our laboratory. The agreement is remarkably good (Table 3.1). Most recently, we have compared conformational stabilities estimated from hydrogen exchange rates measured under native state conditions with those from thermal or solvent denaturation [37]. Again, the conformational stabilities are in good agreement.

There is considerable experimental evidence that proteins are more extensively unfolded in solutions of urea than they are in water (see, for example, Qu et al.

Tab. 3.1. Comparison of $\Delta G(T)$ values from DSC with $\Delta G(H_2O)$ values from UDC.

pH	T (°C)	$T_m{}^a$ (°C)	$\Delta H_m{}^a$ (kcal mol^{-1})	$\Delta G(T)^a$ (kcal mol^{-1})	$\Delta G(H_2O)^b$ (kcal mol^{-1})
2.8	17.1	44.9	79.4	5.5	5.4
2.8	21.1	44.9	79.4	4.9	4.9
2.8	24.9	44.9	79.4	4.3	4.3
2.8	27.8	44.9	79.4	3.7	3.5
2.8	25.0	44.9	79.4	4.5	4.3
3.0	25.0	49.1	82.7	5.1	5.2
3.55	25.0	54.5	91.5	6.7	6.4
4.0	25.0	56.1	94.2	7.2	7.3
5.0	25.0	58.6	99.1	8.1	7.9
6.0	25.0	60.3	100.7	8.5	8.6
7.0	25.0	61.8	102.3	8.9	9.1

[a] The T_m, ΔH_m, and $\Delta C_p = 1.15$ kcal mol^{-1} K^{-1} values are from Pace et al. [36]. They were used in Eq. (6) to calculate the $\Delta G(T)$ values.
[b] The first four $\Delta G(H_2O)$ values are from Pace and Laurents [39], and the last five $\Delta G(H_2O)$ values are interpolated from figure 6 in Pace et al. [56].

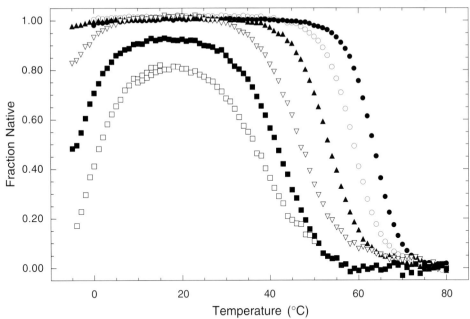

Fig. 3.4. Thermal unfolding curves for *ec*HPr in various urea concentrations at pH 7.0 (10 mM K/P$_i$) as monitored by the change in ellipticity at 222 nm and converted to fraction folded, f_F. The error on any single point is smaller than the size of the symbol. The urea concentrations are 0 M (filled circle), 1 M (open circle), 2 M (filled triangle), 3 M (open triangle), 3.5 M (filled square), and 4 M (open square). From Nicholson and Scholtz [41].

[38]). Thus, it is surprising that conformational stabilities determined under conditions where the denatured state ensembles differ would be the same. This suggests that the denatured state ensemble that exists under physiological conditions is thermodynamically equivalent to the ensembles that exist after thermal or urea denaturation, even though they do not appear to be structurally equivalent. This is supported by studies where both thermal and urea induced unfolding are used to study the conformational stability of a protein.

This approach had its beginnings in a study that showed that results from thermal and urea denaturation could be combined to get a reasonable estimate of ΔC_p [39]. The method was then extended to characterize the stability of the HPr proteins from *Bacillus subtilis* [40] and *Escherichia coli* [41] as a function of both urea concentration and temperature. Some typical results are shown in Figure 3.4. More recently, Felitsky and Record [42] used this approach to characterize the stability of the Lac repressor DNA-binding domain in the most detailed study of the stability of a protein yet published. A related approach has been described by Ibarra-Molera and Sanchez-Ruiz [43].

The excellent agreement between conformational stabilities based on DSC results and the LEM suggests that the LEM is a reliable method for estimating the conformational stability of a protein. The Record group has developed an approach for analyzing the results of urea and GdmCl denaturation studies in terms of preferential interaction coefficients that is firmly based on thermodynamics. This method is called the local-bulk domain model and they show that this analysis "justifies the use of linear extrapolation to estimate ($\leq 5\%$ error) the stability of the native protein in the absence of a denaturation" [21]. The history of this area is summarized in a delightful review titled: "Fifty years of solvent denaturation" by Schellman [44].

Two other methods have been used to analyze urea denaturation curves. They both give estimates of $\Delta G(H_2O)$ that are larger than those based on the LEM. The differences are not large with urea, but they are with GdmCl. One is an extrapolation based on Tanford's model [4] that is discussed in the next section. This method uses ΔG_{tr} values for transfer of the constituent groups of a protein from water to urea solutions. Unfortunately, there is uncertainty in the ΔG_{tr} value to use for a peptide group [38], and this introduces uncertainty into the $\Delta G(H_2O)$ values obtained by this method [45]. Experiments are underway to determine a more reliable value of ΔG_{tr} for a peptide group [7] (Auton and Bolen, personal communication). Makhatadze [46] has shown that the data on which this method is based are consistent with the LEM when urea but not GdnHCl is used as the denaturant.

The other method that can be used to determine $\Delta G(H_2O)$ is the denaturant binding model first used by Aune and Tanford [47]. The plots of log K vs. log [urea] mentioned above show that more urea molecules are bound by the denatured state than by the native state. Also, urea increases the solubility of all of the constituent parts of a protein, and it is possible to account for the enhanced solubility in terms of urea binding, even for leucine side chains (see table XI in Tanford [48]). Thus, it might seem reasonable to try to analyze urea denaturation curves in terms of the denaturant binding model. Generally, all of the urea binding sites are considered identical and independent and $K = 0.1$ is used for the binding constant [30]. However, this is surely not stoichiometric binding and it is unlikely that the binding sites are identical and independent. The enhanced solubility of the model compound data requires "binding constants" in the range 0.05–0.3 [46, 48, 49]. Schellman has shown convincingly that a solvent exchange model is more reasonable than a stoichiometric binding model, and this model is consistent with the LEM. (Schellman's recent article [22] gives references to his earlier work. It is also reviewed in Timasheff's paper [50].)

At first it was puzzling that the LEM worked so well. It seemed likely that there would sometimes be binding sites for urea molecules on folded proteins of at least moderate affinity and this would lead to nonlinearity that would cause the LEM to give erroneous results. Apparently, this rarely occurs. Our understanding of how urea denatures proteins has improved, thanks to the approaches of Timasheff [20], Record [21], and Schellman [22], which all show now why the LEM method works as well as it does.

$\Delta G(H_2O)$ values from GdmCl denaturation studies are less reliable, as pointed out most recently by Makhatadze [46]. This conclusion was reached earlier by Schellman and Gassner [51] who said:

Finally this study as well as a number of others indicates that urea has a number of advantages over guanidinium chloride as far as thermodynamic interactions are concerned. The free energy and enthalpy functions of both proteins and model compounds are more linear; the solutions more ideal; extrapolations to zero concentration are more certain; and least-squares analysis of the data is more stable.

One problem is that the ionic strength cannot be controlled with GdmCl, and this has been shown to affect the results in several studies (see, for example, Ibarra-Molero et al. [52], Monera et al. [53], Santoro and Bolen [54]).

In summary, the agreement between $\Delta G(T)$ values from DSC and thermal denaturation studies and $\Delta G(H_2O)$ values from urea denaturation studies analyzed by the LEM suggests that either method can be used to reliably measure the conformational stability of a protein. In addition, we now have a good understanding of why the LEM works so well.

3.5
m-Values

The thermodynamic cycle in Figure 3.5 shows that the dependence of ΔG on denaturant concentration is determined by the groups in a protein that are exposed to solvent in the denatured state, D, but not in the native state, N[4]. In support of this, Myers et al. [55] showed that there is a good correlation between m-values and the change in accessible surface area on unfolding. Since $\Delta g_{tr,i}$ values are available for

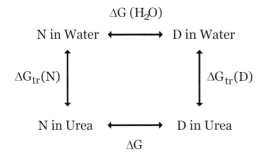

$$\Delta G + \Delta G_{tr}(N) = \Delta G(H_2O) + \Delta G_{tr}(D)$$
$$\Delta G - \Delta G(H_2O) = \Delta G_{tr}(D) - \Delta G_{tr}(N) = \Sigma n_i \alpha_i \Delta g_{tr,i}$$
$$\Delta G = \Delta G(H_2O) + \Sigma n_i \alpha_i \Delta g_{tr,i}$$
$$\Delta G = \Delta G(H_2O) + (\Delta \alpha) \Sigma n_i \Delta g_{tr,i}$$

Fig. 3.5. A thermodynamic cycle connecting protein folding in water and urea. N and D denote the native and denatured states, ΔG and $\Delta G(H_2O)$ are defined by Eq. (4), n_i is the total number of groups of type i in the protein, α_i is fraction of groups of type i that are exposed in D but not N, and $\Delta g_{tr,i}$ is the free energy of transfer of groups of type i from water to a given concentration of urea. $\Delta \alpha$ is the average change in exposure of all of the peptide groups and uncharged side chains in the protein [56].

most of the constituent groups in a protein, the equation at the bottom of Figure 3.5 can be used to calculate the dependence of ΔG on denaturant concentration. The m-value is an experimental measure of the dependence of ΔG on denaturant concentration. Consequently, for proteins that unfold by a two-state mechanism, $\Delta\alpha$ can be estimated by comparing the calculated and measured m-values. Thus, $\Delta\alpha$ is the fraction of buried groups that must become exposed to solvent on unfolding to account for the measured m-value [56]. In the original LEM paper [27], this approach was used to estimate $\Delta\alpha$ for four proteins. It was clear that differences among the $\Delta\alpha$ values for individual proteins revealed differences in the accessibility of the denatured state and that these differences depended to some extent on the number and location of the disulfide bonds in the protein.

Interest in m-values was stimulated by a series of papers from the Shortle laboratory [57, 58]. (A review of these studies was titled: "Staphylococcal nuclease: a showcase of m-value effects." [31].) Their studies of mutants of staphylococcal nuclease (SNase) showed a wide range of m-values. Those with m-values 5% or more greater than wild type were designated m^+ mutants ($\approx$ 25%) and those with m-values 5% or more less than wild type were designated m^- mutants ($\approx$ 50%). Their interpretation was that m^+ mutants unfolded to a greater extent than wild type and vice versa for m^- mutants. These papers were extremely important because they focused attention on the denatured state which had largely been ignored in discussions of protein stability. (Another review from the Shortle lab [58] was titled: "The denatured state (the other half of the folding equation) and its role in protein stability.") However, this interpretation is only straightforward if all of the mutants unfold by a two-state mechanism. Shortle and Meeker [57] chose to assume a two-state mechanism to analyze their results even though the m^- mutants did not appear to unfold by a two-state mechanism. It is now clear that some of the m^- mutants unfold by a three-state mechanism [59–62]. The presence of an intermediate state will generally lower the m-value [63]. One important consequence of this is that the estimates of $\Delta G(H_2O)$ will be too low when a two-state mechanism is assumed and the mechanism is, in fact, three-state. In the case of the m^- mutant V66L + G88V + G79S, a two-state analysis indicates that this mutant is 3.0 kcal mol^{-1} LESS stable than wild type [57], but a three-state analysis indicates that that this mutant is 0.8 kcal mol^{-1} MORE stable than wild type [60]. This is also true for other m^- mutants [64]. This means that the $\Delta(\Delta G)$ values for the m^- mutants are not correct. Since 50% of the SNase mutants are m^-, this calls into question the interpretation of many of the $\Delta(\Delta G)$ values determined for SNase.

In Table 3.2, we have compiled m and $\Delta\alpha$ values for a selection of proteins. In all cases, the folding of the proteins is thought to closely approach a two-state mechanism. The value of $\Delta\alpha$ expected for unfolding a protein depends on the model assumed for the denatured state. If the accessibility of the denatured state is modeled as a tripeptide using the Lee and Richards [65] approach, a value of $\Delta\alpha \approx 0.7$ would be expected. However, if a compact denatured state based on fragments with native-state conformations is assumed, a $\Delta\alpha \approx 0.53$ would be expected [66]. Note in Table 3.2 that most of the $\Delta\alpha$ values are considerably less than 0.53. This suggests that the urea-denatured states of proteins are less completely unfolded than in the hypothetical state that Creamer et al. [66] thought might be a lower limit for

Tab. 3.2. Change in accessibility ($\Delta\alpha$) for urea denaturation calculated from the measured m-values using Tanford's model[a].

Protein	Residues	Disulfides	(Urea)$_{1/2}$	m	$\Delta\alpha$
RNase Sa (0K) (pH 7)[b]	96	1	6.44	0.99	0.32
RNase Sa (0K) (pH 3)[b]	96	1	1.48	1.75	0.41
RNase Sa (3K) (pH 7)[b]	96	1	4.07	0.99	0.29
RNase Sa (3K) (pH 3)[b]	96	1	1.47	2.00	0.46
RNase Sa (5K) (pH 7)[b]	96	1	5.31	1.14	0.36
RNase Sa (5K) (pH 3)[b]	96	1	1.30	2.21	0.50
RNase T1[c]	104	2	5.30	1.21	0.32
RNase A[d]	124	4	6.98	1.40	0.35
Lysozyme (Hen)[c]	129	4	6.80	1.29	0.29
SH3 domain[c]	62	0	3.74	0.77	0.34
Calbindin[c]	75	0	5.30	1.13	0.51
HPr (E. coli)[e]	88	0	4.32	1.14	0.45
Acyl phosphatase[f]	99	0	4.54	1.25	0.39
λ Repressor[c]	102	0	4.42	1.09	0.38
RNase T1[d]	104	0	3.34	1.64	0.38
FKBP[c]	107	0	4.18	1.46	0.42
Thioredoxin[c]	108	0	6.70	1.30	0.26
Barnase[c]	110	0	4.49	1.94	0.50
Che Y[c]	129	0	3.51	1.60	0.36
FABP[c]	131	0	5.43	1.77	0.40
SNase[g]	149	0	2.56	2.36	0.44
ApoMb[h]	153	0	2.05	2.49	0.47
RNase H[c]	155	0	3.88	1.93	0.35
DHFR (E. coli)[c]	159	0	3.10	1.90	0.32
T4 Lysozyme[c]	163	0	6.30	2.00	0.38

[a] Tanford's model is described in Tanford [4] and the details of the approach used to calculate $\Delta\alpha$ are given in Pace et al. [56]. (Urea)$_{1/2}$ values are given in M; m values are given in kcal mol^{-1} M^{-1}.
[b] The (urea)$_{1/2}$ and m-values are from Shaw [82].
[c] The original reference for these proteins is given in Myers et al. [55].
[d] See Pace et al. [56].
[e] See Nicholson and Scholtz [41].
[f] See Taddei et al. [83].
[g] See Shortle [31].
[h] See Hughson and Baldwin [84].

the compactness of the denatured state at the midpoint of a thermal denaturation curve in water. However, because of uncertainty in the ΔG_{tr} value for a peptide group and for some of the side chains, interpreting the absolute values of $\Delta\alpha$ is hazardous. In contrast, differences among the $\Delta\alpha$ values for individual proteins or for the same protein under different conditions are more trustworthy.

For the proteins without disulfide bonds, the $\Delta\alpha$ values in Table 3.2 range from a low of 0.26 for thioredoxin to a high of 0.51 for calbindin. Thus, if these proteins unfold by a two-state mechanism, calbindin unfolds to a much greater extent than thioredoxin. Barnase also unfolds more completely than most of the other proteins in Table 3.2. In support of this, we have shown that the tryptophan and tyrosine side chains of urea-denatured barnase are more accessible to a perturbant than those of several other proteins that have lower $\Delta\alpha$ values [67]. The results in Table 3.2 suggest that urea-denatured proteins are not completely unfolded and that they unfold to different extents.

We were surprised to find that for RNases A and T1 [56] and barnase [67], the m-value increases markedly as the pH is lowered from 7 to 3. This indicates that the denatured states interact more extensively with urea as the pH is lowered, and we suggested that the denatured state ensemble expands at low pH due to electrostatic repulsion among the positive charges. In support of this interpretation, Privalov et al. [68] have shown that the intrinsic viscosity of unfolded proteins increases at low pH. We have recently obtained results with RNase Sa that provide further support for this idea [69].

RNase Sa is an acidic protein with a pI = 3.5 that contains no lysine residues. We have prepared a triple mutant that we call 3K (D1K, D17K, E41K) with a pI = 6.4, and a quintuple mutant that we call 5K (D1K, D17K, D25K, E41K, E74K) with a pI = 10.2. At pH 3, the estimated net charges are +8 for wild-type RNase Sa, +11 for 3K, and +13 for 5K. The m-values for the three proteins at pH 7 and pH 3 are given in Table 3.2. For all three proteins, the m-value is considerably greater at pH 3 than at pH 7, and the $\Delta\alpha$ values increase from 0.32 for wild-type RNase Sa to 0.36 for 3K to 0.50 for 5K. This is consistent with an increase in accessibility to denaturant caused by an expansion of the denatured state due to electrostatic repulsion among the positive charges. As the pH increases from 3 to 7, the carboxyl groups are titrated and both negative and positive charges are present. Now attractive charge–charge interactions are possible, and the decrease in the m-values suggests that the denatured state ensemble is more compact because of attractive Coulombic interactions. In support of this, the $\Delta\alpha$ value at pH 7 is lowest for 3K where the number of positive and negative charges is almost equal that for wild-type RNase Sa, which has an excess of negative charges, or 5K, which has an excess of positive charges. This and other evidence led us to conclude that long-range electrostatic interactions are important in determining the denatured state ensemble [69].

We suggest that when the hydrophobic and hydrogen bonding interactions that stabilize the folded state are disrupted, the unfolded polypeptide chain rearranges to compact conformations that improve long-range electrostatic interactions. These charge–charge interactions in the denatured state will reduce the net contribution of electrostatic interactions to protein stability but will help determine the denatured state ensemble. If this idea is correct, then long-range charge–charge interactions in the denatured state may play an important role in the mechanism of protein folding.

3.6
Concluding Remarks

Protein denaturants are used for a variety of purposes by protein chemists when they need to solubilize or denature proteins. Urea and GdmCl are the denaurants most often used. The LEM has proven to be a reliable method for measuring the conformational stability of a protein or the difference in stability between two proteins that differ slightly in structure. In addition, studies of the *m*-values of proteins, especially SNase, have focused attention on the denatured state and we are slowly gaining a better understanding of the denatured state ensemble.

3.7
Experimental Protocols

3.7.1
How to Choose the Best Denaturant for your Study

Both urea and GdmCl will increase the solubility of the folded and unfolded states of a protein. If the denaturant is just going to be used to increase the solubility or unfold your protein of interest, GdmCl is preferred since it is a more potent denaturant and is chemically stable. Guanidinium thiocyanate is an even more potent denaturant than GdmCl and can be used for the same purposes. (The reason for the greater potency is discussed in Mason et al. [12].)

For quantitative studies of protein stability, we previously preferred results from urea and GdmCl unfolding over thermal unfolding studies for several reasons: the product of unfolding seemed to be better characterized, unfolding was more likely to closely approach a two-state mechanism, and unfolding was more likely to be completely reversible. However, in the cases that have been carefully investigated, the two techniques seem to give estimates of $\Delta G(H_2O)$ that are in good agreement [36, 70]. After getting the equipment set up and learning the procedures, either type of unfolding curve can be determined and analyzed in a single day. Consequently, the safest course is to determine both types of unfolding curves whenever possible. Also, we generally prefer urea over GdmCl because salt effects can be investigated, and sometimes reveal interesting information [46, 51, 71, 72].

3.7.2
How to Prepare Denaturant Solutions

Both urea and GdmCl can be purchased commercially in highly purified forms. However, some lots of these denaturants may contain fluorescent or metallic impurities. Methods are available for checking the purity of GdmCl and for recrystallization if necessary [73]. A procedure for purifying urea has also been described [74]. GdmCl solutions are stable for months, but urea solutions slowly decompose to form cyanate and ammonium ions in a process accelerated at high pH [75]. The cyanate ions can react with amino groups on proteins [76]. Consequently, a fresh urea stock solution should be prepared for each unfolding curve and used within the day.

Table 3.3 summarizes useful information for preparing urea and GdmCl stock solutions. We prepare urea stock solutions by weight, and then check the concentration by refractive index measurements using the equation given in Table 3.3. The preparation of a typical urea stock solution is outlined below. Since GdmCl is

Tab. 3.3. Information for preparing urea and GdmCl stock solutions.

Property	Urea	GdmCl
Molecular weight	60.056	95.533
Solubility (25.0 °C)	10.49 M	8.54 M
d/d_0 [a]	$1 + 0.2658W + 0.0330W^2$	$1 + 0.2710W + 0.0330W^2$
Molarity[b]	$117.66(\Delta N) + 29.753(\Delta N)^2 + 185.56(\Delta N)^3$	$57.147(\Delta N) + 38.68(\Delta N)^2 - 91.60(\Delta N)^3$
Grams of denaturant per gram of water to prepare		
6 M	0.495	1.009
8 M	0.755	1.816
10 M	1.103	–

[a] W is the weight fraction denaturant in the solution, d is the density of the solution and d_0 is density of water [85].
[b] ΔN is the difference in refractive index between the denaturant solution and water (or buffer) at the sodium D line. The equation for urea solutions is based on data from Warren and Gordon [86], and the equation for GdmCl solutions is from Nozaki [73].

quite hygroscopic, it is more difficult to prepare stock solutions by weight. Consequently, the molarity of GdmCl stock solutions is generally determined from the refractive index measurement (step 5 below).

Preparation of a urea stock solution This describes the preparation of ~100 mL of ~10 M urea stock solution containing 30 mM MOPS buffer, pH 7.0. We use a top loading balance with an accuracy of about ±0.02 g.

1. Add ≈60 g of urea to a tared beaker and weigh (59.91 g). Now add 0.69 g of MOPS buffer (sodium salt), 1.8 mL of 1 M HCl and ≈52 mL of distilled water and weigh the solution again (114.65 g).
2. Allow the urea to dissolve and check the pH. If necessary, add a weighed amount of 1 M HCl to adjust to pH 7.0.
3. Prepare 30 mM MOPS buffer, pH 7.0.
4. Determine the refractive index of the urea stock solution (1.4173) and of the buffer (1.3343). Therefore, $\Delta N = 1.4173 - 1.3343 = 0.0830$.
5. Calculate the urea molarity from ΔN using the equation given in Table 3.3: M = 10.08.
6. Calculate the urea molarity based on the recorded weights. The density is calculated with the equation given in Table 3.1: weight fraction area $(W) = 59.91/114.65 = 0.5226$; therefore $d/d_0 = 1.148$. Therefore volume = 114.65/1.148 = 99.88 mL. Therefore urea molarity = 59.91/60.056/.09988 = 9.99 M.

3.7.3
How to Determine Solvent Denaturation Curves

Figure 3.2A shows a typical urea unfolding curve for RNase Sa using CD to follow unfolding. We will refer to the physical parameter used to follow unfolding as y

for the following discussion. The curves can be conveniently divided into three regions: (i) The pretransition region, which shows how y for the folded protein, y_F, depends upon the denaturant. (ii) The transition region, which shows how y varies as unfolding occurs. (iii) The posttransition region, which shows how y for the unfolded protein, y_U, varies with denaturant. All of these regions are important for analyzing unfolding curves. As a minimum, we recommend determining four points in the pre- and posttransition regions, and five points in the transition region. Of course, the more points determined the better defined the curve.

With thermodynamic measurements, it is essential that equilibrium is reached before measurements are made, and that the unfolding reaction is reversible. The time required to reach equilibrium can vary from seconds to days, depending upon the protein and the conditions. For example, the unfolding of RNase T1 requires hours to reach equilibrium at 25 °C, as compared to only minutes for RNase Sa. For any given protein, the time to reach equilibrium is longest at the midpoint of the transition and decreases in both the pre- and posttransition regions. To ensure that equilibrium is reached, y is measured as a function of time to establish the time required to reach equilibrium.

To test the reversibility of unfolding, allow a solution to reach equilibrium in the posttransition region and then by dilution return the solution to the pretransition region and measure y. If the reaction is reversible, the value of y measured after complete unfolding should be within $\pm 5\%$ of that determined directly. For many proteins, unfolding by urea and GdmCl is completely reversible. We have left RNase T1 in 6 M GdmCl for over 3 months and found that the protein will refold completely on dilution.

Proteins containing free sulphydryl groups present special problems. If the protein contains only free –SH groups and no disulfide bonds, then a reducing agent such as 10 mM dithiothreitol (DTT) can be added to ensure that no disulfide bonds form during the experiments. For proteins containing both free –SH groups and disulfide bonds, disulfide interchange can occur and this may lead to irreversibility. Disulfide interchange can be minimized by working at low pH [77].

If the unfolding of the protein of interest requires considerable time to reach equilibrium, then each point shown in Figure 3.2A will have to be determined from a separate solution, prepared volumetrically using the best available pipettes. A fixed volume of protein stock solution is mixed with the appropriate volumes of buffer, and urea or GdmCl stock solutions. The protein and buffer solutions are prepared by standard procedures. After the solutions for measurement have been prepared (see the protocol below), they are incubated until equilibrium is reached at the temperature chosen for determining the unfolding curve. After the measurements have been completed, it is good practise to measure the pH of the solutions in the transition region. If the protein equilibrates quickly, as is the case with RNase Sa, titration methods can be used to determine the denaturation curve. This was the method used to obtain the curve in Figure 3.2A. This protocol is described in detail in Pace [78] and Scholtz [40]. If the amount of protein is limited, instruments have been developed to follow the unfolding of a protein simultaneously with several spectral techniques using a single protein solution [79].

Tab. 3.4. Experimental results from a urea denaturation curve[a].

Urea[b] (mL)	Buffer[c] (mL)	Protein[d] (mL)	Urea[e] (M)	θ_{234} (mdeg)[f]
0.298	2.152	0.050	1.2	18
0.795	1.655	0.050	3.2	18
1.193	1.257	0.050	4.8	16
1.267	1.183	0.050	5.1	14
1.342	1.108	0.050	5.4	12
1.392	1.058	0.050	5.6	10
1.466	0.984	0.050	5.9	7
1.542	0.909	0.050	6.2	4
1.615	0.835	0.050	6.5	1
1.690	0.760	0.050	6.8	−2
1.740	0.710	0.050	7.0	−4
1.789	0.661	0.050	7.2	−4
1.839	0.611	0.050	7.4	−5
2.087	0.363	0.050	8.4	−5
2.286	0.164	0.050	9.2	−6

[a] Only two or three solutions are shown for the pre- and posttransition regions. In the actual experiment, a total of 32 solutions were prepared and measured.
[b] 10.06 M urea stock solution (30 mM MOPS, pH 7.0).
[c] 30 mM MOPS buffer, pH 7.0.
[d] 5.0 mg mL^{-1} RNase Sa E41H stock solution (30 mM MOPS, pH 7.0).
[e] Urea molarity = 10.06 ((mL urea)/(mL urea + mL protein + mL buffer)).
[f] Circular dichroism measurements at 234 nm using an Aviv 62DS spectropolarimeter.

3.7.3.1 Determining a Urea or GdmCl Denaturation Curve

1. Prepare three solutions: a denaturant stock solution, a protein stock solution and a buffer solution.
2. Prepare the sample solutions volumetrically (e.g., with Rainin EDP2 pipettes) in clean, dry test tubes. The solutions made to determine a urea unfolding curve for E41H RNase Sa, along with the resulting CD measurements, are given in Table 3.4.
3. Allow these solutions to equilibrate at the temperature chosen for the experiment until they reach equilibrium. (This is best determined in a separate experiment as described in the text.)
4. Measure the experimental parameter being used to follow unfolding on the solutions in order of increasing denaturant concentration. Leave the cuvette in the spectrophotometer, and do not rinse between samples. Simply remove the old solution carefully with a pasteur pipette with plastic tubing attached to the tip, and then add the next sample. Leaving the cuvette in position improves the quality of the measurements, and the error introduced by the small amount of old solution is negligible.

5. Plot these results to determine if any additional points are needed. If so, prepare the appropriate solutions and make the measurements as described for the original solutions.
6. Measure the pH of the solutions in the transition region.
7. Analyze the results as described below.

3.7.3.2 How to Analyze Urea or GdmCl Denaturant Curves

We assume a two-state folding mechanism for the discussion here (this will be correct for many small globular proteins). Consequently, for any of the points shown in Figure 3.2A, only the folded and unfolded conformations are present at significant concentrations, and $f_F + f_U = 1$, where f_F and f_U represent the fraction protein present in the folded and unfolded conformations, respectively. Thus, the observed value of y at any point will be $y = y_F f_F + y_U f_U$, where y_F and y_U represent the values of y characteristic of the folded and unfolded states, respectively, under the conditions where y is being measured. Combining these equations gives:

$$f_U = (y_F - y)/(y_F - y_U) \tag{A1}$$

The equilibrium constant, K, and the free energy change, ΔG, can be calculated using

$$K = f_U/f_F = f_U/(1 - f_U) = (y_F - y)/(y - y_U) \tag{A2}$$

and

$$\Delta G = -RT \ln K = -RT \ln[(y_F - y)/(y - y_U)] \tag{A3}$$

where R is the gas constant (1.987 cal mol^{-1} K^{-1} or 8.3144 J K^{-1} mol^{-1}) and T is the absolute temperature. Values of y_F and y_U in the transition region are obtained by extrapolating from the pre- and posttransition regions in Figure 3.2A. Usually y_F and y_U are linear functions of denaturant concentration, and a least-squares analysis is used to determine the linear expressions for y_F and y_U.

The calculation of f_U, K, and ΔG from the data in Figure 3.2A is illustrated in Table 3.5. Values of K can be measured most accurately near the midpoints of the curves, and the error becomes substantial for values outside the range 0.1–10. Consequently, we generally only use ΔG values within the range ± 1.5 kcal mol^{-1}. See Eftink and Ionescu [33] and Allan and Pielak [80] for discussions of the problems most frequently encountered in analyzing solvent and thermal denaturation curves.

As discussed above, the simplest method of estimating the conformational stability in the absence of urea, $\Delta G(H_2O)$, is to use the LEM, which typically gives results such as those shown in Figure 3.2B. We strongly recommend that values of $\Delta G(H_2O)$, m, and $[D]_{1/2}$ be given in any study of the unfolding of a protein by

Tab. 3.5. Analysis of a urea denaturation curve[a].

Urea (M)	y	y_F	y_U	f_F	f_U	K	ΔG (kcal mol^{-1})
6.26	9.16	12.01	−10.53	0.87	0.13	0.14	1.15
6.51	8.38	12.08	−10.22	0.83	0.17	0.20	0.95
6.76	7.35	12.16	−9.88	0.78	0.22	0.28	0.76
7.01	6.17	12.23	−9.55	0.72	0.28	0.39	0.56
7.26	4.45	12.30	−9.23	0.64	0.36	0.57	0.33
7.51	2.75	12.38	−8.91	0.55	0.45	0.83	0.11
7.76	1.15	12.45	−8.58	0.46	0.54	1.16	−0.09
8.01	−0.66	12.52	−8.26	0.37	0.63	1.73	−0.33
8.26	−2.16	12.60	−7.94	0.28	0.72	2.55	−0.56
8.51	−3.14	12.67	−7.61	0.22	0.78	3.53	−0.75
8.75	−3.94	12.74	−7.29	0.17	0.83	4.98	−0.95
9.00	−4.55	12.82	−6.96	0.12	0.88	7.16	−1.17

[a] Data points were from the transition region of the unfolding curve shown in Figure 3.2A.

urea or GdmCl. Remember also that if the LEM is used, one can apply Eq. (5) and use nonlinear least squares [29] to conveniently obtain the unfolding parameters.

3.7.4
Determining Differences in Stability

It is frequently of interest to determine differences in conformational stability among proteins that differ slightly in structure. The structural change might be a single change in amino acid sequence achieved through site-directed mutagenesis, or a change in the structure of a side chain resulting from chemical modification. In Table 3.6, we present results from urea unfolding studies of wild-type RNase Sa and a mutant that differs in amino acid sequence by one residue. Two different methods of calculating the differences in stability, $\Delta(\Delta G)$, are illustrated.

Tab. 3.6. Determining differences in conformational stability[a].

Protein	$\Delta G(H_2O)$[b]	$\Delta(\Delta G)$[c]	m[d]	$(Urea)_{1/2}$[e]	$\Delta(Urea)_{1/2}$[f]	$\Delta(\Delta G)$[g]
Wild-type RNase Sa	6.09	–	1170	5.20	–	–
N44 → A RNase Sa	4.12	2.0	1174	3.51	1.69	2.0

[a] Data from Hebert et al. [87].
[b] From Eq. (4), in kcal mol^{-1}.
[c] Difference between the $\Delta G(H_2O)$ values in kcal mol^{-1}.
[d] From Eq. (4), in cal mol^{-1} M^{-1}.
[e] Midpoint of the urea unfolding curve in M.
[f] Difference between the (urea)$_{1/2}$ values in M.
[g] From $\Delta(urea)_{1/2} \times 1172$ (the average of the two m-values) in kcal mol^{-1}.

The midpoints of urea, $(\text{urea})_{1/2}$, and thermal, T_m, unfolding curves can be determined quite accurately and do not depend to a great extent on the unfolding mechanism [30]. In contrast, measures of the steepness of urea unfolding curves, m, cannot be determined as accurately, and deviations from a two-state folding mechanism will generally change these values. Consequently, differences in stability determined by comparing the $\Delta G(\text{H}_2\text{O})$ values can have large errors. However, when comparing completely different proteins or forms of a protein that differ markedly in stability, no other choice is available. This approach is illustrated by the first column of $\Delta(\Delta G)$ values in Table 3.6. There is also danger in drawing conclusions about the conformational stabilities of unrelated proteins based solely on the midpoints of their unfolding curves. For example, lysozyme and myoglobin have similar $\Delta G(\text{H}_2\text{O})$ values at 25 °C, but a much higher concentration of GdmCl is needed to denature lysozyme because the m-value is much smaller. Likewise, lysozyme and cytochrome c are unfolded at about the same temperature, even though lysozyme has a much larger value of $\Delta G(\text{H}_2\text{O})$ at 25 °C [81].

A second approach that can be used to determine differences in stability of single-site mutants is to take the difference between $[D]_{1/2}$ values and multiply this by the average of the m-values. This is illustrated by the $\Delta(\Delta G)$ values in the last column of Table 3.6. The rationale here is that the error in measuring the m-values should generally be greater than any differences resulting from the effect of small changes in structure on the m-value. However, substantial differences in m-values between proteins differing by only one amino acid in sequence have been observed with some proteins, i.e., SNase, so one must use caution when applying this method.

Acknowledgments

This work was supported by grants GM-37039 and GM-52483 from the National Institutes of Health (USA), and grants BE-1060 and BE-1281 from the Robert A. Welch Foundation. We thank many colleagues and coworkers over the years that have made important contributions to the field of protein folding.

References

1 SPIRO, K. (1900). Uber die beeinflussung der eiweisscoagulation durch stickstoffhaltige substanzen. *Z Physiol Chem* 30, 182–199.

2 RAMSDEN, W. (1902). Some new properties of urea. *J Physiol* 28, 23–27.

3 GREENSTEIN, J. P. (1938). Sulfhydryl groups in proteins. *J Biol Chem* 125, 501–513.

4 TANFORD, C. (1964). Isothermal unfolding of globular proteins in aqueous urea solutions. *J Am Chem Soc* 86, 2050–2059.

5 SOPER, A. K., CASTNER, E. W., and LUZAR, A. (2003). Impact of urea on water structure: a clue to its properties as a denaturant? *Biophys Chem* 105, 649–666.

6 Modig, K., Kurian, E., Prendergast, G., and Halle, B. (2003). Water and urea interactions with the native and unfolded forms of a β-barrel protein. *Protein Sci* 12, 2768–2781.

7 Zou, Q., Bennion, B. J., Daggett, V., and Murphy, K. P. (2002). The molecular mechanism of stabilization of proteins by TMAO and its ability to counteract the effects of urea. *J Am Chem Soc* 124, 1192–1202.

8 Tsai, J., Gerstein, M., and Levitt, M. (1996). Keeping the shape but changing the charges: A simulation study of urea and its iso-steric analogs. *J Chem Phys* 104, 9417–9430.

9 Tirado-Rives, J., Orozco, M., and Jorgensen, W. L. (1997). Molecular dynamics simulations of the unfolding of barnase in water and 8 M aqueous urea. *Biochemistry* 36, 7313–7329.

10 Sokolic, F. I. A. (2002). Concentrated aqueous urea solutions: A molecular dynamics study of different models. *J Chem Phys* 116, 1636–1646.

11 Bennion, B. J. and Daggett, V. (2003). The molecular basis for the chemical denaturation of proteins by urea. *Proc Natl Acad Sci USA* 100, 5142–5147.

12 Mason, P. E., Neilson, G. W., Dempsey, C. E., Barnes, A. C., and Cruickshank, J. M. (2003). The hydration structure of guanidinium and thiocyanate ions: implications for protein stability in aqueous solution. *Proc Natl Acad Sci USA* 100, 4557–4561.

13 Muller, N. (1990). A model for the partial reversal of hydrophobic hydration by addition of a urea-like cosolvent. *J Phys Chem* 94, 3856–3859.

14 Laidig, K. E. and Daggett, V. (1996). Testing the modified hydration-shell hydrogen-bond model of hydrophobic effects using molecular dynamics simulation. *J Phys Chem* 100, 5616–5619.

15 Timasheff, S. N. (2002). Protein hydration, thermodynamic binding, and preferential hydration. *Biochemistry* 41, 13473–13482.

16 Karplus, P. A. and Faerman, C. (1996). Ordered water in macro-molecular structure. *Curr Opin Struct Biol* 4, 770–776.

17 Pike, A. C. W. and Acharya, K. R. (1994). A structural basis for the interaction of urea with lysozyme. *Protein Sci* 3, 706–710.

18 Dunbar, J., Yennawar, H. P., Banerjee, S., Luo, J., and Farber, G. K. (1997). The effect of denaturants on protein structure. *Protein Sci* 6, 1727–1733.

19 Makhatadze, G. I. and Privalov, P. L. (1992). Protein interactions with urea and guanidinium chloride. A calorimetric study. *J Mol Biol* 226, 491–505.

20 Timasheff, S. N. and Xie, G. (2003). Preferential interactions of urea with lysozyme and their linkage to protein denaturation. *Biophys Chem* 105, 421–448.

21 Courtenay, E. S., Capp, M. W., Saecker, R. M., and Record, M. T., Jr. (2000). Thermodynamic analysis of interactions between denaturants and protein surface exposed on unfolding: interpretation of urea and guanidinium chloride m-values and their correlation with changes in accessible surface area (ASA) using preferential interaction coefficients and the local-bulk domain model. *Proteins* Suppl 4, 72–85.

22 Schellman, J. A. (2003). Protein stability in mixed solvents: a balance of contact interaction and excluded volume. *Biophys J* 85, 108–125.

23 Goldenberg, D. P. (2003). Computational simulation of the statistical properties of unfolded proteins. *J Mol Biol* 326, 1615–1633.

24 Schellman, J. A. (1987). Selective binding and colvent denaturation. *Biopolymers* 26, 549–559.

25 Alexander, S. S. and Pace, C. N. (1971). A comparison of the denaturation of bovine-lactoglobulins A and B and goat-lactoglobulin. *Biochemistry* 10, 2738–2743.

26 Pace, C. N. and Tanford, C. (1968). Thermodynamics of the unfolding of β-lactoglobulin A in aqueous urea solutions between 5 and 55. *Biochemistry* 7, 198–208.

27 Greene, R. F. J. and Pace, C. N. (1974). Urea and guanidine hydrochloride denaturation of ribonuclease, lysozyme, alpha-chymotrypsin, and beta-lactoglobulin. *J Biol Chem* 249, 5388–5393.
28 Shelton, V. M., Sosnick, T. R., and Pan, T. (1999). Applicability of urea in the thermodynamic analysis of secondary and tertiary RNA folding. *Biochemistry* 38, 16831–16839.
29 Santoro, M. M. and Bolen, D. W. (1988). Unfolding free energy changes determined by the linear extrapolation method. 1. Unfolding of phenylmethane-sulfonyl α-chymotrypsin using different denaturants. *Biochemistry* 27, 8063–8068.
30 Pace, C. N. (1986). Determination and analysis of urea and guanidine hydrochloride denaturation curves. *Methods Enzymol* 131, 266–280.
31 Shortle, D. (1995). Staphylococcal nuclease: a showcase of m-value effects. *Adv Protein Chem* 46, 217–247.
32 Becktel, W. J. and Schellman, J. A. (1987). Protein stability curves. *Biopolymers* 26, 1859–1877.
33 Eftink, M. R. and Ionescu, R. (1997). Thermodynamics of protein unfolding: questions pertinent to testing the validity of the two-state model. *Biophys Chem* 64, 175–197.
34 Yu, Y., Makhatadze, G. I., Pace, C. N., and Privalov, P. L. (1994). Energetics of ribonuclease T1 structure. *Biochemistry* 33, 3312–3319.
35 Hu, C.-Q., Sturtevant, J. M., Thomson, J. A., Erickson, R. E., and Pace, C. N. (1992). Thermodynamics of ribonuclease T1 denaturation. *Biochemistry* 31, 4876–4882.
36 Pace, C. N., Grimsley, G. R., Thomas, S. T., and Makhatadze, G. I. (1999). Heat Capacity Change for Ribonuclease A Folding. *Protein Sci* 8, 1500–1504.
37 Huyghues-Despointes, B. M., Scholtz, J. M., and Pace, C. N. (1999). Protein conformational stabilities can be determined from hydrogen exchange rates. *Nat Struct Biol* 6, 910–912.

38 Qu, Y., Bolen, C. L., and Bolen, D. W. (1998). Osmolyte-driven contraction of a random coil protein. *Proc Natl Acad Sci USA* 95, 9268–9273.
39 Pace, C. N. and Laurents, D. V. (1989). A new method for determining the heat capacity change for protein folding. *Biochemistry* 28, 2520–2525.
40 Scholtz, J. M. (1995). Conformational stability of HPr: the histidine-containing phosphocarrier protein from *Bacillus subtilis*. *Protein Sci* 4, 35–43.
41 Nicholson, E. M. and Scholtz, J. M. (1996). Conformational stability of the *Escherichia coli* HPr protein: test of the linear extrapolation method and a thermodynamic characterization of cold denaturation. *Biochemistry* 35, 11369–11378.
42 Felitsky, D. J. and Record, M. T., Jr. (2003). Thermal and urea-induced unfolding of the marginally stable lac repressor DNA-binding domain: a model system for analysis of solute effects on protein processes. *Biochemistry* 42, 2202–2217.
43 Ibarra-Molero, B. and Sanchez-Ruiz, J. M. (1996). A model-independent, nonlinear extrapolation procdure for the characterization of protein folding energetics from solvent-denaturation data. *Biochemistry* 35, 14689–14702.
44 Schellman, J. A. (2002). Fifty years of solvent denaturation. *Biophys Chem* 96, 91–101.
45 Zhang, O. and Forman-Kay, J. D. (1995). Structural characterization of folded and unfolded states of an SH3 domain in equilibrium in aqueous buffer. *Biochemistry* 34, 6784–6794.
46 Makhatadze, G. I. (1999). Thermodynamics of protein interactions with urea and guanidinium hydrochloride. *J Phys Chem B* 103, 4781–4785.
47 Aune, K. C. and Tanford, C. (1969). Thermodynamics of the denaturation of lysozyme by guanidine hydrochloride II. dependence on denaturant concentration at 25°. *Biochemistry* 8, 4586–4590.
48 Tanford, C. (1970). Protein denaturation. C. Theoretical models

for the mechanism of denaturation. *Adv Protein Chem* 24, 1–95.
49 LIEPINSH, E. and OTTING, G. (1994). Specificity of urea binding to proteins. *J Am Chem Soc* 116, 9670–9674.
50 TIMASHEFF, S. N. (2002). Thermodynamic binding and site occupancy in the light of the Schellman exchange concept. *Biophys Chem* 101–102, 99–111.
51 SCHELLMAN, J. A. and GASSNER, N. C. (1996). The enthalpy of transfer of unfolded proteins into solutions of urea and guanidinium chloride. *Biophys Chem* 59, 259–275.
52 IBARRA-MOLERO, B., LOLADZE, V. V., MAKHATADZE, G. I., and SANCHEZ-RUIZ, J. M. (1999). Thermal versus guanidine-induced unfolding of ubiquitin. An analysis in terms of the contributions from charge–charge interactions to protein stability. *Biochemistry* 38, 8138–8149.
53 MONERA, O. D., KAY, C. M., and HODGES, R. S. (1994). Protein denaturation with guanidine hydrochloride or urea provides a different estimate of stability depending on the contributions of electrostatic interactions. *Protein Sci* 3, 1984–1991.
54 SANTORO, M. M. and BOLEN, D. W. (1992). A test of the linear extrapolation of unfolding free energy changes over an extended denaturant concentration range. *Biochemistry* 31, 4901–4907.
55 MYERS, J. K., PACE, C. N., and SCHOLTZ, J. M. (1995). Denaturant m values and heat capacity changes: relation to changes in accessible surface areas of protein unfolding. *Protein Sci* 4, 2138–2148.
56 PACE, C. N., LAURENTS, D. V., and THOMSON, J. A. (1990). pH dependence of the urea and guanidine hydrochloride denaturation of ribonuclease A and ribonuclease T1. *Biochemistry* 29, 2564–2572.
57 SHORTLE, D. and MEEKER, A. K. (1986). Mutant forms of staphylococcal nuclease with altered patterns of guanidine hydrochloride and urea denaturation. *Proteins* 1, 81–89.

58 SHORTLE, D. (1996). The denatured state (the other half of the folding equation) and its role in protein stability. *FASEB J* 10, 27–34.
59 GITTIS, A. G., W. E. STITES, E. E. LATTMAN. (1993). The phase transition between a compact denatured state and a random coil state in staphylococcal nuclease is first-order. *J Mol Biol* 232, 718–724.
60 CARRA, J. H. and PRIVALOV, P. L. (1995). Energetics of denaturation and m values of staphylococcal nuclease mutants. *Biochemistry* 34, 2034–2041.
61 IONESCU, R. M., M. R. EFTINK. (1997). Global analysis of the acid-induced and urea-induced unfolding of staphylococcal nuclease and two of it variants. *Biochemistry* 35, 1129–1140.
62 BASKAKOV, I., BOLEN, D. W. (1998). Forcing thermodynamically unfolded proteins to fold. *J Biol Chem* 273, 4831–4834.
63 PACE, C. N. (1975). The stability of globular proteins. *CRC Crit Rev Biochem* 3, 1–43.
64 CARRA, J. H., ANDERSON, E. A., and PRIVALOV, P. L. (1994). Three-state thermodynamic analysis of the denaturation of staphylococcal nuclease mutants. *Biochemistry* 33, 10842–10850.
65 LEE, B. and RICHARDS, F. M. (1971). The interpretation of protein structures: estimation of static accessibility. *J Mol Biol* 55, 379–400.
66 CREAMER, T. P., SRINIVASAN, R., and ROSE, G. D. (1997). Modeling unfolded states of proteins and peptides. II. backbone solvent accessibility. *Biochemistry* 36, 2832–2835.
67 PACE, C. N., LAURENTS, D. V., and ERICKSON, R. E. (1992). Urea denaturation of barnase: pH dependence and characterization of the unfolded state. *Biochemistry* 31, 2728–2734.
68 PRIVALOV, P. L., TIKTOPULO, E. I., VENYAMINOV, S., GRIKO YU, V., MAKHATADZE, G. I., and KHECHINASHVILI, N. N. (1989). Heat capacity and conformation of proteins in the denatured state. *J Mol Biol* 205, 737–750.

69. PACE, C. N., ALSTON, R. W., and SHAW, K. L. (2000). Charge–charge interactions influence the denatured state ensemble and contribute to protein stability. *Protein Sci* 9, 1395–1398.
70. THOMSON, J. A., SHIRLEY, B. A., GRIMSLEY, G. R., and PACE, C. N. (1989). Conformational stability and mechanism of folding of ribonuclease T1. *J Biol Chem* 264, 11614–11620.
71. PACE, C. N. and GRIMSLEY, G. R. (1988). Ribonuclease T1 is stabilized by cation and anion binding. *Biochemistry* 27, 3242–3246.
72. YANG, M., FERREON, A. C., and BOLEN, D. W. (2000). Structural thermodynamics of a random coil protein in guanidine hydrochloride. *Proteins* Suppl, 44–49.
73. NOZAKI, Y. (1972). The preparation of guanidine hydrochloride. In *Methods in Enzymology* (HIRS, C. H. W. and TIMASHEFF, S. N., eds), Vol. 26, pp. 43–50. Academic Press, New York.
74. PRAKASH, V., LOUCHEUX, C., SCHEUFELE, S., GORBUNOFF, M. J., and TIMASHEFF, S. N. (1981). Interactions of proteins with solvent components in 8 M urea. *Arch Biochem Biophys* 210, 455–464.
75. HAGEL, P., GERDING, J., FIEGGEN, W., and BLOEMENDAL, H. (1971). Cyanate formation in solutions of urea. I. Calculation of cyanate concentrations at different temperature and pH. *Biochim Biophys Acta* 243, 366–373.
76. STARK, G. R. (1965). Reactions of cyanate with functional groups of proteins. 3. Reactions with amino and carboxyl groups. *Biochemistry* 4, 1030–1036.
77. GRAY, W. R. (1997). Disulfide bonds between cysteine residues. In *Protein Structure: A Practical Approach*, 2nd edn (CREIGHTON, T. E., ed.), pp. 165–186. IRL Press, Oxford.
78. PACE, C. N., HEBERT, E. J., SHAW, K. L. et al. (1998). Conformational stability and thermodynamics of folding of ribonucleases Sa, Sa2 and Sa3. *J Mol Biol* 279, 271–286.
79. SAITO, Y. and WADA, A. (1983). Comparative study of GuHCl denaturation of globular proteins. I. Spectroscopic and chromatographic analysis of the denaturation curves of ribonuclease A, cytochrome c, and pepsinogen. *Biopolymers* 22, 2105–2122.
80. ALLEN, D. L. and PIELAK, G. J. (1998). Baseline length and automated fitting of denaturation data. *Protein Sci* 7, 1262–1263.
81. PACE, C. N. and SCHOLTZ, J. M. (1997). Measuring the conformational stability of a protein. In *Protein Structure: A Practical Approach*, 2nd edn. (CREIGHTON, T. E., ed.), pp. 299–321. IRL Press, Oxford.
82. SHAW, K. L. (2000). Reversing the net charge of ribonuclease sa. Texas A&M University, PhD dissertation. College Station, TX.
83. TADDEI, N., CHITI, F., PAOLI, P. et al. (1999). Thermodynamics and kinetics of folding of common-type acylphosphatase: comparison to the highly homologous muscle isoenzyme. *Biochemistry* 38, 2135–2142.
84. HUGHSON, F. M. and BALDWIN, R. L. (1989). Use of site-directed mutagenesis to destabilize native apomyoglobin relative to folding intermediates. *Biochemistry* 28, 4415–4422.
85. KAWAHARA, K. and TANFORD, C. (1966). Viscosity and density of aqueous solutions of urea and guanidine hydrochloride. *J Biol Chem* 241, 3228–3232.
86. WARREN, J. R. and GORDON, J. A. (1966). On the refractive indices of aqueous solutions of urea. *J Phys Chem* 70, 297–300.
87. HEBERT, E. J., GILETTO, A., SEVCIK, J. et al. (1998). Contribution of a conserved asparagine to the conformational stability of ribonucleases Sa, Ba, and T1. *Biochemistry* 37, 16192–16200.
88. WANG, A., BOLEN, D. W. (1997). A naturally occurring protective system in urea-rich cells: mechanism of osmolyte protection of proteins against urea denaturation. *Biochemistry* 36, 9101–9108.

4
Thermal Unfolding of Proteins Studied by Calorimetry

George I. Makhatadze

4.1
Introduction

Protein folding/unfolding reactions and protein–ligand interactions, like any other chemical reactions, are accompanied by heat effects. At a constant pressure, these heat effects are described by the enthalpy of the process. Temperature-induced changes in the enthalpy of the macroscopic states of a protein are of particular significance. Temperature and enthalpy are coupled extensive and intensive thermodynamic parameters. Thus, the functional relationship between enthalpy and temperature $f = H(T)$ includes all the thermodynamic information on the macroscopic states in the system. This information can be extracted using the formalism of equilibrium thermodynamics.

The enthalpy $H(T)$ function of the system can be determined experimentally from the calorimetric measurements of the heat capacity at constant pressure, $C_p(T)$, of a system over a broad temperature range, as:

$$C_p(T) = \left(\frac{\partial H}{\partial T}\right)_p \quad \text{and thus} \quad H(T) = \int_{T_0}^{T} C_p(T)\,dT + H(T_0) \tag{1}$$

The heat capacity function of a system, $C_p(T)$, is directly measured by differential scanning calorimetry (DSC), making DSC the method of choice to study the conformational stability of biological macromolecules in general and proteins in particular. To study biological macromolecules the DSC instrumentation must satisfy a number of requirements, the most important of which is that measurements must be performed on dilute (<1 mg mL^{-1}/0.1 weight %) protein solutions to minimize intermolecular interactions, and must use a small volume because in most cases only a few milligrams of protein are available. With these requirements satisfied, the partial heat capacity of a protein in aqueous solution is on the order of 0.03–0.5%, which is measured on the background of more than 99.5% heat capacity of solvent. Contemporary DSC instruments [1, 2] are designed to satisfy these extreme sensitivity, reproducibility, and signal stability requirements (Figure 4.1).

Protein Folding Handbook. Part I. Edited by J. Buchner and T. Kiefhaber
Copyright © 2005 WILEY-VCH Verlag GmbH & Co. KGaA, Weinheim
ISBN: 3-527-30784-2

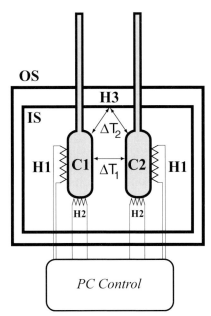

Fig. 4.1. Schematic design of the cell assembly of a contemporary DSC instrument. The matched sample cell (C1) and reference cell (C2) of the total fill type with inlet capillary tubes for filling and cleaning are surrounded by inner (IS) and outer (OS) shields. Two main heating elements (H1) on the cells are supplying electrical power for scanning upward in temperature. The auxiliary heating elements (H2) are activated by a feedback loop to equilibrate the difference in temperatures between cells (i.e., to maintain $\Delta T_1 = 0\ °C$). Additional heaters (H3) on the inner shield (IS) maintain the temperature difference between the cell and the shield $\Delta T_2 = 0\ °C$. The feedback loops are controlled by a PC board.

Extreme sensitivity of the instrumentation implies that particular care should be taken over experimental procedures for sample preparation and routine operation of DSC [3, 4]. Some essential considerations for setting up a DSC experiment, and discussion of critical factors are given in Section 4.7. For those interested in collection of numerical data of thermodynamic parameters such as Gibbs free energy change, enthalpy change, heat capacity change, transition temperature etc., two excellent sources of information are Refs [5] and [6].

4.2
Two-state Unfolding

First we consider in details the simplest case – a two-state model of protein unfolding. Some of the more complex cases will be discussed in Section 4.6.

For a monomolecular two-state process:

$$N \overset{K_U}{\leftrightarrow} U$$

the equilibrium constant is defined by the concentrations of protein in the native and unfolded states at a given temperature:

$$K(T) = \frac{[U]}{[N]} \tag{2}$$

while the fraction of the molecules in the native, F_N, and unfolded, F_U, states is defined as:

$$F_N(T) = \frac{[N]}{[N]+[U]} \quad \text{and} \quad F_U(T) = \frac{[U]}{[N]+[U]} \tag{3}$$

Combining Eqs (2) and (3), the populations of the native and unfolded states are defined by the equilibrium constant as:

$$F_N(T) = \frac{1}{1+K(T)} \quad \text{and} \quad F_U(T) = \frac{K(T)}{1+K(T)} \tag{4}$$

The equilibrium constant of the unfolding reaction, $K(T)$, on the other hand, is related to the Gibbs energy change upon unfolding as:

$$K(T) = \exp\left(-\frac{\Delta G(T)}{RT}\right) \tag{5}$$

The Gibbs energy of unfolding, $\Delta G(T)$, is defined as:

$$\Delta G(T) = \Delta H(T) - T \cdot \Delta S(T) \tag{6}$$

where $\Delta H(T)$ is the enthalpy change and $\Delta S(T)$ is the entropy changes upon protein unfolding. The temperature dependence of both parameters is defined by the heat capacity change upon unfolding, ΔC_p:

$$\Delta C_p = \frac{d\Delta H(T)}{dT} = T \cdot \frac{d\Delta S(T)}{dT} \tag{7}$$

Assuming that ΔC_p is independent of temperature we can write:

$$\Delta H(T) = \Delta H(T_0) + \Delta C_p(T-T_0) \tag{8}$$

$$\Delta S(T) = \Delta S(T_0) + \Delta C_p \cdot \ln\left(\frac{T}{T_0}\right) \tag{9}$$

where $\Delta H(T_0)$ and $\Delta S(T_0)$ are the enthalpy and entropy changes at a reference temperature T_0. For a two-state process it is convenient to set the reference temperature at which fractions for the native, F_N, and unfolded, F_U, protein are equal. Correspondingly, at this temperature, which is usually called the transition temperature, T_m, the equilibrium constant is equal to 1 and thus the Gibbs energy is equal to zero:

$$\Delta G(T_m) = -RT \ln K(T_m) = \Delta H(T_m) - T_m \Delta S(T_m) = 0 \tag{10}$$

which allows the entropy of unfolding at T_m to be expressed as:

$$\Delta S(T_m) = \frac{\Delta H(T_m)}{T_m} \tag{11}$$

and the entropy function as

$$\Delta S(T) = \frac{\Delta H(T_m)}{T_m} + \Delta C_p \cdot \ln\left(\frac{T}{T_m}\right) \tag{12}$$

Thus the temperature dependence of the Gibbs energy can be written as:

$$\Delta G(T) = \frac{T_m - T}{T_m} \cdot \Delta H(T_m) + \Delta C_p \cdot (T - T_m) + T \cdot \Delta C_p \cdot \ln\left(\frac{T_m}{T}\right) \tag{13}$$

The importance of Eq. (13) is that if just three parameters, T_m, $\Delta H(T_m)$, and ΔC_p, are known, all thermodynamic parameters ΔH, ΔS, and ΔG for the two-state process can be calculated for any temperature.

Figure 4.2 shows the temperature dependence of the partial heat capacity, $C_{p,pr}(T)$ of a hypothetical protein. It consists of two terms:

$$C_{p,pr}(T) = C_p^{prg}(T) + \langle C_p(T) \rangle^{exc} \tag{14}$$

where $C_p^{prg}(T)$ is the so-called progress heat capacity (sometimes also called the "chemical baseline") which is defined by the intrinsic heat capacities of protein in different states, and $\langle C_p(T) \rangle^{exc}$ is the excess heat capacity which is caused by the heat absorbed upon protein unfolding. The progress heat capacity for a two-state unfolding can be defined as:

$$C_p^{prg}(T) = F_N \cdot C_{p,N}(T) + F_U \cdot C_{p,U}(T) \tag{15}$$

where $C_{p,N}(T)$ and $C_{p,U}(T)$ are the temperature dependencies of the heat capacities of the native and unfolded states, respectively. It is important to emphasize that the heat capacity of the unfolded state is higher than that of the native state, $C_{p,U} > C_{p,N}$. Extensive studies of the heat capacities of the native and unfolded

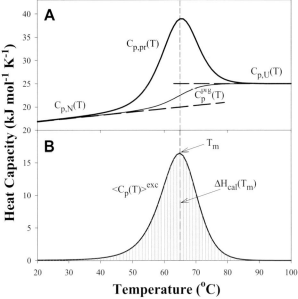

Fig. 4.2. A) Temperature dependence of the partial heat capacity, $C_{p,pr}(T)$, of a hypothetical protein that has the following thermodynamic parameters of unfolding: $T_m = 65$ °C, $\Delta H = 250$ kJ mol^{-1}, $\Delta C_p = 5$ kJ mol^{-1} K^{-1}. $C_{p,N}(T)$ and $C_{p,U}(T)$ are the temperature dependencies of the heat capacities of the native and unfolded states, respectively, and $C_p^{prg}(T)$ is the progress heat capacity (sometimes also called the "chemical baseline"). B) The excess heat capacity, $\langle C_p(T) \rangle^{exc}$, is obtained as a difference between $C_{p,pr}(T)$ and $C_p^{prg}(T)$. The area under $\langle C_p(T) \rangle^{exc}$ is the enthalpy of unfolding, and temperature at which $\langle C_p(T) \rangle^{exc}$ is maximum defines the transition temperature, T_m.

states of small globular proteins have established some general properties [7]. The heat capacity function for the native state appears to be a linear function of temperature with a rather similar slope 0.005–0.008 J K^{-2} g^{-1} when calculated per gram of protein. The absolute values for $C_{p,N}$ vary between 1.2 and 1.80 J K^{-1} g^{-1} at 25 °C. For the unfolded state, the absolute value at 25 °C is always higher than that for the native state ($C_{p,U} > C_{p,N}$) and ranges between 1.85 and 2.2 J K^{-1} g^{-1}. The heat capacity of the unfolded state has nonlinear dependence on temperature. It increases with increase of temperature and reaches a plateau between 60 and 75 °C. Above this temperature $C_{p,U}$ $C_{p,U}$ remains constant with the typical values ranging between 2.1 and 2.5 J K^{-1} g^{-1}.

The excess heat capacity function can thus be calculated as a difference between experimental and progress heat capacities. The excess heat capacity function is very important as it provides the temperature dependence of the fraction of protein in the native and unfolded states:

$$F_N(T) = 1 - F_U(T) = \frac{Q_i(T)}{Q_{tot}} = \frac{\int_0^T \langle C_p(T) \rangle^{exc} \, dT}{\int_0^\infty \langle C_p(T) \rangle^{exc} \, dT} \tag{16}$$

and the temperature dependence of the equilibrium constant

$$K(T) = \frac{\int_T^\infty \langle C_p(T) \rangle^{\text{exc}} \, dT}{\int_0^T \langle C_p(T) \rangle^{\text{exc}} \, dT} \qquad (17)$$

The temperature dependence of the equilibrium constant defines the van't Hoff enthalpy of a two-state process as:

$$\Delta H_{\text{vH}} = -R \cdot \frac{d \ln K(T)}{d(1/T)} \qquad (18)$$

The area under the $\langle C_p(T) \rangle^{\text{exc}}$ is the calorimetric (model independent) enthalpy of unfolding:

$$\Delta H_{\text{cal}} = \int_0^\infty \langle C_p(T) \rangle^{\text{exc}} \, dT \qquad (19)$$

Comparison of the two enthalpies – van't Hoff enthalpy calculated for a two-state transition and model-free calorimetric enthalpy of unfolding – is the most rigorous test of whether protein unfolding is a two-state process [8–11]. If calorimetric enthalpy is equal to van't Hoff enthalpy, $\Delta H_{\text{cal}}(T_m) = \Delta H_{\text{vH}}(T_m)$, the unfolding process is a two-state process. If calorimetric enthalpy is larger than the van't Hoff enthalpy, $\Delta H_{\text{cal}}(T_m) > \Delta H_{\text{vH}}(T_m)$, the unfolding process is more complex, such as being not monomolecular or involving intermediate states (see Section 4.6).

Instead of the van't Hoff enthalpy, it is more practical to use the fitted enthalpy of two-state unfolding. This can be done using the following approach. The excess enthalpy function (relative to the native state) is defined as [2, 8]:

$$\langle \Delta H^{\text{exc}} \rangle = F_U \cdot \Delta H(T) \qquad (20)$$

where $\Delta H(T)$ is the enthalpy difference between native and unfolded states. By definition, the excess heat capacity function is a temperature derivative of the excess enthalpy at constant pressure:

$$\langle \Delta C_p \rangle^{\text{exc}} = \frac{\partial \langle \Delta H^{\text{exc}} \rangle}{\partial T} = \Delta H \cdot \frac{\partial F_U}{\partial T} \qquad (21)$$

Note that the way $\langle C_p(T) \rangle^{\text{exc}}$ is defined by Eq. (14), the temperature dependence of ΔH is in the $C_p^{\text{prg}}(T)$ function. Combining Eqs (4), (8), and (21) it is easy to show that the excess heat capacity $\langle C_p(T) \rangle^{\text{exc}}$ is defined as:

$$\langle C_p(T) \rangle^{\text{exc}} = \frac{\Delta H_{\text{fit}}^2}{R \cdot T^2} \cdot \frac{K}{(1+K)^2} \qquad (22)$$

where the enthalpy function is defined as $\Delta H(T)$:

$$\Delta H(T) = \Delta H_{\text{fit}}(T_m) + \Delta C_p \cdot (T - T_m) \tag{23}$$

The experimental partial molar heat capacity function, $C_{p,\text{pr}}(T)$, is then fitted to the following expression:

$$C_{p,\text{pr}}(T) = F_N(T) \cdot C_{p,N}(T) + \langle C_p(T) \rangle^{\text{exc}} + F_U(T) \cdot C_{p,U}(T) \tag{24}$$

The heat capacity functions for the native and unfolded states can be represented by the linear functions of temperature as:

$$C_{p,N}(T) = A_N \cdot T + B_N \tag{25}$$

$$C_{p,U}(T) = A_U \cdot T + B_U \tag{26}$$

There are seven parameters to fit: $T_m, \Delta H_{\text{fit}}, \Delta C_p, A_N, A_U, B_N,$ and B_U.

4.3
Cold Denaturation

Initial studies of the thermodynamics of protein unfolding [9, 12–15] done on small globular proteins established that equilibrium protein unfolding is a two-state process, i.e., $\Delta H_{\text{vH}} = \Delta H_{\text{cal}}$ and that the protein unfolding is accompanied by a large positive heat capacity change. Direct consequence of large heat capacity changes upon unfolding is that both enthalpy and entropy functions are strongly temperature dependent which in turn leads to a temperature-dependent Gibbs energy function (Figure 4.3). Both the enthalpy and entropy are expected to be zero at temperature T_h and T_s. By definition, the slope of temperature dependence of the ΔG function is equal to $-\Delta S$:

$$\frac{\partial \Delta G(T)}{\partial T} = -\Delta S(T)$$

Thus at $T_s \Delta S = 0$ and ΔG has a maximum at T_s, so that $\Delta G(T_s) = \Delta H(T_s)$. The enthalpy function crosses zero at T_h, which is always lower than T_s (i.e., $T_h < T_s$), however since $T_s - T_h = \Delta H(T_s)/\Delta C_p$, this difference is only a few degrees. There is a simple relationship between T_h and the transition temperature T_m: $T_m - T_h = \Delta H(T_m)/\Delta C_p$.

The fact that ΔG function has a maximum at T_s means that it can cross zero at two different temperatures: once at temperature $T_m > T_s$ and again at $T_{m'} < T_s$. T_m is the temperature of denaturation observed upon heating the solution. $T_{m'}$ is the temperature of protein denaturation upon cooling, the so-called cold denaturation. There is a simple relationship between T_m and $T_{m'}$. The equation $T_{m'} \approx T_m - 2 \cdot \Delta H(T_m)/\Delta C_p$ overestimates the temperature of cold denaturation only by few degrees.

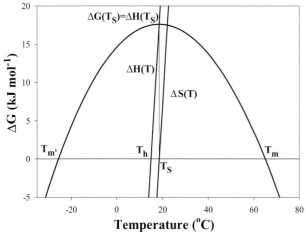

Fig. 4.3. Temperature dependencies of Gibbs energy, ΔG, enthalpy, ΔH, and entropic contribution, $T \cdot \Delta S$, for the same hypothetical protein as in Figure 4.2. Thermodynamics functions were calculated according to equations using the following parameters: $T_m = 65\ °C$, $\Delta H = 250\ kJ\ mol^{-1}$, $\Delta\Delta C_p = 5\ kJ\ mol^{-1}\ K^{-1}$. Other abbreviations: $T_{m'}$ temperature of cold denaturation, T_h temperature at which enthalpy is zero, T_s temperature at which entropy is zero.

Cold denaturation was predicted and later its existence confirmed based on the extrapolation of thermodynamic functions obtained from the study of heat denaturation process [8, 16–19]. Experimental validation of the cold denaturation phenomenon once again demonstrated the power of statistical thermodynamics in the analysis of the conformational properties of proteins.

4.4
Mechanisms of Thermostabilization

From the point of view of applied properties of proteins, two parameters are important – the thermodynamic stability as measured by the Gibbs energy and the thermostability as measured by the transition temperature.

Depending on the relationship between the T_s, T_m and absolute values of ΔG, three extreme models [20–24] account for the increase in ΔG and/or T_m and the relationship between these two parameters (Figure 4.4).

- *Model I* Increase in thermostability is accompanied by little or no change in T_s but with an increase in stability at T_s. This can be achieved by lowering the entropy of unfolding without changes in the enthalpy of unfolding or in ΔC_p.
- *Model II* Increase in thermostability is accompanied by little or no change in T_s and no changes in stability at T_s, but the temperature dependence of ΔG be-

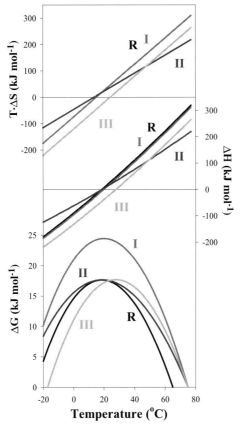

Fig. 4.4. Thermodynamic models for thermostabilization relative to the reference protein. Detailed thermodynamic parameters for the reference protein (R), and models I, II and III are given in Table 4.1.

Tab. 4.1. Thermodynamic parameters used in the simulation of different models for thermostabilization shown in Figure 4.4.

	T_m (°C)	T_s (°C)	ΔC_p (kJ mol^{-1} K^{-1})	$\Delta H(T_m)$ (kJ mol^{-1})	$\Delta G(T_S)$ (kJ mol^{-1})	ΔH(65 °C) (kJ mol^{-1})	$T \cdot \Delta S$(65 °C) (kJ mol^{-1})	$\Delta\Delta G$(65 °C) (kJ mol^{-1})
Reference	65	18	5.00	250	17.6	250	250	0
Model 1	75	19	5.00	300	24.6	250	242	7.9
Model 2	75	18	3.44	212	17.6	177	172	5.6
Model 3	75	27	5.00	254	17.6	204	197	6.6

comes less steep. This can be achieved by lowering ΔC_p, but both enthalpy and entropy will change as well.
- *Model III* Increase in thermostability is accompanied by the increase in T_s but no changes in stability at T_s, or in the temperature dependence of ΔG function. The increase in stability is due to the decrease in both enthalpic and entropic factors, but decrease in $T \cdot \Delta S$ is larger than in ΔH, while ΔC_p remains unchanged.

Of course these three models represent "extreme" cases and the real systems can use combination of all three simultaneously. Nevertheless, understanding the contributions of different factors to the thermodynamics of proteins in terms of $\Delta C_p, \Delta H$, and ΔS allows strategies for rational modulation of protein stability and/or thermostability to be designed. These strategies are discussed in the next section.

4.5
Thermodynamic Dissection of Forces Contributing to Protein Stability

The ultimate goal of the field is to be able to understand the determinants of protein stability in such details that will allow the prediction of the stability profile (i.e., ΔG) of any protein under any conditions such as temperature, pH, ionic strength, etc.

For example, the slope of the ΔG dependence on temperature is $-\Delta S$, which in turn is dependent on temperature and this dependence is defined by ΔC_p. Thus one needs to know not only a value for ΔG at one particular temperature, but also $\Delta S, \Delta H$, and ΔC_p to calculate the $\Delta G(T)$ function.

Another potential complexity arises from the fact that ΔG being the difference between the enthalpic (ΔH) and entropic ($T \cdot \Delta S$) terms, is subject to the so-called enthalpy–entropy compensation phenomenon [25–27]. It has been observed in a number of systems that changes in the enthalpy of a system are often compensated by the changes in the entropy in such a way that the resulting changes in ΔG are very small. One particularly striking example of such compensation is the effect of heavy (D_2O) and ordinary light (H_2O) water on protein stability [28]. Heavy water as a solvent provides probably the smallest possible perturbation to the properties H_2O. The stability (ΔG) of proteins in H_2O is very similar to the stability of proteins in D_2O (Figure 4.5). Such similarity in the ΔG values might lead to the conclusion that D_2O did not perturb the properties of the system relative to H_2O. However, if one looks at the corresponding changes in enthalpy (Figure 4.5) and entropy (not shown), it is clear that D_2O has a profound effect on the system, and decreases the enthalpy of unfolding by $\sim 20\%$. This decrease in enthalpy is accompanied by a decrease in entropy in such a way that the resulting ΔG is largely unchanged.

This result underlines the importance of the analysis of the protein stability in terms of enthalpy, entropy, and heat capacity, so that underlying physicochemical processes are not masked by the enthalpy–entropy compensations. The effect of

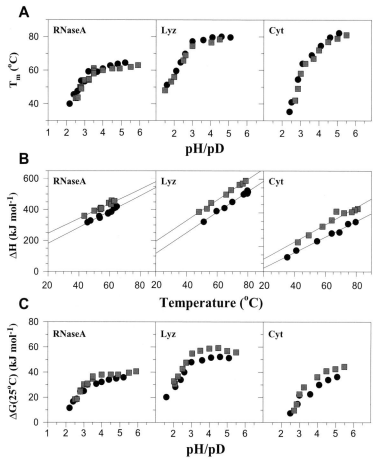

Fig. 4.5. Comparison of the unfolding thermodynamics for three different proteins ribonucleases A (RNaseA), hen-egg white lysozyme (Lyz) and cytochrome c (Cyt), in ordinary (H_2O, squares) and heavy (D_2O, circles) water. A) Dependence of the transition temperature, T_m, on the pH/pD. B) Dependence of the enthalpy of unfolding, ΔH, on the transition temperature, T_m. The slopes which represent the heat capacity change upon unfolding (ΔC_p, kJ K^{-1} mol^{-1}) are 4.8 (d-RNase), 5.2 (h-RNase), 6.4 (d-Lyz), 6.7 (h-Lyz), 4.7 (d-Cyt$c$) and 5.5 (h-Cyt$c$). C) Dependence of the Gibbs energy of unfolding at ΔG (25 °C) on the pH/pD. Reproduced with permission from Ref. [28].

D_2O on protein stability also underlines the importance of the solvent (or interactions of protein groups with the solvent) for protein stability. Even such small perturbation in the properties of solvent such as D_2O versus H_2O produces dramatic changes in the enthalpies and entropies of unfolding. Thus in analysis of the factors important for proteins stability one needs to consider not only the interactions between protein groups but also interactions between protein groups and the solvent. The next three sections will discuss the contribution of different types of interactions to the heat capacity, enthalpy and entropy changes upon protein unfolding.

4.5.1
Heat Capacity Changes, ΔC_p

It is currently believed that the heat capacity changes upon unfolding are entirely defined by the interactions with the solvent of the proteins groups that were buried in the native state but become exposed upon unfolding. Both polar and nonpolar groups seem to contribute although with different sign and magnitude. It appears that nonpolar groups have a large positive contribution to the heat capacity of unfolding, while the polar groups have smaller (for a group of similar size) and negative contribution to ΔC_p. These estimates of ΔC_p for polar and nonpolar groups were obtained from the study of thermodynamics of transfer for model compounds [7, 29–35]. To link the data on model compounds to the results obtained on proteins, the heat capacity changes for model compounds were computed per unit of solvent accessible surface area, and then used as coefficients to calculated the hydration heat capacity for proteins. It has been shown that the ΔC_p values calculated are in a very good agreement with the experimentally determined heat capacity changes for proteins. From this correlation it was concluded that the ΔC_p of protein unfolding is caused exclusively by the hydration and that the disruption of internal interactions in the native state do not contribute to the ΔC_p. This conclusion was in disagreement with the estimates done by Sturtevant [36], who showed that the changes in vibrational entropy of protein upon unfolding should contribute 20–25% to the experimentally observed changes in ΔC_p.

The common belief that ΔC_p is entirely defined by the hydration was questioned by experiments of Loladze et al. [37]. These authors measured experimentally the effects of single amino acid substitutions at fully buried positions of ubiquitin on the ΔC_p of this protein. It was shown that substitution of the fully buried nonpolar residues with the polar ones leads to a decrease in ΔC_p, while the substitutions of fully buried polar residues with nonpolar ones leads to an increase in ΔC_p. These results provided direct experimental support for the model compound data (that indicated difference in the sign for the ΔC_p of hydration for polar and nonpolar groups) to be qualitatively applicable to proteins. Quantitatively, however, the results of substitutions on ΔC_p were not possible to predict based on the model compound data and surface area proportionality. This led to the conclusion that the dynamic properties of the native state also contribute to ΔC_p [37, 38]. At this point quantitative correlation between native state dynamics and ΔC_p awaits experimental validation.

4.5.2
Enthalpy of Unfolding, ΔH

The enthalpy of protein unfolding is positive, which means that enthalpically the native state is favored over the unfolded state. Two major types of interactions can be considered as a possible source of enthalpy: interactions with the solvent and internal interactions between protein groups in the native state.

Interactions with the solvent by the protein groups newly exposed upon unfolding of the native protein structure are expected to be enthalpically unfavorable (enthalpically hydration of the unfolded state is favored over the hydration of the native state). This follows from the data on the transfer of the model compounds from the gas phase into water; for both polar and nonpolar groups, the enthalpy of transfer is negative [28, 32, 34, 48, 39]. The absolute magnitude of the effects for a polar group and a nonpolar group of the same size is very different. It is far more enthalpically unfavorable to transfer a polar group into water that it is for a nonpolar group. This large difference in the enthalpy of hydration of polar and nonpolar groups in proteins was recently demonstrated experimentally by measuring the effects of nonpolar to polar substitutions at fully buried sites on the enthalpy of unfolding [40].

If the enthalpy of dehydration is unfavorable, it is clear that the internal interactions between protein groups in the native state are enthalpically very stabilizing. The internal interactions are the hydrogen bonding and packing interactions between groups in the protein interior. The packing of the protein core is as good as the packing of organic crystals. Calculations show that the packing density in proteins is very similar to that for organic crystals [41, 42]. Melting or sublimation of a crystal is accompanied by large enthalpy, and one can expect that disruption of packing interactions in the protein interior will also lead to large positive enthalpy. Direct evidence for this was obtained by measuring the effects of substitutions that perturb packing in the protein interior [40]. It was shown that cavity creating substitutions in the protein core led to substantial decrease in the enthalpy of unfolding.

Hydrogen bonding also makes a favorable contribution to the protein stability [43, 44]. The net effect of disruption of hydrogen bonding inside the protein is difficult to measure. However, substitutions that disrupt hydrogen bonding with minimal (but not negligible) perturbations in the packing and hydration indicate that there is a significant enthalpic effect associated with the disruption of internal hydrogen bonds [40]. One needs to keep in mind that this effect is for disruption of hydrogen bonds in the protein interior and does not include the effect of hydration. These two effects are very difficult to estimate separately. The combined effect, i.e., enthalpy of disruption of hydrogen bonds in the protein interior followed by exposure (hydration) of hydrogen bonding groups to the solvent is more straightforward for the experimental analysis. Recent advances in calorimetric instrumentation and the development of novel model systems have provided estimates for the enthalpy of hydrogen bonding upon helix–coil transition [45–47]. The measured value is relatively small (~ 3 kJ mol^{-1}) but positive, indicating that the favorable enthalpy of internal hydrogen bonding is larger than the unfavorable enthalpy of hydration.

To summarize, native protein is enthalpically more stable than the unfolded state. Internal interactions (packing and hydrogen bonding) are enthalpically stabilizing, while exposure to the solvent upon unfolding of the buried in the native state groups is enthalpically destabilizing.

4.5.3
Entropy of Unfolding, ΔS

Protein unfolding is accompanied by increase in entropy. This entropy increase is caused by two factors. One factor is the interaction of the newly exposed groups with the solvent in the unfolded state. The other factor is the increase in the number of accessible conformations in the unfolded state relative to the native state. Interactions with the solvent by both polar and nonpolar groups occur with the negative entropy [33, 34, 48]. The molecular details of these effects just start to become clear but in a simplified way they can be described by the increase in order of water molecules around the solute (polar or nonpolar) relative to the bulk water. The molecular mechanism of this ordering is different for polar and nonpolar groups [49]. For a group of the same size, the entropy decrease due to hydration for polar groups appears (based on the model compound study) to be larger than that for the nonpolar groups. Such an effect is expected intuitively, based on consideration that the polar groups will have more ordering effect on the water dipole.

This decrease in entropy due to hydration of protein groups newly exposed to solvent in the unfolded state is overcompensated by the increase (relative to the native state) in the configurational freedom of the polypeptide chain in the unfolded state. There are several different estimates of the magnitude of the configurational entropy in proteins [50–55]. The actual value will probably be strongly context dependent, and will depend on the conformations that can be populated within the unfolded state ensemble.

Overall, hydration of both polar and nonpolar groups (the latter sometimes called hydrophobic effect) entropically favors the native state, while the increase in conformational freedom favors the unfolded state.

On balance, the protein structure is stabilized enthalpically. Large enthalpic stabilization is offset by entropic destabilization, rendering the protein molecule marginally stable [56].

One can only speculate why proteins evolve to be marginally stable. From the biological standpoint, low stability might be advantageous for rapid regulation of protein levels in the cell, because low stability most certainly will facilitate protein degradation by proteases or more complex cellular machineries. Transport of proteins among different cellular compartments involves folding–unfolding and thus low stability will facilitate transport across the membrane [57]. Low stability will probably permit conformational transitions between different functional states, and thus will be advantageous for regulation. Low stability will also have important consequences for the kinetics of folding/unfolding reactions and thus might be important in preventing accumulation of intermediate states that have detrimental properties, such as a tendency to aggregate [57–59].

In discussing protein stability we largely concentrate on the residues that are buried in the native state as a major source for the stability. Indeed, interactions in the core of the native protein and interactions of these residues with the solvent in the unfolded state provide the dominant effects for protein stability. This does not mean, however, that the surface residues do not contribute to stability [60–63]. They are in most cases not dominant factors defining protein energetics, yet are significantly large relative to the overall protein stability (ΔG) and thus they are important for modulating protein stability and, in the case of charge–charge interactions, enhancing protein solubility.

4.6
Multistate Transitions

In 1978 Biltonen and Freire [8] proposed a universal mechanism for equilibrium unfolding of proteins based on standard statistical thermodynamics approach. They considered a general reaction in which a initial native state, N, undergoes conformational transition to the final unfolded state, U, via a series of intermediate states, I_i:

$$N \leftrightarrow I_1 \leftrightarrow I_2 \cdots \leftrightarrow I_n \leftrightarrow U$$

and showed that the excess enthalpy of the systems (enthalpies of all states in the system relative to the initial state) can be written as:

$$\langle \Delta H^{exc} \rangle = \sum_{i=0}^{n} F_i \cdot \Delta H_i \tag{27}$$

where F_i is the fractional population of i-th state and ΔH_i is the enthalpy of the i-th state relative to the initial state. Keeping in mind that since

$$\langle \Delta H^{exc} \rangle = \int_{T_0}^{T} \Delta C_p^{exc} \cdot dT \tag{28}$$

knowledge of the excess heat capacity function provides a unique description of the thermodynamic properties of the system. One practical approximation for the multistate transitions is that when the transitions involve many states, it is convenient to assume that the thermodynamic domains behave in an independent manner, with each individual transition approximated by a two-state transition. In such case the excess heat capacity function can be simply written as:

$$\langle \Delta C_p^{exc} \rangle = \sum_i \frac{(\Delta H_i)^2}{RT^2} \cdot \frac{K_i}{(1+K_i)^2} \tag{29}$$

where K_i is the equilibrium constant for the i-th reaction.

4.6.1
Two-state Dimeric Model

In this model only the native dimer, N_2, and unfolded monomer, U, are significantly populated.

$$N_2 \overset{K}{\leftrightarrow} 2 \cdot U$$

The equilibrium constant is then defined as:

$$K = \frac{[U]^2}{[N_2]} \tag{30}$$

where $[N_2]$ and $[U]$ are the concentrations of native dimers and monomers, respectively. The fraction of protein in these two states can be presented as:

$$[N_2] = \tfrac{1}{2} \cdot P_T \cdot F_{N_2} \quad \text{and} \quad [U] = P_T \cdot F_U \tag{31}$$

where F_{N_2} and F_U are the fractions of native dimer and unfolded monomers, respectively, and P_T, the total protein concentration expressed per mole of monomers is:

$$P_T = 2 \cdot [N_2] + [U] \tag{32}$$

Combining Eqs (30) and (32) allows the equilibrium constant to be expressed as a function of total protein concentration and a fraction of unfolded protein as:

$$K = \frac{2 \cdot P_T \cdot F_U^2}{1 - F_U} \tag{33}$$

Solving a simple quadratic equation for F_U, we arrive at:

$$F_U = \frac{\sqrt{(K^2 + 8 \cdot P_T \cdot K)} - K}{4 \cdot P_T} \tag{34}$$

The excess enthalpy of the system is related to the enthalpy of unfolding as:

$$\langle \Delta H^{\text{exc}} \rangle = F_U \cdot \Delta H(T) \tag{35}$$

which allows calculation of the excess heat capacity function by simple analytical differentiation:

$$\langle \Delta C_p^{\text{exc}} \rangle = \frac{\partial \langle \Delta H^{\text{exc}} \rangle}{\partial T} = \Delta H \cdot \frac{\partial F_U}{\partial T} = \frac{\Delta H(T)^2}{R \cdot T^2} \frac{F_U \cdot (1 - F_U)}{2 - F_U} \qquad (36)$$

Thus the experimental partial molar heat capacity profile can be expressed as:

$$C_{p,\text{pr}}(T) = F_N(T) \cdot C_{p,N}(T) + C_p^{\text{exc}}(T) + F_U(T) \cdot C_{p,U}(T) \qquad (37)$$

where the heat capacity functions for the native and unfolded states are expressed according to Eqs (25–26).

Calculation of the standard thermodynamic quantities such as Gibbs energy, can be done by defining the reference temperature T_m as a temperature at which $F_{N_2} = 0.5$, and thus according to the Eq. (33) the equilibrium constant at this temperature is equal to P_T. The standard Gibbs energy can then be expressed as:

$$\Delta G^0 = \frac{T_m - T}{T_m} \cdot \Delta H(T_m) + \Delta C_p \cdot (T - T_m)$$

$$+ T \cdot \Delta C_p \cdot \ln\left(\frac{T_m}{T}\right) - R \cdot T \cdot \ln K(T_m) \qquad (38)$$

It must be emphasized that the ΔG^0 function in this case is refers to a 1 mol L^{-1} concentration of dimers.

4.6.2
Two-state Multimeric Model

For the reaction of unfolding of homo-oligomeric system that occurs without significant population of intermediate states, i.e.

$$N_n \overset{K}{\leftrightarrow} n \cdot U$$

it is easy to show that the excess heat capacity function can be expressed simply as

$$\langle \Delta C_p^{\text{exc}} \rangle = \frac{\Delta H(T)^2}{R \cdot T^2} \frac{F_U \cdot (1 - F_U)}{n - F_U \cdot (n - 1)} \qquad (39)$$

where n is the order of the reaction. Obviously, for $n = 2$, Eq. (39) is reduced to Eq. (36).

4.6.3
Three-state Dimeric Model

In this model, not only the native dimer, N_2, and the unfolded monomer, U, but also a native monomer, N, are significantly populated.

4.6 Multistate Transitions

$$N_2 \overset{K_d}{\leftrightarrow} 2 \cdot N \overset{K_U}{\leftrightarrow} 2 \cdot U$$

Equilibrium constants of dissociation, K_d, and of unfolding, K_U, are then defined as:

$$K_d = \frac{[N]^2}{[N_2]} \quad \text{and} \quad K_U = \frac{[U]}{[N]} \tag{40}$$

where $[N_2]$, $[N]$, and $[U]$ are the concentrations of the native dimers, the native monomers and the unfolded monomers, respectively. The fraction of protein in these three states can be presented as:

$$[N_2] = \tfrac{1}{2} F_{N_2} \cdot P_T \tag{41}$$

$$[N] = F_N \cdot P_T \tag{42}$$

$$[U] = P_T \cdot F_U \tag{43}$$

where F_{N_2}, F_N, and F_U are the fractions of native dimer, native monomer, and unfolded monomer, respectively, and P_T, the total protein concentration expressed per mole of monomers, is:

$$P_T = 2[N_2] + [N] + [U] \tag{44}$$

Combining Eqs (40)–(44) allows the equilibrium constants to be expressed as a function of total protein concentration and fractions of native and unfolded protein as:

$$K_d = \frac{2 \cdot F_N^2 \cdot P_T}{F_{N_2}} \tag{45}$$

$$K_U = \frac{F_U}{\sqrt{\dfrac{K_d \cdot F_{N_2}}{2 \cdot P_T}}} \tag{46}$$

Keeping in mind that $F_{N_2} + F_N + F_U = 1$ we can combine Eqs (45) and (46) to get:

$$F_{N_2} = 1 - \sqrt{\frac{K_d \cdot F_{N_2}}{2 \cdot P_T}} - K_U \cdot \sqrt{\frac{K_d \cdot F_{N_2}}{2 \cdot P_T}} \tag{47}$$

Substituting $1 + K_U = q$ and solving the quadratic equation we arrive at the following expressions for the fractions of proteins in the native dimer, native monomer, and unfolded monomer:

$$F_{N_2} = \frac{K_d}{8 \cdot P_T} \cdot \left[\sqrt{q^2 + \frac{8 \cdot P_T}{K_d}} - q\right]^2 \tag{48}$$

$$F_N = \frac{K_d}{4 \cdot P_T} \cdot \left[\sqrt{q^2 + \frac{8 \cdot P_T}{K_d}} - q\right] \tag{49}$$

$$F_U = \frac{K_U \cdot K_d}{4 \cdot P_T} \cdot \left[\sqrt{q^2 + \frac{8 \cdot P_T}{K_d}} - q\right] \tag{50}$$

The excess enthalpy function is defined as:

$$\langle \Delta H^{exc} \rangle = \Delta H_d \cdot F_N + (\Delta H_d + \Delta H_U) \cdot F_U \tag{51}$$

which allows calculation of the excess heat capacity function as a temperature derivative of the excess enthalpy. The corresponding Gibbs energies can be expressed as:

$$\Delta G_d^0 = \frac{T_d - T}{T_d} \cdot \Delta H(T_d) + \Delta C_p \cdot (T - T_d) + T \cdot \Delta C_p \cdot \ln\left(\frac{T_d}{T}\right) - R \cdot T \cdot \ln K(T_d) \tag{52}$$

$$\Delta G_U = \frac{T_m - T}{T_m} \cdot \Delta H(T_m) + \Delta C_p \cdot (T - T_m) + T \cdot \Delta C_p \cdot \ln\left(\frac{T_m}{T}\right) \tag{53}$$

where ΔG_d^0 is the standard Gibbs energy of dissociation of native dimer into native monomers, and ΔG_U is the monomolecular Gibbs energy of unfolding of monomers. For convenience the reference temperatures are defined as follows: T_d is defined as a temperature at which $F_{N_2} = F_N^2$, and thus according to the Eq. (45) the equilibrium constant is equal to $2P_T$; T_m is the temperature at which $F_N = F_U$ and thus the equilibrium constant is equal to 1.

4.6.4
Two-state Model with Ligand Binding

In this model, the native state, N, but not the unfolded state, U, can bind a ligand, L. The ligand does not undergo any conformational transition upon binding.

$$NL \overset{K_d}{\leftrightarrow} N + L \overset{K_U}{\leftrightarrow} U + L$$

The equilibrium constants of dissociation, K_d, and of unfolding, K_U, are then defined as:

$$K_d = \frac{[N] \cdot [L]}{[NL]} \quad \text{and} \quad K_U = \frac{[U]}{[N]} \tag{54}$$

where [NL], [N], [U], and [L] are the concentrations of ligated, and unligated native states, unfolded state, and free ligand concentration, respectively. The fraction of protein in these three states can be presented as:

$$[N] = F_N \cdot P_T \tag{55}$$

$$[U] = F_U \cdot P_T \tag{56}$$

$$[NL] = F_{NL} \cdot P_T \tag{57}$$

where the total concentration of the protein, P_T, is

$$P_T = [NL] + [N] + [U] \tag{58}$$

Similarly, the total concentration of the ligand, L_T, can be expressed as:

$$L_T = [NL] + [L] \tag{59}$$

Combining Eqs (54)–(59) allows the equilibrium constants to be expressed as a function of total protein and ligand concentrations and fractions of native and unfolded protein as:

$$K_d = \frac{F_N \cdot (L_T - F_{NL} \cdot P_T)}{F_{NL}} \tag{60}$$

$$K_U = \frac{F_U \cdot (L_T - F_{NL} \cdot P_T)}{K_d \cdot F_{NL}} \tag{61}$$

Keeping in mind that $F_{NL} + F_N + F_U = 1$ we can combine Eqs (60)–(61) and write:

$$F_{NL} + \frac{K_d \cdot F_{NL}}{(L_T - F_{NL} \cdot P_T)} + \frac{K_d \cdot K_U \cdot F_{NL}}{(L_T - F_{NL} \cdot P_T)} = 1 \tag{62}$$

Equation (62) is a simple quadratic equation and can be solved for F_{NL} to give:

$$F_{NL} = \frac{\sqrt{q^2 - 4 \cdot P_T \cdot L_T} - q}{2 \cdot P_T} \tag{63}$$

where $q = P_T + L_T + K_d + K_U K_d$

The fractions of the unligated native and of the unfolded states can be expressed as:

$$F_N = \frac{K_d \cdot (\sqrt{q^2 - 4 \cdot P_T \cdot L_T} - q)}{2 \cdot L_T \cdot P_T - P_T \cdot (\sqrt{q^2 - 4 \cdot P_T \cdot L_T} - q)} \tag{64}$$

$$F_U = \frac{K_U \cdot K_d \cdot (\sqrt{q^2 - 4 \cdot P_T \cdot L_T} - q)}{2 \cdot L_T \cdot P_T - P_T \cdot (\sqrt{q^2 - 4 \cdot P_T \cdot L_T} - q)} \tag{65}$$

The excess enthalpy function is defined as:

$$\langle \Delta H^{exc} \rangle = \Delta H_d \cdot F_N + (\Delta H_d + \Delta H_U) \cdot F_U \tag{66}$$

which allows calculation of the excess heat capacity function as a temperature derivative of the excess enthalpy.

The corresponding Gibbs energies can be expressed as:

$$\Delta G_d^0 = \Delta H(T_0) + \Delta C_p \cdot (T - T_0) - T \cdot \Delta S(T_0)$$
$$- T \cdot \Delta C_p \cdot \ln\left(\frac{T_0}{T}\right) \quad R \cdot T \cdot \ln K_d(T_0) \tag{67}$$

$$\Delta G_U = \frac{T_m - T}{T_m} \cdot \Delta H(T_m) + \Delta C_p \cdot (T - T_m) + T \cdot \Delta C_p \cdot \ln\left(\frac{T_m}{T}\right) \tag{68}$$

where ΔG_d^0 is the standard Gibbs energy of dissociation, ΔG_U is the monomolecular Gibbs energy of unfolding, T_0 is the reference temperature for the dissociation reaction, and T_m is the temperature at which $F_N = F_U$ and thus the equilibrium constant is equal to 1. In practice, the ΔG_U function is determined from the DSC experiments performed in the absence of the ligand, and then is used to fit the DSC profiles obtained in the presence of ligand.

4.6.5
Four-state (Two-domain Protein) Model

In this model, the protein consists of two interacting domains A and B, each of which unfolds as a two state.

$$\begin{array}{ccc} A_N B_N & \xrightleftharpoons{K_1} & A_U B_N \\ {\scriptstyle K_2}\updownarrow & & \updownarrow{\scriptstyle K_3} \\ A_N B_U & \xrightleftharpoons[K_4]{} & A_U B_U \end{array}$$

The equilibrium constants for the four reactions can be expressed through the concentrations of four states, native with both A and B domains folded ($A_N B_N$), domain A folded domain B unfolded ($A_N B_U$), domain A unfolded domain B folded ($A_U B_N$), and both domains unfolded ($A_U B_U$), as

$$K_1 = \frac{[A_U B_N]}{[A_N B_N]} \tag{69}$$

$$K_2 = \frac{[A_N B_U]}{[A_N B_N]} \tag{70}$$

$$K_3 = \frac{[A_U B_U]}{[A_U B_N]} \tag{71}$$

$$K_4 = \frac{[A_U B_U]}{[A_N B_U]} \tag{72}$$

It is important to emphasize that the four equilibrium constants are not fully independent since the following relationship is valid $K_1 \cdot K_3 = K_2 \cdot K_4$.

The fraction of each of the four states can be written as:

$$F_{A_N B_N} = \frac{[A_N B_N]}{[A_N B_N] + [A_U B_N] + [A_N B_U] + [A_U B_U]} \tag{73}$$

$$F_{A_U B_N} = \frac{[A_U B_N]}{[A_N B_N] + [A_U B_N] + [A_N B_U] + [A_U B_U]} \tag{74}$$

$$F_{A_N B_U} = \frac{[A_N B_U]}{[A_N B_N] + [A_U B_N] + [A_N B_U] + [A_U B_U]} \tag{75}$$

$$F_{A_U B_U} = \frac{[A_U B_U]}{[A_N B_N] + [A_U B_N] + [A_N B_U] + [A_U B_U]} \tag{76}$$

Combining Eqs (69)–(72) and (73)–(76) we can relate the fraction of individual states to the corresponding equilibrium constants:

$$F_{A_N B_N} = \frac{1}{1 + K_1 + K_2 + K_3} \tag{77}$$

$$F_{A_U B_N} = \frac{K_1}{1 + K_1 + K_2 + K_1 \cdot K_3} \tag{78}$$

$$F_{A_N B_U} = \frac{K_2}{1 + K_1 + K_2 + K_1 \cdot K_3} \tag{79}$$

$$F_{A_U B_U} = \frac{K_1 \cdot K_3}{1 + K_1 + K_2 + K_1 \cdot K_3} \tag{80}$$

Difference in K_1 and K_4 or K_2 and K_3 defines the energy of interaction between domains, ΔG_{int}.

$$K_1 = \frac{K_4}{K_{\text{int}}} = \exp\left(-\frac{\Delta G_4 - \Delta G_{\text{int}}}{R \cdot T}\right) \tag{81}$$

$$\Delta G_4 = \Delta H(T_A) \cdot \left(1 + \frac{T}{T_A}\right) + \Delta C_{p,A} \cdot (T - T_A - T \cdot \ln(T/T_A)) \tag{82}$$

$$K_2 = \frac{K_3}{K_{\text{int}}} = \exp\left(-\frac{\Delta G_3 - \Delta G_{\text{int}}}{R \cdot T}\right) \tag{83}$$

$$\Delta G_3 = \Delta H(T_B) \cdot \left(1 + \frac{T}{T_B}\right) + \Delta C_{p,B} \cdot (T - T_B - T \cdot \ln(T/T_B)) \tag{84}$$

$$K_{int} = \exp\left(-\frac{\Delta G_{int}}{R \cdot T}\right) \tag{85}$$

$$\Delta G_{int} = \Delta H_{int} - T \cdot \Delta S_{int} \tag{86}$$

The reference temperatures T_A and T_B are defined as the temperatures at which the each of the two intermediate states, $A_U B_N$ and $A_N B_U$ that have one of the domains unfolded awhile the other one folded, is 50% populated. It is important to emphasize that the energy of interaction as defined above also includes possible conformational changes in each domain upon their interactions.

The excess enthalpy of the system can be written as:

$$\langle \Delta H^{exc} \rangle = (\Delta H_A - \Delta H_{int}) \cdot F_{A_U B_N} + (\Delta H_B - \Delta H_{int}) \cdot F_{A_N B_U}$$
$$+ (\Delta H_A + \Delta H_B - \Delta H_{int}) \cdot F_{A_U B_U} \tag{87}$$

and the excess heat capacity function can be obtained by differentiation.

$$\langle C_p^{exc}(T) \rangle = \frac{\partial \langle \Delta H^{exc} \rangle}{\partial T} = (\Delta H_A - \Delta H_{int}) \cdot \frac{dF_{A_U B_N}}{dT} + (\Delta H_B - \Delta H_{int}) \cdot \frac{dF_{A_N B_U}}{dT}$$
$$+ (\Delta H_A + \Delta H_B - \Delta H_{int}) \cdot \frac{dF_{A_U B_U}}{dT} \tag{88}$$

Note that after one of the domains melts, the interaction energy goes to zero. Thus, ΔG_{int} has apparent effect only on a domain with lower T_m.

4.7
Experimental Protocols

4.7.1
How to Prepare for DSC Experiments

The DSC instrument should be turned on prior to the experiment and several (at least three) baseline scans with both cells filled with the buffer should be recorded. The last two profiles must be overlapping. This establishes the thermal history and verifies that the instrument is working within specifications.

The pre-experiment baseline scans should be done at the scan rate that will be used for the protein scan. The heating rate is one of the most important parameters. Several considerations have to be taken into account. First, the higher the scan rate, the higher the sensitivity of the instrument. The increase in sensitivity is linear with respect to the heating rate (i.e., the sensitivity at a heating rate of 2° min^{-1} is twice as high as that at 1° min^{-1}), but the increase in sensitivity does not mean an increase in the signal-to-noise ratio. Second, if the expected transition is very sharp (e.g., within a few degrees) a high heating rate will affect the shape of the heat absorption profile and lead to an error in the determination of all thermodynamic parameters for this transition. This error will be particularly large for the transition temperature. Third, the higher the heating rate the less time the system has to equilibrate. So for slow unfolding/refolding processes it is advisable to use low heating rates.

In practice, for small globular proteins exhibiting reversible temperature induced unfolding, a maximal (on most DSC instruments) heating rate of 2° min^{-1} is acceptable. For large proteins lower heating rates (0.5–1° min^{-1}) are more appropriate. In the case of fibrillar proteins, which usually exhibit very narrow transitions, a heating rate of 0.1–0.25° min^{-1} is most suitable.

The purity of the protein sample is very important. Contamination with other proteins, nucleic acids, or lipids can affect both the shape of the calorimetric profiles and the estimation of the thermodynamic parameters of unfolding through the errors in determination of the sample concentration. PAAG electrophoresis, UV/VIS spectroscopy and mass spectroscopy should be used to validate the purity of the protein sample.

Since calorimetric experiments are performed over a broad temperature range, the pH of the buffer should have a small temperature dependence. For this reason for example, very popular "physiological" Tris buffer is not the best choice. Glycine, sodium/potassium acetate, sodium/potassium phosphate, sodium cacodilate, and sodium citrate are the buffers of choice.

Equilibration of protein with buffer is a vital part of the sample preparation for calorimetric experiments. One should keep in mind that the DSC instrument measures the heat capacity difference between buffer and protein solutions. Thus any small inequality in the buffer composition will result in an incorrect absolute value and an incorrect slope of the heat capacity profile. The best way to achieve equilibrium between buffer and protein solution is dialysis. The protein sample should be dialyzed against at least two changes of buffer. The buffer solutions should be filtered prior to dialysis using a 0.45 μm filter to eliminate all insoluble material.

Following dialysis, all insoluble particles should be carefully removed from the protein solution. This can be achieved by centrifugation at $\sim$12 000 g in a microcentrifuge for $\sim$15–20 min.

Knowledge of the exact protein concentration *after dialysis* is very important. Among all available methods for determination of protein concentration, spectrophotometric methods seem to be most accurate and rapid. This method requires knowledge of the extinction coefficient of the protein at a given wavelength. The extinction coefficient can be calculated from the number of aromatic residues and disulfide bonds in a protein using an empirical equation [64]:

$$\varepsilon_{280\ nm}^{0.1\%,\ 1\ cm} = (5690 \cdot N_{Trp} + 1280 \cdot N_{Tyr} + 120 \cdot N_{SS})/MW \tag{89}$$

where MW is the molecular weight of the protein in daltons. A more elaborate procedure is described by Pace et al. [65] and improves the accuracy of the calculations. For proteins that do not contain any aromatic residues, the method that is based on absorption of peptide backbone gives fast and accurate results. Light scattering in spectrophotometric measurements of protein concentration can be a significant problem, but can be taken into account using a procedure suggested by Winder and Gent [66].

4.7.2
How to Choose Appropriate Conditions

The most frequent problem in DSC experiments is irreversibility of the unfolding transition. This particularly applies to large proteins that have tendencies to aggregate after unfolding. Sensitivity of contemporary DSC instruments provides reliable heat capacity profiles at concentrations as low as 0.3 mg mL^{-1} for proteins of the molecular range 6–17 kDa. The concentration for larger proteins can be even lower. Low protein concentration leads to decreased probability for aggregation. However, even at low protein concentrations, finding the proper solvent conditions is not a simple task. A rule of thumb is that at a pH value close to the isoelectric point of the protein, aggregation after unfolding is highly probable. High concentrations of neutral salts also promote aggregation. In many cases, aggregation can be avoided by keeping the ionic strength of the solvent very low (10–50 mM) and by using pH values a couple of units away (usually below) from the isoelectric point.

4.7.3
Critical Factors in Running DSC Experiments

Two important pieces of information should be established prior to the analysis of the calorimetric profiles. This will help to decide whether the formalism of equilibrium thermodynamics is applicable to the studied protein:

1. Reversibility of the unfolding transition under several conditions should be examined by rescanning the sample. Reversibility depends on the upper temperature of heating. The unfolding transition can be considered reversible if rescanning after the sample has been heated to the temperature just past the completion of the transition leads to the recovery of at least 85–90% of the initial endotherm. If no endotherm is observed after rescanning, the unfolding reaction is irreversible. Thermodynamic analysis for irreversible transitions is possible only in special cases [11, 67].
2. Equilibrium conditions of reversible unfolding should be examined to find the optimal scan rate that permits establishment of equilibrium between native and unfolded states at all temperatures (i.e., scan rate should not be faster than the folding/unfolding rates). This can be examined by comparing the heat capacity profiles obtained on exactly the same sample at two different heating rates, usually 2° min^{-1} and 0.5° min^{-1}. If the difference in the transition temperature exceeds 0.5–0.7 °C, additional experiments at a lower heating rate (i.e., 0.1° min^{-1}) should be performed and compared with a 0.5° min^{-1} experiment. If there is still a difference in the transition temperature all experiments at a given solvent composition should be performed at several different heating rates and the transition temperatures should be extrapolated to a zero heating rate (for more details see Ref. [68]).

References

1. PLOTNIKOV, V. V., BRANDTS, J. M., LIN, L. N., & BRANDTS, J. F. (1997). A new ultrasensitive scanning calorimeter. *Anal Biochem* 250, 237–244.
2. PRIVALOV, G., KAVINA, V., FREIRE, E., & PRIVALOV, P. L. (1995). Precise scanning calorimeter for studying thermal properties of biological macromolecules in dilute solution. *Anal Biochem* 232, 79–85.
3. LOPEZ, M. M. & MAKHATADZE, G. I. (2002). Differential scanning calorimetry. *Methods Mol Biol* 173, 113–119.
4. MAKHATADZE, G. I. (1998). *Measuring protein thermostability by differential scanning calorimetry.* Current Protocols in Protein Chemistry, 2, John Wiley & Sons, New York.
5. PFEIL, W. (1998). *Protein Stability and folding. A collection of thermodynamic data,* Springer, Berlin, Heidelberg, New York.
6. http://www.rtc.riken.go.jp/jouhou/Protherm/protherm.html.
7. MAKHATADZE, G. I. (1998). Heat capacities of amino acids, peptides and proteins. *Biophys Chem* 71, 133–156.
8. BILTONEN, R. L. & FREIRE, E. (1978). Thermodynamic characterization of conformational states of biological macromolecules using differential scanning calorimetry. *CRC Crit Rev Biochem* 5, 85–124.
9. PRIVALOV, P. L. (1979). Stability of proteins: small globular proteins. *Adv Protein Chem* 33, 167–241.
10. BECKTEL, W. J. & SCHELLMAN, J. A. (1987). Protein stability curves. *Biopolymers* 26, 1859–1877.
11. SANCHEZ-RUIZ, J. M., LOPEZ-LACOMBA, J. L., CORTIJO, M., & MATEO, P. L. (1988). Differential scanning calorimetry of the irreversible thermal denaturation of thermolysin. *Biochemistry* 27, 1648–1652.
12. LUMRY, R. & BILTONEN, R. (1966). Validity of the "two-state" hypothesis for conformational transitions of proteins. *Biopolymers* 4, 917–944.
13. JACKSON, W. M. & BRANDTS, J. F. (1970). Thermodynamics of protein denaturation. A calorimetric study of the reversible denaturation of chymotrypsinogen and conclusions regarding the accuracy of the two-state approximation. *Biochemistry* 9, 2294–2301.
14. TSONG, T. Y., HEARN, R. P., WRATHALL, D. P., & STURTEVANT, J. M. (1970). A calorimetric study of thermally induced conformational transitions of ribonuclease A and certain of its derivatives. *Biochemistry* 9, 2666–2677.
15. PRIVALOV, P. L. & KHECHINASHVILI, N. N. (1974). A thermodynamic approach to the problem of stabilization of globular protein structure: a calorimetric study. *J Mol Biol* 86, 665–684.
16. PACE, N. C. & TANFORD, C. (1968). Thermodynamics of the unfolding of beta-lactoglobulin A in aqueous urea solutions between 5 and 55 degrees. *Biochemistry* 7, 198–208.
17. PRIVALOV, P. L., GRIKO YU, V., VENYAMINOV, S., & KUTYSHENKO, V. P. (1986). Cold denaturation of myoglobin. *J Mol Biol* 190, 487–498.
18. PRIVALOV, P. L. (1990). Cold denaturation of proteins. *Crit Rev Biochem Mol Biol* 25, 281–305.
19. GRIKO, Y. V., PRIVALOV, P. L., STURTEVANT, J. M., & VENYAMINOV, S. (1988). Cold denaturation of staphylococcal nuclease. *Proc Natl Acad Sci USA* 85, 3343–3347.
20. BEADLE, B. M., BAASE, W. A., WILSON, D. B., GILKES, N. R., & SHOICHET, B. K. (1999). Comparing the thermodynamic stabilities of a related thermophilic and mesophilic enzyme. *Biochemistry* 38, 2570–2576.
21. HOLLIEN, J. & MARQUSEE, S. (1999). A thermodynamic comparison of mesophilic and thermophilic ribonucleases H. *Biochemistry* 38, 3831–3836.
22. REES, D. C. & ROBERTSON, A. D. (2001). Some thermodynamic

implications for the thermostability of proteins. *Protein Sci* 10, 1187–1194.

23 DEUTSCHMAN, W. A. & DAHLQUIST, F. W. (2001). Thermodynamic basis for the increased thermostability of CheY from the hyperthermophile *Thermotoga maritima*. *Biochemistry* 40, 13107–13113.

24 MAKHATADZE, G. I., LOLADZE, V. V., GRIBENKO, A. V., & LOPEZ, M. M. (2004). Mechanism of thermostabilization in a designed cold shock protein with optimized surface electrostatic interactions. *J Mol Biol* 336, 929–942.

25 LUMRY, R. & RAJENDER, S. (1970). Enthalpy–entropy compensation phenomena in water solutions of proteins and small molecules: a ubiquitous property of water. *Biopolymers* 9, 1125–1227.

26 SHARP, K. (2001). Entropy–enthalpy compensation: fact or artifact? *Protein Sci* 10, 661–667.

27 BEASLEY, J. R., DOYLE, D. F., CHEN, L., COHEN, D. S., FINE, B. R., & PIELAK, G. J. (2002). Searching for quantitative entropy–enthalpy compensation among protein variants. *Proteins* 49, 398–402.

28 MAKHATADZE, G. I., CLORE, G. M., & GRONENBORN, A. M. (1995). Solvent isotope effect and protein stability. *Nat Struct Biol* 2, 852–855.

29 MAKHATADZE, G. I. & PRIVALOV, P. L. (1990). Heat capacity of proteins. I. Partial molar heat capacity of individual amino acid residues in aqueous solution: hydration effect. *J Mol Biol* 213, 375–384.

30 SPOLAR, R. S., HA, J. H., & RECORD, M. T., JR. (1989). Hydrophobic effect in protein folding and other noncovalent processes involving proteins. *Proc Natl Acad Sci USA* 86, 8382–8385.

31 SPOLAR, R. S., LIVINGSTONE, J. R., & RECORD, M. T., JR. (1992). Use of liquid hydrocarbon and amide transfer data to estimate contributions to thermodynamic functions of protein folding from the removal of nonpolar and polar surface from water. *Biochemistry* 31, 3947–3955.

32 BALDWIN, R. L. (1986). Temperature dependence of the hydrophobic interaction in protein folding. *Proc Natl Acad Sci USA* 83, 8069–8072.

33 MAKHATADZE, G. I. & PRIVALOV, P. L. (1995). Energetics of protein structure. *Adv Protein Chem* 47, 307–425.

34 ROBERTSON, A. D. & MURPHY, K. P. (1997). Protein structure and the energetics of protein stability. *Chem Rev* 97, 1251–1268.

35 MYERS, J. K., PACE, C. N., & SCHOLTZ, J. M. (1995). Denaturant m values and heat capacity changes: relation to changes in accessible surface areas of protein unfolding. *Protein Sci* 4, 2138–2148.

36 STURTEVANT, J. M. (1977). Heat capacity and entropy changes in processes involving proteins. *Proc Natl Acad Sci USA* 74, 2236–2240.

37 LOLADZE, V. V., ERMOLENKO, D. N., & MAKHATADZE, G. I. (2001). Heat capacity changes upon burial of polar and nonpolar groups in proteins. *Protein Sci* 10, 1343–1352.

38 COOPER, A., JOHNSON, C. M., LAKEY, J. H., & NOLLMANN, M. (2001). Heat does not come in different colours: entropy-enthalpy compensation, free energy windows, quantum confinement, pressure perturbation calorimetry, solvation and the multiple causes of heat capacity effects in biomolecular interactions. *Biophys Chem* 93, 215–230.

39 LAZARIDIS, T., ARCHONTIS, G., KARPLUS, M. (1995). Enthalpic contribution to protein stability: insights from atom-based calculations and statistical mechanics. *Adv Protein Chem* 47, 231–306.

40 LOLADZE, V. V., ERMOLENKO, D. N., & MAKHATADZE, G. I. (2002). Thermodynamic consequences of burial of polar and nonpolar amino acid residues in the protein interior. *J Mol Biol* 320, 343–357.

41 RICHARDS, F. M. (1977). Areas, volumes, packing and protein structure. *Annu Rev Biophys Bioeng* 6, 151–176.

42 FLEMING, P. J. & RICHARDS, F. M. (2000). Protein packing: dependence

on protein size, secondary structure and amino acid composition. *J Mol Biol* 299, 487–498.
43. PACE, C. N., SHIRLEY, B. A., McNUTT, M., & GAJIWALA, K. (1996). Forces contributing to the conformational stability of proteins. *FASEB J* 10, 75–83.
44. MYERS, J. K. & PACE, C. N. (1996). Hydrogen bonding stabilizes globular proteins. *Biophys J* 71, 2033–2039.
45. SCHOLTZ, J. M., MARQUSEE, S., BALDWIN, R. L., YORK, E. J., STEWART, J. M., SANTORO, M., & BOLEN, D. W. (1991). Calorimetric determination of the enthalpy change for the alpha-helix to coil transition of an alanine peptide in water. *Proc Natl Acad Sci USA* 88, 2854–2858.
46. RICHARDSON, J. M., McMAHON, K. W., MacDONALD, C. C., & MAKHATADZE, G. I. (1999). MEARA sequence repeat of human CstF-64 polyadenylation factor is helical in solution. A spectroscopic and calorimetric study. *Biochemistry* 38, 12869–12875.
47. LOPEZ, M. M., CHIN, D. H., BALDWIN, R. L., & MAKHATADZE, G. I. (2002). The enthalpy of the alanine peptide helix measured by isothermal titration calorimetry using metal-binding to induce helix formation. *Proc Natl Acad Sci USA* 99, 1298–1302.
48. YANG, A. S., SHARP, K. A., & HONIG, B. (1992). Analysis of the heat capacity dependence of protein folding. *J Mol Biol* 227, 889–900.
49. MADAN, B. & SHARP, K. (1999). Changes in water structure induced by a hydrophobic solute probed by simulation of the water hydrogen bond angle and radial distribution functions. *Biophys Chem* 78, 33–41.
50. CREAMER, T. P. & ROSE, G. D. (1992). Side-chain entropy opposes alpha-helix formation but rationalizes experimentally determined helix-forming propensities. *Proc Natl Acad Sci USA* 89, 5937–5941.
51. DOIG, A. J. & STERNBERG, M. J. (1995). Side-chain conformational entropy in protein folding. *Protein Sci* 4, 2247–2251.
52. AMZEL, L. M. (2000). Calculation of entropy changes in biological processes: folding, binding, and oligomerization. *Methods Enzymol* 323, 167–177.
53. PICKETT, S. D. & STERNBERG, M. J. (1993). Empirical scale of side-chain conformational entropy in protein folding. *J Mol Biol* 231, 825–839.
54. STERNBERG, M. J. & CHICKOS, J. S. (1994). Protein side-chain conformational entropy derived from fusion data – comparison with other empirical scales. *Protein Eng* 7, 149–155.
55. BLABER, M., ZHANG, X. J., LINDSTROM, J. D., PEPIOT, S. D., BAASE, W. A., & MATTHEWS, B. W. (1994). Determination of alpha-helix propensity within the context of a folded protein. Sites 44 and 131 in bacteriophage T4 lysozyme. *J Mol Biol* 235, 600–624.
56. DILL, K. A. (1990). Dominant forces in protein folding. *Biochemistry* 29, 7133–7155.
57. MATOUSCHEK, A. (2003). Protein unfolding – an important process in vivo? *Curr Opin Struct Biol* 13, 98–109.
58. DOIG, A. J. & WILLIAMS, D. H. (1992). Why water-soluble, compact, globular proteins have similar specific enthalpies of unfolding at 110 degrees C. *Biochemistry* 31, 9371–9375.
59. SANCHEZ-RUIZ, J. M. (1995). Differential scanning calorimetry of proteins. *Subcell Biochem* 24, 133–176.
60. LOLADZE, V. V., IBARRA-MOLERO, B., SANCHEZ-RUIZ, J. M., & MAKHATADZE, G. I. (1999). Engineering a thermostable protein via optimization of charge–charge interactions on the protein surface. *Biochemistry* 38, 16419–16423.
61. SANCHEZ-RUIZ, J. M. & MAKHATADZE, G. I. (2001). To charge or not to charge? *Trends Biotechnol* 19, 132–135.
62. THOMAS, S. T., LOLADZE, V. V., & MAKHATADZE, G. I. (2001). Hydration of the peptide backbone largely defines the thermodynamic propensity scale of residues at the C' position of the C-capping box of alpha-helices. *Proc Natl Acad Sci USA* 98, 10670–10675.

63 Makhatadze, G. I., Loladze, V. V., Ermolenko, D. N., Chen, X., & Thomas, S. T. (2003). Contribution of surface salt bridges to protein stability: guidelines for protein engineering. *J Mol Biol* 327, 1135–1148.

64 Gill, S. C. & von Hippel, P. H. (1989). Calculation of protein extinction coefficients from amino acid sequence data. *Anal Biochem* 182, 319–326.

65 Pace, C. N., Vajdos, F., Fee, L., Grimsley, G., & Gray, T. (1995). How to measure and predict the molar absorption coefficient of a protein. *Protein Sci* 4, 2411–2423.

66 Winder, A. F. & Gent, W. L. (1971). Correction of light-scattering errors in spectrophotometric protein determinations. *Biopolymers* 10, 1243–1251.

67 Sanchez-Ruiz, J. M. (1992). Theoretical analysis of Lumry-Eyring models in differential scanning calorimetry. *Biophys J* 61, 921–935.

68 Yu, Y., Makhatadze, G. I., Pace, C. N., & Privalov, P. L. (1994). Energetics of ribonuclease T1 structure. *Biochemistry* 33, 3312–3319.

5
Pressure–Temperature Phase Diagrams of Proteins

Wolfgang Doster and Josef Friedrich

5.1
Introduction

One of the most peculiar features of proteins is their marginal stability within a narrow range of thermodynamic conditions. Biologically active structures can be disrupted by increasing or decreasing the temperature, the pressure, the pH or by adding denaturants. By disrupting a structure one can study its architecture and energetics. In this article we focus on the thermodynamic aspects of temperature and pressure denaturation. A variation of temperature inevitably leads to a simultaneous change of entropy and volume through thermal expansion. The advantage of using pressure as a thermodynamic variable is that volume-dependent effects can be isolated from temperature-dependent effects. Just as the increase in temperature drives the system in equilibrium towards states of higher entropy, a pressure increase will bias the ensemble of accessible states towards those with a smaller volume. This is the meaning of le Chatelier's principle. Thus, if the protein has the lower volume in its denatured form, the native structure will become unstable above a critical pressure. Similarly the opposite effect, stabilization of the native state by pressure, is sometimes associated with heat denaturation.

Packing defects in the water-excluding native state, reorganization of the solvent near exposed nonpolar side chains, and electrostriction by newly formed charges act as volume reservoirs which can lower the volume in the unfolded state. While dissociation of oligomeric proteins into subunits occurs at low pressures, i.e., below 200 MPa [1], pressures above 300 MPa are required to denature monomeric proteins. A number of useful review articles on this subject have been published recently (see, for example, Ref. [2]).

The isothermal compressibility of water (0.56/GPa) is quite small. At 11 000 m below sea level and pressures near 100 MPa, the density of water is only 5% higher than at ambient pressure. On the other hand, the density at the freezing transition changes by 9%. Moreover, proteins are 5–10 times less compressible than water. As a result, pressure-induced volume changes in proteins are quite small, typically 0.5% of the total volume at the unfolding transition. For monomeric proteins the difference corresponds approximately to the volume of five water molecules

Protein Folding Handbook. Part I. Edited by J. Buchner and T. Kiefhaber
Copyright © 2005 WILEY-VCH Verlag GmbH & Co. KGaA, Weinheim
ISBN: 3-527-30784-2

(5×18 mL mol^{-1}). Thus, minor volume changes often induce large structural rearrangements, where essentially incompressible matter is displaced. Therefore pressure can be a powerful structural gear. On the other hand, the microscopic interpretation of protein volume changes is still one of the difficult and unresolved questions of the field [3, 4].

Part of the problem arises from the fact that the denatured state itself is not very well defined [5]. It is generally assumed that thermal denaturing leads to a random coil state. As to pressure-induced denaturation and cold denaturation, there are indications that the denatured state is still compact and resembles the molten globule state [5–7]. How such a situation can be reconciled with the simple thermodynamic description of protein stability based on a two-state model has to be investigated in each case.

Below we present a thermodynamic frame for describing and interpreting protein stability phase diagrams. The main features are illustrated using our own results obtained with various optical techniques as well as with neutron scattering.

5.2
Basic Aspects of Phase Diagrams of Proteins and Early Experiments

The most striking thermodynamic properties of proteins are related to the hydrophobic effect. These are

- the large and positive heat capacity increment upon unfolding [8–10] and
- the phenomenon of cold denaturation.

The latter was first predicted by Brandts [11, 12] based on studies of heat denaturation on ribonuclease. The transition temperatures for cold denaturation are usually below the freezing point of most aqueous solutions at ambient pressure. However, under high pressure, water remains liquid down to much lower temperatures (-18 °C at 200 MPa, Figure 5.1).

Thus cold denaturation can be studied conveniently at elevated pressures under mild denaturing conditions, without the need to add denaturants.

The phase diagram of water (Figure 5.1) illustrates the relevance of the pressure–temperature plane for locating states of structural stability and for classifying their thermodynamic properties.

The slope of the phase boundary of liquid water and ice I is negative: Increase in pressure extends the stability range of the liquid phase. The slope of the phase boundary is given by the ratio of the negative entropy change and the increase in specific volume at the freezing transition (Eq. (3)). The expansion ($+9\%$) causes substantial damage when biological material is frozen. In contrast, the slope of the liquid to ice III boundary is large and positive, since the corresponding volume change is small and negative. The application of high pressure to the liquid and then cooling into ice III minimizes the damage by volume changes (-3%) at the freezing transition.

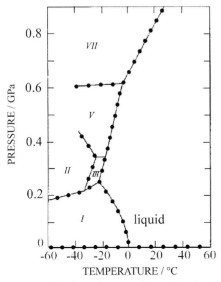

Fig. 5.1. The phase diagram of water: For the boundary from the liquid to hexagonal ice (I) one has $dP/dT = \Delta S/\Delta V < 0$, unlike for the technically relevant modifications, II and III where the volume discontinuity is small.

Proteins, consisting of a few thousand atoms, are mesoscopic objects. Experience and theoretical analysis shows [13] that they can be labeled by well-defined thermodynamic averages, such as the specific heat capacity, the compressibility, and the thermal expansion. The structural reorganization may then be regarded as a change of the macroscopic state of the system. Such changes occur in a highly cooperative manner. Since the native and denatured states of proteins differ not only in heat capacity but also in volume and entropy, the transition between the two states must involve a discontinuity in the first derivatives of the thermodynamic potential. This is the signature of a first-order phase transition. Consequently, the native and denatured states of a single-domain protein may be interpreted as two phases of a macroscopic system, which differ in structural order. For a single protein molecule the transformation can only be abrupt and not gradually. From this point of view a cooperative domain of a protein resembles a crystal, which has, however, a critical size, since it can only exist as a whole. Experimentally, however, one usually deals with protein ensembles.

The properties of the ensemble, which are reflected in the heterogeneity of the characteristic parameters of a protein, such as its energy, structure, compressibility, etc., render some specific features to the denaturing transition compared with phase transitions of thermodynamic systems: The transition is usually rather broad in the variables P or T (e.g., see Figures 5.9 and 5.10). This dispersion of the transition region is directly related to the heterogeneity of the protein ensemble, which has its roots in the fact that a protein, although a large molecule, is not an infinite

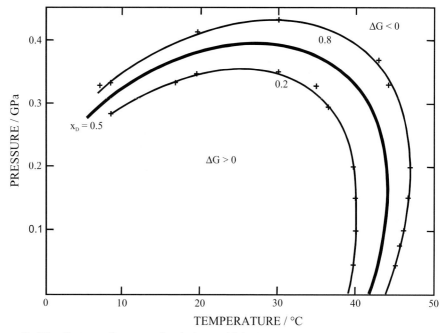

Fig. 5.2. Contours of constant chemical potential $\Delta\mu = \mu_D - \mu_N$ between the denatured (D) and the native (N) state of chymotrypsinogen (after Hawley [14]).

system. The equilibrium constant $K_{DN}(T, P)$ follows from the pressure analog of the van t'Hoff equation with $\partial T \to \partial P$ and $\Delta S \to -\Delta V$:

$$\partial(\ln K_{DN})/\partial P = -\Delta V/RT$$

The transition width, determined by the magnitude of the transition volume ΔV, is finite for a mesoscopic protein molecule, in contrast to the zero width of a macroscopic first-order transition.

Hawley, in 1971, was the first to interpret the curve on which the free energy difference between native and unfolded state vanishes as an elliptical phase boundary. This provided a natural explanation for the so-called re-entrant phase behavior of heat and cold denaturation [14]. Figure 5.2 shows the contours of the Gibbs free energy $\Delta G(P, T)$ versus pressure and temperature for chymotrypsinogen. Note that at moderate pressure levels (< 150 MPa), the temperature of heat denaturation increases slightly with pressure. Hence, applying pressure to the thermally denatured protein may actually drive the protein back into the native state. Upon further increasing the pressure, however, unfolding may occur again. In other words, the denatured phase is re-entered, irrespective of the fact that the pressure is monotonously changed. Phase diagrams with this property are called "re-entrant." A similar behavior is found if the temperature is changed. For instance, if one starts out at sufficiently high pressure levels (200 MPa) in the denatured

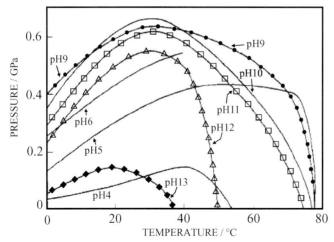

Fig. 5.3. Phase boundaries ($\Delta\mu = 0$) of myoglobin at various pH values, the native state is stable inside the contours, the denatured state outside (after Zipp and Kauzmann [18]).

state and lowers the temperature, one enters the native regime which is, however, left again at even lower temperatures to re-enter the denatured regime again. This latter transition is called "cold denaturation".

Re-entrant phase diagrams of perfect elliptical shape have been observed for liquid crystals by Cladis et al. [15] and by Klug and Whalley [16]. The elliptical shape was analyzed in detail by Clark [16].

In 1973, Zipp and Kauzmann, in a seminal paper, investigated the pressure–temperature–pH phase diagram of myoglobin [18]. They also observed re-entrant behavior, but did not attempt to fit the phase boundaries using ellipses. Depending on pH, the boundaries vary strongly in shape and show nonelliptical distortions, as shown in Figure 5.3.

In recent years, a significant number of studies have suggested diagrams of elliptical shape for protein denaturation or for even more complex units, such as bacteria [19].

The stability phase diagram of proteins and related phenomena have been discussed in several review articles [20, 21, 22].

Our goal is to elucidate the physical basis and thermodynamics of re-entrant protein phase diagrams, the conditions for elliptical shapes, and the limitations of this approach.

5.3
Thermodynamics of Pressure–Temperature Phase Diagrams

The phase equilibrium between the native and the denatured state N ⇌ D is controlled by the difference in the chemical potentials $\Delta\mu = \mu_D - \mu_N$. The ratio of concentrations of the denatured to the native form assuming a dilute solution is given by:

$$K_{DN} = c_D/c_N = \exp(-\Delta\mu/RT) \tag{1}$$

The chemical potential is the driving force that induces transport and transformations of a substance. It is analogous to the electrical potential, which can only induce currents because of charge conservation. Since the concentration of components, water, and protein is fixed (only pressure and temperature are varied), we restrict ourselves to one-component systems. The change $d\mu_i$ with temperature and pressure obeys the Gibbs–Duhem relation for each phase D and N:

$$d\mu_N = -S_N\,dT + V_N\,dP = -RT\,d(\ln c_N/c_0)$$
$$d\mu_D = -S_D\,dT + V_D\,dP = -RT\,d(\ln c_D/c_0) \tag{2}$$

where S_N, S_D and V_N, V_D are the partial molar entropy and volume of the protein in solution in the native and denatured phase, respectively. Since entropy and volume are generally positive quantities, the chemical potential always decreases with increasing temperature, while it increases with increasing pressure, as shown in Figures 5.4 and 5.6. The potentials of two phases, differing in entropy, will thus cross at a particular temperature where a transition to the phase with the lower potential occurs. A similar change in phase will take place with pressure when the volumes of the two phases differ. Thus the more compact phase will be stable at high pressure.

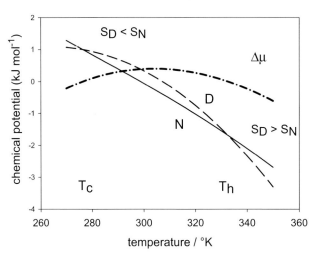

Fig. 5.4. The chemical potentials of the denatured (D) and the native (N) state of myoglobin and the resulting $\Delta\mu$ per mole of amino acid residue. Parameters [25]: $\Delta\mu_{0N} = -0.2$ kJ mol^{-1}, $\Delta\mu_{0D} = 0.2$ kJ mol^{-1}, $\Delta S_C = 14$ J K^{-1} mol^{-1}, $\Delta C_p = 75$ J K^{-1} mol^{-1}. T_H, T_C, and T_M are the temperatures for heat denaturation, for cold denaturation and for maximum stability.

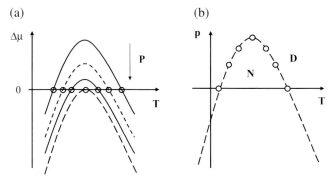

Fig. 5.5. a) Schematic representation of the difference $\Delta\mu = \mu_D - \mu_N$ of the chemical potential as a function of temperature for various pressures as deduced from Figure 5.4. b) The phase boundary as schematically deduced from the various pairs of points with $\Delta\mu = 0$ of Figure 5.5a for negative values of $\Delta V = V_D - V_N$.

For proteins in particular, $S_D > S_N$ at higher temperature because of an excess in configurational entropy of the unfolded chain. Thus μ_D has the more pronounced temperature dependence and will fall below μ_N at T_H, the heat denaturation temperature. For myoglobin it is suggested in Figure 5.4 that the peculiar thermodynamic behavior of proteins arises from the strongly temperature-dependent entropy of the unfolded phase: The slope of $\mu_D(T)$ and thus the entropy $S_D(T)$ decreases with decreasing temperature.

On a microscopic scale this reflects the ordering force of nonpolar groups on surface water molecules. Since the corresponding hydrophobic effect is less pronounced in the native state, S_N is expected to be almost temperature independent. This discrepancy explains why the potential surfaces μ_D and μ_N meet again at T_C, the cold denaturation temperature, as shown in Figure 5.4. The chemical potential difference, $\Delta\mu = \mu_D - \mu_N$, consequently assumes the shape of a convex curve, which is close to an inverted parabola with the maximum at T_M, the temperature of maximum stability (Figure 5.5).

Changes in pressure displace the difference potential surface $\Delta\mu$ vertically, depending on the unfolding volume ΔV as shown in Figure 5.5a.

Since the phase transition occurs at constant chemical potential, $\Delta\mu(P, T) = 0$, or $c_N = c_D$, while pressure and temperature are varied, one derives, from Eq. (2), a differential equation for the phase boundary, the Clausius–Claperon equation:

$$(dP/dT)_{\Delta\mu=0} = \Delta S/\Delta V \qquad (3)$$

where $\Delta S = S_D - S_N$ and $\Delta V = V_D - V_N$ can depend on T and P. The phase boundary (Figure 5.5b) will assume the shape of an inverted parabola if the unfolding volume ΔV is negative and constant in P and T. For positive unfolding volumes a regular parabola, open to the high pressure side, would be obtained excluding denaturation by pressure.

The latent quantities ΔS and ΔV are composed of intrinsic contributions of the protein and those of the hydration shell, while the values of bulk water in both phases cancel.

The entropy difference reflects mainly the increase in configurational entropy of the chain, ΔS_C, and the entropy change ΔS_H in the hydration shell, due to exposure of buried residues. The same reasoning applies to the latent volume. Hence, we can write:

$$\Delta S = \Delta S_C + \Delta S_H$$
$$\Delta V = \Delta V_C + \Delta V_H \tag{4}$$

We argue that the closed phase boundaries displayed in Figures 5.2 and 5.3 are the consequence of the variation of ΔS_H and ΔV_H with temperature and pressure due to the hydrophobic effect. Exposure of nonpolar groups upon unfolding tends to immobilize water molecules, thereby decreasing their entropy. This mechanism is strongly temperature dependent. Its experimental signature is the positive heat capacity increment at constant pressure, $\Delta C_p > 0$, of the denatured relative to the native form. In simple terms, to increase the temperature of the unfolded phase requires additional entropy to continuously melt the immobilized water [8–10, 23, 24]. However, above a certain temperature, T_0, approximately 140 °C, the ordering effect has vanished.

Below T_0, the hydration entropy change, $\Delta S_H(T, T_0)$, is rapidly decreasing with decreasing temperature. Assuming ΔC_p to be approximately independent of temperature, one obtains for the change in "hydration entropy" [25]:

$$\Delta S_H(T) = \int_{T_0}^{T} \frac{\Delta C_p}{T} dT = -\Delta C_p \cdot \ln\left(\frac{T_0}{T}\right) \tag{5}$$

and thus from Eqs (2)–(5):

$$\Delta \mu = \Delta \mu_0 - (T - T_0) \cdot [\Delta S_C - \Delta C_p \ln(T_0/T)] + \Delta V \cdot P \tag{6}$$

The hydration term $\propto \Delta C_p$ is always negative below T_0, stabilizing the denatured form. This is known as the "wedging effect" of water, which softens the native structure. The combined entropy difference ΔS vanishes at the maximum of $\Delta \mu$, at the temperature of maximum stability T_M (Figure 5.4):

$$(\partial \Delta \mu / \partial T)_{T=T_M} = -\Delta S = 0 \tag{7}$$

At T_M the negative hydration entropy ΔS_H just compensates for the positive configurational term ΔS_C:

$$\Delta S_C = \Delta C_p \ln(T_0/T_M) \tag{8}$$

5.3 Thermodynamics of Pressure–Temperature Phase Diagrams

ΔS in Eq. (7) depends on pressure and temperature according to:

$$d\Delta S(P,T) = (\partial \Delta S/\partial T)\,dT + (\partial \Delta S/\partial P)\,dP = (\Delta C_p/T)\cdot dT - \Delta\alpha_P \cdot dP = 0 \quad (9)$$

where we have introduced the volume expansion increment $\Delta\alpha_P = (\partial \Delta V/\partial T)_{P=\text{const}}$.

Equation (9) defines the slope of a boundary $\Delta S(T,P) = $ constant in the P–T plane:

$$(dP/dT)_{\Delta S=\text{const}} = \Delta C_p/(T\cdot \Delta\alpha_P) \quad (10)$$

The special $\Delta S = 0$ line, separating regions of positive and negative transition entropies, is fixed by the condition $(dP/dT)_{\Delta\mu=0} = 0$ as shown in Figure 5.11. Since the slope in Figure 5.11 is positive and since both ΔC_p and ΔV are positive, it follows that $\Delta\alpha_P > 0$. Integrating Eq. (10) yields the effect of pressure on the temperature of maximum stability:

$$T_M = T_{M0} \cdot \exp(P \cdot \Delta\alpha_P/\Delta C_p) \quad (11)$$

For reasonably low pressures one has for the $\Delta S = 0$ line: $T_M \approx T_{M0} \cdot (1 + P\cdot \Delta\alpha_P/\Delta C_p)$. Expanding the entropic part of the chemical potential (Eq. (6)) about $T_M \approx T_{M0}$ yields an approximate parabola in T with negative curvature (Figure 5.5a):

$$\Delta\mu(T,P) = \Delta\mu_M - \tfrac{1}{2}\Delta C_p \cdot (T - T_M)^2/T + \Delta V \cdot P = -RT\ln(c_D/c_N) \quad (12)$$

For $\Delta V < 0$, increasing pressure diminishes the stability range of the native state. If ΔV is constant, one obtains the parabolic pressure–temperature phase diagram, $P(T)_{\Delta\mu=0}$, shown in Figure 5.5b.

Experimental data, like those shown in Figure 5.2, 5.3, and 5.11 suggest a reentrant phase behavior not only as a function of temperature but also as a function of pressure. This leads to a closed or "ellipsoidal" phase boundary with ΔV depending on pressure and temperature. The volume of the denatured form, i.e. the slope of $\mu_D(P)$, in Figure 5.6 increases with decreasing pressure. Thus a second crossing of the D–N potentials may occur at low or even negative pressure denoted by P_L. Thus the unfolded protein is more easily stretched than the compact native state. Moreover, near $P = 0$, a pressure increase will stabilize the native state. In analogy to the temperature (Eq. (7)) we can define a pressure of maximum stability, P_M, according to:

$$(\partial \Delta\mu/\partial P)_{P=P_M} = \Delta V = 0 \quad (13)$$

Combining Eqs (4) and (13) shows that the changes in configurational and hydration volume cancel at $P = P_M$. The small observed unfolding volumes may result from such compensation effects. Equation (13) allows determination of the $\Delta V(P,T) = 0$ line, separating regions of positive and negative volume changes:

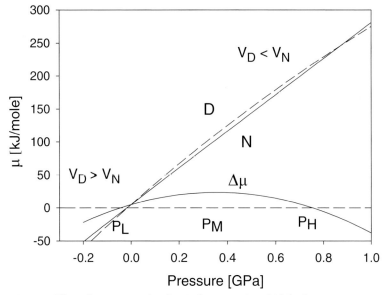

Fig. 5.6. Effect of pressure on the chemical potential $\mu(\div 4)$ for the parameters of cytochrome c at 280 K (Table 5.1 and Eq. (12, 17)). The volume of the denatured state decreases with pressure to below the value of the native protein, which leads to a crossing at high pressure P_H. A second crossing occurs at negative pressure P_L. P_M is the pressure of maximum stability.

$$(\partial \Delta V / \partial T)\, dT + (\partial \Delta V / \partial P)\, dP = \Delta \alpha_P \cdot dT - \Delta \beta_T \cdot dP = 0 \tag{14}$$

thus

$$(dP/dT)_{\Delta V = \text{const}} = \Delta \alpha_P / \Delta \beta_T \tag{15}$$

The $\Delta V = 0$ line follows from the additional condition: $(dP/dT)_{\Delta \mu = 0} = \infty$. For the pressure of maximum stability one obtains:

$$P_M = P_r + (\Delta \alpha_P / \Delta \beta_T)(T - T_r) \tag{16}$$

$\Delta \beta_T$ denotes the increment in the compressibility coefficient: $\Delta \beta_T = -(\partial \Delta V / \partial P)$. T_r denotes an unspecified reference temperature. For second-order phase transitions, sometimes associated with molten globule denaturation [26, 27], there is no discontinuity in the extensive quantities.

Equations (9) and (14) then become the so-called Ehrenfest equations: $(dP/dT)_{\Delta S=0} = (dP/dT)_{\Delta V=0}$

For most proteins cold denaturation occurs at subzero temperatures in the unstable or supercooled region of the water phase diagram. For this reason cold denaturation was formerly considered to be irrelevant to protein science. Similarly, low-pressure denaturation at $P_L(T)$ occurs for most proteins in a large temperature range where pressure is negative. Liquids under tension are unstable, although metastable states can exist, since the creation of a new liquid–gas interface involves the crossing of an energy barrier. The first experiments were performed by Berthelot in 1850 using a spinning capillary [28]. Stretched states of water at pressures as low as −280 bars have been observed using the same method [29]. Unfolding experiments with protein solutions at negative pressures still need to be performed. The folding of unstable proteins such as nuclease conA remains incomplete at positive pressure, while negative pressure may further stabilize the native state [30]. At sufficiently high temperature, low-pressure denaturation may occur in the positive range (Figure 5.11) and in this case it is easily accessible experimentally.

Pressure- and temperature-dependent transition volumes originate from finite increments in the partial compressibility $\Delta\beta_T$ and the volume expansion coefficient $\Delta\alpha_P$. Expanding the latent volume ΔV about a reference point (T_r, P_r) yields:

$$\Delta V(P, T) = \Delta V_r - \Delta\beta_T \cdot (P - P_r) + \Delta\alpha_P \cdot (T - T_r) \qquad (17)$$

The major part of the change in the compressibility upon global transformations of proteins is due to hydration processes [31]. The greater the hydration, the smaller the partial compressibility of protein and hydration shell. This is the rule for protein solutions at normal temperature and pressure. For native proteins the intrinsic compressibility is as low as that of organic solids [32–34]. The outer surface contribution to the measured partial compressibility is quite negative. Thus complete unfolding leads to a considerable decrease in the partial compressibility due to the loss of intramolecular voids and the expansion of the surface area contacting the bulk water [35]. In this low pressure/high temperature regime $\Delta\beta_T$ is negative, while $\Delta\alpha_P$ is mostly positive. This results in a positive unfolding volume (Eq. (17)) as indicated in Figure 5.6. Consequently moderate pressures stabilize the native state. A number of studies suggest that high-pressure denaturation leads to incomplete unfolding and structures resembling those of molten globule states (MG) [5–7, 36]. The N → MG transition is generally accompanied by an increase in compressibility. The tendency towards a positive $\Delta\beta_T$ at higher pressures due to partial unfolding may lead to the observed negative unfolding volumes. Combining Eqs (12) and (17), the phase boundary $\Delta\mu(T, P) = 0$ with respect to a reference point (T_r, P_r) assumes the form of a second-order hyperface:

$$-\tfrac{1}{2}\Delta C_p \cdot (T - T_r)^2/T - \tfrac{1}{2}\Delta\beta_T \cdot (P - P_r)^2 + \Delta\alpha_P \cdot (T - T_r) \cdot (P - P_r)$$
$$- \Delta S_r \cdot (T - T_r) + \Delta V_r \cdot (P - P_r) + \Delta\mu_r = 0 \qquad (18)$$

For approximately constant increments $\Delta C_p, \Delta\beta_T, \Delta\alpha_P$, the phase boundary assumes parabolic ($\Delta\beta_T = 0$), hyperbolic or closed elliptical shapes. The basic condition to obtain an elliptical phase diagram is given by:

$$(\Delta C_p)(\Delta\beta_T) - (\Delta\alpha_P)^2 > 0 \qquad (19)$$

Since experiments show that $\Delta C_p > 0$, Eq. (19) requires the compressibility change to be positive: $\Delta\beta_T > 0$. A hyperbolic shape would imply a negative left-hand side in Eq. (19). Thus far only elliptical diagrams, in some cases with distortions (e.g., Figure 5.3) have been discussed. A complete solution would require data in the range where pressure is negative: the data shown in Figure 5.2 are also consistent with a parabolic shape.

5.4
Measuring Phase Stability Boundaries with Optical Techniques

5.4.1
Fluorescence Experiments with Cytochrome *c*

Protein denaturing processes can be investigated with many techniques. The most convenient ones concerning the determination of the whole stability diagram are spectroscopic techniques. Almost all methods have been used: IR, Raman, NMR, absorption and fluorescence spectroscopy. However, data covering complete phase diagrams are quite rare. Some of them were reviewed above. The various techniques have their specific advantages as well as disadvantages. For instance, with vibrational spectroscopy one obtains information on structural changes such as the weakening of the amide I band [20, 21, 37, 38], with NMR one obtains information on hydrogen exchange and the associated protection factors [6] from which conclusions on structural details of the denatured state can be drawn. However, as a rule no information on the structure disrupting mechanism under pressure is obtained. The situation with absorption and fluorescence spectroscopy is different. In optical spectroscopy it is the spectral position and the linewidth of an electronic transition that is measured. From these quantities it is, as a rule, quite difficult to extract any structural information. On the other hand, one obtains detailed information on the interaction of the electronic states with the respective environment. Since these interactions are known in detail, it is possible to extract from their behavior under pressure and temperature changes information on the mechanism of phase crossing.

Compared with other spectroscopic techniques, fluorescence spectroscopy is also the most sensitive. Hence, it is possible to work at very low concentration levels so that aggregation processes can easily be avoided. In most experiments tryptophan is used as a fluorescent probe molecule. With chromoproteins this is, generally speaking, impossible since fast energy transfer to the chromophore quenches the UV fluorescence to a high degree. In the following we describe fluorescence experiments on a cytochrome *c*-type protein in a glycerol/water matrix [39]. The native heme iron was substituted by Zn in order to make the protein strongly fluorescent in the visible range. As a short notation for this protein we use the abbreviation Zn-Cc.

The set-up for a fluorescence experiment is simple and is shown in Figure 5.7.

The sample is in a temperature- and pressure-controlled diamond anvil cell.

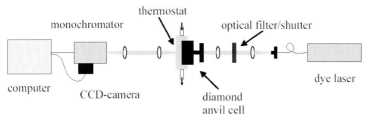

Fig. 5.7. Sketch of a fluorecence experiment for measuring the phase diagram of proteins.

Pressure can be varied up to about 2 GPa. Its magnitude is determined from the pressure shift of the ruby fluorescence, for which reference values can be taken from the literature [40]. The temperature is controlled by a flow thermostat between −20 °C and 100 °C. Excitation is carried out into the Soret band at 420 nm with light from a pulsed dye laser pumped by an excimer laser. The fluorescence is collected in a collinear arrangement, dispersed in a spectrometer and detected via a CCD camera. The quantities of interest for measuring the protein stability phase diagram are the spectral position and the width of the $S_1 \rightarrow S_0$ 00-transition (587 nm, Figure 5.8). Generally speaking the spectral changes for Zn-Cc are rather small. Nevertheless the measured changes of the spectral shifts and widths are accurate within about 3%. Figure 5.8 shows the fluorescence spectrum of Zn-Cc between 550 and 700 nm as it changes with pressure. The sharp line around 695 nm is the fluorescence from ruby.

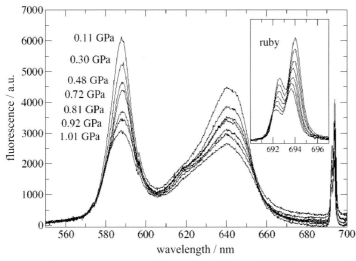

Fig. 5.8. Fluorescence spectrum of Zn-cytochrome c in glycerol/water at ambient temperature as it changes with pressure. The stability diagram was determined from the changes of the first and second moment of the 00-band at 584 nm. The fluorescence from ruby is shown as well. It serves for gauging the pressure.

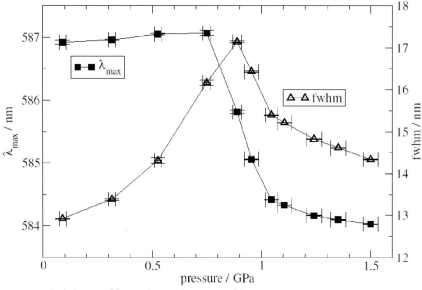

Fig. 5.9. The behavior of first and second moment of the fluorescence under pressure variation at ambient temperature.

5.4.2
Results

Figures 5.9 and 5.10 show typical results for the changes in the first and second moment (position of the maximum and width) of the 00-band under pressure and temperature variation.

If pressure is increased the band responds with a red shift, although a very small one (Figure 5.9). Around 0.75 GPa the red shifting regime changes rather abruptly into a blue shifting regime which levels off beyond 1 GPa. A qualitatively similar behavior is observed for the band shift under a temperature increase: a red shifting phase (in this case more pronounced) is followed by a blue shifting phase (Figure 5.10). Quite interesting is the behavior of the bandwidth: An increase of pressure leads to a rather strong increase in the width (Figure 5.9). This is the usual behavior and just reflects the fact that an increase in density results in an increase in the molecular interactions due to their strong dependence on distance. However, if the pressure is increased beyond about 0.9 GPa, the band all of a sudden narrows. This narrowing levels off beyond about 1 GPa. As can be seen from the figure, the maximum in the bandwidth coincides with the midpoint of the blue shifting regime. The thermal behavior of the bandwidth is different (Figure 5.10): There is no narrowing phase, but there is kind of a kink in the increase of the bandwidth with temperature around 65 °C. Again, this change in the thermal behavior of the width

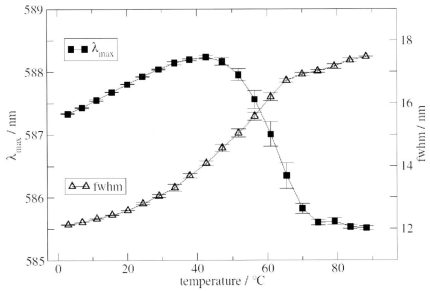

Fig. 5.10. The behavior of the first and second moment of the fluorescence under temperature variation at ambient pressure.

coincides rather well with the midpoint in the thermally induced blue shifting phase.

For measuring the complete phase diagram we performed a series of pressure scans from ambient pressure up to about 1.5 GPa at various temperatures and a series of temperature scans from a few centigrades up to about 90 °C at various pressure levels. The respective pattern of the changes of the fluorescence always showed the sigmoid-like behavior as discussed above. We associate the midpoint of the blue shifting regime with the stability boundary between the native and the denatured state of the protein due to reasons discussed below.

The complete stability diagram is shown in Figure 5.11. The solid line represents an elliptic least square fit to the experimental points. The two solid lines $\Delta S = 0$ and $\Delta V = 0$ cut the stability diagram at points where the volume and the entropy difference change sign (see also Eqs (11) and (16)). For instance, for all data points to the left of the $\Delta S = 0$ line the entropy in the denatured state S_D is smaller than S_N, the entropy in the native state (see below).

Figures 5.12 and 5.13 show specific features of the fluorescence behavior which we want to discuss separately. Figure 5.12 shows the behavior of the first and second moment for cold denaturation at ambient pressure. The point which we want to stress is that the band shift no longer reflects the qualitative change in its behavior. Instead, we observe a blue shift which increases in a nonlinear fashion with decreasing temperature. From such a behavior it is hard to determine a transition

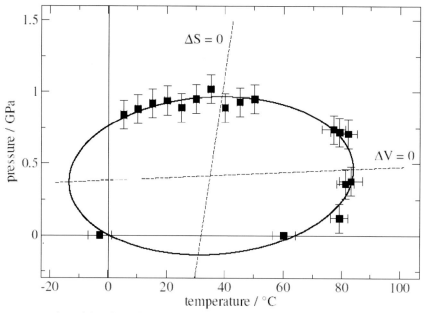

Fig. 5.11. The stability phase diagram of Zn-cytochrome *c*. The solid line is an elliptic least square fit to the data points. The straight lines mark the points where ΔS and ΔV change sign.

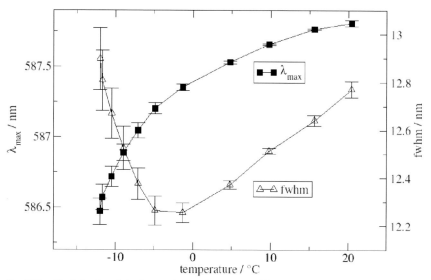

Fig. 5.12. The behavior of the first and second moment of the fluorescence for cold denaturation at ambient pressure.

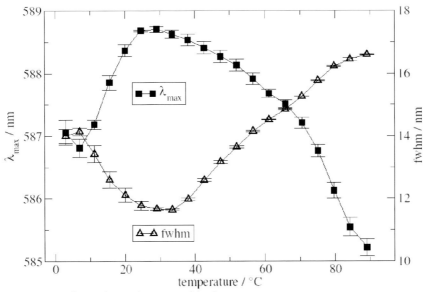

Fig. 5.13. The first and second moment of the fluorescence under a temperature variation at high pressure (0.7 GPa). The pattern shows features characteristic for re-entrant phase crossing.

point. The bandwidth, on the other hand, shows a quite unusual behavior: it narrows with decreasing temperature, but around −3 °C it runs through a minimum and starts broadening upon further decreasing the temperature. The unusual behavior concerns the broadening with decreasing temperature. This broadening obviously originates from an additional state which becomes populated below −3 °C. Naturally, this state must be the denatured state of the protein. Hence, we associate this minimum in the bandwidth with the stability boundary.

The temperature dependence of the moments in Figure 5.13 follows a rather complex pattern. This pattern comes from the re-entrant character of the phase diagram which has been stressed in Section 5.2. At a pressure level of 0.7 GPa, the protein is denatured to a high degree if the temperature is below around 10 °C (see Figure 5.11). Hence, upon increasing the temperature, the protein enters the stable regime. This is most clearly reflected in a narrowing of the bandwidth due to the fact that the denatured state becomes less and less populated At the same time the band maximum undergoes a red shift in line with the observation that the denatured state is blue shifted from the native one. However, around 25 °C the red shifting phase turns into a blue shifting phase, signaling that the protein re-enters the denatured regime. The bandwidth data support this conclusion. Interestingly, the data convey the impression that there are more than just two transitions involved; their nature, however, is not clear.

5.5
What Do We Learn from the Stability Diagram?

5.5.1
Thermodynamics

The eye-catching feature in Figure 5.11 is the nearly perfect elliptic shape of the stability diagram in agreement with the simple model developed in Section 5.3. The elliptic shape implies that the characteristic thermodynamic parameters of the protein, namely the increments of the specific heat capacity, the compressibility and the thermal expansion are rather well-defined material parameters, i.e., do not significantly depend on temperature and pressure. In addition, since the model is based on just two states, we conclude from the elliptic shape that the protein investigated, namely Zn-Cc, can be described as a two-state folder. Applying the Clausius–Clapeyron equation (Eq. (3)) to the elliptic phase boundary, we see that there are distinct points characterized by $(dP/dT)_{\Delta\mu=0} = \infty$ and by $(dP/dT)_{\Delta\mu=0} = 0$. The former condition separates the regime with positive ΔV from the regime with negative ΔV. The latter one separates the regime with positive ΔS from the regime with negative ΔS (Eqs (7) and (13)).

Let us consider thermal denaturation at ambient pressure. ΔS is definitely positive because the entropy of the chain increases upon denaturation and the entropy of the solvent increases as well. Since the slope of the phase boundary is positive, ΔV must be positive, too. However, moving along the phase boundary, we eventually cross the $\Delta V = 0$ line. For all points above that line, ΔV is negative. In other words: application of pressure destabilizes the native state and favors the denatured state. Below the $\Delta V = 0$-line, ΔV is positive. Accordingly, application of pressure favors the native state.

In a similar way, for all points to the right of the $\Delta S = 0$ line, ΔS is positive. A positive entropy change upon denaturation is what one would straightforwardly expect. However, to the left of the $\Delta S = 0$ line, the entropy change is negative, meaning that the entropy in the denatured state is smaller than in the native state. As was discussed above, the negative entropy change is due to an upcoming ordering of the hydration water as the temperature is decreased: The exposure of hydrophobic groups causes the water molecules in the hydration shell to become more and more immobilized due to the formation of a stronger hydrogen network. Right at the point where the $\Delta S = 0$ line cuts the phase diagram, the chain entropy and the entropy of the hydration shell exactly cancel, as we have stressed above.

There is another interesting outcome from the elliptic shape of the phase diagram: Eq. (19) tells us that the change in the compressibility upon denaturation has to have the same sign as the change in the specific heat. The change in the specific heat is positive meaning that the heat capacity is larger in the unfolded state. As outlined above, this is due to the fact that an increase of the temperature in the denatured state requires the melting of the ordered immobilized hydration shell. Accordingly, $\Delta\beta$ has to be positive, meaning that the compressibility in the denatured state is larger than in the native state. As to the absolute values of the thermodynamic parameters which determine the phase diagram, they can be determined only if one of the parameters is known. In the following section we will show how the equilibrium constant can be determined as a function of temperature and pressure and how this can be exploited to extract the thermodynamic parameters on an absolute scale.

5.5.2
Determination of the Equilibrium Constant of Denaturation

The body of information that can be extracted from the thermodynamics of protein denaturation reveals surprising details. However, it should be stressed again that all the modeling is based on two important assumptions, namely that folding and denaturing is described within the frame of just two states, N and D, and that these two states are in thermal equilibrium. Neither of these assumptions is straightforward. It is well known that the number of structural states, even of small proteins is, is extremely large [41–47] and the communication between the various states can be very slow so that the establishment of thermal equilibrium is not always ensured. We understand the two-state approximation on the basis of a concept which we call "state lumping." In simple terms this means that the structural phase space can be partitioned into two areas comprising, on the one hand, all the states in which the protein is functioning and, on the other hand, the area in which the protein is dead. We associate the native state N and the denatured state D with these two areas. In order for the "state lumping" concept to work, the immediate consequence is that all states within the two and between the two areas are in thermal equilibrium. Whether this is true or not can only be proven by the outcome of the experiments.

For Zn-Cc this concept seems to hold sufficiently well. In our experiments, typical waiting times for establishing equilibrium were of the order of 20 minutes. In order to make sure that this time span is reasonable, we measured for some points on the phase boundary also the respective kinetics. However, we also stress that at high pressures and low temperatures (left side of the phase diagram, Figure 5.11) we could not get reasonable data and we attributed this to the fact that equilibrium could not be reached within the experimental time window.

Assuming that equilibrium is established and the two-state approximation holds with sufficient accuracy, we can determine the equilibrium constant as a function of pressure and temperature from the fluorescence experiments and from the fact that the phase diagram has an elliptic shape. We proceed in the following way: The fluorescence intensity $F(\lambda)$ in the transformation range is a superposition from the two states N and D with their respective fluorescence maximum at λ_N and λ_D:

$$F(\lambda) = p_N a_N \exp[-(\lambda - \lambda_N)^2/2\sigma_N^2] + p_D a_D \exp[-(\lambda - \lambda_D)^2/2\sigma_D^2] \qquad (20)$$

where p_N and p_D are the population factors of the two states N and D, a_N and a_D are the respective oscillator strengths. Changes of the temperature or the pressure of the system cause a change in the population factors p_N and p_D which, in turn, leads to a change in the first moment of $F(\lambda)$, that is in the spectral position of the band maximum, λ_{max}. As to Zn-Cc, we can make use of some of its specific properties: First, the chromophore itself is quite rigid, hence, it is not likely affected by pressurizing or heating the sample. Accordingly, it is safe to assume that the oscillator strengths in the native and in the denatured state are not significantly different from each other. Second, the spectral changes induced by varying the temperature or the pressure are rather small compared to the total width of the long wavelength band (Figure 5.8). In addition, the wavelengths λ_N and λ_D are rather close and well within the inhomogeneous band. As a consequence, the exponentials in Eq. (20) are roughly of the same magnitude irrespective of the value of λ. Along these lines of reasoning, λ_{max} is readily determined from the condition $dF/d\lambda = 0$:

$$\lambda_{max} = p_N(T, P)\lambda_N + p_D(T, P)\lambda_D = p_N(T, p)[\lambda_N - \lambda_D] - \lambda_D \tag{21}$$

The last term on the right hand side of Eq. (21) holds because we restrict our evaluation to an effective two-state system. According to the above equation, the band maximum in the transition region is determined by a population weighted average of λ_N and λ_D. Since λ_N and λ_D depend on pressure and temperature themselves, we have to determine the respective edge values in the transition region. How this is done is shown in Figure 5.14 for thermal denaturation at ambient pressure. Similar figures are obtained for any parameter variation. Having λ_{max} as a function of pressure or temperature from the experiment and knowing the respective edge

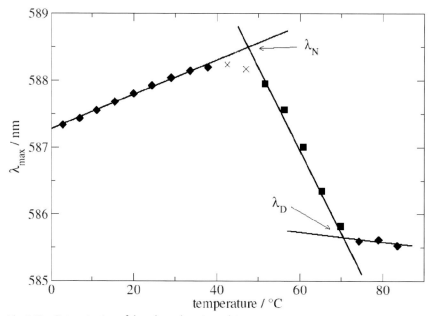

Fig. 5.14. Determination of the edge values λ_N and λ_D.

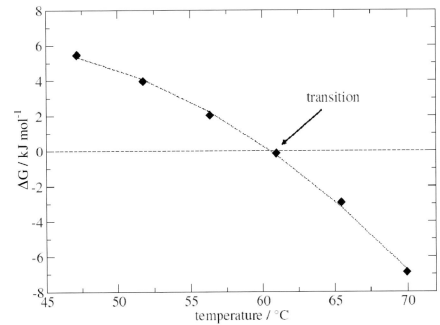

Fig. 5.15. Temperature dependence of the difference of the Gibbs free energy ΔG. (Note that, for a one-component system, $\Delta G/\text{Mol} = \Delta\mu$).

values λ_N and λ_D, the population factors p_N and p_D can readily be evaluated as a function of pressure and temperature using Eq. (21). From the population factors the equilibrium constant K follows immediately:

$$K_{DN}(T, P) = [1 - p_N(T, P)] / p_N(T, P) \tag{22}$$

Once K is known, $\Delta\mu$ (or, equivalently, ΔG mol^{-1}) can be determined as a function of pressure and temperature from Eq. (1). For a fixed pressure, say p_i, $\Delta\mu(T, P_i)$ forms an inverted parabola in T, as was shown in Section 5.3. The same is true for $\Delta\mu(P, T_i)$. So we determined $\Delta\mu(T)$ and $\Delta\mu(P)$ by fitting parabolas to the few data points (Figures 5.15 and 5.16) under the constraints that both branches of these parabolas have to go through the phase boundaries of the stability diagram (Figure 5.11). From the first and second derivative of $\Delta\mu$ with respect to temperature and pressure all the thermodynamic parameters, namely $\Delta V(T, P), \Delta S(T, P)$, $\Delta C_p, \Delta\beta$ and $\Delta\alpha$, can be determined. For Zn-Cc this parameters are listed in Table 5.1. We took ΔC_p from the equilibrium constant because it seems to be the most accurate parameter (Figure 5.15). As a matter of fact our value is rather close to what was measured by Makhatadze and Privalov for unfolding native cytochrome c [10]. All the other parameters are determined from the phase diagram. Note that $\Delta\beta$ has the same sign as ΔC_p, as required for an elliptic shape of the diagram.

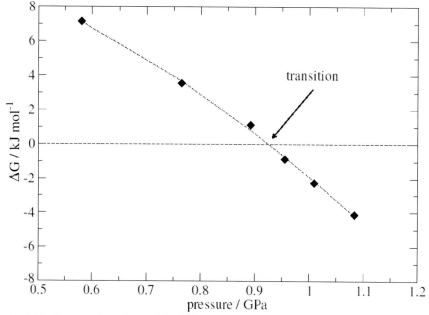

Fig. 5.16. Pressure dependence of the difference in the Gibbs free energy ΔG.

5.5.3
Microscopic Aspects

Having determined the complete set of thermodynamic parameters which govern denaturation of Zn-Cc, we may now proceed with exploring the microscopic driving forces of denaturation. We stress that the pattern in the behavior of the first moment of the fluorescence 00-transition is qualitatively very similar for thermal and pressure denaturation: A red shifting regime upon an increase in both parameters is followed by a blue shifting regime which signals the onset of the transformation range. The solvent shift of an optical transition is mainly determined by

Tab. 5.1. Thermodynamic parameters for the denaturation of Zn-cytochrome c.

Parameters	From the fit
ΔC_p (kJ mol^{-1} K^{-1})	5.87
$\Delta \beta$ (cm^3 mol^{-1} GPa^{-1})	148.1
$\Delta \alpha$ (cm^3 mol^{-1} K^{-1})	0.139
ΔV (0.91 GPa, 298 K) (cm^3 mol^{-1})	−74.6
ΔS (0.91 GPa, 298 K) (kJ mol^{-1} K^{-1})	−0.263
ΔV (0.1 MPa, 333 K) (cm^3 mol^{-1})	65.0
ΔS (0.1 MPa, 333 K) (kJ mol^{-1} K^{-1})	0.530

two types of interactions, namely the dispersive and the higher order electrostatic interaction (dipole–induced dipole). Both types of interaction are very short ranged. They fall off with distance R as R^{-6}. From the pioneering papers by Bayliss, McRae and Liptay [48–50], it is well known that the dispersive interaction is always red shifting because the polarizability in the excited state is higher than in the ground state. However, electrostatic interactions can cause shifts in both directions, to the red as well as to the blue, depending on how the dipole moment of the chromophore changes in the excited state compared with the ground state Accordingly, in the blue shifting regime, the electrostatic interaction of the probe with its environment obviously increases compared with the dispersive interaction. As a consequence, we conclude that polar groups with a sufficiently large dipole moment have to come close to the chromophore to induce this shift.

As to thermal denaturation, such an interpretation seems to fit quite well into the scenario. Baldwin [23], for instance, could show that thermal denaturation of a protein is remarkably well described in analogy to the solvation of liquid hydrocarbons in water. This "oil droplet model" accounts for the temperature dependence of the hydrophobic interaction: The entropic driving force for folding, which comprises the major part of the hydrophobic interaction, decreases as the temperature increases. This driving force is associated with the formation of an ordered structure of the water molecules surrounding the hydrophobic amino acids. At sufficiently high temperature this ordered structure melts away so that the change of the unfolding entropy of the polypeptide chain takes over and the protein attains a random coil-like shape [5]. Since a random coil is an open structure, the chromophore may now be exposed to water molecules. Water molecules have a rather high permanent dipole moment, hence, may become polarized through the electrostatic interaction with the chromophore. Along these lines of reasoning, it seems straightforward that the observed blue shifting regime of the fluorescence in the thermal denaturation of Zn-Cc comes from the water molecules of the solvent. However, as was pointed out by Kauzmann [51], the "oil droplet model" has severe shortcomings, despite its success in explaining the specific features of thermal denaturation. It almost completely fails in explaining the specific features of pressure induced denaturation. Pressure denaturation is governed by the volume change ΔV (Section 5.3). In this respect hydrophobic molecules behave quite differently from proteins: ΔV for protein unfolding is negative above a few hundred MPa (see, for instance, Figure 5.11 and discussions in Sections 5.3 and 5.5.1), whereas it is positive in the same pressure range for transfering hydrophobic molecules into water. There are two possible consequences: Either the "oil droplet model" is completely wrong, or the pressure denatured state is different from the thermally denatured state.

Indeed, Hummer and coworkers could show that the latter possibility is true [52, 53]. On the basis of their calculations they suggested that, as pressure is increased, the tretrahedral network of H-bonds in the solvent becomes more and more frustrated so that the pressure-induced inclusion of water molecules within the hydrophobic core becomes energetically more and more favorable. As a consequence, the contact configuration between hydrophobic molecules is destabilized by pressure whereas the solvent-separated contact configuration (a water molecule between two hydrophobic molecules) is stabilized. In simple words this means that water is pressed into the protein, the protein swells but does not completely unfold into a random coil conformation but rather seems to retain major parts of its globular shape. The important conclusion is that pressure denaturation is based on the presence of water. Larger solvent molecules, such as glycerol, can obviously not penetrate the protein, hence may act as stabilizing elements against pressure [54].

It seems that this view on the microscopic aspects of pressure denaturation is convincingly supported by our experiments: First, we stress again that the general pattern in the variation of the first moment with pressure is very similar to the respective variation with temperature. Since, in the latter case, we attributed this pattern to water molecules approaching the chromophore, it is quite natural to associate the pressure-induced pattern with the same phenomenon. Second, the magnitude of the change in the first moments is about the same for thermally and pressure-induced denaturation. Accordingly, the change in the respective interaction has to be very similar. This means, that, on average, the number and the distances of the additionally interacting water molecules have to be the same.

In order to get additional support for this reasoning we performed pressure tuning hole burning experiments [55]. With these experiments it is possible to estimate the size of an average interaction volume of the chromophore with its environment. We found that the respective radius is about 4–5 Å. Accordingly, application of high pressure forces water molecules from the hydration shell into the interior of the protein where they fill the voids around the chromophore in the heme pocket. High-pressure (0.9 GPa) MD simulations (M Reif and Ch Scharnagl, in preparation) are in full agreement with this view of pressure denaturation.

Summarizing this section we state that fluorescence experiments in combination with other optical experiments and computer simulations [22] provide a detailed insight into the processes involved in pressure-induced protein denaturation.

5.5.4
Structural Features of the Pressure-denatured State

From the discussion above it is evident that pressure denaturation leads to a structural state which is different from the respective one obtained by thermal denaturation. It seems that many structural features of the native protein remain preserved under pressure denaturation. Experimental evidence comes from NMR experiments [6], but also from IR experiments [20, 21, 37, 38].

Figure 5.17 shows another experiment [56], namely a neutron-scattering experiment with myoglobin as a target from which the conservation of native structural elements in the denatured state is directly seen. Myoglobin denatures at pH 7 in the range between 0.3 and 0.4 GPa as judged from the optical absorption spectrum of the heme group. The corresponding changes in the protein structure and at the protein–water interface can be deduced from the neutron structure factor $H(Q)$ of a concentrated solution of myoglobin in D_2O (Figure 5.17). Q denotes the length of the scattering vector. The large maximum around $Q = 2 \text{Å}^{-1}$ arises from O–O cor-

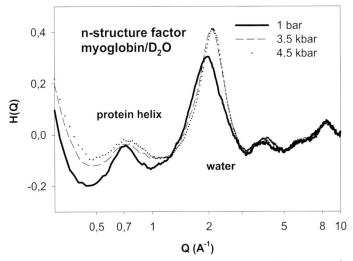

Fig. 5.17. Comparative neutron scattering experiments at ambient pressure and at high pressure. Plotted is the neutron structure factor $H(Q)$ as a function of Q. Data are for myoglobin. The important feature is the partial persistence of the protein helical structure (peak at $Q = 0.7\,\text{Å}^{-1}$) in the pressure-denatured state.

relations of water near the protein interface. This maximum increases with density. The smaller maximum near $Q = 0.7\,\text{Å}^{-1}$ reflects helix-helix correlations. The contrast is provided by the negative scattering length density of the protein hydrogen atoms versus the positive scattering length density of C, N, and O. The most remarkable feature is the persistence of this maximum above the transition at 0.45 GPa, indicating residual secondary structure in the pressure-unfolded form.

5.6
Conclusions and Outlook

Folding and denaturing processes of proteins are extremely complicated due to the complex nature of the protein molecules with their huge manifold of structural states. Hence, it is surprising that some of the characteristic features of equilibrium thermodynamics associated with the folded and denatured state are already revealed on the basis of simple models. Most important in this context is the reduction of the folding and denaturing processes to just two essential states, namely the native state and the denatured state. To reconcile this crude assumption with the large structural phase space, we introduced the concept of "state lumping." Another important approximation concerns the vanishing higher (higher than two) derivatives of the chemical potential, rendering pressure- and temperature-independent state parameters (specific heat, compressibility, thermal expansion) to the protein. The result of these simplifying assumptions is an elliptically shaped stability phase diagram, provided that the sign of the change in the specific heat capacity and in the compressibility is the same.

Our experiments on the stability of a modified cytochrome c in a glycerol/water solvent showed an almost perfect ellipse from which the regimes with positive and negative volume changes as well as positive and negative entropy changes could be deduced. We tried to shed some light on the microscopic aspects responsible for these regimes. A dominant force is the hydrophobic interaction which depends not only on temperature but also on pressure [57]. The characteristic structural features of the pressure-denatured state are different from the respective ones of the thermally denatured state. In both cases, however, water molecules come very close to the chromophore.

As to the open questions in context with the stability phase diagram, we refer to the elliptic shape. So far only closed diagrams have been observed, and in most cases an elliptic shape was an appropriate description. The question comes up how general this observation is and what the microscopic implications are. Admittedly, up to now few experiments have measured the full diagram.

Another problem concerns negative pressure. From an experimental point of view negative pressures as low as −200 bar seem to be feasible. Denaturation under negative pressure, that is under conditions where the relevant interactions are weakened, would add valuable information on pressure-induced denaturing forces.

Finally the role of the solvent has to be addressed in more detail. The solvent may have a dramatic influence on the hydrophobic interaction, the size of the solvent molecules may play an important role in all processes involving pressure, and, last but not least, understanding the influence of the solvent may also shed light on the behavior of membrane proteins.

Acknowledgment

The authors acknowledge financial support from the DFG (FOR 358, A1/2), from the Fonds der Chemischen Industrie (J.F.) and from the Bundesministerium für Bildung und Forschung 03DOE2M1.

Scientific input to this work came from many of our friends. In particular we want to thank J. M. Vanderkooi, H. Lesch, and G. Job and F. Herrmann for the idea to base thermodynamics on entropy and chemical potential.

References

1 G. WEBER, H. DRICKAMMER, Q. Rev. Biophys. **1983**, *16*, 89–112.
2 Special Issue on "Frontiers in high pressure biochemistry and biophysics", Biochim. Biophys. Acta 2002, 1595, Issue 1–2.
3 C. A. ROYER, Biochim. Biophys. Acta **2002**, *1595*, 201–209.
4 M. GERSTEIN, C. GOTHIA, Proc. Natl Acad. Sci. USA **1996**, *93*, 10167–10172.
5 A. FERSHT, Structure and Mechanism in Protein Science, W. H. FREEMAN, New York, 1999.
6 D. P. NASH, J. JONAS, Biochemistry **1997**, *36*, 14375–14383.

7 G. J. A. Vidugiris, J. L. Markley, C. A. Royer, *Biochemistry* **1995**, *34*, 4909–4912.
8 P. L. Privalov, *Adv. Protein Chem.* **1979**, *33*, 167–241.
9 P. L. Privalov, S. J. Gill, *Adv. Protein Chem.* **1988**, *39*, 191–234.
10 G. I. Makhatadze, P. L. Privalov, *Adv. Protein Chem.* **1995**, *47*, 307–425.
11 J. F. Brandts, *J. Am. Chem. Soc.* **1964**, *86*, 4302–4314.
12 J. F. Brandts, *J. Am. Chem. Soc.* **1965**, *87*, 2759–2760.
13 A. Cooper, *Prog. Biophys. Mol. Biol.* **1984**, *44*, 181–214.
14 S. A. Hawley, *Biochemistry* **1971**, *10*, 2436–2442.
15 P. E. Cladis, R. K. Bogardus, W. B. Daniels, G. N. Taylor, *Phys. Rev. Lett.* **1977**, *39*, 720–723.
16 D. D. Klug, E. J. Whalley, *J. Chem. Phys.* **1979**, *71*, 1874–1877.
17 N. A. Clark, *J. Physique* **1979**, *40*, 345–349.
18 A. Zipp, W. Kauzmann, *Biochemistry* **1973**, *12*, 4217–4228.
19 H. Ludwig, W. Scigalla, B. Sojka, in *High Pressure Effects in Molecular Biophysics and Enzymology*, J. L. Markley, D. Northrop, C. A. Royer (Eds), Oxford University Press, Oxford, 1996, 346–363.
20 K. Heremans, L. Smeller, *Biochim. Biophys. Acta* **1998**, *1386*, 353–370.
21 Y. Taniguchi, N. Takeda, in *High Pressure Effects in Molecular Biophysics and Enzymology*, J. L. Markley, D. Northrop, C. A. Royer (Eds), Oxford University Press, Oxford, 1996, 87–95.
22 E. Paci, *Biochim. Biophys. Acta* **2002**, *1595*, 185–200.
23 R. Baldwin, *Proc. Natl Acad. Sci. USA* **1986**, *83*, 8069–8072.
24 R. Baldwin, N. Muller, *Proc. Natl Acad. Sci. USA* **1992**, *89*, 7110–7113.
25 P. L. Privalov, *Biofizika* **1987**, *32*, 742–746.
26 Y. V. Griko, P. Privalov, *J. Mol. Biol.* **1994**, *235*, 1318–1325.
27 Th. Kiefhaber, R. L. Baldwin, *J. Mol. Biol.* **1995**, *252*, 122–132.
28 M. Berthelot, *Ann. Chim.* **1850**, *30*, 232–237.
29 L. J. Briggs, *J. Appl. Phys.* **1950**, *21*, 721–722.
30 M. Eftink, G. Ramsey, in *High Pressure Effects in Molecular Biophysics and Enzymology*, J. Markley, D. Northrop, C. A. Royer (Eds), Oxford University Press, Oxford, 1996, 62.
31 T. V. Chalikian, K. J. Breslauer, *Curr. Opin. Struct. Biol.* **1998**, *8*, 657–664.
32 K. Gekko, H. Noguchi, *J. Phys. Chem.* **1979**, *83*, 2706–2714.
33 B. Gavish, E. Gratton, C. J. Hardy, *Proc. Natl Acad. Sci. USA* **1983**, *80*, 750–754.
34 J. Zollfrank, J. Friedrich, *J. Opt. Soc. Am. B.* **1992**, *9*, 956–961.
35 D. P. Kharakoz, A. P. Sarvazyan, *Biopolymers* **1993**, *33*, 11–25.
36 G. J. A. Vidugiris, C. A. Royer, *Biophys. J.* **1998**, *75*, 463–470.
37 G. Panick, R. Malessa, R. Winter, G. Rapp, K. J. Frye, C. A. Royer, *J. Mol. Biol.* **1998**, *275*, 389–402.
38 H. Herberhold, S. Marchal, R. Lange, C. H. Scheying, R. F. Vogel, R. Winter, *J. Mol. Biol.* **2003**, *330*, 1153–1164.
39 H. Lesch, H. Stadlbauer, J. Friedrich, J. M. Vanderkooi, *Biophys. J.* **2002**, *82*, 1644–1653.
40 M. Eremets, *High Pressure Experimental Methods*, Oxford University Press, Oxford, 1996.
41 J. D. Bryngelson, J. N. Onuchic, N. D. Socchi, P. G. Wolynes, *Proteins Struct. Funct. Genet.* **1995**, *21*, 167–195.
42 H. Frauenfelder, F. Parak, R. D. Young, in *Annu. Rev. Biophys. Chem.* **1988**, *17*, 451–479.
43 H. Frauenfelder, S. G. Sligar, P. G. Wolynes, *Science* **1991**, *254*, 1598–1603.
44 H. Frauenfelder, D. Thorn Leeson, *Nat. Struct. Biol.* **1998**, *5*, 757–759.
45 H. Frauenfelder, *Nat. Struct. Biol.* **1995**, *2*, 821–823.
46 A. E. Garcia, R. Blumenfeld, G. Hummer, J. A. Krummhansl, *Physica D* **1997**, *107*, 225–239.
47 K. A. Dill, H. S. Chan, *Nat. Struct. Biol.* **1997**, *4*, 10–19.
48 N. S. Bayliss, *J. Chem. Phys.* **1950**, *18*, 292–296.

49 N. S. Bayliss, E. G. McRae, *J. Phys. Chem.* **1954**, *58*, 1002–1006.
50 W. Liptay, *Z. Naturforsch.* **1965**, *20a*, 272–289.
51 W. Kauzmann, *Nature* **1987**, *325*, 763–764.
52 G. Hummer, S. Garde, A. E. Garcia, M. E. Paulalitis, L. R. Pratt, *Proc. Natl Acad. Sci. USA* **1998**, *95*, 1552–1555.
53 G. Hummer, S. Garde, A. E. Garcia, M. E. Paulalitis, L. R. Pratt, *J. Phys. Chem.* **1998**, *102*, 10469–10482.
54 A. C. Oliveira, L. P. Gaspart, A. T. Da Poian, J. L. Silva, *J. Mol. Biol.* **1994**, *240*, 184–187.
55 H. Lesch, J. Schlichter, J. Friedrich, J. M. Vanderkooi, *Biophys. J.* **2004**, *86*, 467–472.
56 W. Doster, R. Gebhardt, A. K. Soper, in *Advances in High Pressure Bioscience and Biotechnology*, R. Winter (Ed.), Springer, Berlin, 2003, 29.
57 K. A. Dill, *Biochemistry* **1990**, *29*, 7133–7155.

6
Weak Interactions in Protein Folding: Hydrophobic Free Energy, van der Waals Interactions, Peptide Hydrogen Bonds, and Peptide Solvation

Robert L. Baldwin

6.1
Introduction

Hydrophobic free energy has been widely accepted as a major force driving protein folding [1, 2], although a dispute over its proper definition earlier made this issue controversial. When a hydrocarbon solute is transferred from water to a nonaqueous solvent, or a nonpolar side chain of a protein is buried in its hydrophobic core through folding, the transfer free energy is referred to as hydrophobic free energy. The earlier dispute concerns whether the transfer free energy can be legitimately separated into two parts and the free energy of hydrophobic hydration treated separately from the overall free energy change [3–5]. If the hydrophobic free energy is defined as the entire transfer free energy [5], then there is general agreement that transfer of the nonpolar solute (or side chain) out of water and into a nonaqueous environment drives folding in a major way. A related concern has come forward, however, and scientists increasingly question whether the energetics of forming the hydrophobic core of a protein should be attributed chiefly to packing interactions (van der Waals interactions, or dispersion forces) rather than to burial of nonpolar surface area. This question is closely related to the issue of whether the hydrophobic free energy in protein folding should be modeled by liquid–liquid transfer experiments or by gas–liquid transfer experiments.

The energetic role of peptide hydrogen bonds (H-bonds) was studied as long ago as 1955 [6] but the subject has made slow progress since then, chiefly because of difficulty in determining how water interacts with the peptide group both in the unfolded and folded forms of a protein. Peptide H-bonds are likely to make a significant contribution to the energetics of folding because there are so many of them: about two-thirds of the residues in folded proteins make peptide H-bonds [7]. Peptide backbone solvation can be predicted from electrostatic algorithms but experimental measurements of peptide solvation are limited to amides as models for the peptide group.

This chapter gives a brief historical introduction to the "weak interactions in protein folding" and then discusses current issues. It is not a comprehensive review and only selected references are given. The term "weak interaction" is somewhat misleading because these interactions are chiefly responsible for the folded structures of proteins. The problem of evaluating them quantitatively lies at the heart of the structure prediction problem. Although there are methods such as homology modeling for predicting protein structures that bypass evaluation of the weak interactions, de novo methods of structure prediction generally rely entirely on evaluating them. Thus, the problem of analyzing the weak interactions will continue to be a central focus of protein folding research until it is fully solved.

6.2
Hydrophobic Free Energy, Burial of Nonpolar Surface and van der Waals Interactions

6.2.1
History

The prediction in 1959 by Walter Kauzmann [1] that hydrophobic free energy would prove to be a main factor in protein folding was both a major advance and a remarkable prophecy. No protein structure had been determined in 1959 and the role of hydrophobic free energy in structure formation could not be deduced by examining protein structures. The first protein structure, that of sperm whale myoglobin, was solved at 2 Å resolution only in 1960 [8]. On the other hand, the predicted structure of the α-helix [9] given by Pauling and coworkers in 1951, which was widely accepted, suggested that peptide H-bonds would prove to be the central interaction governing protein folding. Peptide H-bonds satisfied the intuitive belief of protein scientists that the interactions governing protein folding should be bonds with defined bond lengths and angles. This is not a property of hydrophobic free energy.

Kauzmann [1] used the ambitious term "hydrophobic bonds," probably aiming to coax protein scientists into crediting their importance, while Tanford [10] introduced the cautious term "the hydrophobic effect." "Hydrophobic interaction" has often been used because a factor that drives the folding process should be an interaction. However, hydrophobic interaction is also used with a different meaning than removal of nonpolar surface from contact with water, namely the direct interaction of nonpolar side chains with each other. The latter topic is discussed under the heading "van der Waals interactions." The term "hydrophobic free energy" is used here to signify that nonpolar groups help to drive the folding process. Tanford [10] points out that a hydrophobic molecule has both poor solubility in water and good solubility in nonpolar solvents. Thus, mercury is not hydrophobic because it is insoluble in both solvents. Early work leading to the modern view of hydrophobic free energy is summarized by Tanford [11] and a recent discussion by Southall et al. [12] provides a valuable perspective.

6.2.2
Liquid–Liquid Transfer Model

Kauzmann [1] proposed the liquid–liquid transfer model for quantitating hydrophobic free energy. His proposal was straightforward. Hydrophobic molecules prefer to be in a nonpolar environment rather than an aqueous one and the free energy difference corresponding to this preference should be measurable by partitioning hydrocarbons between water and a nonaqueous solvent. Nozaki and Tanford [13] undertook a major program of using the liquid–liquid transfer model to measure the contributions of nonpolar and partially nonpolar side chains to the energetics of folding. They measured the solubilities of amino acids with free α-COO$^-$ and α-NH$_3^+$ groups, while Fauchère and Pliska [14] later studied amino acids with blocked end groups, because ionized end groups interfere with the validity of assuming additive free energies of various groups. They measured partitioning of solutes between water and n-octanol (saturated with water), which is less polar than the two semi-polar solvents, ethanol and dioxane, used by Nozaki and Tanford [13]. Wimley et al. [15] used a pentapeptide host to redetermine the partition coefficients of the amino acid side chains between water and water-saturated n-octanol and they obtained significantly different results from those of Fauchère and Pliska. They emphasize the effect of neighboring side chains ("occlusion") in reducing the exposure of a given side chain to water. Radzicka and Wolfenden [16] studied a completely nonpolar solvent, cyclohexane, and observed that the transfer free energies of hydrocarbons are quite different when cyclohexane is the nonaqueous solvent as compared to n-octanol.

In Figure 6.1 the transfer free energies of model compounds for nonpolar amino

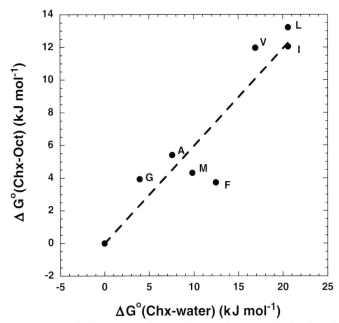

Fig. 6.1. Transfer free energies from cyclohexane to water compared with ones from cyclohexane to water-saturated n-octanol (data from Ref. [16]). The model solutes undergoing transfer represent the amino acid side chains shown on the plot. Note that the transfer free energies between cyclohexane and n-octanol are more than half as large as those from cyclohexane to water.

acid side chains are compared using either cyclohexane or n-octanol as the nonaqueous solvent [16]. If different nonaqueous solvents may be used equally well to model the hydrophobic core of a protein, then the transfer free energies of hydrocarbons from cyclohexane to n-octanol should be small compared with the transfer free energies from cyclohexane to water. Figure 6.1 shows this is not the case: the transfer free energies measured between cyclohexane and n-octanol are more than half as large as the ones between cyclohexane and water. Thus, these results pose the first serious question about the use of the liquid–liquid transfer model: which nonaqueous solvent should be used to model the hydrophobic core and how valid are the results if no single solvent is a reliable model?

6.2.3
Relation between Hydrophobic Free Energy and Molecular Surface Area

A second important step in quantifying hydrophobic free energy was taken when several authors independently observed that the transfer free energy of a nonpolar solute is nearly proportional to the solute's surface area for a homologous series of solutes [17–19]. This observation agrees with the intuitive notion that the transfer free energy of a solute between two immiscible solvents should be proportional to the number of contacts made between solute and solvent (however, see Section 6.2.6). Lee and Richards [20] in 1971 developed an automated algorithm for measuring the water-accessible surface area (ASA) of a solute by rolling a spherical probe, with a radius equivalent to that of a water molecule ($1.4\,\text{Å}$) ($10\,\text{Å} = 1$ nm), over its surface. Their work showed how to make use of the surface area of a solute to analyze its hydrophobicity. Proportionality between transfer free energy and ASA does not apply to model compounds containing polar side chains because polar groups interact strongly and specifically with water.

A plot of transfer free energy versus ASA is shown in Figure 6.2 for linear alkanes. The slope of the line for linear (including branched) hydrocarbons is 31 cal mol^{-1} Å^{-2} (1.30 J mol^{-1} nm^{-2}) when partition coefficients on the mole fraction scale [21] are used. Earlier data for the solubilities of liquid hydrocarbons in water are used to provide the transfer free energies. In Figure 6.2 the transfer free energy is nearly proportional to the ASA of the solute. The line does not pass through (0,0), but deviation from strict proportionality is not surprising for small solutes [22].

Hermann [22] points out that linear hydrocarbons exist in a broad range of configurations in solution and each configuration has a different accessible surface area. He also points out [17] that the transfer free energy arises from a modest difference between the unfavorable work of making a cavity in a liquid and the favorable van der Waals interaction between solute and solvent. Consequently, a moderate change in the van der Waals interaction can cause a large change in the transfer free energy. Tanford [10] analyzes the plot of transfer free energy versus the number of carbon atoms for hydrocarbons of various types, and discusses data for the different slopes of these plots.

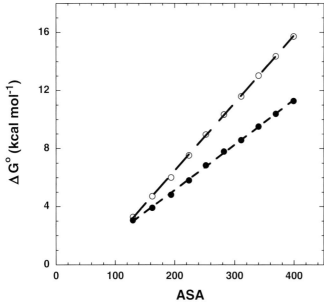

Fig. 6.2. Transfer free energies of linear alkanes (from 1 to 10 carbon atoms) from liquid alkane to water, measured from hydrocarbon solubility in water. They are plotted against water-accessible surface area (ASA in Å^2). Data are from Ref. [21]. The uncorrected transfer free energies (filled circles) refer to the mole fraction scale while the corrected values (open circles) refer to the molarity scale and are corrected for the ratio of molecular volumes, solute/solvent, according to Sharp et al. [21]. The data here and in Figure 6.3 are given in kcal mol^{-1} to conform with the literature on this subject. Note that the plots do not pass through 0,0 and note the larger slope (47 cal mol^{-1} Å^{-2}) of the corrected plot versus the uncorrected plot (31 cal mol^{-1} Å^{-2}).

6.2.4
Quasi-experimental Estimates of the Work of Making a Cavity in Water or in Liquid Alkane

In modern solution chemistry, the solvation free energy of a solute is defined as its transfer free energy from the gas phase into the solvent, when the appropriate standard state concentration (1 M) is used in each phase, as specified by Ben-Naim and Marcus [23]. Gas–liquid transfer free energies are used to analyze the nature of liquid–liquid transfer free energies. The reason for adopting the 1 M standard state concentration in both the gas and liquid phases is to ensure that the density of the solute is the same in both phases at the standard state concentration. Then the transfer free energy gives the free energy change for transferring the solute from a fixed position in the gas phase to a fixed position in the liquid phase [23]. The solute in the gas phase can be treated as an ideal gas [24, 25] and the nonideal behavior of real gases at 1 M concentration can be omitted from consideration.

Modern theories of solvation indicate that the gas–liquid transfer process can be formally divided into two steps of an insertion model of solvation: see discussion by Lee [24, 25] and basic theory by Widom [26]. In step 1 thermal fluctuations create a cavity in the liquid with a size and shape appropriate for containing the solute. The structure of the liquid undergoes reorganization to make the cavity [24]. In step 2 the solute is inserted into the cavity, van der Waals interactions occur between the solute and the solvent, and the solvent structure undergoes further reorganization at the surface of the cavity. Lee [24] determines quasi-experimental values for the entropy and enthalpy changes in the two steps of the insertion model: (1) making a cavity in the liquid, and (2) inserting the solute into the cavity. Experimental transfer data are used for each of five alkanes undergoing transfer from the gas phase to the liquid phase, either to water or to neat liquid alkane. The transfer thermodynamics then are combined with literature estimates for the van der Waals interaction energies, obtained by Jorgensen and coworkers [27] from Monte-Carlo simulations. The results give quasi-experimental estimates of the enthalpy and entropy changes in each step of the insertion model. Table 6.1 gives these values for the free energy cost of making a cavity to contain the alkane solute both in water and in liquid alkane. The 1977 theory of hydrophobic solvation by Pratt and Chandler [28] divides the process of solvation into the two steps of cavity formation and solute insertion, and the authors consider the rules for separating the solvation process into these two steps. The two steps of dissolving an alkane in water have also been simulated by molecular dynamics and the results analyzed by the free energy perturbation method [29].

The following conclusions can be drawn from Lee's data [24, 25]. (1) The work of making a cavity in water is much larger than in liquid alkane and this difference

Tab. 6.1. Quasi-experimental estimates of the free energy of cavity formation and simulation-based results for van der Waals interaction energies between solvent and solute[a].

Hydrocarbon	ΔG_c(water)	ΔG_c(alkane)	E_a(water)	E_a(alkane)
Methane	20.4	9.6	−12.1	−14.7
Ethane	27.7	18.1	−20.1	−26.8
Propane	35.8	23.7	−27.6	−34.4
Isobutane	42.4	24.4	−32.7	−35.1
Neopentane	46.5	27.6	−36.0	−39.6

[a] Values in kJ mol^{-1}. ΔG_c is the free energy cost of making a cavity in the solvent to contain the hydrocarbon solute, from a study by Lee [24]. E_a is the van der Waals interaction energy between solute and solvent; values from Lee [24], based on parameters from a Monte-Carlo simulation study by Jorgensen and coworkers [27]. For water as solvent, conditions are 25 °C, 1 atm; for neat hydrocarbon as solvent, either the temperature or pressure is chosen that will liquefy the hydrocarbon.

is the major factor determining the size of the hydrophobic free energy in liquid–liquid transfer. For example, it costs 46.5 kJ mol^{-1} to make a cavity for neopentane in water but only 27.6 kJ mol^{-1} in liquid neopentane. To understand hydrophobic free energy, it is necessary first of all to understand the free energies of cavity formation in water and in nonpolar liquids. The work of making a cavity in water is large because it depends on the ratio of cavity size to solvent size [30, 31] and water is a small molecule (see Section 6.2.7). It is more difficult to make a cavity of given size by thermal fluctuations if the solvent molecule is small. (2) The work of making a cavity in a liquid is chiefly entropic [24], while the van der Waals interactions between solute and solvent are enthalpic. (3) The van der Waals interaction energy between an alkane solute and water is nearly the same as between the alkane solute and liquid alkane (see Table 6.1). For example, the interaction energies between neopentane and water versus neopentane and liquid neopentane are −36.0 kJ mol^{-1} and −39.6 kJ mol^{-1} [24]. Earlier, Tanford [32] used interfacial tensions to show that the attractive force between water and hydrocarbon is approximately equal, per unit area, to that between hydrocarbon and hydrocarbon. Scaled particle theory predicts well the work of making a cavity either in water or in liquid alkane [24], but it predicts only semi-quantitatively the enthalpy of solvent reorganization for these cavities.

6.2.5
Molecular Dynamics Simulations of the Work of Making Cavities in Water

In 1977 a physico-chemical theory of hydrophobic free energy by Pratt and Chandler [28] already gave good agreement between predicted and observed transfer free energies of linear alkanes, both for gas–liquid and liquid–liquid transfer. Molecular dynamics simulations can be used to obtain the free energy cost of cavity formation in liquids and the results are of much interest because they basically depend only on the specific water model used for the simulations. It should be kept in mind that the physical properties of water used as constraints when constructing water models do not normally include surface tension, and consequently good agreement between the predicted and known surface tension of water is not necessarily to be expected. (For macroscopic cavities, the work of making a cavity equals surface tension times the surface area of the cavity.) In 1982 Berendsen and coworkers [33] determined the free energy of cavity formation in water for cavities of varying size and compared the results to values predicted by scaled particle theory, with reasonable agreement. Remarkably, the comparison with scaled particle theory also gave a value for the surface tension of water close to the known value. Because of the importance of the problem, simulations of cavity formation in water by molecular dynamics continued in other laboratories (see references in [34]). An important result is the development by Hummer and coworkers [34] of an easily used information-theory model to represent the results for water in the cavity size range of interest. Some applications of the information theory model are mentioned below.

6.2.6
Dependence of Transfer Free Energy on the Volume of the Solute

Evidence is discussed in Section 6.2.3 that the transfer free energy is correlated with the surface area (ASA) of the solute. Because it is straightforward to compute ASA [20] from the structure of a peptide or protein, this correlation provides a very useful means of computing the change in hydrophobic free energy that accompanies a particular change in conformation. In recent years, evidence has grown, however, that the transfer free energy of a nonpolar solute depends on its size and shape for reasons that are independent of hydrophobic free energy. In 1990 DeYoung and Dill [35] brought the problem forcibly to the attention of protein chemists by demonstrating that the transfer free energy of benzene from liquid hydrocarbon to water depends on the size and shape of the liquid hydrocarbon molecules. Section 6.2.2 reviews evidence that liquid–liquid transfer free energies depend on the polarity (and perhaps on water content) of the nonaqueous solvent. But in the study by DeYoung and Dill [35] the size and shape of the nonaqueous solvent molecules affect the apparent hydrophobic free energy. A large literature has developed on this subject and recently Chan and Dill [36] have provided a comprehensive review.

Chandler [37] briefly discusses the reason why solvation free energy in water depends on the volume of a sufficiently small nonpolar solute. This dependence can be found in both the information-theory model [34] and the Lum-Chandler-Weeks theory [38] of hydrophobic solvation. Effects of the size and shape of the solute are taken into account in the Pratt and Chandler theory [28].

Stimulated by the results of DeYoung and Dill [35], Sharp and coworkers [21] used a thermodynamic cycle and an ideal gas model to relate the ratio of sizes, solute to solvent, to the transfer free energy for gas–liquid transfer. They conclude that the transfer free energy depends on the ratio of solute/solvent molecular volumes. Their paper has generated much discussion and controversy. In 1994 Lee [39] gave a more general derivation for the transfer free energy, based on statistical mechanics, and considered possible assumptions that will yield the result of Sharp et al. [21].

The Lum-Chandler-Weeks theory of hydrophobic solvation [38] predicts a crossover occurring between the solvation properties of macroscopic and microscopic cavities when the cavity radius is 10 Å. Huang and Chandler [40] point out that the ratio of the work of making a cavity in water to its surface area reaches a plateau value for radii above 10 Å, and this value agrees with the known surface tension of water at various temperatures. On the other hand, the hydrophobic free energy found from hydrocarbon transfer experiments increases slightly with temperature (see Section 6.2.9), implying that the work of making a sufficiently small cavity in water increases with temperature. Chandler [37] explains that these two different outcomes, which depend on solute size, arise naturally from the H-bonding properties of water, because the sheath of water molecules that surrounds a nonpolar solute remains fully H-bonded when the solute is sufficiently small but not when the solute radius exceeds a critical value. Southall and Dill [41] find that a highly simplified model of the water molecule (the "Mercedes-Benz" model), which reproduces several remarkable properties of water, also predicts such a transition from microscopic to macroscopic solvation behavior.

The question of interest to protein chemists is: should a transfer free energy be corrected for the ratio of solute/solvent volumes or not? Figures 6.2 and 6.3 compare the uncorrected with the volume-corrected plots of transfer free energy versus ASA, for both liquid–liquid transfer (Figure 6.2) and gas–liquid transfer (Figure 6.3). Both correlations show good linearity. However, the hydrophobic free energy corresponding to a given ASA value is substantially larger if the volume-corrected transfer free energy is used (see Ref. [21]). Whether the volume correction should be made remains controversial. Scaled particle theory emphasizes the role of surface area in determining the free energy of cavity formation while the information-theory model [34] and the Lum-Chandler-Weeks theory [38] both emphasize the

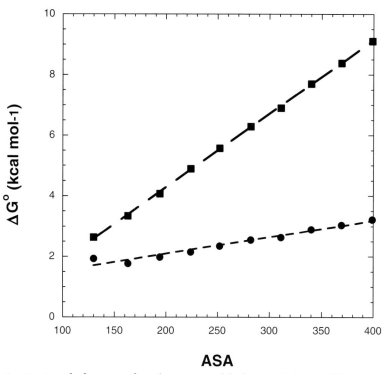

Fig. 6.3. Transfer free energy from the gas phase to liquid water for linear alkanes (from 1 to 10 carbon atoms) plotted against water-accessible surface area (in $Å^2$). Data are from Ref. [21]. The uncorrected transfer free energies (filled circles) refer to the mole fraction scale while the corrected values (filled squares) refer to the molarity scale and are volume-corrected according to Ref. [21]. The slopes of the lines are 5.5 cal mol^{-1} $Å^{-2}$ (uncorrected) and 24 cal mol^{-1} $Å^{-2}$ (corrected).

role of molecular volume. The molecular dynamics simulations of Berendsen and coworkers [33] were interpreted by scaled-particle theory while those of Hummer and coworkers were interpreted by the information-theory model [34], which has much in common with the Lum-Chandler-Weeks theory. Pohorille and Pratt [42] give a detailed discussion of how the scaled particle interpretation [33] may be reconciled with their own analysis.

6.2.7
Molecular Nature of Hydrophobic Free Energy

The molecular nature of hydrophobic free energy has been controversial for a long time [3, 4, 11, 12]. A long-standing proposal, supported by liquid–liquid transfer data at 25 °C [1] and by simulation results [27], is that the arrangement of water molecules in the solvation shell around a dissolved hydrocarbon is entropically unfavorable. Consequently the unfavorable entropy change for dissolving a hydrocarbon in water should provide the driving force for expelling the solute from water. This proposal required modification when it was learned that the hydrophobic free energy found from liquid–liquid transfer is purely entropic only at an isolated temperature near room temperature [43]. Hydrophobic free energy becomes increasingly enthalpy-driven as the temperature increases and it becomes entirely enthalpic upon reaching its maximum value at a temperature above 100 °C (see Section 6.2.9). The characteristic property of hydrophobic free energy that dominates its temperature-dependent behavior is the large positive value of ΔC_p, the difference between the heat capacities of the hydrocarbon in water and in the nonaqueous solvent.

In contradiction to the thesis developed by Kauzmann [1], Privalov and Gill [3, 4] proposed that the hydration shell surrounding a dissolved hydrocarbon tends to stabilize the hydrocarbon in water while the van der Waals interactions between the hydrocarbon solute and the nonaqueous solvent account for the hydrophobicity of the solute. They assume that the van der Waals interaction energy between solute and solvent is large in the nonaqueous solvent compared to water. However, Lee's data (see Table 6.1) show that the large work of making a cavity in water is responsible for the hydrophobic free energy while a hydrocarbon solute makes nearly equal van der Waals interactions with water and with a liquid hydrocarbon [24]. Privalov and Gill coupled two proposals: (1) the van der Waals interactions between nonpolar side chains drive the formation of the hydrophobic core of a protein, and (2) the hydration shell surrounding a dissolved hydrocarbon tends to stabilize it in water. The latter proposal is now widely believed to be incorrect but there is increasing interest in the first proposal.

There are restrictions both on the possible orientations of water molecules in the solvation shell around a hydrocarbon solute and on the hydrogen bonds they form [12]. Models suggesting how these restrictions can explain the large positive values of ΔC_p found for nonpolar molecules in water have been discussed as far back as 1985, in Gill's pioneering study of the problem [44]. Water-containing clathrates of nonpolar molecules surrounded by a single shell of water molecules have been crystallized and their X-ray structures determined [45]. The water molecules form interconnected 5- and 6-membered rings. Water molecules in ice I are oriented tetrahedrally in a lattice and the oxygen atoms form six-membered rings [45].

In 1985 Lee [30] used scaled particle theory to argue that the low solubilities of nonpolar solutes in water, and the magnitude of hydrophobic free energy, depend strongly on the solute/solvent size ratio, and he reviews prior literature on this subject. Rank and Baker [31] confirmed his conclusion with Monte-Carlo simulations of the potential of mean force between two methane molecules in water. The remarkable temperature-dependent properties of hydrophobic free energy (see Section 6.2.9) are determined, however, chiefly by ΔC_p, which depends on the hydrogen bonding properties of water according to most authors (see Ref. [46]).

6.2.8
Simulation of Hydrophobic Clusters

Formation of the hydrophobic core of a protein during folding must proceed by direct interaction between nonpolar side chains. Yet direct interaction between two hydrocarbon molecules in water is known to be extremely weak (compare Ref. [28]). Benzene dimers or complexes between benzene and phenol in water are barely detectable [47]. Raschke, Tsai and Levitt used molecular dynamics to simulate the formation of hydrophobic clusters, starting from a collection of isolated hydrocarbon molecules in water [48]. Their results give an interesting picture of the thermodynamics of the process. The gain in negative free energy from adding a hydrocarbon molecule to a hydrocarbon cluster in water increases with the size of the cluster until limiting behavior is reached for large clusters. The simulations of cluster formation yield a proportionality between transfer free energy and burial of nonpolar surface area that is similar to the one found from liquid–liquid transfer experiments [48]. The simulation results make the important point that a standard molecular force field is able to simulate the thermodynamics of hydrophobic free energy when a hydrophobic cluster is formed in water [48]. Rank and Baker [49] found that solvent-separated hydrocarbon clusters precede the desolvated clusters found in the interior of large hydrocarbon clusters. Thus, a hydrocarbon desolvation barrier may be important in the kinetics of protein folding [49]. In both simulation studies [48, 49], the authors find that the molecular surface area (defined by Richards [50]) is more useful than water-accessible surface area in analyzing cluster formation.

6.2.9
ΔC_p and the Temperature-dependent Thermodynamics of Hydrophobic Free Energy

Although the gas–liquid transfer model is now often used instead of the liquid–liquid transfer model to analyze hydrophobic free energy, nevertheless thermodynamic data for the transfer of liquid hydrocarbons to water are remarkably successful in capturing basic thermodynamic properties of hydrophobic free energy in protein folding. Figure 6.4A shows ΔH versus temperature for the process

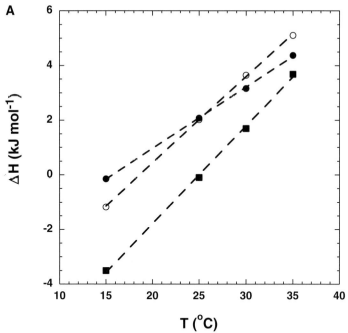

Fig. 6.4. (A) Enthalpy of transfer plotted against temperature for three liquid hydrocarbons (benzene (filled circles), ethylbenzene (open circles) and cyclohexane (filled squares)) undergoing transfer from liquid hydrocarbon to water. Data are from Ref. [51]. Note that the plots are straight lines in this temperature range, where ΔC_p = constant, and ΔH passes through 0 near room temperature.

of transferring three hydrocarbon solutes from neat liquid hydrocarbon to water, taken from the calorimetric study by Gill and coworkers [51]. These results illustrate the large positive values of ΔC_p found when nonpolar molecules are dissolved in water, a property discovered by Edsall [52] in 1935. Although ΔH depends strongly on temperature, ΔC_p (the slope of ΔH versus T) is nearly constant in the temperature range 15–50 °C (see figure 12 of Ref. [3]). ΔC_p decreases perceptibly at temperatures above 50 °C [3, 4].

The following thermodynamic expressions take a simple form when ΔC_p is constant. A strong dependence of ΔH on temperature must be accompanied by a strong dependence of $\Delta S°$ on temperature. When ΔC_p is constant, then:

$$\Delta H(T_2) = \Delta H(T_1) + \Delta C_p(T_2 - T_1) \tag{1}$$

$$\Delta S°(T_2) = \Delta S°(T_1) + \Delta C_p T_2 \ln(T_2/T_1) \tag{2}$$

Figure 6.4B compares ΔH, $T\Delta S°$ and $\Delta G°$ as functions of temperature for the transfer of benzene from liquid benzene to water. (The behavior of a liquid alkane,

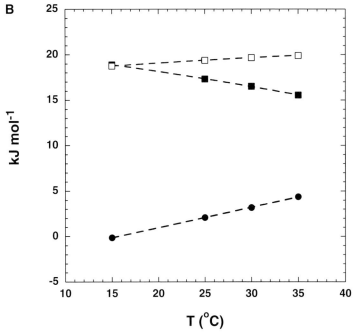

Fig. 6.4. (B) Transfer free energy and the contributions to free energy from ΔH (filled circles) and $T\Delta S$ (filled squares), plotted against temperature, for the transfer of benzene from liquid benzene to water. Data are from Ref. [43]. Note that the relative contribution to ΔG (open squares) from ΔH increases with temperature while the contribution of $T\Delta S$ decreases, and ΔG increases only slightly with temperature. The plots illustrate how entropy–enthalpy compensation affects hydrophobic free energy.

pentane, is fairly similar to that of the aromatic hydrocarbon benzene, see figure 12 of Ref. [3].) Substituting Eqs (1) and (2) into the standard relation

$$\Delta G° = \Delta H - T\Delta S° \qquad (3)$$

gives

$$\Delta G°(T_2) = \Delta G°(T_1) + \Delta C_p(T_2 - T_1) - T_2\Delta C_p \ln(T_2/T_1) \qquad (4)$$

Equation (4) shows that the ΔC_p-induced changes in ΔH and $\Delta S°$ with temperature tend to compensate each other to produce only a small net increase in $-\Delta G°$ as the temperature increases. This property of enthalpy–entropy compensation is one of the most characteristic features of hydrophobic free energy. (ΔG and ΔS depend strongly on the solute concentration and the superscript ° emphasizes that $\Delta G°$ and $\Delta S°$ refer to the standard state concentration.)

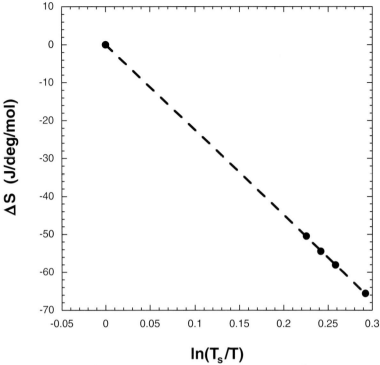

Fig. 6.5. Decrease of $-\Delta S$ towards 0 as temperature (K) approaches T_s (386 K), the temperature at which $\Delta S = 0$. The transfer of benzene from liquid benzene to water is shown. Data are from Ref. [43]. Note the linear decrease in $-\Delta S$ with temperature [43], which depends on using data from the temperature range (15–50 °C) where ΔC_p = constant (see figure 12 of Ref. [3]). T_s = 386 K is an average value for several hydrocarbon solutes [43].

Figures 6.4A and 6.4B illustrate some basic thermodynamic properties of hydrophobic free energy as modeled by liquid–liquid transfer. The enthalpy change ΔH is zero near room temperature (the exact temperature depends on the hydrocarbon) and the transfer process is entropy-driven around room temperature. However, as the temperature increases above 25 °C, the transfer process gradually becomes enthalpy-driven. A surprising property is that of entropy convergence: if ΔC_p = constant, then different hydrocarbons share a convergence temperature at which $\Delta S° = 0$ (386 K or 113 °C) [43]. Data taken between 15 and 35 °C are extrapolated linearly versus $\ln T$ in Figure 6.5, according to Eqn (2). When the gradual decrease in ΔC_p at temperatures above 50 °C [3] is taken into account by using a curved extrapolation, hydrocarbon data for $\Delta S°$ still approach a common value near T_s, the temperature at which the entropy change is 0, which is approximately 140 °C [3, 4]. The property of entropy convergence is predicted both by the information-theory model [53] and scaled particle theory [54].

Privalov [55] discovered in 1979 that values for the specific entropy change on protein unfolding converge near 113 °C when results for some different proteins are extrapolated linearly versus temperature. His observation, taken together with the entropy convergence shown by hydrocarbon solutes, suggests that the hydrophobic entropy change of a protein unfolding reaction might be removed by extrapolating the total entropy change to T_s for hydrocarbon transfer [43]. However, when Robertson and Murphy [56] analyzed unfolding data from several laboratories, they found more scatter in the data and they did not find a common intercept for the specific entropy change on protein unfolding. The extrapolation offers a possible route to determining the change in conformational entropy on unfolding, one of the basic unsolved problems in protein folding. The hydrophobic contribution to the entropy of unfolding is opposite in sign and, at room temperature, comparable in size to the change in conformational entropy [43].

Schellman [57] points out that there are advantages to using $\Delta G/T$ as an index of stability instead of ΔG itself when considering the temperature dependence of hydrophobic free energy. The low solubilities of hydrocarbons in water have minimum values near room temperature, corresponding to the maximum values of $\Delta G/T$ and to $\Delta H = 0$, whereas ΔG itself increases steadily but slowly with temperature until it reaches a maximum value above 100 °C, at a temperature where $\Delta S = 0$.

A standard test of whether the hydrophobic free energy found from liquid–liquid transfer experiments applies to protein folding experiments is to compare the unfolding free energy change caused by a "large to small" mutation with the free energy change predicted from the change in ASA on unfolding. Pace [58] surveyed the literature on this subject (see also Ref. [21]). He concludes there is good agreement provided the proportionality coefficient between $\Delta\Delta G$ and ΔASA is the volume-corrected value 47 cal mol^{-1} Å^{-2} [21]. However, this type of comparison between liquid–liquid transfer and protein folding thermodynamics is complicated by the presence of cavities produced when large-to-small mutations are made [59].

A direct test of the correspondence between liquid–liquid transfer thermodynamics and protein folding is to compare the value of ΔC_p for a protein unfolding experiment with the value predicted from the change in ASA on unfolding. No other factor besides the burial of nonpolar surface area is known to make a significant positive contribution to ΔC_p. This comparison was studied in 1991 by Record and coworkers [60], who found that values of ΔC_p measured in unfolding experiments do agree with the ones predicted from the change in nonpolar ASA on unfolding. The issue became somewhat complicated, however, by contemporary work showing that polar groups (and especially the peptide group) contribute negatively to ΔC_p for unfolding. Makhatadze and Privalov [61] gave extensive model compound data and used them to estimate contributions to the enthalpy and heat capacity changes on unfolding of all groups present in proteins. Freire and coworkers [62] used empirical calibration of protein unfolding data to give the contributions to ΔC_p expected from the changes in polar and nonpolar ASA on unfolding. The issue is seriously complicated, however, by evidence from the study of model compounds that the enthalpies of interaction with water are not related in any simple way to polar ASA: see Table 6.2 and Section 6.3.2. To make progress in analyzing this problem, it is important to get data for the ΔC_p values accompanying unfolding of peptide helices and progress has been reported recently by Richardson and Makhatadze [63].

A long-standing puzzle in comparing the thermodynamics of protein unfolding with those of liquid–liquid transfer has been the unfolding results produced by high pressures [64]. The information-theory model [65] provides new insight into the problem by predicting that water penetrates the hydrophobic cores of proteins during pressure-induced unfolding, as observed experimentally [66], but does not penetrate the hydrophobic cores during thermal unfolding.

6.2.10
Modeling Formation of the Hydrophobic Core from Solvation Free Energy and van der Waals Interactions between Nonpolar Residues

As explained above, solvation free energies of solutes are based on gas–liquid transfer data. They provide an alternative model for the energetics of forming the hydrophobic core of a protein. Compared with the liquid–liquid transfer model, this approach has two major advantages. (1) It considers the van der Waals interactions explicitly, and (2) it avoids the question of which nonaqueous solvent to use for modeling the hydrophobic core. This approach was pioneered by Ooi and Oobatake [67, 68] and by Makhatadze and Privalov [61]. Simonson and Brünger [69] analyze model compound data for gas–liquid transfer. They report that transfer free energies for gas to liquid transfer of cyclic hydrocarbons fall well below the line for linear alkanes when plotted against ASA.

The use of gas–liquid transfer to probe the energetics of folding is illustrated in the cycle shown in Scheme 6.1, in which the native protein (N) is formed by folding the unfolded protein (U) either in the gas phase (g) or in aqueous solution (aq). The cycle is completed by the process of transferring U from aqueous solution to the gas phase and by transferring N from the gas phase back to aqueous solution. Only the nonpolar side chains and nonpolar moieties of polar side chains are considered here. Transfer of the polar peptide NH and CO groups between the gas phase and aqueous solution is considered in Section 6.3.8, together with formation of the peptide H-bonds. The polar moieties of polar side chains are supposed to be fully exposed to water in both U and N, and so their transfer between aqueous solution and the gas phase cancels energetically in steps 1 and 3. The change in conformational entropy when the unfolded polypeptide U folds to the native protein N is supposed to be the same in steps 2 and 4.

In Scheme 6.1, two processes contribute to $\Delta G°_{UN}$, the standard free energy

U (g) $\Leftarrow$(2)$\Rightarrow$ N (g)

$\Updownarrow$ (1) $\Updownarrow$ (3)

U (aq) $\Leftarrow$(4)$\Rightarrow$ N (aq)

Scheme 6.1

change for folding in aqueous solution. Process A is burial (removal from water) of nonpolar side chains (or moieties) in steps 1 and 3, and process B is formation of van der Waals interactions between nonpolar side chains (or moieties) in step 2. Let ΔG_{bur} and ΔG_{vdw} be the contributions to $\Delta G°_{\text{UN}}$ from processes A and B in Eq. (5).

$$\Delta G°_{\text{UN}} = \Delta G_{\text{bur}} + \Delta G_{\text{vdw}} \tag{5}$$

Let ΔASA be the net change in nonpolar solvent-accessible surface area in steps 1 + 3. Evidence is presented above that ΔG_{bur} can be related to ΔASA even for microscopic cavities. Thus ΔG_{bur} can be represented by

$$\Delta G_{\text{bur}} = \beta \Delta \text{ASA} \tag{6}$$

in which β is the proportionality coefficient between $\Delta G°_{\text{solv}}$ and ASA for alkanes in transfer experiments from the vapor phase to aqueous solution. Figure 6.3 shows good linearity between $\Delta G°_{\text{solv}}$ and ASA for several alkanes, and β is reasonably constant. However, the value of β changes substantially (from 5.5 to 24 cal mol^{-1} Å^{-2} [21]) when volume-corrected transfer free energy is used. Evaluation of ΔG_{vdw} depends sensitively on parameters that are difficult to determine experimentally, and there is little discussion in the literature of how mutations cause ΔG_{vdw} to vary. See, however, the discussion of ΔG_{vdw} values for selected proteins by Makhatadze and Privalov [61] and note that the *number* of van der Waals contacts is being discussed in mutational studies of packing [70].

For the process of forming the hydrophobic core, ΔG_{vdw} can be estimated approximately in the following manner. First, consider the empirical relation between $\Delta G°_{\text{UN}}$ and ΔASA given by the liquid–liquid transfer model.

$$\Delta G°_{\text{UN}} \sim B(\Delta \text{ASA}) \tag{7}$$

In Eq. (7), ΔASA is the net value of ASA buried upon folding and B is the proportionality factor between liquid–liquid transfer free energy and ASA. Combining Eqns (5)–(7) gives:

$$\Delta G_{\text{vdw}} \sim (B - \beta)\Delta \text{ASA} \tag{8}$$

The value of $(B - \beta)$ depends on whether or not the volume correction is made to the transfer free energies. If the volume correction is made, then $(B - \beta) = 47 - 25 = 22$ cal mol^{-1} Å^{-2} [21] and ΔG_{vdw} should account for approximately $22/47 = 47\%$ of the free energy change on forming the hydrophobic core. If the volume correction is not made and the mole fraction scale is used for computing transfer free energies, then $(31 - 5)/31$ [21] $= 84\%$, the percentage of $\Delta G°_{\text{UN}}$ that is assigned to ΔG_{vdw}. This argument follows the one given by Havranek and Harbury [71], who used different numbers for B and β from the ones given by Sharp et al. [21].

Data for the melting of solid crystalline alkanes show that ΔH is substantial and favors the solid alkane, but ΔG for melting is small because there is a substantial ΔS that favors the liquid alkane [72]. These results have been used to argue that a liquid hydrocarbon should model satisfactorily the hydrophobic core of a protein [72], and therefore the van der Waals interactions need not be considered explicitly.

6.2.11
Evidence Supporting a Role for van der Waals Interactions in Forming the Hydrophobic Core

An important role for van der Waals interactions in protein folding is suggested by the unusual close-packed arrangement of the side chains in folded proteins. This argument was pointed out in 1971 by Klapper, who compares the protein core to a wax ball rather than an oil drop [73], and in 1977 by Richards, who compares the protein core to an organic crystal [50]. The fraction of void space in folded proteins is quite small and comparable to that in close-packed spheres, whereas there is a sharp contrast between the small void space in folded proteins and the large void space in nonpolar liquids, as pointed out by these authors. In 1994 Chothia and coworkers [74] made a detailed study of the volumes of aliphatic side chains inside proteins and found they are substantially smaller than in aqueous solution. Water molecules are nevertheless rather well packed around hydrocarbons in aqueous solution and hydrocarbons occupy larger volumes in nonpolar solvents than in water, as pointed out by Klapper [73], who found good agreement between experimental data and predictions based on scaled particle theory.

Stites and coworkers have made extensive mutational experiments on the hydrophobic core of staphylococcal nuclease (SNase) [70, 75] including many single, double, triple, and quadruple mutants. The authors measure interaction energies by taking the difference between the $\Delta G°$ values of two single mutants and the double mutant, and so forth. Their overall results suggest that the packing arrangement of nonpolar side chains in the core is an important determinant of SNase stability because favorable interaction energies are associated with certain pairs of residues [75]. They tested this interpretation by constructing seven multiple mutants that were predicted to be hyperstable [75]. The X-ray structures of five mutants were determined. The mutants uniformly show high temperatures for thermal unfolding (T_m). The largest increase in T_m is 22.9 °C and the increase in T_m correlates with the number of van der Waals contacts. The overall results indicate that increased protein stability is correlated with an increased number of van der Waals contacts, and certain pairs of buried side chains make characteristic contributions to the number of van der Waals contacts.

A novel approach to analyzing the thermodynamics of burial of nonpolar side chains in protein folding was undertaken by Varadarajan and coworkers [76, 77]. They measure $\Delta\Delta H$ of folding for mutants with varying extents of burial of nonpolar surface area. Note that $\Delta\Delta H$ should be small near 25 °C for burial of nonpolar surface area, according to liquid–liquid transfer experiments with hydrocarbons (see Figures 6.4A and 6.4B). On the other hand, van der Waals packing interactions are enthalpy-driven and larger values of $\Delta\Delta H$ are expected if the folding thermodynamics are dominated by packing interactions. The authors are able to measure ΔH directly at fixed temperatures by titration calorimetry for their ribonuclease S system (which is composed of two dissociable species, S-peptide + S-protein) by titrating unfolded S-peptide versus folded S-protein. They determine the X-ray structures of the mutant RNase S species and measure ΔH for folding S-peptide at varying temperatures. They find substantial $\Delta\Delta H$ values for the mutants and conclude that van der Waals packing interactions are responsible; see also the study of cavity mutants of T4 lysozyme by Matthews and coworkers [59].

Zhou and Zhou [78, 79] present an interesting analysis of the energetics of burying different side-chain types. They construct a database of mutational experiments on protein folding from the literature and extract the "transfer free energy" contributed by each amino acid side chain to the unfolding free energy: the transfer process is from the folded protein to the unfolded protein in aqueous solution. Then the authors simulate the mutational results by a theoretical analysis based on a potential of mean force, and plot the results as $\Delta\Delta G$ versus ΔASA for each side-chain type. The resulting plots are similar for the various nonpolar amino acids, and the slopes are in the range 25–30 cal mol^{-1} Å^{-2}. Interestingly, the plots for polar amino acids are similar in character but have substantially smaller slopes. The authors conclude that it is energetically favorable to fold polar as well as nonpolar side chains, but folding polar side chains contributes less to stability. There should be an energetic penalty if the polar moiety of a side chain is buried without making an H-bond, because the polar group interacts favorably with water (see below).

6.3
Peptide Solvation and the Peptide Hydrogen Bond

6.3.1
History

Interest in the role of the peptide hydrogen bond in stabilizing protein folding took off when Pauling and Corey proposed structures for the α-helix and the parallel and antiparallel β-sheets, and when Schellman analyzed [6] the factors that should determine the stability of the helix backbone in water. He estimated the net enthalpy change for forming a peptide H-bond in water as -1.5 kcal mol^{-1}, based on heat of dilution data for urea solutions [80], and he concluded that an isolated α-helix backbone should be marginally stable in water and might or might not be detectable [6]. Klotz and Franzen [81] investigated by infrared spectroscopy whether H-bonded dimers of N-methylacetamide are detectable in water and concluded that they have at best marginal stability: see also a later study of the dimerization of δ-valerolactam [82].

Determination of the first X-ray structures of proteins sparked interest in the problem of whether the peptide sequence for a protein helix can form the helix in water. Initial results based on circular dichroism (CD) (or optical rotatory dispersion) spectroscopy were disappointing. In 1968 Epand and Scheraga [83] examined two peptides obtained by cyanogen bromide cleavage from the highly helical protein sperm whale myoglobin and found very low values for the possible helix contents. In 1969 Taniuchi and Anfinsen [84] cleaved staphylococcal nuclease, which has 149 residues, between residues 126 and 127 and found that both fragments 1–126 and 127–149 (each of which has a helix of modest size) were structureless by CD as well as by other physical criteria. In 1969–1971 Klee obtained tantalizing results, suggestive of some helix formation at low temperatures, with the "S-peptide" (residues 1–20) [85] and "C-peptide" (residues 1–13) [86] of ribonuclease A. (The RNase A helix contains residues 3–12.) The helix problem languished for 11 years, perhaps because of the disappointing results found in the laboratories of Anfinsen and Scheraga. In 1982, Bierzynski et al. [87] used NMR as well as CD spectroscopy to reinvestigate the claim of Brown and Klee [86] and confirmed that partial helix formation occurs. Bierzynski et al. found that helix formation is strongly pH dependent and the pK values indicate that both a His residue and a Glu residue are required for helix formation, thus implicating specific side-chain interactions [87]. Several kinds of helix-stabilizing side-chain interactions were later studied, such as salt bridges, amino–aromatic interactions, and charge–helix dipole interactions [88]. The helix formed by C-peptide or S-peptide has basic properties in common with the protein helix of RNase A: both the peptide and protein helices have the same two helix-stabilizing side-chain interactions and the same helix termination signals. Today hundreds of peptides with sequences derived from proteins form partly stable helices in water [89]. The helix termination signals are well-studied [90] and consistent results are found from protein and peptide helices. The AGADIR algorithm of Muñoz and Serrano [89] fits the helix contents of peptides by statistically evaluating the many parameters describing side-chain interactions and helix propensities.

In 1989 Marqusee et al. [91] found that alanine-rich peptides form stable helices in water without the aid of specific side-chain interactions. Of the 20 natural amino acids, only alanine has this ability [92]. Because alanine has only a –CH$_3$ group for a side chain, this result strongly suggests that the helix backbone itself has measurable stability in water and that side chains longer than alanine somehow destabilize the helix. In 1985 Dyson and coworkers used NMR to detect a reverse turn conformation in a nine-residue peptide [93], and later studied the specific sequence requirements for forming this class of turn in peptides [94]. In 1994 Serrano and coworkers [95] found the β-hairpin conformation in a protein-derived peptide, and other β-hairpin sequences were soon found. Thus, H-bonded secondary structures derived from proteins can in general be studied in peptides in aqueous solution.

Determination of the energetic nature of the peptide H-bond was a chief motivation for the pioneering development in 1979 of a molecular force field by Lifson and coworkers [96]. They concluded that the peptide H-bond is represented to a good first approximation by an electrostatic model based on partial charges residing on the peptide CO and NH groups. Their analysis indicates that the peptide

CO and NH groups interact with water as well as with each other. Solvation of the peptide polar groups has been analyzed by using amides such as N-methylacetamide (NMA) as models for the peptide group.

6.3.2
Solvation Free Energies of Amides

In a classic experiment, Wolfenden demonstrated in 1978 that amides interact very strongly with water [97]. He measured the equilibrium distribution of a radioactively labeled amide between aqueous solution and the vapor phase, and calculated its transfer free energy from the vapor phase to aqueous solution. The (hypothetical) standard state concentration of the amide solute is 1 M in the vapor phase as well as in the liquid phase (see Section 6.2.4). Wolfenden and coworkers next measured the transfer free energies of model compounds for amino acid side chains [98]. The ionized forms of the basic and acidic residues are excluded from their study because species containing a full formal charge interact too strongly with water to be measured. Their results show that the interaction of water with amides is stronger than with any of the model compounds representing side chains. Saturated hydrocarbons, which serve as models for the side chains of leucine, isoleucine and valine, are preferentially excluded from aqueous solution and have unfavorable solvation free energies. The basic cause is the large unfavorable work of making a cavity in water, which exceeds the favorable van der Waals interaction energy (see Refs [24, 25] and Section 6.2.4). Figure 6.3 shows a direct proportionality between the transfer free energy $\Delta G°$ and ASA for transfer of alkanes from aqueous solution to the vapor phase.

For polar solutes as well as nonpolar solutes, there is an unfavorable work of making a cavity in water for the solute and a favorable van der Waals interaction energy between the solute and water. These effects must be removed from the observed solvation free energy to obtain the interaction free energy between water and the polar groups. Currently the sum of the cavity and van der Waals terms is approximated by experimental data for the transfer free energies of nonpolar solutes, as suggested by Makhatadze and Privalov [61] and Sitkoff et al. [99]. The observed solvation free energy in Eq. (9) is split into two terms: one for the polar groups ($\Delta G(\text{pol})$) and one for the sum of the cavity and van der Waals terms ($\Delta G(\text{np})$). The value of $\Delta G(\text{np})$ in Eq. (10) is taken from the plot of transfer free energy versus ASA for alkanes and the proportionality coefficient β is taken from this plot (Figure 6.3).

$$\Delta G°(\text{obs}) = \Delta G(\text{pol}) + \Delta G(\text{np}) \quad (9)$$

$$\Delta G(\text{np}) = \beta \text{ASA} \quad (10)$$

Values of $\Delta G(\text{pol})$ are given for four amides in Table 6.2, taken from Avbelj et al. [100]. Although the correction term $\Delta G(\text{np})$ is obtained by the approximate procedure of using experimental data for nonpolar solutes, the correction term is fairly

Tab. 6.2. Solvation free energy data for amides as models for the free peptide group[a].

Quantity[b] (kJ mol^{-1})	Acetamide	N-methyl-acetamide	N,N-dimethyl-acetamide	Propionamide
$\Delta G°$(obs)	−40.58	42.13	−35.73	−39.25
ΔG(np)	8.16	−9.04	9.71	8.87
ΔG(pol)	−48.74	−51.17	−45.44	−48.12
ΔH(obs)	−68.28	−71.42	−69.29	−73.01
ΔH(np)	−19.54	−23.35	−26.15	−22.64
ΔH(pol)	−48.74	−48.07	−43.14	−50.38
ASA(pol) (nm^2)	0.95	0.51	0.32	0.88

[a] Data are from Ref. [100].
[b] $\Delta G°$(obs) is the observed transfer free energy from gas to water. Data are taken from Ref. [100], based on a standard state concentration of 1 M in both the gas and liquid phases and without volume correction. ΔG(np) is the correction for the cavity and van der Waals terms according to Eq. (10) and using the value of β given in Ref. [100] (6.4 cal mol^{-1} Å^{-2}), which is not volume-corrected. ΔG(pol) is the solvation free energy of the polar groups in the amide, found from Eq. (9). The quantities ΔH(obs), ΔH(np) and ΔH(pol) have corresponding meanings and are explained in the text. Values for ΔH(obs) are taken from various calorimetric papers, and references are given in Ref. [100]. ASA(pol) is the value of ASA assigned to the polar groups.

small (9.0 kJ mol^{-1} for N-methylacetamide) compared with the observed solvation free energy (−42.1 kJ mol^{-1}) or the value of ΔG(pol) (−51.2 kJ mol^{-1}).

The value of β (Eq. (6)) used by Sitkoff et al. [99] (to obtain values for ΔG(np) in Eq. (10)) is based on transfer free energies that are not volume-corrected [21, 99]. The resulting values of ΔG(pol) (obtained from Eq. (9) for a database of model compounds) have been used to calibrate the PARSE parameter set of the DelPhi algorithm [99] (see Section 6.3.5). For this reason, the values of ΔG(np) in Table 6.1 are based on transfer free energies that are not volume-corrected.

The enthalpy of interaction between the polar groups and water, ΔH(pol), has been obtained [100] from calorimetric data in the literature after approximating ΔH(np) by experimental data for alkanes, as before.

$$\Delta H(\text{obs}) = \Delta H(\text{pol}) + \Delta H(\text{np}) \qquad (11)$$

The value for ΔH(obs) in Eq. (11) is found by combining results from two calorimetric experiments. The first experiment yields the heat of vaporization of the liquid amide and the second experiment yields the heat of solution of the liquid amide in water. The value of ΔH(np) is found from the heats of solution, plotted against ASA, of gaseous hydrocarbons in water, for which the heats of solution depend chiefly [24] on the favorable van der Waals interaction energy of the hydrocarbon solute with water.

Some interesting conclusions may be drawn from Table 6.2: (1) The four different amides have similar values for ΔG(pol) and also for ΔH(pol). Although there is a threefold difference between the values of polar ASA for acetamide and N,N-dimethylacetamide, their values of $\Delta G(\text{pol})$ differ by less than 10%. Thus, the computationally convenient approximation of assuming proportionality between $\Delta G(\text{pol})$ and polar ASA [101] is not a satisfactory assumption for amides. The polar groups of an amide interact strongly with water even when they are only partly exposed. (2) The value of $\Delta G(\text{pol})$ is nearly equal to the value of $\Delta H(\text{pol})$ for each amide, and so the $T\Delta S(\text{pol})$ term must be small by comparison. This conclusion is surprising if one expects a large entropy change when water interacts with polar groups. It is not surprising, however, after considering the entropies of hydration of crystalline hydrate salts [102] or the prediction [103] made from the Born equation that $\Delta G(\text{pol})$ is almost entirely enthalpic. The basic reason why the entropy term looks small is that the enthalpy term is large. (3) The values of $\Delta G(\text{pol})$ are impressively large for amides. NMA is sometimes taken as a model for studying the interaction between water and the peptide group, and its value of $\Delta G(\text{pol})$ is -51.2 kJ mol^{-1}. This is a huge value for only one peptide group: $-\Delta G°$ for folding a 100-residue protein is typically 40 kJ mol^{-1} or less. Evidently solvation of the peptide group is a major factor in the energetics of folding: see Wolfenden [97] and Makhatadze and Privalov [61].

6.3.3
Test of the Hydrogen-Bond Inventory

Protein chemists often use the H-bond inventory to interpret results of mutational experiments when the experiment involves a change in the number of hydrogen bonds [104, 105]. Experimental data for amides provide a direct test of the H-bond inventory [106]. A key feature of the H-bond inventory is that water (W) is treated as a reactant and the change in the total number of W•••W H-bonds is included in the inventory. When discussing the formation of a peptide H-bond (CO•••HN) in water, the H-bond inventory takes the form of Eq. (12).

$$\text{CO}\bullet\bullet\bullet\text{W} + \text{NH}\bullet\bullet\bullet\text{W} = \text{CO}\bullet\bullet\bullet\text{HN} + \text{W}\bullet\bullet\bullet\text{W} \tag{12}$$

The peptide NH and CO groups are predicted to make one H-bond each to water in the unfolded peptide, and the CO and NH groups are predicted not to interact further with water after the peptide H-bond is made. Gas phase calculations of H-bond energies indicate that all four H-bond types in Eq. (12) have nearly the same energy [107], -25 ± 4 kJ mol^{-1}, and consequently the net enthalpy change predicted for forming a peptide H-bond in water is 0 ± 4 kJ mol^{-1}. The enthalpy of sublimation of ice has been used to give an experimental estimate of -21 kJ mol^{-1} for the enthalpy of the W•••W H-bond [45].

When a dry molecule of amide (Am) in the gas phase becomes hydrated upon solution in water, the H-bond inventory model postulates that the CO and NH groups each make one H-bond to water and one W•••W H-bond is broken:

$$\text{HN}-\text{Am}-\text{CO} + \text{W}\bullet\bullet\bullet\text{W} = \text{W}\bullet\bullet\bullet\text{HN}-\text{Am}-\text{CO}\bullet\bullet\bullet\text{W} \tag{13}$$

Thus, the H-bond inventory predicts a net increase of one H-bond when a dry amide becomes solvated, with a net enthalpy change of -25 ± 4 kJ mol^{-1}. But the experimental enthalpy change for solvating the amide polar groups is ~ -50 kJ mol^{-1} (Table 6.2) and the H-bond inventory fails badly.

6.3.4
The Born Equation

In 1920 Max Born wanted to know why atom-smashing experiments can be visualized in Wilson's cloud chamber [108]: an ion leaves a track of tiny water droplets as it passes through supersaturated water vapor. He computed the work of charging a spherical ion in vacuum and in water, and he treated water as a continuum solvent with dielectric constant D. He gave the favorable change in free energy for transferring the charged ion from vacuum to water as:

$$-\Delta G = (q^2/2r)[1 - (1/D)] \qquad (14)$$

in which q is the charge and r is the radius of the ion [109]. This simple calculation answers Born's question: the free energy change is enormous, of the order of -420 kJ mol^{-1} for a monovalent ion transferred to water. Consequently the ion easily nucleates the formation of water droplets. Inorganic chemists – notably W. M. Latimer – realized that Born's equation provides a useful guide to the behavior of solvation free energies. Note that the solvation free energy predicted by Born's equation is inversely proportional to the ionic radius, but which radius should be used? Ions are strongly solvated and the solvated radius is subject to argument. Rashin and Honig [103] argued that the covalent radius is a logical choice and gives consistent results for different monovalent anions and cations. They also point out that one can obtain the enthalpy of solvation from Born's equation and it predicts that, if the solvent is water with a high dielectric constant (near 80), the solvation free energy is almost entirely enthalpy. Quite recently the solvation free energies of monovalent ions have been analyzed by a force field employing polarizable water molecules [110], with excellent agreement between theory and experiment.

6.3.5
Prediction of Solvation Free Energies of Polar Molecules by an Electrostatic Algorithm

The success of the Born equation in rationalizing solvation free energies of monovalent anions and cations suggests that it should be possible to predict the solvation free energies of polar molecules by using an appropriate electrostatic algorithm plus knowledge of the structure and partial charges of the polar molecule. The problem has a long history. Kirkwood and Westheimer [111] developed a theory in 1938, based on a simple geometrical model, that gives the effect of a low dielectric environment within the molecule on the separation between the two pK values of a dicarboxylic acid. The problem was treated later by an electrostatic algorithm that is free from geometrical assumptions about the shape of the molecule

[112]. The latter treatment includes the effect of electrostatic solvation (the "Born term") as well as the electrostatic interaction between the two charged carboxylate groups.

Various electrostatic algorithms are in current use [99, 113] today, including one based on using Langevin dipoles [113] to treat the polarization of water molecules in the vicinity of charged groups. The focus here is on the DelPhi algorithm of Honig and coworkers [99], because the PARSE parameter set of DelPhi has been calibrated against a database of experimental solvation free energies for small molecules that includes amides [99]. Thus, DelPhi may plausibly be used to predict the solvation free energies of the polar CO and NH groups of peptides in various conformations. There are no adjustable parameters and the predicted values of electrostatic solvation free energy (ESF) may be compared directly with experiments if suitable data are available. The DelPhi algorithm uses a low dielectric ($D = 2$) cavity to represent the shape of the solute while the partial charges of the solute are placed on a finely spaced grid running through the cavity, and the solvent is represented by a uniform dielectric constant of 80 [99]. The results are calculated by Poisson's equation if the solvent does not contain mobile ions, or by the Poisson-Boltzmann equation if it does.

6.3.6
Prediction of the Solvation Free Energies of Peptide Groups in Different Backbone Conformations

A basic prediction about the electrostatic interactions among polar CO and NH groups in the peptide backbone is that the interactions depend strongly on the peptide backbone conformation [114–116]. The peptide group is normally fixed in the trans conformation while the CO and NH dipoles of adjacent peptide units are aligned antiparallel in the extended β-strand conformation and parallel in the α-helix conformation. When adjacent peptide dipoles are parallel they make unfavorable interactions unless the dipoles are placed end to end, when formation of peptide H-bonds occurs. These simple observations have important consequences, as pointed out especially by Brant and Flory [114, 115] and by Avbelj and Moult [116]. Thus, intrachain electrostatic free energy favors the extended β conformation over the compact α conformation and this factor tends to make β the default conformation in unfolded peptides [114]. There is a very large difference, ~20 kJ mol^{-1}, between the local electrostatic free energy of the α- and β-strand conformations when calculated with $D = 1$ [116]. Nucleation of an α-helix is difficult for this electrostatic reason [115] as well as for the commonly cited loss in backbone conformational entropy: the peptide dipoles in the helix nucleus are parallel and make unfavorable electrostatic interactions. When helix propagation begins and additional H-bonded residues are added onto the helical nucleus, the favorable H-bond energy drives helix formation.

Calculations of polar group solvation (ESF) in peptides [100, 121] using DelPhi show that ESF depends strongly on two factors: the access of solvent to the peptide group and the local electrostatic potential in the peptide chain. Nearby side chains

Tab. 6.3. Calculated solvation free energies of peptide groups in different backbone conformations[a].

Structure[b]	ESF (kJ mol^{-1})	Reference
Helix, H-bonded, solvent-exposed	−10.5	100
Helix, not H-bonded, solvent-exposed	−39.8	100
β-hairpin, H-bonded, exposed	−10.5	121
β-hairpin, H-bonded, buried	0	121
β-strand[c], not H-bonded, exposed	−35.6	
Polyproline II[b], not H-bonded, exposed	−38.1	

[a] The solvation free energies of the peptide polar groups (CO, NH) are calculated by DelPhi, as explained in the text (see Section 6.3.5). No adjustable parameters are used except those that describe the structure. The calculations listed here are made for all-alanine peptides and they refer to interior peptide groups, not to N- or C-terminal groups.
[b] The H-bonded, solvent-exposed helix refers to the central residue of a 15-residue helical peptide. The solvent-exposed helix, not H-bonded, refers to the central residue of a five-residue peptide in the α-helical conformation ($\phi, \psi = -65°, -40°$). The β-hairpin (H-bonded, solvent-exposed) refers to typical H-bonded residues in a 15-residue peptide with a β-hairpin conformation taken from a segment of the GB1 structure. The solvent-exposed, extended β-strand, not H-bonded ($\phi, \psi = -120°, 120°$) refers to a nine-residue peptide (F. Avbelj and R. L. Baldwin, to be published). The ESF value given here is more negative than the value previously given [100]. The earlier value by accident had a conformation deviating from ($\phi, \psi = -120°, 120°$) because it was used for an Ala to Val substitution and had a conformation suitable for receiving a Val residue. The polyproline II structure (F. Avbelj and R. L. Baldwin, to be published) refers to the central residue of a nine-residue peptide and has ($\phi, \psi = -70°, 150°$).

hinder access of solvent to the peptide group and reduce the negative ESF value. Consequently, helix and β-structure propensities of the different amino acids should depend on the ESF values of the peptide groups [100, 116–118]. The different helix propensities of the amino acids are often attributed instead to the loss in side-chain entropy when an unfolded peptide forms a helix [119]. Because ESF is almost entirely enthalpic (see Table 6.2), these alternative explanations can be tested by determining if the helix propensity differences are enthalpic or entropic. Temperature-dependence results [120] measured for the nonpolar amino acids (see also Ref. [63]) indicate that the helix propensity differences are largely enthalpic.

Table 6.3 gives some ESF values for alanine peptide groups in different backbone conformations. There are several points of interest: (1) A peptide group in a non-H-bonded alanine peptide has substantially different ESF values in the three major backbone conformations β, α_R, and P_{II} (polyproline II). Consequently any analysis based on group additivity that predicts the overall enthalpy change by assigning constant energetic contributions (independent of backbone conformation) to the polar peptide CO and NH groups cannot be valid. The overall enthalpy change contains contributions from both ESF and the local intrachain electrostatic potential (E_{local}) [116–118], both of which depend on backbone conformation. (2) Because ESF depends on the accessibility of water to a peptide group, the N-terminal and C-terminal peptide groups of an all-alanine peptide have more negative ESF values than interior peptide groups. (3) Replacement of Ala by a larger or more bulky residue such as Val [100] changes the ESF not only at the substitution site but also at neighboring sites (Avbelj and Baldwin, to be published). (4) N-Methylacetamide, whose ΔG_{pol} (or ESF) is -51.2 kJ mol^{-1} (Table 6.2) is a poor model for the interaction with water of a free peptide group. There is a 16 kJ mol^{-1} difference between the ESF of NMA and that of an alanine peptide group in the β-conformation (Tables 6.2 and 6.3).

6.3.7
Predicted Desolvation Penalty for Burial of a Peptide H-bond

ESF calculations for completely buried peptide groups, with no accessibility to water, show that their ESF values fall to zero [121]. On the other hand, DelPhi calculations for H-bonded and solvent-exposed peptide groups in either the α-helical [100] or β-hairpin [121] conformation show that H-bonded peptide groups interact with solvent and have highly significant ESF values, about -10.5 kJ mol^{-1} for alanine peptides. Of this, -8.5 kJ mol^{-1} is assigned to the peptide CO group and -2 kJ mol^{-1} to the peptide NH group [100]. Consequently, there should be a large desolvation penalty (equal to the ESF of the solvent-exposed, H-bonded peptide group) if a solvent-exposed peptide H-bond becomes completely buried [100, 121]. This deduction agrees with the prediction by Honig and coworkers [122] of an even larger desolvation penalty (16 kJ mol^{-1}) for transferring an amide H-bond in a NMA dimer from water to liquid alkane. Two points of interest for the mechanism of protein folding arise from these ESF calculations. First, the ESF values of H-bonded, solvent-exposed peptide groups are predicted to be a major factor stabilizing molten globule folding intermediates [100, 121]. Second, complete burial of H-bonded secondary structure should involve a substantial desolvation penalty [121]. The size of the penalty depends on how solvated each peptide group is when the secondary structure is solvent-exposed. Neighboring side chains larger or more bulky than Ala reduce the negative ESF of the peptide H-bond.

When Ben-Tal et al. [122] studied the transfer of a H-bonded NMA dimer from water to liquid alkane, they found the formation of the H-bonded dimer in water to be stable by -5 kJ mol^{-1}. Their results give the penalty for forming the H-bond in water and then transferring it to liquid alkane as $-5 + 16 = 11$ kJ mol^{-1}. When their predictions are compared with values for an alanine peptide helix, the results are numerically different (not surprising, in view of the structural difference) but qualitatively the same. Ben-Tal et al. omit the loss in translational and rotational entropy for forming the NMA dimer and their ΔG value of -5 kJ mol^{-1} [122] may be compared with the measured ΔH (-4.0 kJ mol^{-1} [123, 124]) for forming the alanine peptide helix.

Myers and Pace [125] discuss whether or not peptide H-bonds stabilize protein folding by considering mutational data for H-bonds made between specific pairs of side chains. The ESF studies considered here emphasize the role of context: solvent-exposed peptide H-bonds should stabilize folding but buried H-bonds should detract from stability.

The ESF perspective provides a simple explanation for why alanine, alone of the 20 amino acids in proteins, forms a stable helix in water. The H-bonded, solvent-exposed peptide group in an alanine helix has the most stabilizing ESF value of any nonpolar amino acid except glycine, and glycine fails to form a helix because of the exceptional flexibility of the glycine peptide linkage.

The interaction energy between water and the peptide group is a critical quantity in predicting the structures of membrane proteins from amino acid sequences [126]. Accurate values are needed for the energy of desolvation both of the free and H-bonded forms of the peptide group. The solvent n-octanol, which has been used extensively to model the interiors of water-soluble proteins (see Sections 6.2.2 and 6.2.3), has also been used as an experimental model for the lipid bilayer environment of membrane proteins, when measuring transfer free energies of peptides between water and a membrane-like solvent. The small value of the transfer free energy found for the free glycine peptide group transferred from water to water-saturated n-octanol (4.8 kJ mol^{-1} [15]) emphasizes the role of the water contained in this solvent, which increases its attraction for the peptide group. Cyclohexane has been used as an alternative solvent that provides a nearly water-free environment when partitioning model compounds [16]. The transfer free energies of amides are much larger in the cyclohexane/water pair than the 4.8 kJ mol^{-1} determined for the peptide group in octanol/water: 25 and 21 kJ mol^{-1} for acetamide and propionamide, respectively [16].

6.3.8
Gas–Liquid Transfer Model

The gas–liquid transfer model has been used in Section 6.2.8 to discuss the respective roles of van der Waals interactions and burial of nonpolar surface area in protein folding (Scheme 6.1). The gas–liquid transfer model can be adapted to discuss the roles of peptide H-bonds and peptide solvation in folding. Scheme 6.2 is written for a simple folding reaction involving only a single peptide H-bond.

Here the conformation of the unfolded form U must be considered because changes in both local electrostatic free energy (E_{local}) and ESF contribute to ΔH

$U(\gamma)(g) \Leftarrow (2) \Rightarrow U(\delta)(g) \Leftarrow (3) \Rightarrow N(g)$

⇅ (1)   ⇅ (7)   ⇅ (4)

$U(\gamma)(aq) \Leftarrow (5) \Rightarrow U(\delta)(aq) \Leftarrow (6) \Rightarrow N(aq)$

Scheme 6.2

and they depend on backbone conformation. In order to form the peptide H-bond, a change in peptide backbone conformation must usually take place that involves substantial changes in E_{local} and ESF [116]. The unfolded form U(γ) normally exists as a complex equilibrium mixture of conformations and U(δ) represents the new conformation needed to make the peptide H-bond.

In principle, Scheme 6.2 may be adapted to predict the enthalpy change for forming an alanine peptide helix, whose experimental value is known [123, 124], by assigning provisional values to the different steps. Step 1, desolvating the free peptide group, has $-$ESF $= 36$ kJ mol^{-1} if the free peptide conformation is an extended β-strand (Table 6.2), and this $-$ESF value gives the contribution to ΔH because ESF and ΔH(pol) are the same within error for amides (Table 6.2). Likewise, solvating the alanine helix in step 4 can be assigned a contribution to ΔH equal to the ESF of the solvent-exposed, H-bonded peptide group, -10.5 kJ mol^{-1} [100]. In step 3, making the peptide H-bond in the gas phase, the H-bond energy has been calculated by quantum mechanics to be -28 kJ mol^{-1} for the H-bonded NMA dimer [122]. Step 3 also contains, however, an additional unknown quantity, the van der Waals interaction energy made in forming the helix backbone. In step 1 there is also an additional problem. The ESF depends not only on the appropriate mixture of backbone conformations (values for α_R, β, and P_{II} are given in Table 6.3), but also on the tendency of a peptide to bend back on itself, which reduces the exposure to solvent of the peptide groups [100]. Likewise, there are unknown – and possibly quite large – changes in E_{local} in both steps 2 and 3. Thus, the puzzle of predicting ΔH even for a simple folding reaction, like that of forming an alanine peptide helix is far from being completely understood.

Nevertheless, the gas–liquid transfer model provides the background for a simple interpretation of the enthalpy of forming the alanine helix. The observed enthalpy of helix formation, ΔH(H-C), can be written as a sum of three terms, and the enthalpy of interaction with water of the helix or of the coil can be approximated by the ESF value, as explained above.

$$\Delta H(\text{H-C}) = \Delta H(\text{hb}) + \Delta H(\text{H-W}) - \Delta H(\text{C-W}) \tag{15}$$

In Eq. (15), H refers to helix, C to coil (the unfolded peptide), W to water and hb to the peptide H-bond. ΔH(H-C) is the observed enthalpy of helix formation in water (-4.0 kJ mol^{-1}, see above). It equals ΔH(hb), the enthalpy of forming the peptide H-bond (-27.6 kJ mol^{-1} [122]), plus ΔH(H-W), the enthalpy of interaction between water and the helix (-10.5 kJ mol^{-1} (see Table 6.3), minus ΔH(C-W), the enthalpy of interaction between water and the coil (unknown, for reasons explained above). The unknown term ΔH(C-W) is found by solving the equation to be $-27.6 - 10.5 + 4.0 = -34.1$ kJ mol^{-1}. This is a reasonable value for the enthalpy of interaction between water and the coil, in view of the ESF values given in Table 6.3 for different unfolded conformations and given also that bending back of the unfolded peptide on itself is likely to reduce the negative ESF value.

Acknowledgments

I thank Franc Avbelj, David Baker, David Chandler, Ken Dill, Pehr Harbury, B.-K. Lee and John Schellman for discussion, and David Chandler, Alan Grossfield, George Makhatadze and Yaoqi Zhou for sending me their papers before publication.

Footnote added in proof

An important paper [127] was overlooked in writing section 6.2.11 on the role of van der Waals interactions in forming the hydrophobic cores of proteins. Loladze, Ermolenko and Makhatadze [127] made a set of large-to-small mutations (e.g., Val → Ala) in ubiquitin and used scanning calorimetry to measure ΔH, $T\Delta S°$ and $\Delta G°$ for mutant unfolding. In agreement with a related study by Varadarajan, Richards and coworkers [76, 77], they find that large-to small mutations are characterized chiefly by unfavorable enthalpy changes, and not by the expected unfavorable entropy changes. Surprisingly, they find favorable entropy changes for these mutants, indicating that some kind of loosening of the structure occurs, perhaps at residues lining the cavity surface. Thus, they conclude that burial of nonpolar surface area stabilizes folding primarily by a favorable enthalpy change and not by the favorable change in $T\Delta S°$ predicted by the liquid–liquid transfer model. The measurements are reported at 50 °C [127], where the favorable free energy change for burial of nonpolar surface area given by the liquid–liquid transfer model (Section [6.2.9]) is approximately 30% enthalpy and 70% $T\Delta S°$. The changes in conformational entropy that occur in large-to-small mutants [127] invalidate testing the relation between buried ASA and $\Delta G°$ for folding unless the enthalpy changes are also measured and analyzed. There is the further problem of a dependence of $\Delta\Delta G°$ on the size of the cavity, discussed by Matthews and coworkers [59]. Makhatadze and coworkers find values of ΔC_p in these experiments that are too small to be determined accurately [128], in agreement with predictions by the liquid–liquid transfer and other models.

References

1 Kauzmann, W. (1959). Factors in interpretation of protein denaturation. *Adv. Protein Chem.* **14**, 1–63.
2 Dill, K. A. (1990). Dominant forces in protein folding. *Biochemistry* **29**, 7133–7155.
3 Privalov, P. L. & Gill, S. J. (1988). Stability of protein structure and hydrophobic interaction. *Adv. Protein Chem.* **39**, 191–234.
4 Privalov, P. L. & Gill, S. J. (1989). The hydrophobic effect: a reappraisal. *Pure Appl. Chem.* **61**, 1097–1104.
5 Dill, K. A. (1990). The meaning of hydrophobicity. *Science* **250**, 297–298.
6 Schellman, J. A. (1955). The stability of hydrogen-bonded peptide structures in aqueous solution. *C.R. Trav. Lab. Carlsberg Sér Chim.* **29**, 230–259.
7 Stickle, D. F., Presta, L. G., Dill,

K. A., & Rose, G. D. (1992). Hydrogen bonding in globular proteins. *J. Mol. Biol.* **226**, 1143–1159.

8 Kendrew, J. C., Dickerson, R. E., Strandberg, B. E. et al. (1960). Structure of myoglobin. A three-dimensional Fourier synthesis at 2 Å resolution. *Nature* **185**, 422–427.

9 Pauling, L., Corey, R. B. & Branson, H. R. (1951). The structure of proteins: two hydrogen-bonded helical configurations of the polypeptide chain. *Proc. Natl Acad. Sci. USA* **37**, 205–211.

10 Tanford, C. (1980). *The Hydrophobic Effect*, 2nd edn. John Wiley & Sons, New York.

11 Tanford, C. (1997). How protein chemists learned about the hydrophobic factor. *Protein Sci.* **6**, 1358–1366.

12 Southall, N. T., Dill, K. A., & Haymet, A. D. J. (2002). A view of the hydrophobic effect. *J. Phys Chem. B* **106**, 521–533.

13 Nozaki, Y. & Tanford, C. (1971). The solubility of amino acids and two glycine peptides in aqueous ethanol and dioxane solutions. *J. Biol. Chem.* **246**, 2211–2217.

14 Fauchère, J.-L. & Pliska, V. (1983). Hydrophobic parameters π of amino-acid side chains from partitioning of N-acetyl-amino-acid amides. *Eur. J. Med. Chem.* **18**, 369–375.

15 Wimley, W. C., Creamer, T. P., & White, S. H. (1996). Solvation energies of amino acid side chains and backbone in a family of host-guest pentapeptides. *Biochemistry* **35**, 5109–5124.

16 Radzicka, A. & Wolfenden, R. (1988). Comparing the polarities of the amino acids: side-chain distribution coefficients between the vapor phase, cyclohexane, 1-octanol and neutral aqueous solution. *Biochemistry* **27**, 1644–1670.

17 Hermann, R. B. (1972). Theory of hydrophobic bonding. II. The correlation of hydrocarbon solubility in water with solvent cavity surface area. *J. Phys. Chem.* **76**, 2754–2759.

18 Chothia, C. (1974). Hydrophobic bonding and accessible surface area in proteins. *Nature* **248**, 338–339.

19 Reynolds, J. A., Gilbert, D. B., & Tanford, C. (1974). Empirical correlation between hydrophobic free energy and aqueous cavity surface area. *Proc. Natl Acad. Sci. USA* **71**, 2925–2927.

20 Lee, B. & Richards, F. M. (1971). The interpretation of protein structures: estimation of static accessibility. *J. Mol. Biol.* **55**, 379–400.

21 Sharp, K. A., Nicholls, A., Friedman, R., & Honig, B. (1991). Extracting hydrophobic free energies from experimental data: relationship to protein folding and theoretical models. *Biochemistry* **30**, 9686–9697.

22 Hermann, R. B. (1977). Use of solvent cavity area and number of packed molecules around a solute in regard to hydrocarbon solubilities and hydrophobic interactions. *Proc. Natl Acad. Sci. USA* **74**, 4144–4145.

23 Ben-Naim, A. & Marcus, Y. (1984). Solvation thermodynamics of nonionic solutes. *J. Chem. Phys.* **81**, 2016–2027.

24 Lee, B. (1991). Solvent reorganization contribution to the transfer thermodynamics of small nonpolar molecules. *Biopolymers* **31**, 993–1008.

25 Lee, B. (1995). Analyzing solvent reorganization and hydrophobicity. *Methods Enzymol.* **259**, 555–576.

26 Widom, B. (1982). Potential-distribution theory and the statistical mechanics of fluids. *J. Phys. Chem.* **86**, 869–872.

27 Jorgensen, W. L., Gao, J., & Ravimohan, C. (1985). Monte Carlo simulations of alkanes in water. Hydration numbers and the hydrophobic effect. *J. Phys. Chem.* **89**, 3470–3473.

28 Pratt, L. R. & Chandler, D. (1977). Theory of the hydrophobic effect. *J. Chem. Phys.* **67**, 3683–3704.

29 Gallichio, E., Kubo, M. M., & Levy, R. M. (2000). Enthalpy-entropy and cavity decomposition of alkane hydration free energies: Numerical results and implications for theories of hydrophobic solvation. *J. Phys. Chem. B* **104**, 6271–6285.

30 LEE, B. (1985). The physical origin of the low solubility of nonpolar solutes in water. *Biopolymers* **24**, 813–823.

31 RANK, J. A. & BAKER, D. (1998). Contributions of solvent-solvent hydrogen bonding and van der Waals interaction to the attraction between methane molecules in water. *Biophys. Chem.* **71**, 199–204.

32 TANFORD, C. (1979). Interfacial free energy and the hydrophobic effect. *Proc. Natl Acad. Sci. USA* **76**, 4175–4176.

33 POSTMA, J. P. M., BERENDSEN, H. J. C., & HAAK, J. R. (1982). Thermodynamics of cavity formation in water: a molecular dynamics study. *Faraday Symp. Chem. Soc.* **17**, 55–67.

34 HUMMER, G., GARDE, S., GARCÍA, A., POHORILLE, A., & PRATT, L. R. (1996). An information theory model of hydrophobic interactions. *Proc. Natl Acad. Sci. USA* **93**, 8951–8955.

35 DEYOUNG, L. R. & DILL, K. A. (1990). Partitioning of nonpolar solutes into bilayers and amorphous n-alkanes. *J. Phys. Chem.* **94**, 801–809.

36 CHAN, H. S. & DILL, K. A. (1997). Solvation: how to obtain microscopic energies from partitioning and solvation experiments. *Annu. Rev. Biophys. Biomol. Struct.* **26**, 425–459.

37 CHANDLER, D. (2004). Hydrophobicity: two faces of water. *Nature (London)* **417**, 491–493.

38 LUM, K., CHANDLER, D., & WEEKS, J. D. (1999). Hydrophobicity at small and large length scales. *J. Phys. Chem. B* **103**, 4570–4577.

39 LEE, B. (1994). Relation between volume correction and the standard state. *Biophys. Chem.* **51**, 263–269.

40 HUANG, D. M. & CHANDLER, D. (2000). Temperature and length scale dependence of hydrophobic effects and their possible implications for protein folding. *Proc. Natl Acad. Sci. USA* **97**, 8324–8327.

41 SOUTHALL, N. T. & DILL, K. A. (2000). The mechanism of hydrophobic solvation depends on solute radius. *J. Phys. Chem. B* **104**, 1326–1331.

42 POHORILLE, A. & PRATT, L. R. (1990). Cavities in molecular liquids and the theory of hydrophobic solubility. *J. Am. Chem. Soc.* **112**, 5066–5074.

43 BALDWIN, R. L. (1986). Temperature dependence of the hydrophobic interaction in protein folding. *Proc. Natl Acad. Sci. USA* **83**, 8069–8072.

44 GILL, S. J., DEC, S. F., OLOFSSON, G., & WADSÖ, I. (1985). Anomalous heat capacity of hydrophobic solvation. *J. Phys. Chem.* **89**, 3758–3761.

45 PAULING, L. (1960). *The Nature of the Chemical Bond*, 3rd edn, pp. 468–472. Cornell University Press, Ithaca, NY.

46 LAZARIDIS, T. (2001). Solvent size versus cohesive energy as the origin of hydrophobicity. *Acc. Chem. Res.* **34**, 931–937.

47 STELLNER, K. L., TUCKER, E. E., & CHRISTIAN, S. D. (1983). Thermodynamic properties of the benzene-phenol dimer in dilute aqueous solution. *J. Sol. Chem.* **12**, 307–313.

48 RASCHKE, T. M., TSAI, J., & LEVITT, M. (2001). Quantification of the hydrophobic interaction by simulations of the aggregation of small hydrophobic solutes in water. *Proc. Natl Acad. Sci. USA* **98**, 5965–5969.

49 RANK, J. A. & BAKER, D. (1997). A desolvation barrier to cluster formation may contribute to the rate-limiting step in protein folding. *Protein Sci.* **6**, 347–354.

50 RICHARDS, F. M. (1977). Areas, volumes. packing and protein structure. *Annu. Rev. Biophys. Bioeng.* **6**, 151–176.

51 GILL, S. J., NICHOLS, N. F., & WADSÖ, I. (1976). Calorimetric determination of enthalpies of solution of slightly soluble liquids. II. Enthalpy of solution of some hydrocarbons in water and their use in establishing the temperature dependence of their solubilities. *J. Chem. Thermodynam.* **8**, 445–452.

52 EDSALL, J. T. (1935). Apparent molal heat capacities of amino acids and other organic compounds. *J. Am. Chem. Soc.* **57**, 1506–1507.

53 GARDE, S., HUMMER, G., GARCÍA, A., PAULAITIS, M. E., & PRATT, L. R. (1996). Origin of entropy convergence

in hydrophobic hydration and protein folding. *Phys. Rev. Lett.* **77**, 4966–4968.
54 GRAZIANO, G. & LEE, B.-K. (2003). Entropy convergence in hydrophobic hydration: a scaled particle theory analysis. *Biophys. Chem.* **105**, 241–250.
55 PRIVALOV, P. L. (1979). Stability of proteins: small globular proteins. *Adv. Protein Chem.* **33**, 167–241.
56 ROBERTSON, A. D. & MURPHY, K. P. (1997). Protein structure and the energetics of protein stability. *Chem. Rev.* **97**, 1251–1267.
57 SCHELLMAN, J. A. (1997). Temperature, stability and the hydrophobic interaction. *Biophys. J.* **73**, 2960–2964.
58 PACE, C. N. (1992). Contribution of the hydrophobic effect to globular protein stability. *J. Mol. Biol.* **226**, 29–35.
59 ERIKSSON, A. E., BAASE, W. A., ZHANG, X.-J. et al. (1992). Response of a protein structure to cavity-creating mutations and its relation to the hydrophobic effect. *Science* **255**, 178–183.
60 LIVINGSTONE, J. R., SPOLAR, R. S., & RECORD, M. T. (1991). Contribution to the thermodynamics of protein folding from the reduction in water-accessible nonpolar surface area. *Biochemistry* **30**, 4237–4244.
61 MAKHATADZE, G. I. & PRIVALOV, P. L. (1993). Contribution of hydration to protein folding. I. The enthalpy of hydration. *J. Mol. Biol.* **232**, 639–659.
62 GÓMEZ, J., HILSER, V. J., XIE, DONG, & FREIRE, E. (1995). The heat capacity of proteins. *Proteins Struct. Funct. Genet.* **22**, 404–412.
63 RICHARDSON, J. M. & MAKHATADZE, G. I. (2003). Temperature dependence of the thermodynamics of the helix-coil transition. *J. Mol. Biol.*, **335**, 1029–1037.
64 KAUZMANN, W. (1987). Thermodynamics of unfolding. *Nature (London)* **325**, 763–764.
65 HUMMER, G., GARDE, S., GARCÍA, A., PAULAITIS, M. E., & PRATT, L. R. (1998). The pressure dependence of hydrophobic interactions is consistent with the observed pressure denaturation of proteins. *Proc. Natl Acad. Sci. USA* **95**, 1552–1555.
66 AKASAKA, K. (2003). Highly fluctuating protein structures revealed by variable-pressure nuclear magnetic resonance. *Biochemistry* **42**, 10875–10885.
67 OOI, T. & OOBATAKE, M. (1988). Effects of hydrated water on protein unfolding. *J. Biochem. (Tokyo)* **103**, 114–120.
68 OOBATAKE, M. & OOI, T. (1992). Hydration and heat stability effects on protein unfolding. *Prog. Biophys. Mol. Biol.* **59**, 237–284.
69 SIMONSON, T. & BRÜNGER, A. T. (1994). Solvation free energies estimated from macroscopic continuum theory: an accuracy assessment. *J. Phys. Chem.* **98**, 4683–4694.
70 CHEN, J., LU, Z., SAKON, J., & STITES, W. E. (2000). Increasing the thermostability of staphylococcal nuclease: implications for the origin of protein thermostability. *J. Mol. Biol.* **303**, 125–130.
71 HAVRANEK, J. J. & HARBURY, P. B. (2003). Automated design of specificity in molecular recognition. *Nature Struct. Biol.* **10**, 45–52.
72 NICHOLLS, A., SHARP, K. A., & HONIG, B. (1991). Protein folding and association: insights from the interfacial and thermodynamic properties of hydrocarbons. *Proteins Struct. Funct. Genet.* **11**, 281–296.
73 KLAPPER, M. H. (1971). On the nature of the protein interior. *Biochim. Biophys. Acta* **229**, 557–566.
74 HARPAZ, Y., GERSTEIN, M., & CHOTHIA, C. (1994). Volume changes on protein folding. *Structure* **2**, 641–659.
75 CHEN, J. & STITES, W. E. (2001). Packing is a key selection factor in the evolution of protein hydrophobic cores. *Biochemistry* **40**, 15280–15289.
76 VARADARAJAN, R., CONNELLY, P. R., STURTEVANT, J. M., & RICHARDS, F. M. (1992). Heat capacity changes for protein-peptide interactions in the ribonuclease S system. *Biochemistry* **31**, 1421–1426.
77 RATNAPARKHI, G. S. & VARADARAJAN, R. (2000). Thermodynamic and structural studies of cavity formation in proteins suggest that loss of

packing interactions rather than the hydrophobic effect dominates the observed energetics. *Biochemistry* **39**, 12365–12374.

78 ZHOU, H. & ZHOU, Y. (2002). Stability scale and atomic solvation parameters extracted from 1023 mutation experiments. *Proteins Struct. Funct. Genet.* **49**, 483–492.

79 ZHOU, H. & ZHOU, Y. (2003). Quantifying the effect of burial of amino acid residues on protein stability. *Proteins Struct. Funct. Genet.* **54**, 315–322.

80 SCHELLMAN, J. A. (1955). The thermodynamics of urea solutions and the heat of formation of the peptide hydrogen bond. *C.R. Trav. Lab. Carlsberg, Sér. Chim.* **29**, 223–229.

81 KLOTZ, I. M. & FRANZEN, J. S. (1962). Hydrogen bonds between model peptide groups in solution. *J. Am. Chem. Soc.* **84**, 3461–3466.

82 SUSI, H., TIMASHEFF, S. N., & ARD, J. S. (1964). Near infrared investigation of interamide hydrogen bonding in aqueous solution. *J. Biol. Chem.* **239**, 3051–3054.

83 EPAND, R. M. & SCHERAGA, H. A. (1968). The influence of long-range interactions on the structure of myoglobin. *Biochemistry* **7**, 2864–2872.

84 TANIUCHI, H. & ANFINSEN, C. B. (1969). An experimental approach to the study of the folding of staphylococcal nuclease. *J. Biol. Chem.* **244**, 3864–3875.

85 KLEE, W. A. (1968). Studies on the conformation of ribonuclease S-peptide. *Biochemistry* **7**, 2731–2736.

86 BROWN, J. E. & KLEE, W. A. (1971). Helix-coil transition of the isolated amino-terminus of ribonuclease. *Biochemistry* **10**, 470–476.

87 BIERZYNSKI, A., KIM, P. S., & BALDWIN, R. L. (1982). A salt bridge stabilizes the helix formed by the isolated C-peptide of RNase A. *Proc. Natl Acad. Sci. USA* **79**, 2470–2474.

88 BALDWIN, R. L. (1995). α-Helix formation by peptides of defined sequence. *Biophys. Chem.* **55**, 127–135.

89 MUÑOZ, V. & SERRANO, L. (1994). Elucidating the folding problem of helical peptides using emprical parameters. *Nature Struct. Biol.* **1**, 399–409.

90 AURORA, R. & ROSE, G. D. (1998). Helix capping. *Protein Sci.* **7**, 21–38.

91 MARQUSEE, M., ROBBINS, V. H., & BALDWIN, R. L. (1989). Unusually stable helix formation in short alanine-based peptides. *Proc. Natl Acad. Sci. USA* **86**, 5286–5290.

92 ROHL, C. A., CHAKRABARTTY, A., & BALDWIN, R. L. (1996). Helix propagation and N-cap propensities of the amino acids measured in alanine-based peptides in 40 volume percent trifluoroethanol. *Protein Sci.* **5**, 2623–2637.

93 DYSON, H. J., CROSS, K. J., HOUGHTEN, R. A., WILSON, I. A., WRIGHT, P. E., & LERNER, R. A. (1985). The immunodominant site of a synthetic immunogen has a conformational preference in water for a type-II reverse turn. *Nature (London)* **318**, 480–483.

94 DYSON, H. J., RANCE, M., HOUGHTEN, R. A., LERNER, R. A., & WRIGHT, P. E. (1988). Folding of immunogenic peptide fragments of proteins in water solution. I. Sequence requirements for the formation of a reverse turn. *J. Mol. Biol.* **201**, 161–200.

95 BLANCO, F. J., RIVAS, G., & SERRANO, L. (1995). A short linear peptide that folds into a stable native β-hairpin in aqueous solution. *Nature Struct. Biol.* **1**, 584–590.

96 LIFSON, S., HAGLER, A. T., & DAUBER, P. (1979). Consistent force field studies of hydrogen-bonded crystals. 1. Carboxylic acids, amides, and the C=O•••H– hydrogen bonds. *J. Am. Chem. Soc.* **101**, 5111–5121.

97 WOLFENDEN, R. (1978). Interaction of the peptide bond with solvent water: a vapor phase analysis. *Biochemistry* **17**, 201–204.

98 WOLFENDEN, R., ANDERSSON, L., CULLIS, P. M., & SOUTHGATE, C. C. B. (1981). Affinities of amino acid side chains for solvent water. *Biochemistry* **20**, 849–855.

99 SITKOFF, D., SHARP, K. A., & HONIG, B. (1994). Accurate calculation of

hydration free energies using macroscopic solvent models. *J. Phys. Chem.* **98**, 1978–1988.

100 AVBELJ, F., LUO, P., & BALDWIN, R. L. (2000). Energetics of the interaction between water and the helical peptide group and its role in determining helix propensities. *Proc. Natl Acad. Sci. USA* **97**, 10786–10791.

101 EISENBERG, D. & McLACHLAN, A. D. (1986). Solvation energy in protein folding and binding. *Nature (London)* **319**, 199–203.

102 DUNITZ, J. D. (1994). The entropic cost of bound water in crystals and biomolecules. *Science* **264**, 670.

103 RASHIN, A. A. & HONIG, B. (1985). Reevaluation of the Born model of ion hydration. *J. Phys. Chem.* **89**, 5588–5593.

104 FERSHT, A. R., SHI, J.-P., KNILL-JONES, J. et al. (1985). Hydrogen bonding and biological specificity analysed by protein engineering. *Nature (London)* **314**, 235–238.

105 FERSHT, A. R. (1987). The hydrogen bond in molecular recognition. *Trends Biochem. Sci.* **12**, 301–304.

106 BALDWIN, R. L. (2003). In search of the energetic role of peptide H-bonds. *J. Biol. Chem.* **278**, 17581–17588.

107 MITCHELL, J. B. O. & PRICE, S. L. (1991). On the relative strengths of amide•••amide and amide•••water hydrogen bonds. *Chem. Phys. Lett.* **180**, 517–523.

108 GALISON, P. (1997). *Image and Logic*, p. 68, University of Chicago Press, Chicago.

109 BORN, M. (1920). Volumen und Hydrationswärme der Ionen. *Z. Physik* **1**, 45–48.

110 GROSSFIELD, A., REN, P., & PONDER, J. W. (2003). Ion solvation thermodynamics from simulation with a polarizable force field. *J. Am. Chem. Soc.* **125**, 15671–15682.

111 KIRKWOOD, J. G. & WESTHEIMER, F. H. (1938). The electrostatic influence of substituents on the dissociation constants of organic acids. *J. Chem. Phys.* **6**, 506–512.

112 RAJASEKARAN, E., JAYARAM, B., & HONIG, B. (1994). Electrostatic interactions in aliphatic dicarboxylic acids: a computational route to the determination of pK shifts. *J. Am. Chem. Soc.* **116**, 8238–8240.

113 FLORIAN, J. & WARSHEL, A. (1997). Langevin dipoles model for ab initio calculations of chemical processes in solution: parameterization and application to hydration free energies of neutral and ionic solutes and conformational analysis in aqueous solution. *J. Phys. Chem. B* **101**, 5583–5595.

114 BRANT, D. A. & FLORY, P. J. (1965). The configuration of random polypeptide chains. II. Theory. *J. Am. Chem. Soc.* **87**, 2791–2800.

115 BRANT, D. A. & FLORY, P. J. (1965). The role of dipole interactions in determining polypeptide configurations. *J. Am. Chem. Soc.* **87**, 663–664.

116 AVBELJ, F. & MOULT, J. (1995). Role of electrostatic screening in determining protein main chain conformational preferences. *Biochemistry* **34**, 755–764.

117 AVBELJ, F. & FELE, L. (1998). Role of main-chain electrostatics, hydrophobic effect and side-chain conformational entropy in determining the secondary structure of proteins. *J. Mol. Biol.* **279**, 665–684.

118 AVBELJ, F. (2000). Amino acid conformational preferences and solvation of polar backbone atoms in peptides and proteins. *J. Mol. Biol.* **300**, 1335–1359.

119 CREAMER, T. P. & ROSE, G. D. (1994). α-Helix-forming propensities in peptides and proteins. *Proteins Struct. Funct. Genet.* **19**, 85–97.

120 LUO, P. & BALDWIN, R. L. (1999). Interaction between water and polar groups of the helix backbone: an important determinant of helix propensities. *Proc. Natl Acad. Sci. USA* **96**, 4930–4935.

121 AVBELJ, F. & BALDWIN, R. L. (2002). Role of backbone solvation in determining thermodynamic β propensities of the amino acids. *Proc. Natl Acad. Sci. USA* **99**, 1309–1313.

122 BEN-TAL, N., SITKOFF, D., TOPOL, I. A., YANG, A.-S., BURT, S. K., & HONIG, B.

(1997). Free energy of amide hydrogen bond formation in vacuum, in water, and in liquid alkane solution. *J. Phys. Chem. B* **101**, 450–457.
123 LOPEZ, M. M., CHIN, D.-H., BALDWIN, R. L., & MAKHATADZE, G. I. (2002). The enthalpy of the alanine peptide helix measured by isothermal titration calorimetry using metal-binding to induce helix formation. *Proc. Natl Acad. Sci. USA* **99**, 1298–1302.
124 GOCH, G., MACIEJCZYK, OLESZCZUK, M., STACHOWIAK, D., MALICKA, J., & BIERZYNSKI, A. (2003). Experimental investigation of initial steps of helix propagation in model peptides. *Biochemistry* **42**, 6840–6847.
125 MYERS, J. K. & PACE, C. N. (1996). Hydrogen bonding stabilizes globular proteins. *Biophys. J.* **71**, 2033–2039.
126 JAYASINGHE, S., HRISTOVA, K., & WHITE, S. H. (2001). Energetics, stability and prediction of transmembrane helices. *J. Mol. Biol.* **312**, 927–934.
127 LOLADZE, V. V., ERMOLENKO, D. N., & MAKHATADZE, G. I. (2002). Thermodynamic consequences of burial of polar and nonpolar amino acid residues in the protein interior. *J. Mol. Biol.* **320**, 343–357.
128 LOLADZE, V. V., ERMOLENKO, D. N., & MAKHATADZE, G. I. (2001). Heat capacity changes upon burial of polar and nonpolar groups in proteins. *Protein Sci.* **10**, 1343–1352.

7
Electrostatics of Proteins: Principles, Models and Applications

Sonja Braun-Sand and Arieh Warshel

7.1
Introduction

The foundation of continuum (classical) electrostatic theory was laid in the early eighteenth century and formulated rigorously by the Maxwell equations. Unfortunately, this formal progress and the availability of analytical solutions of a few simple cases have not provided proper tools for calculations of electrostatic energies in proteins. Here one is faced with electrostatic interactions in microscopically small distances, where the concepts of a dielectric constant are problematic at best. Furthermore, the irregular shape of protein environments made the use of analytical models impractical. Nevertheless, the realization that some aspects of protein action must involve electrostatic interactions led to the emergence of simplified phenomenological models [1, 2] that could provide in some cases a useful insight (see below).

The availability of X-ray structures of proteins and the search for concrete quantitative results led to the emergence of modern, microscopically based studies of electrostatic effects in proteins (see below). These studies and subsequent numerical continuum studies led to the gradual realization that the electrostatic energies provide by far the best structure function correlator for proteins and other macromolecules (Section 7.6). However, the overwhelming impact of macroscopic concepts still leads to major confusions in the field and to oversimplified assumptions. Here the problem can be reduced by focusing on relevant validation studies (Section 7.5).

This review will consider the above issues and try to lead the reader from concepts to models and to applications, pointing out pitfalls and promising directions. General background material can also be found in the following reviews [3–9].

7.2
Historical Perspectives

The basis for all electrostatic theories is Coulomb's law, which expresses the reversible work of bringing two charges (Q_i and Q_j) together by:

Protein Folding Handbook. Part I. Edited by J. Buchner and T. Kiefhaber
Copyright © 2005 WILEY-VCH Verlag GmbH & Co. KGaA, Weinheim
ISBN: 3-527-30784-2

$$\Delta W = 332 Q_i Q_j / r_{ij} \tag{1}$$

Where the free energy is given in kcal mol^{-1}, the distance in Å, and Q in atomic units (au). Manipulations of this led to the general macroscopic equation [10]:

$$\nabla \mathbf{E} = 4\pi\rho \tag{2}$$

where **E** is the macroscopic electric field and ρ is the charge density ($\rho(\mathbf{r}) = \sum_i Q_i(\delta(\mathbf{r} - \mathbf{r}_i))$). Further manipulation, using the fact that the field **E** can be expressed as a gradient of a scalar potential, U, gives:

$$\mathbf{E} = -\nabla U \tag{3}$$

which leads to the Poisson equation:

$$\nabla^2 U(r) = -4\pi\rho(r) \tag{4}$$

Assuming that the effects which are not treated explicitly can be represented by a dielectric constant, ε, leads to:

$$\nabla \varepsilon(r) \nabla U(r) = -4\pi\rho(\mathbf{r}) \tag{5}$$

When ions are presented in the solution around our system and when the ion distribution is described by the Boltzmann distribution we obtain rather some approximation, the linearized Poisson–Boltzmann (PB) equation:

$$\nabla \varepsilon(\mathbf{r}) \nabla U(\mathbf{r}) = -4\pi\rho(\mathbf{r}) + \kappa^2 U \tag{6}$$

where κ is the Debye–Hückel screening parameter. The major problem with these "rigorous" equations is their physical foundation. Not only that the continuum assumptions are very doubtful on a molecular level, but the nature of the dielectric constant in this equation is far from being obvious when applied to heterogeneous environments. Nevertheless, the early history of protein electrostatics has been marked by the emergence of phenomenological models [2, 11]. Perhaps the most influential model was the Tanford Kirkwood (TK) model [2] that describes the protein as a sphere with a uniform dielectric constant. The validity of this model was not obvious since it was formulated before the elucidation of the structure of proteins. However, subsequent analysis by Warshel and coworkers [12] indicates that the model ignores one of the most important physical aspects of protein electrostatics, the self-energy of the charged groups (the energy of forming these charges in the given protein site). This problem was not obvious at the time of the formulation of the TK model, when it was thought that the ionizable groups reside on the surface of the protein). However, this assumption was found to be incorrect in key cases. Nevertheless, the TK model is repeatedly invoked as a physically reasonable model, where only the assumption of a spherical shape is considered as a rough approximation. Interestingly, even today we find cases where the self-energy term is neglected [13]. Another fundamental problem of the early phenomenological models was the selection of the protein dielectric constant. This problem remains a major issue in modern continuum models (see Section 7.4).

The emergence of modern electrostatic models can probably be traced to the work of Warshel and Levitt [14]. In this work it was realized that the only way to reach clear conclusions about enzyme catalysis is to obtain quantitative estimates of the corresponding electrostatic energies. This meant that any concept about the dielectric constant of proteins would generate nonunique conclusions about the origin of enzyme catalysis. The best way to overcome this problem was to abandon the concept of a dielectric constant and to develop a simplified microscopic model that included all feasible microscopic contributions. The solvent was modeled by an explicit grid of Langevin dipoles (LD) and the protein was modeled explicitly, taking into account its permanent and induced dipoles, as well as a limited reorganization (by energy minimization).

The resulting protein dipoles Langevin dipoles (PDLD) model was the first model to capture correctly the physics of protein electrostatics. In fact, the microscopic perspective of the PDLD microscopic model has been the best guarantee against the traps of the continuum concepts (see below).

The introduction of discretized continuum (DC) models [4, 15, 16] and the inherent respect to the authoritative continuum equations (e.g., Eqs (2)–(6)) led to a wide acceptance of DC models. In fact, many workers have no problem accepting the questionable continuum description while questioning more microscopic models (see Section 7.3.2). This trend was not slowed down by the identification of major conceptual problems that were pointed out and resolved by the use of microscopic models [3, 7, 12, 17]. More specifically, early DC models overlooked self-energy, ignored the key role of the protein permanent dipoles [17] and used unjustified dielectric constants [3, 19]. However, the so-called continuum models kept evolving and becoming more and more microscopic and in many respects converging to the PDLD and its semi-macroscopic version, the PDLD/S-LRA model (see below). At present we have a wide range of almost reasonable electrostatic models, but there is still major ignorance about the nature and the meaning of different aspects of these models (e.g., the dielectric constant). Some of these issues will be clarified in Sections 7.3 and 7.4. At any rate, it is important to realize here that almost any conceptual problem of the continuum models have been resolved and clarified by the use of microscopic concepts. Perhaps the best example is the elucidation of the nature of the protein dielectric constant [3, 19].

Early, insightful observations of Perutz [20] and others (see, for example, Refs [21–23]) have indicated that electrostatic interactions must play an important role in biology. However, it was the development of computational methods and their successful applications that led to the realization that electrostatic energies provide the best structure–function correlation for proteins [18]. This has been demonstrated in studies of a wide range of properties that will be considered in Section 7.6. However, in some cases it took major effort to convince the scientific community of the importance of electrostatic contributions to different effects. For example, some workers invoked entropic and dynamic effects as crucial factors, even in cases of charge formation, such as redox and ionization processes. This assumption may have reflected unfamiliarity with the fact that electrostatic free energies involve entropy changes, and that averaging over configurations is an integral part of the proper evaluation of electrostatic free energies rather than an inherent dynamic effect. In some important cases, such as enzyme reactions, it was impossible to assess the importance of electrostatic effects without developing quantitative models. Here the use of macroscopic estimates led repeatedly to the conclusion that electrostatic effects cannot play a major role in catalysis (see, for example, Ref. [24]).

7.3
Electrostatic Models: From Microscopic to Macroscopic Models

Although the progress in studies of electrostatic effects is sometimes taken to be synonymous with that in continuum approaches [25], in fact a wide range of modeling approaches exist (Figure 7.1), each with its own scope and limitations. This includes, of course, the aforementioned difficulty of analyzing continuum results without the use of microscopic models. Thus, it is important to understand the relationship between various approaches, spanning a range from continuum dielectrics to all-atom classical representations and quantum mechanical models. For example, there is a surprising degree of confusion regarding dipolar models that bridge the all-atom and continuum dielectric models. Below we will consider the key classes of electrostatic models and discuss their current scope and limitations.

7.3.1
All-Atom Models

An all-atom model describes a system as a collection of particles that interact via a quantum mechanical potential surface or the corresponding classical force field. Although such approaches can be quite rigorous, adequate convergence may sometimes not be achieved even with nanosecond simulations, particularly for macromolecules. One of the most significant "unresolved" issues in atomistic simulations is the proper representation of an infinite system, which is critical for the proper treatment of long-range electrostatic effects. Periodic boundary conditions, although appropriate in nonpolar systems for nonelectrostatic problems or in molecular crystals [26], cannot produce truly infinite noncrystalline systems without artifacts when long-range electrostatic interactions are significant. It is only now becoming commonly appreciated that the Ewald method for periodic boundary conditions gives divergent solvation energies for the same solute depending on the size of the simulated system [27]. Correction formulas for specific charge distributions [28] do not provide a general solution (see discussion in Ref. [29]). As demonstrated elsewhere [29–31], problems do not exist in simulations that use spherical boundary conditions and utilize the local reaction field (LRF) approach [32].

7.3 Electrostatic Models: From Microscopic to Macroscopic Models

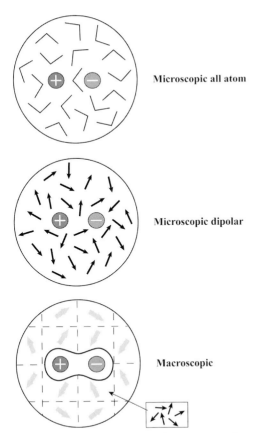

Fig. 7.1. The three main options for representing the solvent in computer simulation approaches. The microscopic model uses detailed all-atom representation and evaluates the average interaction between the solvent residual charges and the solute charges. Such calculations are expensive. The simplified microscopic model replaces the time average dipole of each solvent molecule by a point dipole, while the macroscopic model is based on considering a collection of solvent dipoles in a large volume element as a polarization vector.

Spherical boundary conditions offer an obvious alternative that is more physically consistent than the Ewald methods. There is no spurious periodicity; a spherical region of atomic representation is surrounded by a simplified region. Because the simplified outer region is far from the focal point of the simulation, its effect can be made to be indistinguishable from that of an atomistic representation. The issues involved in constructing proper boundaries for spherical models are logical and obvious. The most important requirement is that the system will maintain proper polarization at and around the boundaries [33]. This requirement, which is far more important than the physically trivial requirement of maintaining proper heat flow by stochastic boundaries [34], has been solved by the surface constraint all-atom solvent (SCAAS) model [35].

It is important to reemphasize here that the problem is not the well-known bulk contribution from the surrounding of the simulation sphere [36], which has been implemented repeatedly in simulation studies (see, for example, Refs [37] and [38]), but the polarization at the microscopic boundaries of the simulation sphere.

With reasonable boundary conditions and proper long-range treatment, one may use all-atom models for evaluation of electrostatic free energies. Here the most obvious approach is the free energy perturbation (FEP) approach [39, 40] which involves a gradual change of the solute charge from zero to its actual value $Q = Q_0$ using a mapping potential of the form (see, for example, Refs [39] and [41]):

$$V_m = V(Q=0)(1-\lambda_m) + V(Q_0)\lambda_m \tag{7}$$

and then evaluating the corresponding free energy change by:

$$\exp\{-\Delta G(\lambda_m \to \lambda_{m+1})\beta\} = \langle \exp\{-(V_{m+1}-V_m)\beta\}\rangle_{V_m} \tag{8}$$

where $\langle\ \rangle_{V_m}$ designates a molecular dynamics (MD) average over V_m and $\beta = 1/k_B T$. A very useful alternative approach is the linear response approximation (LRA) where the free energy is given by [42]:

$$\Delta G(Q=0 \to Q_0) = \tfrac{1}{2}[\langle V(Q_0) - V(Q=0)\rangle_{V(Q_0)}$$
$$+ \langle V(Q_0) - V(Q=0)\rangle_{V(Q=0)}] \tag{9}$$

Although the use of the FEP and LRA approaches is very appealing, it still involves major convergence problems and a relatively small number of systematic studies (see, for example, Refs [39] and [43–45]).

7.3.2
Dipolar Lattice Models and the PDLD Approach

The most natural simplification of all-atom polar models involves the representation of molecules or groups as dipoles. Dipolar models have a long history in the theory of polar solvents and in providing the fundamental microscopic basis of electrostatic theories. Simulations with sticky dipoles [46] and formal studies [47] addressing the origin of the dipolar representation have been recently reported. The nature of the Brownian dipole lattice (BDL) and Langevin dipole (LD) models and their relationship to continuum models has been established [48]. This study indicated that continuum solvation models are the infinite dipole density limit of a more general dipolar representation, and that their linearity is a natural consequence of being at that limit (although the discreteness of dipole lattice models is not equal to that of continuum models [49]). It was also found that the linearity of a continuum model is a direct consequence of being the infinite density limit of a dipolar model. An instructive conclusion of the above studies is the finding that

7.3 Electrostatic Models: From Microscopic to Macroscopic Models

the solvation behavior of the LD model is identical to that of the more rigorous BDL model. This is significant since a version of the LD model, the protein dipoles Langevin dipoles (PDLD) model, has been used extensively to treat electrostatics in macromolecules [3]. PDLD is probably the first to properly treat macromolecular electrostatics.

At this point it might be useful to readdress the criticism of the PDLD model by some workers who have little difficulties accepting macroscopic models as a realistic description of macromolecules. Some of this criticism involved selective citation of incorrect arguments (see discussion in footnote 21 of Ref. [43]). Unfortunately the criticism continues despite the obvious conclusions of repeated rigorous studies of dipolar lattices [49] and related models (for example, see the recent excellent study of Borgis and coworkers [50]). An instructive example is the suggestion [51] that the PDLD model of Ref. [14] did not evaluate self-energies and that this model is in fact macroscopic in nature and that it is physically equivalent to the use of a finite-difference grid within DC methods. These assertions are incorrect both in presentation of facts and in interpretation of the physics of dipolar solvents. It should be clear by now that a grid of dipoles with finite spacing or the equivalent system of polar molecules on a lattice is not equivalent to the numerical grid used in evaluating the electrostatic potential in DC approaches (see the clear analysis of footnote 26 of Ref. [43]).

Another more recent example is the attempt to support the argument that the LD and DC models are equivalent, which was advanced by Ref. [8]. This work stated "the LD model is not only analogous to the continuum model (as stated many times in the literature), but identical to the continuum model in its discrete approximation," continuing "to our knowledge this rigorous equivalence was only pointed out very recently [50]." Not only does this reflect a clear misunderstanding of the conclusions of Ref. [50], but it also overlooks the rigorous finding of Ref. [49], where it was found that the LD and DC models have different structural features even at the *limit* of infinitely small spacing. Only with a regular cubic grid of linearized Langevin dipoles and with a Clausius-Mossotti based local polarizability one can move, at the limit of infinitely small spacing, from the LD to the DC limit.

In considering the difference between the LD and continuum philosophies, it should be realized that the original LD model, with its finite spacing, has always been a simplified microscopic model. Insisting on the clear physical features of the model (i.e., the clear identification of each grid point as a discrete particle) has been the reason for the fact that the PDLD model did not fall into all the traps that plagued continuum studies of proteins. For example, looking at the microscopic interactions led immediately to the explicit considerations of the protein permanent dipoles rather than to the futile attempt of representing them by a low dielectric constant. At any rate, saying that the LD model (or other dipolar model) is equal to a continuum model is like saying that an all-atom model is equal to a continuum model. Finally, if the reader is still confused at this point by the suggestion that the LD model is a continuum model [8, 51], it is useful to point out that macroscopic models involve scaling by a dielectric constant but no such scaling is used in the PDLD model.

7.3.3
The PDLD/S-LRA Model

The PDLD model and other microscopic models do not involve any dielectric constant. While this is a conceptual strength it is also sometimes a practical "weakness" since they involve the balance between very large contributions, which is hard to obtain without a perfect convergence. In some cases it is possible to assume that the balance must exist and to represent this balance by scaling the results with a dielectric constant (see Section 7.4). Such a philosophy leads to semi-macroscopic models, which are obviously more stable (but not necessarily more reliable) than the corresponding microscopic models. A promising way to exploit the stability of the semi-macroscopic models, and yet to keep a clear physical picture, is the semi-macroscopic version of the PDLD model (the PDLD/S model [52]). The PDLD/S model takes the PDLD model and scales the corresponding contributions by assuming that the protein has a dielectric constant ε_p. Now, in order to reduce the unknown factors in ε_p it is useful to move to the PDLD/S-LRA model, where the PDLD/S energy is evaluated within the LRA approximation with an average on the configurations generated by MD simulation of the charged and the uncharged states (see Refs [19, 37, 43]). In this way one uses MD simulations to automatically generate protein configurations for the charged and uncharged forms of the given "solute" and then average these contributions according to Eq. (9). This treatment expresses the free energy of moving a charge from water to the protein active site, which is given by [42]:

$$\Delta\Delta G_{\text{sol}}^{\text{w}\rightarrow\text{p}} = \tfrac{1}{2}[\langle \Delta U^{\text{w}\rightarrow\text{p}}(Q=0 \rightarrow Q=Q_0) \rangle_{Q_0} + \langle \Delta U^{\text{w}\rightarrow\text{p}}(Q=Q_0) \rangle_{Q=0}] \quad (10)$$

where

$$\Delta U^{\text{w}\rightarrow\text{p}} = [\Delta\Delta G_{\text{p}}^{\text{w}}(Q=0 \rightarrow Q=Q_0) - \Delta G_{Q}^{\text{w}}]\left(\frac{1}{\varepsilon_p} - \frac{1}{\varepsilon_w}\right) + \Delta U_{Q\mu}^{\text{p}}(Q=Q_0)\frac{1}{\varepsilon_p} \quad (11)$$

Here ΔG_{Q}^{w} is the solvation energy of the given charge in water, $\Delta\Delta G_{\text{p}}^{\text{w}}$ is the change in the solvation energy of the entire protein (plus its bound charge) upon changing this charge from its actual value (Q_0) to zero. $\Delta U_{Q\mu}$ is the microscopic electrostatic interaction between the charge and its surrounding protein polar groups (for simplicity we consider the case when the interaction with the protein ionizable groups is treated in a separate cycle (see, for example, Ref. [43]) with the effective dielectric constant ε_{eff} of Section 7.3.5. Now we bring here the explicit expression of Eq. (11), since to the best of our knowledge it provides the clearest connection between microscopic quantities such as $\Delta\Delta G_{\text{p}}^{\text{w}}$ and $\Delta U_{Q\mu}$ and semi-macroscopic treatments. In fact, the PDLD/S treatment without the LRA treatment converges to the PB treatment and thus provides the clearest connection between PB treatments and microscopic treatments, allowing one to demonstrate the exact nature of ε_{eff} (see, for example, Ref. [19] and Section 7.4). At any rate, it is important to keep in mind that the LRA considerations of Eq. (10) treat the protein reorganization explicitly, and this reduces the uncertainties about the magnitude of ε_p. It is also important to mention that the recently introduced molecular mechanics PB surface area (MM-PBSA) model [53] is basically an adaptation of the PDLD/S-LRA idea of MD generation of configurations for implicit solvent calculations, while only calculating the average over the configurations generated with the charged solute (the first term of Eq. (9)).

7.3.4
Continuum (Poisson-Boltzmann) and Related Approaches

Continuum solvent models have become a powerful tool for calculations of solvation energies of large solutes in solutions. Here one solves the Poisson equation or related equations [54] numerically [15, 16]. However, these treatments need more than the solvent dielectric, ε_s, to be useful. Solvation energy depends on the chosen solute cavity radius much more strongly than it does on ε_s. In addition to solute size, the cavity radius must absorb the effect of local structure near the solute. Consistent studies [49] showed that the cavity radius depends on the degree of solvent discreteness, even though the solute "size," as defined by an exclusion radius, is invariant. Also, a "cavity-in-a-continuum" description of solvation is physically inconsistent for discrete (i.e., real) solvents, leading to Born radii that depend on purely nonstructural factors [49, 55]. Thus, one cannot treat the cavity size as a transferable parameter that can be used in different dielectrics or in a given dielectric under even moderately different conditions.

PB approaches for studies of protein electrostatics represent the solvent and the protein by two dielectric constants (ε_p and ε_s, respectively) and solve numerically Eqs (5) or (6). Although these approaches became extremely popular, they involve significant conceptual and practical problems. Purely macroscopic (continuum) models are entirely inadequate for proteins because they miss the crucial local polarity defined largely by the net orientations of protein permanent dipoles rather than by a dielectric constant [7, 52]. The effect of these dipoles (sometimes called "back field"; see Ref. [43]) is now widely appreciated. Thus, most current "continuum" models are no longer macroscopic but semi-macroscopic and therefore potentially reliable. Their reliability depends, however, on properly selecting the "dielectric constant" ε_p (see below) for the relevant protein region. Despite recent suggestions that the reliability of continuum methods can be increased by configurational averaging [19], such averaging still does not capture the crucial protein reorganization effect (see Section 7.4).

The popularity of continuum models may be due to the belief that the formulation of the PB equation must reflect some fundamental physics and that the criticism of such models only address minor points that can be "fixed" in one way or another. The discussion of conceptually trivial issues such as the linear approximation of the second term in Eq. (6) often created the feeling that this is the true problem in such models. In fact, the real unsolved problem is the fact that the results depend critically on ε_p, whose value cannot be determined uniquely (see Section 7.4). Another reason for the acceptance of PB models has been the use of improper validation, addressing mainly trivial cases of surface groups where all macroscopic models would give the same results (see Section 7.5). At any rate, PB models (see, for example, Ref. [56]) are becoming more and more microscopic, borrowing ideas from the semi-macroscopic PDLD/S-LRA model.

7.3.5
Effective Dielectric Constant for Charge–charge Interactions and the GB Model

Electrostatic energies in proteins can be formulated in terms of a two-step thermodynamic cycle. The first step involves the transfer of each ionized group from water to its specific protein site where all other groups are neutral, and a second step where the interaction between the ionized groups is turned on [19]. Thus, we can write the energy of moving ionized groups from water to their protein site as:

$$\Delta G = \sum_{i=1}^{N} \Delta G_i + \sum_{i>j} \Delta G_{ij} \tag{12}$$

where the first and second terms correspond to the first and second steps, respectively. ΔG_i, which represents the change in self-energies of the i-th ionized group upon transfer from water to the protein site, can be evaluated by the PDLD/S-LRA approach with a relatively small ε_p. However, the charge–charge interaction terms (the ΔG_{ij}) are best reproduced by using a simple macroscopic Coulomb's-type law:

$$\Delta G_{ij} = 332 Q_i Q_j / r_{ij} \varepsilon_{ij} \tag{13}$$

where ΔG is given in kcal mol^{-1}, r_{ij} in Å and ε_{ij} is a relatively high distance-dependent dielectric constant [3, 12]. The dielectric constant ε_{ij} can be described by (e.g., Ref. [12]):

$$\varepsilon(r_{ij}) = 1 + \varepsilon'[1 - \exp(-\mu r_{ij})] \tag{14}$$

Although the fact that ε_{ij} is large has been supported by mutation experiments (e.g., [57, 58]) and conceptual considerations [3], the use of Eq. (13) is still considered by many as a poor approximation for PB treatments (without realizing that the PB approach depends entirely on an unknown ε_p). Interestingly, the so-called generalized Born (GB) model, whose usefulness is now widely appreciated [59], is basically a combination of Eq. (13) and the Born energy of the individual charges. Thus, the GB model is merely a version of an earlier treatment [3]. More specifically, as pointed out originally by Warshel and coworkers [3, 60], the energy of an ion pair in a uniform dielectric medium can be written as [3]:

$$\Delta G_{ij} = 332 Q_i Q_j / r_{ij} + \Delta G_{\text{sol}} = \Delta G_{\text{sol}}^{\infty} + 332 Q_i Q_j / r_{ij} \varepsilon \tag{15}$$

where the free energies are given in kcal mol^{-1} and the distances in Å. ΔG_{sol} is the solvation energy of the ions and ΔG_{sol}^{∞} is the solvation of the ions at infinite separation. Equation (15) gives (see also Ref. [3]):

$$\Delta G_{sol} = \Delta G_{sol}^{\infty} - (332 Q_i Q_j / r_{ij})(1 - 1/\varepsilon)$$
$$= -166[(Q_i^2/a_i + Q_j^2/a_j) + (2 Q_i Q_j / r_{ij})](1 - 1/\varepsilon) \quad (16)$$

where the a_i and a_j are the Born's radii of the positive and negative ions, respectively. Equation (16) with some empirical modifications leads exactly to the GB treatment [61]. Thus, the widely accepted GB treatment is in fact a glorified Coulomb's law treatment [7]. Nevertheless, it is instructive to note that the GB approximation can also be obtained by assuming a *local* model where the vacuum electric field, $\mathbf{E}_0$, and the displacement vector, $\mathbf{D}$, are identical [62]. This model, which can be considered as a "local Langevin Dipole model" [50] or as a version of the noniterative LD model, allows one to approximate the energy of a collection of charge and to obtain (with some assumptions about the position dependence of the dielectric constant) the GB approximation.

It should be noted that the simple derivations of the GB approximation are valid for homogeneous media with a high dielectric (e.g., the solvent around the protein). This can be problematic when one considers models where the protein has a low dielectric constant and when we are dealing with transfer of charges from water to the protein environment. Fortunately, it is reasonable to use a large ε for the second term in Eq. (15), even for protein interiors (see discussion of ε_{ij} of Eq. (14)). Here again, the main issue is the validity of Eq. (13) and not its legitimization by the seemingly "rigorous" GB formulation. It is also important to realize that a consistent treatment should involve the replacement of ΔG_{sol}^{∞} by the ΔG^+ and ΔG^- of the ions in their specific protein site when ε_p is different from ε_w.

7.4
The Meaning and Use of the Protein Dielectric Constant

As demonstrated in the validation studies of Ref. [19], the value of the optimal dielectric constant depends drastically on the model used. This finding might seem a questionable conclusion to readers who are accustomed to the view that macroscopic models have a universal meaning. This might also look strange to some workers who are experienced with microscopic statistical mechanics, where ε is evaluated in a unique way from the fluctuations of the total dipole moment of the system (see below). Apparently, even now, it is not widely recognized that ε_p has little to do with what is usually considered as the protein dielectric constant. That is, as was pointed out and illustrated by the conceptual analysis of Warshel and Russell [3], the value of the dielectric constant of proteins is entirely dependent on the method used to define this constant and on the model used. Subsequent studies [5, 19, 63] clarified this fact and established its validity. As stated in these studies the dependence of ε_p on the model used can be best realized by considering different limiting cases. That is, when all the interactions are treated explicitly, $\varepsilon = 1$; when all but induced dipoles are included explicitly, $\varepsilon = 2$ [3]; and when the solvent is not included explicitly, $\varepsilon > 40$ (although this is a very bad model). When the protein permanent dipoles are included explicitly but their relaxation (the protein reorganization) and the protein induced dipoles are included implicitly, the value of ε_p is not well defined. In this case ε_p should be between 4 and 6 for dipole–charge interactions and >10 for charge–charge interactions [64, 65].

One may clearly argue that there is only one "proper" dielectric constant, which is the one obtained from the fluctuation of the average dipole moment [6, 63, 66–68]. However this protein dielectric constant ($\varepsilon = \bar{\varepsilon}$) is not a constant because it depends on the site considered [63, 66]. More importantly (see above), $\bar{\varepsilon}$ cannot be equal to ε_p, where the simplest example is the use of explicit models where $\varepsilon_p = 1$. Detailed analysis of this issue is given in Ref. [69]. At any rate, it is still useful to review what has been learned on the nature of $\bar{\varepsilon}$. Apparently it has been found [63] that the value of $\bar{\varepsilon}$ can be significantly larger than the traditional value ($\bar{\varepsilon} \simeq 4$) deduced from measurements of dry proteins and peptide powders [70, 71] or from simulations of the entire protein, rather than a specific region (see, for example, Ref. [63]). In fact, the early finding of $\bar{\varepsilon} = 4$, which was obtained from a gas phase study [71], ignored the large effect of the solvent reaction field (see analysis and discussion in Ref. [63]). The larger than expected value of $\bar{\varepsilon}$ was attributed to the fluctuations of ionized side chains [8, 66, 67]. However, consistent simulations [63] that included the reaction field of the solvent gave $\bar{\varepsilon} \simeq 10$ in and near protein active sites, even in the absence of the fluctuations of the ionized residues. This established that the fluctuations of the protein polar groups and internal water molecules contribute significantly to the increase in $\bar{\varepsilon}$. In view of this finding it might be instructive to address recent suggestions [8] that the calculations of reference [63] did not converge. Apart from the problematic nature of this assertion it is important to realize that the SCAAS model used in Ref. [63] converges much faster than any of the periodic models (and the corresponding very large simulation systems) used in the alternative "converging" studies mentioned in Ref. [8]. In fact the convergence of $\bar{\varepsilon}$ in reference [63] was examined by longer runs and by using alternative formulations. Obviously, there is a wide intuitive support for the idea that $\bar{\varepsilon}$ in protein active sites cannot be large unless it reflects the effect of ionized surface groups. However, the scientific way to challenge the finding of Ref. [63] is to repeat the calculations of $\bar{\varepsilon}$ in the *active site* of trypsin (rather than the whole protein) and to see if the conclusions of Ref. [63] reflect convergence problems. Unfortunately, no such attempt to repeat the calculations of Ref. [63] (or to study $\bar{\varepsilon}$ in other active sites with and without ionizable surface groups) has been reported. Finally, it is important to realize that recent experimental studies [73, 74] provided a clear support to the finding that $\bar{\varepsilon}$ is large in active sites even when they are far from the protein surface.

In view of the above analysis, and despite the obvious interest in the nature of $\bar{\varepsilon}$, it should be recognized that $\bar{\varepsilon}$ is not so relevant to electrostatic energies in proteins. Here one should focus on the energies of charges in their specific environment and the best ways for a reliable and effective evaluation of these energies. Unfortunately, $\bar{\varepsilon}$ does not tell us in a unique way how to determine the ε_p of semi-macroscopic models. For example (see also above), obtaining $\varepsilon_p = 1$ in a fully microscopic model, or $\varepsilon_p = 2$ in a model with implicit induced dipoles has nothing to do with the corresponding $\bar{\varepsilon}$. The best way to realize this point is to think of charges in a water sphere (our hypothetical protein) surrounded by water. In such a model $\bar{\varepsilon} \cong 80$, whereas microscopic models of the same system with implicit induced dipoles will be best described by using $\varepsilon_p = 2$. Of course, one may argue that the entire concept of a dielectric constant is invalid in the heterogeneous environment of proteins [14]. However, because fully microscopic models are still not giving sufficiently precise results, it is very useful to have reasonable estimates using implicit models and in particular semi-macroscopic models. Thus, it is justified to look for optimal ε_p values after realizing, however, what these parameters really mean. Even the use of a consistent semi-macroscopic model such as the PDLD/S-LRA does not guarantee that the corresponding ε_p will approach a universal value. For example, in principal the PDLD/S-LRA model should involve $\varepsilon_p = 2$, because all effects except the effect of the protein-induced dipoles are considered explicitly. However, the configurational sampling by the LRA approach is not perfect. The protein reorganization energy is probably captured to a reasonable extent [69], but the change in water penetration on ionization of charged residues is probably not reproduced in an accurate way. This problem is particularly serious when we have ionized groups in nonpolar sites in the interior of the proteins. In such cases we have significant changes in water penetration and local conformations during the ionization process, (as indicated by the very instructive study of Ref. [75]) but these changes might not be reproduced in standard simulation times.

The fact that ε_{eff} for charge–charge interactions is large has been pointed out by several workers [12, 76, 77]. However, this important observation was not always analyzed with a clear microscopic perspective. Mehler and coworkers [76, 78] argued, correctly, that $\varepsilon_{eff}(r)$ is a sigmoid function with a large value at relatively short distances based on classical studies of $\varepsilon(r)$ in water [79–81]. Similarly, it has been assumed by Jonsson and coworkers [77] that electrostatic interactions in proteins can be described by using a large ε_{eff} for charge–charge interactions. Now, although we completely agree that ε_{eff} is large in many cases and we repeatedly advanced this view [3], we believe that the reasons for this are much more subtle. That is, the finding of a particular behavior of ε in water might be not so relevant to the corresponding behavior in proteins (in principle, ε_{eff} in proteins can be very different from ε_{eff} in water). Here the remarkable reason why ε_{eff} is large, even in proteins, was rationalized in Ref. [3], where it was pointed out that the large ε_{eff} must reflect the compensation of charge–charge interactions and solvation energy in proteins (see, for example, figure 4 in Ref. [64]). It seems to us that the understanding of this crucial compensation effect is essential for the understanding of ε_{eff} in proteins.

Finally, it might be useful to comment on the perception that somehow the macroscopic dielectric of a given region in the protein can still be used to provide a general description of the energetics of changes in the same region. The funda-

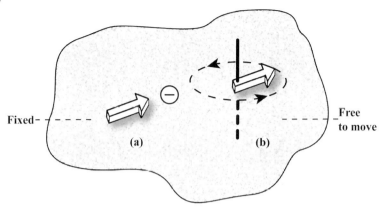

Fig. 7.2. A single macroscopic dielectric constant (which is evaluated by some microscopic prescription) cannot be used to reproduce the energetics of charges in proteins in a unique way. The figure considers a hypothetical site with a relatively fixed and freely rotated dipoles. Using the same ε_p for both dipoles will underestimate the effect of μ_a and overestimate the effect of μ_b.

mental problem with such an assertion is demonstrated in Figure 7.2. This figure considers a typical site with an internal site with two dipolar groups, one fixed (preorganized) and the other free to rotate. Since this is an internal site, the leading term in semi-macroscopic treatments will be the $U_{Q\mu}/\varepsilon_p$ term of Eq. (11). Now, one of the best ways to estimate ε_p is to calculate the reorganization energy (λ) for the given charges [8, 69]. Using this ε_p (for $U_{Q\mu}/\varepsilon_p$) for the interaction between site (a) to the charge will underestimate the corresponding free energy contribution, while using it for site (b) will overestimate this interaction. The same problem will occur if we use $\bar{\varepsilon}$ evaluated from the dipolar fluctuations in the site that includes the dipoles and the charge. Furthermore, if the site involves an ion pair instead of a single ionized group we will need a different ε_p to reproduce the energy of this ion pair.

7.5
Validation Studies

As discussed above, the conceptual and practical problems associated with the proper evaluation of electrostatic energies in proteins are far from trivial. Here the general realization of the problems associated with different models has been slow in part because of the use of improper validation approaches. For example, although it has been realized that pK_a value calculations provide an effective way for validating electrostatic models, [3, 39, 44, 72, 75, 78, 82, 83] most studies have focused mainly on the pK_a values of surface groups. Unfortunately, these pK_a values can be reproduced by many models, including those that are fundamentally incorrect [7]. Thus, the commonly used benchmarks may lead toward unjustified

7.5 Validation Studies

conclusions about the nature of electrostatic effects in proteins. It has been pointed out [7, 19, 75] that ionizable groups in protein interiors should provide a much more discriminating benchmark because these groups reside in a very heterogeneous environment, where the interplay between polar and nonpolar components is crucial. For example, an ionizable group in a true nonpolar environment will have an enormous pK_a shift, and such a shift cannot be reproduced by models that assume a high dielectric in the protein interior. Similarly, a model that describes the protein as a uniform low dielectric medium, without permanent dipoles, will not work in cases where the environment around the ionizable group is polar [3].

The problems with nondiscriminating benchmarks were established in Ref. [19] and are reillustrated in Figure 7.3. This figure considers the pK_a values of acidic and basic groups and correlates the results obtained by the modified TK (MTK) model with the corresponding observed results. The figure gives the impression

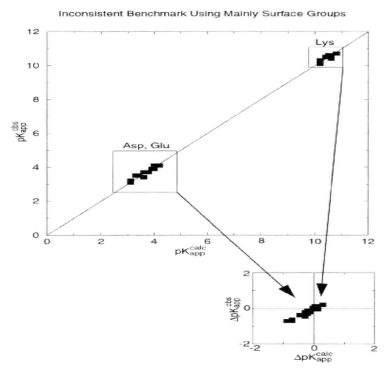

Fig. 7.3. The correlation between the calculated and observed pK_a values gives a nondiscriminative benchmark that only involves surface groups. The figure considers Asp, Glu, and Lys residues on the surfaces of lysozyme and ribonuclease but includes only those residues with $pK_{app}^{obs} < 1$. The calculated pK_a values are evaluated by the MTK(S) model with $\varepsilon_p = 40$ and $\varepsilon_{eff} = 80$. As seen from the figure, we obtain seemingly impressive results for this benchmark, but when the results are displayed as ΔpK_a, it becomes clear that we have simply a poor benchmark where $\Delta pK_a \approx 0$ and any model that produces a small pK_a shift looks like an excellent model.

that the MTK is an excellent model, but the same agreement would be obtained for any model that uses a large ε since most surface groups have very small $\Delta p K_a$ so that Asp and Glu have $pK_a \approx 4$ and all Lys have $pK_a \approx 10$. Thus, as emphasized in the inset in the figure, we are dealing with a very small amount of relevant information and probably the best correlation would have been obtained with the trivial null model when $\Delta p K_a = 0$ for all surface groups.

In view of the above example, it is very important to validate electrostatic models by using discriminative benchmarks (DBM) with $\Delta p K_a > 2$ and to use the same philosophy for other electrostatic properties. The use of a DBM in pK_a calculations [19] has been found to be extremely useful in demonstrating the range of validity and the proper dielectric of different models.

7.6
Systems Studied

7.6.1
Solvation Energies of Small Molecules

Modeling of almost any biological process requires one to know the solvation free energy of the relevant reacting system in aqueous solutions. Thus for example, calculations of binding energies compare the solvation energy of the ligand in its protein site to the corresponding solvation energy in solution. Early attempts to estimate solvation energies (see, for example, Refs [84] and [85]) were based on the use of the Born's or Onsager's model with basically arbitrary cavity radius. The first attempts to move toward quantitative evaluation of solvation energies can be divided into two branches. One direction involves attempts to examine quantum mechanically the interaction between the solute and a single solvent molecule [86, 87]. The other direction that turns out to be more successful involves the realization that quantitative evaluation of solvation free energies requires parameterization of the solute solvent van der Waals interaction in a complete solute solvent system [33] and evaluation of the interaction between the solvent and many (rather than one) solvent molecules.

Although such an empirical approach was considered initially as having "too many parameters" it was realized eventually that having an atom-solvent parameter for each type of the solute atoms is the key requirement in any quantitative semiempirical solvation model. The same number of parameters exists now in more recent continuum solvation models [54, 88, 89] and in all-atom models [35, 40, 41, 90]. At any rate, it took significant time (see Ref. [91]) until the realization that with any solvation model with a given set of solute charges, the most crucial step is the parameterization of the solute–solvent repulsion term (or the corresponding atomic radius).

Now assuming that we have reasonable atomic parameters the next requirement is to obtain converging results with the given modeling method (all-atom solvent simplified solvent model, PB model, GB model, etc.). Here the general progress is quite encouraging, as is apparent from the results reported by many research groups.

FEP calculations have focused on the proper treatment of ionized groups and the role of proper boundary conditions [27, 28, 92]. Recently, the validity of the LRA treatment in solvation calculations was explored [93], and the LRA formalism was used in a quantum mechanical treatment [94]. LD calculations were found to give accurate results when incorporated into quantum mechanical models [95, 96] and similar success was reported in continuum [54, 97] and GB [98] studies. All of these models apparently give similar accuracy once properly parameterized. The deviation from experiments in special cases probably reflects the neglected charge transfer to the solvent.

The fact that the calculations of solvation energies of small molecules in solution give reasonable results does not guarantee reasonable results for the solvation energies of charged (or other) ligands in proteins. This issue will be addressed in subsequent sections. Another interesting and important issue is the solvation energy of the entire macromolecule. Here the options are, in principle, reasonable. One can use the LD model, PB model and GB model. The problem is, however, to get quantitative results. Some attempts will be considered in Section 7.6.10.

7.6.2
Calculation of pK_a Values of Ionizable Residues

The electrostatic energy of ionizable residues in proteins plays a major role in most biological processes including enzymatic reactions, proton pumps, and protein stability. Calculation of pK_a values of ionizable groups in proteins provides a stringent test of different models for electrostatic energies in proteins [3]. Calculations of pK_a values by all-atom FEP approaches have been reported in only a few cases [39, 45]. Recent works include studies of the pK_a of metal-bound water molecules [99] and proton transfer in proteins [100]. All-atom LRA calculations [43, 44] were also reported: the error range of the all-atom models is still somewhat disappointing, although the inclusion of proper long-range treatments and induced dipoles leads to some improvements [43, 45]. Remaining problems might reflect insufficient sampling of protein configurations and solvent penetration. PDLD/S-LRA calculations [19, 43] gave encouraging results with $\varepsilon_p = 4$. Although the LRA treatment accounts at least formally for the protein reorganization, the limit of $\varepsilon_p = 2$ (which would be expected from a "perfect" model with implicitly induced dipoles) has not been reached, since apparently some of this reorganization and/or water penetration is not captured in the computer time used in the simulations.

Semi-macroscopic "continuum" (i.e., PB) models, (which now include the protein permanent dipoles and the self-energy term [12]) also often give reasonable results (e.g., [72, 101, 102]). Yet most PB models do not treat the effect of the protein reorganization explicitly, and the so-called multi-conformation continuum electrostatic (MCCE) model [56] (which has been inspired by the LRA approach) only considers the reorganization of polar side chains instead of performing a proper LRA treatment on all the protein coordinates. Another problem with current PB models is the assumption that the same ε_p should be used in accounting for the effect of the protein dipoles and in providing the screening for charge–charge interactions [19, 43]. This requirement led to inconsistency that can only be demonstrated by the use of discriminating benchmarks for both self-energies and charge–charge interactions. Although it is extremely important to continue to use DBM with large ΔpK_a in attempts to further explore the validity of different electrostatic models, it is important to realize that the trend in pK_a changes can be captured by models that focus more on charge–charge interactions than on the self-energy term or by models that estimate the self-energy term by empirical considerations [78].

It might be useful to point here to the perception that the need for averaging over the protein configurations is the main problem in continuum calculations of pK_a values [103–105]. First, the obvious need for proper configurational averaging in free energy calculations should not be confused, as done in several cases (e.g., [106]), with dynamics effects. Second, and perhaps more importantly, most attempts of configurational averaging involve only one charge state. This corresponds to only one charge state in the LRA treatment (Eq. (9)) and thus to incorrect estimates of the relevant charging free energy. Here it would help to recognize that a proper LRA averaging, rather than a simple average on one charge state, is essential for formally correct pK_a calculations.

Overall, we believe that a more complete consensus about the validity of different models for pK_a calculations will be obtained when the microscopic models start to give stable quantitative results. Here the comparison of the microscopic and semi-macroscopic models (as was done in Ref. [43]) will be particularly useful.

7.6.3
Redox and Electron Transport Processes

Electron transport processes are involved in key energy transduction processes in living systems (most notably photosynthesis). Such processes involve changes in the charges of the donor and acceptor involved, and thus are controlled by the electrostatic energies of the corresponding charges and the reorganization energies involved in the charge transfer process. Here the challenge is to evaluate the redox energies, and the reorganization energies using the relevant protein structure. Probably the first attempt to address this problem was reported by Kassner [107], who represented the protein as a nonpolar sphere. The idea that such a model can be used for analyzing redox properties held up for a significant time (see discussion in Refs [108–115]). However, Warshel and coworkers have shown by using the PDLD model [116, 117] that the evaluation of redox potentials must take into account the protein permanent dipoles, and the penetration of water molecules. The role of the protein permanent dipoles has been most clearly established in subsequent studies of iron–sulfur proteins [118, 119]. Another interesting factor is the effect of ionized groups on redox potentials. The effect of ionized residues can usually be described well using Coulomb's law with a large effective dielectric constant [12, 57, 58].

Microscopic estimates of protein reorganization energies have been reported [57, 69, 120] and used very effectively in studies of the rate constants of biological electron transport. This also includes studies of the nuclear quantum mechanical effect, associated with the fluctuations of the protein polar groups (for review see Ref. [121]). PB studies of redox proteins have progressed significantly since the early studies that considered the protein as a nonpolar sphere (see above). These studies (e.g., [108, 111, 122–124]) started to reflect a gradual recognition of the importance of the protein permanent dipoles, although some confusion still exists (see discussion in Refs [108] and [114]).

Recent studies reported quantum mechanical evaluation of the redox potentials in blue copper proteins [125], but even in this case it was found that the key problem is the convergence issue, and at present semi-macroscopic models give more accurate results than microscopic models.

7.6.4
Ligand Binding

A reliable evaluation of the free energy of ligand binding can potentially play a major role in designing effective drugs against various diseases (e.g., [126]). Here we have interplay between electrostatic, hydrophobic, and steric effects, but accurate estimates of the relevant electrostatic contributions are still crucial.

In principle, it is possible to evaluate binding free energies by performing FEP calculations of the type considered in Eqs (7) and (8), for "mutating" the ligand to "nothing" in water and in the protein active site. This approach, however, encounters major converence problems, and at present the reported results are disappointing except in cases of very small ligands. Alternatively one may study in simple cases the effect of small "mutations" of the given ligand [40], for example replacement of NH_2 by OH. However, when one is interested in the absolute binding of medium-size ligands, it is essential to use simpler approaches. Perhaps the most useful alternative is offered by the LRA approach of Eq. (9), and augmented by estimates of the nonelectrostatic effects. That is, the LRA approach is particularly effective in calculating the electrostatic contribution to the binding energy [42, 127, 128]. With this approximation one can express the binding energy as

$$\Delta G_{bind} = \frac{1}{2}[\langle U^p_{elec,l}\rangle_l + \langle U^p_{elec,l}\rangle_{l'} - \langle U^w_{elec,l}\rangle_l - \langle U^w_{elec,l}\rangle_{l'}] + \Delta G^{nonelec}_{bind} \tag{17}$$

where $U^p_{elec,l}$ is the electrostatic contribution for the interaction between the ligand and its surrounding; p and w designate protein and water, respectively; and l and l' designate the ligand in its actual charge form and the "nonpolar" ligand where all the residual charges are set to zero. In this expression the terms $\langle U_{elec,l} - U_{elec,l'}\rangle$, which are required by Eq. (9), are replaced by $\langle U_{elec,l}\rangle$ since $U_{elec,l'} = 0$. Now, the evaluation of the nonelectrostatic contribution $\Delta G^{nonelec}_{bind}$ is still very challenging, since these contributions might not follow the LRA. A useful option, which was used in Refs [42] and [127], is to evaluate the contribution to the binding free energy from hydrophobic effects, van der Waals, and water penetration. Another powerful option is the so-called linear interaction energy (LIE) approach [93]. This approach adopts the LRA approximation for the electrostatic contribution but neglects the $\langle U_{elec,l} \rangle_{l'}$ terms. The binding energy is then expressed as:

$$\Delta G_{bind} \approx \alpha \lfloor \langle U^p_{elec,l} \rangle_l - \langle U^w_{elec,l} \rangle_l \rfloor + \beta \lfloor \langle U^p_{vdW,l} \rangle_l - \langle U^w_{vdW,l} \rangle_l \rfloor \qquad (18)$$

where α is a constant that is around 1/2 in many cases and β is an empirical parameter that scales the van der Waals (vdW) component of the protein–ligand interaction. A careful analysis of the relationship between the LRA and LIE approaches and the origin of the α and β parameters is given in Refs [127].

More implicit models can, of course, estimate the binding energy. This includes the use of the PDLD/S-LRA model for both the electrostatic and nonelectrostatic components (see Refs [42] and [127]) and the PB approach augmented by estimates of the hydrophobic contributions using the calculated surface area of the ligand (e.g., [128, 129]).

The idea that conformational averaging is useful for calculations of binding energies seems quite obvious, generally understood and widely used [53]. What is much less understood and somehow still ignored by a large part of the computational community is the realization that a proper LRA treatment requires averaging on both the charged (or polar) and nonpolar states of the ligand. The frequently neglected average on the nonpolar state (the second term of Eq. (9)) plays a crucial role in proteins as it reflects the effect of the protein preorganization. The role of this term in binding calculations has been discussed in Ref. [127] and its importance has been illustrated in an impressive way in studies of the fidelity of DNA polymerase [130].

7.6.5
Enzyme Catalysis

Understanding enzyme catalysis is a subject of major practical and fundamental importance [131–134]. The development of combined quantum mechanical/ molecular mechanics (QM/MM) computational models (e.g., [14, 131, 134–140]) that allow enzymatic reactions to be simulated provided a way for quantifying the main factors that contribute to enzyme catalysis. QM/MM studies, including those conducted by the empirical valence bond (EVB) method [132], helped to establish the proposal [141] that the electrostatic effects of preorganized active sites play a major role in stabilizing the transition states of enzymatic reactions [142]. In fact, there is now growing appreciation of this view (e.g., [143, 144]). Simulation approaches that focus on the electrostatic aspects of enzyme catalysis (i.e., the difference between the stabilization in the enzyme and solution) appear to give much more quantitative results than those that focus on the quantum mechanical aspects of the problem but overlook the proper treatment of long-range effects (see discussion in Ref. [145]). In fact, some problems can be effectively treated even by PB approaches (see, for example, Ref. [146]) without considering quantum mechanical issues. Some studies focus on the role of the diffusion step of enzyme catalysis [147, 148]. As pointed out, however (e.g., [149]), these effects are trivial compared with the electrostatic effect of enzymes in actual chemical steps. Interestingly, evaluation of activation free energies of enzymatic reactions appeared to be simpler, in terms of the stability of the corresponding results, than other types of electrostatic calculations such as binding free energies (see discussion in Ref. [150]). This apparent advantage that has been exploited for a long time in EVB studies (see, for example, Ref. [131]) is now being reflected in molecular orbital QM/MM studies [134, 137, 151].

7.6.6
Ion Pairs

Ion pairs play a major role in biological systems and biological processes. This includes the "salt bridges" that control the Bohr effect in hemoglobin [152], ion pair type transition states [132], ion pairs that help in maintaining protein stability (see Section 7.6.9), ion pairs that play crucial roles in signal transduction (Section 7.6.6) and ion pairs that play a major role in photobiological processes [153]. Reliable analysis of the energetics of ion pairs in proteins presents a major challenge and has been the subject of significant conceptual confusions (see discussions in Refs [3] and [12]). The most common confusion is the intuitive assumption that ion pairs are more stable in nonpolar environments rather than less stable in nonpolar environments (see analysis in Refs [154] and [155]). Excellent examples of this confusion are provided even in recent studies (e.g., [156]) and in most of the desolvation models (e.g., [157–159]) which are based on the assumption that enzymes destabilize their ground states by placing them in nonpolar environments. However, as has been demonstrated elsewhere (e.g., [132, 150]) enzymes do not provide nonpolar active sites for ion pair type ground states (they provide a very polar environment that stabilizes the transition state).

In general it is important to keep in mind that ion pairs are not stable in nonpolar sites in proteins. Thus most ion pairs in proteins are located on the surface, and those key ion pairs that are located in protein interiors are always surrounded by polar environments [132]. Forcing ion pairs between amino acids to be in nonpolar environments usually results in the conversion of the pair to its nonpolar form (by a proton transfer from the base to the acid).

The energetics of specific ion pairs in proteins has been studied by microscopic FEP calculations [160, 161] and by semi-macroscopic calculations [19, 37, 52, 162, 163]. However, we only have very few systematic validation studies. An instructive example is the study of Hendsch and Tidor [164] and its reanalysis by Schutz and Warshel [19]. That is, Ref. [164] used a PB approach with a low value of ε_p. This might have led to nonquantitative results. That is, in the case of T4 lysozyme, the calculated energies for $\text{Asp70}^0 \ldots \text{His31}^+ \rightarrow \text{Asp70}^- \ldots \text{His31}^+$ and for $\text{Asp70}^- \ldots \text{His31}^0 \rightarrow \text{Asp70}^- \ldots \text{His31}^+$ were -0.69 kcal mol^{-1} and -0.77 kcal mol^{-1}, respectively. This should be compared with the corresponding observed values (−4.7 and −1.9 kcal mol^{-1}) or the values obtained in Ref. [19] (−4.7 and −2.2 kcal mol^{-1}). Similarly, the estimate of 10 kcal mol^{-1} destabilization of the Glu11-Arg45 ion pair in T4 lysozyme seems to be a significant overestimate relative to the result of ∼3 kcal mol^{-1} obtained by the current PDLD/S-LRA study. It is significant that the Asp70$^-$ ... His31$^+$ ion pair does not appear to destabilize the protein. That is, the energy of moving the ion pair from infinite separation in water to the protein site (relative to the corresponding energy for a nonpolar pair) was found here to be around zero compared with the 3.46 kcal mol^{-1} obtained in Ref. [164]. The origin of difference between the two studies can be traced in part to the estimate of the stabilizing effect of the protein permanent dipoles (the $V_{Q\mu}$ term in Ref. [19] and $\Delta G_{protein}$ in Ref. [164]). The ∼−5 kcal mol^{-1} $V_{Q\mu}$ contribution found by the PDLD/S-LRA calculation corresponds to a contribution of only ∼−0.26 kcal mol^{-1} in Ref. [164]. This might be due to the use of an approach that does not allow polar groups to reorient in response to the charges of the ion pair. It is exactly the stabilizing effect of the protein polar groups that plays a crucial role in stabilizing ion pairs in proteins. Clearly, the estimate of the energetics of ion pairs depends crucially on the dielectric constant used and models that do not consider the protein relaxation explicitly must use larger ε_p than the PDLD/S-LRA model.

Obviously, it is essential to have more systematic validation studies considering internal ion pairs with well-defined structure and energy. In this respect, it is useful to point out that the instructive statistical analysis reported so far (e.g., [165]) may involve similar problems as those mentioned in Section 7.5, since it does not consider a discriminative database; it includes mainly surface groups rather than ion pairs in protein interiors.

7.6.7
Protein–Protein Interactions

Protein–protein interactions play a crucial role in many biological systems, including in signal transduction [166], energy transduction [167], different assembly processes (e.g., [22]) and electron transport [168, 169]. Here again it seems that electrostatic effects are the most important factor in determining the nature and strength of the interaction between protein surfaces.

Calculations of protein–protein interactions are very challenging, particularly because of the large surfaces involved and the large structural changes upon binding. Considering the protein reorganization upon binding is particularly challenging and an LRA study that tried to accomplish this task for the Ras/RAF system [65] did not succeed to obtain quantitative results. However, a much simpler approach that evaluated charge–charge interaction by the effective dielectric constant of Eq. (14) gave encouraging results for the effect of mutations of the binding energy between the two proteins [65].

Recent studies of the binding of Barnase and Barstar [170] also indicate that the results are very sensitive to the dielectric model used. This study identified problems associated with the use of PB with different standard dielectric constant models.

7.6.8
Ion Channels

The control of ion permeation by transmembrane channels underlies many important biological functions (e.g., [171]). Understanding the control of ion selectivity by ion channels is basically a problem in protein electrostatics (e.g., [172, 173]) that turns out to be a truly challenging task. The primary problem is the evaluation of the free energy profile for transferring the given ion from water to the given position in the channel. It is also essential to evaluate the interaction between ions in the channel if the ion current involves a multi-ion process [174]. Early studies of ion channels focused on the energetics of ions in the gramicidine channel [175, 176]. The solution of the structure of the KscA potassium channel [177] provided a model for real biological channels and a major challenge for simulation approaches. Some early studies provided a major overestimate of the barriers for ion transport (e.g., [178, 179]) and the first reasonable results were obtained by the FEP calculations of Åqvist and Luzhkov [180]. These calculations involved the LRF long-range treatment and the SCSSA boundary conditions that probably helped in obtaining reliable results. Microscopic attempts to obtain the selectivity difference between K^+ and Na^+ were also reported [181]. However, these attempts did not evaluate the activation barriers for the two different ions and thus could not be used in evaluating the difference in the corresponding currents. Furthermore, attempts to evaluate the so-called potential of mean force (PMF) for ion penetration, that have the appearance of truly rigorous approaches, have not succeeded to reproduce the actual PMF for moving the ions from water to the channel (see discussion in Ref. [174]).

A promising alternative that appears to be quite effective in capturing the energetics of ion transport through the KcsA channel has been provided by the semi-macroscopic PDLD/S-LRA approach. This approach has provided activation barriers that reproduced the difference between the currents of Na^+ and K^+ [182]. Here it is important to point out that the LRA treatment of the PDLD/S-LRA approach has been crucial for the reproduction of the K^+/Na^+ selectivity. That is, without the LRA or related treatments it is impossible to capture the change in the protein (channel) structure upon change from K^+ to Na^+. This presents a major challenge to PB attempts to reproduce ion selectivity. It is also important to note that the evaluation of ion current by Langevin Dynamics or related approaches requires a very fast evaluation of the relevant electrostatic energies at different configurations of the transferred ions. Here the use of the PDLD/S-LRA approach and the distance dependence of the dielectric constant of Eq. (14) was found to be quite effective [174].

7.6.9
Helix Macrodipoles versus Localized Molecular Dipoles

The idea that the macroscopic dipoles of α-helices provide a critical electrostatic contribution [183, 184] has gained significant popularity and appeared in many proposals. The general acceptance of this idea and the corresponding estimates (see below) is in fact a reflection of a superficial attitude. That is, here we have a case where the idea that microscopic dipoles (e.g., hydrogen bonds and carbonyls) play a major role in protein electrostatics [3] is replaced by a problematic idea that macrodipoles are the source of large electrostatic effects. The main reason for the acceptance of the helix-dipole idea (except the structural appeal of this proposal) has probably been due to the use of incorrect dielectric concepts. That is, estimates of large helix-dipole effects [184–188] involved a major underestimate of the corresponding dielectric constant and the customary tendency to avoid proper validation studies. Almost all attempts to estimate the magnitude of the helix-dipole effect have not been verified by using the same model in calculations of relevant observables (e.g., pK_a shift and enzyme catalysis). The first quantitative estimate of the effect of the helix dipole [189] established that the actual effect is due to the first few microscopic dipoles at the end of the helix and not to the helix macrodipole. It was also predicted that neutralizing the end of the helix by an opposing charge will have a very small effect and this prediction was confirmed experimentally [190].

An interesting recent examples of the need for quantitative considerations of the helix-dipole effect has been provided by the KcsA K^+ channel. The instructive study of Roux and MacKinnon [191] used PB calculations with $\varepsilon_p = 2$ and obtained an extremely large effect of the helix dipoles on the stabilization of the K^+ ion in the central cavity (~ -20 kcal mol^{-1}). However, a recent study [182] that used a proper LRA procedure in the framework of the PDLD/S-LRA approach gave a much smaller effect of the helix macrodipole (Figure 7.4). Basically, the use of $\varepsilon_p = 2$ overestimates the effect of the helix dipole by a factor of 3 and the effect is rather localized on the first few residues.

7.6.10
Folding and Stability

The study of protein folding by simulation approaches dates back to the introduction of the simplified folding model in 1975 [192]. This model, which forms the base for many of the current folding models, demonstrated for the first time that the folding problem is much simpler than previously thought. The simplified folding model and the somewhat less physical model developed by Go [193], which are sometimes classified as off-lattice and on-lattice models, respectively, paved the way for major advances in studies of the folding process and the free energy landscape of this process (e.g., [194–197]). Further progress has been obtained in recent years from all-atom simulation of peptides and small proteins (e.g., [198–200]). Some of these simulations involved the use of explicit solvent models and most involved the use of implicit solvation models ranging from GB type models [198], to the Langevin dipole model [200]. Interesting insights have been provided by the idea of using a simplified folding model as a reference potential for all-atom free energy simulation [200], but the full potential of this approach has not been exploited

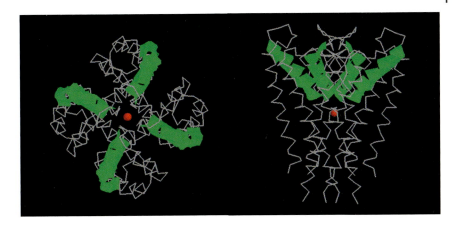

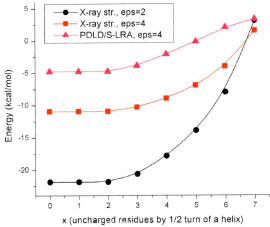

Fig. 7.4. Examination of the effect of the helix dipoles of the KcsA ion channel (upper panel) on a K^+ ion in the central cavity. The lower panel presents the contribution of the residues in the four helices as a function of the dielectric treatment used. It is shown that the use of $\varepsilon_p = 2$ drastically overestimates the contribution of the macrodipoles, which is evaluated more quantitatively with the PDLD/S-LRA treatment.

yet. Another promising direction has been offered by the use of a large amount of distributed computing resources [201].

At present it seems that simplified folding models describe quite well the average behavior of all-atom models (e.g., [200]), although more systematic studies are needed. The use of implicit solvent models also looks promising but here again there is a clear need for detailed validation studies focusing on observed electrostatic energies.

Despite the progress in studies of the folding process, we do not have yet a clear understanding of the factors that determined protein stability. Early insightful considerations [20] argued that electrostatic effects must play a major role in controlling protein stability. Other workers emphasized hydrophobic effects and disulfide bridges, yet no consensus has been reached. Here it might be useful to consider the current situation with regards to the role of electrostatic contributions to thermal stability. Experimental studies of mesophilic, thermophilic, and hyperthermophilic proteins have provided an excellent benchmark for studies of the relationship between electrostatic energy and protein stability (e.g., [202, 203]). In general, the number of charged residues increases in hyperthermophiles, and thus can be considered as a stabilizing factor. However, the effect of ionized residues [164] and polar groups [204] has been thought to lead to destabilization bases on continuum studies (e.g., [164]). A notable example is the study of Hendsch and Tidor [164]. This study concluded that internal salt bridges tend to destabilize proteins, but as discussed in Section 7.6.6, this study did not reproduce the relevant observed energies. On the other hand, some studies (e.g., [205–208]) seem to support the idea that charged residues may help to optimize protein stability. The problem in reaching clear conclusions is closely related to the uncertainty about the value of ε_p in PB and related treatments. Here we have a competition between desolvation penalties, stabilization by local protein dipoles, and charge–charge interactions. The situation is relatively simple when one deals with a hypothetical hydrophobic protein (the original TK model). In this case, as was shown first in Ref. [12], both isolated ions and ion pairs become unstable in the "protein" interiors. The situation becomes, however, much more complex in real proteins where charges are stabilized by polar groups (e.g., [17]). Here the balance between charge–charge interactions and self-energy may depend drastically on the assumed ε_p. As stated in previous sections, it seems to us that different ε_p should be used for charge–charge interactions and self-energies. The above point can be clarified by writing the contribution of the ionizable residues to the folding energy as:

$$\Delta G_{\text{fold}}^{\text{elec}} = \Delta G_f^{\text{elec}} - \Delta G_{uf}^{\text{elec}} = -2.3RT \sum_i Q_i^{(f)}(\text{p}K_{i,\text{int}}^{\text{p}} - \text{pH})$$

$$+ \frac{1}{2} \sum_{i \neq j} Q_i^{(f)} Q_j^{(f)} \left(\frac{332}{r_{ij}^{(f)} \varepsilon_{\text{eff}}(r_{ij}^{(f)})} \right) - \sum_i Q_i^{(uf)}(\text{p}K_{a,i}^{\text{w}} - \text{pH})$$

$$- \frac{1}{2} \sum_{i > j} Q_i^{(uf)} Q_j^{(uf)} \left(\frac{332}{80 r_{ij}^{(uf)}} \right) \quad (19)$$

where f and uf designate, respectively, the folded and unfolded form of the protein, and the Qs are the charges of the ionized groups in their particular configuration. The above expression is based on our previous description of the electrostatic free energies of ionization states in proteins [209] and on the simplified approximation that considered the charges in the unfolded state as being fully solvated by water. If we assume that the same groups are ionized in the folded state, we can get a simpler expression:

$$\Delta G_{\text{fold}}^{\text{elec}} \approx -2.3RT \sum_i Q_i(\text{p}K_{i,\text{int}}^{\text{p}} - \text{p}K_{i,\text{w}}^{\text{w}}) + 166 \sum_{i>j} Q_i Q_j \left[\frac{1}{r_{ij}^{(f)} \varepsilon_{ij}^{(f)}} - \frac{1}{80 r_{ij}^{(\text{uf})}} \right] \quad (20)$$

The first term in Eq. (20) represents the change of self-energy upon moving the given charge from water to the protein active site, and the second term represents the effect of charge–charge interaction. Now, in general the ε_p for the first term and the ε_eff of the second term are quite different, and this makes the analysis complicated. At any rate, our preliminary analysis [210] evaluated the relative folding energies of two proteins expressed from the cold shock protein B (CspB) family, expressed by the mesophilic bacterium *Bacillus subtilis* [211] and the thermophilic bacterium *Bacillus caldolyticus*. This study obtained different trends with different ε_p and ε_eff, and reproduced the increased stability of the thermophilic protein with $\varepsilon_\text{p} = 8$ and $\varepsilon_\text{eff} = 20$. It seems to us, however, that this issue must be subjected to very careful validation studies, focusing in particular on the reproduction of the observed $\text{p}K_\text{a}$ values of the proteins under consideration. Relevant benchmarks can be obtained in part from available experimental studies.

7.7
Concluding Remarks

Electrostatic effects play a major role in almost all biological processes. Thus, accurate evaluation of electrostatic energies is a key to quantitative structure function analysis [18]. Here the key to further progress is expected to depend on wider recognition of the importance of proper validation studies and on the increase in accuracy of microscopic calculations. That is, in many cases the focus on trivial properties of surface groups lead to entirely incorrect views of the validity of different models. Another worrisome direction is the tendency to validate different implicit models by their ability to reproduce PB results. For example, GB calculations are frequently verified by comparing them to the corresponding PB calculations. This implies a belief that the PB results represent the correct electrostatic energies, overlooking the fact that these calculated energies depend critically on an undefined dielectric constant. Consistent validations and developments of reliable models must involve comparisons to relevant experimental results. Proper verification of semi-macroscopic models might also involve a comparison to microscopic results but in many cases these results do not reach proper convergence. Here the wider realization of the importance of long-range effects, boundary conditions and convergence issues are expected to help in increasing the accuracy of microscopic models. The continuing increase in computational resources will also play a key role in providing more accurate electrostatic energies.

Recent studies of electrostatic energies in proteins have repeatedly focused on the formal validations of PB and related approaches, while focusing on the contribution of the solvent around the protein. This focus, however, seems to overlook and sometimes ignore a key issue. That is, the PB and GB approaches provide excellent models for the effect of the solvent around the protein. The problem, however, is the effect of the protein and the proper ε_p. Here we cannot use any formal justification since our results depend drastically on a somewhat arbitrary parameter (i.e., ε_p).

Although this review presents a somewhat critical discussion of different semi-macroscopic approaches, it does not mean that such approaches are not useful. That is, as long as the difficulties of obtaining reliable results by microscopic approaches persist one must resort to semi-macroscopic approaches. Such approaches are also essential in studies of problems that require fast evaluation of the relevant electrostatic energies (e.g., the folding problem). The results obtained by semi-macroscopic models are expected to become more accurate with a better understanding of the protein dielectric "constant".

Acknowledgments

This work was supported by NIH grants GM 40283 and 24492, and NSF grant MCB-0003872. We are grateful to Mitsunori Kato for the work presented in Figure 7.4.

References

1 LINDERSTROM-LANG, K., The ionization of proteins. C. R. Trav. Lab. Carlsberg, 1924; **15**: 29.
2 TANFORD, C. and J. G. KIRKWOOD, Theory of protein titration curves. I. General equations for impenetrable spheres. J. Am. Chem. Soc., 1957; **79**: 5333–5339.
3 WARSHEL, A. and S. T. RUSSELL, Calculations of electrostatic interactions in biological systems and in solutions. Q. Rev. Biophys., 1984; **17**: 283–421.
4 SHARP, K. A. and B. HONIG, Electrostatic interactions in macromolecules: theory and applications. Annu. Rev. Biophys. Biochem., 1990; **19**: 301–332.
5 WARSHEL, A. and J. ÅQVIST, Electrostatic energy and macromolecular function. Annu. Rev. Biophys. Biochem., 1991; **20**: 267–298.
6 NAKAMURA, H., Roles of electrostatic interaction in proteins. Q. Rev. Biophys., 1996; **29**: 1–90.
7 WARSHEL, A. and A. PAPAZYAN, Electrostatic effects in macromolecules: fundamental concepts and practical modeling. Curr. Opin. Struct. Biol., 1998; **8**: 211–217.
8 SIMONSON, T., Electrostatics and dynamics of proteins. Rep. Prog. Phys., 2003; **66**(5): 737–787.
9 DAVIS, M. E. and J. A. McCAMMON, Electrostatics in biomolecular structure and dynamics. Chem. Rev., 1990; **90**: 509–521.
10 JACKSON, J. D., Classical Electrodynamics, 3rd edn. New York: John Wiley & Sons, 1999, 808.
11 LINDERSTROM-LANG, K., On the ionization of proteins. C. R. Trav. Lab. Carlsberg, 1924; **15**: 1.
12 WARSHEL, A., S. T. RUSSELL, and A. K. CHURG, Macroscopic models for studies of electrostatic interactions in proteins: limitations and applicability. Proc. Natl Acad. Sci. USA, 1984; **81**: 4785–4789.
13 CHUNG, S. H. and S. KUYUCAK, Ion channels: recent progress and prospects. Eur. J. Biochem., 2002; **31**: 283–293.
14 WARSHEL, A. and M. LEVITT, Theoretical studies of enzymic reactions:

dielectric, electrostatic and steric stabilization of the carbonium ion in the reaction of lysozyme. *J. Mol. Biol.*, 1976; **103**: 227–249.
15 ORTTUNG, W. H., Polarizability of density of inert-gas atom pairs. 1. *J. Phys. Chem.*, 1985; **89**: 3011–3016.
16 WARWICKER, J. and H. C. WATSON, Calculation of the electric potential in the active site cleft due to alpha-helix dipoles. *J. Mol. Biol.*, 1982; **157**: 671–679.
17 WARSHEL, A., What about protein polarity? *Nature*, 1987; **333**: 15–18.
18 WARSHEL, A., Electrostatic basis of structure-function correlation in proteins. *Acc. Chem. Res.*, 1981; **14**: 284–290.
19 SCHUTZ, C. N. and A. WARSHEL, What are the dielectric "constants" of proteins and how to validate electrostatic models. *Proteins*, 2001; **44**: 400–417.
20 PERUTZ, M. F., Electrostatic effects in proteins. *Science*, 1978; **201**: 1187–1191.
21 MATTHEW, J. B., P. C. WEBER, F. R. SALEMME, and F. M. RICHARDS, Electrostatic orientation during electron-transfer between flavodoxin and cytochrome-c. *Nature*, 1983; **301**(5896): 169–171.
22 BLOOMER, A. C., J. N. CHAMPNESS, G. BRICOGNE, R. STADEN, and A. KLUG, Protein disk of tobacco mosaic-virus at 2.8 Å resolution showing interactions within and between subunits. *Nature*, 1978; **276**(5686): 362–368.
23 VERNON, C. A., Mechanisms of hydrolysis of glycosides and their relevance to enzyme-catalyzed reactions. *Proc. R. Soc. Lond. B Biol.*, 1967; **167**(1009): 389–401.
24 THOMA, J. A., Separation of factors responsible for lysozyme catalysis. *J. Theor. Biol.*, 1974; **44**: 305–317.
25 SCHAEFER, M., M. SOMMER, and M. KARPLUS, pH-dependence of protein stability: absolute electrostatic free energy differences between conformations. *J. Phys. Chem. B*, 1997; **101**: 1663–1683.
26 DUFNER, H., S. M. KAST, J. BRICKMANN, and M. SCHLENKRICH, Ewald summation versus direct summation of shifted-force potentials for the calculation of electrostatic interactions in solids – a quantitative study. *J. Comput. Chem.*, 1997; **18**: 660–675.
27 FIGUEIRIDO, F., G. S. DEL BUONO, and R. M. LEVY, On the finite size corrections to the free energy of ionic hydration. *J. Phys. Chem. B.*, 1997; **101**(29): 5622–5623.
28 HUMMER, G., L. R. PRATT, and A. E. GARCIA, Free energy of ionic hydration. *J. Phys. Chem.*, 1996; **100**: 1206–1215.
29 SHAM, Y. Y. and A. WARSHEL, The surface constrained all atom model provides size independent results in calculations of hydration free energies. *J. Chem. Phys.*, 1998; **109**: 7940–7944.
30 AQVIST, J., Comment on transferability of ion models. *J. Phys. Chem.*, 1994; **98**(33): 8253–8255.
31 AQVIST, J. and T. HANSSON, Analysis of electrostatic potential truncation schemes in simulations of polar solvents. *J. Phys. Chem. B*, 1998; **102**(19): 3837–3840.
32 LEE, F. S. and A. WARSHEL, A local reaction field method for fast evaluation of long-range electrostatic interactions in molecular simulations. *J. Chem. Phys.*, 1992; **97**: 3100–3107.
33 WARSHEL, A., Calculations of chemical processes in solutions. *J. Phys. Chem.*, 1979; **83**: 1640–1650.
34 BROOKS III, C. L. and M. KARPLUS, Deformable stochastic boundaries in molecular dynamics. *J. Chem. Phys.*, 1983; **79**: 6312–6325.
35 KING, G. and A. WARSHEL, A surface constrained all-atom solvent model for effective simulations of polar solutions. *J. Chem. Phys.*, 1989; **91**(6): 3647–3661.
36 KIRKWOOD, J. G., Theory of solutions of molecules containing widely serparated charges with special application to zwitterions. *J. Chem. Phys.*, 1934; **2**: 351–361.
37 LEE, F. S., Z. T. CHU, and A. WARSHEL, Microscopic and semi-microscopic calculations of electrostatic energies in proteins by the POLARIS and ENZYMIX programs. *J. Comput. Chem.*, 1993; **14**: 161–185.

38 ALPER, H. and R. M. LEVY, Dielectric and thermodynamic response of generalized reaction field model for liquid state simulations. *J. Chem. Phys.*, 1993; **99**(12): 9847–9852.

39 WARSHEL, A., F. SUSSMAN, and G. KING, Free energy of charges in solvated proteins: microscopic calculations using a reversible charging process. *Biochemistry*, 1986; **25**: 8368–8372.

40 KOLLMAN, P., Free energy calculations: applications to chemical and biochemical phenomena. *Chem. Rev.*, 1993; **93**: 2395–2417.

41 WARSHEL, A., Dynamics of reactions in polar solvents, semiclassical trajectory studies of electron-transfer and proton-transfer reactions. *J. Phys. Chem.*, 1982; **86**: 2218–2224.

42 LEE, F. S., Z. T. CHU, M. B. BOLGER, and A. WARSHEL, Calculations of antibody-antigen interactions: microscopic and semi-microscopic evaluation of the free energies of binding of phosphorylcholine analogs to McPC603. *Protein Eng.*, 1992; **5**: 215–228.

43 SHAM, Y. Y., Z. T. CHU, and A. WARSHEL, Consistent calculations of pK_a's of ionizable residues in proteins: semi-microscopic and macroscopic approaches. *J. Phys. Chem. B*, 1997; **101**: 4458–4472.

44 DELBUONO, G. S., F. E. FIGUEIRIDO, and R. M. LEVY, Intrinsic pK(a)s of ionizable residues in proteins – an explicit solvent calculation for lysozyme. *Proteins*, 1994; **20**(1): 85–97.

45 SAITO, M., Molecular dynamics free-energy study of a protein in solution with all degrees of freedom and long-range Coulomb interactions. *J. Phys. Chem.*, 1995; **99**: 17043–17048.

46 LIU, Y. and T. ICHIYE, The static dielectric-constant of the soft sticky dipole model of liquid water – Monte Carlo simulation. *Chem. Phys. Lett.*, 1996; **256**: 334–340.

47 COALSON, R. D. and A. DUNCAN, Statistical mechanics of a multipolar gas: a lattice field theory approach. *J. Phys. Chem.*, 1996; **100**: 2612–2620.

48 PAPAZYAN, A. and A. WARSHEL, Continuum and dipole-lattice models of solvation. *J. Phys. Chem.*, 1997; **101**: 11254–11264.

49 PAPAZYAN, A. and A. WARSHEL, Effect of solvent discreteness on solvation. *J. Phys. Chem. B*, 1998; **102**: 5248–5357.

50 HADUONG, T., S. PHAN, M. MARCHI, and D. BORGIS, Electrostatics on particles: phenomenological and orientational density functional theory approach. *J. Chem. Phys.*, 2002; **117**(2): 541–556.

51 SCHAEFER, M. and M. KARPLUS, A comprehensive analytical treatment of continuum electrostatics. *J. Phys. Chem.*, 1996; **100**: 1578–1599.

52 WARSHEL, A., G. NARAY-SZABO, F. SUSSMAN, and J.-K. HWANG, How do serine proteases really work? *Biochemistry*, 1989; **28**: 3629–3673.

53 KOLLMAN, P. A., I. MASSOVA, C. REYES, B. et al., Calculating structures and free energies of complex molecules: combining molecular mechanics and continuum models. *Acc. Chem. Res.*, 2000; **33**: 889–897.

54 TOMASI, J., B. MENNUCCI, R. CAMMI, and M. COSSI, *Quantum Mechanical Models for Reactions in Solution*, in *Computational Approaches to Biochemical Reactivity* (G. NARAY-SZABO, and A. WARSHEL, Eds.). Dordrecht: Kluwer Academic Publishers, 1997, 1–102.

55 PAPAZYAN, A. and A. WARSHEL, A stringent test of the cavity concept in continuum dielectrics. *J. Chem. Phys.*, 1997; **107**(19): 7975–7978.

56 GEORGESCU, R. E., E. G., ALEXOV, and M. R. GUNNER, Combining conformational flexibility and continuum electrostatics for calculating pK(a)s in proteins. *Biophys. J.*, 2002; **83**(4): 1731–1748.

57 ALDEN, R. G., W. W. PARSON, Z. T. CHU, and A. WARSHEL, Calculations of electrostatic energies in photosynthetic reaction centers. *J. Am. Chem. Soc.*, 1995; **117**: 12284–12298.

58 JOHNSON, E. T. and W. W. PARSON, Electrostatic interactions in an integral membrane protein. *Biochemistry*, 2002; **41**: 6483–6494.

59 BASHFORD, D. and D. A. CASE, Generalized born models of

macromolecular solvation effects. *Annu. Rev. Phys. Chem.*, 2000; **51**: 129–152.
60. LUZHKOV, V. and A. WARSHEL, Microscopic models for quantum mechanical calculations of chemical processes in solutions: LD/AMPAC and SCAAS/AMPAC calculations of solvation energies. *J. Comput. Chem.*, 1992; **13**: 199–213.
61. STILL, W. C., A. TEMPCZYK, R. C. HAWLEY, and T. HENDRICKSON, Semianalytical treatment of solvation for molecular mechanics and dynamics. *J. Am. Chem. Soc.*, 1990; **112**: 6127–6129.
62. BORGIS, D., N. LEVY, and M. MARCHI, Computing the electrostatic free-energy of complex molecules: The variational Coulomb field approximation. *J. Chem. Phys.*, 2003; **119**(6): 3516–3528.
63. KING, G., F. S. LEE, and A. WARSHEL, Microscopic simulations of macroscopic dielectric constants of solvated proteins. *J. Chem. Phys.*, 1991; **95**: 4366–4377.
64. SHAM, Y. Y., I. MUEGGE, and A. WARSHEL, The effect of protein relaxation on charge-charge interactions and dielectric constants of proteins. *Biophys. J.*, 1998; **74**: 1744–1753.
65. MUEGGE, I., T. SCHWEINS, R. LANGEN, and A. WARSHEL, Electrostatic control of GTP and GDP binding in the oncoprotein p21 ras. *Structure*, 1996; **4**: 475–489.
66. SIMONSON, T. and C. L. BROOKS, Charge screening and the dielectric constant of proteins: Insights from molecular dynamics. *J. Am. Chem. Soc.*, 1996; **118**: 8452–8458.
67. PITERA, J. W., M. FALTA, and W. F. VAN GUNSTEREN, Dielectric Properties of Proteins from Simulation: The Effects of Solvent, Ligands, pH, and Temperature. *Biophys. J.*, 2001; **80**: 2546–2555.
68. SIMONSON, T., D. PERAHIA, and A. T. BRUNGER, Microscopic theory of the dielectric properties of proteins. *Biophys. J.*, 1991; **59**: 670–690.
69. MUEGGE, I., P. X. QI, A. J. WAND, Z. T. CHU, and A. WARSHEL, The reorganization energy of cytochrome c revisited. *J. Phys. Chem. B*, 1997; **101**: 825–836.
70. HARVEY, S. and P. HOEKSTRA, Dielectric relaxation spectra of water adsorbed on lysozyme. *J. Phys. Chem.*, 1972; **76**: 2987–2994.
71. GILSON, M. and B. HONIG, The dielectric constant of a folded protein. *Biopolymers*, 1986; **25**: 2097–2119.
72. ANTOSIEWICZ, J., J. A. MCCAMMON, and M. K. GILSON, Prediction of pH-dependent properties of proteins. *J. Mol. Biol.*, 1994; **238**: 415–436.
73. COHEN, B. E., T. B. MCANANEY, E. S. PARK, Y. N. JAN, S. G. BOXER, and L. Y. JAN, Probing protein electrostatics with a synthetic fluorescent amino acid. *Science*, 2002; **296**(5573): 1700–1703.
74. BORMAN, S., Fluorescent probe for proteins: nonnatural amino acid used to investigate electrostatic properties within proteins. *Chem. Eng. News*, 2002; **80**(24): 30.
75. GARCIA-MORENO, B., J. J. DWYER, A. G. GITTIS, E. E. LATTMAN, D. S. SPENCER, and W. E. STITES, Experimental measurement of the effective dielectric in the hydrophobic core of a protein. *Biophys. Chem.*, 1997; **64**: 211–224.
76. MEHLER, E. L. and G. EICHELE, Electrostatic effects in water-accessible regions of proteins. *Biochemistry*, 1984; **23**: 3887–3891.
77. PENFOLD, R., J. WARWICKER, and B. JONSSON, Electrostatic models for calcium binding proteins. *J. Phys. Chem. B*, 1998; **108**: 8599–8610.
78. MEHLER, E. L. and F. GUARNIERI, A self-consistent, microscopic modulated screened Coulomb potential approximation to calculate pH-dependent electrostatic effects in proteins. *Biophys. J.*, 1999; **75**: 3–22.
79. DEBYE, P. J. W. and L. PAULING, The inter-ionic attraction theory of ionized solutes. IV. The influence of variation of dielectric constant on the limiting law for small concentrations. *J. Am. Chem. Soc.*, 1925; **47**: 2129–2134.
80. WEBB, T. J., The free energy of hydration of ions and the electro-

striction of the solvent. *J. Am. Chem. Soc.*, 1926; **48**: 2589–2603.
81 HILL, T. L., The electrostatic contribution to hindered rotation in certain ions and dipolar ions in solution iii. *J. Chem. Phys.*, 1944; **12**: 56–61.
82 RUSSELL, S. T. and A. WARSHEL, Calculations of electrostatic energies in proteins; the energetics of ionized groups in bovine pancreatic trypsin inhibitor. *J. Mol. Biol.*, 1985; **185**: 389–404.
83 HONIG, B., SHARP, K., SAMPOGNA, R., GUNNER, M. R., and YANG, A. S., On the calculation of pKas in proteins. *Proteins*, 1993; **15**: 252–265.
84 KOSOWER, E. M., *Introduction to Physical Organic Chemistry*, 1968, Wiley, New York.
85 MATAGA, N. and T. KUBOTA, *Molecular Interactions and Electronic Spectra.*, 1970, M. DEKKAR, New York.
86 PULLMAN, A. and B. PULLMAN, New paths in the molecular orbital approach to solvation in biological molecules. *Q. Rev. Biophys.*, 1975; **7**: 506–566.
87 JORGENSEN, W. L., Ab initio molecular orbital study of the geometric properties and protonation of alkyl chloride. *J. Am. Chem. Soc.*, 1978; **100**: 1057–1061.
88 CRAMER, C. J. and D. G. TRUHLAR, General parameterized SCF model for free energies of solvation in aqueous solution. *J. Am. Chem. Soc.*, 1991; **113**: 8305–8311.
89 HONIG, B., K. SHARP, and A.-S. YANG, Macroscopic models of aqueous solutions: biological and chemical applications. *J. Phys. Chem.*, 1993; **97**: 1101–1109.
90 ÅQVIST, J., Ion-water interaction potentials derived from free energy perturbation simulations. *J. Chem. Phys.*, 1990; **94**: 8021–8024.
91 WARSHEL, A. and Z. T. CHU, Calculations of solvation free energies in chemistry and biology, in *ACS Symposium Series: Structure and Reactivity in Aqueous Solution. Characterization of Chemical and Biological Systems* (D. G. TRUHLAR, Ed.). Washington, DC: ACS, 1994, p. 72–93.
92 FIGUEIRIDO, F., G. S. DELBUONO, and R. M. LEVY, On finite-size effects in computer simulations using the Ewald potential. *J. Chem. Phys.*, 1995; **103**: 6133–6142.
93 ÅQVIST, J. and T. HANSSON, On the validity of electrostatic linear response in polar solvents. *J. Phys. Chem.*, 1996; **100**: 9512–9521.
94 OROZCO, M. and F. J. LUQUE, Generalized linear-response approximation in discrete methods. *Chem. Phys. Lett.*, 1997; **265**: 473–480.
95 FLORIÁN, J. and A. WARSHEL, Langevin dipoles model for ab initio calculations of chemical processes in solution: parameterization and application to hydration free energies of neutral and ionic solutes, and conformational analysis in aqueous solution. *J. Phys. Chem. B*, 1997; **101**(28): 5583–5595.
96 MALCOLM, N. O. J. and J. J. W. MCDOUALL, Assessment of the Langevin dipoles solvation model for Hartree-Fock wavefunctions. *Theochem-J. Mol. Struct.*, 1996; **336**: 1–9.
97 MARTEN, B., K. KIM, C. CORTIS, R. A. et al., New model for calculation of solvation free energies: correction of self-consistent reaction field continuum dielectric theory for short-range hydrogen bonding effects. *J. Phys. Chem.*, 1996; **100**: 11775–11788.
98 HAWKINS, G. D., C. J. CRAMER, and D. G. TRUHLAR, Parametrized models of aqueous free-energies of solvation based on pairwise descreening of solute atomic charges from a dielectric medium. *J. Phys. Chem.*, 1996; **100**: 19824–19839.
99 FOTHERGILL, M., M. F. GOODMAN, J. PETRUSKA, and A. WARSHEL, Structure-energy analysis of the role of metal ions in phosphodiester bond hydrolysis by DNA polymerase I. *J. Am. Chem. Soc.*, 1995; **117**(47): 11619–11627.
100 ÅQVIST, J. and M. FOTHERGILL, Computer simulation of the

triosephosphate isomerase catalyzed reaction. *J. Biol. Chem.*, 1996; **271**: 10010–10016.

101 BEROZA, P. and D. R. FREDKIN, Calculation of amino acid pK(a)s in a protein from a continuum electrostatic model: Method and sensitivity analysis. *J. Comput. Chem.*, 1996; **17**(10): 1229–1244.

102 KOUMANOV, A., H. RUTERJANS, and A. KARSHIKOFF, Continuum electrostatic analysis of irregular ionization and proton allocation in proteins. *Proteins*, 2002; **46**(1): 85–96.

103 YOU, T. J. and D. BASHFORD, Conformation and hydrogen ion titration of proteins: a continuum electrostatic model with conformational flexibility. *Biophys. J.*, 1995; **69**: 1721–1733.

104 VAN VLIJMEN, H. W. T., M. SCHAEFER, and M. KARPLUS, Improving the accuracy of protein pK(a) calculations: Conformational averaging versus the average structure. *Proteins*, 1998; **33**(2): 145–158.

105 ZHOU, H. X. and M. VIJAYAKUMAR, Modeling of protein conformational fluctuations in pK(a) predictions. *J. Mol. Biol.*, 1997; **267**(4): 1002–1011.

106 WENDOLOSKI, J. J. and J. B. MATTHEW, Molecular-dynamics effects on protein electrostatics. *Proteins*, 1989; **5**(4): 313–321.

107 KASSNER, R. J., Effects of non polar environments on the redox potentials of heme complexes. *Proc. Natl Acad. Sci. USA*, 1972; **69**: 2263–2267.

108 ZHOU, H.-X., Control of reduction potential by protein matrix: lesson from a spherical protein model. *J. Biol. Inorg. Chem.*, 1997; **2**: 109–113.

109 BERTINI, I., G. GORI-SAVELLINI, and C. LUCHINAT, Are unit charges always negligible? *J. Biol. Inorg. Chem.*, 1997; **2**: 114–118.

110 MAUK, A. G. and G. R. MOORE, Control of metalloprotein redox potentials: what does site-directed mutagenesis of hemoproteins tell us? *J. Biol. Inorg. Chem.*, 1997; **2**: 119–125.

111 GUNNER, M. R., E. ALEXOV, E. TORRES, and S. LIPOVACA, The importance of the protein in controlling the electro-chemistry of heme metalloproteins: methods of calculation and analysis. *J. Biol. Inorg. Chem.*, 1997; **2**: 126–134.

112 NARAY-SZABO, G., Electrostatic modulation of electron transfer in the active site of heme peroxidases. *J. Biol. Inorg. Chem.*, 1997; **2**: 135–138.

113 ARMSTRONG, F. A., Evaluations of reduction potential data in relation to coupling, kinetics and function. *J. Biol. Inorg. Chem.*, 1997; **2**: 139–142.

114 WARSHEL, A., A. PAPAZYAN, and I. MUEGGE, Microscopic and semimacroscopic redox calculations: what can and cannot be learned from continuum models. *J. Biol. Inorg. Chem.*, 1997; **2**: 143–152.

115 ROGERS, N. K. and G. R. MOORE, On the energetics of conformational-changes and pH dependent redox behavior of electron-transfer proteins. *FEBS Lett.*, 1988; **228**(1): 69–73.

116 CHURG, A. K., R. M. WEISS, A. WARSHEL, and T. TAKANO, On the action of cytochrome c: correlating geometry changes upon oxidation with activation energies of electron transfer. *J. Phys. Chem.*, 1983; **87**: 1683–1694.

117 CHURG, A. K. and A. WARSHEL, Control of redox potential of cytochrome c and microscopic dielectric effects in proteins. *Biochemistry*, 1986; **25**: 1675–1681.

118 STEPHENS, P. J., D. R. JOLLIE, and A. WARSHEL, Protein control of redox potentials of iron-sulfur proteins. *Chem. Rev.*, 1996; **96**: 2491–2513.

119 SWARTZ, P. D., B. W. BECK, and T. ICHIYE, Structural origins of redox potentials in Fe–S proteins-electrostatic potentials of crystal structures. *Biophys. J.*, 1996; **71**: 2958–2969.

120 YELLE, R. B. and T. ICHIYE, Solvation free-energy reaction curves for electron-transfer in aqueous solution – theory and simulation. *J. Phys. Chem. B*, 1997; **101**: 4127–4135.

121 WARSHEL, A. and W. W. PARSON, Dynamics of biochemical and biophysical reactions: insight from computer simluations. *Q. Rev. Biophys.*, 2001; **34**: 563–670.

122 RABENSTEIN, B., G. M. ULLMANN, and

E. W. KNAPP, Energetics of electron-transfer and protonation reactions of the quinones in the photosynthetic reaction-center of *Rhodopseudomonas viridis*. *Biochemistry*, 1998; **37**(8): 2488–2495.

123 NOODLEMAN, L., T. LOVELL, T. Q. LIU, F. HIMO, and R. A. TORRES, Insights into properties and energetics of iron-sulfur proteins from simple clusters to nitrogenase. *Curr. Opin. Chem. Biol.*, 2002; **6**(2): 259–273.

124 TEIXEIRA, V. H., C. M. SOARES, and A. M. BAPTISTA, Studies of the reduction and protonation behavior of tetraheme cytochromes using atomic detail. *J. Biol. Inorg. Chem.*, 2002; **7**(1–2): 200–216.

125 OLSSON, M. H. M., G. HONG, and A. WARSHEL, Frozen density functional free energy simulations of redox proteins: computational studies of the reduction potential of plastocyanin and rusticyanin. *J. Am. Chem. Soc.*, 2003; **125**: 5025–5039.

126 MUEGGE, I. and M. RAREY, Small molecule docking and scoring, in *Reviews in Computational Chemistry* (D. B. BOYD, Ed.). New York: Wiley-VCH, John Wiley and Sons, 2001, 1–60.

127 SHAM, Y. Y., Z. T. CHU, H. TAO, and A. WARSHEL, Examining methods for calculations of binding free energies: LRA, LIE, PDLD-LRA, and PDLD/S-LRA calculations of ligands binding to an HIV protease. *Proteins*, 2000; **39**: 393–407.

128 MADURA, J. D., Y. NAKAJIMA, R. M. HAMILTON, A. WIERZBICKI, and A. WARSHEL, Calculations of the electrostatic free energy contributions to the binding free energy of sulfonamides to carbonic anhydrase. *Struct. Chem.*, 1996; **7**: 131–137.

129 FROLOFF, N., A. WINDEMUTH, and B. HONIG, On the calculation of binding free energies using continuum methods – application to MHC class-I protein–peptide interactions. *Protein Sci.*, 1997; **6**(6): 1293–1301.

130 FLORIAN, J., M. F. GOODMAN, and A. WARSHEL, Theoretical investigation of the binding free energies and key substrate-recognition components of the replication fidelity of human DNA polymerase β. *J. Phys. Chem. B*, 2002; **106**: 5739–5753.

131 WARSHEL, A., Computer simulations of enzyme catalysis: methods, progress, and insights. *Annu. Rev. Biophys. Biomech.*, 2003. **32**: 425–443.

132 WARSHEL, A., *Computer Modeling of Chemical Reactions in Enzymes and Solutions*. New York: John Wiley & Sons, 1991, 236.

133 FERSHT, A., *Structure and Mechanism in Protein Science. A Guide to Enzyme Catalysis and Protein Folding*. New York: W.H. Freeman and Co., 1999, 631.

134 FIELD, M., Stimulating enzyme reactions: challenges and perspectives. *J. Comput. Chem.*, 2002; **23**: 48–58.

135 BASH, P. A., M. J. FIELD, R. C. DAVENPORT, G. A. PETSKO, D. RINGE, and M. KARPLUS, Computer simulation and analysis of the reaction pathway of triosephosphate isomerase. *Biochemistry*, 1991; **30**: 5826–5832.

136 HARTSOUGH, D. S. and K. M. MERZ, JR., Dynamic force field models: molecular dynamics simulations of human carbonic anhydrase II using a quantum mechanical/molecular mechanical coupled potential. *J. Phys. Chem.*, 1995; **99**: 11266–11275.

137 ALHAMBRA, C., J. GAO, J. C. CORCHADO, J. VILLÀ, and D. G. TRUHLAR, Quantum mechanical dynamical effects in an enzyme-catalyzed proton transfer reaction. *J. Am. Chem. Soc.*, 1999; **121**: 2253–2258.

138 MARTÍ, S., J. ANDRÉS, V. MOLINER, E. SILLA, I. TUNON, and J. BERTRAN, Transition structure selectivity in enzyme catalysis: a QM/MM study of chorismate mutase. *Theor. Chem. Acc.*, 2001; **105**: 207–212.

139 MULHOLLAND, A. J., G. H. GRANT, and W. G. RICHARDS, Computer modelling of enzyme catalysed reaction mechanisms. *Protein Eng.*, 1993; **6**: 133–147.

140 SINGH, U. C. and P. A. KOLLMAN, A combined ab initio quantum mechanical and molecular mechanical method for carrying out simulations on complex molecular systems:

applications to the $CH_3Cl + Cl^-$ exchange reaction and gas phase protonation of polyethers. *J. Comput. Chem.*, 1986; **7**(6): 718–730.
141 WARSHEL, A., Energetics of enzyme catalysis. *Proc. Natl Acad. Sci. USA*, 1978; **75**: 5250–5254.
142 WARSHEL, A., Electrostatic origin of the catalytic power of enzymes and the role of preorganized active sites. *J. Biol. Chem.*, 1998; **273**: 27035–27038.
143 ROCA, M., S. MARTI, J. ANDRES et al., Theoretical modeling of enzyme catalytic power: Analysis of "cratic" and electrostatic factors in catechol O-methyltransferase. *J. Am. Chem. Soc.*, 2003; **125**(25): 7726–7737.
144 CANNON, W. R. and S. J. BENKOVIC, Solvation, reorganization energy, and biological catalysis. *J. Biol. Chem.*, 1998; **273**(41): 26257–26260.
145 NARAY-SZABO, G., M. FUXREITER, and A. WARSHEL, Electrostatic basis of enzyme catalysis, in *Computational Approaches to Biochemical Reactivity*. Dordrecht: Kluwer Academic, 1997, 237–294.
146 VAN BEEK, J., R. CALLENDER, and M. R. GUNNER, The contribution of electrostatic and van der Waals interactions to the stereospecificity of the reaction catalyzed by lactate-dehydrogenase. *Biophys. J.*, 1997; **72**: 619–626.
147 FISHER, C. L., D. E. CABELLI, R. A. HALLEWELL, P. BEROZA, E. D. GETZOFF, and J. A. TAINER, Computational, pulse-radiolytic, and structural investigations of lysine-136 and its role in the electrostatic triad of human Cu, Zn superoxide dismutase. *Proteins*, 1997; **29**: 103–112.
148 ANTOSIEWICZ, J., J. A. MCCAMMON, S. T. WLODEK, and M. K. GILSON, Simulation of charge-mutant acetylcholinesterases. *Biochemistry*, 1995; **34**: 4211–4219.
149 FUXREITER, M. and A. WARSHEL, Origin of the catalytic power of acetylcholinesterase: computer simulation studies. *J. Am. Chem. Soc.*, 1998; **120**: 183–194.
150 WARSHEL, A., J. VILLÀ, M. ŠTRAJBL, and J. FLORIÁN, Remarkable rate enhancement of orotidine 5′-monophosphate decarboxylase is due to transition state stabilization rather than ground state destabilization. *Biochemistry*, 2000; **39**: 14728–14738.
151 CUI, Q., M. ELSTNER, E. KAXIRAS, T. FRAUENHEIM, and M. KARPLUS, A QM/MM implementation of the self-consistent charge density functional tight binding (SCC-DFTB) method. *J. Phys. Chem. B*, 2001; **105**: 569–585.
152 PERUTZ, M. F., J. V. KILMARTIN, K. NISHIKURA, J. H. FOGG, P. J. G. BUTLER, and H. S. ROLLEMA, Identification of residues contributing to the Bohr effect of human-hemoglobin. *J. Mol. Biol.*, 1980; **138**(3): 649–670.
153 WARSHEL, A. and N. BARBOY, Energy storage and reaction pathways in the first step of the vision process. *J. Am. Chem. Soc.*, 1982; **104**: 1469.
154 WARSHEL, A., Charge stabilization mechanism in the visual and purple membrane pigments. *Proc. Natl Acad. Sci. USA*, 1978; **75**: 2558.
155 WARSHEL, A., Calculations of enzymic reactions: calculations of pK_a, proton transfer reactions, and general acid catalysis reactions in enzymes. *Biochemistry*, 1981; **20**: 3167–3177.
156 JORDAN, F., H. LI, and A. BROWN, Remarkable stabilization of zwitterionic intermediates may account for a billion-fold rate acceleration by thiamin diphosphate-dependent decarboxylases. *Biochemistry*, 1999; **38**(20): 6369–6373.
157 DEWAR, M. J. S. and D. M. STORCH, Alternative view of enzyme reactions. *Proc. Natl Acad. Sci. USA*, 1985; **82**: 2225–2229.
158 CROSBY, J., R. STONE, and G. E. LIENHARD, Mechanisms of thiamine-catalyzed reactions. Decarboxylation of 2-(1-carboxy-1-hydroxyethyl)-3,4-dimethylthiazolium chloride. *J. Am. Chem. Soc.*, 1970; **92**: 2891–2900.
159 LEE, J. K. and K. N. HOUK, A proficient enzyme revisited: the predicted mechanism for orotidine monophosphate decarboxylase. *Science*, 1997; **276**: 942–945.
160 HWANG, J.-K. and A. WARSHEL, Why ion pair reversal by protein

engineering is unlikely to succeed. *Nature*, 1988; **334**: 270.

161 CUTLER, R. L., A. M. DAVIES, S. CREIGHTON et al., Role of arginine-38 in regulation of the cytochrome c oxidation-reduction equilibrium. *Biochemistry*, 1989; **28**: 3188–3197.

162 BARRIL, X., C. ALEMAN, M. OROZCO, and F. J. LUQUE, Salt bridge interactions: stability of the ionic and neutral complexes in the gas phase, in solution, and in proteins. *Proteins*, 1998; **32**(1): 67–79.

163 WARSHEL, A. (Ed.), *Computational Approaches to Biochemical Reactivity. Understanding Chemical Reactivity*. Dordrecht: Kluwer Academic Publishers, 1997, 379.

164 HENDSCH, Z. S. and B. TIDOR, Do salt bridges stabilize proteins – a continuum electrostatic analysis. *Protein Sci.*, 1994; **3**(2): 211–226.

165 KUMAR, S. and R. NUSSINOV, Relationship between ion pair geometries and electrostatic strengths in proteins. *Biophys. J.*, 2002; **83**(3): 1595–1612.

166 NASSAR, N., G. HORN, C. HERRMANN, A. SCHERER, F. MCCORMICK, and A. WITTINGHOFER, The 2.2 Å crystal structure of the ras-binding domain of the serine/threonine kinase c-Raf1 in complex with Rap1A and a GTP analogue. *Nature*, 1995; **375**: 554–560.

167 MENZ, R. I., J. E. WALKER, and A. G. W. LESLIE, Structure of bovine mitochondrial F1-ATPase with nucleotide bound to all three catalytic sites: implications for the mechanism of rotary catalysis. *Cell*, 2001; **106**: 331–341.

168 FLOCK, D. and V. HELMS, Protein–protein docking of electron transfer complexes: cytochrome c oxidase and cytochrome c. *Proteins*, 2002; **47**(1): 75–85.

169 VICTOR, B. L., J. B. VICENTE, R. RODRIGUES et al., Docking and electron transfer studies between rubredoxin and rubredoxin: oxygen oxidoreductase. *J. Biol. Inorg. Chem.*, 2003; **8**(4): 475–488.

170 DONG, F., M. VIJAYAKUMAR, and H.-X. ZHOU, Comparison of Calculation and Experiment Implicates Significant Electrostatic Contributions to the Binding Stability of Barnase and Barstar. *Biophys. J.*, 2003; **85**: 49–60.

171 HILLE, B., *Ion Channels of Excitable Membranes*, 3rd edn. Sunderland, MA: Sinauer Associates, 2001, 814.

172 EISENMAN, G. and R. HORN, Ionic selectivity revisted: the role of kinetic and equilibrium processes in ion permeation through channels. *J. Membr. Biol.*, 1983; **50**: 1025–1034.

173 EISENMAN, G. and O. ALVAREZ, Structure and function of channels and channelogs as studied by computational chemistry. *J. Membr. Biol.*, 1991; **119**(2): 109–132.

174 BURYKIN, A., C. N. SCHUTZ, J. VILLA, and A. WARSHEL, Simulations of ion current in realistic models of ion channels: the KcsA potassium channel. *Proteins*, 2002; **47**: 265–280.

175 ÅQVIST, J. and A. WARSHEL, Energetics of ion permeation through membrane channels. Solvation of Na^+ by gramicidin A. *Biophys. J.*, 1989; **56**: 171–182.

176 JORDAN, P. C., Microscopic aproaches to ion transport through transmembrane channels. The model system gramicidin. *J. Phys. Chem.*, 1987; **91**: 6582–6591.

177 MORALS-CABRAL, J. H., Y. ZHOU, and R. MACKINNON, Energetic optimization of ion conduction rate by the K^+ selectivity filter. *Nature*, 2001; **414**: 37–42.

178 SHRIVASTAVA, I. H. and M. S. P. SANSOM, Simulations of ion permeation through a potassium channel: molecular dynamics of KcsA in a phospholipid bilayer. *Biophys. J.*, 2000; **78**: 557–570.

179 ALLEN, T. W., S. KUYUCAK, and S. H. CHUNG, Molecular dynamics study of the KcsA potassium channel. *Biophys. J.*, 1999; **77**: 2502–2516.

180 ÅQVIST, J. and V. LUZHKOV, Ion permeation mechanism of the potassium channel. *Nature*, 2000; **404**: 881–884.

181 LUZHKOV, V. and J. AQVIST, K^+/Na^+ selectivity of the KcsA potassium

181 channel from microscopic free energy perturbation calculations. *Biochim. Biophys. Acta Protein Struct. Mol.*, 2001; **36446**: 1–9.

182 BURYKIN, A., M. KATO, and A. WARSHEL, Exploring the origin of the ion selectivity of the KcsA potassium channel. *Proteins*, 2003; **52**: 412–426.

183 WADA, A., The alpha-helix as an electric macro-dipole. *Adv. Biophys.*, 1976; **9**: 1–63.

184 HOL, W. G. J., P. T. V. DUIJNEN, and H. J. C. BERENDSON, Alpha helix dipole and the properties of proteins. *Nature*, 1978; **273**: 443–446.

185 ROUX, B., S. BERNECHE, and W. IM, Ion channels, permeation and electrostatics: insight into the function of KcsA. *Biochemistry*, 2000; **39**(44): 13295–13306.

186 DAGGETT, V. D., P. A. KOLLMAN, and I. D. KUNTZ, Free-energy perturbation calculations of charge interactions with the helix dipole. *Chem. Scripta*, 1989; **29A**: 205–215.

187 GILSON, M. and B. HONIG, The energetics of charge–charge interactions in proteins. *Proteins*, 1988; **3**: 32–52.

188 VANDUIJNEN, P. T., B. T. THOLE, and W. G. J. HOL, Role of the active-site helix in papain, an abinitio molecular-orbital study. *Biophys. Chem.*, 1979; **9**(3): 273–280.

189 ÅQVIST, J., H. LUECKE, F. A. QUIOCHO, and A. WARSHEL, Dipoles localized at helix termini of proteins stabilize charges. *Proc. Natl Acad. Sci. USA*, 1991; **88**(5): 2026–2030.

190 LODI, P. J. and J. R. KNOWLES, Direct evidence for the exploitation of an alpha-helix in the catalytic mechanism of triosephosphate isomerase. *Biochemistry*, 1993; **32**(16): 4338–4343.

191 ROUX, B. and R. MACKINNON, The cavity and pore helices the KcsA K$^+$ channel: Electrostatic stabilization of monovalent cations. *Science*, 1999; **285**: 100–102.

192 LEVITT, M. and A. WARSHEL, Computer simulation of protein folding. *Nature*, 1975; **253**(5494): 694–698.

193 GO, N., Theoretical-studies of protein folding. *Annu. Rev. Biophys. Biol.*, 1983; **12**: 183–210.

194 ONUCHIC, J. N., P. G. WOLYNES, Z. LUTHEY-SCHULTEN, and N. D. SOCCI, Toward an outline of the topography of a realistic protein-folding funnel. *Proc. Natl Acad. Sci. USA*, 1995; **92**: 3626–3630.

195 GODZIK, A., J. SKOLNICK, and A. KOLINSKI, Simulations of the folding pathway of triose phosphate isomerase-type-alpha/beta barrel proteins. *Proc. Natl Acad. Sci. USA*, 1992; **89**(7): 2629–2633.

196 KARANICOLAS, J. and C. L. BROOKS, The structural basis for biphasic kinetics in the folding of the WW domain from a formin-binding protein: Lessons for protein design? *Proc. Natl Acad. Sci. USA*, 2003; **100**(7): 3954–3959.

197 DILL, K. A., Dominant forces in protein folding. *Biochemistry*, 1990; **29**: 7133–7155.

198 SIMMERLING, C., B. STROCKBINE, and A. E. ROITBERG, All-atom structure prediction and folding simulations of a stable protein. *J. Am. Chem. Soc.*, 2002; **124**(38): 11258–11259.

199 SNOW, C. D., N. NGUYEN, V. S. PANDE, and M. GRUEBELE, Absolute comparison of simulated and experimental protein-folding dynamics. *Nature*, 2002; **420**(6911): 102–106.

200 FAN, Z. Z., J. K. HWANG, and A. WARSHEL, Using simplified protein representation as a reference potential for all-atom calculations of folding free energies. *Theor. Chem. Acc.*, 1999; **103**: 77–80.

201 SHIRTS, M. and V. S. PANDE, Computing – Screen savers of the world unite! *Science*, 2000. **290**(5498): 1903–1904.

202 HOLLIEN, J. and S. MARQUSEE, Structural distribution of stability in a thermophilic enzyme. *Proc. Natl Acad. Sci. USA*, 1999; **96**(24): 13674–13678.

203 USHER, K. C., A. F. A. DE LA CRUZ, F. W. DAHLQUIST, R. V. SWANSON, M. I. SIMON, and S. J. REMINGTON, Crystal structures of CheY from *Thermotoga maritima* do not support conventional

explanations for the structural basis of enhanced thermostability. *Protein Sci.*, 1998; **7**(2): 403–412.

204 HONIG, B. and A. S. YANG, Free-energy balance in protein-folding. *Adv. Protein Chem.* 1995; **46**: 27–58.

205 XIAO, L. and B. HONIG, Electrostatic contributions to the stability of hyperthermophilic proteins. *J. Mol. Biol.*, 1999; **289**(5): 1435–1444.

206 IBARRA-MOLERO, B. and J. M. SANCHEZ-RUIZ, Genetic algorithm to design stabilizing surface-charge distributions in proteins. *J. Phys. Chem. B*, 2002; **106**(26): 6609–6613.

207 GILETTO, A. and C. N. PACE, Buried, charged, non-ion-paired aspartic acid 76 contributes favorably to the conformational stability of ribonuclease T-1. Biochemistry, 1999; **38**(40): 13379–13384.

208 GRIMSLEY, G. R., K. L. SHAW, L. R. FEE et al., Increasing protein stability by altering long-range coulombic interactions. *Protein Sci.*, 1999; **8**(9): 1843–1849.

209 WARSHEL, A., Conversion of light energy to electrostatic energy in the proton pump of *Halobacterium halobium*. Photochem. Photobiol., 1979; **30**: 285–290.

210 SCHUTZ, C. N. and A. WARSHEL, unpublished results.

211 WILLIMSKY, G., H. BANG, G. FISCHER, and M. A. MARAHIEL, Characterization of Cspb, a *Bacillus subtilis* inducible cold shock gene affecting cell viability at low temperatures. *J. Bacteriol.*, 1992; **174**(20): 6326–6335.

8
Protein Conformational Transitions as Seen from the Solvent: Magnetic Relaxation Dispersion Studies of Water, Co-solvent, and Denaturant Interactions with Nonnative Proteins

Bertil Halle, Vladimir P. Denisov, Kristofer Modig, and Monika Davidovic

8.1
The Role of the Solvent in Protein Folding and Stability

During the course of three billion years of evolution, proteins have adapted to their aqueous environments by exploiting the unusual physical properties of water [1] in many ways. The solvent therefore play a central role in the noncovalent interactions responsible for the unique three-dimensional structures of native globular proteins. The most important example of this is the hydrophobic effect [2, 3], which provides the main driving force for protein folding (see Chapter 6). The thermodynamic stability of native proteins, conferred by numerous but weak noncovalent interactions, is marginal [4]. Even a small variation in external conditions can therefore cause denaturation, that is, a partial or complete loss of the native polypeptide conformation. Proteins can be denatured by a variety of perturbations [5]: thermal (see Chapter 4), pressure (see Chapter 5), electrostatic (see Chapter 7), or solvent composition (see Chapters 3 and 24). Because each perturbation alters the balance of stabilizing forces in a different way, a given protein may adopt a variety of nonnative states with distinct properties.

A fundamental understanding of protein folding and stability, including the mechanism of action of denaturants and co-solvents, must be based on studies of the structure, solvation, and energetics of nonnative proteins at a level of detail comparable with what has been achieved for native proteins [6]. Because of their complexity, nonnative proteins need to be examined from different vantage points using several techniques. Most experimental studies have focused on the properties of the polypeptide chain, such as its degree of compactness and secondary structure content. Inferences about solvation have usually been indirect – where the peptide chain is not, there is solvent – or have relied on uncertain premises. The current view of nonnative protein solvation derives largely from calorimetric (see Chapter 4), volumetric, and other macroscopic measurements, the interpretation of which is highly model dependent.

Computer simulations can, in principle, give a detailed picture of nonnative protein solvation (see Chapter 32), but suffer from two serious limitations. First,

Protein Folding Handbook. Part I. Edited by J. Buchner and T. Kiefhaber
Copyright © 2005 WILEY-VCH Verlag GmbH & Co. KGaA, Weinheim
ISBN: 3-527-30784-2

because molecular dynamics trajectories cannot (yet) be extended to the time scales where proteins unfold, the experimentally studied nonnative equilibrium states of proteins are inaccessible. Second, protein conformational equilibria are governed by small free energy differences involving hundreds of noncovalent interaction sites. Even minor inaccuracies in the (mostly empirical) force field model can thus produce unacceptably large accumulated errors. At the other extreme, qualitative insights into protein folding are sought from "minimalist" models. Because such models describe the solvent implicitly or, at best, in a highly idealized manner, they cannot be expected to capture the subtle thermodynamics of real proteins.

As one of the few methods that directly probes water molecules interacting with proteins, water ^{17}O magnetic relaxation dispersion (MRD) [7, 8] has been used extensively to study both the internal and surface hydration of native proteins [9–11]. In recent years, the MRD method has also been applied to nonnative proteins, yielding new information about the interaction of water (^{17}O and ^{2}H MRD) and co-solvent (^{2}H and ^{19}F MRD) molecules with proteins denatured by heat, cold, acid, urea, guanidinium chloride, or trifluoroethanol. In the following, we summarize the results of these MRD studies, which, in most cases, indicate that nonnative proteins are more structured and less solvent exposed than commonly believed. Relevant aspects of MRD methodology and data analysis are described at the end of this chapter. More comprehensive and technical accounts of biomolecular MRD are available [7, 8].

8.2
Information Content of Magnetic Relaxation Dispersion

MRD investigations of protein solutions usually entail measurements of the longitudinal relaxation rate, R_1, for a nuclear isotope, such as ^{2}H, ^{17}O, or ^{19}F, belonging to a solvent molecular species. These R_1 measurements are performed as a function of the resonance frequency, v_0, which is determined by the strength of the applied static magnetic field. A data set, $R_1(v_0)$, covering two or more frequency decades is referred to as a dispersion profile. The MRD profile can provide quantitative information about several aspects of native and nonnative proteins, such as (i) long-lived association of water, denaturant, and co-solvent molecules with the protein, (ii) global exposure of the protein to water and other solvent components, (iii) the hydrodynamic volume of the protein, and (iv) side chain flexibility (via order parameters of labile hydrogens). Unlike high-resolution NMR spectroscopy (see Chapter 21), the MRD method is equally applicable to native and nonnative proteins, to small and large proteins, and to high and low (even subzero) temperatures (or other high-viscosity samples).

In the MRD context, a "long-lived association" usually means a residence time longer than 1 ns. The origin of this operational definition is that a correlation time of 1 ns produces a dispersion centered around 100 MHz, which is the highest resonance frequency for low-γ nuclei such as ^{2}H and ^{17}O achievable with present-day superconducting NMR magnets. Fortuitously, the 1 ns residence time also happens to be a convenient dividing line between solvent molecules buried in internal cavities, which typically have residence times in the range 10^{-8}–10^{-4} s at room temperature [9, 10], and solvent molecules interacting with the external protein surface, the vast majority of which have residence times in the range 10^{-11}–10^{-10} s at room temperature [9–11]. The MRD profile yields the quantity $N_\beta S_\beta^2$, the product of the number of long-lived solvent molecules, N_β, and their mean-square orientational order parameter S_β^2, a number in the range 0–1.

Nearly all globular proteins contain buried water molecules in the native state, on average one per 25 amino acid residues [12]. These internal water molecules are conserved to the same extent as the amino acid sequence and must therefore be essential for function [13]. Not all protein cavities of sufficient size are occupied by water molecules. To compensate for the favorable intermolecular interactions in bulk water, the cavity must allow for at least two (more commonly, three or four) hydrogen bonds with the internal water molecule. (Two exceptions to this "rule" would be a large cavity occupied by an internally hydrogen-bonded water cluster and a cavity with a large electric field, produced by nearby ionized side chains.) However, the long residence time of internal water molecules is not a consequence of these hydrogen bonds (which are also present in the bulk solvent), but result from trapping by the protein structure. The quantity $N_\beta S_\beta^2$ thus reflects the structural integrity of the protein. For a completely unfolded protein in a random-coil conformation, we expect that $N_\beta S_\beta^2 = 0$. If the locations of the buried water molecules are known in the native protein (from the crystal structure), then an observed difference in $N_\beta S_\beta^2$ between native and nonnative forms gives an indication of the structural integrity at the corresponding locations in the protein. Similar considerations apply to other solvent species.

The MRD profile also yields a measure of the global solvent exposure of the protein in the form of the quantity $N_\alpha \rho_\alpha$, the product of the number of dynamically perturbed, but short-lived (< 1 ns), solvent molecules, N_α, and their rotational retardation factor, $\rho_\alpha = \tau_\alpha/\tau_{\text{bulk}} - 1$. Because the dynamic perturbation of the solvent is short-ranged, only solvent molecules in direct interaction with the protein surface are significantly perturbed [10]. The correlation time τ_α can therefore be interpreted as the mean rotational correlation time for solvent molecules in contact with the protein surface. Technically, τ_α is the integral of the time correlation function for the second-rank Legendre polynomial [7]. However, because ρ_α involves the ratio $\tau_\alpha/\tau_{\text{bulk}}$, it is independent of the rank and can be compared directly with results obtained by other methods. The number of solvent molecules of species S in contact with the protein surface can be expressed as $N_\alpha^S = N_S \theta_S$, where N_S is the total number of S sites (that can be occupied simultaneously) on the protein surface and θ_S is the mean occupancy of these sites by S molecules. The number of sites can be estimated from geometric considerations as $N_S = A_P^S/a_S$, where A_P^S is the surface area of the protein accessible to S molecules and a_S is the mean surface area occupied by one S molecule at the protein surface. In the case of water (S = W), A_P^W is the usual solvent-accessible surface area computed with a probe radius of 1.4 Å and a_W is approximately 15 Å^2. For a 15 kDa protein, N_W is about 500.

The MRD-derived quantity $N_\alpha \rho_\alpha$ can thus provide information about the global solvent exposure of the protein in different states. If water is the only solvent component, then $\theta_W = 1$ and N_α^W is proportional to the water-accessible surface area A_P^W. Consequently, N_α^W should increase when the protein unfolds. However, the MRD profile does not monitor solvent accessibility directly; an observed change in $N_\alpha \rho_\alpha$ may also be due to a variation in the retardation factor ρ_α. For native proteins, the average τ_α is dominated by a relatively small number of water molecules located in surface pockets, where the geometric constraints prevent the cooperative motions that are responsible for the fast rotational and translational dynamics in bulk water [9–11]. If these surface pockets are disrupted in the unfolded protein, ρ_α will be smaller than for the native protein. On the other hand, water molecules that penetrate a (partially) unfolded protein, perhaps in the form of hydrogen-bonded water chains or small clusters, may be more strongly motionally retarded than water molecules at the fully exposed (convex) surface of the native protein. Because of this interpretational ambiguity, the quantity $N_\alpha \rho_\alpha$ can only provide bounds on the solvent exposure (based on an assumption about ρ_α) or on the rotational retardation (based on an assumption about N_α). On the other hand, penetrating solvent molecules may become long-lived, in which case they contribute to the parameter $N_\beta S_\beta^2$.

For solvent-denatured proteins, we must also deal with the complication of preferential solvation [14, 15]. In other words, N_α^S is determined not only by the solvent-accessible area of the protein, but also by the competition for this area by water and co-solvent molecules. The occupancy θ_S, which describes this competition, can be estimated from the known activities of both solvent species and the (usually unknown) co-solvent binding constant (which may differ between native and nonnative states).

The third parameter provided by an MRD profile is the correlation time τ_β, which is the characteristic time for orientational randomization of long-lived solvent molecules trapped within the protein. In general, this randomization occurs by two parallel and independent processes: rotational diffusion of the protein, with rotational correlation time τ_R, and escape of the solvent molecule from the protein, with mean residence time τ_S (the inverse of the dissociation rate constant). Mathematically, this is expressed as $\tau_\beta = \tau_R \tau_S/(\tau_R + \tau_S)$. Therefore, if $\tau_S \gg \tau_R$, as is often the case, the observed correlation time τ_β can be identified with the rotational correlation time τ_R of the protein. For native globular proteins, τ_R is usually described by the Stokes-Einstein-Debye equation, $\tau_R = \eta_0 V_H/(k_B T)$, where η_0 is the viscosity of the solvent (not the solution!) and V_H is an effective hydrodynamic volume. If the protein were a smooth sphere that did not perturb the solvent ($\rho_\alpha = 0$), then V_H would simply be the protein volume. However, a real protein sweeps out a larger volume when it rotates than does a compact sphere of the same volume. Furthermore, it perturbs the motion of adjacent solvent molecules. The former effect can be handled by molecular hydrodynamics, where the hydrodynamic equations are solved for the actual protein structure, specified in atomic detail [16]. The second effect can be incorporated by assigning a higher viscosity to the hydration layer [17]. An empirical approach [17], which predicts τ_R with similar accuracy, is to set

$V_H = 2V_0$, where V_0 is the protein volume obtained from the partial specific volume, v_P, that is, $V_0 = v_P M_P/N_A$.

MRD exploits trapped solvent molecules as intrinsic probes of protein rotational diffusion. It would thus appear that τ_R cannot be determined for an unfolded protein without long-lived solvent molecules. However, water ^{2}H MRD monitors not only water molecules but also the labile hydrogens in the protein that exchange sufficiently rapidly (usually meaning residence times < 1 ms) with water hydrogens. Under most conditions (pH and temperature being the critical variables), the labile hydrogen contribution is sufficient to produce a ^{2}H dispersion even in the absence of long-lived water molecules. For native proteins, τ_R can be determined by several other techniques besides MRD, for example, ^{15}N spin relaxation and fluorescence depolarization.

Under conditions where $\tau_\beta = \tau_R$, which is always the case for the labile-hydrogen contribution to the ^{2}H dispersion, the correlation time yields the hydrodynamic volume of the protein. However, for extensively unfolded proteins, V_H is an apparent volume that cannot be related to protein structure in a simple way. The Stokes-Einstein-Debye relation between τ_R and V_H is only valid for a rigid and nearly spherical protein. In the random-coil limit, the rotational dynamics would have to be described by a distribution of correlation times, as in the Rouse-Zimm theory [18]. Because a flexible protein exhibits internal rotational modes with shorter correlation times than for the rigid protein, the apparent V_H does not necessarily increase when the protein expands. This complication limits the diagnostic value of τ_R. Another confounding factor is that unfolding leads to increased exposure of nonpolar groups, which may cause the denatured protein to self-associate. However, these complications notwithstanding, the correlation time τ_β always provides a lower bound on the residence time τ_S for the long-lived solvent molecules responsible for the dispersion. As such, it defines the time scale on which the local structural integrity of the protein is probed.

8.3
Thermal Perturbations

8.3.1
Heat Denaturation

The first water ^{2}H and ^{17}O MRD study of heat denaturation examined bovine pancreatic ribonuclease A (RNase A) [19], which is known to undergo a reversible and cooperative thermal unfolding at neutral or acidic pH. However, some studies have indicated that the conformational transition may not be truly two-state. The thermally denatured state of RNase A appears to be relatively compact and is further unfolded on reduction of the four disulfide bonds.

MRD profiles have been recorded at seven temperatures in the interval 4–65 °C at pH 2.0 and 4.0 (all pH values quoted in this chapter are uncorrected for H$_2$O/

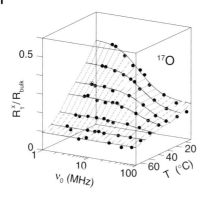

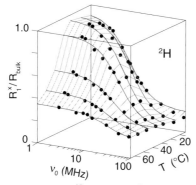

Fig. 8.1. Water ^{17}O (top) and ^{2}H (bottom) MRD profiles from 3.8 mM solution of ribonuclease A in D$_2$O at pH 2.0 and seven temperatures from 4 to 65 °C [19]. The vertical axis shows the excess relaxation rate $R_1^x = R_1 - R_{bulk}$, normalized by R_{bulk} to remove most of the trivial (viscosity-related) temperature dependence. The ^{17}O dispersion curves are three-parameter fits according to Eq. (8).

D$_2$O isotope effects). Figure 8.1 shows the MRD data obtained at pH 2, where the transition midpoint T_m is 31 °C according to CD measurements. For the native protein, observed at low temperatures, the MRD profiles yield $N_\beta S_\beta^2 = 2.3 \pm 0.1$, which requires at least three long-lived water molecules (since $S_\beta^2 \leq 1$). The crystal structure of RNase A reveals six water molecules partly buried in surface pockets, but no deeply buried water molecules. The ^{17}O MRD profile from the native protein must be due to some or all of these six water molecules. The absence of ^{17}O dispersion at high temperatures indicates a loss of persistent native structure at the locations of these hydration sites.

If the conformational transition involves only two states, then $N_\beta S_\beta^2$ yields the fraction native protein at any temperature (see Eq. 17). Figure 8.2a compares this with the native-state fraction derived from CD measurements. The MRD and CD results agree on the width (ΔT) of the transition, but T_m is 5–10 °C higher in the

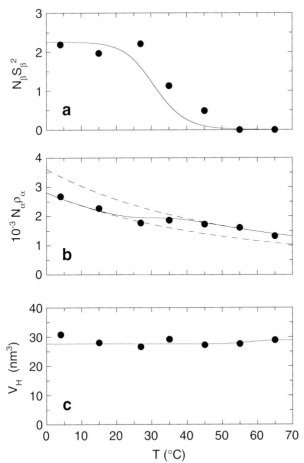

Fig. 8.2. Temperature dependence of MRD parameters derived from water ^{17}O and ^{2}H MRD profiles from 3.8 mM solution of ribonuclease A in D$_2$O: a) $N_\beta S_\beta^2$ from $\alpha(^{17}$O) at pH 2.0; b) $N_\beta S_\beta^2$ from $\beta(^{17}$O) at pH 2.0; and c) V_H from τ_β for the labile-hydrogen contribution to the ^{2}H dispersion at pH 4.0 [19]. The curves correspond to a two-state transition with transition midpoint and width as determined by CD measurements at 222 and 275 nm.

MRD data. This shift indicates that the transition is not truly two-state and that the first step does not affect the tertiary structure that maintains the 3–6 long-lived hydration sites. On the other hand, the MRD data rule out a scenario with an intermediate state extensively permeated by long-lived water molecules.

Figure 8.2b shows the surface hydration parameter $N_\alpha \rho_\alpha$ derived from the ^{17}O MRD profiles. The solid curve is a fit based on the two-state model with T_m and ΔT fixed at the values determined by CD measurements. Since the rotational retardation factor ρ_α involves a ratio of correlation times that may have different activation energies, it is expected to vary with temperature (dashed curves in Figure

8.2b). The fit in Figure 8.2b yields an activation enthalpy of 12 ± 2 kJ mol^{-1} for $\tau_\alpha/\tau_{\text{bulk}}$ ($= \rho_\alpha + 1$), in the range found for the hydration of small organic molecules. For the native state, the fit yields $N_\alpha \rho_\alpha = (1.8 \pm 0.1) \times 10^3$ at 27 °C. The number of water molecules in contact with the native protein surface can be estimated as $N_\alpha = A_P^W/a_W \approx 460$ (see Section 8.2), whereby $\rho_\alpha = 3.9 \pm 0.2$. This value is similar to what has been found for many other native proteins (at 27 °C), but is about twice as large as the ρ_α value obtained for amino acids and other small organic molecules [9–11]. The larger ρ_α value for proteins can be understood if τ_α, which typically is an average over some 500 water molecules in contact with the protein surface, has a sizable contribution from a small number of water molecules with a relatively large dynamic retardation. If ρ_α has the same temperature dependence in the native and denatured states, then the MRD data yield $N_\alpha(U)\rho_\alpha(U)/N_\alpha(N)\rho_\alpha(N) = 1.3 \pm 0.2$. Although this result does not allow us to infer how much the individual factors N_α and ρ_α differ between the two states, it restricts the possible scenarios.

Calorimetric studies showing that the heat capacity of thermally denatured proteins is essentially the same as that of the constituent amino acid side chains (see Chapter 4) have been taken to imply that thermally denatured RNase A is extensively unfolded and that the nonpolar side chains forming the native hydrophobic core are fully hydrated [20, 21]. Changes in the apparent molar volume and adiabatic compressibility on thermal denaturation of RNase A have also been attributed to extensive exposure of nonpolar side chains [22]. For complete unfolding (to an extended β conformation), the solvent-accessible area of RNase A increases by a factor 2.4 (see Chapter 20), but the topological constraints imposed by the four disulfide bonds reduce this factor to 1.9 [23]. To be consistent with a 1.9-fold area increase, the MRD results require that $\rho_\alpha = 2.6$ in the denatured state, 33% less than for the native state. A smaller rotational retardation in the denatured state would be expected if the strongly retarded surface sites in the native state are absent. However, extensive solvent penetration of the partly unfolded protein may well increase ρ_α.

The ^{2}H MRD profiles in Figure 8.1 also reflect the thermal unfolding transition, but differ qualitatively from the ^{17}O profiles in two respects. First, a sizable ^{2}H dispersion remains even at 65 °C, where the native state has negligible population. Second, the ^{2}H dispersion amplitude increases markedly with temperature outside the transition region. These are the hallmarks of a labile hydrogen contribution. A detailed analysis of the ^{2}H MRD profiles confirms that they are dominated by rapidly exchanging hydrogens in the 11 carboxyl and 31 hydroxyl groups of RNase A. This analysis also indicates that the orientational order parameters of the O–D bonds in these groups are strongly reduced in the transition, as expected from the greater conformational flexibility in the unfolded state [19].

The correlation time τ_β associated with the labile hydrogen contribution to the ^{2}H dispersion can be identified with the rotational correlation time τ_R of RNase A, which therefore can be determined also for the denatured state. To remove the trivial (via η/T) temperature dependence in τ_R, we consider the variation of the hydrodynamic volume V_H as the protein unfolds. Figure 8.2c shows V_H derived in this way from ^{2}H MRD data at pH 4.0, where $T_m = 58$ °C according to CD measurements. Thus all temperatures but the two highest refer to the native state, where $V_H = 27.6 \pm 1.0$ nm^3, in reasonable agreement with the value 30.5 nm^3 determined by ^{15}N relaxation at pH 6.4, 25 °C (and sixfold lower protein concentration) [24]. Remarkably, V_H is not significantly different at 65 °C, where the protein is 94% denatured according to CD. Presumably, this invariance is a result of compensating effects on the rotational dynamics of polypeptide expansion and flexibility (see Section 8.2).

In summary, the MRD data show that the 3–6 long-lived hydration sites in native RNase A are disrupted in the thermally denatured state. While the denatured protein lacks structural integrity on the 10 ns time scale at these sites, it is much more compact (less solvent exposed) than the maximally unfolded disulfide-intact protein. This picture is consistent with other observations, indicating substantial residual structure in the heat-denatured state of RNase A.

8.3.2
Cold Denaturation

For most proteins, the temperature-dependent free energy difference, $\Delta G(T)$, between denatured and native states exhibits negative curvature at all (accessible) temperatures, with maximum thermal stability at or near ambient temperature. Such a parabolic stability profile is predicted by macroscopic thermodynamics if the isobaric heat capacity difference, ΔC_P, is positive and temperature-independent. This simplified analysis also predicts that $\Delta G(T) = 0$, not only at the heat denaturation midpoint T_m, but also at a lower temperature, T'_m, identified as the midpoint of an exothermic cold denaturation [25]. Heat denaturation is readily understood as the result of thermal excitation of the numerous disordered configurations of the polypeptide chain. While the molecular details of cold denaturation remain obscure, it is usually attributed to a weakening of the hydrophobic effect caused by the temperature-dependent structure of bulk water [25]. If this notion is correct, cold denaturation might hold the key to understanding the major driving force for protein folding.

Studies of cold denaturation are hampered by the fact that, for most proteins, T'_m is predicted to be well below the freezing point of water. This obstacle can be circumvented by increasing T'_m and/or preventing ice formation. The most common strategy is to expose the protein to a second (nonthermal) perturbation, typically a few kilobars of hydrostatic pressure or a moderate denaturant concentration, which lowers the $\Delta G(T)$ curve, thereby raising T'_m and lowering T_m. Both pressure elevation and co-solvent addition also depress the freezing point of the water (to -19 °C at 2 kbar). However, this approach has a serious drawback: even if the secondary perturbation does not denature the protein at ambient temperature, the relative importance of the two simultaneous perturbations is difficult to establish. Thus, for example, the term pressure-assisted cold denaturation merely indicates that temperature was the experimental variable, but does not imply that the low temperature is more important than the high pressure in determining the properties of the resulting nonnative state. In other words, it is important to distinguish denaturation occurring at low temperature from denaturation caused by the low temperature.

To study cold denaturation per se for proteins with $T'_m < 0$ °C, the aqueous solvent must be maintained in a metastable, supercooled state. By careful water treatment and judicious choice of (small) sample containers to eliminate ice nucleation sites, water can be supercooled to about −20 °C [26]. A more robust method is to subdivide the protein solution into micrometer-sized emulsion droplets, the vast majority of which remain unfrozen down to the homogeneous nucleation temperature of −38 °C (for H_2O) [27]. While sample emulsification interferes with most scattering and spectroscopic measurements, it poses little or no problem for MRD studies.

Cold denaturation of bovine β-lactoglobulin (βLG) has been observed by several methods, using denaturants [28–30] or elevated pressure [31] to bring T'_m into a more accessible range. This 162-residue β-barrel protein is dimeric at physiological pH, but is essentially monomeric at pH < 3 in the absence of salt [32]. Remarkably, βLG is more stable towards urea denaturation at pH 2 than at neutral pH. Calorimetric, ^1H NMR and CD studies of βLG at pH 2 in the presence of 4 M urea (which raises T'_m to about 20 °C) indicate that cold denaturation involves an intermediate with disrupted tertiary structure, which unfolds extensively at temperatures close to 0 °C [28, 29]. However, SAXS, hydrogen exchange and heteronuclear NMR measurements at 0 °C (pH 2.5, 4 M urea) indicate a rather compact structure (27% increase of the radius of gyration) with a residual, native-like β-hairpin structure (stabilized by one of the two disulfide bonds) [30]. All these studies, as well as the MRD studies to be discussed, refer to isoform A of βLG.

Crystal structures of βLG identify two water molecules trapped in small cavities. In addition, the β barrel contains a large, elongated, hydrophobic cavity that can bind retinol, fatty acids and other nonpolar ligands. One crystal structure indicates that the ligand-free cavity is occupied by a hydrogen-bonded string of five water molecules [33]. Consequently, we expect that βLG contains at least two, but possibly more, long-lived water molecules. This structure-based prediction is consistent with ^{17}O MRD studies of native βLG (0.9 mM, pH 2.5) at 10 °C, yielding $N_\beta S_\beta^2 = 2.2 \pm 0.2$ [34]. In addition, the MRD results reveal several water molecules with a correlation time of 2–3 ns, tentatively assigned to partially buried hydration sites in small surface pockets and/or in the large binding pocket.

Lowering of the temperature to −1 °C does not affect $N_\beta S_\beta^2$ significantly (see Figure 8.3), indicating that the native protein structure remains intact at this temperature. However, addition of 4 M urea virtually eliminates the ^{17}O dispersion at −1 °C, as expected from earlier studies of urea-assisted cold denaturation [28–30]. The MRD result, $N_\beta S_\beta^2 = 0.3 \pm 0.2$, indicates a native fraction on the order of 10% under these conditions. In the absence of urea and at temperatures down to −1 °C, the rotational correlation time τ_R of βLG, derived from the ^2H MRD profile (with a large contribution from labile hydrogens), agrees quantitatively (after η/T scaling) with expectations based on hydrodynamic modeling [17] and ^{15}N relaxation [35]. This provides further evidence for a native structure under these conditions. For the denatured protein in 4 M urea at −1 °C, the ^2H correlation time τ_β is 30% shorter than the τ_R expected for the native protein, even though SAXS measurements indicate a twofold volume expansion in the denatured state [30]. This

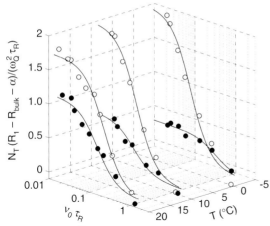

Fig. 8.3. Water ^{17}O MRD profiles from 0.9 mM solutions of β-lactoglobulin A (pH 2.5, 52% D$_2$O) at 19.4, 10.2 and -1.1 °C [34]. The solvent is salt-free water (open circles) or 4 M urea (solid circles). The vertical axis shows the β-dispersion part of the relaxation rate, $R_1 - R_{bulk} - \alpha$, normalized by $\omega_Q^2 \tau_R / N_T$ to remove the trivial dependence on viscosity and temperature. This reduced quantity approaches $N_\beta S_\beta^2$ at low frequencies. The curves are three-parameter fits according to Eq. (8).

shortening of the correlation time shows that the denatured protein does not undergo rigid body rotational diffusion. Rather, it exhibits internal reorientational motions on a range of time scales (shorter than the rigid body τ_R), as expected for labile hydrogens in unfolded polypeptide segments.

To examine whether βLG can be cold denatured in the absence of urea, water ^{2}H and ^{17}O MRD profiles have been recorded on emulsified samples down to -20 °C [34]. Control experiments yield identical MRD profiles before and after emulsification, justifying the neglect of βLG interactions with the interface of the 10 µm diameter emulsion droplets. The MRD data show no evidence of a cold denaturation, but indicate that the native structure persists down to at least -20 °C (see Figure 8.4). Indeed, after normalization to remove the trivial temperature dependence via τ_R, the ^{17}O MRD profiles exhibit little variation from 27 °C to -20 °C. The ^{2}H profiles do have reduced amplitude at low temperatures, but this can be accounted for by the slower labile hydrogen exchange.

In summary, MRD shows that βLG is extensively denatured in 4 M urea at -1 °C, in agreement with previous studies. However, even though the conventional thermodynamic analysis indicates that the midpoint temperature for cold denaturation of βLG in the absence of urea should be about -20 °C, the MRD data indicate that the native protein structure persists down to this temperature. This suggests that the standard thermodynamic analysis is incomplete and that the observed denaturation at low temperature in the presence of urea may not be driven primarily by the temperature dependence of the water structure, which is the usual explanation of cold denaturation.

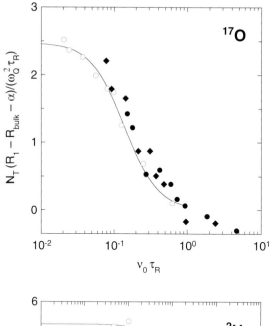

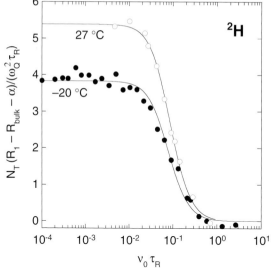

Fig. 8.4. Water ^{17}O (top) and ^{2}H (bottom) MRD profiles from emulsified 0.8 mM solutions of β-lactoglobulin A (pH 2.7, 50–100% D$_2$O) at 27 °C (open circles), −10 °C (solid diamonds) and −20 °C (solid circles) [34]. The relaxation rate has been reduced as in Figure 8.3. The curves are three-parameter fits (to the 27 °C data in the case of ^{17}O) according to Eq. (8).

8.4
Electrostatic Perturbations

Ionizable side chains contribute to protein stability via their mutual Coulomb interactions and via ionic solvation effects, that is, the charge-dependent part of side-chain interactions with the polarizable environment, including water, co-solvents, salt and buffer ions, and the rest of the protein (see Chapter 7). This contribution to protein stability can be manipulated by changing the net charge of side chains, by means of pH titration or site-directed mutagenesis, or by adding salt to the solvent. In contrast to thermal or solvent denaturation, pH titration or mutation of an ionizable side chain is a localized perturbation. However, because the Coulomb interaction is long-ranged, the effect of the perturbation is not necessarily local.

Whereas some proteins remain in the native state even at extreme pH values, others undergo a cooperative transition to a compact globular state with disordered ("molten") side chains, but with native-like secondary structure and backbone fold (see Chapter 23). This so-called molten globule (MG) state is thought to be an equilibrium analog of a universal kinetic intermediate on the protein folding pathway [36]. If the native protein is regarded as an (irregular) crystal, one would expect the MG state to be favored by high temperature. However, in most reported studies, the MG state has been induced by nonthermal perturbations. In particular, many proteins adopt MG-like states when some or all carboxylate groups are neutralized.

MG proteins are remarkably compact. Typically, the radius of gyration is only 10–30% larger than for the native state [36]. Nevertheless, if the MG is modeled as a uniformly expanded version of the native protein, this corresponds to a volume expansion factor of 1.3–2.2. Even for a small protein, the inferred difference in volume between the native and MG states translates into several hundred water molecules. However, it is not clear whether the additional volume of the MG protein is, in fact, occupied by water. This issue, which is of vital importance for the understanding of the MG state, has been addressed by water ^{17}O MRD studies of the acid MG states of three proteins [37].

Among these proteins is α-lactalbumin (αLA), which at pH 2 adopts a conformation that has come to be regarded as the paradigmatic molten globule [36]. The crystal structure of native (human) αLA shows six potentially long-lived water molecules. Two of these coordinate the bound Ca^{2+} ion, residing in a cavity between the α and β domains that traps three additional water molecules. The ^{17}O MRD profile (see Figure 8.5) from native (bovine) αLA yields $N_\beta S_\beta^2 = 3.4 \pm 0.2$, indicating a residence time between 10 ns and a few microseconds (at 27 °C) for at least four, and perhaps all six, of these water molecules. For the MG state at pH 2, $N_\beta S_\beta^2 = 1.8 \pm 0.2$, corresponding to at least two long-lived water molecules. This observation demonstrates that the structural integrity of the MG state is sufficient to trap at least two long-lived water molecules. Because the Ca^{2+} ion is no longer bound at pH 2, the difference in $N_\beta S_\beta^2$ between the native and MG states might be fully accounted for by the loss of the two calcium-coordinated water molecules. However, since the MRD data do not reveal water locations, we cannot say if the long-lived water molecules in the MG state correspond to native hydration sites.

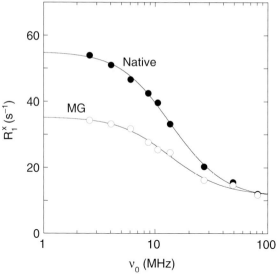

Fig. 8.5. Water ^{17}O MRD profiles from 1.9 mM solutions of bovine α-lactalbumin (D$_2$O, 27 °C) at pH 8.4 (native) and 2.0 (molten globule, MG) [37]. The vertical axis shows the excess relaxation rate $R_1^x = R_1 - R_{bulk}$. The curves are three-parameter fits according to Eq. (8). The presence of the MG state at pH 2.0 was verified by the characteristic CD profile and loss of the upfield methyl resonances in the ^{1}H NMR spectrum.

The convergence at high frequencies of the native and MG profiles (see Figure 8.5) implies that the two states have very similar surface hydration parameter. Indeed, the fits yield $N_\alpha \rho_\alpha = (1.8 \pm 0.1) \times 10^3$ for the native state and $(1.9 \pm 0.1) \times 10^3$ for the MG state. From the solvent-accessible area of the native protein, we estimate $N_\alpha = 450$ and thus $\rho_\alpha = 4.0 \pm 0.3$, a typical value for native globular proteins. Low-temperature MRD studies of protein hydration and ^{17}O relaxation studies of small-solute hydration suggest that $\rho_\alpha = 1-2$ for the vast majority of the water molecules in contact with the protein surface [11]. The ρ_α values of 4–5 typically found for native proteins must therefore be dominated by a small subset of hydration sites with strongly rotationally retarded water molecules. If these special hydration sites are abolished in the MG state, ρ_α would be substantially smaller than for the native protein. In this case, the invariance of $N_\alpha \rho_\alpha$ observed for αLA must be accidental, the reduction of ρ_α being precisely canceled by an increase of N_α due to a greater solvent exposure in the MG state. This additional solvent exposure could reflect extensive water penetration of the MG protein. This scenario would be consistent with a strictly geometric interpretation of the 35–60% reported increase of the hydrodynamic volume for the MG state, leading to the conclusion that 270 water molecules penetrate the interior of the αLA molten globule [38]. However, later measurements have indicated a significantly smaller expansion of the αLA MG (see below). Moreover, as the hydrodynamic volume reflects the viscous energy dissipation associated with protein motion, it depends, not only on protein geometry, but also on water dynamics at the protein surface, as manifested in ρ_α and in the local viscosity [17].

The observed invariance of $N_\alpha \rho_\alpha$ may also be interpreted, in a more restrictive way, as resulting from the invariance of each of the quantities N_α and ρ_α. In this interpretation, the expansion of the MG must be explained without invoking water penetration. Recent reports indicate that the αLA MG is even more compact than previously thought. From NMR diffusion measurements a volume expansion of 20 ± 4% [39] or 22 ± 5% [40] has been reported and the fluorescence anisotropy decay of the three tryptophan residues indicates even smaller values of 12–18% [41]. These results should be compared with the typical 15% expansion on melting of crystals of small organic molecules [42]. Considering also that native proteins are typically 4% more densely packed than small-molecule crystals [42], a 20% expansion of αLA upon "melting" of the side chains seems plausible. In this view, which also appears to be consistent with calorimetric data [43, 44], the weakening of the attractive van der Waals interactions in the protein core is compensated by the configurational entropy associated with alternative side chain packings without the need to invoke a significant amount of water penetration.

Although the individual factors N_α and ρ_α in the product $N_\alpha \rho_\alpha$ cannot be determined from MRD data alone, the interpretational ambiguity for αLA might be resolved by investigating other proteins. If the $N_\alpha \rho_\alpha$ invariance observed for αLA is an accidental result of large compensating changes in N_α and ρ_α, then it is unlikely to be a universal phenomenon. On the other hand, if the generic MG state has an expanded and dynamic but dry core and a native-like surface, then neither N_α nor ρ_α would change much in the native to MG transition and their product should be nearly invariant for most MG proteins. This latter scenario is supported by the MRD data available so far [37]. Thus, for human carbonic anhydrase (CA) in form II (homologous with the bovine B form), the ^{17}O MRD profiles yield $N_\alpha \rho_\alpha = (7.0 \pm 0.4) \times 10^3$ for the native state at pH 9.0 and $(6.9 \pm 0.3) \times 10^3$ for the MG state at pH 3.0 [37]. With $N_\alpha = 770$ estimated from the solvent-accessible area of the native protein, this yields a rotational retardation factor ρ_α of 9.1 ± 0.5, twice as large as for most other native proteins. This large value is attributed to the unusual prevalence in CA of hydrated surface pockets and, possibly, to internal motions of the large number of long-lived buried water molecules ($N_\beta S_\beta^2 = 8–9$ for both the native and the MG state). The observed state invariance of the large $N_\alpha \rho_\alpha$ value (for a protein of this size) suggests that strongly motionally retarded water molecules occupy surface pockets also in the MG state of CA.

A different picture emerged from a ^{1}H NMR study of (bovine) CA in the native (pH 7.2), MG (pH 3.9) and urea-denatured (pH 7.2) states [45]. Magnetization transfer between water and protein protons was observed on the 100 ms time scale in the MG state, but not in the native or denatured states. This finding was taken as evidence for a much greater hydration (due to extensive water penetration) for the MG state than for the other two states. However, a more plausible explanation is that the observed magnetization transfer is relayed by the roughly 50 hydroxyl and carboxyl protons present in CA at pH 3.9, which exchange with water protons

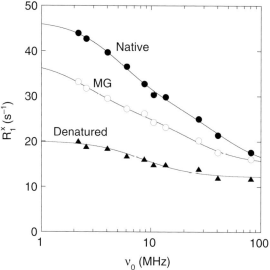

Fig. 8.6. Water ^{17}O MRD profiles from 2.1 mM solutions of equine apomyoglobin (D$_2$O, 27 °C) at pH 5.9 (native), 4.2 (molten globule, MG) and 2.2 (denatured) [37]. The vertical axis shows the excess relaxation rate $R_1^x = R_1 - R_{bulk}$. The curves are three- or five- parameter fits according to Eq. (8) or its bilorentzian generalization. The presence of the MG state at pH 4.2 was verified by the characteristic CD profile and loss of the upfield methyl resonances in the ^{1}H NMR spectrum.

within 100 ms at pH 3.9, but are either titrated (carboxyl) or exchange too slowly (hydroxyl) at pH 7.2 [46].

Also for apomyoglobin (apoMb), the ^{17}O MRD profiles for the native (pH 5.9) and MG (pH 4.2) states converge at high frequencies (see Figure 8.6), yielding $N_\alpha \rho_\alpha = (2.1 \pm 0.4) \times 10^3$ for the native state (corresponding to $\rho_\alpha = 4.0 \pm 0.8$) and $(2.2 \pm 0.2) \times 10^3$ for the MG state [37]. The MG state of apoMb has a high helix content and native-like fold [47], but, with about 70% larger hydrodynamic volume than native apoMb, it is less compact than the αLA MG. A uniform volume expansion by 70% can hardly be achieved without water penetration. However, the observed 20% increase in hydrodynamic radius [48] and radius of gyration [49] could be produced by a smaller expansion of the core (as for αLA) and some protrusion of polypeptide segments from fraying helices and termini [49]. Because water does not penetrate the core and water in contact with protruding polypeptide segments should not be much rotationally retarded ($\rho_\alpha = 1$–2, as for small molecules), this picture of the MG structure is compatible with the observed $N_\alpha \rho_\alpha$ invariance.

Myoglobin has an exceptionally large amount of internal cavities, some of which contain disordered water molecules [50]. The exchange of these buried water molecules among internal hydration sites may be responsible for the observed high-frequency dispersion with a correlation time of a few nanoseconds. The MRD profiles in Figure 8.6 were thus fitted with a bilorentzian dispersion function. It is apparent from the MRD profiles that the number of long-lived water molecules is comparable in the native (at least three) and MG (at least two) states.

At pH 2 in the absence of salt, apoMb is extensively unfolded with 50% larger radius of gyration than in the native state [49]. This contrast with αLA, which forms an MG under these conditions, presumably due to the conformational constraints imposed by the four disulfide bonds (there are no disulfide bonds in apoMb). Another factor contributing to this difference may be the 80% higher surface charge density (ratio of net charge to solvent-accessible area) in apoMb. The ^{17}O MRD profile from apoMb at pH 2 (see Figure 8.6) exhibits a small dispersion ($N_\beta S_\beta^2 = 0.3 \pm 0.1$), probably due to a small fraction of coexisting MG state. Even though the acid-denatured state should be much more solvent exposed (larger N_α) than the native or MG states, it has a smaller $N_\alpha \rho_\alpha$ of $(1.8 \pm 0.1) \times 10^3$. This implies that, on average, water penetrating the unfolded protein is considerably less rotationally retarded (smaller ρ_α) than water at the surface of the native or MG proteins. With $N_\alpha \approx 1300$ computed from the solvent-accessible area of the fully extended polypeptide chain [51] we obtain a lower bound of $\rho_\alpha \approx 1.4$, similar to the rotational retardation for small organic molecules [11]. Because the polypeptide chain cannot be fully solvent exposed even in the denatured state of apoMb, ρ_α should be somewhat larger, but still substantially smaller than the value of 4–5 typical for native proteins. The 15% reduction in $N_\alpha \rho_\alpha$ on acid denaturation of apoMb contrast with the 30% increase on heat denaturation of RNase A at pH 2 (see Section 8.3.1). This difference may result from the four disulfide bonds in RNase A, which force the denatured protein to adopt more compact conformations, where the penetrating water molecules are more strongly rotationally retarded.

In summary, ^{17}O MRD data from three proteins show that the MG state has the same value for the surface hydration parameter $N_\alpha \rho_\alpha$ as in the native state. The possibility that, for all three proteins, this invariance is the result of an accidental near-perfect compensation of a large increase in N_α (due to extensive water penetration of the MG structure) and a corresponding decrease in ρ_α seems unlikely because the native (and, presumably, MG) structures of the proteins are very different: apoMb is mainly α-helical, CA mainly β-sheet, and αLA contains one α and one β domain. Furthermore, the rotational retardation factor ρ_α for native CA is a factor 2.3 larger than for the other two proteins. An accidental compensation would therefore require an exceptionally large water penetration of the MG state of CA. For these reasons, we believe that the MRD data support the picture of a dry molten globule, with little or no water penetration of the protein core and with native-like surface hydration. This view is further supported by the finding that the MG and native states have a comparable number of long-lived water molecules trapped within the protein structure.

To the extent that the MG state plays the role of a kinetic folding intermediate, the MRD results also provide new insights into the mechanism of protein folding. Notably, they indicate that the cooperative transition from the MG to the native state is not accompanied (or driven) by water expulsion. Consequently, this transition only involves the locking in of the side chains in the native conformation, with the consequent entropy loss being compensated mainly by more favorable van der

Waals interactions in the more densely packed protein core. Furthermore, the presence of some long-lived water molecules in the MG state suggests that some or all of the internal water molecules that stabilize the native protein by extending the hydrogen bond framework are already in place when the native tertiary structure is fully formed. This underscores the important role of internal water molecules, not only in stabilizing the native protein, but also in shaping the energy landscape that governs the folding process.

8.5
Solvent Perturbations

Early work suggested that whereas thermally denatured proteins tend to be relatively compact and may retain some native-like structure, proteins exposed to high concentrations of denaturants like urea or guanidinium chloride (and disulfide reduction) are essentially structureless random coils [5]. This view of solvent denatured proteins has received support from SAXS measurements, showing that, for many proteins, solvent denaturation roughly doubles the radius of gyration. The picture of the denatured state as a fully unfolded, and thus fully solvent-exposed, polypeptide chain is also implicit in the widely used solution transfer approach to protein stability [23]. However, because a theta solvent does not exist for a heteropolymer, it is clear that the random-coil model is, at best, an approximation. In fact, a growing body of experimental evidence for the survival of significant amounts of residual structure under the harshest denaturing conditions [52, 53] suggests that the random-coil model is, in many cases, far from reality.

The phenomenon of solvent denaturation raises two fundamental questions: (i) What is the structure (in a statistical sense) of the denatured protein? and (ii) What is the mechanism whereby denaturants destabilize native proteins? Because protein stability results from the free energy difference between native and denatured states, the second problem cannot be fully solved without an answer to the first question [54]. While detailed information about polypeptide conformation in denatured states is beginning to emerge, our knowledge about the other side of the coin – the solvent – remains incomplete and indirect. Solvent denaturation is obviously a result of altered protein–solvent interactions, but there is little consensus about the precise thermodynamic and structural role of the solvent. At the high denaturant concentrations commonly used, nearly all water molecules interact directly with the denaturant. Solvent denaturation thus interferes directly with the main driving force for protein folding – the hydrophobic effect [2–5]. Moreover, a mixed solvent in contact with a protein is, in general, spatially inhomogeneous. In other words, the protein may exhibit preferential solvation, preferential solvent penetration, and differential denaturant "binding" in the native and denatured states [14, 15]. Through these complex phenomena, the structural and mechanistic aspects of solvent denaturation are inextricably linked. To make progress, we need to take a closer look at the solvent.

8.5.1
Denaturation Induced by Urea

Despite the widespread use of urea in studies of protein stability and folding thermodynamics [2, 5, 23] (see Chapter 3), the molecular mechanism whereby urea unfolds proteins has not been established. In particular, it is not clear whether urea acts directly by binding to the polypeptide, indirectly by perturbing solvent-mediated hydrophobic interactions, or by a combination of these mechanisms. The direct mechanism is made plausible by the structural similarity between urea and the peptide group, suggesting that urea–peptide interactions, like peptide–peptide interactions, can compete favorably with water–peptide interactions. If this is the case, then solvent denaturation can be driven simply by the exposure of more binding sites in the denatured protein [55]. The indirect mechanism is supported by the observation that urea enhances the solubility of not-too-small nonpolar solutes or groups [56, 57] and, by implication, weakens the hydrophobic stabilization of the folded protein.

The available structural data on urea–protein interactions are limited and, with few exceptions [58], are restricted to native proteins [59–61]. Explicit solvent simulations cannot access the time scales on which solvent-induced protein unfolding takes place and have only provided information about urea–protein interactions in the native state or in partially unfolded states at very high temperatures [62, 63]. The MRD method is an ideal tool for investigating solvent denaturation. By exploiting different nuclear isotopes, water–protein and denaturant–protein interactions can be monitored simultaneously across the unfolding transition. So far, the MRD method has been used to study urea denaturation of two proteins, β-lactoglobulin (see Section 8.3.2) and intestinal fatty acid-binding protein (I-FABP) [64].

I-FABP is a 15 kDa cytoplasmic protein with a β clam structure composed of 10 antiparallel strands that enclose a very large (500–1000 Å^3) internal cavity. Lipids are thought to enter the cavity via a small "portal" lined by two short α-helices. In the apo form, the cavity is occupied by 20–25 water molecules [65]. The internal and external hydration of native I-FABP in both apo and holo forms have been characterized in detail by ^{17}O and ^{2}H MRD [66, 67]. A three-parameter fit (see Eq. (8)) to the low-frequency part (β dispersion) of the ^{17}O MRD profile at 27 °C yields $N_\beta S_\beta^2 = 2.4 \pm 0.3$ and $\tau_\beta = 6.8 \pm 0.5$ ns [64, 66]. The correlation time τ_β agrees quantitatively with the (η/T scaled) rotational correlation time τ_R of native I-FABP as determined by ^{15}N NMR relaxation [68] and fluorescence depolarization [69].

The MRD profile also exhibits a high-frequency dispersion (labeled γ), which is only partly sampled at 27 °C. The γ dispersion is produced by the 20–25 water molecules in the binding cavity, which exchange among internal hydration sites on a time scale of 1 ns. This intra-cavity exchange has also been seen in molecular simulations [70, 71]. Because these water molecules remain in the cavity for periods longer than τ_R, they also contribute to the β dispersion. Therefore [67], $N_\beta S_\beta^2 = N_I S_I^2 + N_C S_C^2 A_C^2$, where the subscripts I and C refer to singly buried water

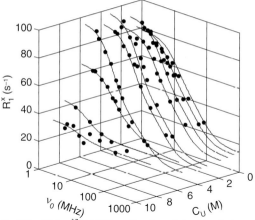

Fig. 8.7. Water ^{17}O MRD profiles from 2.4 mM solution of apo I-FABP at pH 7.0 (52% D_2O, 27 °C) and 10 different urea concentrations in the range 0–8.6 M [64]. The vertical axis shows the excess relaxation rate $R_1^x = R_1 - R_{bulk}$.

molecules and water molecules in the large binding cavity, respectively. The local, root-mean-square order parameters for these two classes of hydration sites are denoted by S_I and S_C, while the anisotropy parameter A_C is, roughly speaking, a measure of the nonsphericity of the cavity shape [67]. A detailed analysis [65–67] indicates that the β dispersion has contributions from the 20–25 cavity water molecules as well as from an isolated water molecule (known as W135) buried near a hydrophobic cluster at the turn between β-strands D and E. This internal water molecule is conserved across the family of lipid-binding proteins and must therefore contribute importantly to the stability of the native protein [72]. Because the hydrophobic cluster at the D–E turn forms early on the folding pathway, W135 can be used as an MRD marker for this (un)folding event.

Figure 8.7 shows ^{17}O MRD profiles from the apo form (without bound fatty acid) of I-FABP at 10 different urea concentrations, $C_U = 0$–8.6 M [64]. An examination of the dispersion parameters shows that the maximum seen in Figure 8.7 is due to an increase of $N_\beta S_\beta^2$ by nearly one unit in the range 0–3 M urea, where CD data indicate that the protein is fully native (see Figure 8.8a). Whereas the CD data are well described by a two-state model, the MRD data thus signal the presence of an intermediate state. This result conforms with the detection, by ^{1}H–^{15}N HSQC NMR, of an intermediate with maximum population in the range of 2.0–3.5 M urea [73].

The MRD data yield a significantly higher denaturation midpoint ($C_{1/2} = 6.5$ M) than the CD data (5.1 M). Again, this is consistent with ^{1}H–^{15}N HSQC NMR spectra [73], demonstrating that native-like structural elements persist up to 6.5 M urea, where CD and fluorescence data suggest that the protein is fully denatured. The observation of a substantial ^{17}O dispersion at 6.5 M urea implies that the residual protein structure is sufficiently permanent to trap water molecules for pe-

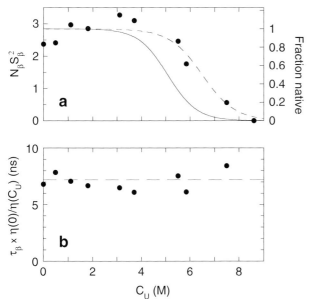

Fig. 8.8. Urea concentration dependence of MRD parameters for apo I-FABP, derived from the ^{17}O MRD profiles in Figure 8.7 [64]. a) $N_\beta S_\beta^2$ and the two-state fit (dashed curve). The solid curve and the right-hand axis refer to the apparent fraction native protein, derived from 216 and 222 nm CD data (12 μM I-FABP) and a fit based on the standard two-state linear free energy model. b) The correlation time τ_β, viscosity-corrected to remove the effect of urea on the hydrodynamic friction. The dashed line corresponds to the independently determined rotational correlation time of native I-FABP.

riods longer than 10 ns. This residual structure may be related to the equilibrium folding intermediate detected at $C_U = 4$–7 M by ^{19}F NMR on fluorinated Trp82 [74], the backbone NH of which donates a hydrogen bond to the long-lived internal water molecule (W135) in the D–E turn. The vanishing of $N_\beta S_\beta^2$ at 8.6 M urea (see Figure 8.8a) therefore indicates that the binding cavity has collapsed and that the hydrophobic cluster at the D–E turn has disintegrated.

In contrast to $N_\beta S_\beta^2$, the correlation time τ_β, when corrected for the variation in solvent viscosity, is virtually independent of urea concentration (see Figure 8.8b). In the absence of urea, τ_β can be identified with the rotational correlation τ_R of the protein. The viscosity-corrected τ_β should therefore reflect any global changes in protein structure upon denaturation. However, the disappearance of the ^{17}O dispersion at 8.6 M urea shows that there are no long-lived water molecules in the fully denatured protein. For a two-state denaturation, the frequency-dependent part of R_1 would thus be entirely due to the native protein fraction (see Eq. (17)). The invariance of τ_β in Figure 8.8b then indicates that the overall structure of the native state is essentially independent of urea concentration. This conclusion is consistent with studies of BPTI [60] and hen lysozyme [59, 61], showing that the native protein structure is virtually unaffected by high urea concentrations. On the other hand, MRD and other data reveal intermediate states in the urea denaturation of I-FABP. The observed invariance of τ_β thus also requires the hydrodynamic volume of the intermediate species to be similar to the native state.

The third piece of information obtained from a three-parameter fit (see Eq. (8)) to (the low-frequency part of) the ^{17}O MRD profile is the surface hydration parameter $N_\alpha \rho_\alpha$. For native I-FABP, the interpretation of this quantity is complicated by water exchange among hydration sites in the large binding cavity, which contributes to $N_\alpha \rho_\alpha$ to roughly the same extent as the rotational retardation of some 460 water molecules in contact with the external protein surface. However, in the denatured state of I-FABP, present at 8.6 M urea, the binding cavity has collapsed. Under these conditions, $N_\alpha \rho_\alpha$ is governed by solvent exposure, preferential solvation, and rotational retardation. While it is impossible to separate these effects, some conclusions can be drawn with the aid of independent data. Thus, if the urea-binding constant K_U is assumed to lie in the range 0.05–0.2 M^{-1}, we obtain $N_S \rho_S = (4.6–9.4) \times 10^3$ (see Section 8.7.2.6; the subscript S refers to the external protein surface). If the urea denatured state of I-FABP resembled a fully solvent-exposed polypeptide chain, we would expect $\rho_S \approx 1.3$ [11, 37] and $N_S \approx 1250$ (based on the solvent-accessible area of the fully unfolded protein). The $N_S \rho_S$ value predicted for the fully unfolded polypeptide chain is thus a factor 3–6 smaller than the experimental result. Because N_S cannot exceed the estimate for the unfolded chain, it follows that the rotational retardation factor ρ_S for urea denatured I-FABP is substantially larger than expected for a fully unfolded structure. This is expected to be the case if the denatured protein is penetrated by water strings or clusters that interact simultaneously with several polypeptide segments. In other words, the MRD data indicate that denatured I-FABP is considerably more compact than a random coil even in 8.6 M urea. This conclusion is in line with recent reports of residual native-like structure in 8 M urea for staphylococcal nuclease [52] and hen lysozyme [53].

Because the expected value of $N_S \rho_S$ is similar for the native ($460 \times 4.0 = 1.8 \times 10^3$) and fully unfolded ($1250 \times 1.3 = 1.6 \times 10^3$) states, it is clear that this quantity is not related to protein structure in a simple way. Nevertheless, one can easily imagine denatured protein structures where $N_S \rho_S$ would have much smaller or larger values. Experimental $N_S \rho_S$ values thus provide valuable information, even if a unique structural interpretation is precluded. For example, if the relative change in $N_S \rho_S$ on going from the native to the denatured state varies substantially among different proteins (of similar size) or for different denaturation agents, then we can expect a corresponding variation in the properties of the denatured states. For the two urea denatured proteins that have been investigated by MRD, βLG and I-FABP, the relative change in $N_S \rho_S$ is 100–300%. This may be contrasted with the much smaller changes of 30% for heat denatured RNase A (see Section 8.3.1) and -15% for acid-denatured apoMb (see Section 8.4). While the molecular basis of these results remains to be clarified, the very much larger change in $N_S \rho_S$ for urea-denatured proteins can hardly be explained by more extensive unfolding or solvent exposure (larger N_S). More likely, the dramatic increase in $N_S \rho_S$ reflects a stronger rotational retardation (larger ρ_S) of water molecules that penetrate urea denatured proteins. Ultimately, this must be related to the perturbations involved. Thermal and electrostatic perturbations can be viewed as general driving forces for protein expansion, where a configurational entropy gain or a reduction of Coulomb repulsion pays the free energy price for penetration by weakly interacting (with small ρ_S) water molecules. Solvent perturbations act more locally, for example, by replacing intramolecular hydrogen bonds by protein–solvent hydrogen bonds. Solvent molecules will thus penetrate the protein only if they can interact favorably with the polypeptide (leading to a large ρ_S). To elaborate this view, we need to know the relative affinity of water and urea for the interior of the unfolded protein as well as for the surface of the native protein.

Information about urea–protein interactions can be obtained from ^{2}H MRD. When the solvent contains urea and D_2O, hydrogen exchange distributes the ^{2}H nuclei uniformly among water and urea molecules. The ^{2}H magnetization therefore reports on both species. Separate water and urea resonance peaks are observed only at high magnetic fields and neutral pH, where water–urea hydrogen exchange is slow on the chemical shift time scale. Nevertheless, because the exchange remains in the slow to intermediate regime on the relaxation time scale, the individual water and urea ^{2}H relaxation rates can be determined also at low fields from a quantitative analysis of the biexponential ^{2}H magnetization recovery [64].

Whereas the water ^{2}H and ^{17}O MRD profiles yield a similar picture of I-FABP hydration, the urea ^{2}H MRD profile (see Figure 8.9) provides new information

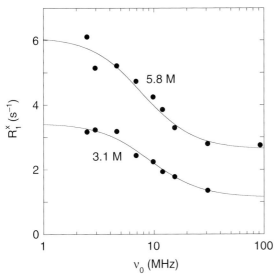

Fig. 8.9. Urea ^{2}H MRD profiles from 2.3 mM solutions of apo I-FABP at pH 7.0 (52% D_2O, 27 °C) and two different urea concentrations [64]. The vertical axis shows the excess urea ^{2}H relaxation rate $R_1^x = R_1 - R_{\text{bulk}}$, normalized to the same urea/protein mole ratio ($N_T^U = 1470$, corresponding to $C_U = 3.1$ M) to remove the trivial dependence on urea concentration. The curves are three-parameter fits according to Eq. (8).

about denaturant–protein interactions. As for the water dispersions, the urea correlation time τ_β can be identified with the tumbling time τ_R of the protein. In principle, the species giving rise to the urea dispersion could be either bound urea or labile hydrogens in the protein. However, at pH 7 hydrogen exchange between protein and urea, whether direct or water mediated, is slow on the relaxation time scale [60, 64]. Consequently, the urea ^{2}H dispersion, observed at all investigated urea concentrations from 3.1 to 7.5 M, provides direct evidence for urea binding to I-FABP with a residence time between 10 ns and 200 μs.

The fits to the urea ^{2}H profiles yield $N_\beta S_\beta^2$ values in the range 0.5–1.0. These dispersions might result from urea molecules trapped in the binding cavity, but then $N_\beta S_\beta^2$ should decrease with increasing urea concentration and vanish at $C_U = 7.5$ M, where the CD data indicate that the cavity is disrupted (see Figure 8.8a). Since $N_\beta S_\beta^2$ actually increases (slightly) with C_U, this possibility can be ruled out. It can therefore be concluded that both the native and denatured forms of I-FABP contain at least one specific urea-binding site. This finding raises the possibility that strong urea binding contributes significantly to the unfolding thermodynamics, thus compromising the linear extrapolation method widely used to determine the stability of the native protein in the absence of urea [23].

The high urea concentrations needed to denature proteins implies that weak binding to many sites is involved. Information about such interactions is contained in the MRD parameter $N_\alpha \rho_\alpha$, which can be attributed to urea molecules in short-lived (< 1 ns) association with the external protein surface. Taking into account exchange averaging over coexisting native and denatured proteins (see Eq. (17)) as well as preferential solvation effects (see Eq. (18)), it can be shown [64] that the $N_\alpha \rho_\alpha$ results are consistent with urea-binding constants in the range 0.05–0.2 M^{-1} and a similar rotational retardation for urea and water in the denatured protein.

To summarize, the urea ^{2}H and water ^{17}O MRD data support a picture of the denatured state where much of the polypeptide chain participates in clusters that are more compact and more ordered than a random coil but nevertheless are penetrated by large numbers of water and urea molecules. These solvent-permeated clusters must be sufficiently compact to allow side chains from different polypeptide segments to come into hydrophobic contact, while, at the same time, permitting solvent molecules to interact favorably with peptide groups and with charged and polar side chains. The exceptional hydrogen-bonding capacity and small size of water and urea molecules are likely to be essential attributes in this regard. In such clusters, many water and urea molecules will simultaneously interact with more than one polypeptide segment and their rotational motions will therefore be more strongly retarded than at the surface of the native protein. While the hydrogen-bonding capacity per unit volume is similar for water and urea, the 2.5-fold larger volume of urea reduces the entropic penalty for confining a certain volume of solvent to a cluster. The energetics and dynamics of solvent included in clusters is expected to differ considerably from solvent at the surface of the native protein. This view is supported by the slow water and urea rotation in the denatured state, as deduced from the MRD data.

8.5.2
Denaturation Induced by Guanidinium Chloride

Guanidinium chloride (GdmCl) is among the most potent denaturants and, at high concentrations, is thought to induce extensive unfolding of most proteins [5]. Like urea and water, GdmCl has a high hydrogen-bonding potential, but it is also a strong electrolyte. GdmCl thus introduces an electrostatic perturbation by screening Coulomb interactions. Being a salt, GdmCl should have a weaker affinity than urea for the protein interior, which is less polarizable than the bulk solvent. Consequently, GdmCl may denature proteins by a different mechanism than urea, and the resulting denatured states may exhibit qualitative differences.

Water ^{17}O MRD studies of GdmCl denaturation have been carried out on four proteins: bovine α-lactalbumin (αLA), hen lysozyme (HEWL), ribonuclease A (RNase A), and human carbonic anhydrase [37]. The first three of these proteins are of similar size (about 14 kDa) and they are all stabilized by four disulfide bonds. In fact, HEWL is structurally homologous with αLA, but lacks the strongly bound Ca^{2+} ion. According to the crystal structures, all three proteins have 6–7 potentially long-lived water molecules, but in αLA two of these coordinate the Ca^{2+} ion and in RNase A none of the water molecules is deeply buried. In accordance with these structure-based predictions, the ^{17}O MRD profiles for the native proteins (see Figure 8.10) show that $N_\beta S_\beta^2$ is largest for αLA, 3.4 ± 0.2, decreasing to 2.7 ± 0.4 for HEWL and 2.3 ± 0.3 for RNase A. The amplitude of the dispersion steps differs more than $N_\beta S_\beta^2$ because the correlation time τ_β shows the same trend among the proteins. In particular, the relatively short residence time of the buried water molecules in RNase A reduces τ_β by a factor of 2 compared with the other two proteins (see Section 8.3.1). The surface hydration parameter $N_\alpha \rho_\alpha$ has similar values for the three native proteins, yielding rotational retardation factors ρ_α in the typical range 4.0–4.5.

Surprisingly, all three proteins exhibit dispersions at high GdmCl concentrations, where conventional methods indicate that they are extensively unfolded. The residual $N_\beta S_\beta^2$ is 0.4 ± 0.1 for αLA and HEWL and 0.2 ± 0.1 for RNase A, and the correlation time τ_β is about 5 ns (the difference from the native state is barely significant). These findings demonstrate unequivocally that the extensively solvent-denatured states of these proteins trap at least one, and more if $S_\beta^2 \ll 1$, long-lived water molecules. The residence time of these water molecules, that is, the time spent within the protein before they escape into the bulk solvent, must be longer than 5 ns but shorter than 1–10 μs (the upper bound depends on the value of S_β^2).

For all three proteins, the dispersion disappears when the four disulfide bonds are broken by dithiothreitol reduction (see Figure 8.10). This indicates that the long-lived water molecules in the denatured proteins are trapped by persistent structures stabilized by the disulfide bonds. In accord with this view, no ^{17}O dispersion was observed for the solvent denatured state (4 M GdmCl) of human carbonic anhydrase (form II), which lacks disulfide bonds [37]. The MRD results are also consistent with NMR diffusion measurements on bovine αLA at 5 °C, showing

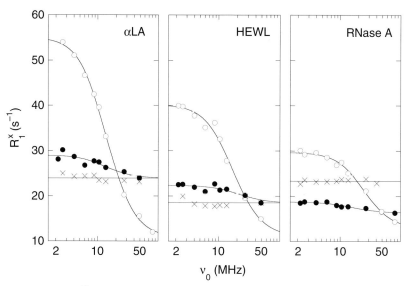

Fig. 8.10. Water ^{17}O MRD profiles from native and denatured states of bovine α-lactalbumin (1.9 mM αLA, pH 8.4 native, pH 7.0 denatured), hen lysozyme (7.0 mM HEWL, pH 4.4 native, pH 5.0 denatured), and ribonuclease A (3.8 mM RNase A, pH 4.0), all at 27 °C [37]. The MRD profiles refer to the native (open circles), denatured (filled circles) and reduced-denatured (crosses) protein. The solvent was D$_2$O and the GdmCl concentration for the denatured states was 4.4 M (αLA), 7.0 M (HEWL), or 4.0 M (RNase A). The vertical axis shows the excess relaxation rate $R_1^x = R_1 - R_{bulk}$, scaled to the same water/protein mole ratio ($N_T = 28\,500$) for all three proteins. The curves are three-parameter fits according to Eq. (8).

that the hydrodynamic volume of the native protein expands by 120 ± 10% in the disulfide-intact denatured state (6 M GdmCl) and by 380 ± 20% on disulfide reduction (without GdmCl) [40].

As judged by the ^{17}O MRD data in Figure 8.10, the reduced–denatured states of the three proteins are very similar with regard to hydration and are therefore likely to have similar configurational statistics. (The smaller excess relaxation for HEWL can be attributed to the higher GdmCl concentration. When HEWL and αLA are both studied at 7 M GdmCl, nearly identical MRD profiles are obtained for the denatured states.) However, the disulfide-intact reduced states are markedly different. For the structurally homologous αLA and HEWL, disulfide reduction has virtually no effect on the surface hydration parameter $N_\alpha \rho_\alpha$ (see Figure 8.10). This is an unexpected result, because disulfide reduction is accompanied by a large expansion of the hydrodynamic volume of αLA [40]. Because the invariance of $N_\alpha \rho_\alpha$ under disulfide reduction has only been observed for two structurally related proteins, and is not seen for RNase A, it may result from an accidental compensation of a sizeable increase in N_α by a corresponding decrease in ρ_α. For RNase A, $N_\alpha \rho_\alpha$ increases by 42 ± 2% when the four disulfide bonds are broken (see Figure 8.10). The MRD

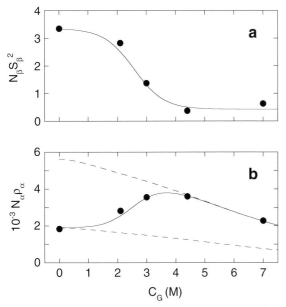

Fig. 8.11. Guanidinium chloride concentration dependence of a) $N_\beta S_\beta^2$ and b) $N_\alpha \rho_\alpha$, derived from ^{17}O MRD profiles for 1.9 mM disulfide-intact α-lactalbumin at pH 7 and 27 °C [37]. The solid curves resulted from a simultaneous two-state fit based on Eqs (17) and (18) and the standard linear free energy model. The dashed curves show the effect on $N_\alpha \rho_\alpha$ for the native and denatured states of replacing water by GdmCl.

data thus indicate that the hydration of the disulfide-intact denatured state is more native-like for RNase A than for the other two proteins (smaller N_α and/or ρ_α). This is consistent with reports of substantial residual helix content and aromatic side chain clustering in GdmCl denatured RNase A.

For a quantitative analysis of MRD data from solvent-denatured proteins, it is necessary to take preferential solvation effects into account (see Section 8.7.2.6). This requires MRD data to be recorded at several denaturant concentrations. In the case of GdmCl, this has been done for αLA (see Figure 8.11) [37]. The analysis of these data shows that the variation of the MRD parameters $N_\beta S_\beta^2$ and $N_\alpha \rho_\alpha$ with the GdmCl concentration is well described by the standard two-state linear free energy model [23]. The coincidence of the $N_\beta S_\beta^2$ and $N_\alpha \rho_\alpha$ transitions shows that the release of the long-lived water molecules from the native protein occurs in the same cooperative unfolding event as the influx of short-lived water molecules into the expanded denatured state. The apparent two-state behavior also shows that the MG intermediate, which is maximally populated at GdmCl concentrations of 2–3 M, cannot have a much larger number (N_β) of long-lived water molecules than the native state. The fit to the combined $N_\beta S_\beta^2$ and $N_\alpha \rho_\alpha$ data yields a midpoint GdmCl concentration of 2.6 ± 0.2 M and an m-value of 4.8 ± 1.7 kJ mol^{-1} M^{-1}. The close agreement of these values with those derived from the far-UV CD measurements

[75] indicates that the major hydration changes upon GdmCl denaturation are correlated with the disruption of secondary structure.

The fit in Figure 8.11 yields an average GdmCl binding constant $K_G = 0.16 \pm 0.07$ M^{-1}, in good agreement with calorimetrically determined values for HEWL and RNase A [76, 77]. For the denatured state (see the upper dashed curve in Figure 8.11b), we obtain $N_S \rho_S = 5.6 \times 10^3$. This value is a factor 3.8 larger than expected for the fully hydrated polypeptide chain (based on the maximum solvent-accessible area and $\rho_S = 1.3$). Water molecules interacting with the GdmCl denatured state of αLA are thus much more dynamically perturbed (larger ρ_S) than water molecules in contact with a fully exposed polypeptide chain. This suggests that a substantial fraction of the water molecules that penetrate the denatured protein interact strongly with several polypeptide segments and, therefore, that the structure of the denatured state contains relatively compact domains. The finding that $N_S \rho_S$ is unaffected by reduction of the four disulfide bonds in αLA (see Figure 8.10) implies that the compact domains are not due to topological constraints, although the loss of the residual dispersion suggests that these domains become more dynamic. As viewed from the solvent, the GdmCl denatured state of αLA appears to be similar to the urea-denatured state of I-FABP (see Section 8.5.1).

In summary, the MRD data show that loss of specific internal water sites and influx of external solvent are concomitant with disintegration of secondary structure (as seen by CD), consistent with the view that disruption of α-helices and β-sheets is promoted by water acting as a competitive hydrogen bond partner [78]. The large rotational retardation inferred for the denatured state suggests that water penetrating the GdmCl denatured protein differs substantially from the hydration shell of a fully exposed polypeptide chain. Taken together with the residual dispersion from the denatured state of all three investigated proteins, this suggests that even strongly solvent-denatured proteins contain relatively compact domains. This is consistent with the finding that mutations can exert their destabilizing effects directly on the denatured state [54]. Furthermore, for αLA and HEWL, disulfide bond cleavage appears to affect the flexibility of the denatured protein more than its compactness. The MRD data thus suggest that, even under extremely denaturing conditions, proteins are far from the idealized random-coil state.

8.5.3
Conformational Transitions Induced by Co-solvents

Fluorinated alcohols, such as 2,2,2-trifluoroethanol (TFE), are known to stabilize regular secondary structure, in particular α-helices, in peptides and to trigger a cooperative transition to an open helical structure in many proteins [79] (see Chapter 24). Apparently, TFE allows amino acid residues to manifest their intrinsic helical propensity, which may be suppressed by nonlocal interactions in the native protein [79–81]. Bovine βLG has been widely used as a model for studies of the alcohol-induced β → α transition in proteins (see Chapter 24). The 48% α-helix content predicted theoretically greatly exceeds the 7% found in the crystal structure of βLG [82], consistent with the finding that isolated βLG fragments may adopt nonnative helical structure in aqueous solution and invariably do so in the presence of TFE [83, 84]. βLG undergoes a β → α transition in the range 15–20% TFE, with the helix content increasing from 7% in the native protein (N) to 60–80% in the helical state (H) at 30% TFE. Several studies indicate that the TFE-induced structural transformation is best described as a three-state equilibrium, N ↔ I ↔ H, with the intermediate state (I) being most populated around 15–20% TFE [85, 86]. (Section 8.3.2, which deals with urea and cold denaturation of βLG, contains further references to the structure and hydration of native βLG.)

Although the effects of TFE on peptides and proteins have been thoroughly studied, no consensus has emerged about the underlying molecular mechanisms. Most studies have focused on peptides, where the relevant equilibrium involves unstructured and helical forms: U ↔ H. One may then ask whether TFE shifts the equilibrium by stabilizing the H form or by destabilizing the U form, or both. It is frequently assumed that the peptide is preferentially solvated by TFE, that is, that TFE accumulates near the peptide–solvent interface. Direct evidence for such preferential solvation has recently come from intermolecular NOE measurements and molecular simulations [87–89]. Whether the principal effect of TFE is to strengthen intrapeptide hydrogen bonds (because TFE is a less polar solvent than water and a less potent hydrogen bond competitor) or to weaken hydrophobic interactions among side chains (by displacing water and modifying its structure) is less clear.

Because the TFE-induced N ↔ H equilibrium in proteins does not involve the unfolded state, the thermodynamics and mechanism may differ considerably from the peptide case. While CD and high-resolution NMR studies have elucidated the dependence of protein secondary structure on TFE concentration, little direct information is available about the critical role of protein solvation in TFE/water mixtures. Molecular dynamics simulations on the time scale of the N → H transition are not yet feasible and the only reported intermolecular NOE study (on hen egg white lysozyme at pH 2) detected TFE binding in the active site but did not provide quantitative results on preferential solvation or solvent penetration [90].

The MRD method has been used to monitor the interactions of water and TFE with βLG in native and TFE-induced nonnative states at pH 2.4 (where βLG is monomeric) [91]. Water ^{2}H and ^{17}O MRD profiles measured at 4 °C (see Figure 8.12) are bilorentzian, that is, a γ dispersion has to be added to Eq. (8), with the exception of the ^{17}O profile at 30% TFE, which only exhibits a γ dispersion. In the absence of TFE, the ^{2}H and ^{17}O profiles yield essentially the same correlation times: $\tau_\beta = 16$–17 ns, as expected for rotational diffusion of monomeric βLG [35], and $\tau_\gamma = 3$–4 ns, which reflects exchange of (partly) buried water molecules. The ^{17}O dispersion amplitude parameters are $N_\beta S_\beta^2 = 2.2 \pm 0.8$, consistent with the presence of two fully buried water molecules in all crystal structures of βLG, and $N_\gamma S_\gamma^2 = 9 \pm 6$, attributed to water molecules in deep surface pockets and/or in the large binding cavity. The surface hydration parameter $N_\alpha \rho_\alpha$ corresponds to $\rho_\alpha = 5.3 \pm 1.8$, slightly above the typical range. The ^{2}H MRD data contain a substantial contribution from the hydroxyl group of TFE and from about 50 carboxyl and hydroxyl hydrogens in βLG, in fast exchange with water hydrogens at pH 2.4. The ^{2}H

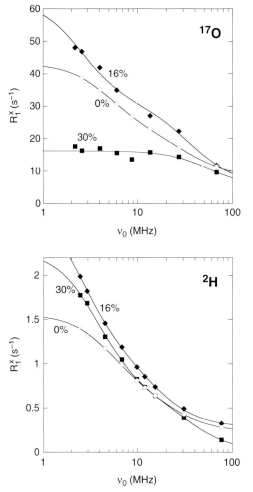

Fig. 8.12. Water ^{17}O (top) and ^{2}H (bottom) MRD profiles from 0.5 mM solutions of β-lactoglobulin A (25–50% D$_2$O, 4 °C, pH 2.4) at the indicated TFE concentrations [91]. The vertical axis shows the excess relaxation rate $R_1^x = R_1 - R_{bulk}$, scaled to the same water/protein mole ratio ($N_T^W = 103\,600$) for all samples. The curves are three-parameter (^{17}O, 30% TFE) or five-parameter (all other profiles) fits according to Eq. (8) or its bilorentzian generalization.

profile can thus provide the tumbling time (and hydrodynamic volume) of the protein even in the absence of long-lived water molecules.

Analysis of the 4 °C data in Figure 8.12 shows that the effect of adding 16% TFE is essentially to increase the viscosity (thereby increasing τ_β), whereas $N_\beta S_\beta^2$ is hardly affected. This indicates that βLG retains a compact native-like structure at 16% TFE and 4 °C, consistent with the finding that the ^{1}H–^{15}N heteronuclear single-quantum coherence (HSQC) NMR spectrum of βLG in 15% TFE is close to that of native βLG at temperatures below 15 °C [85]. However, at 27 °C, 16% TFE should be close to the midpoint of the N ↔ H transition, where the I state is highly populated [82, 83, 86]. The correlation time, $\tau_\beta = 24 \pm 4$ ns, then exceeds the rotational correlation time, $\tau_R = 14.6$ ns, predicted for the native protein at this solvent viscosity. This implies that the hydrodynamic volume of the protein has increased by $65 \pm 30\%$, consistent with the prevalent view of the TFE-induced helical state as a relatively open assembly of weakly interacting helical segments [82]. A substantial structural change is also signaled by a fourfold reduction of $N_\beta S_\beta^2 (^{17}O)$, indicating that the long-lived hydration sites are largely disrupted.

At 30% TFE, the protein should be fully converted to the H state at 27 °C [82, 83, 86]. Consistent with this expectation, τ_β is about twice as long as the τ_R predicted for the native protein, implying a correspondingly large expansion of the protein. A progressive expansion of βLG in going from 0 to 16 to 30% is also indicated by the continued decrease of $N_\beta S_\beta^2$ for both nuclei. The small value $N_\beta S_\beta^2 (^{17}O) = 0.14 \pm 0.05$ presumably reflects several weakly ordered water molecules that penetrate the open H state. At 4 °C, however, the ^{17}O profile only exhibits a γ dispersion (see Figure 8.12), indicating that the nonnative state in 30% TFE undergoes significant structural change between 4 and 27 °C.

TFE–protein interactions can be monitored by ^{19}F MRD (see Section 8.7.2.4). The ^{19}F profiles in Figure 8.13, recorded at 27 °C, demonstrate long-lived binding of TFE to βLG at 7% and 21% TFE. The correlation time τ_β of the Lorentzian dispersion is 7 ± 1 ns at 7% TFE and 4 ± 1 ns at 21% TFE, several-fold shorter than the τ_β values derived from the corresponding ^2H profiles. This implies (see Eq. (4))

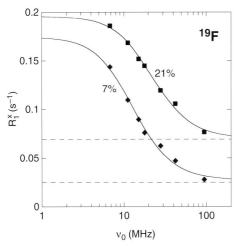

Fig. 8.13. Trifluoroethanol ^{19}F MRD profiles from 1.0 mM solutions of β-lactoglobulin A (50% D_2O, 4 °C, pH 2.4) at the indicated TFE concentrations [91]. The vertical axis shows the excess relaxation rate $R_1^x = R_1 - R_{bulk}$, scaled to the same TFE/protein mole ratio ($N_T^{TFE} = 970$) for both samples. The curves are three-parameter fits according to Eq. (8) and the dashed lines indicate the high-frequency (α) plateau.

that the residence time of TFE molecules bound to βLG is in the range 5–10 ns at 27 °C. The ^{19}F parameter $N_\beta S_\beta^2$ is 4.5 and 6 at the two concentrations, indicating that βLG contains at least five such long-lived TFE molecules. (Since S_β^2 may be well below its rigid-binding limit of 1, the number N_β might be an order of magnitude larger.)

According to CD and high-resolution NMR studies [82, 83, 85, 86], βLG retains its native structure at 7% TFE, whereas at 21% TFE the I and H states are present at roughly equal populations without any coexisting native protein. The ^{19}F MRD results thus suggest that $N_\beta S_\beta^2$ is nearly the same in the native and TFE-induced states. However, the TFE-binding sites are not necessarily the same in the different protein states. In fact, judging by the strong variation of the ^{2}H and ^{17}O profiles with TFE concentration, this is highly improbable. A more likely scenario is that the native state has a relatively small number of long-lived binding sites with large S_β^2, whereas the I and H states are penetrated by a considerably larger number of less ordered TFE molecules.

As for water, TFE residence times in the nanosecond range cannot be rationalized solely in terms of favorable hydrogen bonds and other noncovalent interactions. These interactions are also present in the bulk solvent, where molecular motions take place on a picosecond time scale. Instead, long residence times result from trapping of solvent molecules within the protein structure. In the native protein at 7% TFE, the most likely sites for long-lived TFE molecules are the small surface pockets and large binding cavity seen in the crystal structures. If the water molecules that occupy these pockets in the absence of TFE have residence times of order 100 ps, TFE binding should be accompanied by a reduction of $N_\alpha \rho_\alpha$ in the ^{17}O profile. Although ^{17}O data at 7% TFE are not available, such a reduction is indeed seen at 16% TFE.

Addition of 30% TFE reduces $N_\alpha \rho_\alpha(^{17}\text{O})$ by a factor 4 (at both temperatures), even though the expanded state of βLG should have a larger solvent-accessible area than the native state. If this is due to replacement of water by TFE at the protein surface, it should be accompanied by an increase in $N_\alpha \rho_\alpha(^{19}\text{F})$. This is indeed observed: $N_\alpha \rho_\alpha(^{19}\text{F})$ increases from 150 ± 20 at 7% TFE to 350 ± 90 at 21% TFE. Because TFE is a poor hydrogen bond acceptor and interacts only weakly with peptides, we expect ρ_α to be smaller for TFE than for water. The $N_\alpha \rho_\alpha(^{19}\text{F})$ values are therefore consistent with several hundred TFE molecules at the protein surface. On purely geometrical grounds, we estimate that about 225 TFE molecules are needed to cover the surface of native βLG. In the expanded state, this number would be larger. On the other hand, if water and TFE were uniformly distributed throughout the solvent, then only about 30 TFE molecules would make contact with the (native) protein surface at 21% TFE. To account for $N_\alpha \rho_\alpha(^{19}\text{F}) = 350$, we would then need a retardation factor of about 10, which is clearly unrealistic. The ^{19}F MRD data thus indicate that βLG has a strong preference for solvation by TFE. This is in accord with the two- to threefold TFE enrichment in the surface layer of peptides in 30% TFE seen in several simulations [87–89].

In summary, the transformations N → I → H at 27 °C are found to be accompanied by a progressive expansion of the protein and loss of specific long-lived hydration sites. The observation of ^{17}O and ^{19}F dispersions when the protein is in the H state demonstrates long-lived association of water (> 40 ns) and TFE (5–10 ns) with the expanded protein. These long residence times indicate that solvent molecules penetrate the protein matrix. The number of long-lived internal TFE molecules is $\sim 5/S_\beta^2$, with S_β^2 in the range 0–1. In the N $\rightarrow$ H transition, the surface hydration parameter decreases by a factor 4, while the corresponding TFE parameter increases by a factor 2 between 7 and 21% TFE. The MRD data are consistent with a strong accumulation of TFE at the surface as well as in the interior of the protein. The ^{19}F MRD data also demonstrate long-lived binding of several TFE molecules in the native protein, presumably by displacement of water molecules in deep surface pockets. At 4 °C, βLG is much less affected by TFE. The protein remains in the native state in the presence of 16% TFE, but adopts a nonnative structure at 30% TFE. This nonnative structure is not penetrated by long-lived water molecules, but presumably contains internal TFE molecules.

8.6
Outlook

During the past 5 years, the MRD method has been used to study proteins in a variety of nonnative states induced by thermal, electrostatic, and solvent perturbations. By focusing on the solvent, the MRD method provides a unique perspective on protein folding and stability and a valuable complement to techniques that probe the conformation and dynamics of the polypeptide chain. While the quantitative interpretation of MRD data is not always straightforward, the underlying theoretical framework of nuclear spin relaxation is rigorous. In general, the MRD method yields clear-cut information about long-lived protein–solvent interactions, while inferences about the more transient surface solvation tend to be more model dependent. Given the high complexity of the investigated systems, one cannot hope that a single technique will provide all the answers. MRD applications to nonnative proteins are still in an early exploratory phase and further studies are needed to establish the generality of the (sometimes unexpected) results obtained so far. MRD studies of nonnative proteins should not only be extended to a larger set of proteins, they can also be used to examine other types of denaturants and co-solvents as well as different perturbations, including pressure-denatured proteins.

8.7
Experimental Protocols and Data Analysis

8.7.1
Experimental Methodology

The term magnetic relaxation dispersion (MRD) refers to the frequency dependence of spin relaxation rates arising from thermal molecular motions. In a typical

MRD experiment, the longitudinal spin relaxation time, T_1, is measured as a function of the resonance frequency, ν_0, and the plot of the relaxation rate, $R_1 = 1/T_1$, versus ν_0 is known as the dispersion profile. The MRD profile resolves the total fluctuation amplitude of the so-called spin-lattice coupling (the interaction that couples the nuclear spin to the molecular degrees of freedom) into contributions from motions on different time scales. This distribution of motional frequencies, called the spectral density function, is the essential information provided by spin relaxation experiments on liquid samples [92]. In most MRD experiments, the probing frequency is controlled by the strength of the static magnetic field applied during the evolution period. The MRD method provides direct access to the spectral density over a wide frequency range: 2–3 decades can be covered with multiple-field measurements (MF-MRD), while 4–6 decades can be accessed with the field-cycling technique (FC-MRD). This allows molecular motions on a wide range of time scales to be studied. At the low (detection) fields used in most MRD studies, only relatively abundant low-molecular-weight species, such as water, co-solvents, and salt ions, can be studied [8].

8.7.1.1 Multiple-field MRD

In bulk water at room temperature, the dispersion is centered at 45 GHz. However, current NMR magnet technology only provides access to frequencies below 1 GHz. In aqueous solutions of globular proteins with molecular weights in the range 5–50 kDa, on the other hand, long-lived (> 1 ns) interactions of water molecules or co-solvents with the protein give rise to a dispersion in the range 1–100 MHz, which is accessible with conventional NMR spectrometers. Typically, the dispersion profile is the result of relaxation measurements at three or four magnetic fields in the range 2.35–14.1 T, using several NMR spectrometers equipped with fixed-field cryomagnets, combined with measurements at 5–10 fields in the range 0.1–2 T, using a dedicated MRD instrument: a tunable, multi-channel NMR spectrometer equipped with an iron-core electromagnet and a variable external field-frequency lock.

To determine the three parameters defining a Lorentzian dispersion, accurate relaxation data covering approximately two frequency decades are needed. By maintaining a high (> 100) signal-to-noise ratio at all fields, the longitudinal relaxation rate can usually be determined with an accuracy of 0.5% using the 180–τ–90 inversion recovery pulse sequence with standard phase cycling and carefully calibrated pulse length. An accurate and robust procedure for temperature calibration and control is vital for reducing systematic errors in MRD experiments. Parallel measurements on a reference sample with a known temperature-dependent (but frequency-independent) relaxation rate is a reliable temperature check.

8.7.1.2 Field-cycling MRD

For large proteins or subzero temperatures (as in studies of cold denaturation), the dispersion profile is shifted to frequencies below the accessible range of MF-MRD. At these low frequencies, MRD studies can only be performed with the field-cycling (FC) technique [93, 94]. FC-MRD differs fundamentally from MF-MRD in that the static field is varied during the course of an individual relaxation experiment. The principal limitations of FC-MRD are the relatively low maximum attainable field and the time required to switch the field, factors that have so far precluded FC-MRD studies of fast-relaxing nuclei. Due to their different strengths and weaknesses, the MF-MRD and FC-MRD techniques complement each other.

The initial nonequilibrium polarization required for a longitudinal relaxation experiment can be created either by manipulating the spin state populations with a coherent radiofrequency magnetic field without altering the spin energy levels, as in an MF-MRD experiment at a fixed magnetic field, or by changing the level spacing without altering the populations, as in the cyclic field variation performed in FC-MRD. In either case, the nonequilibrium state should preferably be established (by the radiofrequency pulse or the static field switch) in a time that is short compared with the longitudinal relaxation time T_1. Field switching can be accomplished mechanically or electronically. The mechanical approach, where the sample is pneumatically shuttled between two magnetic fields, has the advantage that the detection field can be provided by a high-field cryomagnet, with consequent improvements in sensitivity and resolution. However, the long shuttling time (typically on the order of 100 ms) limits the mechanical approach to samples with relatively slow relaxation. With electronic switching of the magnet current, the maximum field is usually in range 0.5–2 T, but the field switching rate may be 100 T s^{-1} or higher, allowing measurements of T_1 values in the millisecond range.

8.7.1.3 Choice of Nuclear Isotope

For MRD studies of protein hydration, three stable nuclides are available: ^{1}H, ^{2}H, and ^{17}O (disregarding the radioactive ^{3}H isotope). The first MRD studies of protein solutions employed the ^{1}H isotope, which has the highest sensitivity and the widest accessible frequency range. These advantages, however, are offset by serious drawbacks. Most importantly, ^{1}H relaxation generally contains a potentially confounding contribution from labile protein protons in rapid exchange with water protons. Due to proton exchange catalysis, this contribution is more pronounced away from neutral pH and in the presence of buffers. In the past, exchange-averaged labile proton contributions to ^{1}H MRD data have frequently been unjustifiably ignored. Despite having two orders of magnitude lower receptivity and one order of magnitude faster relaxation than ^{1}H, the ^{2}H isotope can be used for water MRD studies (both MF and FC). Due to the shorter intrinsic relaxation time, ^{2}H is less susceptible than ^{1}H to labile hydrogen exchange averaging. Nevertheless, rapidly exchanging protein deuterons can dominate water ^{2}H relaxation at low and high pH values and can never be neglected a priori.

Unlike the hydrogen isotopes, water ^{17}O does not exchange with protein atoms on the relaxation time scale and therefore reports exclusively on water molecules. An observed water ^{17}O dispersion in the 1–100 MHz range is therefore indisputable evidence for water molecules in long-lived association with the protein. Due to the low natural abundance of the ^{17}O nuclide, ca. 20 mM in water, isotope-enriched water is essential for accurate MRD work. ^{17}O-enriched H$_2$O and D$_2$O is commercially available at enrichment levels up to at least 60%. The fast quadrupolar relaxation of ^{17}O allows for efficient signal averaging and, in combination with the small magnetic moment, makes ^{17}O much less susceptible than ^{1}H to paramagnetic impurities.

For MRD studies of denaturants and co-solvents, the nuclides ^{2}H and ^{19}F have been used so far. The ^{2}H isotope is particularly convenient when the studied molecule can be deuterated at nonlabile positions, as with DMSO-d$_6$ in H$_2$O [95]. However, even species with labile deuterons can be studied, for example, urea-d$_4$ in D$_2$O [64], provided that the (pH- and temperature-dependent) hydrogen exchange with water is not too fast. Fluorinated co-solvents, like trifluoroethanol [91], can be studied by ^{19}F MRD. Like ^{1}H, the ^{19}F nuclide has a predominantly dipolar relaxation mechanism. Because T_1 is long (several seconds) for these spin-1/2 nuclei, paramagnetic O$_2$ makes a significant contribution to the dispersion profile. This contribution can be eliminated by O$_2$ purging of the protein solution with Ar or N$_2$ gas.

8.7.2
Data Analysis

8.7.2.1 Exchange Averaging

In MRD studies of protein solutions, the observed nucleus usually resides in a mobile solvent molecule that can diffuse between the bulk solvent region and one or more sites on (or in) the protein. Spin relaxation is then influenced by molecular motions at two levels. At the level of spin dynamics, translational diffusion transfers magnetization between microenvironments with different local relaxation rates. If sufficiently fast, such diffusional exchange leads to spatial averaging of the local relaxation rates. At the level of time correlation functions, molecular rotation (and translation, in the case of intermolecular couplings) averages out the anisotropic nuclear coupling and thus determines the local spin relaxation rates.

The theoretical framework for analyzing relaxation data from nuclei exchanging between two or more states or regions is well established. The simplest description of exchange averaging of the longitudinal relaxation rate in a protein solution yields for the excess relaxation rate, $R_1^x(\omega_0)$, that is, the relaxation rate $R_1(\omega_0)$ measured on the protein solution minus the rate R_{bulk} measured on a protein-free reference sample,

$$R_1^x(\omega_0) \equiv R_1(\omega_0) - R_{\text{bulk}} = f_\alpha(R_\alpha - R_{\text{bulk}}) + \sum_k \frac{f_{\beta k}}{\tau_k + 1/R_{1\beta k}(\omega_0)} \quad (1)$$

where $\omega_0 = 2\pi\nu_0$ is the resonance frequency in angular frequency units. The first (α) term in Eq. (1) refers to nuclei residing in molecules that have so short-lived encounters with the protein that the local relaxation rate is frequency independent throughout the investigated frequency range. The second (β) term is due to nuclei in molecules that associate with the protein for sufficiently long periods (usually meaning > 1 ns) that their local relaxation rate is frequency dependent within the investigated frequency range. Such long-lived encounters generally involve well-defined sites, each characterized by a mean residence time τ_k and a local longitudinal relaxation rate $R_{1k}(\omega_0)$. At any time, a fraction $f_\beta = \sum_k f_{\beta k}$ of the nuclei reside in such sites. The simple form of the second term in Eq. (1) is valid provided that $f_{\beta k} \ll 1$ and $R_{1\beta k} \gg R_{\text{bulk}}$. This is nearly always the case in MRD studies of protein solutions.

8.7.2.2 Spectral Density Function

Spin relaxation of nuclides with spin quantum number $I \geq 1$, like $^2\text{H}(I=1)$ and $^{17}\text{O}(I=5/2)$, is induced by the coupling of the nuclear electric quadrupole moment with fluctuating electric field gradients [92]. The local relaxation rate, $R_{1\beta k}(\omega_0)$, can then be expressed as

$$R_{1\beta k}(\omega_0) = 0.2 J_{\beta k}(\omega_0) + 0.8 J_{\beta k}(2\omega_0) \tag{2}$$

This expression is exact for ^2H and an excellent approximation for ^{17}O. The spectral density function, $J(\omega)$, is the cosine transform of the time autocorrelation function for the fluctuating electric field gradient. For a small molecule rigidly bound to a protein undergoing spherical-top rotational diffusion, it has the Lorentzian form

$$J_{\beta k}(\omega) = \omega_{Qk}^2 \frac{\tau_{\beta k}}{1 + (\omega \tau_{\beta k})^2} \tag{3}$$

The effective correlation time $\tau_{\beta k}$ is determined by the residence time τ_k in site k and the rotational correlation time τ_R of the biomolecule according to

$$\frac{1}{\tau_{\beta k}} = \frac{1}{\tau_R} + \frac{1}{\tau_k} \tag{4}$$

The quadrupole frequency ω_Q depends on the spin quantum number, the quadrupole coupling constant, and the asymmetry parameter of the electric field gradient tensor. For ^2H and ^{17}O in water, $\omega_Q = 8.7 \times 10^5 \, \text{s}^{-1}$ and $7.6 \times 10^6 \, \text{s}^{-1}$, respectively [7]. Equation (3) is readily generalized to nonspherical proteins with symmetric-top rather than spherical-top rotational diffusion. However, for globular proteins, the effect of anisotropic rotational diffusion on the shape of the relaxation dispersion is usually negligible.

In general, small molecules are not rigidly attached to the protein on the time scale (τ_R) of protein tumbling, but undergo restricted local rotational motions on time scales short compared with τ_R. If these local motions are much faster than the global motion (with correlation time $\tau_{\beta k}$) and the highest investigated MRD frequency, then the generalization of Eq. (3) is [96]

$$J_{\beta k}(\omega) = \omega_{Qk}^2 \left\{ (1 - S_{\beta k}^2) \tau_{\beta k}^{\text{loc}} + S_{\beta k}^2 \frac{\tau_{\beta k}}{1 + (\omega \tau_{\beta k})^2} \right\} \tag{5}$$

where $\tau_{\beta k}^{\text{loc}}$ is an effective correlation time for the local restricted rotation. Furthermore, $S_{\beta k}$ is the generalized second-rank orientational order parameter for site k, which has a maximum value of unity for a rigidly attached molecule. When this spectral density is substituted into Eq. (2), we obtain

$$R_{1\beta k}(\omega_0) = R_{\beta k}^{\text{loc}} + b_k \tau_{\beta k} \left[\frac{0.2}{1 + (\omega_0 \tau_{\beta k})^2} + \frac{0.8}{1 + (2\omega_0 \tau_{\beta k})^2} \right] \tag{6}$$

where $R_{\beta k}^{\text{loc}} = \omega_{Qk}^2 (1 - S_{\beta k}^2) \tau_{\beta k}^{\text{loc}}$ and $b_k = (\omega_{Qk} S_{\beta k})^2$. Using this expression for $R_{1\beta k}(\omega_0)$, we can manipulate Eq. (1) into the usual fast-exchange form, but with renormalized parameters:

$$R_1^x(\omega_0) = \alpha + \sum_k f_{\beta k} b_k' \tau_{\beta k}' \left[\frac{0.2}{1 + (\omega_0 \tau_{\beta k}')^2} + \frac{0.8}{1 + (2\omega_0 \tau_{\beta k}')^2} \right] \tag{7}$$

where $\alpha = f_\alpha (R_\alpha - R_{\text{bulk}}) + \sum_k f_{\beta k} R_{\beta k}^{\text{loc}}$ and we have assumed that $f_\beta \ll 1$ and $R_{\beta k}^{\text{loc}} \tau_k \ll 1$. Furthermore, we have introduced the renormalized parameters $b_k' = b_k / [1 + R_{1\beta k}(0)\tau_k]^{1/2}$ and $\tau_{\beta k}' = \tau_{\beta k} / [1 + R_{1\beta k}(0)\tau_k]^{1/2}$. Equation (7) shows that a given site produces a Lorentzian dispersion profile even outside the fast-exchange regime. With increasing residence time τ_k, the dispersion shifts to higher frequency ($\tau_{\beta k}' < \tau_{\beta k}$) and the dispersion step is reduced in magnitude ($b_k' < b_k$), but the shape remains Lorentzian (to an excellent approximation). However, if several sites with different effective correlation times $\tau_{\beta k}$ contribute, then the dispersion profile will be stretched out over a wider frequency range.

If all sites have the same effective correlation time $\tau_{\beta k}$, which is the case if $\tau_k \gg \tau_R$ for all sites, and if also all sites are in the fast-exchange regime, meaning that $R_{1\beta k}(0)\tau_k \ll 1$, then Eq. (7) reduces to

$$R_1^x(\omega_0) = \alpha + \beta \tau_\beta \left[\frac{0.2}{1 + (\omega_0 \tau_\beta)^2} + \frac{0.8}{1 + (2\omega_0 \tau_\beta)^2} \right] \tag{8}$$

where $\beta = \sum_k f_{\beta k} b_k$. The dispersion profile is then fully characterized by the three parameters α, β, and τ_β. In the fast-exchange regime, sites of a particular type k contribute to the β parameter a quantity $f_{\beta k} b_k = N_{\beta k} (\omega_{Qk} S_{\beta k})^2 / N_T$, where $N_{\beta k}$ is the number of small (water or co-solvent) molecules bound to k-sites and N_T is the total number of such molecules, both counted per protein molecule. Usually, ω_{Qk} can be taken from solid-state data, while N_T is obtained from the sample composition (which requires an accurate determination of the protein concentration). The dispersion amplitude parameter β thus provides the molecular quantity

$$N_\beta S_\beta^2 = \frac{\beta N_T}{\omega_Q^2} \tag{9}$$

Here, $N_\beta = \sum_k N_{\beta k}$ is the total number of small molecules in long-lived (> 1 ns) association with the protein and S_β is their root-mean-square order parameter. Because $0 \leq S_\beta^2 \leq 1$, the quantity $N_\beta S_\beta^2$ furnishes a lower bound on the number N_β.

The number N_α of perturbed but short-lived solvent molecules is usually very much larger than the number N_β of long-lived molecules. We can then neglect the local-motion contribution to α from the long-lived molecules. The high-frequency amplitude parameter α thus provides the molecular quantity

$$N_\alpha \rho_\alpha = \frac{\alpha N_T}{R_{bulk}} \qquad (10)$$

where $\rho_\alpha = R_\alpha/R_{bulk} - 1 = \tau_\alpha/\tau_{bulk} - 1$ is a measure of the increase of the rotational correlation time τ_α for the N_α perturbed molecules relative to the rotational correlation time τ_{bulk} of the same molecule in the bulk solvent. The quantity ρ_α, known as the rotational retardation factor, is an average over all N_α short-lived solvation sites. The dynamic solvent perturbation is short-ranged [10], so that ρ_α differs significantly from zero only for molecules that interact directly with the surface. If the solvent is pure water, the number N_α can therefore be estimated by dividing the solvent-accessible surface area of the protein by the mean area occupied by a water molecule, usually taken as 15 Å² (see Section 8.2).

8.7.2.3 Residence Time

The MRD profile separates contributions from interaction sites with long and short residence times τ_k. The observation of a dispersion step demonstrates unambiguously that a fraction of the nuclei reside in sites with correlation times on the order of the inverse dispersion frequency. To contribute maximally to the relaxation dispersion, a site k must have a residence time τ_k much longer than the rotational correlation time τ_R of the biomolecule, but much shorter than the zero-frequency local relaxation time, $1/R_{1\beta k}(0)$. If $R_{1\beta k}(0) \gg R_{\beta k}^{loc}$, as is usually the case, these conditions are expressed by the inequalities

$$\tau_R \ll \tau_k \ll \frac{1}{b_k \tau_R} \qquad (11)$$

which may be said to define the MRD residence time window. Of course, sites that do not obey these inequalities may still contribute to the dispersion, but do so with less than the maximum contribution $f_{\beta k} b_k \tau_R$. If $R_{\beta k}^{loc}$ is neglected, violation of the right-hand (fast-exchange) inequality in Eq. (11) reduces the effective dispersion amplitude parameter and the effective correlation time by the relative amounts

$$\frac{\beta_k'}{\beta_k} = \left[\frac{1 + \tau_R/\tau_k}{1 + \tau_R/\tau_k + b_k \tau_R \tau_k}\right]^{1/2} \qquad (12)$$

$$\frac{\tau_{\beta k}'}{\tau_R} = \frac{1}{[(1 + \tau_R/\tau_k)(1 + \tau_R/\tau_k + b_k \tau_R \tau_k)]^{1/2}} \qquad (13)$$

For residence times on the central plateau of the MRD window, only lower and upper bounds on τ_k can be established, as expressed by the inequalities in Eq. (11). On the wide flanks of the MRD window, however, τ_k can be quantitatively determined. On the short-τ_k flank, this requires independent information about τ_R. If the τ_β value deduced from the MRD profile is much smaller than τ_R, then it can be directly identified with the residence time τ_k without the need for an accurate estimate of τ_R. In contrast, determination of residence times comparable to τ_R requires knowledge of the precise value of τ_R. This can be obtained from ^{1}H or ^{2}H MRD data, which generally contain contributions from labile protein hydrogens (with $\tau_k \gg \tau_R$).

Longer residence times can be determined on the long-τ_k flank of the MRD window. According to Eqs (12) and (13), τ_k can be obtained from either β'_k or $\tau'_{\beta k}$, since $\beta'_k/\beta_k = \tau'_{\beta k}/\tau_R = (1 + b_k \tau_R \tau_k)^{-1/2}$ in this regime. To determine τ_k at a given temperature, τ_R and b_k must be known. From a variable-temperature MRD study, b_k (assumed to be temperature independent) and the activation enthalpy of τ_k can be determined if τ_R is taken to scale as η/T (as expected for rotational diffusion). If τ_R is known at one temperature, τ_k is obtained at all investigated temperatures [97].

8.7.2.4 ^{19}F Relaxation

The ^{19}F nuclei in a partially fluorinated co-solvent molecule, such as trifluoroethanol (TFE), are relaxed by fluctuating magnetic dipole–dipole interactions involving ^{19}F–^{19}F and ^{19}F–^{1}H pairs. In place of Eq. (2), we then have [91, 92]

$$R_{1\beta k}(\omega_0) = 0.2 J_{\beta k}^{FF}(\omega_F) + 0.8 J_{\beta k}^{FF}(2\omega_F) + 0.1 J_{\beta k}^{FH}(\omega_H - \omega_F) + 0.3 J_{\beta k}^{FH}(\omega_F)$$
$$+ 0.6 J_{\beta k}^{FH}(\omega_H + \omega_F) \tag{14}$$

where ω_F and ω_H are the ^{19}F and ^{1}H angular resonance frequencies. The dipolar spectral density functions $J_{\beta k}^{FF}(\omega)$ and $J_{\beta k}^{FH}(\omega)$ are given by Eq. (5), but with the quadrupole frequency ω_Q replaced by the dipole-coupling frequencies ω_{FF} and ω_{FH}. For the TFE molecule, the ^{19}F amplitude parameters α and $\beta = \beta_{FF} + \beta_{FH}$ are given by [91]

$$\alpha = \frac{R_{\text{bulk}}}{N_T^{\text{TFE}}} N_\alpha \rho_\alpha \tag{15}$$

$$\beta_{FF} = 2 \times \frac{3}{2} \frac{\omega_{FF}^2}{N_T^{\text{TFE}}} N_\beta S_\beta^2 \tag{16a}$$

$$\beta_{FH} = 2 \frac{\omega_{FH}^2}{N_T^{\text{TFE}}} N_\beta S_\beta^2 \tag{16b}$$

where N_T^{TFE} is the TFE/protein mole ratio and $N_\alpha, \rho_\alpha, N_\beta$, and S_β now refer to TFE rather than to water. The factor 2 in Eq. (16) appears because each ^{19}F nucleus is dipole coupled to two other ^{19}F nuclei in the CF$_3$ group or to two protons in the adjacent CH$_2$ group. The factor 3/2 in Eq. (16a) reflects cross-relaxation within a homonuclear pair of dipole-coupled nuclei. The dipole-coupling frequencies ω_{FF} and ω_{FH} in Eq. (16) refer to bound TFE molecules that tumble with the protein and are averaged over the much faster internal rotation of the CF$_3$ group. From the molecular geometry of TFE, one obtains $\omega_{FF} = -3.39 \times 10^4$ rad s^{-1} and $\omega_{FH} = 2.39 \times 10^4$ rad s^{-1}.

At high magnetic fields, it is necessary to include a contribution to ^{19}F relaxation from rotational modulation of the anisotropic ^{19}F shielding tensor and at high temperatures there may also be a contribution from the so-called spin-rotation mechanism [92]. Under certain conditions it may also be necessary to include the effect of intermolecular dipole–dipole couplings.

8.7.2.5 Coexisting Protein Species

Under conditions where significant populations of two or more protein states are present simultaneously, the relaxation theory must be generalized. Typically, the observed solvent molecules exchange rapidly (on the relaxation time scale) between solvation sites on the protein (in different states) and the bulk solvent. The observed (excess) relaxation rate is then simply a population-weighted average. For example, for coexisting native (N) and unfolded (U) states:

$$R_1^x = x_N R_1^x(N) + (1 - x_N) R_1^x(U) \tag{17}$$

where x_N is the fraction native protein. It follows from Eq. (8), that the high-frequency amplitude parameter α may be expressed as a population-weighted average in the same way. In principle, the observed (β) dispersion is a population-weighted superposition of dispersions from the native and unfolded states. In practice, the rotational correlation times (τ_R) of the native and (partially) unfolded states are often sufficiently similar that the dispersion profile can be accurately described by a single Lorentzian. Also the amplitude parameter β can then be expressed as a population-weighted average. Of course, if the long-lived solvent sites have been lost in the unfolded state, then only the native state contributes to the β parameter.

8.7.2.6 Preferential Solvation

In MRD studies of solvent-induced denaturation, the variation of the native protein fraction x_N with the denaturant concentration C_D may be obtained from the usual assumption that the free energy of unfolding varies linearly with C_D [23]. It then follows that $x_N = 1/(1 + K_U)$, with the unfolding constant $K_U = \exp[m(C_D - C_{1/2})/(RT)]$, where the parameters $C_{1/2}$ and m characterize the midpoint and slope, respectively, of the N → U transition [23].

The denaturant concentration not only controls the N ↔ U equilibrium; it also affects the number of water (N_α^W) and denaturant (N_α^D) molecules in contact with the protein surface (in each state). The molecular parameters $N_\alpha^W \rho_\alpha^W$ and $N_\alpha^D \rho_\alpha^D$ derived from the MRD parameters α^W and α^D by means of Eq. (10) can be decomposed into contributions from individual protein states as in Eq. (17), but now also the numbers N_α^W and N_α^D depend on C_D. This dependence can be handled by the solvent-exchange model [14], where denaturant binding to the protein surface is described thermodynamically as a one-to-one exchange with water. The denaturant occupancy averaged over all binding sites is then given by

$$\theta_D = \frac{K_D a_D^c}{a_W^x + K_D a_D^c} \tag{18}$$

where K_D is an effective denaturant binding constant (with units M^{-1}), a_D^c is the denaturant activity on the molarity scale, and a_W^x is the water activity on the mole fraction scale. We then write $N_\alpha^D(N) = \mathcal{N}_D(N)\theta_D(N)$ and $N_\alpha^W(N) = \mathcal{N}_W(N)[1 - \theta_D(N)]$ along with analogous expressions for the unfolded (U) state. Here, $\mathcal{N}_D(N)$ and $\mathcal{N}_W(N)$ are the number of sites on the surface of the native protein that can be occupied by denaturant and water molecules, respectively. Because a denaturant molecule is larger than a water molecule, we allow these numbers to be different. This may be regarded as an ad hoc generalization of the one-to-one solvent exchange model.

References

1 EISENBERG, D. and KAUZMANN, W. (1969). *The Structure and Properties of Water*. Clarendon, Oxford.
2 KAUZMANN, W. (1959). Some factors in the interpretation of protein denaturation. *Adv. Protein Chem.* **14**, 1–63.
3 TANFORD, C. (1980). *The Hydrophobic Effect*, 2nd edn, Wiley, New York.
4 DILL, K. A. (1990). Dominant forces in protein folding. *Biochemistry* **29**, 7133–7155.
5 TANFORD, C. (1968). Protein denaturation. Part A. Characterization of the denatured state. *Adv. Protein Chem.* **23**, 121–282.
6 SHORTLE, D. (1996). Structural analysis of non-native states of proteins by NMR methods. *Curr. Opin. Struct. Biol.* **6**, 24–30.
7 HALLE, B., DENISOV, V. P., and VENU, K. (1999). Multinuclear relaxation dispersion studies of protein hydration. In *Biological Magnetic Resonance* (KRISHNA, N. R. and BERLINER, L. J., eds), Vol. 17, pp. 419–484, Kluwer/Plenum, New York.
8 HALLE, B. and DENISOV, V. P. (2001). Magnetic relaxation dispersion studies of biomolecular solutions. *Methods Enzymol.* **338**, 178–201.
9 DENISOV, V. P. and HALLE, B. (1996). Protein hydration dynamics in aqueous solution. *Faraday Discuss.* **103**, 227–244.
10 HALLE, B. (1998). Water in biological systems: the NMR picture. In *Hydration Processes in Biology* (BELLISENT-FUNEL, M.-C., ed.), pp. 233–249, IOS Press, Dordrecht.
11 MODIG, K., LIEPINSH, E., OTTING, G., and HALLE, B. (2004). Dynamics of protein and peptide hydration. *J. Am. Chem. Soc.* **126**, 102–114.
12 WILLIAMS, M. A., GOODFELLOW, J. M., and THORNTON, J. M. (1994). Buried waters and internal cavities in monomeric proteins. *Protein Sci.* **3**, 1224–1235.
13 BAKER, E. N. (1995). Solvent interactions with proteins as revealed by X-ray crystallographic studies. In *Protein–Solvent Interactions* (GREGORY, R. B., ed.), pp. 143–189, M. Dekker, New York.
14 SCHELLMAN, J. A. (1994). The thermodynamics of solvent exchange. *Biopolymers* **34**, 1015–1026.
15 TIMASHEFF, S. N. (2002). Protein hydration, thermodynamic binding, and preferential hydration. *Biochemistry* **41**, 13473–13482.
16 GARCIA DE LA TORRE, J., HUERTAS, M. L., and CARRASCO, B. (2000). Calculation of hydrodynamic properties of globular proteins from their atomic-level structure. *Biophys. J.* **78**, 719–730.
17 HALLE, B. and DAVIDOVIC, M. (2003). Biomolecular hydration: from water dynamics to hydrodynamics. *Proc. Natl Acad. Sci. USA* **100**, 12135–12140.
18 DOI, M. and EDWARDS, S. F. (1986). *The Theory of Polymer Dynamics*. Clarendon, Oxford.

19 Denisov, V. P. and Halle, B. (1998). Thermal denaturation of ribonuclease A characterized by water ^{17}O and ^{2}H magnetic relaxation dispersion. *Biochemistry* **37**, 9595–9604.

20 Privalov, P. L., Tiktopulo, E. I., Venyaminov et al. (1989). Heat capacity and conformation of proteins in the denatured state. *J. Mol. Biol.* **205**, 737–750.

21 Privalov, P. L. and Makhatadze, G. I. (1990). Heat capacity of proteins. II. Partial molar heat capacity of the unfolded polypeptide chain of proteins: protein unfolding effects. *J. Mol. Biol.* **213**, 385–391.

22 Tamura, Y. and Gekko, K. (1995). Compactness of thermally and chemically denatured ribonuclease-A as revealed by volume and compressibility. *Biochemistry* **34**, 1878–1884.

23 Myers, J. K., Pace, C. N., and Scholtz, J. M. (1995). Denaturant *m* values and heat capacity changes: relation to changes in accessible surface areas of protein unfolding. *Protein Sci.* **4**, 2138–2148.

24 Cole, R. and Loria, J. P. (2002). Evidence for flexibility in the function of ribonuclease A. *Biochemistry* **41**, 6072–6081.

25 Privalov, P. L. (1990). Cold denaturation of proteins. *Crit. Rev. Biochem. Mol. Biol.* **25**, 281–305.

26 Sabelko, J., Ervin, J., and Gruebele, M. (1998). Cold-denatured ensemble of apomyoglobin: implications for the early steps of folding. *J. Phys. Chem. B* **102**, 1806–1819.

27 Rasmussen, D. H. and MacKenzie, A. P. (1973). Clustering in supercooled water. *J. Chem. Phys.* **59**, 5003–5013.

28 Griko, Yu. V. and Privalov, P. L. (1992). Calorimetric study of the heat and cold denaturation of β-lactoglobulin. *Biochemistry* **31**, 8810–8815.

29 Griko, Yu. V. and Kutyshenko, V. P. (1994). Differences in the processes of β-lactoglobulin cold and heat denaturations. *Biophys J.* **67**, 356–363.

30 Katou, H., Hoshino, M., Kamikubo, H., Batt, C. A., and Goto, Y. (2001). Native-like β-hairpin retained in the cold-denatured state of bovine β-lactoglobulin. *J. Mol. Biol.* **310**, 471–484.

31 Valente-Mesquita, V. L., Botelho, M. M., and Ferreira, S. T. (1998). Pressure-induced subunit dissociation and unfolding of dimeric β-lactoglobulin. *Biophys. J.* **75**, 471–476.

32 Sakurai, K., Oobatake, M., and Goto, Y. (2001). Salt-dependent monomer-dimer equilibrium of bovine β-lactoglobulin at pH 3. *Protein Sci.* **10**, 2325–2335.

33 Jameson, G. B., Adams, J. J., and Creamer, L. K. (2002). Flexibility, functionality and hydrophobicity of bovine β-lactoglobulin. *Int. Dairy J.* **12**, 319–329.

34 Davidovic, M. and Halle, B., manuscript in preparation.

35 Uhrinova, S., Smith, M. H., Jameson, G. B., Uhrin, D., Sawyer, L., and Barlow, P. N. (2000). Structural changes accompanying pH-induced dissociation of the β-lactoglobulin dimer. *Biochemistry* **39**, 3565–3574.

36 Arai, M. and Kuwajima, K. (2000). Role of the molten globule state in protein folding. *Adv. Protein Chem.* **53**, 209–282.

37 Denisov, V. P., Jonsson, B.-H., and Halle, B. (1999). Hydration of denatured and molten globule proteins. *Nature Struct. Biol.* **6**, 253–260.

38 Kharakoz, D. P. and Bychkova, V. E. (1997). Molten globule of human α-lactalbumin: hydration, density, and compressibility of the interior. *Biochemistry* **36**, 1882–1890.

39 Redfield, C., Schulman, B. A., Milhollen, M. A., Kim, P. S., and Dobson, C. M. (1999). α-Lactalbumin forms a compact molten globule in the absence of disulfide bonds. *Nature Struct. Biol.* **6**, 948–952.

40 Balbach, J. (2000). Compaction during protein folding studied by real-time NMR diffusion experiments. *J. Am. Chem. Soc.* **122**, 5887–5888.

41 Chakraborty, S., Ittah, V., Bai, P., Luo, L., Haas, E., and Peng, Z. (2001). Structure and dynamics of the α-lactalbumin molten globule:

fluorescence studies using proteins containing a single tryptophan residue. *Biochemistry* **40**, 7228–7238.

42 HARPAZ, Y., GERSTEIN, M., and CHOTHIA, C. (1994). Volume changes on protein folding. *Structure* **2**, 641–649.

43 GRIKO, YU. V., FREIRE, E., and PRIVALOV, P. L. (1994). Energetics of the α-lactalbumin states: a calorimetric and statistical thermodynamic study. *Biochemistry* **33**, 1889–1899.

44 GRIKO, YU. V. (2000). Energetic basis of structural stability in the molten globule state: α-lactalbumin. *J. Mol. Biol.* **297**, 1259–1268.

45 KUTYSHENKO, V. P. and CORTIJO, M. (2000). Water–protein interactions in the molten-globule state of carbonic anhydrase b: an NMR spin-diffusion study. *Protein Sci.* **9**, 1540–1547.

46 LIEPINSH, E., OTTING, G., and WÜTHRICH, K. (1992). NMR spectroscopy of hydroxyl protons in aqueous solutions of peptides and proteins. *J. Biomol. NMR* **2**, 447–465.

47 ELIEZER, D., YAO, J., DYSON, H. J., and WRIGHT, P. E. (1998). Structural and dynamic characterization of partially folded states of apomyoglobin and implications for protein folding. *Nature Struct. Biol.* **5**, 148–155.

48 GAST, K., DAMASCHUN, H., MISSEL-WITZ, R., MÜLLER-FROHNE, M., ZIRWER, D., and DAMASCHUN, G. (1994). Compactness of protein molten globules: temperature-induced structural changes of the apomyoglobin folding intermediate. *Eur. Biophys. J.* **23**, 297–305.

49 KATAOKA, M., NISHII, I., FUJISAWA, T., UEKI, T., TOKUNAGA, F., and GOTO, Y. (1995). Structural characterization of the molten globule and native states of apomyoglobin by solution X-ray scattering. *J. Mol. Biol.* **249**, 215–228.

50 OSTERMAN, A., TANAKA, I., ENGLER, N., NIIMURA, N., and PARAK, F. G. (2002). Hydrogen and deuterium in myoglobin as seen by a neutron structure determination at 1.5 Å resolution. *Biophys. Chem.* **95**, 183–193.

51 CREAMER, T. P., SRINIVASAN, R., and ROSE, G. D. (1995). Modeling unfolded states of peptides and proteins. *Biochemistry* **34**, 16245–16250.

52 SHORTLE, D. and ACKERMAN, M. S. (2001). Persistence of native-like topology in a denatured protein in 8 M urea. *Science* **293**, 487–489.

53 KLEIN-SEETHARAMAN, J., OIKAWA, M., GRIMSHAW, S. B., et al. (2002). Long-range interactions within a nonnative protein. *Science* **295**, 1719–1722.

54 SHORTLE, D. (1996). The denatured state (the other half of the folding equation) and its role in protein stability. *FASEB J.* **10**, 27–34.

55 SCHELLMAN, J. A. (1987). Selective binding and solvent denaturation. *Biopolymers* **26**, 549–559.

56 WETLAUFER, D. B., MALIK, S. K., STOLLER, L., and COFFIN, R. L. (1964). Nonpolar group participation in the denaturation of proteins by urea and guanidinium salts. Model compound studies. *J. Am. Chem. Soc.* **86**, 508–514.

57 SHIMIZU, S. and CHAN, H. S. (2002). Origins of protein denatured state compactness and hydrophobic clustering in aqueous urea: Inferences from nonpolar potentials of mean force. *Proteins* **49**, 560–566.

58 DÖTSCH, V. (1996). Characterization of protein–solvent interactions with NMR-spectroscopy: The role of urea in the unfolding of proteins. *Pharm. Acta Helv.* **71**, 87–96.

59 LUMB, K. J. and DOBSON, C. M. (1992). ^{1}H nuclear magnetic resonance studies of the interaction of urea with hen lysozyme. *J. Mol. Biol.* **227**, 9–14.

60 LIEPINSH, E. and OTTING, G. (1994). Specificity of urea binding to proteins. *J. Am. Chem. Soc.* **116**, 9670–9674.

61 PIKE, A. C. W. and ACHARYA, K. R. (1994). A structural basis for the interaction of urea with lysozyme. *Protein Sci.* **3**, 706–710.

62 TIRADO-RIVES, J., OROZCO, M., and JORGENSEN, W. L. (1997). Molecular dynamics simulations of the unfolding of barnase in water and 8 M aqueous urea. *Biochemistry* **36**, 7313–7329.

63 CAFLISCH, A. and KARPLUS, M. (1999).

Structural details of urea binding to barnase: a molecular dynamics analysis. *Structure* **7**, 477–488.

64 MODIG, K., KURIAN, E., PRENDERGAST, F. G., and HALLE, B. (2003). Water and urea interactions with the native and unfolded forms of a β-barrel protein. *Protein Sci.* **12**, 2768–2781.

65 SCAPIN, G., GORDON, J. I., and SACCHETTINI, J. C. (1992). Refinement of the structure of recombinant rat intestinal fatty acid-binding apoprotein at 1.2-Å resolution. *J. Biol. Chem.* **267**, 4253–4269.

66 WIESNER, S., KURIAN, E., PRENDERGAST, F. G., and HALLE, B. (1999). Water molecules in the binding cavity of intestinal fatty acid binding protein: Dynamic characterization by water ^{17}O and ^{2}H magnetic relaxation dispersion. *J. Mol. Biol.* **286**, 233–246.

67 MODIG, K., RADEMACHER, M., LÜCKE, C., and HALLE, B. (2003). Water dynamics in the large cavity of three lipid-binding proteins monitored by ^{17}O magnetic relaxation dispersion. *J. Mol. Biol.* **332**, 965–977.

68 HODSDON, M. E. and CISTOLA, D. P. (1997). Ligand binding alters the backbone mobility of intestinal fatty acid-binding protein as monitored by ^{15}N NMR relaxation and ^{1}H exchange. *Biochemistry* **36**, 2278–2290.

69 FROLOV, A. and SCHROEDER, F. (1997). Time-resolved fluorescence of intestinal and liver fatty acid binding proteins: Role of fatty acyl CoA and fatty acid. *Biochemistry* **36**, 505–517.

70 LIKIC, V. A. and PRENDERGAST, F. G. (2001). Dynamics of internal waters in fatty acid binding protein: Computer simulations and comparison with experiments. *Proteins* **43**, 65–72.

71 BAKOWIES, D. and VAN GUNSTEREN, W. F. (2002). Simulations of apo and holo-fatty acid binding protein: Structure and dynamics of protein, ligand and internal water. *J. Mol. Biol.* **315**, 713–736.

72 LIKIC, V. A., JURANIC, N., MACURA, S., and PRENDERGAST, F. G. (2000). A "structural" water molecule in the family of fatty acid binding proteins. *Protein Sci.* **9**, 497–504.

73 HODSDON, M. E. and FRIEDEN, C. (2001). Intestinal fatty acid binding protein: the folding mechanism as determined by NMR studies. *Biochemistry* **40**, 732–742.

74 ROPSON, I. J. and FRIEDEN, C. (1992). Dynamic NMR spectral analysis and protein folding: identification of a highly populated folding intermediate of rat intestinal fatty acid-binding protein by ^{19}F NMR. *Proc. Natl Acad. Sci. USA* **89**, 7222–7226.

75 KUWAJIMA, K., NITTA, K., YONEYAMA, M., and SUGAI, S. (1976). Three-state denaturation of α-lactalbumin by guanidine hydrochloride. *J. Mol. Biol.* **106**, 359–373.

76 MAKHATADZE, G. I. and PRIVALOV, P. L. (1992). Protein interactions with urea and guanidinium chloride. A calorimetric study. *J. Mol. Biol.* **226**, 491–505.

77 SCHELLMAN, J. A. and GASSNER, N. C. (1996). The enthalpy of transfer of unfolded proteins into solutions of urea and guanidinium chloride. *Biophys. Chem.* **59**, 259–275.

78 BARRON, L. D., HECHT, L., and WILSON, G. (1997). The lubricant of life: a proposal that solvent water promotes extremely fast conformational fluctuations in mobile heteropolypeptide structure. *Biochemistry* **36**, 13143–13147.

79 BUCK, M. (1998). Trifluoroethanol and colleagues: cosolvents come of age. Recent studies with peptides and proteins. *Q. Rev. Biophys.* **31**, 297–355.

80 THOMAS, P. D. and DILL, K. A. (1993). Local and nonlocal interactions in globular proteins and mechanisms of alcohol denaturation. *Protein Sci.* **2**, 2050–2065.

81 JASANOFF, A. and FERSHT, A. R. (1994). Quantitative determination of helical propensities from trifluoroethanol titration curves. *Biochemistry* **33**, 2129–2135.

82 SHIRAKI, K., NISHIKAWA, K., and GOTO, Y. (1995). Trifluoroethanol-induced stabilization of the α-helical structure of β-lactoglobulin: implica-

tion for non-hierarchical protein folding. *J. Mol. Biol.* **245**, 180–194.
83 HAMADA, D., KURODA, Y., TANAKA, T., and GOTO, Y. (1995). High helical propensity of the peptide fragments derived from β-lactoglobulin, a predominantly β-sheet protein. *J. Mol. Biol.* **254**, 737–746.
84 KURODA, Y., HAMADA, D., TANAKA, T., and GOTO, Y. (1996). High helicity of peptide fragments corresponding to β-strand regions of β-lactoglobulin observed by 2D NMR spectroscopy. *Folding Des.* **1**, 255–263.
85 KUWATA, K., HOSHINO, M., ERA, S., BATT, C. A., and GOTO, Y. (1998). α → β transition of β-lactoglobulin as evidenced by heteronuclear NMR. *J. Mol. Biol.* **283**, 731–739.
86 MENDIETA, J., FOLQUÉ, H., and TAULER, R. (1999). Two-phase induction of the nonnative α-helical form of β-lactoglobulin in the presence of trifluoroethanol. *Biophys. J.* **76**, 451–457.
87 FIORONI, M., DÍAZ, M. D., BURGER, K., and BERGER, S. (2002). Solvation phenomena of a tetrapeptide in water/trifluoroethanol and water/ethanol mixtures: a diffusion NMR, intermolecular NOE, and molecular dynamics study. *J. Am. Chem. Soc.* **124**, 7737–7744.
88 DÍAZ, M. D., FIORONI, M., BURGER, K., and BERGER, S. (2002). Evidence of complete hydrophobic coating of bombesin by trifluoroethanol in aqueous solution: an NMR spectroscopic and molecular dynamics study. *Chem. Eur. J.* **8**, 1663–1669.
89 ROCCATANO, D., COLOMBO, G., FIORONI, M., and MARK, A. E. (2002). Mechanism by which 2,2,2-trifluoroethanol/water mixtures stabilize secondary-structure formation in peptides: a molecular dynamics study. *Proc. Natl Acad. Sci. USA* **99**, 12179–12184.
90 MARTINEZ, D. and GERIG, J. T. (2001). Intermolecular ^{1}H{^{19}F} NOEs in studies of fluoroalcohol-induced conformations of peptides and proteins. *J. Magn. Reson.* **152**, 269–275.
91 KUMAR, S., MODIG, K., and HALLE, B. (2003). Trifluoroethanol-induced β → α transition in β-lactoglobulin: hydration and cosolvent binding studied by ^{2}H, ^{17}O, and ^{19}F magnetic relaxation dispersion. *Biochemistry* **42**, 13708–13716.
92 ABRAGAM, A. (1961). *The Principles of Nuclear Magnetism.* Clarendon, Oxford.
93 NOACK, F. (1986). NMR field-cycling spectroscopy: principles and applications. *Prog. NMR Spectrosc.* **18**, 171–276.
94 ANOARDO, E., GALLI, G., and FERRANTE, G. (2001). Fast field-cycling NMR: applications and instrumentation. *Appl. Magn. Reson.* **20**, 365–404.
95 JÓHANNESSON, H., DENISOV, V. P., and HALLE, B. (1997). Dimethyl sulfoxide binding to globular proteins: a nuclear magnetic relaxation dispersion study. *Protein Sci.* **6**, 1756–1763.
96 HALLE, B. and WENNERSTRÖM, H. (1981). Interpretation of magnetic resonance data from water in heterogeneous systems. *J. Chem. Phys.* **75**, 1928–1943.
97 DENISOV, V. P., PETERS, J., HÖRLEIN, H. D., and HALLE, B. (1996). Using buried water molecules to explore the energy landscape of proteins. *Nature Struct. Biol.* **3**, 505–509.

9
Stability and Design of α-Helices

Andrew J. Doig, Neil Errington, and Teuku M. Iqbalsyah

9.1
Introduction

Proteins are built of regular local folds of the polypeptide chain called secondary structure. The α-helix was first described by Pauling, Corey, and Branson in 1950 [1], and their model was quickly supported by X-ray analysis of hemoglobin [2]. Irrefutable proof of the existence of the α-helix came with the first protein crystal structure of myoglobin, where most secondary structure is helical [3]. α-Helices were subsequently found in nearly all globular proteins. It is the most abundant secondary structure, with $\approx 30\%$ of residues found in α-helices [4]. In this review, we discuss structural features of the helix and their study in peptides. Some earlier reviews on this field are in Refs [5–11].

9.2
Structure of the α-Helix

A helix combines a linear translation with an orthogonal circular rotation. In the α-helix the linear translation is a rise of 5.4 Å per turn of the helix and a circular rotation is 3.6 residues per turn. Side chains spaced $i, i+3$, $i, i+4$, and $i, i+7$ are therefore close in space and interactions between them can affect helix stability. Spacings of $i, i+2$, $i, i+5$, and $i, i+6$ place the side chain pairs on opposite faces of the helix, avoiding any interaction. The helix is primarily stabilized by $i, i+4$ hydrogen bonds between backbone amide groups.

The conformation of a polypeptide can be described by the backbone dihedral angles ϕ and ψ. Most ϕ, ψ combinations are sterically excluded, leaving only the broad β region and narrower α region. One reason why the α-helix is so stable is that a succession of the sterically allowed α, ϕ, ψ angles naturally position the backbone NH and CO groups towards each other for hydrogen bond formation. It is possible that a succession of the most stable conformation of an isolated residue in a polymer with alternative functional groups could point two hydrogen bond donors or acceptors towards each other, making secondary structure formation un-

Protein Folding Handbook. Part I. Edited by J. Buchner and T. Kiefhaber
Copyright © 2005 WILEY-VCH Verlag GmbH & Co. KGaA, Weinheim
ISBN: 3-527-30784-2

favourable. One reason why polypeptides may have been selected as the polymer of choice for building functional molecules is that the sterically most stable conformations also give strong hydrogen bonds.

The residues at the N-terminus of the α-helix are called N'-N-cap-N1-N2-N3-N4 etc., where the N-cap is the residue with nonhelical ϕ, ψ angles immediately preceding the N-terminus of an α-helix and N1 is the first residue with helical ϕ, ψ angles [12]. The C-terminal residues are similarly called C4-C3-C2-C1-C-cap-C' etc. The N1, N2, N3, C1, C2, and C3 residues are unique because their amide groups participate in $i, i + 4$ backbone–backbone hydrogen bonds using either only their CO (at the N-terminus) or NH (at the C-terminus) groups. The need for these groups to form hydrogen bonds has powerful effects on helix structure and stability [13]. The following rules are generally observed for N-capping in α-helices: Thr and Ser N-cap side chains adopt the *gauche*$^-$ rotamer, hydrogen bond to the N3 NH and have ψ restricted to $164 \pm 8°$. Asp and Asn N-cap side chains either adopt the *gauche*$^-$ rotamer and hydrogen bond to the N3 NH with $\psi = 172 \pm 10°$, or adopt the *trans* rotamer and hydrogen bond to both the N2 and N3 NH groups with $\psi = 107 \pm 19°$. With all other N-caps, the side chain is found in the gauche$^+$ rotamer so that the side chain does not interact unfavorably with the N-terminus by blocking solvation and ψ is unrestricted. An $i, i + 3$ hydrogen bond from N3 NH to the N-cap backbone C=O is more likely to form at the N-terminus when an unfavorable N-cap is present [14].

Bonds between sp^3 hybridized atoms show preferences for dihedral angles of $+60°$ (*gauche*$^-$), $-60°$ (*gauche*$^+$) or $180°$ (*trans*). Within amino acid side chains the preferences for these rotamers varies between secondary structures [15–18]. Within the α-helix, rotamer preferences vary greatly between different positions of N-cap, N1, N2, N3, and interior [14, 19]. At helix interior positions, the *gauche*$^+$ rotamer is usually most abundant (64%), followed by *trans* (33%) with *gauche*$^-$ very rare (3%), though considerable variations are seen. For example, Ser and Thr are *gauche*$^-$ 19% and 14% of the time, respectively, as their hydroxyl groups can hydrogen bond to the helix backbone in this conformation. The β-branched side chains Val, Ile, and Thr are the most restricted with *gauche*$^+$ very common.

9.2.1
Capping Motifs

The amide NH groups at the helix N-terminus are satisfied predominantly by side chain hydrogen bond acceptors. In contrast, carbonyl CO groups at the C-terminus are satisfied primarily by backbone NH groups from the sequence following the helix [13]. The presence of such interactions would therefore stabilize helices. A well-known interaction is helix capping, defined as specific patterns found at or near the ends of helices [12, 20–24]. The patterns involve hydrogen bonding and hydrophobic interactions as summarized in Table 9.1.

A common pattern of capping at the helix N-terminus is the capping box. Here, the side chain of the N-cap forms a hydrogen bond with the backbone of N3 and,

Tab. 9.1. The most common capping motifs at α-helix termini.

Sequence	Related position	Designation	Interactions	References
N-capping				
p-XXp	N-cap → N3	Capping box	Reciprocal H-bonds between residues at N-cap and N3, where S/T at N-cap and E at N3 are the most frequently observed residues	24, 25
hp-XXph	N' → N4	Expanded capping box/Hydrophobic staple	Reciprocal H-bonds between residues at N-cap and N3 accompanied by hydrophobic interactions between residues at N' and N4	26, 298
hP-XXVh	N' → N4	Pro-box motif	N-cap Pro residues are usually associated with Ile and Leu at N', Val at N3 and a hydrophobic residue at N4	28
C-capping				
hXp-XGh	C3 → C''	Schellman motif	H-bonds between C'' NH and the C3 CO at and between C' NH and C2 CO, respectively accompanied by hydrophobic interaction between C3 and C''	31, 223
hXp-XGp	C2 → C''	α_L motifs	As The Schellman motif but H-bond between C' NH and C3 CO where C'' polar	32
X-Pro	C-cap → C'	Pro-capping motif	A stabilizing electrostatic interaction of the residues at positions C-cap and Pro at C' with the helix macrodipole	34

p, polar amino acids; h, hydrophobic amino acids; X, any amino acids;
P, proline; G, glycine.

reciprocally, the side chain of N3 forms a hydrogen bond with the backbone of the N-cap [25]. The definition of the capping box was expanded by Seale et al. [26] to include an associated hydrophobic interaction between residues N' and N4 and is also known as a "hydrophobic staple" [27]. A variant of the capping box motif, termed the "big" box with an observed hydrophobic interaction between nonpolar side chain groups in residues N4 and N'' (not N') [26].

The Pro-box motif involves three hydrophobic residues and a Pro residue at the N-cap [28]. The N-cap Pro residue is usually associated with Ile or Leu at position N'', Val at position N3, and a hydrophobic residue at position N4. This motif is arguably very destabilizing since it discourages helix elongation due to the low N-cap preference of Pro [29] and the poor helix propagation parameter value of Val [30]. This suggests that the Pro-box motif will not specially contribute to protein stability but to the specificity of its fold.

The two primary capping motifs found at helix C-termini are the Schellman and the α_L motifs [31–33]. The Schellman motif is defined by a doubly hydrogen-bonded pattern between backbone partners, consisting of hydrogen bonds between the amide NH at C″ and the carbonyl CO at C3 and between the amide NH at C′ and the carbonyl CO at C2, respectively. The associated hydrophobic interaction is between C3 and C″. In a Schellman motif, polar residues are highly favored at the C1 position and the C′ residue is typically glycine.

If C″ is polar, the alternative α_L motif is observed. The α_L motif is defined by a hydrogen bond between the amide NH at C′ and the carbonyl CO at C3. As in the Schellman motif, the C′ residue is typically glycine, which adopts a positive value of ϕ dihedral angle. However, the hydrophobic interaction in an α_L is heterogeneous, occurring between C3 and any of several residues external to the helix ($C^{3\prime}$, $C^{4\prime}$, or $C^{5\prime}$) [32].

A statistical analysis of the protein crystal structures database detected another possible C-cap local motif, designated Pro C-capping [34]. Certain combinations of X-Pro pairs, in which residue X is the C-cap and the Pro is at position C′, are more abundant than expected. The aliphatic (Ile, Val, and Leu) and aromatic residues (Phe, Tyr, and Trp), together with Asn, His, and Cys are the most favored residues accompanying Pro at position C′.

A notable difference between the N- and C-terminal motifs is that at the N-terminus, helix geometry favors side chain-to-backbone hydrogen bonding and selects for compatible polar residues [14, 19]. Accordingly, the N-terminus promotes selectivity in all polar positions, especially N-cap and N3 in the capping box. In contrast, at the C-terminus, side chain-to-backbone hydrogen bonding is disfavored. Backbone hydrogen bonds are satisfied instead by posthelical backbone groups. The C-terminus need only select for C′ residues that can adopt positive values of the backbone dihedral angle ϕ, most notably Gly [32].

9.2.2
Metal Binding

In general, the stabilization of folded forms can be achieved by reducing the entropy by cross-linking, such as via metal ions. Another way to stabilize helix conformations, especially in short peptides, is to introduce an artificial nucleation site composed of a few residues fixed in a helical conformation. For example, the calcium-binding loop from EF-hand proteins saturated with a lanthanide ion promotes a rigid short helical conformation at its C-terminus region [35].

Metal ions play a key structural and functional role in many proteins. There is therefore interest in using peptide models containing metal complexes. In stabilizing proteins, the rule of thumb in metal–ligand binding is that "hard metals" prefer "hard ligands". For example Ca and Mg prefer ligands with oxygen as the coordinating atoms (Asp, Glu) [36]. In contrast, soft metals, such as Cu and Zn, bind mostly to His, Cys, and Trp ligands and sometimes indirectly via water molecules [37].

In studies of helical peptide models, soft ligands have been mostly used. In the presence of Cd ions, a synthetic peptide containing Cys-His ligands $i, i+4$ apart at the C-terminal region promoted helicity from 54% to 90%. The helicity of a similar peptide containing His-His ligands increased by up to 90% as a result of Cu and Zn binding [38]. The addition of a *cis*-Ru(III) ion to a 6-mer peptide, Ac-AHAAAHA-NH$_2$, changed the peptide conformation from random coil to 37% helix [39]. An 11-residue peptide was converted from random coil to 80% helix content by the addition of Cd ions, although the ligands used were not natural amino acids but aminodiacetic acids [40]. As(III) stabilizes helices when bound to Cys side chains spaced $i, i+4$ by -0.7 to -1.0 kcal mol^{-1} [41].

9.2.3
The 3$_{10}$-Helix

3$_{10}$-Helices are stabilized by $i, i+3$ hydrogen bonds instead of the $i, i+4$ bonds found in α-helices, making the cylinder of the 3$_{10}$-helix narrower than α and the hydrogen bonds nonlinear. About 3–4% of residues in crystal structures are in 3$_{10}$-helices [4, 42], making the 3$_{10}$-helix the fourth most common type of secondary structure in proteins after α-helices, β-sheets, and reverse turns. Most 3$_{10}$-helices are short, only 3 or 4 residues long, compared with a mean of 10 residues in α-helices [4]. 3$_{10}$-Helices are commonly found as N- or C-terminal extensions to an α-helix [4, 43, 44]: Strong amino acid preferences have been observed for different locations within the interiors [42] and N- and C-caps [14] of 3$_{10}$-helices in crystal structures.

The 3$_{10}$-helix is being recognized as of increasing importance in isolated peptides and as a possible intermediate in α-helix formation [45, 46]. It has been proposed that peptides made of the standard 20 L-amino acids can form 3$_{10}$-helices [47], or at least populate a significant amount of 3$_{10}$-helix at their N- and C-termini. The fraction of 3$_{10}$-helix present in a helical peptide is strongly sequence dependent [48, 49]. 3$_{10}$-helix formation can be induced by the introduction of a $C_{\alpha,\alpha}$-disubstituted α-amino acid, of which α-aminoisobutyric acid (AIB) is the prototype. For most amino acids, the α-helical geometry ($\phi = -57°$, $\psi = -70°$) is of lower energy than the 3$_{10}$ geometry ($\phi = -49°$, $\psi = -26°$). There is no disallowed region between the α and 3$_{10}$ conformations in the Ramachandran plot, and a peptide can therefore be gradually transformed from one helix to the other [50].

9.2.4
The π-Helix

In contrast to the widely occurring α- and 3$_{10}$-helices, the π-helix is extremely rare. The π-helix is unfavorable for three reasons: its dihedral angles are energetically unfavorable relative to the α-helix [51, 52], its three-dimensional structure has a 1 Å hole down the center that is too narrow for access by a water molecule, resulting in the loss of van der Waals interactions, and a higher number of residues (four) must be correctly oriented before the first $i, i+5$ hydrogen bond is formed, making helix initiation more entropically unfavorable than for α- or 3$_{10}$-helices. The π-helix may be of more than theoretical interest [53–57].

9.3
Design of Peptide Helices

The first protein crystal structures showed an abundance of α-helices, leading to speculation whether peptide fragments of the helical sequences could be stable in isolation. Since isolated helices lack hydrophobic cores, this question could be rephrased to ask whether the amide–amide hydrogen bond is strong enough to oppose the loss of conformational entropy arising from restricting the peptide into a helical structure [58]. Early estimates of these terms suggested that hydrogen bonds were not strong enough and this was confirmed by peptide studies of helices from myoglobin [59] and staphylococcal nuclease [60], which found no helix formation. Early estimates of helix/coil parameters from a host–guest system (see Section 9.3.1) suggested that polypeptides would need to be hundreds of residues long to form stable helices. Hence, the earliest work on peptide helices was on long homopolymers of Glu or Lys, which show coil-to-helix transitions on changing the pH from charged to neutral. The neutral polypeptides are metastable and prone to aggregation, ultimately to β-sheet amyloid [61].

A conflicting result was found by Brown and Klee in 1971 [62], who reported that the C-peptide of ribonuclease A, which contains the first 13 residues of the protein and which forms a helix in the protein, was helical at 0 °C. This observation was not followed up for 10 years until extensive work on the sequence features responsible for helix formation in this peptide, and in the larger S-peptide, was performed by Baldwin and coworkers [63–73], while NMR studies by Rico and coworkers precisely defined the helical structure [74–82]. Some important features responsible for the helicity of the C-peptide that emerged from this work included an $i, i+4$ Phe-His interaction, an $i, i+3$ Glu-His salt bridge, an unusual $i, i+8$ Glu-Arg salt bridge that spanned two turns of the helix and a helix termination signal at Met. The stabilizing effects of salt bridges were inferred from pH titrations and circular dichroism (CD), where a decrease in helicity was seen when a residue participating in a salt bridge was neutralized. Quantification of these features in terms of a free energy was not possible at that time, as helix/coil theory that included side-chain interactions and termination signals (caps) had not been developed.

Perhaps the most interesting result from the C- and S-peptide work was the importance of Ala. The replacement of interior helical residues with Ala was stabilizing, indicating that a major reason why this helix was folded in isolation was the presence of three successive alanines from positions 4–6. This led to the successful design of isolated, monomeric helical peptides in aqueous solution, first containing several salt bridges and a high alanine content, based on $(EAAAK)_n$ [83–84] and then a simple sequence with a high alanine content solubilized by several lysines [85]. These "AK peptides" are based on the sequence $(AAKAA)_n$, where n is typically 2–5. The Lys side chains are spaced $i, i+5$ so they are on opposite faces of the helix, giving no charge repulsion and may be substituted with Arg or Gln to give a neutral peptide. Hundreds of AK peptides have been studied, giving most of the results on helix stability in peptides (see below). The Alanines in the

(EAAAK)$_n$ type peptides may be removed entirely; E$_4$K$_4$ peptides, with sequences based on (EEEEKKKK)$_n$ or EAK patterns are also helical, stabilized by large numbers of salt bridges [8, 86–88].

9.3.1
Host–Guest Studies

Extensive work from the Scheraga group has obtained helix/coil parameters using a host–guest method. Long random copolymers were synthesized of a water-soluble, nonionic guest (poly[N^5-(3-hydroxypropyl)-L-glutamine] (PHPG) or poly[N^5-(4-hydroxybutyl)-L-glutamine] (PHBG)), together with a low (10–50%) content of the guest residue. Using the s and σ Zimm-Bragg helix/coil parameters (see below) for the host homopolymer, it was possible to calculate those for the guest using helix/coil theory as a function of temperature. The results from the host–guest work are in disagreement with most of those from short peptides of fixed sequence (see below).

9.3.2
Helix Lengths

Helix formation in peptides is cooperative, with a nucleation penalty. Helix stability therefore tends to increase with length, in homopolymers at least. As the length of a homopolymer increases, the mean fraction helix will level off below 100% as long helices tend to break in two. In heteropolymers, observed lengths are highly sequence dependent. As helices are at best marginally stable in monomeric peptides in aqueous solution, they are readily terminated by the introduction of a strong capping residue or a residue with a low intrinsic helical preference.

The length distribution of helices in peptides is very different to proteins [4]. Most helices are short, with 5–14 residues most common. There is a general trend for a decrease in frequency as the length increases beyond 13 residues. Helix lengths longer than 25 are rare. This is a consequence of the organization of proteins into domains of similar size, rather than showing different rules for stability; helices do not extend beyond the boundaries of the domain and so terminate. There is also a preference to have close to an integral number of turns so that their N- and C-caps are on the same side of the helix [89].

9.3.3
The Helix Dipole

The secondary amide group in a protein backbone is polarized with the oxygen negatively charged and hydrogen positively charged. In a helix the amides are all oriented in the same direction with the positive hydrogens pointing to the N-terminus and negative oxygens pointing to the C-terminus. This can be regarded as giving a positive charge at the helix N-terminus and a negative charge at the helix C-terminus [90–92]. In general, therefore, negatively charged groups are stabilizing at the N-terminus and positive at the C-terminus, as shown by numerous titrations that measure helix content as a function of pH and amino acid preferences for helix terminal positions. An alternative interpretation of these results is that favored side chains are those that can make hydrogen bonds to the free amide NH groups at N1, N2, and N3 or free CO groups at C1, C2, and C3 [93]. Charged groups can form stronger hydrogen bonds than neutral groups, thus providing an alternative rationalization of the pH titration results. These hypotheses are not mutually exclusive, as a charged side chain can also function as a hydrogen bond acceptor or donor. A free energy simulation of Tidor [94] suggested that helix-stabilizing interactions, as a result of a Tyr → Asp substitution at an N-cap site, arose from hydrogen-bonding interactions from its direct hydrogen-bonding partners, and from more distant electrostatic interactions with groups within the first two turns of the helix.

Measurements of the amino acid preferences for the N-cap, N1, N2, and N3 positions in the helix allow a comparison to be made of the relative importance of helix-dipole and hydrogen-bonding interactions [29, 95–97]. The helix-dipole model implies that the side chains most favored at the helix N-terminus are those with a negative charge while positive charges are disfavored. A pure hydrogen-bonding model implies that favored side chains are those that can make hydrogen bonds to the free amide NH groups at N1, N2, and N3. In general, the N-cap results suggest that hydrogen bonding is more important than helix-dipole interactions; the best N-caps are Asn, Asp, Ser, and Thr [29], which can accept hydrogen bonds from the N2 and N3 NH groups [14]. Glu has only a moderate N-cap preference despite its negative charge. In contrast, N1, N2, and N3 results suggest that helix-dipole interactions are more important. The contrasting results between the different helix N-terminal positions can be rationalized by considering the geometry of the hydrogen bonds. N-cap hydrogen bonds are close to linear [14] and so are strong, while N1 and N2 hydrogen bonds are close to $90°$ [19], making them much weaker. Helix-dipole effects are likely to be present at all sites, as also shown by every pH titration where a more negative side chain is favored over a more positive side chain, but this can be overwhelmed by strong hydrogen bonds, as at the N-cap. In the absence of strong hydrogen bonds, helix-dipole effects dominate. Hydrogen bonds can therefore make a substantial contribution to protein stability, but only if their geometry is close to linear (as it is for backbone to backbone hydrogen bonds in α-helices and β-sheets).

9.3.4
Acetylation and Amidation

A simple, yet effective, way to increase the helicity of a peptide is to acetylate its N-terminus [22, 98]. This is readily done with acetic anhydride/pyridine, after completion of a peptide synthesis, but before cleavage of the peptide from the resin and deprotection of the side chains. Acetylation removes the positive charge that is present at the helix terminus at low or neutral pH; this would interact unfavorably with the positive helix dipole and free N-terminal NH groups. The extra CO group from the acetyl group can form an additional hydrogen bond to the NH group, putting the acetyl at the N-cap position. This has a strong stabilizing effect by approximately 1.0 kcal mol^{-1} compared with alanine [29, 30, 99]. The acetyl group is one of the best N-caps. We found only Asn and Asp to be more stabilizing [29].

Amidation of the peptide C-terminus is achieved by using different types of resin in solid-phase peptide synthesis, resulting in the replacement of COO$^-$ with CONH$_2$. This is similar structurally to acetylation: the helix is extended by one hydrogen bond and an unfavorable charge–charge repulsion with the helix dipole is removed. The energetic benefit of amidation is rather smaller, however, with the amide group being no better than Ala and in the middle if the C-cap residues are ranked in order of stabilization effect [29]. As most helical peptides studied to date are both acetylated and amidated, and acetylation is more stabilizing than amidation, the helicity of peptides is generally skewed so that residues near the N-terminus are more helical than those near the C-terminus. Acetylation and amidation are often also beneficial in removing a pH titration that can obscure other effects.

9.3.5
Side Chain Spacings

Side chains in the helix are spaced at 3.6 residues per turn of the helix. Side chains spaced at $i, i+3$, $i, i+4$, and $i, i+7$ are therefore close in space and interactions between them can affect helix stability. Spacings of $i, i+2$, $i, i+5$, and $i, i+6$ place the side chain pairs on opposite faces of the helix, avoiding any interaction. Care must therefore be taken to avoid unwanted interactions when designing helices. For example, we studied these peptides to determine the energetics of the Phe-Met interaction [100]:

F7M12 Ac-YGAAKA<u>F</u>AAKA<u>M</u>AAKAA-NH$_2$
F8M12 Ac-YGAAKAA<u>F</u>AKA<u>M</u>AAKAA-NH$_2$

The control peptide F7M12 has Phe and Met spaced $i, i+5$ where they cannot interact; F8M12 moves the Phe one place towards the C-terminus so that Phe and Met are spaced $i, i+4$ where they can form a noncovalent interaction. The increase in helix content was used to derive the Phe-Met interaction energy. These sequences are not ideal, however. F7M12 contains a Phe-Lys $i, i+3$ interaction that is replaced by a Lys-Phe $i, i+3$ interaction in F8M12. If either of these are stabilizing or nonstabilizing the Phe-Met $i, i+4$ energy will be in error. While this problem is likely to be small, in this case at least, our more recent designs try to minimize such potential problems. For example, the following peptides, designed to measure Ile-Lys $i, i+4$ energies, do not introduce or remove an unwanted $i, i+3$ spacing [101]:

IKc Ac-AK̲IAAAAK̲IAAAAK̲AKAGY-NH$_2$
IKi Ac-AKA̲IAAAK̲AI̲AAAK̲AKAGY-NH$_2$

9.3.6
Solubility

Lack of peptide solubility can be a problem. Peptides designed to be helices can even become highly insoluble amyloid with a high β-sheet content. We have occasionally found that alanine-based peptides designed to be helical instead form amyloid, though this may only take place after several years storage in solution. The measurement of interactions in a helix will be compromised by peptide oligomerization so it is generally essential to check that the peptides are monomeric. This can be done rigorously by sedimentation equilibrium, which determines the oligomeric state of a molecule in solution. This is difficult, however, with the short peptides often used, as their molecular weights are at the lower limit for this technique. A simpler method is to check a spectroscopic technique that depends on peptide structure, most obviously CD, as a function of concentration. If the signal depends linearly on peptide concentration across a large range, including that used to study the peptide structure, it is safe to assume that the peptide is monomeric. For example, if helicity measurements are made at 10 µM, CD spectra can be acquired from 5 µM to 100 µM. An oligomer that does not change state, such as a coiled-coil, across the concentration range cannot be excluded, however. Light scattering can detect aggregation. A monomeric peptide should have a flat baseline in a UV spectrum outside the range of any chromophores in the peptide. In stock solutions of a peptide with a single tyrosine isolated from the helix region by Gly should have $A_{300}/A_{275} < 0.02$ and $A_{250} < A_{275} < 0.2$ [102].

Consideration of solubility is essential when designing helical peptides. While Ala has the highest helix propensity and would provide an ideal theoretical background for substitutions, poly(Ala) is insoluble. Solubility can be achieved most easily by including polar side chains spaced $i, i + 5$ in the sequence where they cannot interact. Lys, Arg, and Gln are used most often for this purpose. Gln may be preferred if unwanted interactions with charged Lys or Arg may be a problem, but some AQ peptides lack sufficient solubility and AQ peptides are less helical. We have found that it is not easy to predict peptide solubility. For example, 20 peptides with sequence Ac-XAAAAQAAAAQAAGY-NH$_2$, where X is all the amino acids encoded by the genetic code, were all soluble [95]. However, when the homologous peptide Ac-AAAAAAQAAAAQAAAAQGY-NH$_2$ was synthesized to study the N2 position it tended to aggregate. The Gln side chains were therefore replaced with Lys, giving no further problems [96].

The spacing of side chains in the helix is best visualized with a helical wheel, to ensure that the designed helix does not have a nonpolar face that may lead to dimerization. The following web pages provide useful resources for this:

http://www.site.uottawa.ca/~turcotte/resources/HelixWheel/
http://marqusee9.berkeley.edu/kael/helical.htm

9.3.7
Concentration Determination

An accurate measurement of helix content depends on an accurate spectroscopic measurement and, equally importantly, peptide concentration. This is usually achieved by including a Tyr side chain at one end of the peptide. The extinction coefficient of Tyr at 275 nm is 1450 M^{-1} cm^{-1} [103]. If Trp is present, measurements at 281 nm can be used where the extinction coefficient of Trp is 5690 M^{-1} cm^{-1} and Tyr 1250 M^{-1} cm^{-1} [104]. Phe absorbance is negligible at this wavelength. These UV absorbances are ideally made in 6 M GuHCl, pH 7.0, 25 °C, though we have found very little variation with solvent so measurements in water are identical within error. The main source of error is in pipetting small volumes; this is typically around 2%. Pipetting larger volumes with well-maintained fixed volume pipettors can help minimize this error, which may well be the largest when measuring a percentage helix content by CD.

Though the inclusion of aromatic residues is required for concentration determination, this can have the unwanted side effect of perturbing a CD spectrum, leading to an inaccurate measure of helix content. A simple solution to this is to separate the terminal Tyr from the rest of the sequence by one or more Gly residues [105]. If the aromatic residues must be included within the helical region, the CD spectrum should be corrected to remove this perturbation (see later).

9.3.8
Design of Peptides to Measure Helix Parameters

Numerous peptides have been studied to measure the forces responsible for helix formation (see below). In general, one or more peptides are synthesized that contain the interaction of interest, while all other terms that can contribute to helix stability are known. The helix content of the peptide is measured and the statistical weight of the interaction is varied until predictions from helix/coil theory match experiment. The weight of the interaction (and hence its free energy, as $-RT \ln(\text{weight})$) is then known. In practice it is wise to also synthesize a control peptide that lacks the interaction, but is otherwise very similar. The helix content of this peptide should be predictable from helix/coil theory using known parameters. For example, Table 9.2 gives sequences of control and interaction peptides used to measure Phe-Lys $i, i+4$ side-chain interaction energies [106]. The FK control peptide has no $i, i+4$ side-chain interactions and all the helix/coil weights for these residues are known. The helicity predicted by SCINT2 [30] for this peptide is in close agreement with experiment. This is an important confirmation that the helix/coil model and the parameters used are accurate. The discrepancy between prediction and experiment for the FK interaction peptide is because this peptide contains two Phe-Lys $i, i+4$ side-chain interactions whose weights are assumed to be 1 (i.e., $\Delta G = 0$). If the Phe-Lys weight is changed to 1.3, the prediction agrees with experiment. Hence ΔG (Phe-Lys) $= -RT \ln 1.3 = -0.14$ kcal mol^{-1}.

It is important to minimize the error in determining helix/coil parameters. Errors can be calculated by assuming an error in measurement of % helix (± 3% is

Tab. 9.2. Peptides designed to measure the energetics of the Phe-Lys interaction.

Peptide name	Sequence	Experimental helicity	Predicted helicity (no side-chain interaction)	Phe-Lys weight to fit experiment	ΔG (Phe-Lys) kcal mol^{-1}
FK control	Ac-AAAKFAAAAKFAAAAKAKAGY-NH$_2$	35.4%	35.6	N/A	N/A
FK interaction	Ac-AAAKAFAAAKAFAAAAKAKAGY-NH$_2$	41.1%	34.6	1.3	−0.14

reasonable) and refitting the results across this experimental error range. For the case above, ΔG (Phe-Lys) was calculated for 38.1% and 44.1% helix, giving a range of -0.05 to -0.18 kcal mol^{-1}. This is a remarkably small error range, resulting from the high sensitivity of helix content to a small change in helix stability. This sensitivity can be maximized by including multiple identical interactions; here the FK interaction peptide has two Phe-Lys interactions, for example. The helix contents of the control and interaction peptides should be close to 50%. This may be difficult to achieve if the residues being used have low helix preferences. The best way to maximize sensitivity is to calculate it in advance using possible sequences, helix/coil theory, an error of $\pm 3\%$ and guessing a sensible value for the interaction energy. The interaction of interest can be placed at various positions in the peptide, its length can be changed by adding further AAKAA sequences, terminal residues such as a Tyr can be moved, and solubilizing side chains can be changed. Peptide design can be lengthy, considering sensitivity and solubility, but is a valuable process. The complex nature of the helix/coil equilibrium with frayed conformations highly populated means that considerable variations can be seen for apparently small changes, such as moving an interaction towards one terminus of a sequence.

9.3.9
Helix Templates

A major penalty to helix formation is the loss of entropy arising from the requirement to fix three consecutive residues to form the first hydrogen bond of the helix. Following this nucleation, propagation is much more favored as only a single residue need be restricted to form each additional hydrogen bond. A way to avoid this barrier is to synthesize a template molecule that facilitates helix initiation, by fixing hydrogen bond acceptors or donors in the correct orientation for a peptide to bond in a helical geometry. The ideal template nucleates a helix with an identical geometry to a real helix. Kemp's group applied this strategy and synthesized a proline-like template that nucleated helices when a peptide chain was covalently attached to a carboxyl group (Figure 9.1a) [107–111]. The template is in an equilibrium between *cis* and *trans* isomers of the proline-like part of the molecule. Only the *trans* isomer can bond to the helix so helix content is determined by measuring the *cis/trans* ratio by NMR. Bartlett et al. reported on a hexahydroindol-4-one template (Figure 9.1b) [112] that induced 49–77% helicity at 0 °C, depending on the method of determination, in an appended hexameric peptide. Several other templates were less successful and could only induce helicity in organic solvents [113–115]. Their syntheses are often lengthy and difficult, partly due to the challenging requirement of orienting several dipoles to act as hydrogen bond acceptors or donors.

9.3.10
Design of 3$_{10}$-Helices

Peptides can be induced to form 3$_{10}$-helix by the incorporation of a significant amount of disubstituted C$_\alpha$ amino acids, of which the simplest is α-aminoisobuty-

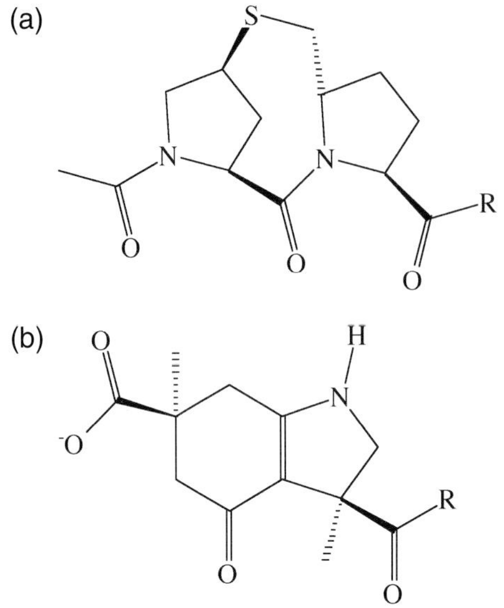

Fig. 9.1. Helix templates. a) Kemp, b) Bartlett.

ric acid (Aib) [116–119]. The presence of steric interactions from the two methyl groups on the α-carbon in AIB results in the 3_{10} geometry being energetically favored over the α. Peptides rich in $C_{\alpha,\alpha}$-disubstituted α-amino acids are readily crystallized and many of their structures have been solved (reviewed in Refs [50] and [120]). Aib is conformationally restricted so that shorter Aib-based helices are more stable than Ala-based α-helices [121]. Many examples of helix stabilization or 3_{10}-helix formation by Aib have been reported. The α/3_{10} equilibrium has been studied in peptides. Yokum et al. [122] synthesized a peptide composed of $C_{\alpha,\alpha}$-disubstituted α-amino acids that forms mixed 3_{10}-/α-helices in mixed aqueous/organic solvents. Millhauser and coworkers have studied the 3_{10}/α equilibrium in peptides by electron spin resonance (ESR) and nuclear magnetic resonance (NMR) [47, 49, 123, 124] and argued that mixed 3_{10}-/α-helices are common in polyalanine-based helices. Yoder et al. [125] studied a series of L-(αMe)-Val homopolymers and showed that their peptides can form coil, 3_{10}-helix, or α-helix depending on concentration, peptide length, and solvent. Kennedy et al. [126] used Fourier transform infrared (FTIR) spectroscopy to monitor 3_{10}- and α-helix formation in poly(Aib)-based peptides. Hungerford et al. [127] synthesized peptides that showed an α to 3_{10} transition upon heating. 3_{10}-Helices have also been proposed as thermodynamic intermediates where helical peptides of moderate stability exist as a mixture of α- and 3_{10}-structures [45, 128].

Guidelines for the inclusion of the 20 natural amino acids in 3_{10}-helices can be taken from their propensities in proteins, both at interior positions [42] and at N-caps [14]. These are similar, but not identical to α preferences. Side-chain interactions in the 3_{10}-helix are spaced $i, i + 3$, with $i, i + 4$ on opposite sides of the helix.

This offers scope to preferentially stabilize 3_{10} over α, by including stabilizing $i, i + 3$ interactions and $i, i + 3$ repulsions. As the 3_{10}- and α-helix structures are so similar, with only a small change in backbone dihedral angle, peptides designed to form 3_{10}-helix are likely to form a mixture of 3_{10} and α, with a central α segment and 3_{10} at the helix termini common. This complex equilibrium, and the spectroscopic similarity of α and 3_{10} makes the analysis of these peptides difficult.

9.3.11
Design of π-helices

To our knowledge, no peptide has yet been made that forms π-helix. There are 4.4 side chains per turn of the π-helix, so $i, i + 5$ interactions may weakly stabilize π-helix over α. Given the other strongly destabilizing features of the π-helix, however, we doubt whether this effect will ever be strong enough. We expect that no peptide composed of the 20 natural amino acids will ever fold to a monomeric π-helix in aqueous solution. We are willing to back up this prediction, but in view of the outcome of the Paracelsus challenge [129, 130], we are taking the advice of George Rose [131] and offering only an "I Met the π-Helix Challenge" T-shirt as a prize.

9.4
Helix Coil Theory

Peptides that form helices in solution do not show a simple two-state equilibrium between a fully folded and fully unfolded structure. Instead they form a complex mixture of all helix, all coil or, most frequently, central helices with frayed coil ends. In order to interpret experiments on helical peptides and make theoretical predictions on helices it is therefore essential to use a helix/coil theory that considers every possible location of a helix within a sequence. The first wave of work on helix/coil theory was in the late 1950s and early 1960s and this has been reviewed in detail by Poland and Scheraga [132]. In 1992, Qian and Schellman [133] reviewed current understanding of helix/coil theories. Our review covered the extensive development of helix/coil theory since this date [134].

The simplest way to analyze the helix/coil equilibrium, still occasionally seen, is the two-state model where the equilibrium is assumed to be between a 100% helix conformation and 100% coil. This is incorrect and its use gives serious errors. This is because helical peptides are generally most often found in partly helical conformations, often with a central helix and frayed, disordered ends, rather than in the fully folded or fully unfolded states.

9.4.1
Zimm-Bragg Model

The two major types of helix/coil model are (i) those that count hydrogen bonds, principally Zimm-Bragg (ZB) [135] and (ii) those that consider residue conforma-

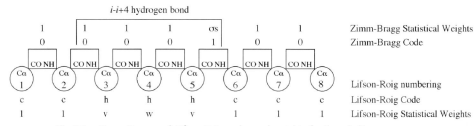

Fig. 9.2. Zimm-Bragg and Lifson-Roig codes and weights for the α-helix.

tions, principally Lifson-Roig [136]. In the ZB theory the units being considered are peptide groups and they are classified on the basis of whether their NH groups participate in hydrogen bonds within the helix. The ZB coding is shown in Figure 9.2. A unit is given a code of 1 (e.g., peptide unit 5 in Figure 9.2) if its NH group forms a hydrogen bond and 0 otherwise. The first hydrogen-bonded unit proceeding from the N-terminus has a statistical weight of σs, successive hydrogen bonded units have weights of s and nonhydrogen-bonded units have weights of 1. The s-value is a propagation parameter and σ is an initiation parameter. The most fundamental feature of the thermodynamics of the helix/coil transition is that the initiation of a new helix is much more difficult than the propagation of an existing helix. This is because three residues need to be fixed in a helical geometry to form the first hydrogen bond while adding an additional hydrogen bond to an existing helix require that only one residue is fixed. These properties are thus captured in the ZB model by having σ smaller than s. The statistical weight of a homopolymeric helix of a N hydrogen bonds is σs^{N-1}. The cost of initiation, σ, is thus paid only once for each helix while extending the helix simply multiplies its weight by one additional s-value for each extra hydrogen bond.

9.4.2
Lifson-Roig Model

In the LR model each residue is assigned a conformation of helix (h) or coil (c), depending on whether it has helical ϕ, ψ angles. Every conformation of a peptide of N residues can therefore be written as a string of N c's or h's, giving 2^N conformations in total. Residues are assigned statistical weights depending on their conformations and the conformations of surrounding residues. A residue in an h conformation with an h on either side has a weight of w. This can be thought of as an equilibrium constant between the helix interior and the coil. Coil residues are used as a reference and have a weight of 1. In order to form an $i, i+4$ hydrogen bond in a helix three successive residues need to be fixed in a helical conformation. M consecutive helical residues will therefore have M−2 hydrogen bonds. The two residues at the helix termini (i.e., those in the center of chh or hhc conformations) are therefore assigned weights of v (Figure 9.2). The ratio of w to v gives the approximately the effect of hydrogen bonding (1.7:0.036 for Ala [30] or $-RT \ln(1.7/0.036) = -2.1$ kcal mol^{-1}). A helical homopolymer segment of M residues has a weight of $v^2 w^{M-2}$ and a population in the equilibrium of $v^2 w^{M-2}$ divided by the sum of the weights of every conformation (i.e., the partition function). In this way the population of every conformation is calculated and all properties of the helix/coil equilibrium evaluated. The LR model is easier to handle conceptually for heteropolymers since the w and v parameters are assigned to individual residues. The substitution of one amino acid at a certain position thus changes the w- and v-values at that position. In the ZB model the initiation parameter σ is associated with several residues and s with a peptide group, rather than a residue. It is therefore easier to use the LR model when making substitutions. Indeed, most recent work has been based on this model.

A further difference is that the ZB model assigns weights of zero to all conformations that contain a chc or chhc sequence. This excludes a very large number of conformations that contain a residue with helical ϕ, ψ angles but with no hydrogen bond. In LR theory, these are all considered. The ZB and LR weights are related by the following formulae [133]: $s = w/(1+v)$; $\sigma = v^2/(1+v)^4$.

The complete helix/coil equilibrium is handled by determining the statistical weight for every possible conformation that contains a helix plus a reference weight of 1 for the coil conformation. Each conformation considered in the helix/coil equilibrium is given a statistical weight. This indicates the stability of that conformation, with the higher the weight, the more probable the conformation. Weights are defined relative to the all coil conformation, which is given a weight of 1. The statistical weight of a conformation can thus be regarded as an equilibrium constant relative to the coil; a weight >1 indicates the conformation is more stable than coil, <1 means less stable and 1 means equally stable. The population of each conformation is given by the statistical weight of that conformation divided by the sum of the statistical weights for every conformation (i.e., the partition function). Thus the greater the statistical weight, the more stable the conformation.

The key to using helix/coil theory is the partition function. All the properties of a system at equilibrium are contained within the partition function, which makes it very valuable. Partition functions are extremely powerful concepts in statistical thermodynamics since they include all properties of an equilibrium. Any property of the equilibrium can be extracted from the partition function by applying the appropriate mathematical function. This is analogous to quantum mechanics where any property can be determined from a system by applying an operator function to a wavefunction. In this case the properties could be the mean number of hydrogen bonds, the mean helix length, the probability that each residue is within a helix etc. In particular, the mean number of residues with a weight x is given by $\partial \ln Z / \partial \ln x$. Circular dichroism is commonly used to give the mean helix content of a helical peptide, namely the fraction of residues that have a weight of w. LR-based models can thus be related to experimental data by equating the measured mean helix content to $(\partial \ln Z / \partial \ln x)/N$, where N is the number of residues in the peptide. Statistical weights can be regarded as equilibrium constants for the equilibrium between coil and the structure (as the reference coil weight is defined as 1). They can therefore be converted to free energies as $-RT \ln(\text{weight})$.

Equations for partition functions are determined as follows: Write a table of the conformations and assigned statistical weights for residues. For the Lifson-Roig model, the weights of a residue depend on its conformation and the weights of its two neighboring residues as follows:

Conformation	Weight of central residue in triplet
ccc	1
cch	1
chc	v
chh	v
hcc	1
hch	1
hhc	v
hhh	w

The same information is rewritten in the form of a matrix (M):

$$M = \begin{array}{c} \\ \bar{hh} \\ \bar{hc} \\ \bar{ch} \\ \bar{cc} \end{array} \begin{array}{c} \bar{hh} \quad \bar{hc} \quad \bar{ch} \quad \bar{cc} \\ \left(\begin{array}{cccc} w & v & 0 & 0 \\ 0 & 0 & 1 & 1 \\ v & v & 0 & 0 \\ 0 & 0 & 1 & 1 \end{array} \right) \end{array}$$

For each triplet, the state of the left-most residue is shown at the start of each row in the matrix. The state of the right-most residue in each triplet is shown at the end of the top of each column. The state of the center residue of each triplet is shown as the barred residue in both the rows and columns. When the states of the center residues differ (i.e., one is c while the other is h), the entry in the matrix is zero. Otherwise, the matrix gives the weight, taken from the table. This apparently simple change in presentation now (amazingly) allows the generation of the partition function (Z) for a polypeptide of N residues as:

$$Z = (0 \quad 0 \quad 1 \quad 1) M^N \begin{pmatrix} 0 \\ 1 \\ 1 \\ 1 \end{pmatrix}$$

The end vectors are present to ensure that the first and last residues in the peptide cannot have a w-weighting. A w-weighting indicates that the residue is between two residues that hydrogen bond to each other and this is impossible for terminal residues. Further extensions to helix/coil theory are dealt with in the same way, by defining residue weight assignments, rewriting as a matrix and determining end vectors. The work may be made easier by combining any identical columns. For example, the Lifson-Roig matrix above has identical third and fourth columns so can be simplified to a 3×3 matrix:

$$M = \begin{array}{c} \\ h\bar{h} \\ h\bar{c} \\ c(h \cup c) \end{array} \begin{array}{ccc} h\bar{h} & h\bar{c} & \bar{c}(h \cup c) \\ \begin{pmatrix} w & v & 0 \\ 0 & 0 & 1 \\ v & v & 1 \end{pmatrix} \end{array}$$

Here, $h \cup c$ shows the residues being combined. The matrix entries for the combined positions are the entries for the noncombined entries, with zero being discounted if combined with a nonzero entry. New end vectors are now required, as their lengths must be the same as the order of the square matrix.

9.4.3
The Unfolded State and Polyproline II Helix

The treatment of peptide conformations is based on Flory's isolated-pair hypothesis [137]. This states that while ϕ and ψ for a residue are strongly interdependent, giving preferred areas in a Ramachandran plot, each ϕ, ψ pair is independent of the ϕ, ψ angles of its neighbors. Pappu et al. examined the isolated-pair hypothesis in detail by exhaustively enumerating the conformations of poly(Ala) chains [138]. Each residue was considered to populate 14 mesostates, defined by ranges of ϕ, ψ values. By considering all 14^N mesostate strings, all conformations were considered for up to seven alanines. The number of allowed conformations was found to be considerably fewer than the maximum, thus showing that the isolated-pair hypothesis is invalid. The chains mostly populate extended or helical conformations as many partly helical conformations are sterically disallowed. Such effects are not included in helix/coil theories, thus presenting a considerable challenge for the future. Helix/coil theories assign the same weight (1) to every coil residue; steric exclusion means that these should vary and be lower than 1 in many cases.

The polyproline II helix may well be an important conformation for unfolded proteins [139–148]. In particular, denatured alanine-rich peptides may form polyproline II helix [143, 148, 149]. It may therefore be valid to consider residues in helical peptides to be in three possible states (helix, coil, or polyproline II), rather than two (helix or coil). No current helix/coil model takes this into account. A scale of amino acid preferences for the polyproline II helix has recently been published [150].

9.4.4
Single Sequence Approximation

Since helix nucleation is difficult, conformations with multiple helical segments are expected to be rare in short peptides. In the one-, or single-, helical sequence approximation, peptide conformations containing more than one helical segment are assumed not to be populated and are excluded from the partition function (i.e., assigned statistical weights of zero). As peptide length increases, the approximation is no longer valid since multiple helical segments can be long enough to overcome the initiation penalty. The single sequence approximation will also break down when a sequence with a high preference for a helix terminus is within the middle of the chain. The error from using the single sequence approximation will therefore show a wide variation with sequence and could be potentially serious if a sequence has a high preference to populate more than one helix simultaneously. Conformations with two or more helices may also often include helix–helix tertiary interactions that are ignored in all helix/coil models.

9.4.5
N- and C-Caps

In the original LR model weights are assigned to residues in the center of hhh triplets (a weight of w for propagation) or in the center of chh or hhc triplets (a weight of v for initiation). Residues in all other triplets have weights of 1. LR-based models have been extended by assigning weights to additional conformations. N-capping can therefore be included in LR theory be assigning a weight of n to the central residue in a cch triplet [99]. Similarly, the C-cap is the first residue in a nonhelical conformation (c) at the C-terminus of a helix. C-cap weights (c-values) are assigned to central residues in hcc triplets. Application of the model to experimental data where the C-terminal amino acid of a helical peptide was varied, allowed the determination of the c-values and hence free energies of C-capping (as $-RT \ln c$) of all the amino acids [29].

A problem with the original definitions of the capping weights above is that they apply to isolated h or hh conformations that are best regarded as part of the random coil. A helical hydrogen bond can only form when a minimum of three consecutive h residues are present. Andersen and Tong [151] and Rohl et al. [9] therefore changed the definition of the N-cap to apply only to the c residue in a chhh quartet. The N-cap residues in chc or chhc conformations have weights of zero in the Andersen-Tong model and 1 in the Rohl et al. model.

9.4.6
Capping Boxes

The N-terminal capping box [25] includes a side chain–backbone hydrogen bond from N3 to the N-cap $(i, i-3)$. This is included in the LR model by assigning a weight of w^*r to the chhh conformation, where r is the weight for the Ser backbone to Glu side chain bond [9].

9.4.7
Side-chain Interactions

As helices have 3.6 residues per turn, side chains spaced $i, i+3$ or $i, i+4$ are close in space. Side-chain interactions are thus possible when four or five consecutive residues are in a helix. They are included in the LR-based model by giving a weight of $w*q$ to hhhh quartets and $w*p$ to hhhhh quintets. The side-chain interaction is between the first and last side chains in these groups; the w weight is maintained to keep the equivalence between the number of residues with a w weighting and the number of backbone helix hydrogen bonds [100].

Scholtz et al. [152] used a model based on the one-helical-sequence approximation of the LR model to quantitatively analyze salt bridge interactions in alanine-based peptides. Only a single interaction between residues of any spacing was considered, though this was appropriate for the sequences they studied. Shalongo and Stellwagen [153] also proposed incorporating side-chain interaction energies into the LR model, using a clever recursive algorithm.

9.4.8
N1, N2, and N3 Preferences

The helix N-terminus shows significantly different residue frequencies for the N-cap, N1, N2, N3, and helix interior positions [12, 19, 154, 155]. A complete theory for the helix should therefore include distinct preferences for the N1, N2, and N3 positions. In the original LR model, the N1 and C1 residues are both assigned the same weight, v. Shalongo and Stellwagen [153] separated these as v_N and v_C. Andersen and Tong [151] did the same and derived complete scales for these parameters from fitting experimental data, though some values were tentative. The helix initiation penalty is v_N*v_C and so v_N and v_C values are all small (≈ 0.04).

We added weights for the N1, N2, and N3 ($n1$, $n2$, and $n3$) positions as follows [156]: The $n1$ value is assigned to a helical residue immediately following a coil residue. The penalty for helix initiation is now $n1.v$, instead of v^2, as v remains the C1 weight. An N2 helical residue is assigned a weight of $n2.w$, instead of w. The weight w is maintained in order to keep the useful definition of the number of residues with a w weighting being equal to the number of residues with an $i, i+4$ main chain–main chain hydrogen bond. The $n2$ value is an adjustment to the weight of an N2 residue that takes into account the structures that can be adopted by side chains uniquely at this position. Similarly, an N3 residue is now assigned the weight $n3.w$, instead of w.

9.4.9
Helix Dipole

Helix-dipole effects were added to the LR model by Scholtz et al. [152], though they used the one-sequence approximation so that only one or no dipoles in total are present. In LR models helix-dipole effects are subsumed within other energies. For example, N-cap, N1, N2, and N3 energies will include a contribution from the helix-dipole interaction so the energy of interaction of charged groups at this position with the dipole should not be counted in addition.

9.4.10
3_{10}- and π-Helices

The Lifson-Roig formalism can easily be adapted to describe helices of other cooperative lengths [157]. The fundamental difference between a 3_{10}-helix and an α-helix is that the 3_{10}-helix has an $i, i + 3$ hydrogen-bonding pattern rather than the $i, i + 4$ pattern characteristic of the α-helix. For a given number of units in helical conformations, a 3_{10}-helix will consequently have one more hydrogen bond than an α-helix. To include this difference in the 3_{10}-helix theory, one of the α-helical-initiating residues (i.e., the central unit of either the $h_\alpha h_\alpha c$ or the $ch_\alpha h_\alpha$ triplet) must become a 3_{10}-helix-propagating residue. We arbitrarily chose to assign the propagating statistical weight, w_τ, to the central unit of the $h_\tau h_\tau c$ triplet such that helix propagating unit i is associated with the hydrogen bond formed between the CO of peptide $i - 2$ and the NH of peptide $i + 1$. The remainder of the statistical weights applicable to the α-helix/coil theory are maintained.

The models described above for the α-helix/coil and 3_{10}-helix/coil transitions can be combined to describe an equilibrium including pure α-helices, pure 3_{10}-helices, and mixed α-/3_{10}-helices [157]. In this model, three conformational states are possible, 3_{10}-helical (h_τ), α-helical (h_α), and coil (c). The pure 3_{10}-helix and the mixed α-/3_{10}-helix models were subsequently extended to include side-chain interactions [158]. Sheinerman and Brooks [128] independently produced a model for the α/3_{10}/coil equilibrium, based on the ZB formalism, rather than the LR model. They similarly extend the classification of conformations from α/coil to α/3_{10}/coil.

In a π-helix, formation of an $i, i + 5$ hydrogen bond requires that four units be constrained to the π-helical conformation, h_π. The π subscript designates the conformation and weights describing the π-helix, whose dihedral angles are distinct from α- and 3_{10}-helices. Assigning statistical weights to individual units requires consideration of the conformations of the unit itself and its three nearest neighbors [157]. The initiating statistical weight, v_π, is assigned to a helical unit when one or more of its two N-terminal and nearest C-terminal neighbors are in the coil conformation. The definition of helix-initiating units as the two N-terminal and one C-terminal units of each helical stretch is again arbitrary. Units in a π-helical conformation with three helical neighbors are assigned the propagating statistical weight, w_π. A π-helix propagating residue, i, is thus associated with the hydrogen bond between the NH of residue $i + 2$ and the CO of residue $i - 3$.

9.4.11
AGADIR

AGADIR is an LR-based helix/coil model developed by Serrano, Muñoz and coworkers. The original model [159] included parameters for helix propensities excluding backbone hydrogen bonds (attributed to conformational entropy), backbone hydrogen bond enthalpy, side-chain interactions, and a term for coil weights at the end of helical sequences (i.e., caps). The single-sequence approximation was used. The original partition function assumed that many helical conformations did not exist, as all conformations in which the residue of interest is not part of a helix were excluded [100, 159]. These were corrected in a later version, AGADIRms, which considers all possible conformations [160]. If AGADIR and LR models are both applied to the same data, to determine a side-chain interaction energy, for example, the results are similar, showing that the models are now not significantly different [160, 161]. The treatment of the helix/coil equilibrium differs in a number of respects from the ZB and LR models and these have been discussed in detail in by Muñoz and Serrano [160]. The minimal helix length in AGADIR is four residues in an h conformation, rather than three. The effect of this assumption is to exclude all helices which contain a single hydrogen bond; only helices with two or more hydrogen bonds are allowed. In practice, this probably makes little difference as chhhc conformations are usually unfavorable and hence have low populations. Early versions of AGADIR considered that residues following an acetyl at the N-terminus or preceding an amide at the C-terminus were always in a c conformation; this was changed to allow these to be helical [162].

The latest version of AGADIR, AGADIR1s-2 [162], includes terms for electrostatics [162], the helix dipole [162, 163], pH dependence [163], temperature [163], ionic strength [162], N1, N2, and N3 preferences [164], and capping motifs such as the capping box, hydrophobic staple, Schellman motif, and Pro-capping motif [162]. The free energy of a helical segment, $\Delta G_{\text{helical-segment}}$, is given by
$$\Delta G_{\text{helical-segment}} = \Delta G_{\text{Int}} + \Delta G_{\text{Hbond}} + \Delta G_{\text{SD}} + \Delta G_{\text{dipole}} + \Delta G_{\text{nonH}} + \Delta G_{\text{electrost}},$$
which are terms for the energy required to fix a residue in helical angles (with separate terms for N1, N2, N3, and N4), backbone hydrogen bonding, side-chain interactions excluding those between charged groups, capping, and helix-dipole interactions, respectively. Electrostatic interactions are calculated with Coulomb's equation. Helix-dipole interactions were all electrostatic interactions between the helix dipole or free N- and C-termini and groups in the helix. Interactions of the helix dipole with charged groups located outside the helical segment were also included. pH-dependence calculations considered a different parameter set for charged and uncharged side chains and their pK_a values. The single sequence approximation (see above) is used again, unlike in AGADIRms. This means that it must not be used for full protein sequences, though this has been done, even if they do not have any tertiary interactions.

AGADIR is at present the only model that can give a prediction of helix content for any peptide sequence, thus making it very useful. It can also predict NMR chemical shifts and coupling constants. In order to do this it must include estimates of all the terms that contribute to helix stability, notably the 400 possible $i, i+4$ side-chain interactions. Since only a few of these interactions have been measured accurately, the terms used cannot be precise. Further determination of energetic contributions to helix stability is therefore still needed.

9.4.12
Lomize-Mosberg Model

Lomize and Mosberg also developed a thermodynamic model for calculating the stability of helices in solution [165]. Interestingly, they extended it to consider helices in micelles or a uniform nonpolar droplet to model a protein core environment. Helix stability in water is calculated as the sum of main chain interactions, which is the free energy change for transferring Ala from coil to helix, the difference in energy when replacing an Ala with another residue, hydrogen bonding and electrostatic interactions between polar side chains and hydrophobic side-chain interactions. An entropic nucleation penalty of two residues per helix is included. Different energies are included for N-cap, N1–N3, C1–C3, C-cap, hydrophobic staples, Schellman motifs, and polar side-chain interactions, based on known empirical data at the time (1996). Hydrophobic interactions were calculated from decreases in nonpolar surface area when they are brought in contact. Helix stability in micelles or nonpolar droplets are found by calculating the stability in water then adding a transfer energy to the nonpolar environment.

9.4.13
Extension of the Zimm-Bragg Model

Following the discovery of short peptides that form isolated helices in aqueous solution, Vásquez and Scheraga extended the ZB model to include helix-dipole and side-chain interactions [166]. The model is very general as it can include interactions of any spacing within a single helix. It was applied to determine $i, i + 4$ and $i, i + 8$ interactions. Long-range interactions, beyond the scope of LR models, can thus be included. Roberts [167] and Gans et al. [88] also refined the ZB model to include side-chain interactions.

9.4.14
Availability of Helix/Coil Programs

Some Lifson-Roig based helix/coil models are available at http://www.bi.umist.ac.uk/users/mjfajdg/HC.htm. We recommend SCINT2 for peptides with side-chain interactions (freely available from us). If no side-chain interactions are present, CAPHELIX is identical and simpler to run. If N1, N2, or N3 preferences are important in your sequence N1N2N3 can be used. If the peptide contains side-chain interactions that have not been well characterized (which is the case for the great majority of sequences) AGADIR can be run at http://www.embl-heidelberg.de/Services/serrano/agadir/agadir-start.html. In our experience, SCINT2 or CAPHELIX are more accurate at predicting helix contents of alanine-based control peptides than AGADIR (see Refs [101] and [106] for examples), though AGADIR can be applied to many more sequences. The Andersen version of an LR-based helix/coil is at http://faculty.washington.edu/nielshan/helix/.

9.5
Forces Affecting α-Helix Stability

9.5.1
Helix Interior

Since the advent of crystal structures of proteins it had been noticed that some amino acids appeared frequently in α-helices and others less frequently [168, 169].

For example alanine and leucine are abundant, whereas proline and glycine appear rarely. As more of this kind of information became available a helix propensity scale was derived [170] and eventually allowed prediction of the location of α-helices (and other structures) in folded proteins from their sequence [171].

Different approaches have been used in order to determine the helical propensity or preference of individual amino acids. Scheraga and coworkers used a host–guest strategy (see Section 9.3.1) to derive values for the helical preference of various amino acid residues [172, 173]. This has been carried out for all 20 naturally occurring amino acids [174]. This work has been criticized as the host side chains can interact with each other [175]. The introduction of a guest residue thus removes host–host interactions and replaces them with PHBG–guest or PHPG–guest side-chain interactions that may obscure the intrinsic helix propensities. We believe that helix propensities are best evaluated in an alanine background, where no side-chain interactions can affect the helix stability.

Rohl et al. [30] used many alanine-based peptides with the general sequences Ac-$(AAKAA)_m$Y-NH_2 (or with Q instead of K) to measure interior helix propensities. Substitutions in the helix interior and subsequent measures of helicity using CD spectroscopy in both water and 40% (v/v) trifluoroethanol (TFE) allowed both the calculation of the Lifson-Roig w parameter and stabilization energy for all 20 amino acids (see Table 9.3). Kallenbach and coworkers also used synthetic peptides

Tab. 9.3. Helix propagation propensities and free energies of amino acids in water.

Amino acid	$\Delta G°$(helix) (kcal mol^{-1})	w value
Ala	−0.27	1.70
Arg$^+$	−0.052	1.14
Leu	0.095	0.87
Met	0.25	0.65
Lys$^+$	0.019	1.00
Gln	0.28	0.62
Glu	0.21	0.70
Ile	0.44	0.46
Trp	0.69	0.29
Ser	0.52	0.40
Tyr	0.42	0.48
Phe	0.73	0.27
Val	0.77	0.25
His	0.57	0.36
Asn	0.69	0.29
Thr	0.95	0.18
Cys	0.64	0.32
Asp	0.52	0.40
Gly	1.7	0.048
Pro	>3.8	<0.001

From Ref. [30].

of the form succinyl-YSEEEEKAKKAXAEEAEKKKK-NH$_2$ where substitutions at X allowed determination of helix stabilising energies for common amino acids [87]. Stellwagen and coworkers made substitutions in position 9 of Ac-Y(EAAAK)$_3$A-NH$_2$ [84]. These agree well with the alanine-based peptide work described previously [30, 176].

Other groups have investigated helical propensities and stabilization using whole protein methods. Blaber et al. [177–179] used mutagenesis in the helices of phage T4 lysozyme to study the structural effects of substitutions of amino acids. With the exception of substituting proline, they found that no substitutions significantly distorted the helix backbone. $\Delta\Delta G$ values correlated well (71–93%) with model peptide studies and with studies on the frequency of amino acid occurrence in protein structures [171]. Fersht and coworkers used a similar method with barnase to study the effect of replacing Ala32 of the second helix in this protein with the other 19 naturally common amino acids [180]. They used reversible urea denaturation to measure free energies of unfolding.

O'Neil and DeGrado [181] used substitutions into an α-helical two-stranded coiled-coil system to deduce helix-forming tendencies of common amino acids, through the design of a peptide that forms a noncovalent α-helical dimer, which is in equilibrium with a randomly coiled monomeric state. The α-helices in the dimer contain a single solvent-exposed site that is surrounded by small, neutral amino acid side chains. Each of the commonly occurring amino acids was substituted into this guest site, and the resulting equilibrium constants for the monomer–dimer equilibrium were determined to provide a list of free energy difference ($\Delta\Delta G^0$) values. Again these values show good agreement with those of other groups working in different model systems.

In 1998 Pace and Scholtz [182] gathered information from many different sources and derived a scale for the propensity of each amino acid in the helix interior. This is summarized in Table 9.4. The values are in $\Delta(\Delta G)$ relative to alanine as zero. Alanine was taken as zero as it is generally (though not universally) agreed that this amino acid has the highest helical propensity. Shown for comparison in Table 9.5 are the s values from the Zimm-Bragg model as derived by the Scheraga group [174]. Both tables are ranked in terms of descending helical propensity.

As can be seen from Tables 9.4 and 9.5, the two series do not agree on even the relative helical propensities of the amino acid residues. The data summarized by Pace and Scholtz shows alanine having the highest helix propensity and all other residues having lower values (a positive $\Delta(\Delta G)$ value relative to Ala). Both series show proline and glycine to have the lowest helical propensity (highest $\Delta(\Delta G)$ value or lowest s value. The most controversial of these differences over the years has been that of alanine. Host–guest analysis showing alanine to be effectively helix-neutral has been supported by data from some other groups, notably the templated helices of Kemp and coworkers [107]. Several efforts have been made to try to explain this discrepancy, including implication of the charged groups used to solubilize the alanine-based peptides [183, 184], but little reconciliation has been achieved. The use of template-nucleated helices has been criticized by Rohl et al. [185] who argued that the low apparent helix propensity of alanine is a conse-

Tab. 9.4. Summary of other experimental helix propensities (relative to alanine).

Amino acid	Helix propensity ($\Delta\Delta G$) (kcal mol^{-1})
Ala	0.00
Arg$^+$	0.21
Leu	0.21
Met	0.24
Lys$^+$	0.26
Gln	0.39
Glu	0.40
Ile	0.41
Trp	0.49
Ser	0.50
Tyr	0.53
Phe	0.54
Val	0.61
His	0.61
Asn	0.65
Thr	0.66
Cys	0.68
Asp	0.69
Gly	1.00
Pro	3.16

From Ref. [182].

quence of properties of the template–helix junction. However, recent work by Kemp and coworkers [186] may at last settle the controversy. Using templates to investigate the helix-forming tendency of polyalanine, these workers extended the length of the polyalanine beyond the previous limit of six residues. Below six residues, both this group and Scheraga's had low helix propensities for alanine (see above) but when the limit of six was exceeded, a dramatic increase in helix propensity was observed. For chains with less than six alanines, $w = 1.03$, in agreement with both Kemp and Scheraga's earlier experimental results. For chains with 6–9 alanines, $w = 1.15$ and for more than 10 Alanines, w is 1.26. This indicates that there is a length-dependent term in the helicity of polyalanine and that the charged groups are not having the effect previously ascribed to them. These values for longer polyalanine sequences are also much more consistent with values published by other groups.

9.5.2
Caps

Serrano and Fersht [20] explored the capping preferences at the N-cap by mutating Thr residues at the N-cap of two helices in barnase. They found that negatively

Tab. 9.5. Helical propensity values from host–guest studies.

Amino acid	s (20 °C)
Glu	1.35
Met	1.2
Leu	1.14
Ile	1.14
Trp	1.11
Phe	1.09
Ala	1.07
Arg+	1.03
Tyr	1.02
Cys	0.99
Gln	0.98
Val	0.95
Lys+	0.94
His	0.85
Thr	0.82
Asn	0.78
Asp	0.78
Ser	0.76
Gly	0.59
Pro	0.19

From Ref. [174].

charged residues were favored at the N-cap with a rank order of Asp > Thr > Glu > Ser > Asn > Gly > Gln > Ala > Val. Interestingly, their result conflicted with the statistical survey result that Asn is one of the most frequently found N-caps in proteins [12]. Experimentally Asn destabilized the helices by 1.3 kcal mol^{-1} relative to Thr. The rank order of amino acids N-cap preferences in T4 lysozyme was found to be Thr > Ser > Asn > Asp > Val = Ala > Gly [21]. They suggested that Asn can be inherently as good an N-cap as Ser or Thr, but it requires a change in backbone dihedral angles of N-cap residues, which might be altered in native proteins as the results of tertiary contacts. Indeed Asn is the most stabilizing residue at N-cap in a peptide model in the absence of tertiary contacts and other side-chain interactions (see below).

The Kallenbach group [187] substituted several amino acids at the N-cap position in peptide models in the presence of a capping box. They found that Ser and Arg are the most stabilizing residues with $\Delta\Delta G$ relative to Ala of −0.74 and −0.58 kcal mol^{-1}, respectively, whilst Gly and Ala are less stabilizing. The results are in agreement with the results of Forood et al. [23], who found that the trend in α-helix inducing ability at the N-cap is Asp > Asn > Ser > Glu > Gln > Ala. A more comprehensive work to determine the preferences for all 20 amino acids at the N-cap position used peptides with a sequence of NH$_2$-XAKAAAAKAAAAKAAGY-CONH$_2$ [22, 29, 99]. N-Capping free energies ranged from Asn (best) to Gln (worst) (Table 9.6).

Tab. 9.6. Amino acid propensities at N- and C-terminal positions of the helix.

Residue	$\Delta\Delta G$ relative to Ala for transition from coil to the position (kcal mol^{-1})												
	N-cap	N1		N2		N3		C3	C2	C1	C-cap		C'
	29[a]	95[b]	164;188[c]	96[d]	164;188[e]	97[f]	164;188[g]	189[h]	190[i]	189[j]	189[k]	29[l]	299[m]
A	0	0	0	0	0	0	0	0	0	0	0	0	0
C°												0.2	
C⁻	−1.4	1.0		0.9									
D°		0.5		0.7								0.2	0.3
D⁻	−1.6	0		−0.2		1.1							
E°		1.0		−0.2								−0.4	0.3
E⁻	−0.7	0.1		−0.4		0.6						−0.5	
F	−0.7	1.4		0.9		1.3			0.6				0.1
G	−1.2	1.0	0.7		0.4		0.8	2.1	1.0	0.6	0.4	0.1	−1.1
H°	−0.7	0.7		0.8		2.6							
H⁺												−0.2	−0.9
I	−0.5	0.5	0.4	0.6	0.5	0.7	0.5	0.2	0.2	0.4	0.5		1.5
K⁺	0.1	0.7		0.9		0.9						−0.1	−0.1
L	−0.7	0.4	0.2	0.5	0.5	0.8	0.4		0.1			−0.1	
M	−0.3	0.5	0.1	0.7	0.3	0.7	0.4		0.1			−0.3	0.1
N	−1.7		0.6	1.7	0.7		0.7	0.5	0.7	0.4	0.3	0.1	−0.4
P	−0.4	0.6	0.5										1.2
Q	2.5	0.5	0.3	0.5	0.3	1.2	0.2	0.2	−0.02	0.2	0.05	−0.5	−0.1
R⁺	−0.1	0.7		0.8								−0.4	−0.2
S	−1.2	0.4	0.4	0.7	0.5	1.1	0.6	0.6	0.5	0.7	0.5	0.8	0.3
T	−0.7	0.5	0.5	0.5	0.5	1.2	0.6	0.8	0.6	0.5	0.8		1.1
V	−0.1	0.6		0.5	0.4			0.4	0.3	0.4	0.7	0.6	0.9	1.6
W	−1.3	0.4		0.8		4.0							0.7
Y	−0.9					1.2						−2.2	

[a] NH$_2$-XAKAAAAKAAAAKAAGY-CONH$_2$
[b] CH$_3$CO-XAAAAQAAAAQAAGY-CONH$_2$
[c] CH$_3$CO-XAAAAAAARAAARGG Y-NH$_2$
[d] CH$_3$CO-AXAAAAKAAAAKAAGY-CONH$_2$
[e] CH$_3$CO-AXAAAAAARAAARGGY-NH$_2$
[f] CH$_3$CO-AAXAAAAKAAAAKAGY-CONH$_2$
[g] CH$_3$CO-AAXAAAAARAAARGGY-NH$_2$
[h] NH$_2$-YGGSAKEAAARAAAAXAA-CONH$_2$
[i] Substitution of residue 32 (C2 position) of α-helix of ubiquitin.
[j] NH$_2$-YGGSAKEAAARAAAAAXA-CONH$_2$
[k] NH$_2$-YGGSAKEAAARAAAAAAX-CONH$_2$
[l] CH$_3$CO-YGAAKAAAAKAAAAKAX-COOH
[m] Substitution of residue 35 (C' position) of α-helix of ubiquitin.

We have used a similar approach using peptide models to probe the preferences at N1 [95], N2 [96], and N3 [97] using peptides with sequences of CH$_3$CO-XAAAA-QAAAAQAAGY-CONH$_2$, CH$_3$CO-AXAAAAKAAAAKAAGY-CONH$_2$ and CH$_3$CO-AAXAAAAKAAAAKAGY-CONH$_2$, respectively. The results have given N1, N2, and N3 preferences for most amino acids for these positions (Table 9.6) and these agree well with preferences seen in protein structures, with the interesting exception of Pro at N1. Petukhov et al. similarly obtained N1, N2, and N3 preferences for nonpolar and uncharged polar residues by applying AGADIR to experimental helical peptide data, and found almost identical results [164, 188]. The complete sequences of peptides used can be seen in the table footnote. In general, at N1, N2, and N3, Asp and Glu as well as Ala are preferred, presumably because negative side chains interact favorably with the helix dipole or NH groups while Ala has the strongest interior helix preference.

Although it is also unique in terms of the presence of unsatisfied backbone hydrogen bonds, the C-terminal region is less explored experimentally. The C-terminus of the α-helix tends to fray more than the N-terminus, making C-terminal measurements less accurate. Preferences at the C-cap position differ from those at the N-cap. At the N-terminus, the helix geometry favors side chain-to-backbone hydrogen bonding, so polar residues are preferred [14, 19]. At the C-terminus unsatisfied backbone hydrogen bonds are fulfilled by interactions with backbone groups upstream of the helix. Zhou et al. [48] found that Asn is the most favored residue at the C-cap followed by Gln > Ser~Ala > Gly~Thr. Forood et al. [23] tested a limited number of amino acids at the C-terminus (C1) finding a rank order of Arg >

Lys > Ala. Doig and Baldwin [29] determined the C-capping preferences for all 20 amino acids in α-helical peptides. The thermodynamic propensities of some amino acids at C′, C-cap, C1, C2, and C3 are also included in Table 9.6 [189, 190].

9.5.3
Phosphorylation

Phosphoserine is destabilizing compared with serine at interior helix positions [191, 192]. We investigated the effect of placing phosphoserine at the N-cap, N1, N2, N3, and interior position in alanine-based α-helical peptides, studying both the −1 and −2 phosphoserine charge states [193]. Phosphoserine stabilizes at the N-terminal positions by as much as 2.3 kcal mol^{-1}, while it destabilizes in the helix interior by 1.2 kcal mol^{-1}, relative to serine. The rank order of free energies relative to serine at each position is N2 > N3 > N1 > N-cap > interior. Moreover, −2 phosphoserine is the most preferred residue known at each of these N-terminal positions. Experimental pK_a values for the −1 to −2 phosphoserine transition are in the order N2 < N-Cap < N1 < N3 < interior. Phosphoserine can form stabilizing salt bridges to arginine [192].

9.5.4
Noncovalent Side-chain Interactions

Many studies have been performed on the stabilizing effects of interactions between amino acid side chains in α-helices. These studies have identified a number of types of interaction that stabilize the helix including salt bridges [83, 86, 88, 152, 194–198], hydrogen bonds [152, 198–200], hydrophobic interactions [100, 201–

203], basic/aromatic interactions [106, 204], and polar/nonpolar interactions [101]. The stabilizing energies of many pairs in these categories have been measured, though some have only been analyzed qualitatively. As described earlier, residue side chains spaced $i, i+3$ and $i, i+4$ are on the same face of the α-helix, though it is the $i, i+4$ spacing that receives most attention in the literature, as these are stronger. A summary of stabilizing energies for side-chain interactions is given in Table 9.7. We give only those that have been measured in helical peptides with the side-chain interaction energies determined by applying helix/coil theory. Almost all are attractive, with the sole exception of the Lys-Lys repulsion.

9.5.5
Covalent Side-chain interactions

Lactam (amide) bonds formed between NH_3^+ and CO_2^- side chains can stabilize a helix, acting in a similar way to disulfide bridges in a protein by constraining the side chains to be close, reducing the entropy of nonhelical states [205]. Lactam bridges between Lys-Asp, Lys-Glu, and Glu-Orn spaced $i, i+4$ have been introduced into analogs of human growth hormone releasing factor [206], and proved to be stabilizing with Lys-Asp most effective. The same Lys-Asp $i, i+4$ lactam was stabilizing in other helical peptide systems [207–210], while Lys-Glu $i, i+4$ lactam bridges were less effective [211]. Two overlapping Lys-Asp lactams were even more stabilizing [212]. The effect of the ring size formed by the lactam was investigated by replacing Lys with ornithine or (S)-diaminopropionic acid. A ring size of 21 or 22 atoms was most stabilizing (a Lys-Asp $i, i+4$ lactam is 20 atoms) [206]. Lactams between side chains spaced $i, i+7$ [213, 214] or $i, i+3$ [214]; [215], spanning two or one turns of the helix have also been reported. $i, i+7$ disulfide bonds have been introduced into alanine-based peptides, using (D)- and (L)-2-amino-6-mercaptohexanoic acid derivatives [216]. Strongly stabilizing effects were observed.

Some interesting recent work has shown that helix formation can be reversibly photoregulated. Two cysteine residues are cross-linked by an azobenzene derivative which can be photoisomerized from *trans* to *cis*, causing a large increase or decrease in the helix content of the peptide, depending on its spacing [217–219].

9.5.6
Capping Motifs

Although the N-terminal capping box sequence stabilizes helices by inhibiting N-terminal fraying, it does not necessarily promote elongation unless accompanied by favorable hydrophobic interactions as in a "hydrophobic staple" motif [220, 221]. The nature of the capping box stabilizing effect thus not only arises from reciprocal hydrogen bonds between compatible residues, but also from local interactions between side chains, helix macrodipole-charged residue interactions and solvation [222].

Despite statistical analyses revealing that Schellman motifs are observed more frequently that expected at the helix C-terminus, this motif populates only transi-

9 Stability and Design of α-Helices

Tab. 9.7. Summary of side-chain interaction energies from literature.

Interaction	$\Delta\Delta G$ (kcal mol^{-1})	Reference
Ile-Lys $(i, i+4)$	−0.22	101
Val-Lys $(i, i+4)$	−0.25	101
Ile-Arg $(i, i+4)$	−0.22	101
Phe-Met $(i, i+4)$	−0.8	100
Met-Phe $(i, i+4)$	−0.5	100
Gln-Asn $(i, i+4)$	−0.5	200
Asn-Gln $(i, i+4)$	−0.1	200
Phe-Lys $(i, i+4)$	−0.14	106
Lys-Phe $(i, i+4)$	−0.10	106
Phe-Arg $(i, i+4)$	−0.18	106
Phe-Orn $(i, i+4)$	−0.4	204
Arg-Phe $(i, i+4)$	−0.1	106
Tyr-Lys $(i, i+4)$	−0.22	106
Glu-Phe $(i, i+4)$	−0.5	300
Asp-Lys $(i, i+3)$	−0.12	227
Asp-Lys $(i, i+4)$	−0.24	227
Asp-His $(i, i+3)$	>−0.63	301
Asp-His $(i, i+4)$	>−0.63	301
Asp-Arg $(i, i+3)$	−0.8	302
Glu-His $(i, i+3)$	−0.23	227
Glu-His $(i, i+4)$	−0.10	227
Glu-Lys $(i, i+3)$	−0.38	152
Glu-Lys $(i, i+4)$	−0.44	152
Phe-His $(i, i+4)$	−1.27	198
Phe-Met $(i, i+4)$	−0.7	203
His-Asp $(i, i+3)$	−0.53	198
His-Asp $(i, i+4)$	−2.38	241
His-Glu $(i, i+3)$	−0.45	227
His-Glu $(i, i+4)$	−0.54	227
Lys-Asp $(i, i+3)$	−0.4	227
Lys-Asp $(i, i+4)$	−0.58	227
Lys-Glu $(i, i+3)$	−0.38	227
Lys-Glu $(i, i+4)$	−0.46	227
Lys-Lys $(i, i+4)$	+0.17	196
Leu-Tyr $(i, i+3)$	−0.44	153
Leu-Tyr $(i, i+4)$	−0.65	153
Met-Phe $(i, i+4)$	−0.37	203
Gln-Asp $(i, i+4)$	−0.97	199
Gln-Glu $(i, i+4)$	−0.31	152
Trp-Arg $(i, i+4)$	−0.4	300
Trp-His $(i, i+4)$	−0.8	161
Tyr-Leu $(i, i+3)$	−0.02	153
Tyr-Leu $(i, i+4)$	−0.44	153
Tyr-Val $(i, i+3)$	−0.13	153
Tyr-Val $(i, i+4)$	−0.31	153
Arg $(i, i+4)$ Glu $(i, i+4)$ Arg	−1.5	303
Arg $(i, i+3)$ Glu $(i, i+3)$ Arg	−1.0	303
Arg $(i, i+3)$ Glu $(i, i+4)$ Arg	−0.3	303
Arg $(i, i+4)$ Glu $(i, i+3)$ Arg	−0.1	303
Phosphoserine-Arg $(i, i+4)$	−0.45	192

ently in aqueous solution but it is formed in 30% TFE [223]. This might be due to the C-terminus being very frayed and the increase of helical content contributed from this motif is small. Energetically this motif is not very favorable due to the entropic cost of fixing a Gly residue at the position C'. The Schellman motif is believed to be a consequence of helix formation and does not involve α-helix nucleation [224]. The α_L motif seems to be more stable than the alternative Schellman motif [221].

9.5.7
Ionic Strength

Electrostatic interactions between charged side chains and the helix macrodipole can stabilize the helix [92, 102, 225]. The interactions are potentially quite strong, but are alleviated by the screening effects of water, ions, and nearby protein atoms. In theory, increasing ionic strength of the solvent (up to 1.0 M) should stabilize the helix through interactions with α-helix dipole moments by shifting the equilibrium between α-helix and random coil, which has a random orientation of the peptide dipoles [226]. The energetics of the interaction between fully charged ion pairs can be diminished by added salt and completely screened at 2.5 M NaCl [197, 227]. In peptides containing side chain-to-side-chain interactions, the effect of ion pairs and charge/helix-dipole interactions cannot be clearly separated. There are, however, indications that the interactions of charged residues with the helix macrodipole are less affected than those between charged side chains [227, 228]. In coiled-coil peptides, salt also affects hydrophobic residues by strengthening their interactions at the coiled-coil interface. This can be explained through alterations of the peptide–water interactions at high salt concentration. However, this requires a strong kosmotropic anion to accompany the screening cation [229].

9.5.8
Temperature

Thermal unfolding experiments show that the helix unfolds with increasing temperature [230–232]. There is no sign of cold denaturation, as seen with proteins. Enthalpy and entropy changes for the helix/coil transition are difficult to determine as the helix/coil transition is very broad, precluding accurate determination of high- and low-temperature baselines by calorimetry [230]. Nevertheless, isothermal titration calorimetric studies of a series of peptides that form helix when binding a nucleating La^{3+}, find ΔH for helix formation to be -1.0 kcal mol^{-1} [135, 233], in good agreement with the earlier work.

9.5.9
Trifluoroethanol

Peptides with sequences of helices in proteins usually show low helix contents in water. An answer to this problem is to add TFE (2,2,2-trifluoroethanol) to induce helix formation [234–237]. For many peptides, the concentration of TFE used to increase the helix content is only up to 40% [234–236, 238, 239]. TFE may act by shielding CO and NH groups from the water solvent while leading to hydrogen bond formation between them. The conformational equilibrium thus shifts toward more compact structures, such as the α-helical conformation [184]. The mechanism involves interaction between TFE and water with several interpretations. One view suggests that TFE indirectly disrupts the solvent shell on α-helices [240, 241]. Another view proposes that TFE destabilizes the unfolded species and thereby indirectly enhances the kinetics and thermodynamics of folding of the coiled coil [242]. A more compromising view suggests that TFE forms clusters in water solution, which at lower concentration pulls the water molecules from the surface of proteins. At higher concentration, TFE clusters associate with appropriate hydrophobic side chains reducing their conformational entropy and switch the conformation at TFE concentration > 40% [243].

The propagation propensities of all amino acids increase variably in 40% TFE relative to water. The propagation propensities of the nonpolar amino acids increase greatly in 40% TFE whilst other amino acids propensity increase uniformly. However, glycine and proline are strong helix breakers in both in water and 40% TFE solvents [30]. In addition, 40% TFE dramatically alters electrostatic (and polar) interactions and increases the dependence of helix propensities on the sequence [244].

9.5.10
pK_a Values

Evaluation of pK_a values of titrable amino acids in a peptide sequence can be used to analyze the strength of the possible interaction they form in water. pK_a shifts of charged residues at the helix termini are significant because they can potentially interact with unsatisfied hydrogen bonds of the NH groups and CO group at the N-terminus and C-terminus, respectively, or the helix dipole. The pK_a values can be measured accurately from the change in ellipticity across a broad range of pH. The asymptotic values of the ellipticities for the different protonation states are fitted to a Henderson-Hasselbach equation to calculate the pK_a.

In general, the pK_a values of Glu and His at N1–N3 are normal compared with those in model compounds. In contrast, Asp and Cys have shifted pK_a to lower values [95, 96, 102, 152, 225, 245–248]. An exception for negatively charged residues at the N-cap is that they have a lower pK_a [29]. This may be because side chains at the N-cap can form strong hydrogen bonds to NH groups of N2 and N3, while the bonds formed by side chains at N1, N2, and N3 are much weaker [14, 19].

The negatively charged residues at higher pH destabilize helices when at the C-cap [29]. The increased pK_a may result from an unfavorable electrostatic interaction with the C-terminal dipole or partial negative charges on the terminal CO groups.

9.5.11
Relevance to Proteins

Many of the features studied in peptide helices are also applicable to proteins and can be used to rationally modify protein stability or to design new helical proteins. Helices in proteins are often found on the surface with one face exposed to solvent and the other buried in the protein core. Helix propensities and side-chain interactions measured in peptides are thus directly applicable to the solvent-exposed face. Substitutions at buried positions are much more complex and tertiary interactions also make major contributions to stability. Tertiary interactions at helix termini are rare; nearly all side-chain interactions are local [14]. Preferences for capping sites and the first and last turn of the helix are therefore applicable to most protein helices. The feature of protein helices of amphiphilicity, reflected in possession of a hydrophobic moment [249], is irrelevant to monomeric isolated helices.

9.6
Experimental Protocols and Strategies

9.6.1
Solid Phase Peptide Synthesis (SPPS) Based on the Fmoc Strategy

Solid phase synthesis was first described by Merrifield [250]. It depends on the attachment of the C-terminal amino acid residue to a solid resin, the stepwise addition of protected amino acids to a growing peptide chain covalently bound to the resin and finally the cleavage of the completed peptide from the solid support. This cyclical process is repeated until the chain assembly process is completed (Figure 9.3).

To select the appropriate resin for the synthetic target, several key features of the linker have to be taken into consideration, for example racemization-free synthetic attachment of the first amino acid residue and resulting C-terminal functionality [251]. A key choice is whether to have a C-terminal amide or carboxyl group, as this is determined by the resin.

The Fmoc (9-fluorenylmethoxycarbonyl) group is the α-amino-protecting group, which is widely used in solid state peptide synthesis for its adaptability to linkers, resin, and cleavage chemistries [252, 253]. Other amino-protecting groups, e.g., BOC (*t*-butyloxycarbonyl), are also commonly used, but only the strategy using Fmoc will be covered here. The procedure can be done either manually or automatically in a peptide synthesizer. Here we describe the procedure of manual peptide synthesis using Fmoc chemistry on a 0.1 mmol scale. There are many other methods in peptide synthesis using different procedures and chemicals [251].

9.6.1.1 Equipment and Reagents
- A sintered glass funnel and a vacuum flask connected to water vacuum pump.
- A high speed centrifuge and centrifuge tubes.
- Freeze dryer with suitable jar.

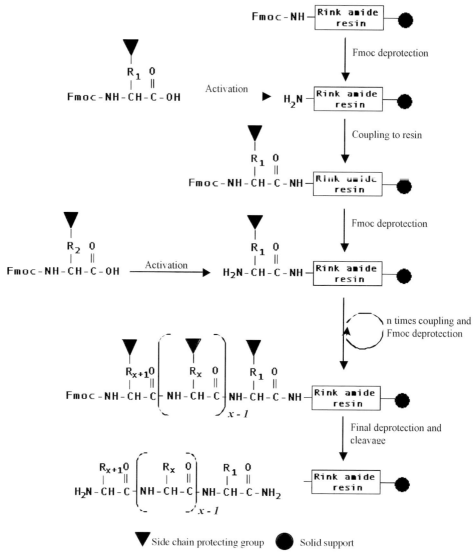

Fig. 9.3. Schematic diagram of peptide chain assembly.

- Rink amide resin 100–200 mesh (other resins can be used depending on the C-terminal functionality required, for example Wang resin is used to give C-terminal carboxyl).
- Deprotecting agent: 20% piperidine in DMF (dimethylformamide).
- Activating/coupling agents:
    −0.45 M HBTU (2-(1H-benzotriazole-1-yl)-1,1,3,3-tetramethyluronium hexafluoro phosphate) dissolved in a solution of HOBt (1-hydroxy-1H-benzotriazole) in DMF. HATU, HCTU or TBTU are popular alternatives
    −2 M DIEA (N,N,diisoprophylethylamine)

- NMP (N-methylpyrrolidone)
- DCM (dichloromethane)
- Diethyl ether
- Pyridine
- Acetic anhydride
- TFA (trifluoroacetic acid)
- TIS (triisopropyl silane) as scavengers (the choice of scavenger depends on peptide sequence)
- Fmoc-protected amino acids. Certain side chains contain functional groups that can react with free amine groups during the formation of the amide bond. Therefore, it is important to mask the functional groups of the amino acid side chain. The following side chain protecting groups are compatible with TFA cleavage (see Section 9.6.1.4):

Fmoc-Arg(Boc)	Fmoc-Cys(Acm)	Fmoc-Lys(Boc)
Fmoc-Arg(Mtr)	Fmoc-Cys(Trt)	Fmoc-Lys(Fmoc)
Fmoc-Arg(Pbf)	Fmoc-Gln(Tmob)	Fmoc-Lys(Mtt)
Fmoc-Arg(Pmc)	Fmoc-Gln(Trt)	Fmoc-Ser(tBu)
Fmoc-Asn(Tmob)	Fmoc-Glu(OtBu)	Fmoc-Thr(tBu)
Fmoc-Asn(Trt)	Fmoc-His(Boc)	Fmoc-Tyr(tBu)
Fmoc-Asp(OtBu)	Fmoc-His(Trt)	

9.6.1.2 Fmoc Deprotection and Coupling

1. Place Fmoc-protected Rink amide resin in a clean sintered glass funnel. For a resin with a loading 0.67 mmol g^{-1} use:

$$\frac{0.1 \text{ mmol}}{0.67 \text{ mmol}} = 0.15 \text{ g of resin}$$

 (using a 0.1 mmol scale).

2. Block the bottom end of the funnel with a stopper (e.g., use a tiny rubber bung). Immerse the resin in NMP for several minutes to make it swell. Remove the stopper temporarily; drain the solution from the funnel by applying suction with water vacuum pump.
3. Add 2 mL of 20% piperidine to the resin, leave it for 20 min, swirl occasionally to deprotect the base labile α-amino Fmoc group of carboxyamidated C-terminus Rink amide resin.
4. Drain the solution from the funnel by applying suction with water vacuum pump. Then wash unprotected resin with NMP several times. Make sure the resin is very clean from the previous reagent.
5. In a vial, dissolve Fmoc-protected amino acids in 2 mL of 0.45 M HBTU/HOBt in DMF, mix for about 5 min. Then add 1 mL of 2 M DIEA. The reagents will

activate the carboxyl group of the amino acid and couple it to the resin. The first amino acid to use is the one at the C-terminus. Use a 10× molar excess of the amino acid compared with the resin, for example 0.300 g of protected amino acid with MW of 300 g mol^{-1} to give 10 mmol of amino acid.
6. Transfer the solution to unprotected resin in the funnel. Leave it for 0.5–3 h, depending on the difficulty of coupling (note: β-branched amino acids sometimes need longer coupling time, i.e., overnight). Swirl the mixture frequently.
7. Repeat step 4.
8. If necessary, do a Kaiser test (see below), a quantitative ninhydrin procedure [254, 255], on a small amount of the peptide resin to check the completeness of the coupling process. Wash the beads with methanol before conducting the Kaiser test.

Continue cyclical process (repeat steps 3–8) of Fmoc deprotection and coupling until the last residue left with a free N-terminus.

9.6.1.3 Kaiser Test
The principle of the Kaiser test is that free amines react with ninhydrin to form a violet colored-compound (Ruhemann's purple). The following procedure is only applicable for Fmoc chemistry.

1. Add 2–3 mL methanol and 2–3 drops acetic acid to a test tube, and dry using a vacuum evaporator.
2. Add a few resin beads to the test tube.
3. Add 4 drops phenol/ethanol (80 g in 20 mL), 8 drops KCN/pyridine (1 mM diluted with 100 mL), and 4 drops ninhydrin/ethanol (500 mg in 10 mL). Do the same to a test tube without resin as a blank.
4. Incubate at 100 °C for 5 min.
5. After heating is completed, remove the tube and immediately add 4.8 mL of 60% ethanol for a final volume of 5 mL, vortex and let the beads settle to the bottom of the tube.
6. Pipette the sample into cuvette and read the absorbance of the sample (including the blank) at 570 nm. Use 60% ethanol to zero the spectrophotometer.
7. Calculate the μmol g^{-1} of amine as follows:

$$\mu\text{mol g}^{-1} = \frac{[(\text{Abs}_{\text{sample}} - \text{Abs}_{\text{blank}}) \times \text{dilution (mL)}]10^6}{\text{Extinction coefficient} \times \text{sample weight (mg)}}$$

where, dilution = 5 mL and extinction coefficient = 15 000 M^{-1} cm^{-1}
8. Calculate the percent coupling as follows:

$$\% \text{ coupled} = \frac{[1 - \text{amine (}\mu\text{mol g}^{-1})]}{10^3 \times \text{substitution (mmol g}^{-1})} \times 100$$

Substitution can be calculated as follows (for Fmoc-Asp(OtBu) on 0.67 mmol g^{-1} Rink amide resin):

$$\frac{1000 \text{ mmol mol}^{-1}}{0.67 \text{ mmol g}^{-1}} = 1492 \text{ g mol}^{-1}$$

Fmoc-Asp(OtBu) = 411 g mol^{-1}

$-H_2O = -18$ g mol^{-1}

Total = 1885 g mol^{-1}

$$\text{Substitution} = \frac{1000 \text{ mmol mol}^{-1}}{1885 \text{ g mol}^{-1}} = 0.53 \text{ mmol g}^{-1}$$

For an easier qualitative observation, the beads remain white and the solution yellow after heating (step 4) indicates a complete reaction. In contrast, a dark blue color on the beads and in the solution indicates an incomplete reaction (positive test). If free amino groups remain, the resin should be subjected again to coupling reaction (repeat Fmoc deprotection and coupling procedure using the same amino acid) or acetylation (see Section 9.6.1.4) to increase the purity of the final product.

9.6.1.4 Acetylation and Cleavage

1. Wash the peptide resin with about 10 mL each of DMF and DCM and further wash with methanol. Apply some suction using a water vacuum pump to dry the peptide resin.
2. When needed (otherwise go to step 4), the free N-terminus of the peptide is acetylated by adding acetylation solution consisting of 2.3 mL DMF, 0.1 mL acetic anhydride, and 0.12 mL pyridine to the peptide resin and leave for 30 min. Swirl occasionally.
3. Remove the acetylation solution from the peptide resin by applying some suction, wash the peptide resin three times each with about 10 mL of DMF and DCM.
4. In a tube, suspend the peptide resin in 5 mL cleavage solution containing 4.75 mL TFA, 125 µL H$_2$O and 125 µL TIS and leave for at least 3 h with an occasional swirl. The later serves as scavengers to protect the peptide from carbocations of amino acid side chain-protecting groups generated during cleavage, which can lead to side reactions. The exact choice of scavengers depends on the amino acids present.
5. Place the peptide resin back into the sintered glass funnel (use clean flask), apply some suction and collect the filtrate. Wash the resin twice with 5 mL of DCM and pool the filtrate.
6. Reduce the volume of the filtrate containing the peptide to about 0.5–1.0 mL by blowing compressed air/nitrogen onto the filtrate.

9.6.1.5 Peptide Precipitation

1. Drop concentrated filtrate into about 30 mL of dry, ice-cold diethyl ether in centrifuge tube. Keep the solution overnight at $-20\ °C$ for the precipitate to develop.
2. Centrifuge the resulting precipitate at 10 000 rpm for 10 min at $4\ °C$.
3. Wash the pellet three times with 30 mL of fresh dry, ice-cold diethyl ether and repeat step 2.
4. Evaporate the ether from the pellet by placing the centrifuge tube in warm water.
5. Dissolve dried pellet containing crude peptide in 5 mL of water, freeze it in dry ice and freeze dry overnight. Crude peptide is ready for HPLC purification.

9.6.2
Peptide Purification

9.6.2.1 Equipment and Reagents
- HPLC system equipped with C18 reverse phase analytical columns (250×10 mm, 5 µm).
- Acetonitrile containing 0.1% TFA.
- Deionized water containing 0.1% TFA.

9.6.2.2 Method

1. Wash the column thoroughly by running through 95% acetonitrile for at least 20 min.
2. When using a UV detector, set the detector wavelength at 214 nm.
3. Equilibrate the column with 5% H_2O for 10 min before the purification run.
4. Load the crude peptide sample dissolved in deionized water and start a linear gradient run. The amount of peptide loaded depends on its concentration. A typical run usually consist of two eluents which are mixed using a linear gradient with a flow rate which will give a 0.5% to 1.0% change per minute.
5. From the chromatogram obtained, optimize the conditions for further purification. This allows the change in flow rate and eluents composition across the elution time. If necessary, the collected fractions can be further purified using isocratic conditions.[1]
6. Inject sample again and start a new linear gradient run with optimized conditions.

---

1) For example, if a peptide elutes out at 30% aqueous acetonitrile (0.1% TFA) (from linear gradient run), a lower concentration of acetonitrile is chosen as the eluent for isocratic run (e.g., 28% (0.1% TFA) aqueous acetonitrile). The peptide peak will come out 4–5 min after the solvent peak under isocratic conditions.

7. Pool fractions of the suspected peptide[2] and analyze by mass spectrometry.[3]
8. When it is confirmed by the mass spectrometry results, collect the desired fractions several times, followed by lyophilization and storage at $-20\,°C$ for further use.[4]

9.6.3
Circular Dichroism

Circular dichroism (CD) is a very useful technique for obtaining secondary structural information on proteins and peptides. This area has been extensively reviewed in the past two decades in relation to both folding and secondary structure determination. Greenfield [256] has reviewed methods of obtaining structural information from CD data. This covers the advantages and pitfalls of each technique and experimental considerations, such as how the wavelength range of data acquisition affects the precision of conformation determination, how precisely must the protein concentration be determined for each method to give reliable answers and what computer resources are necessary to use each method. Manning [257] has reviewed various applications of CD spectroscopy to the study of proteins including analysis of denaturation curves, calculation of secondary structure composition, observation of changes in both secondary and tertiary structure and characterization of folding intermediates. The excellent book *Circular Dichroism and the Conformational Analysis of Biomolecules* [258] covers a wide range of applications of CD in detail, including a review of CD applied to α-helices by Kallenbach and coworkers [8].

In CD, the sample is illuminated with left and right circularly polarized light and the difference in absorbance of each component of the light is measured. Typically this is expressed as change in absorbance (ΔA) or as the ellipticity of the beam exiting the sample (ε, in millidegrees). This signal is proportional to the concentration of the sample, the optical path length of the cell or cuvette and the proportion of the sample exhibiting chiral behavior.

2) The crude peptide obtained from the synthesis will contain many by-products which are a result of deletion or truncated peptides as well as side products stemming from cleaved side chains or oxidation during the cleavage and deprotection process.
3) The peptides mass can be verified by low resolution, electrospray ionization (ESI) or MALDI mass spectrometry. ESI allows the peptide sample, which is dissolved in acetonitrile/water (0.1% TFA) after HPLC purification, to be directly analyzed without further treatment. For small molecules (< 2000 Da) ESI typically generates singly or doubly charged gaseous ions, while for large molecules (> 2000 Da) the ESI process typically gives rise to a series of multiply charged species.
4) It is recommended to first dissolve the peptide in sterile distilled or deionized water. Sonication can be applied if necessary to increase the rate of dissolution. If the peptide is still insoluble (e.g., peptides containing multiple hydrophobic amino acid residues), addition of a small amount of dilute (approximately 10%) acetic acid (for basic peptides) or aqueous ammonia (for acidic peptides) can facilitate dissolution of the peptide. For long-term storage of peptides, lyophilization is highly recommended. Lyophilized peptides can be stored for years at temperatures of $-20\,°C$ or lower with little or no degradation. Peptides in solution are much less stable. Peptides containing methionine, cysteine, or tryptophan residues can have limited storage time in solution due to oxidation. These peptides should be dissolved in oxygen-free solvents or stored with a reducing agent such as DTT.

Two conditions are necessary for obtaining a circular dichroism signal: The sample must be optically active or chiral and there must be a chromophore near to the chiral center.

9.6.4
Acquisition of Spectra

There are several considerations when obtaining spectra, some instrumental, others concerned with sample preparation etc.

9.6.4.1 Instrumental Considerations
Nitrogen flushing
This is essential for several reasons. The xenon lamps used in CD spectrometers have a quartz envelope so if oxygen is present when running, a lot of ozone can be generated. This has health and safety implications for users and can also be severely detrimental to mirror surfaces in the instrument. Oxygen also absorbs UV light below 195 nm. For these reasons, a flow of at least 3 L min^{-1} of clean dry nitrogen should be used at higher wavelengths, with increased flow if the wavelength is to approach 180 nm. Below this level (not available in most lab instruments) a flow of up to 50 L min^{-1} may be necessary. The flow should be started several minutes before the spectra are to be collected. When opening the sample holder compartment, keep the lid open for only a short time to help with flushing.

Scanning parameters
The criteria for selecting scanning parameters for CD spectrophotometers are essentially those for more common UV-visible spectrophotometers. There are five basic selections to be made:

1. Response time. The signal-to-noise ratio of the system is in proportion to the integration time so the easiest way to improve the quality of spectra is to increase the response time. When measuring model helical peptides response times in the region of 1 s are common.
2. Bandwidth or slit width. Setting the width of the slits can also be used to decrease noise level in spectra. The width should be as large as possible but comparable to the natural bandwidth of the bands to be scanned. Scanning parameters for α-helices would typically have bandwidths in the 1 nm region.
3. Scanning speed. The scanning speed selection is important, as speeds that are too high will tend to distort the obtained spectra. The maximum scanning speed is obtained from the bandwidth:response time ratio, so with a 1 nm slit and 1 s response time the maximum speed would be 1 nm s^{-1} or 60 nm min^{-1}. Slower scan speeds will give better quality spectra but obviously take longer times.
4. Data density/pitch. This is the measurement interval or number of data points per nm wavelength. This has no effect on noise in the spectrum, but it is advisable to choose a large number of data points if there will be postaccumulation processing such as curve fitting or filtering to reduce noise.

5. Averaging or accumulation. A further way to reduce the signal-to-noise ratio in spectra is to accumulate several spectra and average them. Noise will reduce as the square root of the number of accumulations, e.g., four accumulations will reduce noise by a factor of two, nine by a factor of three. The main effect is on short-term random noise. Long-term noise such as temperature drifts will not be compensated for in this way.

9.6.5
Data Manipulation and Analysis

CD data are usually expressed as an ellipticity (θ) or delta absorbance (ΔA). These are easily interconvertable as $\Delta A = \theta/32980$. Ellipticity is also converted to molar ellipticity ($[\theta]$) as follows:

$$[\theta] = \theta/(10 \times C \times l)$$

$$[\theta] = \theta M/(1000 \times c \times l)$$

Here θ is ellipticity in millidegrees, C is molar concentration (mol dm^{-3}), l is cell path length (cm), M is molecular weight (g mol^{-1}), and c is weight concentration (g mL^{-1}). $[\theta]$ is expressed in deg cm^{-2} dmol^{-1}. With peptides and proteins the mean residue molar ellipticity is often used. This replaces the C in the above equation with C_r, where $C_r = (1000 \times n \times c_g)/M_r$. Here M_r is the molecular weight of the species, n is the number of peptide bonds, and c_g is the macromolecular concentration (g mL^{-1}).

An example of a peptide helix CD spectrum is shown in Figure 9.4. To obtain the fraction of the peptide/protein that has α-helical secondary structure the mean residue molar ellipticity at 222 nm is most commonly used, as there is a strong helical minimum here and the coil signal is very small. There are several equations that

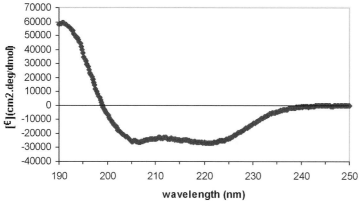

Fig. 9.4. Typical α-helical spectrum from CD spectroscopy. Measured using a Jasco J-810 spectrophotometer at 0.1 °C, data resolution 0.2 nm, Bandwidth 1 nm, scan speed 20 nm/min with 4 accumulations (averages). Model peptide – at 90 μM in 1 mm path length cell. Sequence: Ac-AKAAAAKAAAASAAAAKAAGY-NH$_2$.

can be used for this purpose. The equation below is a long-established equation for this purpose.

$$f_H = (\theta_{OBS} - \theta_C)/(\theta_H - \theta_C)$$

Here f_H is the average fraction of helix content, θ_C is the mean residue ellipticity when the peptide is 100% coil, θ_H is the mean residue ellipticity when the peptide is 100% helical and θ_{OBS} is the observed mean residue molar ellipticity from CD measurements. θ_C and θ_H can be calculated from empirical formulae below derived by Luo and Baldwin [259] where T is temperature (°C) and N_r is the chain length in residues. The value of 3 in the length correction $(1-3/N_r)$ is the number of backbone CO groups in a carboxyamidated peptide that are not hydrogen bonded.

$$\theta_C = 2220 - 53T$$

$$\theta_H = (-44000 + 250T)(1 - 3/N_r)$$

As can be seen from the equations above, accurate determination of protein/peptide concentration is essential for determination of secondary structure content from CD measurements. Any errors in this value will propagate throughout the calculation. Colorimetric assays such as those of Bradford, Lowry, and the BCA (bicinchonic acid) are not accurate enough for this purpose. Spectroscopic (UV absorbance) methods are much better if a reasonable chromophore is present as they are rapid and nondestructive of sample. However, the absorbance maxima and extinction coefficient of chromophores can be environmentally sensitive and so for example, a Trp residue in helical or β-sheet conformation may give a different reading to one in random coil conformation. Extinction coefficients usually quoted are for residues in free solution or fully solvent exposed, therefore for the best accuracy, concentration should be corrected for this, usually involving serial dilution in 6 M guanidinium chloride and comparison with the same dilutions in water (see Section 9.3.7).

A further complication with spectrum measurement and helix content evaluation is the presence of aromatic residues within the helix. In our own model peptides we usually employ tyrosine as a concentration marker but place a glycine (commonly known as a helix-breaker residue) between this and the helical region of the peptide [105]. Aromatic side chains can lead to greatly altered estimates of helix content from θ_{222} measurements. Tyrosine, depending upon the rotamer present, can alter the θ_{222} by up to 4000 deg cm^{-2} dmol^{-1} [106]. The effect on CD measurements of aromatic side chains has long been known [105, 260, 261] and is due to the contributions of electronic transitions in the aromatic groups [262] and favorable interactions with the backbone. Several methods have been proposed and used for correcting CD spectra for this effect. One of the most common is to compute the expected spectrum using the matrix method of Bayley, Nielsen, and Schellman [263]. In this the band intensities are derived directly from transition dipole moments (electronic and magnetic) via the rotational strength of each excited state of the peptide. With model peptides the nonaromatic residues are commonly modeled as alanine to simplify the calculations (see, e.g., Andrew et al. [106]).

Other methods for estimation of secondary structure content involve deconvolution of entire CD spectra. These methods are based upon data from high-quality structure files for which secondary structure content is accurately known or use iterative fitting methods.

Dichroprot (available at http://dicroprot-pbil.ibcp.fr/Documentation_dicroprot.html) is a commonly used program using deconvolution/fitting analysis. It is based mainly upon the work of Sreerama and Woody [264]. They derived a self-consistent procedure fitting sums of spectra of different secondary structures to the experimental spectrum to obtain estimates of secondary structure content.

9.6.5.1 Protocol for CD Measurement of Helix Content

1. Cell path length and sample concentration. This combination should be chosen to give a maximum absorbance of roughly 0.8, this includes all species in the solution, i.e., solvent, buffer salts, etc. as well as the protein/peptide in question. A good starting point in our hands is a peptide concentration of roughly 20 µM with a path length of 1 mm.
2. Baseline measurement for correction. During the same work session as the spectrum for the peptide/protein solution is measured a blank spectrum should be taken with the same cell containing the solvent, buffer salts, etc. at the same concentrations (preferably dialyzed to equilibrium against the peptide solution). Ideally this would give a ΔA value of zero across the spectrum but practically this is rarely the case and so correction is needed by subtraction of this spectrum from the experimental one(s).
3. Measurement time, slit width and data resolution and averaging. These parameters should be set according to the criteria outlined above so as to reduce noise and prevent distortion of the signal. In our experience scan rates of 20 nm min^{-1} with a slit width of 1 nm, response time of 1 s, data resolution of 0.2 nm and 4–6 averages give spectra of good quality for analysis.
4. Spectral range. To check for helical character a spectrum from 250 or 260 nm to 190 or 180 nm is sufficient. α-Helices give a characteristic spectrum (Figure 9.4) with minima at around 222 and 208 nm and a maximum around 190 nm.

9.6.6
Aggregation Test for Helical Peptides

9.6.6.1 Equipment and Reagents
- Circular dichroism spectrometer. See Figure 9.4 text for CD parameters.
- Quartz cuvette
- Peptide with a range of concentration between 5 µM and 500 µM in an appropriate buffer. Notes: Avoid using a buffer that absorbs at far UV wavelength. Typical buffer used is phosphate buffer 5–10 mM containing NaCl 5–10 mM.

9.6.6.2 Method
1. Record UV CD spectra of the buffer.
2. Record UV CD spectra of each peptide.
3. Subtract the average buffer reading from the average peptides readings.
4. Convert the unit of ellipticity (mdeg) to mean residual ellipticity.
5. Aggregation does not occur if the mean residual ellipticities of the peptides are independent of concentration.

9.6.7
Vibrational Circular Dichroism

Circular dichroism can also be performed in the vibrational region of the spectrum. In vibrational CD (VCD) the characteristic transitions of the amide group are resolved well from most other transition types [265]. VCD has important properties due to the local character of vibrational excitations. Overlap with aromatic modes is avoided as aromatic side chains are locally nonchiral and do not couple strongly to amides. Vibrational couplings are also short-range in nature, mostly to the next residue and dipole moments are not strong. This leads to more local structure sampling than in conventional CD [258, 266]. For peptides, empirical correlations between secondary structures and bandshapes were first developed over 20 years ago, mainly for peptides in nonaqueous solvents. A key determination was that the amide band VCD depended on peptide chain conformation and not on other functional groups. The main bandshape criterion was the helical chirality of the backbone [267]. Proteins with a high α-helical content tend to have negative signals in the amide II band with maximum negativity at around 1515–1520 cm^{-1}. As β-sheet content increases a three-featured pattern $(+ - +)$ appears in the amide I band and amide II intensity drops [268].

VCD equipment is now available commercially [269] though most measurements have been performed with modified dispersive or FTIR absorbance spectrometers. The commercial instruments give high-quality data for nonaqueous solutions of small molecules but aqueous solution work is highly demanding, especially in determining baselines and related artifacts [265]. Improvements are continually being made in these instruments in terms of polarization modulation and birefringence effects [269, 270] and so data continue to improve.

9.6.8
NMR Spectroscopy

The full theory and practise of NMR spectroscopy is well beyond the scope of this review. Many reviews of this technique and its application to secondary structure determination exist already. Secondary structure determination by NMR techniques does not require a full three-dimensional structural analysis as does X-ray crystallography. Knowledge of the amide and α proton chemical shifts are in principle all that is necessary, although if this information is available it is likely that nearly complete assignments of side chain protons are also available. While obtaining the sequential resonance assignments is a laborious task, the NMR method is perhaps the most powerful method of secondary structure determination without a crystal structure. Unlike CD and other kinds of spectroscopy, NMR secondary structure determination does not give an average secondary structure content but assigns secondary structures to different parts of the peptide chain.

Several parameters from NMR are needed for unambiguous structure assignments. A mixture of coupling constants, nuclear Overhauser effect (NOE) distances, amide proton exchange rates and chemical shifts may be necessary.

Coupling constants are through-bond or scalar (often called J-coupling) quantities, J-couplings between pairs of protons separated by three or fewer covalent bonds can be measured. The value of a three-bond J-coupling constant contains information about the intervening torsion angle. Unfortunately in general, torsion angles cannot be unambiguously determined from such information since as many as four different torsion angle values correlate with a single coupling constant value. Similar relationships can be determined between the three-bond coupling constant between the α proton and the β proton(s) yielding information on the value of the side chain dihedral angle $\chi 1$. Constraints on the dihedral angles ϕ and $\chi 1$ are important structural parameters in the determination of protein three-dimensional structures by NMR.

The three-bond coupling constant between the intraresidual α and amide protons ($^3 J H \alpha H N$) is the most useful for secondary structure determinations as it can be directly related to the backbone dihedral angle ϕ (Table 9.8). Unfortunately there is no three-bond proton coupling that can be related to the angle ψ.

9.6.8.1 Nuclear Overhauser Effect

The other major source of structural information comes from through space dipole–dipole coupling between two protons. This is the nuclear Overhauser effect or NOE. The intensity of an NOE is proportional to the inverse of the sixth power of the distance separating the two protons and is usually observed if two protons are separated by <5 Å. The NOE is therefore a sensitive probe of short intramolecular distances. They are categorized according to the location of the two protons involved in the interaction. Intraresidual NOEs are between protons within the same amino acid residue whereas sequential, medium, and long-range NOEs are between protons on residues sequentially adjacent, separated by up to three resi-

Tab. 9.8. Typical coupling constants and ϕ angles for different secondary structures.

Secondary structure	ϕ (degrees)	$^3 J H \alpha H N$ (Hz)
Right-handed α-helix	−57	3.9
Right-handed 3/10-helix	−60	4.2
Antiparallel β-sheet	−139	8.9
Parallel β-sheet	−119	9.7
Left-handed α-helix	57	6.9

dues, and separated by four or more residues in the polypeptide sequence. A network of these short interproton distances form the backbone of three-dimensional structure determination by NMR.

A number of short (< 5 Å) distances are fairly unique to secondary structural elements. For example, α-helices are characterized by short distances between certain protons on sequentially neighboring residues (e.g., between backbone amide protons, dNN, as well as between β protons of residue i and the amide protons of residue $i + 1$, dBN). Helical conformations result in short distances between the α proton of residue i and the amide proton of residues $i + 3$ and to a lesser extent $i + 4$ and $i + 2$. These $i + 2, i + 3$, and $i + 4$ NOEs are collectively referred to as medium range NOEs while NOEs connecting residues separated by more than five residues are referred to as long-range NOEs. Extended conformations (e.g., β-strands), on the other hand, are characterized by short sequential, dAN, distances. The formation of sheets also results in short distances between protons on adjacent strands (e.g., dAA and dAN).

9.6.8.2 Amide Proton Exchange Rates

More information on secondary structure can be obtained from the rate of exchange of the amide proton. The regular hydrogen-bonded secondary structures "protect" amide protons involved in them as evidenced by their significantly reduced amide proton exchange rates with the solvent (H_2O). Although nearly all polypeptide amide protons are involved in hydrogen bonds in a globular protein [43], those in regular secondary structures appear to be longer lived. For example, after placing a lyophilized sample of BPTI into D_2O, many amide protons are completely replaced with deuterons within 1 h. Over the next several hours, the amide protons in the N-terminal and then the C-terminal helix also completely exchange. However, some amide protons participating in the central antiparallel sheet are still present after several months. Rohl and Baldwin [271] compared CD and amide exchange measurements as tools for studying helix content in peptides as well as helix/coil transition parameters, over a broad range of helix contents. They showed that helix/coil transition theory did fit both CD and exchange data independently. However, the helix contents measured by exchange are larger than those measured by CD. To bring the techniques into agreement they showed that the assumption that the intrinsic chemical exchange rate in the helix is the same as the exchange rate measured for short unstructured model peptides was incorrect. Modifying this assumption allowed the CD and NH exchange data to be described by the same set of helix parameters and also indicated that the intrinsic exchange rate in the presence of helical structure is reduced approximately 17% relative to the rates measured in unstructured models.

9.6.8.3 ^{13}C NMR

Spera and Bax [272] have used ^{13}C NMR to study secondary structure assignment, helix distribution, capping parameters, and thermal transitions of helical peptide systems. Spera and Bax derived an empirical relationship between backbone conformation and in combination with high-resolution X-ray structures for which ^{13}C chemical shift assignments were available. Stellwagen and coworkers [273–275] have used the same C_α, C_β ^{13}C chemical shifts to study thermal transitions in model peptides to compare Lifson-Roig theory predictions of helicity with experimental values. They showed that mean helix content as measured by CD coincided with that measured by ^{13}C NMR using the mean of all individual fitted transitions of the carbonyl residues. Using the carbonyl, C_α or C_β chemical shifts for a particular residue all gave the same melting temperature and width of thermal transition, although their chemical shift changes upon melting differed in both magnitude and sign. This suggests that each residue acts as a cooperative unit in helix formation. They also indicated that there are more interactions than simply backbone hydrogen bonds at the helix termini, as these appeared less frayed than Lifson-Roig theory would predict.

Detailed treatments of using these parameters are well established in the literature and it is beyond the scope of this current article to treat them in further detail [272, 276, 277].

9.6.9
Fourier Transform Infrared Spectroscopy

Another increasingly popular method for secondary structure determination is Fourier transform infrared spectroscopy (FTIR). There are many descriptions of the basis of IR spectroscopy in the literature [278, 279] and so we will not expand upon that area here.

With proteins/peptides there are nine characteristic absorbance bands produced by the amide functionality. These are named amides A, B, I, II ... VII. Of these, amides I and II are the major bands of interest. The amide I band (between 1600 and 1700 cm^{-1}) is mainly associated with the C=O stretching vibration (70–85%) and is directly related to the backbone conformation. Amide II results from the N–H bending vibration (40–60%) and from the C–N stretching vibration (18–40%). Amide II, a more complex, conformationally sensitive band, is found in the 1510 and 1580 cm^{-1} region. Unfortunately the contribution of side chains also occurs in the same region as the amide I and II bands. Only nine of the 20 common amino acids in proteins have significant absorbance in this region and these are Asp, Asn, Glu, Gln, Lys, Arg, Tyr, Phe and His. The contributions of these side chains must be accounted for before using the amide I and II bands for secondary structure determination. Fortunately this has been well investigated by Venyaminov and Kalnin [280].

9.6.9.1 Secondary Structure
A large number of synthetic polypeptides have been used for the characterization of infrared spectra for proteins with a defined secondary structure content. For example, polylysine adopts random, β-sheet or α-helical structures depending upon conditions of temperature and pH of the solution [281]. Experimental and theoretical work on a large number of synthetic polypeptides has provided insights on the variability of the frequencies for each structure. The large amount of data published by Krimm and Bandekar [282] give an insight in the nature of the amide bond.

For α-helical structures the mean frequency was found to be 1652 cm^{-1} for the amide I and 1548 cm^{-1} for the amide II absorption [283]. The half-width of the α-helix bands depends on the stability of the helix. For the most stable helices, the half-width of about 15 cm^{-1} corresponds with a helix/coil transition free energy of more than 300 cal mol^{-1}. Other α-helices display half-widths of 38 cm^{-1} and helix/coil transition free energies of about 90 cal mol^{-1}. The pioneering work of Byler and Susi [284] on spectral deconvolution has allowed much more detailed analysis of FTIR spectra in terms of secondary structure assignment.

There is a wealth of information that can be used to derive structural information by analyzing the shape and position of bands in the amide I region of the spectrum. The presence of a number of amide I band frequencies have been correlated with the presence of α-helical, antiparallel and parallel β-sheets and random coil structures [285–287]. It is now well accepted that absorbance in the range from 1650 to 1658 cm^{-1} is generally associated with the presence of α-helix in aqueous environments. Precise interpretation of bands in this region are difficult because there is significant overlap of the helical structures with random structures. One way to resolve this issue is to exchange the hydrogen from the peptide N–H with deuterium. If the protein contains a significant amount of random structure, the H–D exchange will result in a large shift in the position of the random structure (will now absorb at around 1646 cm^{-1}) and only a minor change in the position of the helical band. Beta-sheet vibrations have been shown to absorb between 1640 and 1620 cm^{-1} (though there is considerable disagreement on assignments for parallel and antiparallel β-sheets). Absorbances centered at 1670 cm^{-1} have been assigned to β-turns, or simply "turns" by some investigators.

9.6.10
Raman Spectroscopy and Raman Optical Activity

Raman spectroscopy is based on the Raman effect, which is the inelastic scattering of photons by molecules. The Indian physicist, C. V. Raman, discovered the effect in 1928 [288]. The Raman effect comprises a very small fraction, about 1 in 10^7, of the incident photons. In Raman scattering, the energies of the incident and scattered photons are different. Raman spectroscopy is another vibrational mode technique as the incoming photon promotes a vibrational oscillation from the ground state to a virtual, nonstationary state or, if the vibration is already in the virtual state, it can cause relaxation to the ground state. Raman spectroscopy comes in various guises such as difference, Fourier transform and resonance Raman spectroscopy, details of which can be found elsewhere [289, 290]. As with FTIR spectroscopy, we are measuring and assigning vibration bands with these techniques and it is the amide bands from peptide bonds that give most information regarding secondary structure (see Tables 9.9 and 9.10 for details). It is the precise positions of the amide I and III bands that are sensitive to secondary structure [291]. Correlations between these frequencies are widely noted in the literature (see, for example, Ref. [292]). Table 9.10 gives some examples of helix-related bands.

The amide III vibration and the C$_\alpha$H bending vibration in UV Raman resonance spectroscopy are very sensitive to the amide backbone configuration [293]. The var-

Tab. 9.9. Ranges for wavenumbers and vibration assignments for amide vibrations (from Ref. 297).

Band name	Wavenumber (cm^{-1})	Approximate description
Amide A	3250–3300	N–H stretching
Amide I	1630–1700	C=O stretching
Amide II	1510–1570	N–H deformation
Amide III	1230–1330	N–H/C–H deformation
Amide IV	630–750	O=C–N deformation
Amide V	700–750	N–H out-of-plane deformation
Amide VI	~600	C=O out-of-plane deformation

From Ref. [297].

ious contributing signals are mixed when ψ is around 120°, i.e., for "random coil" and β-sheet conformations, but are unmixed at ψ values around 60° (α-helix conformation). This may allow the extraction of amide ψ angles from UV resonance Raman spectra [293].

Raman optical activity (ROA) measures vibrational optical activity. This is due to the small difference in the intensity of Raman scattering from chiral molecules in right and left circularly polarized incident light [142]. The technique is routinely applicable to biomolecules in aqueous environments but requires relatively large amounts of sample (typically 20–100 mg mL^{-1}) [142]. It can provide new information about solution dynamics as well as structure which is complementary to that obtained from more conventional optical spectroscopic techniques described above. Reviews of the theory of ROA can be found in the literature [294, 295] and are beyond the scope of this review.

ROA gives useful information regarding protein/peptide secondary structure in aqueous solution as the signals in the spectrum tend to be dominated by bands from the peptide backbone. Additionally it is sensitive to dynamic aspects of the structure and so can give information on mobile regions and nonnative states. Side chain bands are also present in the spectra though much less prominently. The extended amide III band [296, 297] is very useful in ROA as it is very sensitive to geometry, being composed of coupling between N–H and C$_\alpha$-H deformations. A good description and review of secondary structure assignments from ROE spectra is given by Barron et al. [142].

Tab. 9.10. Raman amide I and III vibrations related to α-helical structure.

Location (cm^{-1})	Assignment	information
1655 ± 5	Amide C=O stretch	α-helix in water
1632	Amide C=O stretch	α-helix in D$_2$O
1661 ± 3	Amide C=O stretch	α-helix in D$_2$O
>1275	Amide III (weak)	α-helix (nothing below 1275 cm^{-1})

9.6.11
pH Titrations

9.6.11.1 Equipment and Reagents
- Circular dichroism spectrometer
- Quartz cuvette with a path length of 1 cm equipped with a magnetic bar
- A pH meter with a small-sized electrode
- A magnetic stirrer
- Peptide at a concentration such that the aggregation does not occur (typically 20–30 µM) dissolved in an appropriate buffer. Notes: Avoid using buffers that absorb at far UV wavelengths. Typical buffer used in pH titration contains 10 mM NaCl, 1 mM sodium phosphate, 1 mM sodium borate, and 1 mM sodium citrate.
- Concentrated HCl and NaOH

9.6.11.2 Method
1. Record UV CD spectra of the buffer at 222 nm.
2. Record UV CD spectra of the peptide at the starting pH. Typical parameters of CD measurement are as in Aggregation test protocol.
3. Record UV CD spectra after pH adjustments that are made by the addition of either NaOH or HCl. Record volume of acid or alkali added.
4. Subtract the average buffer reading from the average peptides readings.
5. Convert the unit of ellipticity (mdeg) to mean residual ellipticity. Note the concentration will decrease as acid or alkali is added.
6. Plot $[\theta]_{222}$ versus pH.
7. Fit the data to a Henderson-Hasselbach equation depending on the number of titrable groups in the peptide. For one apparent pK_a, the value is given by:

$$[\theta]_{222nm} = [\theta]_{222, \text{high\_pH}} \times \left(1 - \frac{1}{1 + 10^{\text{pH} - \text{p}Ka}}\right) + [\theta]_{222, \text{low\_pH}} \times \left(\frac{1}{1 + 10^{\text{pH} - \text{p}Ka}}\right)$$

where $[\theta]_{222, \text{highpH}}$ and $[\theta]_{222, \text{lowpH}}$ are the molar ellipticities measured at 222 nm at the titration end points at high and low pH.

The equation for two different pK_a values is:

$$[\theta]_{222nm} = [\theta]_{222, \text{mid\_pH}} \times \left(1 - \frac{1}{1 + 10^{\text{pH} - \text{p}Ka1}}\right) + [\theta]_{222, \text{low\_pH}} \times \left(\frac{1}{1 + 10^{\text{pH} - \text{p}Ka1}}\right)$$

$$+ [\theta]_{222, (\text{mid\_pH} - \text{high\_pH})} \times \left(1 - \frac{1}{1 + 10^{\text{pH} - \text{p}Ka2}}\right)$$

where, pK_a1 and pK_a2 are the pK_a values measured for the acid–base equilibrium at low and high pH, respectively. $[\theta]_{222, \text{highpH}}$, $[\theta]_{222, \text{midpH}}$, and $[\theta]_{222, \text{lowpH}}$ are the molar ellipticities measured at 222 nm at the titration end points at high, mid, and low pH. $[\theta]_{222\text{mid\_pH} - \text{high\_pH}}$ is the change in molar ellipticity associated with pK_a2.

Acknowledgments

We thank all our coworkers in this field, namely Avi Chakrabartty, Carol Rohl, Buzz Baldwin, Tod Klingler, Ben Stapley, Jim Andrew, Eleri Hughes, Simon Penel, Duncan Cochran, Nicoleta Kokkoni, Jia Ke Sun, Jim Warwicker, Gareth Jones, and Jonathan Hirst. Current work in our lab on helices is supported by the Wellcome Trust (grant references 057318 and 065106). TMI thanks the Indonesian government for a scholarship.

References

1. Pauling, L., Corey, R. B., and Branson, H. R. (1951). The structure of proteins: two hydrogen-bonded helical configurations of the polypeptide chain. *Proc. Natl Acad. Sci. USA* 37, 205–211.
2. Perutz, M. F. (1951). New X-ray evidence on the configuration of polypeptide chains. *Nature* 167, 1053–1054.
3. Kendrew, J. C., Dickerson, R. E., Strandberg, B. E. et al. (1960). Structure of myoglobin. *Nature* 185, 422–427.
4. Barlow, D. J. and Thornton, J. M. (1988). Helix geometry in proteins. *J. Mol. Biol.* 201, 601–19.
5. Scholtz, J. M. and Baldwin, R. L. (1992). The mechanism of α-helix formation by peptides. *Annu. Rev. Biophys. Biomol. Struct.* 21, 95–118.
6. Baldwin, R. L. (1995). α-helix formation by peptides of defined sequence. *Biophys. Chem.* 55, 127–35.
7. Chakrabartty, A. and Baldwin, R. L. (1995). Stability of α-helices. *Adv. Protein Chem.* 46, 141–76.
8. Kallenbach, N. R., Lyu, P., and Zhou, H. (1996). CD spectroscopy and the helix-coil transition in peptides and polypeptides. In *Circular Dichroism and the Conformational Analysis of Biomolecules* G. D. Fasman, editor, Plenum Press, New York, 201–59.
9. Rohl, C. A. and Baldwin, R. L. (1998). Deciphering rules of helix stability in peptides. *Meth. Enzymol.* 295, 1–26.
10. Andrews, M. J. I. and Tabor, A. B. (1999). Forming stable helical peptides using natural and artificial amino acids. *Tetrahedron* 55, 11711–43.
11. Serrano, L. (2000). The relationship between sequence and structure in elementary folding units. *Adv. Protein Chem.* 53, 49–85.
12. Richardson, J. S. and Richardson, D. C. (1988). Amino acid preferences for specific locations at the ends of α helices. *Science* 240, 1648–52.
13. Presta, L. G. and Rose, G. D. (1988). Helix signals in proteins. *Science* 240, 1632–41.
14. Doig, A. J., MacArthur, M. W., Stapley, B. J., and Thornton, J. M. (1997). Structures of N-termini of helices in proteins. *Protein Sci.* 6, 147–55.
15. MacGregor, M. J., Islam, S. A., and Sternberg, M. J. E. (1987). Analysis of the relationship between side-chain conformation and secondary structure in globular proteins. *J. Mol. Biol.* 198, 295–310.
16. Swindells, M. B., MacArthur, M. W., and Thornton, J. M. (1995). Intrinsic ϕ, ψ propensities of amino acids derived from the coil regions of known structures. *Nat. Struct. Biol.* 2, 596–603.
17. Dunbrack, R. and Karplus, M. (1994). Conformational analysis of the backbone-dependent rotamer preferences of protein side-chains. *Nat. Struct. Biol.* 1, 334–9.
18. Dunbrack, R. L. (2002). Rotamer libraries in the 21(st) century. *Curr. Opin. Struct. Biol.* 12, 431–40.

19 PENEL, S., HUGHES, E., and DOIG, A. J. (1999). Side-chain structures in the first turn of the α-helix. *J. Mol. Biol.* 287, 127–43.

20 SERRANO, L. and FERSHT, A. R. (1989). Capping and α-helix stability. *Nature* 342, 296–9.

21 BELL, J. A., BECKTEL, W. J., SAUER, U., BAASE, W. A., and MATTHEWS, B. W. (1992). Dissection of helix capping in T4 lysozyme by structural and thermodynamic analysis of six amino acid substitutions at Thr 59. *Biochemistry* 31, 3590–6.

22 CHAKRABARTTY, A., DOIG, A. J., and BALDWIN, R. L. (1993). Helix capping propensities in peptides parallel those in proteins. *Proc. Natl Acad. Sci. USA* 90, 11332–6.

23 FOROOD, B., FELICIANO, E. J., and NAMBIAR, K. P. (1993). Stabilization of α-helical structures in short peptides via end capping. *Proc. Natl Acad. Sci. USA* 90, 838–42.

24 DASGUPTA, S. and BELL, J. A. (1993). Design of helix ends. Amino acid preferences, hydrogen bonding and electrostatic interactions. *Int. J. Pept. Protein Res.* 41, 499–511.

25 HARPER, E. T. and ROSE, G. D. (1993). Helix stop signals in proteins and peptides: the capping box. *Biochemistry* 32, 7605–9.

26 SEALE, J. W., SRINIVASAN, R., and ROSE, G. D. (1994). Sequence determinants of the capping box, a stabilizing motif at the N-termini of α-helices. *Protein Sci.* 3, 1741–5.

27 MUÑOZ, V. and SERRANO, L. (1995). The hydrophobic-staple motif and a role for loop residues in α-helix stability and protein folding. *Nat. Struct. Biol.* 2, 380–385.

28 VIGUERA, A. R. and SERRANO, L. (1999). Stable proline box motif at the N-terminal end of α-helices. *Protein Sci.* 8, 1733–42.

29 DOIG, A. J. and BALDWIN, R. L. (1995). N- and C-capping preferences for all 20 amino acids in α-helical peptides. *Protein Sci.* 4, 1325–36.

30 ROHL, C. A., CHAKRABARTTY, A., and BALDWIN, R. L. (1996). Helix propagation and N-cap propensities of the amino acids measured in alanine-based peptides in 40 volume percent trifluoroethanol. *Protein Sci.* 5, 2623–37.

31 SCHELLMAN, C. (1980). The αL conformation at the ends of helices. *Protein Folding* 53–61.

32 AURORA, R., SRINIVASAN, R., and ROSE, G. D. (1994). Rules for α-helix termination by glycine. *Science* 264, 1126–30.

33 AURORA, R. and ROSE, G. D. (1998). Helix capping. *Protein Sci.* 7, 21–38.

34 PRIETO, J. and SERRANO, L. (1997). C-capping and helix stability: the Pro C-capping motif. *J. Mol. Biol.* 274, 276–88.

35 SIEDLECKA, M., GOCH, G., EJCHART, A., STICHT, H., and BIERZYNSKI, A. (1999). α-Helix nucleation by calcium-binding peptide loop. *Proc. Natl Acad. Sci. USA* 96, 903–8.

36 JERNIGAN, R., RAGHUNATHAN, G., and BAHAR, I. (1994). Characterisation of interactions and metal-ion binding-sites in proteins. *Curr. Opin. Struct. Biol.* 4, 256–63.

37 ALBERTS, I. L., NADASSY, K., and WODAK, S. J. (1998). Analyisis of zinc binding sites in protein crystal structures. *Protein Sci.* 7, 1700–16.

38 GHADIRI, M. R. and CHOI, C. (1990). Secondary structure nucleation in peptides – transition-metal ion stabilized α-helices. *J. Am. Chem. Soc.* 112, 1630–2.

39 KISE, K. J. and BOWLER, B. E. (2002). Induction of helical structure in a heptapeptide with a metal cross-link: Modification of the Lifson-Roig Helix-Coil theory to account for covalent cross-links. *Biochemistry* 41, 15826–37.

40 RUAN, F., CHEN, Y., and HOPKINS, P. B. (1990). Metal ion enhanced helicity in synthetic peptides containing unnatural, metal-ligating residues. *J. Am. Chem. Soc.* 112, 9403–4.

41 CLINE, D. J., THORPE, C., and SCHNEIDER, J. P. (2003). Effects of As(III) binding on α-helical structure. *J. Am. Chem. Soc.* 125, 2923–9.

42 KARPEN, M. E., DE HASET, P. L., and NEET, K. E. (1992). Differences in the

amino acid distributions of 3_{10}-helices and α-helices. *Protein Sci.* 1, 1333–42.

43 BAKER, E. N. and HUBBARD, R. E. (1984). Hydrogen bonding in globular proteins. *Prog. Biophys. Mol. Biol.* 44, 97–179.

44 NÉMETHY, G., PHILLIPS, D. C., LEACH, S. J., and SCHERAGA, H. A. (1967). A second right-handed helical structure with the parameters of the Pauling-Corey α-helix. *Nature* 214, 363–5.

45 MILLHAUSER, G. L. (1995). Views of helical peptides – a proposal for the position of 3_{10}-helix along the thermodynamic folding pathway. *Biochemistry* 34, 3872–7.

46 BOLIN, K. A. and MILLHAUSER, G. L. (1999). α and 3_{10}: The split personality of polypeptide helices. *Acc. Chem. Res.* 32, 1027–33.

47 MIICK, S. M., MARTINEZ, G. V., FIORI, W. R., TODD, A. P., and MILLHAUSER, G. L. (1992). Short alanine-based peptides may form 3_{10}-helices and not α-helices in aqueous solution. *Nature* 359, 653–5.

48 ZHOU, H. X. X., LYU, P. C. C., WEMMER, D. E., and KALLENBACH, N. R. (1994). Structure of C-terminal α-helix cap in a synthetic peptide. *J. Am. Chem. Soc.* 116, 1139–40.

49 FIORI, W. R. and MILLHAUSER, G. L. (1995). Exploring the peptide 3_{10}-helix/α-helix equilibrium with double label electron spin resonance. *Biopolymers* 37, 243–50.

50 TONIOLO, C. and BENEDETTI, E. (1991). The polypeptide 3_{10}-helix. *Trends Biochem. Sci.* 16, 350–3.

51 RAMACHANDRAN, G. N. and SASISEKHARAN, V. (1968). Conformation of polypeptides and proteins. *Adv. Protein Chem.* 23, 283–437.

52 Low, B. W. and GRENVILLE-WELLS, H. J. (1953). Generalized mathematical relations for polypeptide chain helixes. The coordinates for the π helix. *Proc. Natl Acad. Sci. USA* 39, 785–801.

53 SHIRLEY, W. A. and BROOKS, C. L. (1997). Curious structure in "canonical" alanine based peptides. *Proteins: Struct. Funct. Genet.* 28, 59–71.

54 LEE, K. H., BENSON, D. R., and KUCZERA, K. (2000). Transitions from α to π helix observed in molecular dynamics simulations of synthetic peptides. *Biochemistry* 39, 13737–47.

55 WEAVER, T. M. (2000). The π-helix translates structure into function. *Protein Sci.* 9, 201–6.

56 MORGAN, D. M., LYNN, D. G., MILLER-AUER, H., and MEREDITH, S. C. (2001). A designed Zn2+-binding amphiphilic polypeptide: Energetic consequences of π-helicity. *Biochemistry* 40, 14020–9.

57 FODJE, M. N. and AL-KARADAGHI, S. (2002). Occurrence, conformational features and amino acid propensities for the π-helix. *Protein Eng.* 15, 353–8.

58 BALDWIN, R. L. (2003). In search of the energetic role of peptide hydrogen bonds. *J. Biol. Chem.* 278, 17581–8.

59 EPAND, R. M. and SCHERAGA, H. A. (1968). *Biochemistry* 7, 2864–72.

60 TANIUCHI, J. H. and ANFINSEN, C. B. (1969). *J. Biol. Chem.* 244, 2864–72.

61 SPEK, E. J., GONG, Y., and KALLENBACH, N. R. (1995). Intermolecular interactions in a helical oligo- and poly(L-glutamic acid) at acidic pH. *J. Am. Chem. Soc.* 117, 10773–4.

62 BROWN, J. E. and KLEE, W. A. (1971). *Biochemistry* 10, 470–6.

63 BIERZYNSKI, A., KIM, P. S., and BALDWIN, R. L. (1982). A salt bridge stabilizes the helix formed by isolated C-peptide of ribonuclease A. *Proc. Natl Acad. Sci. USA* 79, 2470–4.

64 KIM, P. S., BIERZYNSKI, A., and BALDWIN, R. L. (1982). A competing salt-bridge suppresses helix formation by the isolated C-peptide carboxylate of Ribonuclease A. *J. Mol. Biol.* 162, 187–99.

65 KIM, P. S. and BALDWIN, R. L. (1984). A helix stop signal in the isolated S-peptide of ribonuclease A. *Nature* 307, 329–34.

66 SHOEMAKER, K. R., KIM, P. S., BREMS, D. N. et al. (1985). Nature of the charged-group effect on the stability of the C-peptide helix. *Proc. Natl Acad. Sci. USA* 82, 2349–53.

67 STREHLOW, K. G. and BALDWIN, R. L. (1989). Effect of the substitution Ala-

Gly at each of 5 residue positions in the C-peptide helix. *Biochemistry* 28, 2130–3.

68 OSTERHOUT, J. J., BALDWIN, R. L., YORK, E. J., STEWART, J. M., DYSON, H. J., and WRIGHT, P. E. (1989). H-1 NMR studies of the solution conformations of an analog of the C-peptide of ribonuclease-A. *Biochemistry* 29, 7059–64.

69 SHOEMAKER, K. R., FAIRMAN, R., SCHULTZ, D. A. et al. (1990). Side-chain interactions in the C-peptide helix – Phe8–His12+. *Biopolymers* 29, 1–11.

70 FAIRMAN, R., SHOEMAKER, K. R., YORK, E. J., STEWART, J. M., and BALDWIN, R. L. (1990). The Glu2– Arg10+ side-chain interaction in the C-peptide helix of ribonuclease A. *Biophys. Chem.* 37, 107–19.

71 STREHLOW, K. G., ROBERTSON, A. D., and BALDWIN, R. L. (1991). Proline for alanine substitutions in the C-peptide helix of ribonuclease A. *Biochemistry* 30, 5810–14.

72 FAIRMAN, R., ARMSTRONG, K. M., SHOEMAKER, K. R., YORK, E. J., STEWART, J. M., and BALDWIN, R. L. (1991). Position effect on apparent helical propensities in the C-peptide helix. *J. Mol. Biol.* 221, 1395–401.

73 BLANCO, F. J., JIMENEZ, M. A., RICO, M., SANTORO, J., HERRANZ, J., and NIETO, J. L. (1992). The homologous angiogenin and ribonuclease N-terminal fragments fold into very similar helices when isolated. *Biochem. Biophys. Res. Commun.* 182, 1491–8.

74 RICO, M., NIETO, J. L., SANTORO, J., BERMEJO, F. J., and HERRANZ, J. (1983). H1-NMR parameters of the N-terminal 13-residue C-peptide of ribonuclease in aqueous-solution. *Org. Magn. Res.* 21, 555–63.

75 GALLEGO, E., HERRANZ, J., NIETO, J. L., RICO, M., and SANTORO, J. (1983). H1-NMR parameters of the N-terminal 19-residue S-peptide of ribonuclease in aqueous-solution. *Int. J. Pept. Protein Res.* 21, 242–53.

76 RICO, M., NIETO, J. L., SANTORO, J., BERMEJO, F. J., HERRANZ, J., and GALLEGO, E. (1983). Low temperature H1-NMR evidence of the folding of isolated ribonuclease S-peptide. *FEBS Lett.* 162, 314–19.

77 RICO, M., GALLEGO, E., SANTORO, J., BERMEJO, F. J., NIETO, J. L., and HERRANZ, J. (1984). On the fundamental role of the Glu2– ... Arg10+ salt bridge in the folding of isolated ribonuclease A S-peptide. *Biochem. Biophys. Res. Commun.* 123, 757–63.

78 NIETO, J. L., RICO, M., JIMENEZ, M. A., HERRANZ, J., and SANTORO, J. (1985). Amide H1-NMR study of the folding of ribonuclease C-peptide. *Int. J. Biol. Macromol.* 7, 66–70.

79 NIETO, J. L., RICO, M., SANTORO, J., and BERMEJO, J. (1985). NH resonances of ribonuclease S-peptide in aqueous solution – low temperature NMR study. *Int. J. Pept. Protein Res.* 25, 47–55.

80 SANTORO, J., RICO, M., NIETO, J. L., BERMEJO, J., HERRANZ, J., and GALLEGO, E. (1986). C13 NMR spectral assignment of ribonuclease S-peptide – Some new structural information about its low temperature folding. *J. Mol. Struct.* 141, 243–8.

81 RICO, M., HERRANZ, J., BERMEJO, J. et al. (1986). Quantitative interpretation of the helix coil transition in RNase A S-peptide. *J. Mol. Struct.* 143, 439–44.

82 RICO, M., SANTORO, J., BERMEJO, J. et al. (1986). Thermodynamic parameters for the helix coil thermal transition of ribonuclease S-peptide and derivatives from H1 NMR data. *Biopolymers* 25, 1031–53.

83 MARQUSEE, S. and BALDWIN, R. L. (1987). Helix stabilization by Glu– ... Lys+ salt bridges in short peptides of de novo design. *Proc. Natl Acad. Sci. USA* 84, 8898–902.

84 PARK, S. H., SHALONGO, W., and STELLWAGEN, E. (1993). Residue helix parameters obtained from dichroic analysis of peptides of defined sequence. *Biochemistry* 32, 7048–53.

85 MARQUSEE, S., ROBBINS, V. H., and BALDWIN, R. L. (1989). Unusually stable helix formation in short alanine based peptides. *Proc. Natl Acad. Sci. USA* 86, 5286–90.

86 LYU, P. C., MARKY, L. A., and KALLEN-

86 Bach, N. R. (1989). The role of ion-pairs in α-helix stability – 2 new designed helical peptides. *J. Am. Chem. Soc.* 111, 2733–34.

87 Lyu, P. C., Liff, M. I., Marky, L. A., and Kallenbach, N. R. (1990). Side-chain contributions to the stability of α-helical structure in peptides. *Science* 250, 669–73.

88 Gans, P. J., Lyu, P. C., Manning, P. C., Woody, R. W., and Kallenbach, N. R. (1991). The helix-coil transition in heterogeneous peptides with specific side chain interactions: Theory and comparison with circular dichroism. *Biopolymers* 31, 1605–14.

89 Penel, S., Morrison, R. G., Mortishire-Smith, R. J., and Doig, A. J. (1999). Periodicity in α-helix lengths and C-capping preferences. *J. Mol. Biol.* 293, 1211–19.

90 Wada, A. (1976). The α-helix as an electric macro-dipole. *Adv. Biophys.* 9, 1–63.

91 Hol, W. G. J., van Duijnen, P. T., and Berendsen, H. J. C. (1978). The α-helix dipole and the properties of proteins. *Nature* 273, 443–6.

92 Aqvist, J., Luecke, H., Quiocho, F. A., and Warshel, A. (1991). Dipoles located at helix termini of proteins stabilize charges. *Proc. Natl Acad. Sci. USA* 88, 2026–30.

93 Zhukovsky, E. A., Mulkerrin, M. G., and Presta, L. G. (1994). Contribution to global protein stabilization of the N-capping box in human growth hormone. *Biochemistry* 33, 9856–64.

94 Tidor, B. (1994). Helix-capping interaction in λ Cro protein: a free energy simulation analysis. *Proteins: Struct. Funct. Genet.* 19, 310–23.

95 Cochran, D. A. E., Penel, S., and Doig, A. J. (2001). Contribution of the N1 amino acid residue to the stability of the α-helix. *Protein Sci.* 10, 463–70.

96 Cochran, D. A. E. and Doig, A. J. (2001). Effects of the N2 residue on the stability of the α-helix for all 20 amino acids. *Protein Sci.* 10, 1305–11.

97 Iqbalsyah, T. M. and Doig, A. J. (in press). Effect of the N3 residue on the stability of the α-helix. *Protein Sci.* 13, 32–39.

98 Decatur, S. M. (2000). IR spectroscopy of isotope-labeled helical peptides: Probing the effect of N-acetylation on helix stability. *Biopolymers* 180–5.

99 Doig, A. J., Chakrabartty, A., Klingler, T. M., and Baldwin, R. L. (1994). Determination of free energies of N-capping in α-helices by modification of the Lifson-Roig helix-coil theory to include N- and C-capping. *Biochemistry* 33, 3396–403.

100 Stapley, B. J., Rohl, C. A., and Doig, A. J. (1995). Addition of side chain interactions to modified Lifson-Roig helix-coil theory: application to energetics of phenylalanine-methionine interactions. *Protein Sci.* 4, 2383–91.

101 Andrew, C. D., Penel, S., Jones, G. R., and Doig, A. J. (2001). Stabilising non-polar/polar side chain interactions in the α-helix. *Proteins: Struct. Funct. Genet.* 45, 449–55.

102 Huyghues-Despointes, B. M., Scholtz, J. M., and Baldwin, R. L. (1993). Effect of a single aspartate on helix stability at different positions in a neutral alanine-based peptide. *Protein Sci.* 2, 1604–11.

103 Brandts, J. R. and Kaplan, K. J. (1973). Derivative spectroscopy applied to tyrosyl chromophores. Studies on ribonuclease, lima bean inhibitor, and pancreatic trypsin inhibitor. *Biochemistry* 10, 470–6.

104 Edelhoch, H. (1967). Spectroscopic determination of tryptophan and tyrosine in proteins. *Biochemistry* 6, 1948–54.

105 Chakrabartty, A., Kortemme, T., Padmanabhan, S., and Baldwin, R. L. (1993). Aromatic side-chain contribution to far-ultraviolet circular dichroism of helical peptides and its effect on measurement of helix propensities. *Biochemistry* 32, 5560–5.

106 Andrew, C. D., Bhattacharjee, S., Kokkoni, N., Hirst, J. D., Jones, G. R., and Doig, A. J. (2002). Stabilising interactions between aromatic and basic side chains in α-helical peptides and proteins. Tyrosine effects on helix circular dichroism. *J. Am. Chem. Soc.* 124, 12706–14.

107 KEMP, D. S., BOYD, J. G., and MUENDEL, C. C. (1991). The helical s-constant for alanine in water derived from template-nucleated helices. *Nature* 352, 451–4.

108 KEMP, D. S., ALLEN, T. J., and OSLICK, S. L. (1995). The energetics of helix formation by short templated peptides in aqueous solution. 1. Characterization of the reporting helical template Ac-HE1(1). *J. Am. Chem. Soc.* 117, 6641–57.

109 GROEBKE, K., RENOLD, P., TSANG, K. Y., ALLEN, T. J., McCLURE, K. F., and KEMP, D. S. (1996). Template-nucleated alanine-lysine helices are stabilized by position-dependent interactions between the lysine side chain and the helix barrel. *Proc. Natl Acad. Sci. USA* 93, 4025–9.

110 KEMP, D. S., OSLICK, S. L., and ALLEN, T. J. (1996). The structure and energetics of helix formation by short templated peptides in aqueous solution. 3. Calculation of the helical propagation constant s from the template stability constants t/c for Ac-Hel(1)-Ala(n)-OH, n = 1–6. *J. Am. Chem. Soc.* 118, 4249–55.

111 KEMP, D. S., ALLEN, T. J., OSLICK, S. L., and BOYD, J. G. (1996). The structure and energetics of helix formation by short templated peptides in aqueous solution. 2. Characterization of the helical structure of Ac-Hel(1)-Ala(6)-OH. *J. Am. Chem. Soc.* 118, 4240–8.

112 AUSTIN, R. E., MAPLESTONE, R. A., SEFLER, A. M. et al. (1997). Template for stabilization of a peptide α-helix: Synthesis and evaluation of conformational effects by circular dichroism and NMR. *J. Am. Chem. Soc.* 119, 6461–72.

113 ARRHENIUS, T. and SATTHERTHWAIT, A. C. (1990). *Peptides: Chemistry, Structure and Biology*: Proceedings of the 11th American Peptide Symposium, ESCOM, Leiden.

114 MULLER, K., OBRECHT, D., KNIERZINGER, A. et al. (1993). *Perspectives in Medicinal Chemistry*, pp. 513–531, Verlag Chemie, Weinheim.

115 GANI, D., LEWIS, A., RUTHERFORD, T. et al. (1998). Design, synthesis, structure and properties of an α-helix cap template derived from N-[(2S)-2-chloropropionyl]-(2S)-Pro-(2R)-Ala-(2S,4S)-4-thioPro-OMe which initiates α-helical structures. *Tetrahedron* 54, 15793–819.

116 ALEMAN, C. (1997). Conformational properties of α-amino acids disubstituted at the α-carbon. *J. Phys. Chem.* 101, 5046–50.

117 SACCA, B., FORMAGGIO, F., CRISMA, M., TONIOLO, C., and GENNARO, R. (1997). Linear oligopeptides. 401. In search of a peptide 3₁₀-helix in water. *Gazz. Chim. Ital.* 127, 495–500.

118 TANAKA, M. (2002). Design and conformation of peptides containing α,α-disubstituted α-amino acids. *J. Synth. Org. Chem. Japan* 60, 125–36.

119 LANCELOT, N., ELBAYED, K., RAYA, J. et al. (2003). Characterization of the 3_{10}-helix in model peptides by HRMAS NMR spectroscopy. *Chem. Eur. J.* 9, 1317–23.

120 KARLE, I. L. and BALARAM, P. (1990). Structural characteristics of α-helical peptide molecules containing Aib residues. *Biochemistry* 29, 6747–56.

121 BANERJEE, R. and BASU, G. (2002). A short Aib/Ala-based peptide helix is as stable as an Ala-based peptide helix double its length. *ChemBiochem* 3, 1263–6.

122 YOKUM, T. S., GAUTHIER, T. J., HAMMER, R. P., and McLAUGHLIN, M. L. (1997). Solvent effects on the 3_{10}-/α-helix equilibrium in short amphipathic peptides rich in α,α-disubstituted amino acids. *J. Am. Chem. Soc.* 119, 1167–8.

123 MILLHAUSER, G. L., STENLAND, C. J., HANSON, P., BOLIN, K. A., and VAN DE VEN, F. J. M. (1997). Estimating the relative populations of 3_{10}-helix and α-helix in Ala-rich peptides. A hydrogen exchange and high field NMR study. *J. Mol. Biol.* 267, 963–74.

124 FIORI, W. R., LUNDBERG, K. M., and MILLHAUSER, G. L. (1994). A single carboxy-terminal arginine determines the amino-terminal helix conformation of an alanine-based peptide. *Nat. Struct. Biol.* 1, 374–7.

125 YODER, G., POLESE, A., SILVA, R. A.

G. D. et al. (1997). Conformation characterization of terminally blocked L-(αMe)Val homopeptides using vibrational and electronic circular dichroism. 3_{10}-Helical stabilization by peptide-peptide interaction. *J. Am. Chem. Soc.* 119, 10278–85.

126 KENNEDY, D. F., CRISMA, M., TONIOLO, C., and CHAPMAN, D. (1991). Studies of peptides forming 3_{10}- and α-helices and β-bend ribbon structures in organic solution and in model biomembranes by Fourier transform infrared spectroscopy. *Biochemistry* 30, 6541–8.

127 HUNGERFORD, G., MARTINEZ-INSUA, M., BIRCH, D. J. S., and MOORE, B. D. (1996). A reversible transition between an α-helix and a 3_{10}-helix in a fluorescence-labelled peptide. *Angew. Chem. Int. Ed. Engl.* 35, 326–9.

128 SHEINERMAN, F. B. and BROOKS, C. L. (1995). 3_{10}-Helices in peptides and proteins as studied by modified Zimm-Bragg theory. *J. Am. Chem. Soc.* 117, 10098–103.

129 ROSE, G. D. and CREAMER, T. P. (1994). Protein-folding – predicting predicting. *Proteins: Struct. Funct. Genet.* 19, 1–3.

130 DALAL, S., BALASUBRAMANIAN, S., and REGAN, L. (1997). Protein alchemy: Changing β-sheet into α-helix. *Nat. Struct. Biol.* 4, 548–52.

131 ROSE, G. D. (1997). Protein folding and the paracelsus challenge. *Nat. Struct. Biol.* 4, 512–14.

132 POLAND, D. and SCHERAGA, H. A. (1970). *Theory of Helix-Coil Transitions in Biopolymers*. Academic Press, New York and London.

133 QIAN, H. and SCHELLMAN, J. A. (1992). Helix/coil theories: A comparative study for finite length polypeptides. *J. Phys. Chem.* 96, 3987–94.

134 DOIG, A. J. (2002). Recent advances in helix-coil theory. *Biophys. Chem.* 101–102, 281–93.

135 ZIMM, B. H. and BRAGG, J. K. (1959). Theory of the phase transition between helix and random coil in polypeptide chains. *J. Chem. Phys.* 31, 526–35.

136 LIFSON, S. and ROIG, A. (1961). On the theory of the helix-coil transition in polypeptides. *J. Chem. Phys.* 34, 1963–74.

137 FLORY, P. J. (1969). *Statistical Mechanics of Chain Molecules*. Wiley, New York.

138 PAPPU, R. V., SRINIVASAN, R., and ROSE, G. D. (2000). The Flory isolated-pair hypothesis is not valid for polypeptide chains: Implications for protein folding. *Proc. Natl Acad. Sci. USA* 97, 12565–70.

139 TIFFANY, M. L. and KRIMM, S. (1968). New chain conformations of poly(glutamic acid) and polylysine. *Biopolymers* 6, 1379–82.

140 KRIMM, S. and TIFFANY, M. L. (1974). The circular dichroism spectrum and structure of unordered polypeptides and proteins. *Israel J. Chem.* 12, 189–200.

141 WOODY, R. W. (1992). Circular dichroism and conformations of unordered polypeptides. *Adv. Biophys. Chem.* 2, 37–79.

142 BARRON, L. D., HECHT, L., BLANCH, E. W., and BELL, A. F. (2000). Solution structure and dynamics of biomolecules from Raman optical activity. *Prog. Biophys. Mol. Biol.* 73, 1–49.

143 BLANCH, E. W., MOROZOVA-ROCHE, L. A., COCHRAN, D. A. E., DOIG, A. J., HECHT, L., and BARRON, L. D. (2000). Is polyproline II helix the killer conformation? A Raman optical activity study of the amyloidogenic prefibrillar intermediate of human lysozyme. *J. Mol. Biol.* 301, 553–63.

144 SMYTH, E., SYME, C. D., BLANCH, E. W., HECHT, L., VASAK, M., and BARRON, L. D. (2000). Solution structure of native proteins irregular folds from Raman optical activity. *Biopolymers* 58, 138–51.

145 SYME, C. D., BLANCH, E. W., HOLT, C. et al. (2002). A Raman optical activity study of the rheomorphism in caseins, synucleins and tau. *Eur. J. Biochem.* 269, 148–56.

146 RUCKER, A. L. and CREAMER, T. P. (2002). Polyproline II helical structure in protein unfolded states: Lysine peptides revisited. *Protein Sci.* 11, 980–5.

147 SHI, Z. S., WOODY, R. W., and KALLENBACH, N. R. (2002). Is polyproline II a major backbone conformation in unfolded proteins? *Adv. Protein Chem.* 62, 163–240.

148 PAPPU, R. V. and ROSE, G. D. (2002). A simple model for polyproline II structure in unfolded states of alanine-based peptides. *Protein Sci.* 11, 2437–55.

149 SHI, Z., OLSON, C. A., ROSE, G. D., BALDWIN, R. L., and KALLENBACH, N. R. (2002). Polyproline II structure in a sequence of seven alanine residues. *Proc. Natl Acad. Sci. USA* 99, 9190–5.

150 RUCKER, A. L., PAGER, C. T., CAMPBELL, M. N., QUALIS, J. E., and CREAMER, T. P. (2003). Host-guest scale of left-handed polyproline II helix formation. *Proteins: Struct. Funct. Genet.* 53, 68–75.

151 ANDERSEN, N. H. and TONG, H. (1997). Empirical parameterization of a model for predicting peptide helix/coil equilibrium populations. *Protein Sci.* 6, 1920–36.

152 SCHOLTZ, J. M., QIAN, H., ROBBINS, V. H., and BALDWIN, R. L. (1993). The energetics of ion-pair and hydrogen-bonding interactions in a helical peptide. *Biochemistry* 32, 9668–76.

153 SHALONGO, W. and STELLWAGEN, E. (1995). Incorporation of pairwise interactions into the Lifson-Roig model for helix prediction. *Protein Sci.* 4, 1161–6.

154 ARGOS, P. and PALAU, J. (1982). Amino acid distribution in protein secondary structures. *Int. J. Pept. Protein Res.* 19, 380–93.

155 KUMAR, S. and BANSAL, M. (1998). Dissecting α-helices: position-specific analysis of α-helices in globular proteins. *Proteins* 31, 460–76.

156 SUN, J. K., PENEL, S., and DOIG, A. J. (2000). Determination of α-helix N1 energies by addition of N1, N2 and N3 preferences to helix-coil theory. *Protein Sci.* 9, 750–4.

157 ROHL, C. A. and DOIG, A. J. (1996). Models for the 3_{10}-helix/coil, π-helix/coil, and α-helix/3_{10}-helix/coil transitions in isolated peptides. *Protein Sci.* 5, 1687–96.

158 SUN, J. K. and DOIG, A. J. (1998). Addition of side-chain interactions to 3_{10}-helix/coil and α-helix/3_{10}-helix/coil theory. *Protein Sci.* 7, 2374–83.

159 MUÑOZ, V. and SERRANO, L. (1994). Elucidating the folding problem of helical peptides using empirical parameters. *Nat. Struct. Biol.* 1, 399–409.

160 MUÑOZ, V. and SERRANO, L. (1997). Development of the multiple sequence approximation within the AGADIR model of α-helix formation: comparison with Zimm-Bragg and Litson-Roig formalisms. *Biopolymers* 41, 495–509.

161 FERNÁNDEZ-RECIO, J., VÁSQUEZ, A., CIVERA, C., SEVILLA, P., and SANCHO, J. (1997). The tryptophan/histidine interaction in α-helices. *J. Mol. Biol.* 267, 184–97.

162 LACROIX, E., VIGUERA, A. R., and SERRANO, L. (1998). Elucidating the folding problem of α-helices: local motifs, long-range electrostatics, ionic-strength dependence and prediction of NMR parameters. *J. Mol. Biol.* 284, 173–91.

163 MUÑOZ, V. and SERRANO, L. (1995). Elucidating the folding problem of helical peptides using empirical parameters. II. Helix macrodipole effects and rational modification of the helical content of natural peptides. *J. Mol. Biol.* 245, 275–96.

164 PETUKHOV, M., MUÑOZ, V., YUMOTO, N., YOSHIKAWA, S., and SERRANO, L. (1998). Position dependence of non-polar amino acid intrinsic helical propensities. *J. Mol. Biol.* 278, 279–89.

165 LOMIZE, A. L. and MOSBERG, H. I. (1997). Thermodynamic model of secondary structure for α-helical peptides and proteins. *Biopolymers* 42, 239–69.

166 VÁSQUEZ, M. and SCHERAGA, H. A. (1988). Effect of sequence-specific interactions on the stability of helical conformations in polypeptides. *Biopolymers* 27, 41–58.

167 ROBERTS, C. H. (1990). A hierarchical nesting approach to describe the stability of α helices with side chain interactions. *Biopolymers* 30, 335–47.

168 GUZZO, A. V. (1965). The influence of amino acid sequence on protein structure. *Biophys. J.* 5, 809–22.

169 DAVIES, D. R. (1964). A correlation between amino acid composition and protein structure. *J. Mol. Biol.* 9, 605–9.

170 CHOU, P. Y. and FASMAN, G. D. (1974). Conformational parameters for amino acids in helical, b-sheet and random coil regions calculated from proteins. *Biochemistry* 13, 211–21.

171 CHOU, P. Y. and FASMAN, G. D. (1978). Empirical predictions of protein conformation. *Annu. Rev. Biochem.* 47, 251–76.

172 VON DREELE, P. H., POLAND, D., and SCHERAGA, H. A. (1971). Helix-coil stability constants for the naturally occurring amino acids in water. I. Properties of copolymers and approximate theories. *Macromolecules* 4, 396–407.

173 VON DREELE, P. H., LOTAN, N., ANANTHANARAYANAN, V. S., ANDREATTA, R. H., POLAND, D., and SCHERAGA, H. A. (1971). Helix-coil stability constants for the naturally occurring amino acids in water. II. Characterization of the host polymers and application of the host-guest technique to random poly-(hydroxypropylglutamine-co-hydroxybutylglutamine). *Macromolecules* 4, 408–17.

174 WOJCIK, J., ALTMANN, K. H., and SCHERAGA, H. A. (1990). Helix-coil stability constants for the naturally occurring amino acids in water. XXIV. Half-cysteine parameters from random poly-(hydroxybutylglutamine-co-S-methylthio-L-cysteine). *Biopolymers* 30, 121–34.

175 PADMANABHAN, S., YORK, E. J., GERA, L., STEWART, J. M., and BALDWIN, R. L. (1994). Helix-forming tendencies of amino acids in short (hydroxybutyl)-L-glutamine peptides: An evaluation of the contradictory results from host-guest studies and short alanine-based peptides. *Biochemistry* 33, 8604–9.

176 CHAKRABARTTY, A., KORTEMME, T., and BALDWIN, R. L. (1994). Helix propensities of the amino acids measured in alanine-based peptides without helix-stabilizing side-chain interactions. *Protein Sci.* 3, 843–52.

177 BLABER, M., ZHANG, X.-J., and MATTHEWS, B. W. (1993). Structural basis of amino acid α-helix propensity. *Science* 260, 1637–40.

178 BLABER, M. W., BAASE, W. A., GASSNER, N., and MATTHEWS, B. W. (1993). Structural basis of amino acid α-helix propensity. *Science* 260, 1637–40.

179 BLABER, M., ZHANG, X. J., LINDSTROM, J. D., PEPIOT, S. D., BAASE, W. A., and MATTHEWS, B. W. (1994). Determination of α-helix propensity within the context of a folded protein: sites 44 and 131 in bacteriophage-T4 lysozyme. *J. Mol. Biol.* 235, 600–24.

180 HOROVITZ, A., MATTHEWS, J. M., and FERSHT, A. R. (1992). α-Helix stability in proteins. II. Factors that influence stability at an internal position. *J. Mol. Biol.* 227, 560–8.

181 O'NEIL, K. T. and DEGRADO, W. F. (1990). A thermodynamic scale for the helix-forming tendencies of the commonly occurring amino acids. *Science* 250, 646–51.

182 PACE, C. N. and SCHOLTZ, J. M. (1998). A helix propensity scale based on experimental studies of peptides and proteins. *Biophys. J.* 75, 422–7.

183 VILA, J., WILLIAMS, R. L., GRANT, J. A., WOJCIK, J., and SCHERAGA, H. A. (1992). The intrinsic helix-forming tendency of L-alanine. *Proc. Natl Acad. Sci. USA* 89, 7821–5.

184 VILA, J. A., RIPOLL, D. R., and SCHERAGA, H. A. (2000). Physical reasons for the unusual α-helix stabilization afforded by charged or neutral polar residues in alanine-rich peptides. *Proc. Natl Acad. Sci. USA* 97, 13075–9.

185 ROHL, C. A., FIORI, W., and BALDWIN, R. L. (1999). Alanine is helix-stabilizing in both template-nucleated and standard peptide helices. *Proc. Natl Acad. Sci. USA* 96, 3682–7.

186 KENNEDY, R. J., TSANG, K.-Y., and KEMP, D. S. (2002). Consistent helicities from CD and template t/c data for N-templated polyalanines:

progress toward resolution of the alanine helicity problem. *J. Am. Chem. Soc.* 124, 934–44.

187 LYU, P. C., ZHOU, H. X. X., JELVEH, N., WEMMER, D. E., and KALLENBACH, N. R. (1992). Position-dependent stabilizing effects in α-helices – N-terminal capping in synthetic model peptides. *J. Am. Chem. Soc.* 114, 6560–2.

188 PETUKHOV, M., UEGAKI, K., YUMOTO, N., YOSHIKAWA, S., and SERRANO, L. (1999). Position dependence of amino acid intrinsic helical propensities II: Non-charged polar residues: Ser, Thr, Asn, and Gln. *Protein Sci.* 8, 2144–50.

189 PETUKHOV, M., UEGAKI, K., YUMOTO, N., and SERRANO, L. (2002). Amino acid intrinsic α-helical propensities III: Positional dependence at several positions of C-terminus. *Protein Sci.* 11, 766–77.

190 ERMOLENKO, D. N., RICHARDSON, J. M., and MAKHATADZE, G. I. (2003). Noncharged amino acid residues at the solvent-exposed positions in the middle and at the C terminus of the α-helix have the same helical propensity. *Protein Sci.* 12, 1169–76.

191 SZALIK, L., MOITRA, J., KRYLOV, D., and VINSON, C. (1997). Phosphorylation destabilizes α-helices. *Nat. Struct. Biol.* 4, 112–14.

192 LIEHR, S. and CHENAULT, H. K. (1999). A comparison of the α-helix forming propensities and hydrogen bonding properties of serine phosphate and α-amino-γ-phosphonobutyric acid. *Bioorg. Med. Chem. Lett.* 9, 2759–62.

193 ANDREW, C. D., WARWICKER, J., JONES, G. R., and DOIG, A. J. (2002). Effect of phosphorylation on α-helix stability as a function of position. *Biochemistry* 41, 1897–905.

194 HOROVITZ, A., SERRANO, L., AVRON, B., BYCROFT, M., and FERSHT, A. R. (1990). Strength and co-operativity of contributions of surface salt bridges to protein stability. *J. Mol. Biol.* 216, 1031–44.

195 MERUTKA, G. and STELLWAGEN, E. (1991). Effect of amino acid ion pairs on peptide helicity. *Biochemistry* 30, 1591–4.

196 STELLWAGEN, E., PARK, S.-H., SHALONGO, W., and JAIN, A. (1992). The contribution of residue ion pairs to the helical stability of a model peptide. *Biopolymers* 32, 1193–200.

197 HUYGHUES-DESPOINTES, B. M., SCHOLTZ, J. M., and BALDWIN, R. L. (1993). Helical peptides with three pairs of Asp-Arg and Glu-Arg residues in different orientations and spacings. *Protein Sci.* 2, 80–5.

198 HUYGHUES-DESPOINTES, B. M. and BALDWIN, R. L. (1997). Ion-pair and charged hydrogen-bond interactions between histidine and aspartate in a peptide helix. *Biochemistry* 36, 1965–70.

199 HUYGHUES-DESPOINTES, B. M., KLINGLER, T. M., and BALDWIN, R. L. (1995). Measuring the strength of side-chain hydrogen bonds in peptide helices: the Gln.Asp $(i, i+4)$ interaction. *Biochemistry* 34, 13267–71.

200 STAPLEY, B. J. and DOIG, A. J. (1997). Hydrogen bonding interactions between Glutamine and Asparagine in α-helical peptides. *J. Mol. Biol.* 272, 465–73.

201 PADMANABHAN, S. and BALDWIN, R. L. (1994). Tests for helix-stabilizing interactions between various nonpolar side chains in alanine-based peptides. *Protein Sci.* 3, 1992–7.

202 PADMANABHAN, S. and BALDWIN, R. L. (1994). Helix-stabilizing interaction between tyrosine and leucine or valine when the spacing is $i, i+4$. *J. Mol. Biol.* 241, 706–13.

203 VIGUERA, A. R. and SERRANO, L. (1995). Side-chain interactions between sulfur-containing amino acids and phenylalanine in α-helices. *Biochemistry* 34, 8771–9.

204 TSOU, L. K., TATKO, C. D., and WATERS, M. L. (2002). Simple cation-p interaction between a phenyl ring and a protonated amine stabilizes an α-helix in water. *J. Am. Chem. Soc.* 124, 14917–21.

205 TAYLOR, J. W. (2002). The synthesis and study of side-chain lactam-bridged peptides. *Biopolymers* 66, 49–75.

206 CAMPBELL, R. M., BONGERS, J., and FELXI, A. M. (1995). Rational design, synthesis, and biological evaluation of novel growth-hormone releasing-factor analogs. *Biopolymers* 37, 67–8.

207 CHOREV, M., ROUBINI, E., MCKEE, R. L. et al. (1991). Cyclic parathyroid-hormone related protein antagonists – lysine 13 to aspartic acid 17 [I to (I + 40] side-chain to side-chain lactamization. *Biochemistry* 30, 5968–74.

208 OSAPAY, G. and TAYLOR, J. W. (1992). Multicyclic polypeptide model compounds. 2. Synthesis and conformational properties of a highly α-helical uncosapeptide constrained by 3 side-chain to side-chain lactam bridges. *J. Am. Chem. Soc.* 114, 6966–73.

209 BOUVIER, M. and TAYLOR, J. W. (1992). Probing the functional conformation of Neuropeptide-T through the design and study of cyclic analogs. *J. Med. Chem.* 35, 1145–55.

210 KAPURNIOTU, A. and TAYLOR, J. W. (1995). Structural and conformational requirements for human calcitonin activity – design, synthesis, and study of lactam-bridged analogs. *J. Med. Chem.* 38, 836–47.

211 OSAPAY, G. and TAYLOR, J. W. (1990). Multicyclic polypeptide model compounds. 1. Synthesis of a tricyclic amphiphilic α-helical peptide using an oxime resin, segment-condensation approach. *J. Am. Chem. Soc.* 112, 6046–51.

212 BRACKEN, C., GULYAS, J., TAYLOR, J. W., and BAUM, J. (1994). Synthesis and nuclear-magnetic-resonance structure determination of an α-helical, bicyclic, lactam-bridged hexapeptide. *J. Am. Chem. Soc.* 116, 6431–2.

213 CHEN, S. T., CHEN, H. J., YU, H. M., and WANG, K. T. (1993). Facile synthesis of a short peptide with a side-chain-constrained structure. *J. Chem. Res.* (S) 6, 228–9.

214 ZHANG, W. T., and TAYLOR, J. W. (1996). Efficient solid-phase synthesis of peptides with tripodal side-chain bridges and optimization of the solvent conditions for solid-phase cyclizations. *Tetrahedron Lett.* 37, 2173–6.

215 LUO, P. Z., BRADDOCK, D. T., SUBRAMANIAN, R. M., MEREDITH, S. C., and LYNN, D. G. (1994). Structural and thermodynamic characterization of a bioactive peptide model of apolipoprotein-E–side-chain lactam bridges to constrain the conformation. *Biochemistry* 33, 12367–77.

216 JACKSON, D. Y., KING, D. S., CHMIELEWSKI, J., SINGH, S., and SCHULTZ, P. G. (1991). General approach to the synthesis of short α-helical peptides. *J. Am. Chem. Soc.* 113, 9391–2.

217 KUMITA, J. R., SMART, O. S., and WOOLLEY, G. A. (2000). Photo-control of helix content in a short peptide. *Proc. Natl Acad. Sci. USA* 97, 3803–8.

218 FLINT, D. G., KUMITA, J. R., SMART, O. S., and WOLLEY, G. A. (2002). Using an azobenzene cross-linker to either increase or decrease peptide helix content upon trans-to-cis photoisomerization. *Chem. Biol.* 9, 391–7.

219 KUMITA, J. R., FLINT, D. G., SMART, O. S., and WOOLLEY, G. A. (2002). Photo-control of peptide helix content by an azobenzene cross-linker: steric interactions with underlying residues are not critical. *Protein Eng.* 15, 561–9.

220 JIMÉNEZ, M. A., MUÑOZ, V., RICO, M., and SERRANO, L. (1994). Helix stop and start signals in peptides and proteins. The capping box does not necessarily prevent helix elongation. *J. Mol. Biol.* 242, 487–96.

221 KALLENBACH, N. R. and GONG, Y. X. (1999). C-terminal capping motifs in model helical peptides. *Bioorg. Med. Chem.* 7, 143–51.

222 PETUKHOV, M., YUMOTO, N., MURASE, S., ONMURA, R., and YOSHIKAWA, S. (1996). Factors that affect the stabilization of α-helices in short peptides by a capping box. *Biochemistry* 35, 387–97.

223 VIGUERA, A. R. and SERRANO, L. (1995). Experimental analysis of the Schellman motif. *J. Mol. Biol.* 251, 150–60.

224 Gong, Y., Zhou, H. X., Guo, M., and Kallenbach, N. R. (1995). Structural analysis of the N- and C-termini in a peptide with consensus sequence. *Protein Sci.* 4, 1446–56.

225 Armstrong, K. M. and Baldwin, R. L. (1993). Charged histidine affects α-helix stability at all positions in the helix by interacting with the backbone charges. *Proc. Natl Acad. Sci. USA* 90, 11337–40.

226 Scholtz, J. M., York, E. J., Stewart, J. M., and Baldwin, R. L. (1991). A neutral water-soluble, α-helical peptide: The effect of ionic strength on the helix-coil equilibrium. *J. Am. Chem. Soc.* 113, 5102–4.

227 Smith, J. S. and Scholtz, J. M. (1998). Energetics of polar side-chain interactions in helical peptides: Salt effects on ion pairs and hydrogen bonds. *Biochemistry* 37, 33–40.

228 Lockhart, D. J. and Kim, P. S. (1993). Electrostatic screening of charge and dipole interactions with the helix backbone. *Science* 260, 198–202.

229 Jelesarov, I., Durr, E., Thomas, R. M., and Bosshard, H. B. (1998). Salt effects on hydrophobic interaction and charge screening in the folding of a negatively charged peptide to a coiled coil (leucine zipper). *Biochemistry* 37, 7539–50.

230 Scholtz, J. M., Marqusee, S., Baldwin, R. L. et al. (1991). Calorimetric determination of the enthalpy change for the α-helix to coil transition of an alanine peptide in water. *Proc. Natl Acad. Sci. USA* 88, 2854–8.

231 Yoder, G., Pancoska, P., and Keiderling, T. A. (1997). Characterization of alanine-rich peptides, Ac-(AAKAA)n-GY-NH2 (n) 1–4), using vibrational circular dichroism and fourier transform infrared conformational determination and thermal unfolding. *Biochemistry* 36, 5123–33.

232 Huang, C. Y., Klemke, J. W., Getahun, Z., DeGrado, W. F., and Gai, F. (2001). Temperature-dependent helix-coil transition of an alanine based peptide. *J. Am. Chem. Soc.* 123, 9235–8.

233 Goch, G., Maciejczyk, M., Oleszczuk, M., Stachowiak, D., Malicka, J., and Bierzynski, A. (2003). Experimental investigation of initial steps of helix propagation in model peptides. *Biochemistry* 42, 6840–7.

234 Nelson, J. W. and N. R., K. (1986). Stabilization of the ribonuclease S-peptide α-helix by trifluoroethanol. *Proteins: Struct. Funct. Genet.* 1, 211–17.

235 Nelson, J. W. and N. R., K. (1989). Persistence of the α-helix stop signal in the S-peptide in trifluoroethanol solutions. *Biochemistry* 28, 5256–61.

236 Sonnichsen, F. D., Van Eyk, J. E., Hodges, R. S., and Sykes, B. D. (1992). Effect of trifluoroethanol on protein secondary structure: an NMR and CD study using a synthetic actin peptide. *Biochemistry* 31, 8790–8.

237 Waterhous, D. V. and Johnson, W. C. (1994). Importance of environment in determining secondary structure in proteins. *Biochemistry* 33, 2121–8.

238 Jasanoff, A. and Fersht, A. R. (1984). Quantitative determination of helical propensities. Analysis of data from trifluoroethanol titration curves. *Biochemistry* 33, 2129–35.

239 Albert, J. S. and Hamilton, A. D. (1995). Stabilization of helical domains in short peptides using hydrophobic interactions. *Biochemistry* 34, 984–90.

240 Walgers, R., Lee, T. C., and Cammers-Goodwin, A. (1998). An indirect chaotropic mechanism for the stabilization of helix conformation of peptides in aqueous trifluoroethanol and hexafluoro-2-propanol. *J. Am. Chem. Soc.* 120, 5073–9.

241 Luo, P. and Baldwin, R. L. (1999). Interaction between water and polar groups of the helix backbone: An important determinant of helix propensities. *Proc. Natl Acad. Sci. USA* 96, 4930–5.

242 Kentsis, A. and Sosnick, T. R. (1998). Trifluoroethanol promotes helix formation by destabilizing backbone

exposure: Desolvation rather than native hydrogen bonding defines the kinetic pathway of dimeric coiled coil folding. *Biochemistry* 37, 14613–22.

243 REIERSEN, H. and REES, A. R. (2000). Trifluoroethanol may form a solvent matrix for assisted hydrophobic interactions between peptide sidechains. *Protein Eng.* 13, 739–43.

244 MYERS, J. K., PACE, N., and SCHOLTZ, J. M. (1998). Trifluoroethanol effects on helix propensity and electrostatic interactions in the helical peptide from ribonuclease T1. *Protein Sci.* 7, 383–8.

245 NOZAKI, Y. and TANFORD, C. (1967). Intrinsic dissociation constants of aspartyl and glutamyl carboxyl groups. *J. Biol. Chem.* 242, 4731–5.

246 KYTE, J. (1995). *Structure in Protein Chemistry*, Garland Publishing, New York and London.

247 KORTEMME, T. and CREIGHTON, T. E. (1995). Ionisation of cysteine residues at the termini of model α-helical peptides. Relevance to unusual thiol pKa values in proteins of the thioredoxin family. *J. Mol. Biol.* 253, 799–812.

248 MIRANDA, J. J. (2003). Position-dependent interactions between cysteine and the helix dipole. *Protein Sci.* 12, 73–81.

249 EISENBERG, D., WEISS, R. M., and TERWILLIGER, T. C. (1982). The helical hydrophobic moment: a measure of the amphiphilicity of a helix. *Nature* 299, 371–4.

250 MERRIFIELD, R. B. (1963). Solid phase peptide synthesis. I. The synthesis of a tetrapeptide. *J. Am. Chem. Soc.* 85, 2149–54.

251 WHITE, P. D. and CHAN, W. C. (2000). Fmoc solid phase peptide synthesis. A practical approach (HAMES, B. D., ed.), pp. 9–40. Oxford University Press, Oxford.

252 CARPINO, L. A. and HAN, G. Y. (1970). The 9-fluorenylmethoxycarbonyl function, a new base-sensitive amino-protecting group. *J. Am. Chem. Soc.* 92, 5748–9.

253 CARPINO, L. A. and HAN, G. Y. (1972). The 9-fluorenylmethoxycarbonyl amino-protecting group. *J. Org. Chem.* 37, 3404–9.

254 KAISER, E., COLESCOTT, R. L., BOSSINGER, C. D., and COOK, P. I. (1970). Color test for detection of free terminal amino groups in the solid-phase synthesis of peptide. *Anal. Biochem.* 34, 595–8.

255 SARIN, V. K., KENT, S. B. H., TAM, J. P., and MERRIFIELD, R. B. (1981). Quantitative monitoring of solid-phase peptide synthesis by the ninhydrin reaction. *Anal. Biochem.* 117, 147–57.

256 GREENFIELD, N. J. (1996). Methods to estimate the conformation of proteins and polypeptides from circular dichroism data. *Biochemistry* 235, 1–10.

257 MANNING, M. C. (1993). ACS Symposium Series, Vol. 516, pp. 33–52, OUP, New York.

258 FASMAN, G. D. (1996). *Circular Dichroism and the Conformational Analysis of Biomolecules*. Plenum Press, New York.

259 LUO, P. and BALDWIN, R. L. (1997). Mechanism of helix induction by trifluoroethanol: a framework for extrapolating the helix-forming properties of peptides from trifluoroethanol/water mixtures back to water. *Biochemistry* 36, 8413–21.

260 STRICKLAND, E. H. (1974). Thermodynamics and dynamics of histidine-binding protein, the water-soluble receptor of histidine permease. *CRC Crit. Rev. Biochem.* 2, 113–74.

261 MANNING, M. C. and WOODY, R. W. (1989). Theoretical study of the contribution of aromatic side chains to the circular dichroism of basic bovine pancreatic trypsin inhibitor. *Biochemistry* 28, 8609–13.

262 WOODY, R. W., DUNKER, A. K., and FASMAN, G. D. (1996). In *Circular Dichroism and the Conformational Analysis of Biomolecules* (FASMAN, G. D., ed.), pp. 109–157. Plenum Press, New York.

263 BAYLEY, P. M., NIELSEN, E. B., and SCHELLMAN, J. A. (1969). The rotatory properties of molecules containing two peptide groups. *J. Phys. Chem.* 73, 228–43.

264 Sreerama, N. and Woody, R. W. (1993). A self-consistent method for the analysis of protein secondary structure from circular dichroism. *Anal. Biochem.* 209, 32–44.

265 Keiderling, T. A. (2002). Protein and peptide secondary structure and conformational determination with vibrational circular dichroism. *Curr. Opin. Struct. Biol.* 6, 682–8.

266 Keiderling, T. A. (2002). In *Circular Dichroism: Principles and Applications* (Berova, N., Nakanishi, K., and Woody, R. A., eds), pp. 621–666. Wiley-VCH, New York.

267 Tanaka, T., Inoue, K., Kodama, T., Kyogoku, Y., Hayakawa, T., and Sugeta, H. (2001). Conformational study on poly[γ-(α-phenethyl)-L-glutamate] using vibrational circular dichroism spectroscopy. *Biopolymers* 62, 228–34.

268 Baumruk, V., Pancoska, P., and Keiderling, T. A. (1996). Predictions of secondary structure using statistical analyses of electronic and vibrational circular dichroism and Fourier transform infrared spectra of proteins in H_2O. *J. Mol. Biol.* 259, 774–91.

269 Nafie, L. A. (2000). Dual polarization modulation: a real-time, spectral multiplex separation of circular dichroism from linear birefringence spectral intensities. *Appl. Spectros.* 54, 1634–45.

270 Hilario, J., Drapcho, D., Curbelo, R., and Keiderling, T. A. (2001). Polarization modulation Fourier transform infrared spectroscopy with digital signal processing: comparison of vibrational circular dichroism method. *Appl. Spectros.* 55, 1435–47.

271 Rohl, C. A. and Baldwin, R. L. (1997). Comparison of NH exchange and circular dichroism as techniques for measuring the parameters of the helix-coil transition in peptides. *Biochemistry* 36, 8435–42.

272 Spera, S. and Bax, A. (1991). Empirical correlation between protein backbone conformation and Cα and Cβ ^{13}C nuclear magnetic resonance chemical shifts. *J. Am. Chem. Soc.* 113, 5490–2.

273 Shalongo, W., Dugad, L., and Stellwagen, E. (1994). Distribution of helicity within the model peptide Acetyl(AAQAA)$_3$ amide. *J. Am. Chem. Soc.* 116, 8288–93.

274 Shalongo, W., Dugad, L., and Stellwagen, E. (1994). Analysis of the thermal transitions of a model helical peptides using C-13 NMR. *J. Am. Chem. Soc.* 116, 2500–7.

275 Park, S.-H., Shalongo, W., and Stellwagen, E. (1998). Analysis of N-terminal Capping using carbonyl-carbon chemical shift measurements. *Proteins: Struct. Funct. Genet.* 33, 167–76.

276 Wuthrich, K. (1986). *NMR of Proteins and Nucleic Acids*, John Wiley and Sons, New York.

277 Kilby, P. M., van Eldick, L. J., and Roberts, G. C. K. (1995). Nuclear magnetic resonance assignments and secondary structure of bovine S100β protein. *FEBS Lett.* 363, 90–6.

278 Campbell, I. D. and Dwek, R. A. (1984). *Biological Spectroscopy*. Benjamin Cummings, Menlo Park, CA.

279 Brey, W. S. (1984). *Physical Chemistry and its Biological Applications*. Academic Press, New York.

280 Venyaminov, S.-Y. and Kalnin, N. N. (1990). Quantitative IR Spectrophotometry of peptide compounds in water (H_2O) solutions. 2. Amide absorption-bands of polypeptides and fibrous proteins in α-coil, β-coil, and random coil conformations. *Biopolymers* 30, 1243–57.

281 Susi, H., Timasheff, S. N., and Stevens, L. (1967). Infra-red spectrum and protein conformations in aqueous solutions. *J. Biol. Chem.* 242, 5460–6.

282 Krimm, S. and Bandekar, J. (1986). Vibrational spectroscopy and conformation of peptides, polypeptides and protein. *Adv. Protein Chem.* 38, 181–364.

283 Nevskaya, N. A. and Chirgadze, Y. N. (1976). Infrared spectra and resonance interactions of amide-I and II vibration of α-helix. *Biopolymers* 15, 637–48.

284 Byler, D. M. and Susi, H. (1986).

Examination of the secondary structure of proteins by deconvolved FTIR spectra. *Biopolymers* 25, 469–87.

285 YANG, W.-J., GRIFFITHS, P. R., BYLER, D. M., and SUSI, H. (1985). Protein Conformation by Infra-red spectroscopy – Resolution enhancement by Fourier self deconvolution. *Appl. Spectros.* 39, 282–7.

286 SUSI, H. and BYLER, D. M. (1988). Fourier deconvolution of the Amide-I Raman band of proteins as related to conformation. *Appl. Spectros.* 42, 819–25.

287 KATO, K., MATSUI, T., and TANAKA, S. (1987). Quantitative estimation of α-helix coil content in bovine sreum albumin by Fourier transform-infrared spectroscopy. *Appl. Spectros.* 41, 861–5.

288 RAMAN, C. V. and KRISHNAN, K. S. (1928). A new type of secondary radiation. *Nature* 121, 501–2.

289 HENDRA, P. J., JONES, C., and WARNES, G. (1991). *Fourier Transform Raman Spectroscopy*. Ellis Horwood, Chichester.

290 CALLENDER, R., DENG, H., and GILMANSHIN, R. (1998). Raman difference studies of protein structure and folding, enzymatic catalysis and ligand binding. *J. Raman Spectros.* 29, 15.

291 WITHNALL, R. (2001). Protein-Ligand Interactions: structure and spectroscopy (HARDING, S. E. and CHOWDHRY, B. Z., eds). Oxford University Press, Oxford.

292 PETICOLAS, W. L., ed. (1995). *Vibrational Spectra: Principles and Aplications with Emphasis on Optical Activity*. Vol. 246. *Methods in Enzymology* (SAUER, K., ed.). Academic Press, New York.

293 ASHER, S. A., IANOUL, A., MIX, G., BOYDEN, M. N., KARNOUP, A., DIEM, M., and SCHWEITZER-STENNER, R. (2001). Dihedral ψ angle dependence of the amide III vibration: A uniquely sensitive UV resonance Raman secondary structural probe. *J. Am. Chem. Soc.* 123, 11775–81.

294 POLAVARAPU, P. L. (1998). *Vibrational Spectra: Principles and Applications with Emphasis on Optical Activity*. Elsevier, Amsterdam.

295 NAFIE, L. A. (1997). Infrared and Raman vibrational optical activity: Theoretical and experimental aspect. *Annu. Rev. Phys. Chem.* 48, 357–86.

296 FORD, S. J., WEN, Z. Q., HECHT, L., and BARRON, L. D. (1994). Vibrational Raman optical-activity of alanyl peptide oligomers – a new perspective on aqueous-solution conformation. *Biopolymers* 34, 303–13.

297 DIEM, M. (1993). *Introduction to Modern Vibrational Spectroscopy*, Wiley, New York.

298 MUÑOZ, V. and SERRANO, L. (1995). Analysis of $i, i + 5$ and $i, i + 8$ hydrophobic interactions in a helical model peptide bearing the hydrophobic staple motif. *Biochemistry* 34, 15301–6.

299 THOMAS, S. T., LOLADZE, V. V., and MAKHATADZE, G. I. (2001). Hydration of the peptide backbone largely defines the thermodynamic propensity scale of residues at the C$'$ position of the C-capping box of α-helices. *Proc. Natl Acad. Sci. USA* 98, 10670–5.

300 SHI, Z., OLSON, C. A., BELL, A. J., and KALLENBACH, N. R. (2002). Non-classical helix-stabilizing interactions: C–H ... O H-bonding between Phe and Glu side chains in α-helical peptides. *Biophys. Chem.* 101–2, 267–79.

301 LUO, R., DAVID, L., HUNG, H., DEVANEY, J., and GILSON, M. K. (1999). Strength of solvent-exposed salt-bridges. *J. Phys. Chem. B* 103, 366–80.

302 MARQUSEE, S. and SAUER, R. T. (1994). Contributions of a hydrogen bond/salt bridge network to the stability of secondary and tertiary structure in lambda repressor. *Protein Sci.* 3, 2217–25.

303 SHI, Z., OLSON, C. A., BELL, A. J., and KALLENBACH, N. R. (2001). Stabilization of α-helix structure by polar side-chain interactions: Complex salt bridges, cation π interactions and C–H ... O–H bonds. *Biopolymers* 60, 366–80.

10
Design and Stability of Peptide β-sheets

Mark S. Searle

10.1
Introduction

The pathway by which the polypeptide chain assembles from the unfolded, "disordered" state to the final active folded protein has been the subject of intense investigation. Hierarchical models of protein folding emphasize the importance of local interactions in restricting the conformational space of the polypeptide chain in the search for the native state. These nuclei of structure promote interactions between different parts of the sequence leading ultimately to a cooperative rate-limiting step from which the native state emerges [1–4]. Designed peptides that fold autonomously in water (α-helices and β-sheets) have proved extremely valuable in probing the relationship between local sequence information and folded conformation (the stereochemical code) in the absence of the tertiary interactions found in the native state of proteins. This has allowed intrinsic secondary structure propensities to be investigated in isolation, and enabled the nature and strength of the weak interactions relevant to a wide range of molecular recognition phenomena in chemistry and biology to be put on a quantitative footing.

While the literature is rich in studies of α-helical peptides [5–8], water-soluble, nonaggregating monomeric β-sheets have emerged relatively recently [9–11]. For reasons of design and chemical synthesis, these are almost exclusively antiparallel β-sheets, although others have used nonnatural linkers to engineer parallel strand alignments [12–14]. Here we focus on contiguous antiparallel β-sheet systems. Autonomously folding β-hairpin motifs, consisting of two antiparallel β-strands linked by a reverse β-turn (Figure 10.1), represent the simplest systems for probing weak interactions in β-sheet folding and assembly, although more recently a number of three- and four-stranded β-sheet structures have been described. From this growing body of data, key factors have come into focus that are important in rational design. The following will be considered: the role of the β-turn in promoting and stabilizing antiparallel β-sheet formation; the role of intrinsic backbone ϕ, ψ propensities in preorganizing the extended conformation of the polypeptide chain; the role of cooperativity in β-sheet folding and stability and in the propagation of

Protein Folding Handbook. Part I. Edited by J. Buchner and T. Kiefhaber
Copyright © 2005 WILEY-VCH Verlag GmbH & Co. KGaA, Weinheim
ISBN: 3-527-30784-2

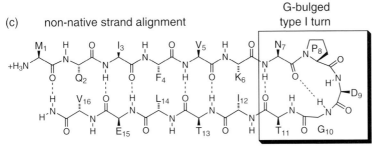

Fig. 10.1. a) Beta-strand alignment and interstrand hydrogen bonds in an antiparallel β-hairpin peptide; R groups represent amino acid side chains, main chain φ and ψ angles are shown. N-terminal β-hairpin of ubiquitin in which the native TLTGK turn sequence has been replaced by the sequence NPDG. b) Native strand alignment giving rise to a type I NPDG turn. c) Nonnative strand alignment with a G-bulged type I turn (NPDGT).

multistranded β-sheets, and quantitative approaches to estimating the energetics of β-sheet stability and folding.

10.2
β-Hairpins Derived from Native Protein Sequences

The early focus on peptides excised from native protein structures provided the first insights into autonomously folding β-hairpins, preceding the more rational approach to β-sheet design. Peptides derived from tendamistat [15], B1 domain of protein G [16, 17], ubiquitin [18, 19], and ferredoxin [20] showed that these sequences could exist in the monomeric form without aggregating, but that in most cases they showed a very limited tendency to fold in the absence of tertiary contacts. The use of organic co-solvents appeared to induce native-like conformation [17, 18, 20]. The study by Blanco et al. [20] of a peptide derived from the B1 domain of protein G provided the first example of native-like folding in water of a fully native peptide sequence. In contrast, the studies of hairpins isolated from the proteins ubiquitin and ferredoxin, which are structurally homologous to the G B1 domain (all form an α/β-roll fold) showed these to be unfolded in water [18, 20]. More recent studies of the native ubiquitin peptide have now revealed evidence for a small population of the folded state in water [21], while a β-turn mutation has been shown to significantly enhance folding [22].

The apparent lack of evidence for folding of the peptide derived from residues 15–23 of tendamistat (YQSWRYSQA) [15], and from the N-terminal sequence of ubiquitin (MQIFVKTLTGKTITLEV) [19] led to partial redesign of the sequence to enhance folding through modification to the β-turn sequence by introducing an NPDG type I turn, which is the most common type I turn sequence in proteins. Thus, in the former peptide SWRY was replaced by NPDG [15], and in the latter the G-bulged type I turn TLTGK was replaced by NPDG [19]. By introducing this tight two-residue loop across PD it was envisaged that both hairpins would be stabilized. This was certainly the case, however, the most striking observation was that both peptides folded into nonnative conformations with a three-residue G-bulged type I turn (PDG) reestablished across the turn (Figure 10.1b and c). These initial studies in rational redesign led to strikingly irrational results, prompting a much more systematic approach to β-hairpin design. The above results revealed that the β-turn sequence, which dictates the preferred backbone geometry, appears to be an important factor in dictating the β-strand alignment. From the protein folding viewpoint, as demonstrated by the redesigned ubiquitin hairpin sequence, it is evident that one important role of the native turn sequence may be to preclude the formation of nonnative conformations that may be incompatible with formation of the native state.

10.3
Role of β-turns in Nucleating β-hairpin Folding

The systematic classification of β-turns in proteins reveals a wide variety of geometries and sizes of loop [22–24]. For the design of β-hairpins, the emphasis has been on incorporation of the smallest turn sequence possible to limit the entropic destabilization effects. While two-residue type I and type II turns are generally common in β-turns, the backbone conformation (ϕ and ψ torsion angles) of a two-residue type I or type II turn results in a local left-handed twist, which is not compatible with the right-handed twist found in protein β-sheets. Consequently, these turns are less commonly found between contiguous antiparallel β-strands. However, the diastereomeric type I′ and II′ turns have the ϕ and ψ angles complementary to the right-handed twisted orientation of the β-strands. These conclusions appear to rationalize, at least in part, the above observations with β-turn modifications introduced into the β-hairpins of tendamistat and ubiquitin. The introduced NPDG type I turn is not compatible with the right-handed twist of the β-strands resulting in a refolding to a more flexible G-bulged type I turn.

The work of Gellman et al. [25] has shown the importance of backbone ϕ, ψ angle preferences for the residues in the turn sequence by comparing the stability of a number of β-hairpin peptides derived from the ubiquitin sequence (MQIFVKSXXKTITLVKV) containing either XX = L-Pro-Xaa or D-Pro-Xaa. Replacing L-Pro with D-Pro switches the twist from left-handed (type I or II) to right-handed (type I′ or II′), making the latter compatible with the right-handed twist of the two β-strands. NMR data (Hα shifts and long-range nuclear Overhauser effects (NOEs)) indicate that each of the D-Pro containing peptides showed a significant degree of folding, whereas the L-Pro analogs appeared to be unfolded. Similar conclusions were drawn from studies of a series of 12-mers containing XG turn sequences, with X = L-Pro or D-Pro [26].

A number of natural L-amino acids are commonly found in the α_L region of conformational space and are compatible with the type I′ or II′ turn conformation. Statistical analyses from a number of groups have identified Xaa-Gly as a favored type I′ turn. A number of studies have used the Asn-Gly sequence to design β-hairpin motifs and have demonstrated a high population of the folded structure with the required turn conformation and strand alignment [27–32]. The work by de Alba et al. [28] identified two possible conformations of the peptide ITSYNGKTYGR. The NOE data appear to be compatible with rapidly interconverting conformations involving an YNGK type I′ turn and a NGKT type II′ turn, each giving a distinct pattern of cross-strand NOEs. Ramirez-Alvarado et al. [27] also described an NG-containing 12-mer (RGITVXGKTYGR; X = N), which they subsequently extended to a series of hairpins to examine the correlation between hairpin stability and the database frequency of occurrence of residues in position X [29]. Using nuclear magnetic resonance (NMR) and circular dichroism (CD) measurements they concluded that X = Asn > Asp > Gly > Ala > Ser in promoting hairpin folding, in agreement with the intrinsic ϕ, ψ preferences of these residues. Despite extensive analysis of NG type I′ turns in the protein database (PDB), it is still not entirely clear why Asn at the first position is so effective in promoting turn formation. There is no evidence for specific side chain to main chain hydrogen bonds that might stabilise the desired backbone conformation, although specific solvation effects cannot be ruled out [30].

Evidence that the NG turn is able to nucleate folding in the absence of cross-strand interactions was demonstrated in a truncated analog of one designed 16-residue β-hairpin sequence KKYTVSINGKKITVSI (β1 in Figure 10.2), in which the sequence was shortened to SINGKKITVSI, lacking the N-terminal six residues [30]. Evidence from NOE data showed that the turn was significantly populated with interactions observed between Ser6 Hα ↔ Lys11 Hα and Ile7 NH ↔ Lys10

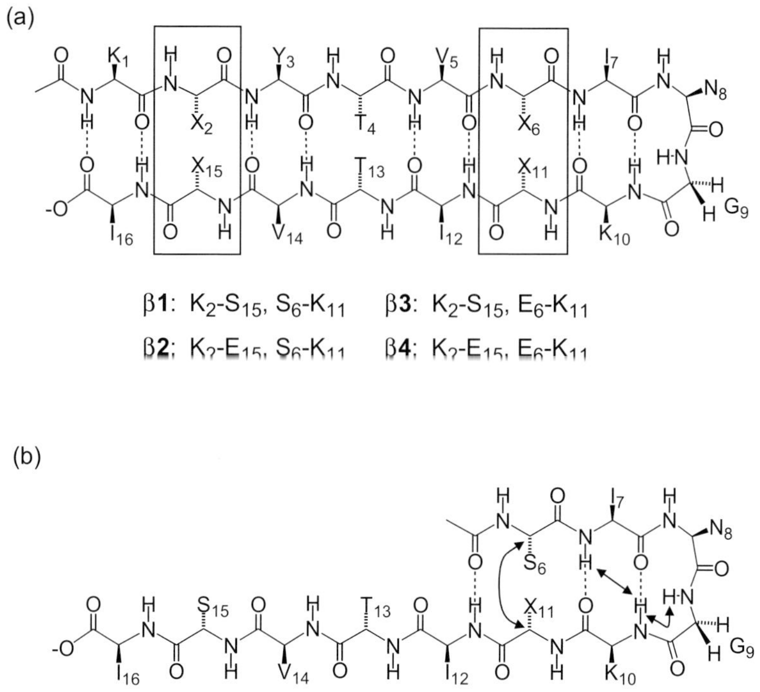

(a)

β1: K_2-S_{15}, S_6-K_{11} β3: K_2-S_{15}, E_6-K_{11}
β2: K_2-E_{15}, S_6-K_{11} β4: K_2-E_{15}, E_6-K_{11}

(b)

Fig. 10.2. Structure of the 16-residue β-hairpin sequence β1, and mutants β2, β3 and β4 in which cross-strand salt bridges (Lys-Glu) have been introduced at position X_2-X_{15} and X_6-X_{11}. In b), hairpin β1 has been truncated by removing the N-terminal hydrophobic residues (1–5). The resulting peptide still shows the ability to fold around the turn sequence as evident from medium range NOEs (indicated by arrows) that show that the INGK sequence is adopting a type I' NG turn.

NH (Figure 10.2b) that are only compatible with a folded type I' turn around NG. Two destabilized β-hairpin mutants (KKYTVSINGKKITKSK with electrostatic repulsion between the N- and C-terminal Lys residues, and KKATASINGKKITVSI with the loss of key hydrophobic residues in one strand) showed no evidence from NMR chemical shift data for cross-strand interactions; however, careful examination of NOE data revealed evidence for NG turn nucleation [30]. Titration with organic co-solvent showed both peptides to fold significantly, indicating that the turn sequence probably already predisposes the peptide to form a β-hairpin but that favorable cross-strand interactions are required for stability.

The work of de Alba et al. convincingly illustrated this principle in a series of six hairpin sequences (10-mers) where strand residues were conserved but turn sequences varied [28]. Using a number of NMR criteria they were able to show that changes in turn sequence could result in a variety of turn conformations including two residue 2:2 turns, 3:5 turns and 4:4 turns (see earlier nomenclature

[22–24]), with different pairings of amino acid side chains. As with the earlier examples cited with the turn modification described for hairpins derived from tendamistat and ubiquitin, the bulged-type I turn (3:5 turn) appears to be an intrinsically stable turn with the necessary right-handed twist. Together these data strongly support a model for hairpin folding in which the turn sequence strongly dictates its preferred conformation, and that strand alignment, cross-strand interactions and subsequently conformational stability are dictated by the specificity of the turn.

10.4
Intrinsic ϕ, ψ Propensities of Amino Acids

Statistical analyses of high-resolution structures in the PDB have provided significant insights into residue-specific intrinsic backbone ϕ, ψ preferences in polypeptide chains. A novel approach presented by Swindells et al., was to determine ϕ, ψ propensities of different residues in nonregular regions of protein structure where backbone geometry is free of interactions associated with regular hydrogen bonded β-sheet or α-helical secondary structure [33]. The striking observation is that in this context ϕ and ψ angles (see Figure 10.3) are far from randomly distributed, and that most occupy regions of Ramachandran space associated with regular secondary structure. The observed ϕ, ψ distributions for individual residues has been taken as representative of those found in denatured states of proteins providing the basis of a "random coil" state from which residue-specific NMR parameters ($^{3}J_{NH-H}\alpha$ and NOE intensities) can be derived as a reference state for folding studies [34, 35]. While β-propensity is found to vary significantly from one residue to the next, context-dependent effects also appear to play an important part [36].

While V, I, F, and Y, for example, have a high intrinsic preference to be in the β-region of the Ramachandran plot where steric interactions with flanking residues are minimized, their conformation is relatively insensitive to the nature of the flanking residues. In contrast, small or unbranched side chains have a higher preference for the α-helical conformation, however, this preference can be significantly modulated by its neighbors through a combination of steric and hydrophobic interactions, as well as both repulsive and attractive electrostatic interactions. General effects of flanking residues (grouped as α-like or β-like, reflecting ϕ, ψ propensities) on the central residue of a XXX triplet are illustrated in Figure 10.4a as an average over all residues at the central position. More specific effects are also shown for Ser, Val, and Lys (Figure 10.4b–d). With Ser, for example, having bulky flanking residues either side with high β-propensity (denoted $\beta S\beta$, where β could be V, I, F or Y), significantly increases the β-propensity of the Ser residue to minimize the steric repulsion between the two bulky neighboring residues [36]. Thus, intrinsic structural propensities appear to be highly context dependent.

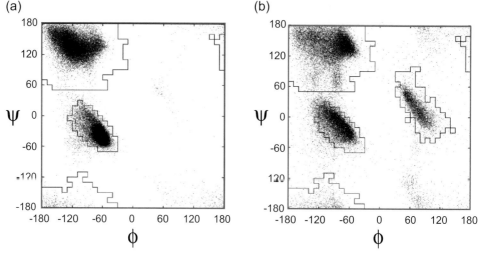

Fig. 10.3. Ramachandran plots of residue backbone ϕ and ψ angles taken from a database of 512 high-resolution protein X-ray structures showing: a) residues in regular β-sheet (ϕ, ψ: $-120°, 120°$) and α-helix (ϕ, ψ: $-60°, 60°$), and b) residues in the irregular coil regions of the same structures (ϕ, ψ: $60°, 0°$ is the α_L region of conformational space mainly occupied by Gly). The distribution in b) shows that residues have a natural propensity to occupy the α and β regions of conformational space even when they are not involved in regular protein secondary structure. Taken from Ref. [36].

This statistical framework has been extended to a number of experimental systems to examine the extent to which isolated β-strand sequences (in the absence of secondary structure interactions) are predisposed by the primary sequence to adopt an extended β-like conformation. The isolated 8-mer (GKKITVSI), corresponding to the C-terminal β-strand of the hairpin **β1** (Figure 10.2; KKYTVSINGK-KITVSI), was examined by NMR analysis of $^3J_{NH-H\alpha}$ values and backbone NOE intensities. Surprisingly, many of these parameters are similar to those for the folded hairpin despite the monomeric nature of the 8-mer [32, 36]. In an analogous study of the C-terminal strand of the ubiquitin hairpin described above [21, 37], similarly large deviations of coupling constants and NOE intensities from random coil values suggested that the isolated β-strands are partially preorganized into an extended conformation supporting a model for hairpin folding which may not require a significant further organisation of the peptide backbone, a factor that may contribute significantly to hairpin stability. Several studies of denatured states of proteins have also highlighted the influence of neighboring residues in modulating main chain conformational preferences [38–41], and the importance of residual structure in the unfolded state in guiding the conformational search to the native state [42, 43].

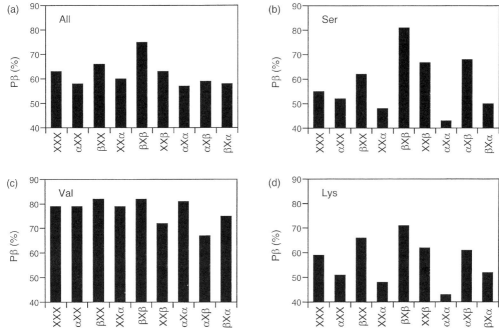

Fig. 10.4. Effects of neighboring residues on residue β-propensities within the triplets XϕX calculated from the data in Figure 10.3b. The β-propensity of residue ϕ is a measure of the number of times a particular residues is found in the β-region of the Ramachandran plot as a fraction of the total distribution between α- and β-space $[\beta/(\alpha+\beta)]$. The context dependence of the β-propensity is estimated by considering the nature of the neighboring residue (X) (X = any residue, α is a residue that prefers the α-helical region – Asp, Glu, Lys, or Ser, β is a residue that prefers the β-sheet region – Ile, Val, Phe, or Tyr). The effects of neighboring residues on the average β-propensity is shown in (a), specific effects on Ser (b), Val (c) and Lys (d) are also shown. While Val is relatively insensitive to the nature of the flanking residues, the smaller Ser residue can be forced to adopt a higher β-propensity if it has bulky neighbors. Thus, the intrinsic β-propensity of a particular residue is highly context dependent. Taken in part from Ref. [36].

10.5
Side-chain Interactions and β-hairpin Stability

There is a strong case, at least in the context of isolated peptide fragments, that the origin of the specificity of β-hairpin folding is largely dictated by the conformational preferences of the turn sequence. However, the stability of the folded state has been attributed to interstrand hydrogen bonding and/or hydrophobic interactions, though which dominates is still a matter of debate. Ramirez-Alvarado et al. [27] reported that the population of the folded state of the hairpin RGITVNGK-TYGR was significantly diminished by replacing residues on the N-terminal strand, and then the C-terminal strand, by Ala. The loss of stability was attributed firstly to a reduction in hydrophobic surface burial, but also due to the intrinsically lower β-propensity of Ala, the latter contributing through an adverse conformational entropy term. To compensate for this de Alba et al. [44] described a family of hairpins derived from the sequence IYSNSDGTWT. The effects of residue substitutions in the first three positions was examined while maintaining the overall β-character of the two strands. Several favorable cross-strand pair-wise interactions were identified that were apparent in earlier, and more recent, PDB analysis of β-sheet interactions [45, 46]. For example, Thr-Thr and Tyr-Thr cross-strand pairs produced stabilizing interactions, whilst Ile-Thr and Ile-Trp had a destabilizing effect. Other studies of Ala substitution in one β-strand have similarly highlighted hydrophobic burial as a key factor in conformational stability [30], while the observation of large numbers of side chain NOEs have been used as evidence for hydrophobic stabilization in water (see below) [19, 25–27, 30].

10.5.1
Aromatic Clusters Stabilize β-hairpins

The first example of a natural β-hairpin sequence that folded autonomously in water (residues 41–56 of the B1 domain of protein G) identified an aromatic-rich cluster of residues that appears to impart considerable stability [17]. The interstrand pairing of Trp/Val and Tyr/Phe has subsequently been exploited in the design of a number of model hairpin systems, in particular to examine the relationship between the position of this stabilizing cluster and the β-turn sequence. The separation between the loop sequence and cluster strongly influences stability and the extent of participation in β-sheet forming interactions (Figure 10.5a). In an isomeric family of 20-mers (peptides **1**, **2**, and **3**), all of which contain exactly the same residue composition, the most stable hairpin is that in which the smallest cost in conformational entropy is paid to bring the cluster together, i.e., where the cluster is closest to the loop sequence [47]. This arrangement results in the largest Hα chemical shift deviations, indicating a well-formed core; however, the terminal residues show a lower propensity to fold. Thus, there exists a strong interplay between the two key stabilizing components of the hairpin. A statistical model was developed to rationalize the experimental observations and estimate the free energy change versus the number of peptide hydrogen bonds formed (Figure 10.5b).

It is a well-known phenomenon that α-helical peptides become more stable as the length increases, reflecting the fact that while helix nucleation is energetically unfavorable, the propagation step has a small net increase in stability. In β-hairpin systems, Stanger et al. [48] suggest that this may not be the case. There is some evidence for an increase in stability as strands lengthen from five to seven residues, however, further extension (to nine) does not lead to a further stability increment, suggesting that there may be an intrinsic limit to strand length. Since the choice of

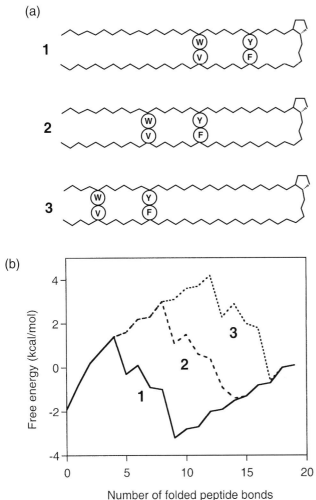

Fig. 10.5. Interplay between hydrophobic cluster and turn position in a family of isomeric β-hairpins. The distance of the WVYF cluster from the turn is shown in (a) for peptides **1**, **2**, and **3**. A statistical mechanical model was used to estimate the free energy of folding for the three peptides according to the number of peptide hydrogen bonds formed and the relative position of the stabilizing hydrophobic cluster (b). Thus, peptide **1**, which has the hydrophobic cluster closest to the turn, has the smallest energetic cost of forming the cluster which then significantly stabilizes this core motif, but leads to fraying of the N- and C-termini. In contrast, in the case of peptide **3**, there is a large energy penalty in ordering the peptide backbone to bring the residues of the cluster together. Overall the core hairpin is less stable but more residues are involved in ordered structure. Adapted from Ref. [47].

Tab. 10.1. trpzip β-hairpin peptides, sequences and stability.

β-hairpin	Sequence	Turn type	T_m	ΔH (kJ mol^{-1})
trpzip1	SWTWEGNKWTWK	(Type II' turn)	323	−45.4
trpzip2	SWTWENGKWTWK	(Type I' turn)	345	−70.6
trpzip3	SWTWE^{D}PNKWTWK	(Type II' turn)	352	−54.8

sequence extension in this study was limited to an all Thr extension or an alternating Ser-Thr (ST)$_n$ extension the conclusions should be viewed with caution. Since cross-strand Ser/Ser and Thr/Thr pairings do not bury very much hydrophobic surface area, it is not surprising perhaps that cross-strand side-chain interactions may only just compensate for the entropic cost of organizing the peptide backbone. Studies with other more favorable pairings may be enlightening.

Undoubtedly the most successfully designed structural motif to date has been the tryptophan zipper (trpzip) motif [49], whose stability exceeds substantially all those already described. The design is based around stabilizing nonhydrogen-bonded cross-strand Trp-Trp pairs (Table 10.1).

The most successful designs involved two such Trp-Trp pairs which NMR structural analysis reveals are interdigitated in a zipper-like manner stabilized through face–face offset π-stacking giving a compact structure (Figure 10.6). This arrangement of the indole rings results in a pronounced signature in the CD spectrum with intense exciton-coupling bands at 215 and 229 nm indicative of interactions between aromatic chromophores in a highly chiral environment (see Figure 10.7). The high sensitivity of the CD bands permits thermal denaturation curves to be determined with high sensitivity, in contrast to other β-sheet systems where only small changes in the CD spectrum at 216 nm are evident, resulting in poor signal-to-noise. The thermal unfolding curves are sigmoidal and reversible, fitting to a two-state folding model (Figure 10.7).

The trpzip hairpins 1, 2, and 3 show high T_m values (see Table 10.1) with folding strongly enthalpy driven. Changing the turn sequence (GN versus NG versus D-PN) has a significant impact on stability. The unfolding curve for trpzip2 with the NG type I' turn gives the most cooperative unfolding transition and appears to be the most stable of the three at room temperature, despite other studies that suggest that the unnatural D-PN (type II') turn is the most stabilizing [25]. Clearly, context-dependent factors are at work. Also, the strongly enthalpy-driven signature for folding is different to that reported previously for other systems [32], reflecting differences in the nature of the stabilizing interactions which in the case of the trpzip peptides involve π–π stacking interactions. Despite the fact that the trpzip hairpins represent a highly stabilized motif, no such examples have been found in the protein structure database. The authors suggest that on steric grounds it may be difficult to accommodate the trpzip motif within a multistranded β-sheet or through packing against other structural elements.

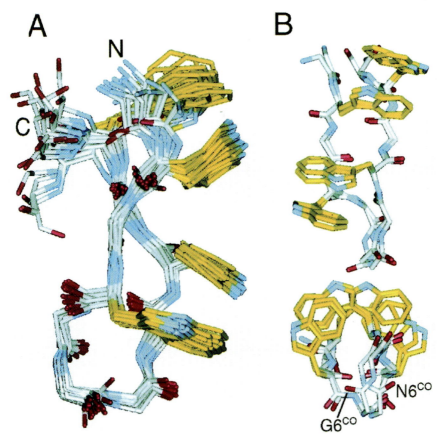

Fig. 10.6. NMR structures of trpzips 1 and 2. A) Ensemble of 20 structures of trpzip1 (residues 2–11) showing the relative orientations of the indole rings. B) Overlay of trpzip 1 and 2 aligned to the peptide backbone of residues 2–5 and 8–11 (top), and rotated by 90° for the end-on view of the indole rings. The backbone carbonyl of residue 6 is labeled to illustrate the different turn geometries of the two hairpins (type II′ and type I′). Taken from Ref. [49].

10.5.2
Salt Bridges Enhance Hairpin Stability

The high abundance of salt bridges on the surface of hyperthermophilic proteins [50, 51], together with examples of rational enhancement of protein stability through redesign of surface charge, has strongly implicated ionic interactions as a stabilizing force [52, 53]. In model peptides that are only weakly folded in water, the relative importance of ionic versus hydrophobic interactions in stabilizing local

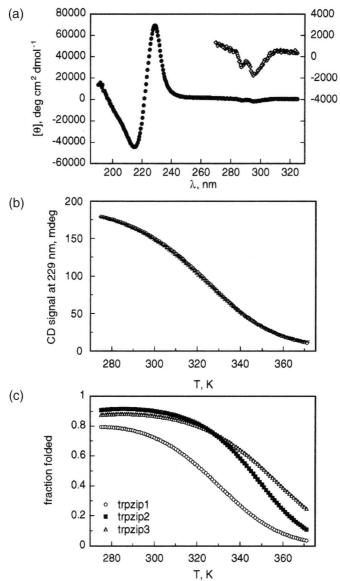

Fig. 10.7. Folding data for trpzips 1 to 3 determined from near- and far-UV CD spectra. a) CD spectrum of trpzip 1, with inset of near-UV CD showing buried aromatic residues (10-fold expansion). b) Reversible thermal denaturation of trpzip 1 monitored by CD at 229 nm (unfolding and refolding curves overlayed). c) Temperature dependence of folding for trpzips 1 to 3 plotted as fraction folded. Taken from Ref. [49].

secondary structure has been less well investigated in terms of a detailed quantitative description. To this end, peptide *β*1 (Figure 10.2) was mutated to introduce Lys-Glu salt bridges at two positions within the hairpin involving substitution of Ser → Glu: one salt bridge is postioned adjacent to the *β*-turn, while the other involves residues close to the N- and C-termini (see Figure 10.2) [54]. Using NMR chemical shift data from Hα resonances we have estimated the net contribution to stability from these two interactions. Although the contribution to stability in each case is small, Hα chemical shifts are extremely sensitive to small shifts in the population of the folded state, as evident from the $\Delta\delta$Hα data in Figure 10.8. On this basis, the individual contributions of these two interactions was estimated (K2-E15, peptide *β*2, and E6-K11, peptide *β*3) compared with their K2-S15 and S6-K11 counterparts (peptide *β*1) and found to be similar (-1.2 and -1.3 kJ mol^{-1} at 298K). When the two salt bridges are introduced simultaneously into the hairpin sequence (*β*4), the energetic contribution of the two interactions together (-3.6 kJ mol^{-1}) is significantly greater than the sum of the individual interactions. This effect is readily apparent from the large increase in $\Delta\delta$Hα values in Figure 10.8, indicating that the contribution of a given interaction appears to depend on the relative stability of the system in which the interaction is being measured. Similar observations have been reported using a disulfide cyclized *β*-hairpin scaffold [55]. The indication is that the strength of the interaction appears to depend on the degree of preorganization of the *β*-hairpin template that pays varying degrees of the entropic cost in bringing the pairs of side chains together.

This is more clearly illustrated by the schematic representation shown in Figure 10.9a that shows the thermodynamic cycle indicating the effects on stability of the introduction of each mutation. Thus, the energetic contribution of the S15/E15 mutation could be measured either from *β*1 ↔ *β*2 or from *β*3 ↔ *β*4. The values obtained are quite different, the latter suggesting that the interaction is more favorable (-1.2 versus -2.3 kJ mol^{-1}). This appears to correlate with the fact that the stability of *β*3 is greater than *β*1 and that the degree of preorganization of the former determines the energetic contribution of the interaction. The same principle is evident when determining the energetics of the E6-K11 interaction. This can be estimated from *β*1 ↔ *β*3 or *β*2 ↔ *β*4. Again, the energetics are quite different (-1.3 versus -2.4 kJ mol^{-1}) with the larger contribution from the *β*2 ↔ *β*4 pair, where the intrinsic stability of the *β*2 reference state is higher.

The NMR data show that the folded conformation of *β*4 is highly populated (> 70%) giving rise to an abundance of cross-strand NOEs (Figure 10.9b) [55]. On the basis of 173 restraints (NOEs and torsion angle restraints from $^3J_{NH-H\alpha}$ values) we have calculated a family of structures compatible with the NMR data (Figure 10.9c). The large number of van der Waals contacts between hydrophobic residues (V, I, and Y) evident from the NOE data leads us to conclude that hydrophobic interactions still provide the overall driving force for folding with structure and stability further consolidated by additional Coulombic interactions from the salt bridges. CD melting curves for hairpin *β*4 in water and various concentrations of methanol (Figure 10.10) show the same characteristics described above from temperature-dependent NMR data for *β*1: a shallow melting curve for *β*4 in water

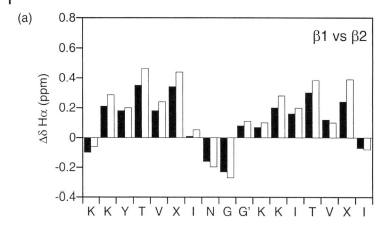
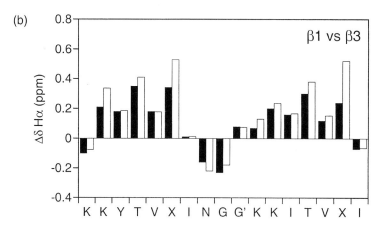
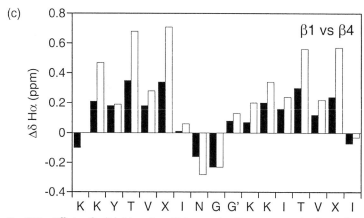

Fig. 10.8. Effects of salt bridges (Lys-Glu) on β-hairpin stability. Hα chemical shift deviation from random coil values (ΔδHα values) for β-hairpin peptides **β2** to **β4** compared with those of the reference hairpin peptide **β1** containing Lys-Ser cross-strand pairs at the X_2-X_{15} and X_6-X_{11} mutation sites (see Figure 10.2). a) **β1** versus **β2**; b) **β1** versus **β3**, and c) **β1** versus **β4**. All data at 298K and pH 5.5. Taken from Ref. [54].

(a)
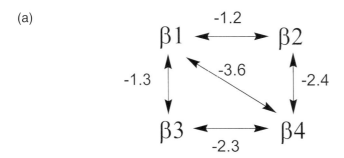

(b)
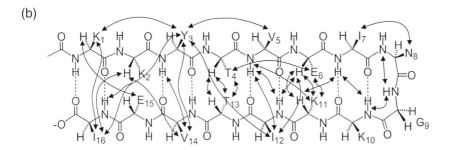

(c)
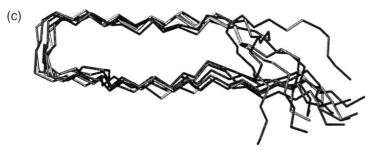

Fig. 10.9. a) Thermodynamic cycle showing the context-dependent energetic contribution to hairpin stability of each electrostatic interaction in the β1 to β4 family of hairpins (Figure 10.2) by comparing relative hairpin stabilities at 298K determined from NMR chemical shift data; ΔΔG values are shown for Lys-Glu interactions at pH 5.5. The stabilizing contribution of each salt bridge is context dependent, showing some degree of cooperativity between the two mutation sites according to the degree of preorganization of the hairpin structure. b) Shematic illustration of the abundance of cross-strand NOEs in hairpin peptide β4. c) Family of five NMR structures of hairpin β4 showing the peptide backbone alignment, with some fraying at the N- and C-termini. Taken from Ref. [54].

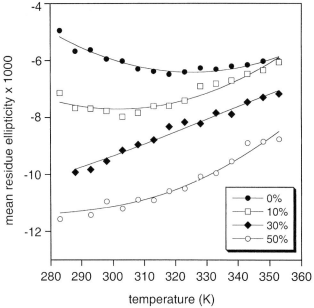

Fig. 10.10. CD melting curves for *β*4 at various concentrations (% v/v) of MeOH, as indicated. The nonlinear least squares fit is shown in each case from which thermodynamic data for folding were determined (see Ref. [54]). The change in heat capacity on folding in 30% and 50% aqueous methanol is assumed to be zero; there is evidence for cold-denaturation in water and 10% methanol. Taken from Ref. [54].

with evidence for cold denaturation is consistent with the hydrophobic interaction again providing the dominant driving force for folding ($\Delta H = +11.9$ kJ mol^{-1} and $\Delta S = +38$ J K^{-1} mol^{-1}), while folding becomes strongly enthalpy driven in 50% aqueous methanol ($\Delta H = -39.8$ kJ mol^{-1} and $\Delta S = -106$ J K^{-1} mol^{-1}).

10.6
Cooperative Interactions in *β*-sheet Peptides: Kinetic Barriers to Folding

Folding kinetics for a *β*-hairpin derived from the C-terminus of the B1 domain of protein G have been described from measurements of tryptophan fluorescence following laser-induced temperature-jump [56, 57]. Kinetic analysis of this peptide (and a dansylated analogue) reveals a single exponential relaxation process with time constant 3.7 ± 0.3 μs. The data indicate a single kinetic barrier separating folded and unfolded states, consistent with a two-state model for folding. Subsequently, the authors developed a statistical mechanical model based on these observations, describing the stability in terms of a minimal numbers of parameters: loss of conformational entropy, backbone stabilization by hydrogen bonding and formation of a stabilizing hydrophobic cluster between three key residues. This model seems sufficient to reproduce all of the features observed experimentally, with a rough, funnel-like energy landscape dominated by two global minima representing the folded and unfolded states. The formation of the hydrophobic cluster appears to be a key folding event. Nucleation by the turn seems most likely, consistent with experimental measurements of loop formation on the timescale of ~1 μs [58]. However, simulation studies by others suggest that folding may proceed by hydrophobic collapse followed by rearrangement to form the hydrophobic cluster, with hydrogen bonds then propagating outward from the cluster in both directions [59]. Such a model does not appear to require a turn-based nucleation event.

10.7
Quantitative Analysis of Peptide Folding

Quantitative analysis of the population of folded β-sheet structures in solution still presents a challenge, largely as a consequence of uncertainties in limiting spectroscopic parameters for the fully folded state. Far UV-CD has been considered to be unreliable as a consequence of the complicating influence of the β-turn conformation and possibly aromatic residues, where present [10]. Added to this, the CD spectrum of β-sheet is intrinsically weak compared with α-helical secondary structure. The trpzip peptides [49] represent the exception to the rule, as discussed above. The use of NMR parameters (Hα chemical shifts, $^3J_{NH-H\alpha}$ values and NOE intensities) to quantify folded populations has been discussed [9–11, 32, 60]. NMR offers the advantage that several independent parameters can be used in quantitative analysis to provide a consensus picture of the folded state. There still appear to be significant discrepancies between CD analysis and NMR, with peptides that appear to be significantly folded by NMR giving rise to a largely random coil CD spectrum. One interpretation of this observation is that in aqueous solution the peptides fold as a collapsed state with an ill-defined hydrogen bonding network dominated by side-chain interactions. Interestingly, in many cases the addition of organic co-solvents changes the CD spectrum dramatically. The interpretation of co-solvent-induced folding is also subject to some uncertainty. Does trifluorethanol or methanol actually significantly perturb the equilibrium between the folded and unfolded states (induce folding), or do these solvents exert their influence by changing the nature of the folded state such as to stabilize interstrand hydrogen bonding interactions without significantly changing the folded population? The latter hypothesis would appear to more readily account for the observation of solvent-induced effects on the CD spectrum, and finds some support from studies of cyclic β-hairpin analogs where the folded population is fixed, but whose CD spectrum undergoes large solvent-induced changes [60].

The use of cyclic β-hairpin analogs has been exploited in a number of studies to generate a fully folded NMR reference state for comparison with the folding of acyclic analogs [61, 62]. Backbone cyclization through amide bond formation or through a disulfide bridge seem to work equally well. Such an approach has been used effectively to measure the thermodynamics of folding of a short hairpin carrying a motif of aromatic residues [62]. The cyclic analogs show a much higher stability than their acyclic counterparts, including significant protection from amide H/D exchange due to enforced interstrand hydrogen bonding. While peptide cyclization seems a worthwhile approach to defining the fully folded state, there may also be some caveats to this approach that have not been tested. When conformational constraints (β-turns) are imposed at both ends of the structure this may affect the intrinsic twist of the two β-strands, resulting in a more pronounced twist in the cyclic analog than in the acyclic hairpin. Further, the latitude for conformational dynamics in the fully folded state will also be different. Both of these factors are likely to influence Hα shifts chemical shifts to some degree.

10.8
Thermodynamics of β-hairpin Folding

The number of β-hairpin model systems is expanding rapidly with a greater focus now on quantitative analysis and thermodynamic characterization. The thermodynamic signature for the folding of **β1** (Figure 10.2) presents an insight into the nature of the stabilizing weak interactions in various solvent milieu [30, 32]. **β1** exhibits the property of cold denaturation, with a maximum in the stability curve occurring at 298K as judged by changes in Hα chemical shift (Figure 10.11a). Such pronounced curvature is clear evidence for entropy-driven folding accompanied by a significant change in heat capacity. Both of these thermodynamic signatures are hallmarks of the hydrophobic effect contributing strongly to hairpin stability. More recent studies by Tatko and Waters [63] have also shown that cold denaturation effects can be observed for simple model peptides, consistent with substantial changes in heat capacity on folding. Further examination of folding in the presence of methanol co-solvent shows that the signature changes such that folding becomes strongly enthalpy driven, and that in 50% aqueous methanol the temperature–stability profile is indicative of the absence of any significant contribution of the hydrophobic effect to folding [32]. The population of the folded state appears to be enhanced by methanol, reflecting similar observations in helical peptides where the phenomenon has been attributed to the effects of the co-solvent destabilizing the unfolded peptide chain so promoting hydrogen bonding interactions in the folded state.

It is unlikely that the folded state of a model β-hairpin peptide resembles a β-sheet in a native protein, with the former sampling a much larger number of conformations of similar energy stabilized by a fluctuating ensemble of transient interactions. IR analysis of the amide I band of **β1** does not show significant differences in the region expected for β-sheet formation (≈ 1630 cm^{-1}) from data on a nonhydrogen-bonded short reference peptide [64]. However, this band does appear under aggregating conditions (Figure 10.11b). In other cases, IR spectral fea-

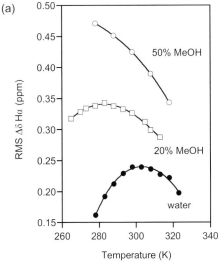 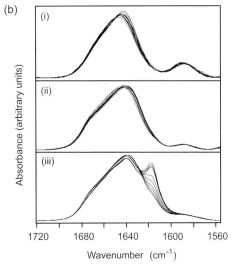

Fig. 10.11. a) Temperature-dependent stability of the hairpin peptide β1 (see Figure 10.2) at pH 5.5 in water, 20% methanol and 50% methanol. Temperature is plotted against the RMS value for the deviation of Hα chemical shifts from random coil values, assuming a two-state folding model. In water the peptide shows "hot" and "cold" denaturation, but in 50% methanol the folded population increases at low temperature. The best fits to the three sets of data are shown; in water folding is slightly entropy-driven with a large negative $\Delta C°_p$ value, while in 50% methanol folding is strongly enthalpy-driven with $\Delta C°_p$ close to zero. In 20% methanol the values are in between. (Reproduced with permission from the *Journal of the American Chemical Society* [32]). b) Variable-temperature FTIR spectra of i) 2 mM aqueous solution of the 8-mer peptide GKKITVSI (C-terminal β-strand of hairpin β1); ii) 2 mM solution of β-hairpin peptide β1; iii) 10 mM solution of β1, all in the temperature range 278–330 K and in D$_2$O solution, pH 5.0 (uncorrected). In iii) the band that appears around 1620 cm^{-1} is formed irreversibly indicative of peptide aggregation. Reproduced from Ref. [64].

tures characteristic of β-hairpins have been shown to form reversibly at relatively high peptide concentrations [65]. Similarly, β-hairpins that appear to be well folded on the basis of various NMR criteria seem to be weakly folded by CD analysis [60]. This discrepancy has been attributed largely to weak interstrand hydrogen bonding interactions in the folded state. Indeed, molecular dynamics simulations using ensemble-averaging approaches or time-averaged NOEs tend to de-emphasize the role of hydrogen bonding between the peptide backbone of the two strands, but emphasize the role of hydrophobic side chains interactions [30, 66, 67]. In all cases described, side chain NOEs across the β-strands support such interactions, however, cross-strand backbone NOEs only imply the possibility of hydrogen bonds since weakly populated folded states lead to only small NH ↔ ND protection factors in water.

In studies (both calorimetric and NMR) of the folding of the C-terminal hairpin from the B1 domain of protein G, enthalpy-driven folding is observed in water

[68]. In contrast with the above data, where the stabilizing hydrophobic interactions involved aliphatic side chains, here an aromatic cluster is responsible for folding. A strongly enthalpy-driven transition is also apparent for the trpzip motifs described above [49] all of which is consistent with π–π interactions stabilizing β-hairpin structures through fundamentally electrostatic interactions rather than through only solvophobic effects. Studies of a designed β-hairpin system, also carrying the same motif of three aromatic residues, demonstrate qualitatively similiar enthalpy-driven folding [62], while the thermodynamics of the N-terminal hairpin component of a designed three-stranded antiparallel β-sheet enables similar conclusions to be drawn (see further below) [69]. The difference in the thermodynamic signature for aliphatic versus aromatic side-chain interactions has been suggested to have its origins in enthalpy–entropy compensation effects, such that enthalpy-driven interactions may be fundamentally a consequence of tighter interfacial interactions, giving rise to stronger electrostatic (enthalpic) interactions [62]. More recent work by Tatko and Waters [70] probing the energetics of aromatic–aliphatic interactions in hairpins suggests that there is a preference for the self-association of aromatics over aromatic–aliphatic interactions, and that the unique nature of this interaction may impart some degree of sequence selectivity. The limited data available from such systems suggest that the thermodynamic driving force for β-hairpin folding is highly dependent on the nature of the side-chain interactions involved. Further quantitative analysis of peptide folding is required to substantiate these hypotheses.

10.9
Multistranded Antiparallel β-sheet Peptides

The natural extension of the earlier studies on β-hairpin peptides was to design three- and four-stranded antiparallel β-sheets using the design principles already discussed, focusing on the importance of turn sequence in defining stability and strand alignment, and employing motifs of interacting side chains already identified to impart stability. An overriding question concerns the extent to which cooperative interactions perpendicular to the strand direction are important in stabilizing these structures (see Figure 10.12a); in other words, how good is a preorganized β-hairpin motif at templating the interaction of a third strand. Several studies have attempted to address this important question. The earliest study described a 24-residue peptide incorporating two NG turns (KKFTLSINGKKY-TISNGKTYITGR) that showed little evidence for folding in water but was significantly stabilised in aqueous methanol solutions [71, 72]. By comparison with a 16-residue β-hairpin analog consisting of the same C-terminal sequence (GKKY-TISNGKTYITGR), it was possible to show that the Hα shift perturbations for the C-terminal β-hairpin were greater in the presence of the interactions of the third strand. Subsequent design strategies, incorporating the D-Pro-Gly loop, together with other Asn-Gly-containing turn sequences have illustrated that it is possible to design structures that fold in water [69, 73–75], some of which show a degree of

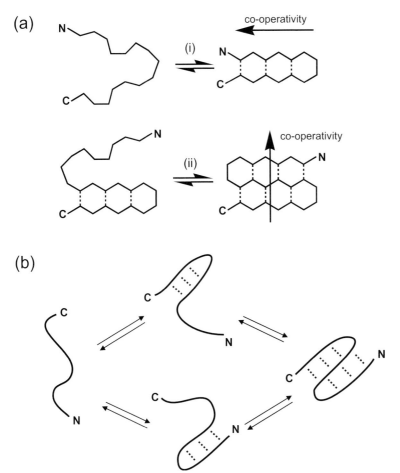

Fig. 10.12. a) Models for the folding of β-hairpin and three-stranded β-sheet peptides illustrating the possibility of cooperative interactions being propagated parallel i) and perpendicular ii) to the β-strand direction. b) Four-state model for the folding of a three-stranded antiparallel β-sheet peptide showing the presence at equilibrium of the unfolded state and partially folded states in which the C-terminal β-hairpin is formed but the N-terminus is disordered, the N-terminal β-hairpin is formed but the C-terminus is disordered, and the fully folded state. The preformed hairpins can act as templates for the folding of the third strand.

cooperative stabilization between different strands. Other studies have reported peptides that fold to three- (and four-) stranded β-sheets in organic solvents [76], but a quantitative dissection of stabilizing interactions between β-strands has not been presented. Redesign of the three-stranded Betanova, following earlier work [74], reported the use of an automated approach using the algorithm PERLA (protein engineering rotamer library algorithm) to define stabilizing and destabilizing single and multiple-residue mutations producing incremental increases in stability over the original design of up to ~4 kJ mol^{-1} [77]. Schenck and Gellman [73] demonstrated cooperative interactions using their D-ProGly to L-ProGly switch, the latter destabilizing one hairpin component selectively. From chemical shift analysis they showed that the individual β-hairpins are cooperatively stabilized by the presence of the third strand. De Alba et al. [75] also reported a designed β-sheet system, but were unable to find convincing evidence for cooperative stabilization of either hairpin by the third strand.

It is clear that folding is not a highly cooperative process in these peptide systems. Thus, the folded three-stranded β-sheet is more than likely in equilibrium with populations of the individual hairpins and the unfolded state (see Figure 10.12b). One quantitative study of cooperative interactions between the strands of a three-stranded β-sheet [69] was based on a designed system incorporating a previously studied β-hairpin with the third strand capable of forming a stabilizing motif of aromatic residues (Figure 10.13a), similar to that already described [17, 62]. The earlier study of the isolated C-terminal β-hairpin showed cold denaturation [30], approximating to two-state unfolding. In the designed three-stranded system, the same hairpin component shows the same cold denaturation profile, however, the N-terminal hairpin (sharing a common central strand) showed increased folding at low temperature. While the first process is characterized by entropy-driven folding, the latter is enthalpy-driven (Figure 10.13b). Clearly, two different thermodynamic profiles are not consistent with a single two-state folding model, but the data could be rationalized in terms of a four-state model in which the individual hairpins with an unfolded tail are also populated (Figure 10.12b). Examination of the folding of the isolated C-terminal hairpin, and comparison of the data with that of the three-stranded analog, shows good evidence that the C-terminal hairpin is cooperatively stabilized by the interaction of the third strand, even though overall folding is not highly cooperative. The data in Figure 10.13b show the temperature dependence of the splitting of the Gly Hα resonances in the two NG turns. Larger values indicate a higher folded population. The splitting for G17 is greater in the three-stranded sheet than for the isolated C-terminal hairpin, while many Hα resonances are further downfield shifted than in the isolated hairpin. The temperature dependence of the stability shows the C-terminal hairpin in both cases to undergo cold denaturation. The N-terminal hairpin carrying the aromatic motif of residues increases in population at low temperature; fitting the data shows the former to be entropy driven and the latter enthalpy driven [69]. Entropic factors seem to be the likely explanation for the small cooperative stabilization effect on the C-terminal hairpin (< 2 kJ mol^{-1}), with each hairpin providing a possible template against which the third strand can interact. With one strand preorganized, the entropic cost of association of an additional strand is largely confined to the associating strand [73, 78]. The nature of the folded state is unlikely to compare with that of a β-sheet in a native protein, more likely, hydrophobic contacts between side chains stabilize a collapsed conformation where interstrand hydrogen bonds may play a minor stabilizing role. These "loosely" defined interactions between side chains,

(a)
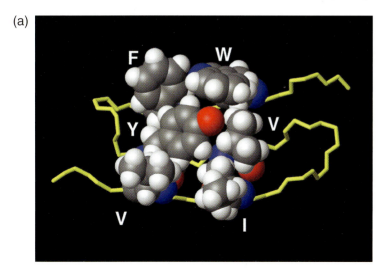

(b)
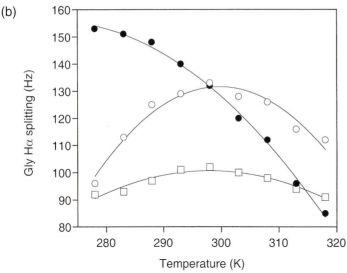

Fig. 10.13. a) Mean NMR structure of the fully folded state of the three-stranded antiparallel β-sheet structure of sequence KGEWTFVNG⁹KYTVSING¹⁷KKITVSI showing the core cluster of hydrophobic residues (underlined). b) Temperature-dependent stability profiles for the various β-hairpin components of the three-stranded sheet using the Hα splitting of Gly9 and Gly17 in the two β-turns: Gly9 (filled circles), Gly17 (open circles). Gly17 (open squares) in the isolated C-terminal hairpin peptide (G⁹KYTVSING¹⁷KKITVSI) is also shown. A larger Gly Hα splitting indicates a higher population of the folded hairpin. Different profiles for Gly9 and Gly17 in the three-stranded sheet show that the peptide cannot be folding via a simple two-state model involving only random coil and fully folded peptide. The data have been fitted assuming the four-state model shown in Figure 10.12b. Reproduced from Ref. [69].

(a)

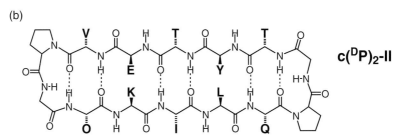

(b)

Fig. 10.14. Structure of the four-stranded β-sheet peptide ᴅPᴅPᴅP-cc (a), and the cyclic reference peptide c(ᴅP)₂-II (b) used for estimating limiting values for the fully folded state of the C-terminal hairpin of ᴅPᴅPᴅP-cc. Adapted from the work of Ref. [78].

rather than a native-like "crystalline" array of hydrogen bonds, may explain why cooperative interactions have only a small effect on overall stability because only a small energy barrier separates the folded from partially or fully unfolded states.

Several of the above studies have attempted to address the issue of whether cooperative interactions are propagated orthogonal to the strand direction as the number of β-strands increases. Gellman and coworkers have examined the influence of strand number on antiparallel β-sheet stability in designed three- and four-stranded structures (see Figure 10.14), using the ᴅ-Pro-Gly β-turn sequence to define β-turn position and strand length. The results are not dissimilar to those described above; a third strand stabilizes an existing two-stranded β-sheet by $\sim$2 kJ mol^{-1}, while a fourth strand seems to confer a similar small stability increment [79].

10.10
Concluding Remarks: Weak Interactions and Stabilization of Peptide β-Sheets

In contrast to the highly cooperative folding behavior characteristic of native globular proteins, simple model β-sheet systems, which lack defining tertiary interactions, do not possess this characteristic. The limited number of three- and four-stranded β-sheet peptides so far described confirms this, although some evidence for cooperative interactions between β-strands have been presented and rationalized on the basis of the entropic benefits associated with adding an additional β-strand to a preformed template. In all cases to date, it seems that designed three- and four-stranded β-sheet structures are in equilibrium with their partially folded β-hairpin components. This is an indication that the energy landscape is relatively flat unlike that for most native proteins where there is a significant energy difference between the single native structure and other conformations. It is interesting to look at examples of the smallest known β-sheet proteins and make comparisons with the designed systems described above.

The WW domains [80–82] form a single folded motif consisting of three strands of antiparallel β-sheet but with well-defined tertiary interactions arising from folding back of the N- and C-termini to form a small compact hydrophobic core. This defining feature appears to allow the WW domains to fold co-operatively despite their small size (some as small as 35 residues) (Figure 10.15). Apart from the desire to be able to design molecules to order with specific tailored properties, the model β-sheet systems described have enabled considerable progress to be made in testing our understanding of the basic design principles of β-sheets and fundamental aspects of weak interactions.

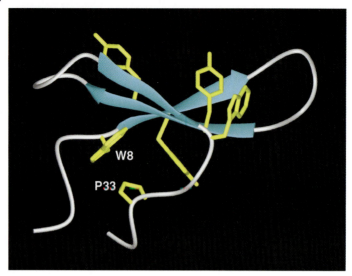

Fig. 10.15. NMR structure of the WW domain of the formin binding protein (PDB code: 1EOI), consisting of a three-stranded antiparallel β-sheet motif; the side chains of conserved residues are shown with tertiary contacts evident between W8 and P33.

References

1 Honig, B. *J. Mol. Biol.* **1999**; *293*, 283–93.
2 Brockwell, D. J., Smith, D. A., Radford, S. E. *Curr. Opin. Struct. Biol.* **2000**; *10*, 16–25.
3 Baldwin, R. L., Rose, G. *Trends Biochem. Sci.* **1999**; *24*, 26–76.
4 Baldwin, R. L., Rose, G. *Trends Biochem. Sci.* **1999**; *24*, 77–83.
5 Chakrabartty, A., Baldwin, R. *Adv. Protein Chem.* **1995**; *46*, 141–76.
6 Munoz, V., Serrano, L. *J. Mol. Biol.* **1995**; *245(3)*, 275–96.
7 O'Neil, K. T., Degrado, W. F. *Science* **1990**; *250(4981)*, 646–51.
8 Hill, R., Degrado, W. *J. Am. Chem. Soc.* **1998**; *120(6)*, 1138–45.
9 Gellman, S. H. *Curr. Opin. Chem. Biol.* **1998**; *2*, 717–25.
10 Ramirez-Alvarado, M., Kortemme, T., Blanco, F. J., Serrano, L. *Bio-org. Med. Chem.* **1999**; *7*, 93–103.
11 Searle, M. S. *J. Chem. Soc. Perkin Trans.* **2001**; *2*, 1011–20.
12 Kemp, D. S., Bowen, B. R., Meundel, C. C. *J. Org. Chem.* **1990**; *55*, 4650–7.
13 Nowick, J. S., Smith, E. M., Pairish, M. *Chem. Soc. Rev.* **1996**; 401–15.
14 Diaz, H., Tsang, K. Y., Choo, D., Espina, J. R., Kelly, J. W. *J. Am. Chem. Soc.* **1993**; *115*, 3790–1.
15 Blanco, F. J., Jimenez, M. A., Herranz, J., Rico, M., Santoro, J., Nieto, J. L. *J. Am. Chem. Soc.* **1993**; *115*, 5887–9.
16 Blanco, F. J., Rivas, G., Serrano, L. *Nat. Struct. Biol.* **1994**; *1*, 584–90.
17 Blanco, F. J., Jimenez, M. A., Pineda, A., Rico, M., Santoro, J., Nieto, J. L. *Biochemistry* **1994**; *33*, 6004–14.
18 Cox, J. P. L., Evans, P. A., Packman, L. C., Williams, D. H., Woolfson, D. N. *J. Mol. Biol.* **1993**; *234*, 483–92.
19 Searle, M. S., Williams, D. H., Packman, L. C. *Nat. Struct. Biol.* **1995**; *2*, 999–1006.

20 Searle, M. S., Zerella, R., Williams, D. H., Packman, L. C. *Protein Eng.* **1996**; *9*, 559–65.
21 Zerella, R., Evans, P. A., Ionides, J. M. C. et al. *Protein Sci.* **1999**; *8*, 1320–31.
22 Zerella, R., Chen, P. Y., Evans, P. A., Raine, A., Williams, D. H. *Protein Sci.* **2000**; *9*, 2142–50.
23 Sibanda, B. L., Blundell, T. L., Thornton, J. M. *J. Mol. Biol.* **1989**; *206*, 759–77.
24 Hutchinson, E. G., Thornton, J. M. *Protein Sci.* **1994**; *3*, 2207–16.
25 Haque, T. S., Gellman, S. H. *J. Am. Chem. Soc.* **1997**; *119*, 2303–4.
26 Stanger, H. E., Gellman, S. H. *J. Am. Chem. Soc.* **1998**; *120*, 4236–7.
27 Ramirez-Alvarado, M., Blanco, F. J., Serrano, L. *Nat. Struct. Biol.* **1996**; *3*, 604–12.
28 De Alba, E., Jimenez, M. A., Rico, M. *J. Am. Chem. Soc.* **1997**; *119*, 175–83.
29 Ramirez-Alvarado, M., Blanco, F. J., Niemann, H., Serrano, L. *J. Mol. Biol.* **1997**; *273*, 898–912.
30 Griffiths-Jones, S. R., Maynard, A. J., Searle, M. S. *J. Mol. Biol.* **1999**; *292*, 1051–69.
31 Maynard, A. J., Searle, M. S. *J. Chem. Soc. Chem. Commun.* **1997**; 1297–8.
32 Maynard, A. J., Sharman, G. J., Searle, M. S. *J. Am. Chem. Soc.* **1998**; *120*, 1996–2007.
33 Swindells, M. B., Macarthur, M. W., Thornton, J. M. *Nat. Struct. Biol.* **1995**; *2*, 596–603.
34 Fiebig, K. M., Schwalbe, H., Buck, M., Smith, L. J., Dobson, C. M. *J. Phys. Chem.* **1996**; *100*, 2661–6.
35 Smith, L. J., Bolin, K. A., Schwalbe, H., Macarthur, M. W., Thornton, J. M., Dobson, C. M. *J. Mol. Biol.* **1996**; *255*, 494–506.
36 Griffiths-Jones, S. R., Sharman, G. J., Maynard, A. J., Searle, M. S. *J. Mol. Biol.* **1998**; *284*, 1597–609.
37 Jourdan, M., Griffiths-Jones, S. R., Searle, M. S. *Eur. J. Biochem.* **2000**; *267*, 3539–48.
38 Neri, D., Billeter, M., Wider, G., Wuthrich, K. *Science* **1992**; *257*, 1559–63.
39 Frank, M., Clore, G. M., Gronenborn, A. *Protein Sci.* **1995**; *4*, 2605–15.
40 Buckler, D., Haas, E., Sheraga, H. *Biochemistry* **1995**; *34*, 15965–78.
41 Shortle, D. *FASEB J.* **1996**; *10*, 27–34.
42 Klein-Seetharaman, J., Oikawa, M., Grimshaw, S. B. et al. *Science* **2002**; *295*, 1719–22.
43 Shortle, D., Ackerman, M. S. *Science* **2001**; *293*, 487–9.
44 De Alba, E., Rico, M., Jimenez, M. A. *Protein Sci.* **1997**; *6*, 2548–60.
45 Wouters, M. A., Curmi, P. M. G. *Protein Struct. Funct. Genet.* **1995**; *22*, 119–31.
46 Hutchinson, E. G., Sessions, R. B., Thornton, J. M., Woolfson, D. N. *Protein Sci.* **1998**; *7*, 2287–300.
47 Espinosa, J. F., Munoz, V., Gellman, S. H. *J. Mol. Biol.* **2001**; *306*, 397–402.
48 Stanger, H. E., Syud, F. A., Espinosa, J. F., Giriat, I., Muir, T., Gellman, S. H. *Proc. Natl Acad. Sci. USA* **2001**; *98*, 12105–20.
49 Cochran, A. G., Skelton, N. J., Starovasnik, M. A. *Proc. Natl Acad. Sci. USA* **2001**; *98*, 5578–83.
50 Karshikoff, A., Ladenstein, R. *Trends Biochem. Sci.* **2001**; *26*, 550–6.
51 Kumar, S., Nussinov, R. *Cell. Mol. Life Sci.* **2001**, *58*, 1216–33.
52 Takano, K., Tsuchimori, K., Yamagata, Y., Yutani, K. *Biochemistry* **2000**, *39*, 12375–81.
53 Loladze, V. V., Ibarra-Molero, B., Sanchez-Ruiz, J., Makhatadze, G. I. *Biochemistry* **1999**, *38*, 16419–23.
54 Ciani, B., Jourdan, M., Searle, M. S. *J. Am. Chem. Soc.* **2003**; *125*, 9038–47.
55 Russell, S. J., Blandl, T., Skelton, N. J., Cochran, A. G. *J. Am. Chem. Soc.* **2003**; *125*, 388–95.
56 Munoz, V., Thompson, P. A., Hofrichter, J., Eaton, W. A. *Nature* **1997**; *390*, 196–199.
57 Munoz, V., Henry, E. R., Hofrichter, J., Eaton, W. A. *Proc. Natl Acad. Sci. USA* **1998**; *95*, 5872–9.
58 Hagen, S. J., Hofrichter, J., Szabo, A., Eaton, W. A. *Proc. Natl Acad. Sci. USA* **1996**; *93*, 11615–17.

59 Dinner, A. R., Lazaridis, T., Karplus, M. *Proc. Natl Acad. Sci. USA* **1999**; *96*, 9068–73.

60 Lacroix, E., Kortemme, T., Lopez de la Paz, M., Serrano, L. *Curr. Opin. Struct. Biol.* 1999; 487–93.

61 Syud, F. A., Espinosa, J. F., Gellman, S. H. *J. Am. Chem. Soc.* **1999**; *121*, 11577–8.

62 Espinosa, J. F., Gellman, S. H. *Angew. Chem.* **2000**; *39*, 2330–3.

63 Tatko, C. D., Waters, M. L. *J. Am. Chem. Soc.* **2004**; *126*, 2018–2034.

64 Colley, C. S., Griffiths-Jones, S. R., George, M. W., Searle, M. S. *J. Chem. Soc. Chem. Commun.* 2000; 593–4.

65 Hilario, J., Kubelka, J., Keiderling, T. A. *J. Am. Chem. Soc.* **2003**; *125*, 7562–7574.

66 Ma, B., Nussinov, R. *J. Mol. Biol.* **2000**; *296*, 1091–104.

67 Wang, H., Sung, S.-S. *Biopolymers* **1999**; *50*, 763–76.

68 Honda, S., Kobayashi, N., Munekata, E. *J. Mol. Biol.* **2000**; *295*, 269–78.

69 Griffiths-Jones, Searle, M. S. *J. Am. Chem. Soc.* **2000**; *122*, 8350–6.

70 Tatko, C. D., Waters, M. L. *J. Am. Chem. Soc.* **2002**; *124*, 9372–3.

71 Sharman, G. J., Searle, M. S. *J. Chem. Soc. Chem. Commun.* 1997; 1955–6.

72 Sharman, G. J., Searle, M. S. *J. Am. Chem. Soc.* **1998**; *120*, 5291–300.

73 Schenck, H. L., Gellman, S. H. *J. Am. Chem. Soc.* **1998**; *120*, 4869–70.

74 Kortemme, T., Ramirez-Alvarado, M., Serrano, L. *Science* **1998**; *281*, 253–6.

75 De Alba, E., Santoro, J., Rico, M., Jimenez, M. A. *Protein Sci.* **1999**; *8*, 854–65.

76 Das, C., Raghothama, S., Balaram, P. *J. Am. Chem. Soc.* **1998**; *120*, 5812–13; Das, C., Raghothama, S., Balaram, P. *Chem. Commun.* **1999**, 967–8.

77 Lopez de la Paz, M., Lacroix, E., Ramirez-Alvarado, M., Serrano, L. *J. Mol. Biol.* **2001**; *312*, 229–46.

78 Finkelstein, A. V. *Proteins* **1991**; *9*, 23–7.

79 Syud, F. A., Stanger, H. E., Schenck Mortell, H. et al. *J. Mol. Biol.* **2003**; *326*, 553–68.

80 Macias, M. J., Gervais, V., Civera, C., Oschkinat, H. *Nat. Struct. Biol.* **2000**; *7*, 375–9.

81 Koepf, E. K., Petrassi, H. M., Ratnaswamy, G., Huff, M. E., Sudol, M., Kelly, J. W. *Biochemistry* **1999**; *38*, 14338–51.

82 Kanelis, V., Rotin, D., Forman-Kay, J. D. *Nat. Struct. Biol.* **2001**; *8*, 407–12.

11
Predicting Free Energy Changes of Mutations in Proteins

Raphael Guerois, Joaquim Mendes, and Luis Serrano

The ability to predict with good accuracy the effect of mutations on protein stability or complex formation is needed in order to design new proteins, as well as to understand the effect of single nucleotide polymorphisms on human health. In the following sections we will summarize what is known and can be used for this purpose. We will start by describing the forces that act on proteins, an understanding of which is a prerequisite to understanding the following section dealing with prediction methods. Finally we will briefly summarize other possible unexpected results of making mutants, such as changes in "in vivo" stability or inducing protein aggregation.

11.1
Physical Forces that Determine Protein Conformational Stability

In this section, we will briefly discuss the physical forces that determine the conformational stability of proteins. This section is intended to serve as a basis for the understanding of the following sections, and no effort has been made to comprehensively cover the relevant literature. However, many excellent reviews that thoroughly cover this topic have been written, several of which are cited in the titles to the various subsections for the interested reader; more specific references are given in the text.

11.1.1
Protein Conformational Stability [1]

Under physiological conditions, proteins exist in a biologically active state (the native (N) state), which is in dynamic equilibrium with a biologically inactive state (the denatured (D) state). The relative thermodynamic stability of the N state with respect to the D state is measured by the folding free energy: $\Delta G_{fold} = G_N - G_D$. Although G_N and G_D are individually large numbers, ΔG_{fold} is always small – typically on the order of 5–20 kcal mol^{-1}. This cancellation of large values is due to enthalpy–entropy compensation. The N state of proteins is, therefore, only marginally stable with respect to the D state, this stability being determined by a delicate balance between the forces that stabilize the N state and those that stabilize the D state.

11.1.2
Structures of the N and D States [2–6]

A detailed understanding of protein conformational stability requires that the structures of all microstates involved be known. Since the N and D states contribute on the same footing to ΔG_{fold}, the structures of both states are, in principle, required.

The N state of proteins consists of a compact structure in which the polypeptide chain folds up upon itself. This structure is frequently described as a single, unique conformation, which is a feature typical of proteins, compared with synthetic polymers. It is this feature that allows the structure of the N state to be determined experimentally with high accuracy, using methods such as X-ray crystallography and nuclear magnetic resonance (NMR) spectroscopy. However, the N state corresponds more accurately to an ensemble of structurally very similar conformations, which manifests itself experimentally as flexibility. Two levels of flexibility can be distinguished [7]: vibrational flexibility, corresponding to structural fluctuations within the same energy well on the energy hypersurface, and conformational flexibility, corresponding to structural transitions between different energy wells. Vibrational flexibility can be observed, for example, in the atomic B-factors of high-resolution X-ray models, whereas conformational flexibility can be observed in these models in the alternative rotamer conformations adopted predominantly by surface side chains.

In contrast to the N state, little is known structurally about the D state of proteins. Proteins can be denatured by a variety of means, including extremes of high or low temperatures, extremes of pH, addition of organic solvents, and addition of chemical denaturants. However, the denatured state relevant to determining protein stability is that which exists in equilibrium with the N state under physiological conditions. It has been extremely difficult to observe this state experimentally, since it exists at very low concentration relative to the N state. Initially, the D state was conceived of as an ensemble of structurally very diverse, totally unfolded conformations of the polypeptide chain, which are highly solvated. These conformations were believed to be completely random and, therefore, the ensemble would be devoid of any structural features, i.e., a random coil state. This conception was based on data from a variety of experimental methods that indicate that almost all tertiary and secondary structure present in the N state is disrupted upon denaturation (probed by fluorescence, UV absorption, and near-UV circular dichroism), and that the polypeptide chain expands to a size approximately equal to that expected for a random coil (probed by viscometry, gel filtration, and small angle X-ray scattering). However, several inconsistencies with the random coil model of the D state appeared: (i) significant amounts of secondary structure can be detected in the D state of some proteins using far-UV circular dichroism even under relatively strong denaturing conditions, (ii) the D state of some proteins is only slightly expanded relative to the N state and much more compact than would be expected for a random coil, (iii) some single-point mutations can drastically alter the compactness of the D state, and (iv) the free energy change arising from some point mutations cannot be rationalized when a random coil model is assumed.

In relation to point (iv), point mutants have been found for which the folding enthalpy is reduced to less than 50% of the wild-type value and the folding entropy increases to a value that almost perfectly compensates the enthalpy change, such that there is practically no difference in ΔG_{fold} of the mutant and wild-type proteins [8]. But by definition, there are no side chain–side chain interactions in a random coil state and, therefore, the change in free energy of the D state accompanying mutation depends only on the identity of the mutated residue in the mutant and wild-type proteins, and not on the particular amino acid sequence; as such, under the random coil assumption, the change in stability produced by a mutation can be rationalized based solely on the structure of the N state. However, to explain the changes in enthalpy and entropy in these mutants solely in terms of effects on the native state requires (1) more than half of the chemical bonds stabilizing the wild type N state to be broken in the mutant N state, and (2) an enormous increase in the vibrational entropy of the mutant N state in order to avoid raising its free energy. Crystallographic studies of the N state of many mutant proteins have invariably shown only modest changes in atomic coordinates and thermal B-factors, making this interpretation highly implausible.

A wealth of studies over the past decade or so have made considerable steps forward in elucidating the structure of the D state. These studies are based on a variety of NMR experiments. In contrast to other spectroscopic and hydrodynamic studies mentioned above, which yield data that correspond to a global property of the protein, NMR spectroscopy provides information about individual residues. The most important of these studies have directly observed the D state under physiological conditions. This has been accomplished by destabilizing the N state relative to the D state, such that both exist in similar concentrations or the D state predominates. The destabilization has been achieved by particular pH conditions [9, 10], a single-point mutation [11, 12], or the deletion of residues at the N- and C-termini [5, 13–15]. The conclusion of these studies is that considerable amounts of secondary structure exist in the D state. This structure can be either native-like or nonnative-like. It is not persistent but, rather, fluctuating. It can exist up to high concentrations of chemical denaturants, but eventually melts out at very high concentrations. Recent studies using paramagnetic probes or residual dipolar couplings have found that, in addition to local secondary structure, a considerable amount of nonlocal structure exists in the D state. For some proteins, the D state appears to adopt the same overall topology of the N state. This nonlocal structure persists even at very high concentrations of denaturant [16]. In the scope of protein engineering it was shown that the specific destabilization of nonnative residual structure in the denatured state could bring about a dramatic stabilization in an SH3 domain [17].

The current conception of the D state under physiological conditions is that of a highly compact ensemble with considerable local and nonlocal structure, which, for some proteins, may already be poised to adopt the N state. The recent systematic survey of 21 well-characterized proteins based on kinetics and protein engineering data also strongly suggests that a majority of proteins studied so far follow such behavior [18].

11.1.3
Studies Aimed at Understanding the Physical Forces that Determine Protein Conformational Stability [1, 2, 8, 19–26]

A large number of studies have been undertaken to assess the relative importance of different physical forces in stabilizing the N and D states of proteins. Two types of study have been performed: (1) studies in which ΔG_{fold} is decomposed into the respective ΔH_{fold} and ΔS_{fold} components and these in turn into various component forces, and (2) studies in which the change in ΔG_{fold} produced by point mutations is decomposed into component forces. These studies provided many important insights and improved the understanding of protein conformational stability considerably. However, some of their conclusions are still very controversial. The controversy arises from several factors: (i) the large enthalpy–entropy compensation accompanying folding leads to a very small ΔG_{fold}, the consequence of which is that a small error in the estimation of a particular force can lead to a stabilizing force being interpreted as being destabilizing, or vice versa, (ii) the impossibility of uniquely decomposing experimental thermodynamic quantities into intramolecular and solvation contributions, (iii) the limited knowledge of the structural details of the D state and the lack of adequate models for it, and (iv) the considerable plasticity of the N state, i.e., its ability to readjust structurally to accommodate point mutations. In spite of the controversy, since ΔG_{fold} is always small, even forces that contribute only a few kcal mol^{-1} make a large contribution to ΔG_{fold} and, thus, tip the balance in favor of the N state or of the D state. Therefore, as a general conclusion that has emerged mainly from mutant studies, each protein uses a different strategy to tweak this stability of the N state, taking advantage of small increases in stability afforded by a particular type of force.

11.1.4
Forces Determining Conformational Stability [1, 2, 8, 19–27]

In their natural environment, most globular proteins are immersed in an aqueous medium. As such, to understand the conformational stability of proteins it does not suffice to consider the intramolecular interactions of the polypeptide chain in the N and D states. One must also consider the intermolecular interactions between the polypeptide chain in each of these states with the water molecules of the aqueous medium, as well as the alterations in the interactions of the water molecules among themselves induced by the presence of the protein. These latter interactions, which contain both enthalpic and entropic components, are usually all collected together in the solvation free energy. Apart from the intramolecular interaction energy and the solvation free energy, one must also take into account the entropy of the N and D states when attempting to understand protein conformational stability.

11.1.5
Intramolecular Interactions

The intramolecular interactions that stabilize the N and D states of a protein are predominantly nonbonded forces. Bonded forces are essential for maintaining the covalent structure of the polypeptide chain, but are less important in determining stability.

11.1.5.1 van der Waals Interactions
The core of proteins in the N state is very densely packed – as dense as in organic crystals. As such, intramolecular van der Waals interactions are very important in stabilizing this state. Since the D state is significantly expanded with respect to the N state, intramolecular van der Waals interactions should be much fewer and weaker because it is a short-ranged interaction and, therefore, should play a much less important role in stabilizing this state.

11.1.5.2 Electrostatic Interactions

Charge–charge interactions Electrostatic interactions between charged groups in proteins can only contribute significantly to ΔG_{fold} if they are nonlocal, i.e., between groups on residues significantly separated in sequence. For neighboring residues, the free energy should be very similar in both the N and D states because charged groups are usually highly exposed in the N state as well.

Salt bridges form in the N state between groups of opposite charge that are in direct contact and simultaneously establish hydrogen bonds with each other. It is frequently argued that buried salt bridges cannot stabilize the N state because the energy gained by the electrostatic interaction, even though high due to the interior of the protein having a low dielectric constant, cannot compensate for the dehydration of two charged groups. However, recent calculations suggest that salt bridges with favourable geometry are likely to be stabilizing anywhere in the protein [28]. On the other hand, it is frequently argued that salt bridge formation at the surface of the protein cannot contribute much to the stability of the N state, because the energy gained by the electrostatic interaction in the aqueous medium of high dielectric constant cannot compensate for the loss of side chain entropy involved in fixing the side chains in the correct orientation for the salt bridge to form. However, if most of the entropy of the side chains is already lost prior to the formation of the salt bridge due to packing with other side chains at the surface, then salt bridge formation does not lead to an entropy loss and is considerably stabilizing.

On a similar note, entropic cooperativity can occur when networks of surface salt bridges are formed. Thus, although formation of the first salt bridge requires the loss of entropy of the two side chains involved, formation of a second salt bridge from a third residue to the already bridged pair requires only loss of the entropy of one side chain. Salt bridge networks are very common in hyperthermophilic proteins, and are thought to be important in stabilizing the N state of these proteins [1].

Long-ranged electrostatic interactions between groups not directly involved in a salt bridge with each other can also contribute to stabilize or destabilize the N state relative to the D state. Thus, too many charges of the same type at the surface of the protein destabilize the N state with respect to the D state, because in the more expanded D state the like charges are further apart and repel each other less strongly. On the other hand, some studies indicate that the distribution of charges at the surface of the protein can be optimized such that the overall long-ranged electrostatic interactions stabilize the N state [29–31].

It is frequently difficult to predict the effect of mutations that aim at increasing protein stability by optimizing electrostatic interactions among charged groups on the surface of the N state of the protein. In fact, the stability increase can be considerably smaller than predicted and, in some cases, a decrease in stability is observed when an increase was predicted. These results suggest that favorable charge–charge interactions occur in the D state, and that the free energy of this state may be decreased even more than that of the N state by the mutations in charged residues that had aimed at improving the electrostatics of the N state [32].

Hydrogen bonding The N state of proteins is extensively hydrogen bonded and intramolecular hydrogen bonding is an important interaction in stabilizing this state. In the D state, intramolecular hydrogen bonds are not persistent and, therefore, probably do not contribute significantly to the stability of this state. Most hydrogen bonds in proteins involve strong hydrogen bond donors and acceptors, and are of type N–H•••O and O–H•••O. However, there is some structural evidence that the α-hydrogen of the polypeptide backbone can participate in C–H•••O hydrogen bonds, which, although somewhat weaker, could make a significant contribution to the stability of the N state [33, 34].

Dipole interactions Molecular mechanic calculations indicate that intramolecular dipole–dipole interactions between polar groups not directly hydrogen bonded to each other can contribute substantially to stabilizing the N state of proteins [22]. These interactions were also found to represent a major stabilizing interaction in the D state, using an extended chain conformation as a model for this state [22]. It is possible that this would also be the case if a more realistic model of the D state were used, since this state is currently thought to contain considerable local and nonlocal structures similar to that in the N state (see above).

Intramolecular charge–dipole interactions can also contribute significantly to stabilizing the N state of proteins. Thus, mutating a residue at the C-terminus of an α-helix to a positively charged amino acid, or mutating a residue at the N-terminus of an α-helix to a negatively charged amino acid, is found to stabilize the

N state of the protein through charge–dipole interactions between the charged residues and the net dipole moment of the helix. Presumably, in these cases, the helix is not formed or only partially formed in the D state and, therefore, the charge–dipole interactions either do not exist or are much weaker in this state.

Aromatic interactions The aromatic groups of aromatic residues have a net electric quadrupole moment. This arises from the electronic distribution being concentrated above and below the aromatic plane, leading to a net negative charge above and below this plane and a net positive charge within the plane and on the aromatic hydrogen atoms. The quadrupole moment is responsible for the particular interactions in which aromatic groups are involved.

Aromatic groups in direct contact can interact with each other through aromatic–aromatic interactions. Two stabilizing geometries are found in proteins: in one, the hydrogen atoms at the edge of one group point towards the center of the other, in the other the two aromatic planes are stacked on top of each other parallel but off-centered. Intramolecular aromatic–aromatic interactions can be considerably strong and can stabilize the N state of proteins significantly. In fact, aromatic residues exist most frequently in the core of native proteins and are in contact with each other more frequently than random.

Aromatic groups can also be involved in cation–π interactions within proteins [35]. In this interaction, a positive ion is in close proximity to the negative partial charge at the center of the aromatic ring, thus interacting favorably with the quadrupole moment. Intramolecular cation–π interactions can contribute considerably to the stability of the N state of proteins.

The aromatic hydrogen atoms can also interact with the sulfur atom of Cys and Met residues through a kind of weak hydrogen bond as has been experimentally shown [36].

Aromatic groups can also participate in a variety of hydrogen bonds of differing strengths [37].

11.1.5.3 Conformational Strain

Conformational strain is a consequence of bonded forces. Presumably, in the D state of proteins strained conformations do not exist because it is expanded and very flexible. Conformational strain can, however, occur in the tightly packed N state of proteins when: (i) a torsion angle or a bond angle deviates significantly from its equilibrium value, (ii) the side chain rotamer selected in the N state is not the one of lowest local energy, and (iii) there is overlap between atoms leading to steric repulsion. The first two of these are coined χ-strain and rotamer strain by Penel and Doig [38], and can make a considerable contribution to destabilizing the N state of proteins. These two types of strain can totally counterbalance the stabilizing effect of a disulfide bridge if this is under strain in the N state. As for the third type of strain, mutations that attempt to fit too large a residue into an internal cavity of a protein can result in so much atomic overlap that the protein does not fold. In less dramatic cases, it can lead to a significant decrease in the stability of the N state.

11.1.6
Solvation

The effects of solvation on protein conformational stability can be quantified through the solvation free energy of the N and the D states, defined as the free energy of transferring the polypeptide chain in each of these states from vacuum into aqueous solution. The free energies of transfer of model compounds indicate that the solvation of polar and charged groups in proteins is strongly favorable, that of aromatic groups is slightly favorable, and that of aliphatic groups is unfavorable. The favorable solvation of polar and charged groups is due to a negative enthalpy associated with favorable van der Waals interactions, hydrogen bond formation, and dipole-dipole or charge-dipole interactions with the solvent molecules. The slightly favorable solvation of aromatic groups is also enthalpic in origin and arises from favorable van der Waals and weak hydrogen bonds to the solvent molecules. The unfavorable solvation of aliphatic groups is due to the hydrophobic effect. This is a complex effect, which is still not completely understood. At room temperature, it is predominantly entropic in nature, involving the ordering of water molecules around the solute, but at other temperatures has both entropic and enthalpic contributions. The solvation data from these model compounds indicates that the solvation of polar and aromatic groups stabilizes the D state more than the N state, because in the D state such groups are more exposed to solvent. In contrast, solvation of aliphatic groups strongly destabilizes the D state with respect to the N state, because a large fraction of these groups is buried in the protein core in the N state but considerably exposed in the D state.

11.1.7
Intramolecular Interactions and Solvation Taken Together

A change in a thermodynamic quantity of folding, such as ΔG_{fold}, ΔH_{fold}, or ΔS_{fold}, resulting from a point mutation cannot be decomposed experimentally into the effects of intramolecular interactions and those of solvation. Therefore, it is common to assess the importance of the combined effects of intramolecular interactions and solvation on protein stability, of the different classes of chemical groups. Aliphatic groups stabilize the N state and destabilize the D state because they form more and stronger intramolecular van der Waals interactions and are excluded from solvent in the core of the protein in the N state. It is frequently argued that the stabilization afforded by the intramolecular van der Waals interactions in the N state is almost exactly offset by the intermolecular van der Waals interactions of the polypeptide chain with water in the D state (contained in the solvation free energy). Therefore, van der Waals interactions as a whole would not contribute to protein conformational stability, and the stability afforded by aliphatic groups is due to the destabilizing effect of hydrophobic solvation in the D state. However, the enthalpy of dissolution of organic crystals in water (as a model of the unfolding of the densely packed protein core) is unfavorable.

Privalov and coworkers have interpreted this result as indicating that the van der Waals interactions in the crystal are stronger than between the organic molecule and water, as a consequence of the much less dense, open structure of water, which leads to fewer and weaker contacts than in the crystal [39]. They extend this explanation to protein stability, arguing that the intramolecular van der Waals interactions in the N state are stronger than the intermolecular van der Waals interactions with water in the D state and, therefore, van der Waals interactions are an important force in stabilizing the N state. However, based on molecular dynamics simulations, Karplus and coworkers suggested a different interpretation of the dissolution data [22]. They find that the intramolecular van der Waals interaction energy in the crystal is approximately the same as the intermolecular van der Waals energy of interaction between the organic molecule and water. Therefore, they argue that van der Waals interactions hardly contribute to protein conformational stability, and that the unfavorable enthalpy of dissolution results from changes in water structure associated with cavity formation.

Although, the net contribution of aliphatic groups is to destabilize the D state through the hydrophobic effect and to stabilize the N state through van der Waals interactions, in some mutants an inverse hydrophobic effect has been observed for very exposed hydrophobic residues in the N state [40]. This has been interpreted as resulting from the hydrophobic residues being more buried in the D state than in the N state.

Aromatic groups probably stabilize the N state more than they do the D state, because they are most frequently buried in the N state establishing favorable intramolecular van der Waals and aromatic interactions, which strongly outweigh the slightly favorable solvation when exposed in the D state. In fact, the very unfavorable free energies of transfer of aromatic compounds from the pure liquid into water, in spite of their favorable solvation free energy, is thought to result from the much more favorable aromatic–aromatic interactions in the pure liquid.

There is still controversy on the contribution of polar groups to protein conformational stability. Some authors defend that the favorable solvation of these groups when exposed to solvent in the D state almost exactly cancels the favorable intramolecular hydrogen bond and dipolar interactions they establish in the N state, and therefore their net contribution to ΔG_{fold} is practically zero. Other authors argue that the intramolecular interactions of the polar groups in the N state outweigh their favorable solvation in the D state, and conclude that polar group interactions contribute about the same to ΔG_{fold} as do the interactions of aliphatic and aromatic groups. Whatever its net contribution to protein stability, hydrogen bonding is always important for achieving a specific fold, since any unsatisfied hydrogen bond donor or acceptor group in the N state of the protein largely destabilizes this state with respect to the D state, in which such groups establish hydrogen bonds with water molecules.

11.1.8
Entropy

The entropy of the D state of proteins is considerably larger than that of the N state, because it consists of a structurally much more diverse ensemble of conformations. As such, entropy stabilizes the D state with respect to the N state. One may consider the entropy of a polypeptide chain to be composed of backbone entropy and side chain entropy, although they are coupled.

Mutants that change the entropy of the backbone in the D state can affect conformational stability substantially. Thus, Xaa-Gly mutations lead to an increased flexibility of the D state, which corresponds to an increase in entropy and, consequently a decrease in G_D. Since the mutation does not affect the entropy of the N state to the same degree, G_N is approximately constant. Therefore, ΔG_{fold} decreases in absolute value, representing a decrease in the stability of the N state with respect to the D state. Xaa-Pro mutations induce exactly the opposite effect. Thus, they lead to a decreased flexibility of the D state, with a consequent decrease in entropy and associated increase in G_D. ΔG_{fold} increases in absolute value, representing an increase in the stability of the N state with respect to the D state. Xaa-Cys mutations designed to form additional disulfide bridges can also drastically reduce the entropy of the D state, thus enhancing stability [41]. However, the disulfide bridge must have the correct geometry in the N state, otherwise conformational strain can totally offset the gain in stabilization afforded by the reduction in entropy of the D state.

The change in side chain entropy upon folding is thought to be a major factor determining α-helix and β-sheet propensities of the various amino acids. The conformational freedom of side chains is lost to a differing degree upon folding from the D to the N state. In the D state, side chains are considerably flexible and can probably access most rotamer states. However, particular interactions in the D state can limit the conformational freedom of side chains [42]. In the N state, totally buried side chains and side chains involved in salt bridges, for example, have very little conformational freedom. However, totally exposed residues may maintain most of the freedom they had in the D state, whereas partially exposed residues may have an intermediate flexibility. Although it is frequently assumed, when interpreting protein conformational stability, that side chain entropy is totally lost upon folding, the differing degrees of side chain flexibility in the N state as well as in the D state should be considered to gain a detailed understanding of the effect of side chain entropy on stability. Clearly, this will only be totally feasible with the advent of more realistic models of the D state.

11.1.9
Cavity Formation

The effect of cavity-forming mutations on protein conformational stability can be difficult to interpret, since it involves a simultaneous change of several interactions. If a hydrophobic cavity that is accessible to solvent is formed, there will be a reduction in the number of intramolecular van der Waals interactions and the exposure of a hydrophobic surface to solvent. All other factors held constant, such mutations will, in general, be destabilizing. If a polar cavity is formed, it is frequently observed that structural water molecules fill the cavity in the mutant. If the group deleted from the wild-type protein was polar, the water molecule(s) may substitute for its interactions and there may be no change in stability. However, if the deleted group was hydrophobic, the exposure of polar groups to solvent, which in the wild-type protein had unsatisfied hydrogen bonding, and the removal of an exposed hydrophobic group will, in general, lead to a stabilization of the mutant.

11.1.10
Summary

Many different forces participate in determining the final stability of a protein, which results from the sum of these forces in the native and denatured states. In general terms, entropy favors the denatured state, while desolvation of apolar groups, van der Waals interactions, hydrogen bonds, and electrostatic interactions favor the native state. Experimentally it has been very difficult to assign absolute values to all these terms, since any mutation will affect many of these terms simultaneously. Thus, breaking a buried H-bond by mutating a Ser into Ala not only breaks the H-bond but also changes the solvation properties of the denatured state, produces a cavity in the native state, and reduces the van der Waals interactions in this state. The native structure can relax, thereby increasing backbone entropy locally, while decreasing side chain entropy by eliminating one degree of freedom of the Ser residue. As a consequence, when protein scientists speak of the H-bond contribution, or of the hydrophobic contribution of an amino acid group, they are not speaking in pure physical terms, but rather are just referring to the main contributor to the interactions being analyzed. Studies over many different proteins and mutants show that all forces contribute to protein stability and that different proteins could use different combinations of forces to achieve stability. In fact, and contrary to what many groups have postulated, it has been shown that it is easier to stabilize a protein through substitutions on the surface than in the protein core [43–45]. A good example of how regular folded structures can be achieved without using hydrophobic amino acids is the formation of β-sheet amyloid fibrils by poly-Gln polypeptides [46]. In this case without any doubt the H-bonding and van der Waals interactions dominate over the hydrophobic contribution.

11.2
Methods for the Prediction of the Effect of Point Mutations on in vitro Protein Stability

11.2.1
General Considerations on Protein Plasticity upon Mutation

Over the past 20 years the protein engineering method was used as an efficient means to probe the chemistry underlying protein structure and interactions. The large amount of thermodynamic data obtained experimentally offers an appealing support for the development and assessment of predictive methods that estimate the effect of point mutation on protein stability. On average, single deletion or conservative point mutations can alter the stability of a protein by 10–20% of its total stability (the upper limit is around 50% of the total stability) and is supposed not to cause too drastic changes in the three-dimensional structure. The observation that the structure is little affected by single deletion or conservative point mutations has been obtained from extensive works of different groups that have systematically solved the structure of the mutated protein [47–52]. The mutation of buried residues is expected to be most deleterious to protein structure and is discussed briefly here. As regards mutants in which a chemical group is deleted, slight adjustments usually partially accommodate changes in side-chain volume, but only to a limited degree [53–55]. In fact, it appears that full compensations are unlikely because it is difficult to reconstitute the equivalent set of interactions that were present in the wild-type structure (Figure 11.1). In the case of a new group inserted in a protein core, the changes in structure tend to accommodate the introduced side chains by rigid-body displacements of groups of linked atoms, achieved through relatively small changes in torsion angles (Figure 11.2). It is rare, for instance, that a side chain close to the site of substitution changes to a different rotamer [56].

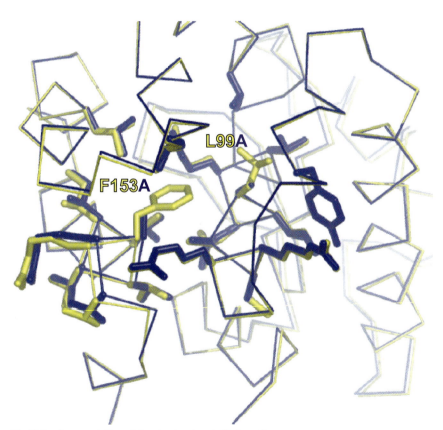

Fig. 11.1. Representation of the structural perturbation of side-chains in the core of the T4 lysozyme upon deletion of chemical groups. The wild-type (PDB: 4LZM) is in yellow and the mutant F153A and L99A (PDB: 1L89) is in blue.

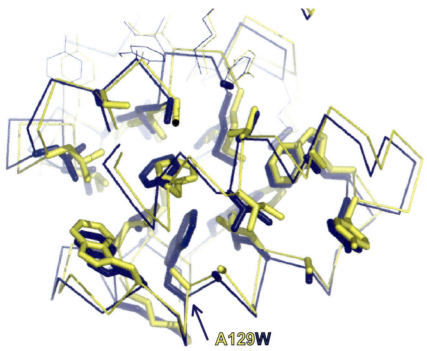

Fig. 11.2. Representation of the structural perturbation of side-chains in the core of the T4 lysozyme upon insertion of chemical groups. The wild-type (PDB: 4LZM) is in yellow and the mutant A129W (PDB: 1QTD) is in blue.

In a sense these observations are consistent with the fact that proteins in their native state are in a deep energy well. Alternative structures are higher in energy and not populated as long as the destabilization effect is limited. Only when several side chains are mutated at one time are wider rearrangements observed in the set of interactions that stabilize protein structure [55, 57]. In cases of extreme perturbation (such as the insertion of amino acids inside a helix), experimental observations reveal an amazing plasticity of proteins that can successfully change and adapt their structures (Figure 11.3, see p. 358) [58]. Such modifications are still far from being predicted by computational methods. Yet, for the majority of the point mutants considered by the predictive algorithms, the assumption that the structure does not undergo large changes is a reasonable one, although some exceptions have also been shown [59].

11.2.2
Predictive Strategies

A wide range of different strategies exist to account for the effect of point mutations on protein stability [60]. The methods can be divided into three classes: (i) statistically based methods (SEEF) that rely on the analysis of either sequence or structure databases to transform probabilities into energies, (ii) protein engineering based methods (EEEF), which are empirically based methods that take advantage of thermodynamic data obtained on protein mutants, and (iii) physically based methods (PEEF) that rely on the derivation of energy terms from model compounds and of rigorous treatment of thermodynamics to calculate free energy changes upon mutation. In the following we consider each of the three approaches, analyzing the bases and limits of their current predictive power and attempting to draw out the critical features that should be implemented for future progress.

11.2.3
Methods

11.2.3.1 From Sequence and Multiple Sequence Alignment Analysis

Several works suggest that the information obtained from multiple sequence alignments could be converted into knowledge-based potentials helpful for guiding protein engineering. Indeed, sequence profiles include in an implicit manner a combination of important protein traits, such as stability, structural specificity, and solubility. For instance, protein engineering experiments based the predictive approach solely on multiple sequence alignments to achieve a successful stabilization of several protein folds [61, 62]. Looking at the reason for the success of this strategy, it was observed that in 25% of the cases the increase in stability could be explained by an enhanced propensity of the substituted residue for the local backbone conformation at the mutated site [61]. This conclusion was reached based on the predictions of local structures using the I-sites library [63]. It would be interesting to use libraries of local structures such as the I-sites library [63] or strategies such as those used in AGADIR [64] to account implicitly for the energetic properties of local structures that cannot be described adequately by the basic α or β secondary structure classes. This may also be an interesting way to calculate the energy of the unfolded state when some residual structure is likely to be formed. However, used as a unique source of information, multiple sequence alignments are unlikely to provide quantitative estimates of the effect of a mutation on protein stability. Yet, if the structure of the protein target is unknown they become the only useful method for rationally modifying the sequence (and might be useful to improve the sample properties as shown in the field of structural genomics [65]).

11.2.3.2 Statistical Analysis of the Structure Databases

Statistical potentials derived from the protein structure database have been widely used in the field of protein structure prediction. As mentioned by Lazaridis and Karplus [66], they have the advantage of being quick to compute, of accounting implicitly for complex effects that may be difficult to consider otherwise, and of perhaps being less sensitive to errors or atom displacement in the protein structure. In the case of protein design however, pairwise statistical potentials for residue–residue interactions may not be adequate for the structural stringency that is required [67, 68]. An elegant method has been developed that may overcome this limitation [69]. A four-body statistical potential, SNAPP, was developed based on the Delaunay tesselation of protein structure. The protein structure is divided into tetrahedra whose edges represent all nearest neighbors of side chain centroids. Quite strong correlations were obtained in the prediction of point mutations of hydrophobic core residues. Yet as discussed further, prediction of the effect of point mutations on buried hydrophobic residues is not the most difficult to achieve. Simple counting methods were shown to perform quite well also [70]. The tessellation strategy might be more difficult to extend to surface accessible positions and a four-body potential method might be less successful for polar residues.

Statistical potentials based on the linear combination of database-derived potentials such as the PoPMuSiC algorithm have also been shown to have good predictive power, especially at fully exposed locations in proteins [71–73]. This program (accessible at http://babylone.ulb.ac.be/popmusic/) was successfully used to guide the engineering of an α_1-antitrypsin protein [74] and helped to identify that cation–π interactions are likely to play a key role in the ability of proteins to undergo domain swapping [75]. The latter is an interesting example of how database-derived potentials can reveal the role of specific interaction types that would have not been detected with atomic potentials that do not explicitly integrate this type of interaction.

11.2.3.3 Helix/Coil Transition Model

The evolution of the algorithms based on the helix/coil transition model illustrates beautifully the extent to which the field of prediction could benefit from protein engineering data. Because of its apparent simplicity, the helix structure has long been recognized to be an excellent model for understanding the fundamentals of protein structure and folding and for probing the effects of mutations on protein stability. The fact the helices are mainly stabilized through local interactions prompted many groups to tackle the analysis of small helical peptides whose sequences could be extensively varied. Short monomeric peptides present less context dependence than proteins. As mentioned in the previous section, changes in local interactions are also likely candidates to contribute to the stability of the native state and the denatured ensemble. Such systems have been successfully used to dissect the contribution of local interactions to helix stability (for reviews see Refs [76] and [77]), and have then been used to analyze β-hairpin and β-sheet formation (for a review see Ref. [77]).

Nowadays spectroscopic analysis, either by circular dichroism (CD) or nuclear magnetic resonance (NMR), has been carried out for a large number of short peptides encompassing helices of natural proteins. While short-range interactions are not enough to determine a single definite helix structure in a peptide, they do determine helix propensities, experimentally observed as helical populations in peptides that are different for every sequence and for each residue in the peptide. Thus, accurate predictions of helix stability require a statistical mechanics approach in which all the possible helical conformations in a peptide and all the energy contributions are taken into account. Its simplest version, postulated by Zimm and Bragg [78], used equilibrium constants characteristic of each amino acid to represent the nucleation and elongation of helical segments. Later versions of the helix/coil transition theory algorithms include detailed interaction terms such as capping interactions, side chain–side chain interactions, $i, i+3$ and $i, i+4$ electrostatic effects, interaction of charged groups with the helix dipole, etc. (for reviews see Refs

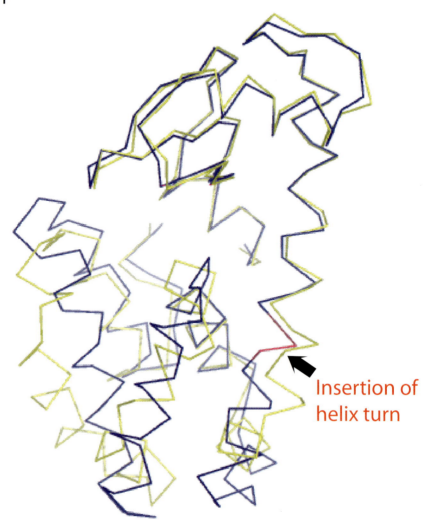

Fig. 11.3. Representation of the structural perturbation induced by the insertion of a triplet of amino acids inside a helix of the T4 lysozyme. The wild-type (PDB: 4LZM) is in yellow and the mutant (PDB: 1L74) is in blue with the inserted region in red.

[76, 77, 79, 80]). These terms have been introduced, either to follow experimental data [64, 76, 81–86], or to correspond to the statistical analysis of the protein database [84]. The latest version of one of these algorithms, AGADIR1s-2, includes: local motifs, a position dependence of the helical propensities for some of the 20 amino acids, an electrostatic model that takes into consideration all electrostatic interactions up to 12 residues in distance in the helix and random coil conformations and the effect of ionic strength [64]. This algorithm predicts with an overall standard deviation value of 6.6% (maximum helix is 100%), the CD helical content of 778 peptides (223 correspond to wild-type and modified protein fragments), as well as the conformational shifts of the C H protons, and the ^{13}C and ^{15}N J coupling values (web access at http://www.embl-heidelberg.de/Services/serrano/agadir/agadir-start.html). The important point for our discussion here is that these methods, originally developed to predict the properties of peptides, could also be used to predict the effect of mutation on protein stability. Several experimental results are now confirming that the use of AGADIR could help understanding or designing the effect of mutations on protein stability [43–45].

The development of an AGADIR-like program for the quantitative prediction of β-hairpin stability has been somewhat less successful than in the case of helices. Several nonaggregating peptides exhibiting a large amount of β-hairpin structure were studied in the 1990s and allowed experimental analysis of the specific interactions both at the turn regions and in the strands [87–91]. In general, a good correlation was found between the experimental data and the statistical analysis of the protein database. As more experimental information is available, the description of hairpin formation in a beta/coil transition algorithm might be undertaken. However, due to the complexity of the interactions present in β-sheets, a residue-based empirical description such as the one used in AGADIR is unlikely to be successful. For example, the energetic contribution of a pairwise interstrand side-chain interaction appears to be dependent on the face of the β-sheet on which it is located, and intrinsic propensities are likely to display a significant position dependence along the strands as well as in the different turn positions, and will vary for the different turn types. Furthermore parallel β-sheets will differ in their properties from the antiparallel β-sheets. Taken together, these different contributions would yield a huge number of parameters to be experimentally determined for any empirical model of β-sheet formation.

As suggested in the next section, a rational approach to quantitatively predict the effect of mutations on β-hairpin and β-sheet stability will be more successful if, in addition to the residue based description used in AGADIR, an atomic-based description is included. Simple peptide models adopting β-hairpin and conformations hence constitute crucial systems for testing and refining such algorithms.

11.2.3.4 Physicochemical Method Based on Protein Engineering Experiments

The physicochemical reasoning methods originate from the rationalization of results obtained from a large quantity of protein engineering data. They have been designed from the assembly of empirical rules that are likely to reflect most of the stability variations in proteins observed upon mutation. The resolution of more than 100 crystal structures of point mutants and the creation of databases gathering all available mutant information such as the Protherm [92] or ASEdb [93] databases constituted decisive steps in pushing forward a large-scale validation of these predictive strategies. We have identified four independent methods recently published that tackled the challenge of developing tools for predicting $\Delta\Delta G$ upon any type of mutation: the SPMP [94, 95], the FOLD-X energy function (FOLDEF) [96] (accessible on the web at http://fold-x.embl-heidelberg.de), the Kortemme et al. method [97], and the Lomize et al. method [98].

The aim of these methods is to estimate free energies from a single structural conformation. This means that they not only consider the potential energies associated with the interactions but also integrate effects such as the protein conformational entropy or the solvation. Because the exploration of the conformational space is limited in these methods, much of these effects have to be accounted for in an implicit manner. Also, the structure of the denatured state is not taken into account explicitly, rather interactions with close neighbours are neglected. Hence, in these methods, an initial fitting procedure is required to adjust the weights of the various explicit and implicit contributions to protein stability. In the training procedure, the structures of the mutants were either modeled from the wild-type structure (FOLDEF, Kortemme et al.) or obtained from experimentally determined X-ray structures (SPMP, Lomize et al.). The fact that most mutations are conservative (involving the deletion or the substitution of chemical groups) prevents large modeling issues.

Table 11.1 summarizes the major factors included in the methods and briefly presents the way the energetic contributions were implemented. A global agreement is reached between all the methods regarding the minimal set of terms that are to be considered. For each term, however, depending on the method, a variable number of parameters were used in the fit. Apart from the work by Lomize et al., in which the different interactions were independently fitted with respect to the atom types (18 parameters all in all), the total number of adjustable parameters is around 8.

A major feature of these methods is that the amplitudes of the various terms are scaled with respect to the solvent proximity. The scaling factor is calculated either from the accessible surface area or the density of protein atoms, which are two highly correlated parameters. In the case of the entropy terms (derived from statistical analyses of the structures database) the scaling accounts implicitly for the fact that, at the surface, side chains or backbone loops are more flexible. In the FOLD-X algorithm, this conformational entropy is also dependent on the existence of hydrogen bond interactions that are likely to freeze the conformations.

The contributions of the solvation effects (derived or consistent with model compound analysis) appear to be approximated in a reasonable manner by the use of the scaling factors. Penalties associated with the burial of polar groups represent the largest contribution to the destabilization of the protein tested in these methods. The implicit treatment of solvation effects is, however, questioned to account for the effect of so-called structural water (or water bridges), which are water molecules making more than two hydrogen bonds with the protein atoms. In the SPMP and FOLD-X methods, an explicit treatment of water bridges was found absolutely required for predicting the $\Delta\Delta G$ of mutants in which a water bridge had been altered. The SPMP method detailed the effect of water even further by considering explicitly the entire water shell network surrounding the protein [99]. This method relied on the X-ray structure of the mutants obtained at very low temperature so that the thermal fluctuations do not impede the observation of water molecules network around the protein. When the structure of the mutant is not known, a fast predictive method for positioning the water molecules is required. The one proposed in [100] was for instance adapted in the FOLD-X method. It is important to note that in their study, Kortemme et al. suggested that the method they used for hydrogen bond calculation allowed an implicit treatment of the effect of these structural water molecules [97].

The role and the way to compute the effects of hydrogen bonds in protein stability has been a matter of debate for some years [101, 102]. In all the methods based on protein engineering analysis, an explicit consideration of the hydrogen bond with distance and angular constraints was used. Other recent studies also suggest that, in order to score a unique structural conformation, the angular and distance constraints inherent to the formation of hydrogen bonds are better modeled by a specific expression than by a global term accounting for all the electrostatic interactions [103, 104].

Different databases were used to test these potentials, which makes a ranking comparison between the methods rather difficult. For the FOLDEF, SPMP, and Kortemme et al. methods a precision ranging from 0.7 to 1 kcal mol^{-1} of standard deviation between prediction and experiments was obtained. The fourth method (Lomize et al.) produced very low rmsd errors between theoretical and experimental (0.4 kcal mol^{-1} and 0.57 kcal mol^{-1} for the training database or the blind test database respectively). Yet the mutant databases used in the latter studies were limited to fully buried residues (test carried out on solvent accessible residues led to mean unsigned errors of 1.4 kcal mol^{-1}) and the blind test database was restricted to 10 point mutants.

Interestingly, each method has specifically focused on at least one term of the energy function: (i) the SPMP method has developed specific analyses for explicit water molecules, (ii) the FOLD-X focused on the way entropy penalties are applied, (iii) the Kortemme et al. method derived a distance and angular dependent expression for hydrogen bonds, and (iv) the Lomidze et al. method used a specific Lennard Jones potential for each atom type interaction. A combination of these specific developments may push forward the current limits of the predictive algorithms. The way hydrogen bonds strength and charge–charge electrostatic interactions are modulated with solvent accessibility and with long-range interactions is probably one of the lines of improvement that can still be achieved in these methods. Yet, to be precise enough, this improvement will require the inclusion of side chain or backbone conformational exploration as was already partly included in the Kortemme et al. method.

An interesting test case to those questions has been proposed in the study of a surface salt bridge K11/E34 in ubiquitin and of the mutant in the reverse orientation E11/K34 [31]. It is clearly shown that long-range interactions contribute to the strength of these salt bridges and that a simple model can predict the effect of the surrounding on this type of interaction. Another track for improvement is related to the selection of the mutants tested in the fitting procedure. The fitting of the adjustable parameters tends to prioritize mutants with a larger $\Delta\Delta G$. Contexts or interactions that involve smaller $\Delta\Delta G$ or interactions that are little represented in the

Tab. 11.1. Summary of the implementation of the energetic factors included in four different methods for predicting free energy changes upon mutation.

	SPMP	FOLD-X	Kortemme et al.	Lomize et al.
Number of adjustable parameters	9	8	8	18
Training database	54 mut. with X-ray structures	339 mut.	743 mut.	106 mut. with X-ray structures
Blind test database	56 mut. with X-ray structures	667 mut. + 82 in complexes	371 mut. + 233 in complexes	10 mut. with X-ray structures
Excluded mutations	High B-factor residues	None	None	Water-accessible residues
Accessibility estimation	ASA based	Protein atomic density from the atom volumic contact parameter	Protein atomic density	ASA based
Van der Waals	None	Set by the atom volumic contact (atomic density)	Lennard-Jones with two parameters for attractive and repulsive terms	Lennard-Jones like. Nine terms depending on atom types
Solvation	ASA-based model (2 parameters)	Derived from average of model compounds analysis	Lazaridis, and Karplus solvation model	ASA-based model (five parameters)
Explicit hydrogen bond	Distance-dependent value. Three terms depend on protein or water interactions	Constant value. Fitted	Angular, distance, and solvent-accessible dependence. Different terms for backbone and side chains. Four parameters	10–12 Lennard-Jones like potential. Three terms depending on atom types
Electrostatics except H-bonds	None	Coulombic dielectric constant varying with atomic density	Coulombic dielectric constant varying with interatomic distance	None
Backbone entropy	Two-states propensity α/β	Propensities from Ramachandran plot statistics	Propensities from Ramachandran plot statistics	None

Side chain entropy	Yes. From statistical table	Yes. From statistical table	None	Yes. From statistical analysis (α/β differences)
Explicit water	Yes from X-ray	Yes. Position of water bridges predicted	None	None
Cavity	Yes	None	None	None
Other specificities		Helix capping term Helix dipole term	Allows side chain flexibility (Monte-Carlo)	Explicit account of torsion strains in side chains

database may tend to be underoptimized. For instance, in the case of mutations made at the surface of a protein, the amplitudes of $\Delta\Delta G$ are on average smaller than those for the buried ones. Also, mutations which alter the cap of a helix or the partners of a structural water are relatively rare, but their corresponding $\Delta\Delta G$ can be quite large. In all those cases, the fitting procedure is likely to limit the optimization of the parameters. A strategy to improve the current algorithms might be to restrict the databases to mutants that specifically address certain classes of interactions, such as the interaction with water bridges or the very solvent exposed ones.

11.2.3.5 Methods Based only on the Basic Principles of Physics and Thermodynamics

These methods use a combination of basic principles of physics and thermodynamics to reproduce with the highest fidelity possible the complex balance of interactions occurring in proteins. With respect to the more empirical methods described above, they restrict as far as possible the number of implicit terms they have to consider in the calculation. The great interest of such approaches is that the origin of any perturbation can then be analyzed and understood in the greatest detail. In principle, the free energy change between two states can be determined by evaluating the partition function in each state. In molecular systems as complex as proteins, however, one cannot sample thoroughly the conformational space to obtain accurate free energies. One of the strategies used instead is to break the change into many smaller steps, so that the free energy change for each step is evaluated by perturbation or thermodynamic integration methods [105–107].

In this method, the ΔG between a state A and B can be expressed with respect to the Hamiltonian of each state, H_A and H_B by defining a λ factor so that

$$H(\lambda) = \lambda H_B + (1 - \lambda) H_A \tag{1}$$

as follows:

1. Using the free energy perturbation method,

$$\Delta G = G_B - G_A = \sum_{\lambda=0}^{1} -RT \ln \langle e^{-\Delta H'/RT} \rangle_\lambda, \quad \text{where } \Delta H' = H_{\lambda+d\lambda} - H_\lambda \tag{2}$$

2. Using the thermodynamic integration method,

$$\Delta G = \int_{\lambda=0}^{\lambda=1} \left\langle \frac{\partial H}{\partial \lambda} \right\rangle_\lambda d\lambda \tag{3}$$

This methodology has been applied to a wide range of free energy calculations dedicated to the study of enzymatic reactions, conformational changes, binding of proteins, nucleic acids or small molecules [108], and also of the effects of mutations on protein stability. Compared with other fields of application, the estimation of protein stability is relatively difficult. Indeed, contrary to the analysis of binding between macromolecules, the estimation of the ΔG of folding requires the energy calculation of a highly flexible and still poorly characterized state, the denatured state ensemble.

Since the early 1990s, molecular dynamics strategies coupled to free energy calculations have been used to analyze the stability changes of about 30 different point mutations in experimentally studied proteins [109–119]. The agreements between experimental and predictions have been good in the large majority of the cases observed, although significant discrepancies could also be observed [120]. The importance of the denatured state in the free energy calculation has been highlighted in many of the most recent calculations. It supports the experimental observation of the existence of residual structure, often with native like properties, in the denatured state of protein as reported in the first section. The systematic calculations carried out on 10 different mutants of chymotrypsin inhibitor 2 (CI2), with various conditions for the representation of the denatured protein emphasize the importance of considering a proper denatured state structure ensemble [118]. This analysis illustrates that when the denatured state is taken into account in the calculation, the predictions are much more successful (correlation going from 0.68 to 0.82). Besides, depending on the model, either an extended tripeptide or a more realistic full-length simulated structure of the denatured state, clear improvements in the simulations could be obtained. Hence, in the case of the hydrophobic mutations considered there, the quality of the prediction was shown to depend on the model of the denatured state.

Similarly, the works by Kitao's group [114, 116] showed a better agreement between experimental values and theory when they changed their reference state from a pentapeptide with an extended conformation to a more realistic pentapeptide model with the native like conformation.

As stated above, the advantage of molecular dynamics simulations or alternative methods that are based on the basic principle of physics is that they allow a detailed inspection of the origin of the stabilization or destabilization. Yet, the way free energies can be divided into energy (such as Lennard Jones, electrostatic, etc.) or residue terms is not a trivial issue. Several theories regarding the path dependence of the free energy components have been advanced [107, 121–125]. Under controlled conditions and proper choice of the λ coupling parameter it seems possible to extract some meaningful information. Unexpected results were observed, for instance, in the case of mutations involving apolar residues. The reason for the loss of stability in the V57A mutant in CI2 or the L56F mutant in T4 lysozyme could be interpreted as a larger variation in the electrostatic component. These particular mutants were not well predicted by the empirical-based methods and this fact emphasizes the advantage of mixing various approaches to get further insight in the origin of protein stabilization or destabilization.

Because it is based on molecular dynamics simulation, the free energy calculation method can account for the change in structure provoked by a mutation and does not depend on the assumption that the structure is unchanged upon mutation. Yet compared with the less sophisticated empirical methods the computer time required to run a calculation on one mutant is several orders of magnitude larger. The most exhaustive study is still limited to eight to ten mutants. It would, however, be of great interest to get results from the calculation of a larger number of point mutants so that a statistical estimation of the error made by these methods could be more easily made. To speed up and overcome the sampling problems mixed strategies using Monte-Carlo sampling coupled to molecular dynamics have been proposed [117].

It is also important to mention alternative strategies using optimized methods for extensive sampling of the side chain conformational space. In this case however the backbone is considered rigid. Instead of extracting free energies from molecular dynamic simulation, these methods account for the conformational entropy component by an extensive sampling of side chain conformations. Different levels of sophistication in the force field can be included in these methods [126–128]. For instance the method in Ref. [128] included an implicit solvation term that allowed the calculation of a large number of point mutants not only at solvent-buried positions but also at solvent-exposed ones. This line of research is, to our view, one of the most promising for the prediction of the effect of point mutation on protein stability since it combines a high sampling quality and a force field that can depend both on physical and statistical terms to achieve the highest prediction rate.

11.3
Mutation Effects on in vivo Stability

A frequently forgotten aspect of proteins is that in general they need to perform their function in vivo. The cell has evolved a series of mechanisms to ensure the removal of proteins that have already carried out their function or of polypeptide chains that are faulty at synthesis or become misfolded. Thus there are some cryptic signals in proteins that are recognized by a cell and determine the half-life of a protein or target for destruction. Knowing about these signals is important since they could determine the outcome of an experiment (i.e., yield, protein activity, etc.).

In general these signals could be classified as follows: (1) N-terminal rules, (2) C-terminal rules, and (3) PEST sequences. However, it is possible that many other spurious signals exist and result in rapid degradation of an otherwise stable protein.

11.3.1
The N-terminal Rule

The N-end rule relates the in vivo half-life of a protein to the identity of its N-terminal residue. Similar but distinct versions of the N-end rule operate in all organisms examined, from mammals to fungi and bacteria. In eukaryotes, the N-end rule pathway is a part of the ubiquitin system. In *Escherichia coli*, three genes

Tab. 11.2. The N-end rule in *E. coli* and *Saccharomyces cerevisiae*.

Residue X in X-gal	In vivo half-life of X-βgal (min)	
	In *E. coli*	In *S. cerevisiae*
Arg	2	2
Lys	2	3
Phe	2	3
Leu	2	3
Trp	2	3
Tyr	2	10
His	>600	3
Ile	>600	30
Asp	>600	3
Glu	>600	30
Asn	>600	3
Gln	>600	10
Cys	>600	>1200
Ala	>600	>1200
Ser	>600	>1200
Thr	>600	>1200
Gly	>600	>1200
Val	>600	>1200
Pro	?	?
Met	>600	>1200

Approximate in vivo half-lives of X-β-galactosidase (βgal) proteins in *E. coli* at 36 °C [130] and in *S. cerevisiae* at 30 °C ([131, 132]). A question mark at Pro indicates its uncertain status in the N-end rule (see text).

From Ref. [129].

have been identified, *clpA*, *clpP*, and *aat*, as part of the degradation machinery involved in the N-terminal rule. In Table 11.2 we show data for the half-life of a protein having different N-terminal amino acids. For a review see Ref. [129].

11.3.2
The C-terminal Rule

In *E. coli* it has been found that when the ribosome encounters a problem in protein synthesis it attaches a 11 amino acid peptide (the ssrA degradation tag) to the C-terminus of the stalled protein chain and then the protein is released and targeted for degradation by the ClpX system [133]. Consequently it has been found that proteins containing an unstructured C-terminal segment with a hydrophobic amino acid at the C-terminus could be targeted for degradation by the ClpX machinery and that the efficiency of the degradation is related to the hydrophobic composition of the last five amino acids [134].

11.3.3
PEST Signals

Polypeptide regions rich in proline (P), glutamic acid (E), serine (S), and threonine (T) (PEST) target intracellular proteins for destruction [135]. Attachment of PEST sequences to proteins results in the fusion protein being degraded from 2- to almost 40-fold faster than the parental molecule [136].

11.4
Mutation Effects on Aggregation

When making a mutation or designing a protein, a frequently forgotten aspect is that the folded protein is in equilibrium with the denatured state and more important that the denatured state, or folding intermediates, can lead to aggregation. It is important at this point to make the distinction with the random process of precipitation, which is related to issues of solubility and does not appear to alter the structure of the proteins, as is seen for example by salting out using ammonium sulfate salts.

Protein aggregation has long been thought of as an unspecific process caused by the formation of nonnative contacts between protein folding intermediates. This view was supported by the wide variation of morphologies of aggregates that were observed by techniques such as electron microscopy and atomic force microscopy. Spectroscopically, two major types of aggregates are observed: the ones that involve the formation of β-sheet [137] and the ones that retain the native spectrum [138], although in a few cases helical aggregates can also be found. Both principal forms of initial aggregates can ultimately be converted to amyloid-like fibers that are invariably rich in cross-beta structure [137]. It seems therefore that nearly all proteins when aggregating will end up in a universal cross-beta structure but that the mechanism of getting there is determined by the competition between retaining the native structure and forming cross-β contacts.

Mutations that favor protein aggregation can be classified into those that destabilize the protein and increase the concentration of aggregation prone species (denatured and intermediate conformations) [139], or those that increase per se the tendency of an amino acid sequence to aggregate [140]. Analysis of the effect of different mutations on the aggregation tendency of unfolded Acp and other peptides [141] has illustrated the importance of secondary structure propensity, hydrophobicity, and charges in the process.

Recently, a statistical mechanics algorithm, TANGO, based on simple physicochemical principles of β-sheet formation extended by the assumption that the core regions of an aggregate are fully buried, has been developed to predict protein aggregating regions. TANGO predicts with surprisingly good accuracy the regions experimentally described to be involved in the aggregation of 176 peptides of over 20 proteins. The predictive capacities of TANGO are further illustrated by two examples: the prediction of the aggregation propensities of Aβ1–40 and Aβ1–42 and in several disease-related mutations of the Alzheimer's β-peptide as well as the prediction of the aggregation profile of human acyl phosphatase. Thus, by capturing the energetics of structural parameters observed to contribute to protein aggregation and taking into account competing conformations, like α-helix and β-turn formation, it is possible to identify protein regions susceptible of promoting protein aggregation. The success of TANGO shows that the underlying mechanism of cross-β formation aggregates is universal [142]. It is expected that improvements of this or similar algorithms developed by other groups in conjunction with other software used to measure the impact of mutations on protein stability will allow us to design new proteins taking into account all the states and interactions.

References

1 VIEILLE, C. and ZEIKUS, G. J. (2001). Hyperthermophilic enzymes: sources, uses, and molecular mechanisms for thermostability. *Microbiol Mol Biol Rev* 65, 1–43.
2 DILL, K. A. (1990). Dominant forces in protein folding. *Biochemistry* 29, 7133–55.
3 DILL, K. A. and SHORTLE, D. (1991). Denatured states of proteins. *Annu Rev Biochem* 60, 795–825.
4 SHORTLE, D. (1996). The denatured state (the other half of the folding equation) and its role in protein stability. *FASEB J* 10, 27–34.
5 SHORTLE, D. (2002). The expanded denatured state: an ensemble of conformations trapped in a locally encoded topological space. *Adv Protein Chem* 62, 1–23.
6 MILLETT, I. S., DONIACH, S., and PLAXCO, K. W. (2002). Toward a taxonomy of the denatured state: small angle scattering studies of unfolded proteins. *Adv Protein Chem* 62, 241–62.
7 KARPLUS, M., ICHIYE, T., and PETTITT, B. M. (1987). Configurational entropy of native proteins. *Biophys J* 52, 1083–5.
8 STURTEVANT, J. M. (1994). The thermodynamic effects of protein mutations. *Curr Opin Struct Biol* 4, 69–78.
9 KORTEMME, T., KELLY, M. J., KAY, L. E., FORMAN-KAY, J., and SERRANO, L. (2000). Similarities between the spectrin SH3 domain denatured state and its folding transition state. *J Mol Biol* 297, 1217–29.
10 LIETZOW, M. A., JAMIN, M., JANE DYSON, H. J., and WRIGHT, P. E. (2002). Mapping long-range contacts in a highly unfolded protein. *J Mol Biol* 322, 655–62.
11 KLEIN-SEETHARAMAN, J., OIKAWA, M., GRIMSHAW, S. B. et al. (2002). Long-range interactions within a nonnative protein. *Science* 295, 1719–22.
12 MAYOR, U., GROSSMANN, J. G., FOSTER, N. W., FREUND, S. M., and FERSHT, A. R. (2003). The denatured state of Engrailed Homeodomain under denaturing and native conditions. *J Mol Biol* 333, 977–91.
13 ACKERMAN, M. S. and SHORTLE, D. (2002). Molecular alignment of denatured states of staphylococcal nuclease with strained polyacrylamide gels and surfactant liquid crystalline phases. *Biochemistry* 41, 3089–95.
14 MOK, Y. K., KAY, C. M., KAY, L. E., and FORMAN-KAY, J. (1999). NOE data demonstrating a compact unfolded state for an SH3 domain under non-denaturing conditions. *J Mol Biol* 289, 619–38.
15 CHOY, W. Y. and FORMAN-KAY, J. D. (2001). Calculation of ensembles of structures representing the unfolded state of an SH3 domain. *J Mol Biol* 308, 1011–32.

16 SHORTLE, D. and ACKERMAN, M. S. (2001). Persistence of native-like topology in a denatured protein in 8 M urea. *Science* 293, 487–9.

17 MOK, Y. K., ELISSEEVA, E. L., DAVIDSON, A. R., and FORMAN-KAY, J. D. (2001). Dramatic stabilization of an SH3 domain by a single substitution: roles of the folded and unfolded states. *J Mol Biol* 307, 913–28.

18 SANCHEZ, I. E. and KIEFHABER, T. (2003). Hammond behavior versus ground state effects in protein folding: evidence for narrow free energy barriers and residual structure in unfolded states. *J Mol Biol* 327, 867–84.

19 CREIGHTON, T. E. (1991). Stability of folded conformations. *Curr Opin Struct Biol* 1, 5–16.

20 MATTHEWS, B. W. (1991). Mutational analysis of protein stability. *Curr Opin Struct Biol* 1, 17–21.

21 FERSHT, A. R. and SERRANO, L. (1993). Principles of protein stability derived from protein engineering experiments. *Curr Opin Struct Biol* 3, 75–83.

22 LAZARIDIS, T., ARCHONTIS, G., and KARPLUS, M. (1995). Enthalpic contribution to protein stability: insights from atom-based calculations and statistical mechanics. *Adv Protein Chem* 47, 231–306.

23 MAKHATADZE, G. I. and PRIVALOV, P. L. (1995). Energetics of protein structure. *Adv Protein Chem* 47, 307–425.

24 HONIG, B. and YANG, A. S. (1995). Free energy balance in protein folding. *Adv Protein Chem* 46, 27–58.

25 JAENICKE, R., SCHURIG, H., BEAUCAMP, N., and OSTENDORP, R. (1996). Structure and stability of hyperstable proteins: glycolytic enzymes from hyperthermophilic bacterium *Thermotoga maritima*. *Adv Protein Chem* 48, 181–269.

26 PACE, C. N., SHIRLEY, B. A., MCNUTT, M., and GAJIWALA, K. (1996). Forces contributing to the conformational stability of proteins. *FASEB J* 10, 75–83.

27 CREIGHTON, T. E. (1993). *Proteins: Structures and Molecular Properties*, 2nd edn. W. H. Freeman and Co., New York.

28 KUMAR, S. and NUSSINOV, R. (1999). Salt bridge stability in monomeric proteins. *J Mol Biol* 293, 1241–55.

29 MARTIN, A., KATHER, I., and SCHMID, F. X. (2002). Origins of the high stability of an in vitro-selected cold-shock protein. *J Mol Biol* 318, 1341–9.

30 TORREZ, M., SCHULTEHENRICH, M., and LIVESAY, D. R. (2003). Conferring thermostability to mesophilic proteins through optimized electrostatic surfaces. *Biophys J* 85, 2845–53.

31 MAKHATADZE, G. I., LOLADZE, V. V., ERMOLENKO, D. N., CHEN, X., and THOMAS, S. T. (2003). Contribution of surface salt bridges to protein stability: guidelines for protein engineering. *J Mol Biol* 327, 1135–48.

32 PACE, C. N., ALSTON, R. W., and SHAW, K. L. (2000). Charge-charge interactions influence the denatured state ensemble and contribute to protein stability. *Protein Sci* 9, 1395–8.

33 SCHEINER, S., KAR, T., and GU, Y. (2001). Strength of the Calpha H•••O hydrogen bond of amino acid residues. *J Biol Chem* 276, 9832–7.

34 CHAMBERLAIN, A. K. and BOWIE, J. U. (2002). Evaluation of C–H•••O hydrogen bonds in native and misfolded proteins. *J Mol Biol* 322, 497–503.

35 PLETNEVA, E. V., LAEDERACH, A. T., FULTON, D. B., and KOSTIC, N. M. (2001). The role of cation-pi interactions in biomolecular association. Design of peptides favoring interactions between cationic and aromatic amino acid side chains. *J Am Chem Soc* 123, 6232–45.

36 VIGUERA, A. R. and SERRANO, L. (1995). Side-chain interactions between sulfur-containing amino acids and phenylalanine in alpha-helices. *Biochemistry* 34, 8771–9.

37 SCHEINER, S., KAR, T., and PATTANAYAK, J. (2002). Comparison of various types of hydrogen bonds involving aromatic amino acids. *J Am Chem Soc* 124, 13257–64.

38 PENEL, S. and DOIG, A. J. (2001). Rotamer strain energy in protein

helices – quantification of a major force opposing protein folding. *J Mol Biol* 305, 961–8.

39 MAKHATADZE, G. I. and PRIVALOV, P. L. (1995). Energetics of protein structure. *Adv Protein Chem* 47, 307–425.

40 BOWLER, B. E., MAY, K., ZARAGOZA, T., YORK, P., DONG, A., and CAUGHEY, W. S. (1993). Destabilizing effects of replacing a surface lysine of cytochrome c with aromatic amino acids: implications for the denatured state. *Biochemistry* 32, 183–90.

41 MATSUMURA, M. and MATTHEWS, B. W. (1991). Stabilization of functional proteins by introduction of multiple disulfide bonds. *Methods Enzymol* 202, 336–56.

42 CHOY, W. Y., SHORTLE, D., and KAY, L. E. (2003). Side chain dynamics in unfolded protein states: an NMR based 2H spin relaxation study of delta131delta. *J Am Chem Soc* 125, 1748–58.

43 MUNOZ, V., CRONET, P., LOPEZ-HERNANDEZ, E., and SERRANO, L. (1996). Analysis of the effect of local interactions on protein stability. *Fold Des* 1, 167–78.

44 VILLEGAS, V., ZURDO, J., FILIMONOV, V. V., AVILES, F. X., DOBSON, C. M., and SERRANO, L. (2000). Protein engineering as a strategy to avoid formation of amyloid fibrils. *Protein Sci* 9, 1700–8.

45 TADDEI, N., CHITI, F., FIASCHI, T. et al. (2000). Stabilisation of alpha-helices by site-directed mutagenesis reveals the importance of secondary structure in the transition state for acylphosphatase folding. *J Mol Biol* 300, 633–47.

46 SCHERZINGER, E., LURZ, R., TURMAINE, M. et al. (1997). Huntingtin-encoded polyglutamine expansions form amyloid-like protein aggregates in vitro and in vivo. *Cell* 90, 549–58.

47 TAKANO, K., SCHOLTZ, J. M., SACCHETTINI, J. C., and PACE, C. N. (2003). The contribution of polar group burial to protein stability is strongly context-dependent. *J Biol Chem* 278, 31790–5.

48 TAKANO, K., YAMAGATA, Y., and YUTANI, K. (2001). Contribution of polar groups in the interior of a protein to the conformational stability. *Biochemistry* 40, 4853–8.

49 TAKANO, K., YAMAGATA, Y., and YUTANI, K. (2003). Buried water molecules contribute to the conformational stability of a protein. *Protein Eng* 16, 5–9.

50 TAKANO, K., YAMAGATA, Y., FUNAHASHI, J., HIOKI, Y., KURAMITSU, S., and YUTANI, K. (1999). Contribution of intra- and intermolecular hydrogen bonds to the conformational stability of human lysozyme. *Biochemistry* 38, 12698–708.

51 YAMAGATA, Y., KUBOTA, M., SUMIKAWA, Y., FUNAHASHI, J., TAKANO, K., FUJII, S., and YUTANI, K. (1998). Contribution of hydrogen bonds to the conformational stability of human lysozyme: calorimetry and X-ray analysis of six tyrosine → phenylalanine mutants. *Biochemistry* 37, 9355–62.

52 MATTHEWS, B. W. (1995). Studies on protein stability with T4 lysozyme. *Adv Protein Chem* 46, 249–78.

53 TAKANO, K., YAMAGATA, Y., FUJII, S., and YUTANI, K. (1997). Contribution of the hydrophobic effect to the stability of human lysozyme: calorimetric studies and X-ray structural analyses of the nine valine to alanine mutants. *Biochemistry* 36, 688–98.

54 ERIKSSON, A. E., BAASE, W. A., ZHANG et al. (1992). Response of a protein structure to cavity-creating mutations and its relation to the hydrophobic effect. *Science* 255, 178–83.

55 BALDWIN, E., XU, J., HAJISEYEDJAVADI, O., BAASE, W. A., and MATTHEWS, B. W. (1996). Thermodynamic and structural compensation in "size-switch" core repacking variants of bacteriophage T4 lysozyme. *J Mol Biol* 259, 542–59.

56 LIU, R., BAASE, W. A., and MATTHEWS, B. W. (2000). The introduction of strain and its effects on the structure and stability of T4 lysozyme. *J Mol Biol* 295, 127–45.

57 GASSNER, N. C., BAASE, W. A., and MATTHEWS, B. W. (1996). A test of the "jigsaw puzzle" model for protein folding by multiple methionine substitutions within the core of T4 lysozyme. *Proc Natl Acad Sci USA* 93, 12155–8.

58 VETTER, I. R., BAASE, W. A., HEINZ, D. W., XIONG, J. P., SNOW, S., and MATTHEWS, B. W. (1996). Protein structural plasticity exemplified by insertion and deletion mutants in T4 lysozyme. *Protein Sci* 5, 2399–415.

59 CONSONNI, R., SANTOMO, L., FUSI, P., TORTORA, P., and ZETTA, L. (1999). A single-point mutation in the extreme heat- and pressure-resistant sso7d protein from sulfolobus solfataricus leads to a major rearrangement of the hydrophobic core. *Biochemistry* 38, 12709–17.

60 MENDES, J., GUEROIS, R., and SERRANO, L. (2002). Energy estimation in protein design. *Curr Opin Struct Biol* 12, 441–6.

61 RATH, A. and DAVIDSON, A. R. (2000). The design of a hyperstable mutant of the Abp1p SH3 domain by sequence alignment analysis. *Protein Sci* 9, 2457–69.

62 WANG, Q., BUCKLE, A. M., and FERSHT, A. R. (2000). Stabilization of GroEL minichaperones by core and surface mutations. *J Mol Biol* 298, 917–26.

63 BYSTROFF, C. and BAKER, D. (1998). Prediction of local structure in proteins using a library of sequence-structure motifs. *J Mol Biol* 281, 565–77.

64 LACROIX, E., VIGUERA, A. R., and SERRANO, L. (1998). Elucidating the folding problem of alpha-helices: local motifs, long-range electrostatics, ionic-strength dependence and prediction of NMR parameters. *J Mol Biol* 284, 173–91.

65 PEDELACQ, J. D., PILTCH, E., LIONG, E. C. et al. (2002). Engineering soluble proteins for structural genomics. *Nat Biotechnol* 20, 927–32.

66 LAZARIDIS, T. and KARPLUS, M. (2000). Effective energy functions for protein structure prediction. *Curr Opin Struct Biol* 10, 139–45.

67 BETANCOURT, M. R. and THIRUMALAI, D. (1999). Pair potentials for protein folding: choice of reference states and sensitivity of predicted native states to variations in the interaction schemes. *Protein Sci* 8, 361–9.

68 THOMAS, P. D. and DILL, K. A. (1996). An iterative method for extracting energy-like quantities from protein structures. *Proc Natl Acad Sci USA* 93, 11628–33.

69 CARTER, C. W., JR., LEFEBVRE, B. C., CAMMER, S. A., TROPSHA, A., and EDGELL, M. H. (2001). Four-body potentials reveal protein-specific correlations to stability changes caused by hydrophobic core mutations. *J Mol Biol* 311, 625–38.

70 SERRANO, L., KELLIS, J. T., JR., CANN, P., MATOUSCHEK, A., and FERSHT, A. R. (1992). The folding of an enzyme. II. Substructure of barnase and the contribution of different interactions to protein stability. *J Mol Biol* 224, 783–804.

71 GILIS, D. and ROOMAN, M. (2000). PoPMuSiC, an algorithm for predicting protein mutant stability changes: application to prion proteins. *Protein Eng* 13, 849–56.

72 GILIS, D. and ROOMAN, M. (1997). Predicting protein stability changes upon mutation using database-derived potentials: solvent accessibility determines the importance of local versus non-local interactions along the sequence. *J Mol Biol* 272, 276–90.

73 GILIS, D. and ROOMAN, M. (1996). Stability changes upon mutation of solvent-accessible residues in proteins evaluated by database-derived potentials. *J Mol Biol* 257, 1112–26.

74 GILIS, D., MCLENNAN, H. R., DEHOUCK, Y., CABRITA, L. D., ROOMAN, M., and BOTTOMLEY, S. P. (2003). In vitro and in silico design of alpha1-antitrypsin mutants with different conformational stabilities. *J Mol Biol* 325, 581–9.

75 DEHOUCK, Y., BIOT, C., GILIS, D., KWASIGROCH, J. M., and ROOMAN, M. (2003). Sequence-structure signals of

3D domain swapping in proteins. *J Mol Biol* 330, 1215–25.
76. CHAKRABARTTY, A. and BALDWIN, R. L. (1995). Stability of alpha-helices. *Adv Protein Chem* 46, 141–76.
77. SERRANO, L. (2000). The relationship between sequence and structure in elementary folding units. *Adv Protein Chem* 53, 49–85.
78. ZIMM, B. H. and BRAGG, J. K. (1959). Theory of the phase transition between helix and random coil in polypeptide chains. *J Chem Phys* 31, 526–535.
79. MUNOZ, V. and SERRANO, L. (1995). Helix design, prediction and stability. *Curr Opin Biotechnol* 6, 382–6.
80. DOIG, A. J. (2002). Recent advances in helix-coil theory. *Biophys Chem* 101–102, 281–93.
81. LIFSON, S. and ROIG, A. (1961). On the helix-coil transition in polypeptides. *J Chem Phys* 34, 1963–74.
82. LOMIZE, A. L. and MOSBERG, H. I. (1997). Thermodynamic model of secondary structure for alpha-helical peptides and proteins. *Biopolymers* 42, 239–69.
83. ANDERSEN, N. H. and TONG, H. (1997). Empirical parameterization of a model for predicting peptide helix/coil equilibrium populations. *Protein Sci* 6, 1920–36.
84. MISRA, G. P. and WONG, C. F. (1997). Predicting helical segments in proteins by a helix-coil transition theory with parameters derived from a structural database of proteins. *Proteins* 28, 344–59.
85. MUNOZ, V. and SERRANO, L. (1994). Elucidating the folding problem of helical peptides using empirical parameters. *Nat Struct Biol* 1, 399–409.
86. MUNOZ, V. and SERRANO, L. (1995). Analysis of $i, i+5$ and $i, i+8$ hydrophobic interactions in a helical model peptide bearing the hydrophobic staple motif. *Biochemistry* 34, 15301–6.
87. RAMIREZ-ALVARADO, M., BLANCO, F. J., and SERRANO, L. (1996). De novo design and structural analysis of a model beta-hairpin peptide system. *Nat Struct Biol* 3, 604–12.
88. DE ALBA, E., RICO, M., and JIMENEZ, M. A. (1997). Cross-strand side-chain interactions versus turn conformation in beta-hairpins. *Protein Sci* 6, 2548–60.
89. RAMIREZ-ALVARADO, M., BLANCO, F. J., NIEMANN, H., and SERRANO, L. (1997). Role of beta-turn residues in beta-hairpin formation and stability in designed peptides. *J Mol Biol* 273, 898–912.
90. DE ALBA, E., RICO, M., and JIMENEZ, M. A. (1999). The turn sequence directs beta-strand alignment in designed beta-hairpins. *Protein Sci* 8, 2234–44.
91. LACROIX, E., KORTEMME, T., LOPEZ DE LA PAZ, M., and SERRANO, L. (1999). The design of linear peptides that fold as monomeric beta-sheet structures. *Curr Opin Struct Biol* 9, 487–93.
92. GROMIHA, M. M., UEDAIRA, H., AN, J., SELVARAJ, S., PRABAKARAN, P., and SARAI, A. (2002). ProTherm, thermodynamic database for proteins and mutants: developments in version 3.0. *Nucleic Acids Res* 30, 301–2.
93. THORN, K. S. and BOGAN, A. A. (2001). ASEdb: a database of alanine mutations and their effects on the free energy of binding in protein interactions. *Bioinformatics* 17, 284–5.
94. TAKANO, K., OTA, M., OGASAHARA, K., YAMAGATA, Y., NISHIKAWA, K., and YUTANI, K. (1999). Experimental verification of the 'stability profile of mutant protein' (SPMP) data using mutant human lysozymes. *Protein Eng* 12, 663–72.
95. FUNAHASHI, J., TAKANO, K., and YUTANI, K. (2001). Are the parameters of various stabilization factors estimated from mutant human lysozymes compatible with other proteins? *Protein Eng* 14, 127–34.
96. GUEROIS, R., NIELSEN, J. E., and SERRANO, L. (2002). Predicting changes in the stability of proteins and protein complexes: a study of more than 1000 mutations. *J Mol Biol* 320, 369–87.
97. KORTEMME, T. and BAKER, D. (2002).

A simple physical model for binding energy hot spots in protein-protein complexes. *Proc Natl Acad Sci USA* 99, 14116–21.

98 LOMIZE, A. L., REIBARKH, M. Y., and POGOZHEVA, I. D. (2002). Interatomic potentials and solvation parameters from protein engineering data for buried residues. *Protein Sci* 11, 1984–2000.

99 FUNAHASHI, J., TAKANO, K., YAMAGATA, Y., and YUTANI, K. (2002). Positive contribution of hydration structure on the surface of human lysozyme to the conformational stability. *J Biol Chem* 277, 21792–800.

100 PITT, W. R. and GOODFELLOW, J. M. (1991). Modelling of solvent positions around polar groups in proteins. *Protein Eng* 4, 531–7.

101 HONIG, B. and YANG, A. S. (1995). Free energy balance in protein folding. *Adv Protein Chem* 46, 27–58.

102 MYERS, J. K. and PACE, C. N. (1996). Hydrogen bonding stabilizes globular proteins. *Biophys J* 71, 2033–9.

103 KORTEMME, T., MOROZOV, A. V., and BAKER, D. (2003). An orientation-dependent hydrogen bonding potential improves prediction of specificity and structure for proteins and protein–protein complexes. *J Mol Biol* 326, 1239–59.

104 MOROZOV, A. V., KORTEMME, T., and BAKER, D. (2003). Evaluation of models of electrostatic interactions in proteins. *J Phys Chem B* 107, 2075–90.

105 BEVERIDGE, D. L. and DICAPUA, F. M. (1989). Free energy via molecular simulation: applications to chemical and biomolecular systems. *Annu Rev Biophys Biophys Chem* 18, 431–92.

106 KOLLMAN, P. A. (1993). Free energy calculations: applications to chemical and biochemical phenomena. *Chem Rev* 93, 2395–417.

107 BRADY, G. P. and SHARP, K. A. (1995). Decomposition of interaction free energies in proteins and other complex systems. *J Mol Biol* 254, 77–85.

108 WANG, W., DONINI, O., REYES, C. M., and KOLLMAN, P. A. (2001). Biomolecular simulations: recent developments in force fields, simulations of enzyme catalysis, protein–ligand, protein–protein, and protein–nucleic acid noncovalent interactions. *Annu Rev Biophys Biomol Struct* 30, 211–43.

109 TIDOR, B. and KARPLUS, M. (1991). Simulation analysis of the stability mutant R96H of T4 lysozyme. *Biochemistry* 30, 3217–28.

110 PREVOST, M., WODAK, S. J., TIDOR, B., and KARPLUS, M. (1991). Contribution of the hydrophobic effect to protein stability: analysis based on simulations of the Ile-96----Ala mutation in barnase. *Proc Natl Acad Sci USA* 88, 10880–4.

111 SNEDDON, S. F. and TOBIAS, D. J. (1992). The role of packing interactions in stabilizing folded proteins. *Biochemistry* 31, 2842–6.

112 YAMAOTSU, N., MORIGUCHI, I., KOLLMAN, P. A., and HIRONO, S. (1993). Molecular dynamics study of the stability of staphylococcal nuclease mutants: component analysis of the free energy difference of denaturation. *Biochim Biophys Acta* 1163, 81–8.

113 YAMAOTSU, N., MORIGUCHI, I., and HIRONO, S. (1993). Estimation of stabilities of staphylococcal nuclease mutants (Met32 → Ala and Met32 → Leu) using molecular dynamics/free energy perturbation. *Biochim Biophys Acta* 1203, 243–50.

114 SUGITA, Y., KITAO, A., and GO, N. (1998). Computational analysis of thermal stability: effect of Ile → Val mutations in human lysozyme. *Fold Des* 3, 173–81.

115 SUGITA, Y. and KITAO, A. (1998). Improved protein free energy calculation by more accurate treatment of nonbonded energy: application to chymotrypsin inhibitor 2, V57A. *Proteins* 30, 388–400.

116 SUGITA, Y. and KITAO, A. (1998). Dependence of protein stability on the structure of the denatured state: free energy calculations of I56V mutation in human lysozyme. *Biophys J* 75, 2178–87.

117 PITERA, J. W. and KOLLMAN, P. A. (2000). Exhaustive mutagenesis in

silico: multicoordinate free energy calculations on proteins and peptides. *Proteins* 41, 385–97.
118. PAN, Y. and DAGGETT, V. (2001). Direct comparison of experimental and calculated folding free energies for hydrophobic deletion mutants of chymotrypsin inhibitor 2: free energy perturbation calculations using transition and denatured states from molecular dynamics simulations of unfolding. *Biochemistry* 40, 2723–31.
119. FUNAHASHI, J., SUGITA, Y., KITAO, A., and YUTANI, K. (2003). How can free energy component analysis explain the difference in protein stability caused by amino acid substitutions? Effect of three hydrophobic mutations at the 56th residue on the stability of human lysozyme. *Protein Eng* 16, 665–671.
120. SPENCER, D. S. and STITES, W. E. (1996). The M32L substitution of staphylococcal nuclease: disagreement between theoretical prediction and experimental protein stability. *J Mol Biol* 257, 497–9.
121. MARK, A. E. and VAN GUNSTEREN, W. F. (1994). Decomposition of the free energy of a system in terms of specific interactions. Implications for theoretical and experimental studies. *J Mol Biol* 240, 167–76.
122. SMITH, P. A. and VAN GUNSTEREN, W. F. (1994). When are free energy components meaningful? *J Phys Chem* 98, 13735–40.
123. BORESCH, S., ARCHONTIS, G., and KARPLUS, M. (1994). Free energy simulations: the meaning of the individual contributions from a component analysis. *Proteins* 20, 25–33.
124. BORESCH, S. and KARPLUS, M. (1995). The meaning of component analysis: decomposition of the free energy in terms of specific interactions. *J Mol Biol* 254, 801–7.
125. BRADY, G. P., SZABO, A., and SHARP, K. A. (1996). On the decomposition of free energies. *J Mol Biol* 263, 123–5.
126. LEE, C. and LEVITT, M. (1991). Accurate prediction of the stability and activity effects of site-directed mutagenesis on a protein core. *Nature* 352, 448–51.
127. LEE, C. (1994). Predicting protein mutant energetics by self-consistent ensemble optimization. *J Mol Biol* 236, 918–39.
128. MENDES, J., BAPTISTA, A. M., CARRONDO, M. A., and SOARES, C. M. (2001). Implicit solvation in the self-consistent mean field theory method: sidechain modelling and prediction of folding free energies of protein mutants. *J Comput Aided Mol Des* 15, 721–40.
129. VARSHAVSKY, A. (1996). The N-end rule: functions, mysteries, uses. *Proc Natl Acad Sci USA* 93, 12142–9.
130. TOBIAS, J. W., SHRADER, T. E., ROCAP, G., and VARSHAVSKY, A. (1991). The N-end rule in bacteria. *Science* 254, 1374–7.
131. BACHMAIR, A., FINLEY, D., and VARSHAVSKY, A. (1986). In vivo half-life of a protein is a function of its amino-terminal residue. *Science* 234, 179–86.
132. BACHMAIR, A. and VARSHAVSKY, A. (1989). The degradation signal in a short-lived protein. *Cell* 56, 1019–32.
133. FLYNN, J. M., LEVCHENKO, I., SEIDEL, M., WICKNER, S. H., SAUER, R. T., and BAKER, T. A. (2001). Overlapping recognition determinants within the ssrA degradation tag allow modulation of proteolysis. *Proc Natl Acad Sci USA* 98, 10584–9.
134. FLYNN, J. M., NEHER, S. B., KIM, Y. I., SAUER, R. T., and BAKER, T. A. (2003). Proteomic discovery of cellular substrates of the ClpXP protease reveals five classes of ClpX-recognition signals. *Mol Cell* 11, 671–83.
135. ROGERS, S., WELLS, R., and RECHSTEINER, M. (1986). Amino acid sequences common to rapidly degraded proteins: the PEST hypothesis. *Science* 234, 364–8.
136. LOETSCHER, P., PRATT, G., and RECHSTEINER, M. (1991). The C terminus of mouse ornithine decarboxylase confers rapid degradation on dihydrofolate reductase. Support for the pest hypothesis. *J Biol Chem* 266, 11213–20.

137 Dobson, C. M. (2002). Getting out of shape. *Nature* 418, 729–30.

138 Bousset, L., Thomson, N. H., Radford, S. E., and Melki, R. (2002). The yeast prion Ure2p retains its native alpha-helical conformation upon assembly into protein fibrils in vitro. *EMBO J* 21, 2903–11.

139 Booth, D. R., Sunde, M., Bellotti, V. et al. (1997). Instability, unfolding and aggregation of human lysozyme variants underlying amyloid fibrillogenesis. *Nature* 385, 787–93.

140 Chiti, F., Taddei, N., Bucciantini, M., White, P., Ramponi, G., and Dobson, C. M. (2000). Mutational analysis of the propensity for amyloid formation by a globular protein. *EMBO J* 19, 1441–9.

141 Chiti, F., Stefani, M., Taddei, N., Ramponi, G., and Dobson, C. M. (2003). Rationalization of the effects of mutations on peptide and protein aggregation rates. *Nature* 424, 805–8.

142 Fernandez-Escamilla, A. M., Rousseau, F., Schymkowitz, J., and Serrano, L. (2004). Prediction of sequence-dependent and mutational effects on the aggregation of peptides and proteins. *Nat Biotechnol* 22, 1302–1306.

Part I
2 Dynamics and Mechanisms of
Protein Folding Reactions

12.1
Kinetic Mechanisms in Protein Folding

Annett Bachmann and Thomas Kiefhaber

12.1.1
Introduction

The understanding of protein folding landscapes requires the characterization of local saddle points (transition states) and minima (intermediates) on the free energy surface between the ensemble of unfolded states and the native protein. This chapter discusses experimental approaches to analyze kinetic data for protein folding reactions, to identify folding intermediates and to assign their role in the folding process. Chapter 12.2 will describe methods used to obtain thermodynamic and structural information on transition states. The methods discussed in these chapters mainly use concepts from classical reaction kinetics and from physical organic chemistry to analyze the dynamics of protein folding and unfolding. It could be argued that these methods may not be applicable to the protein folding process, which represents a conformational search on a complex multidimensional free energy surface. However, results from a variety of different studies suggest that efficient folding proceeds through partially folded intermediates and that protein folding in solution encounters major free energy barriers. In addition, protein folding transition states are usually structurally well-defined local maxima along different testable reaction coordinates [1] (see Chapter 12.2). This suggests that we can indeed derive information on protein folding barriers using the classical concepts.

12.1.2
Analysis of Protein Folding Reactions using Simple Kinetic Models

The aim of kinetic studies is trying to find the minimal model capable of describing the experimental data by ruling out as many alternative models as possible. It is thereby assumed that the mechanism involves a finite number of kinetic species that are separated by energy barriers significantly larger than the thermal energy ($>5k_BT$), but each kinetic species may consist of different conformations in rapid equilibrium. Thus, the transitions between large ensembles of states can be treated

Protein Folding Handbook. Part I. Edited by J. Buchner and T. Kiefhaber
Copyright © 2005 WILEY-VCH Verlag GmbH & Co. KGaA, Weinheim
ISBN: 3-527-30784-2

using classical reaction kinetics and the interconversion between two species, X_i and X_j, can be described with microscopic rate constants for the forward (k_{ij}) and reverse (k_{ji}) reactions

$$X_i \underset{k_{ji}}{\overset{k_{ij}}{\rightleftharpoons}} X_j \quad (1)$$

12.1.2.1
General Treatment of Kinetic Data

Experimental studies on the mechanism of protein folding have focused mainly on monomeric single domain proteins as model systems. Folding of oligomeric proteins and multidomain proteins is discussed in Chapter 27 and Chapter 2 in Part II. Folding kinetics of monomeric proteins have the major advantage that all observed reactions are of first order. Consequently, the time-dependent change in intensity of an observed signal (P) can be represented as a sum of n exponentials with observable rate constants (λ_i) and corresponding amplitudes (A_i)

$$P_t - P_\infty = \sum_{i=1}^{n} A_i \cdot e^{-\lambda_i t} \quad (2)$$

where P_t is the observed intensity at time t and P_∞ is the intensity at $t \to \infty$. The apparent rate constants, λ_i, are functions of the microscopic rate constants, k_{ij}, which depend on external parameters like temperature, pressure, and denaturant concentration. The amplitudes A_i additionally depend on the initial concentrations of the kinetic species (see Protocols, Section 12.1.6). Generally, any kinetic mechanism with n different species connected by first-order reactions results in $n-1$ observable rate constants. Applied to protein folding with fixed initial (U) and final (N) state the observation of $n-1$ apparent rate constants thus indicates the presence of $n-2$ transiently populated intermediate states. A more detailed treatment of the analysis of kinetic data in protein folding is given in Refs [2] and [3] and general treatments of kinetic mechanisms are given in Refs [4] and [5].

12.1.2.2
Two-state Protein Folding

Experimental investigations have revealed simple two-state folding with single exponential folding and unfolding kinetics for more than 30 small proteins (Figure 12.1.1; for an overview see Ref. [6]):

$$U \underset{k_u}{\overset{k_f}{\rightleftharpoons}} N \quad (3)$$

where k_f and k_u are the microscopic rate constants for the folding and unfolding reactions, respectively. The single observable macroscopic rate constant (λ) for this

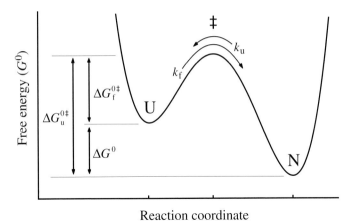

Fig. 12.1.1. Free energy profile for a hypothetical two-state protein folding reaction. The ground states U and N are separated by a free energy barrier with the transition state as point along the reaction coordinate with the highest free energy. The activation free energies for passing the barrier from U or N, $\Delta G_f^{o\ddagger}$ and $\Delta G_u^{o\ddagger}$, respectively, are reflected in the microscopic rate constants k_f or k_u (see Eq. (8)).

mechanism is readily derived as

$$\lambda = k_f + k_u \tag{4}$$

The equilibrium of the reaction is determined by the flows from the native state N to the unfolded state U and vice versa:

$$k_f \cdot [U]_{eq} = k_u \cdot [N]_{eq} \tag{5}$$

and the equilibrium constant (K) for folding is

$$K := \frac{[N]_{eq}}{[U]_{eq}} = \frac{k_f}{k_u} \tag{6}$$

A straightforward parameter that can be inferred from two-state behavior is the free energy for folding ($\Delta G°$) which is connected by the van't Hoff relation with the equilibrium constant (K)

$$\Delta G° = -RT \cdot \ln(K) = -RT \cdot \ln\left(\frac{[N]_{eq}}{[U]_{eq}}\right) = -RT \cdot \ln\left(\frac{k_f}{k_u}\right) \tag{7}$$

Thus, protein stability can be measured in two different ways, either by equilibrium methods or by kinetic measurements. The agreement between the free energies for folding inferred from equilibrium and kinetic measurements is commonly used as a test for two-state folding (see Figure 12.1.2).

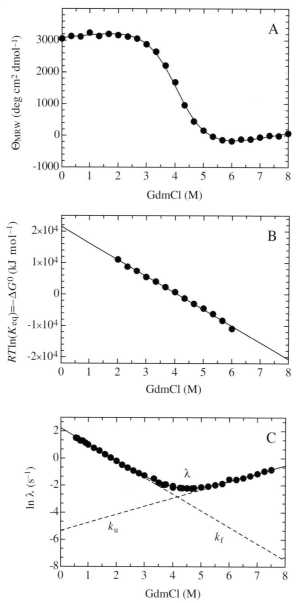

Fig. 12.1.2. Equilibrium and kinetic data for two-state tendamistat folding and unfolding at pH 2.0. A) GdmCl-induced equilibrium transition of wild-type tendamistat at pH 2.0 measured by far-UV circular dichroism. B) Plot of $RT \ln K_{eq}$ for the data in the transition region result in a linear dependency on the denaturant concentration. The slope corresponds to $-m_{eq}$ (see Eq. (11)). C) GdmCl dependence of the logarithm of the single observable folding rate constant ($\lambda = k_f + k_u$) gives a V-shaped profile commonly termed a chevron plot. This reveals a linear dependence between [GdmCl] and the $\ln(k_{f,u})$ as indicated by the dashed lines. The values for K_{eq} and m_{eq} derived from kinetic (panel C) and equilibrium data (panel B) are identical, which is a strong support for a two-state folding mechanism.

The same formalism can be applied to obtain information on the free energy of activation ($\Delta G^{\circ\ddagger}$) for a reaction using transition state theory

$$k = k_0 \cdot e^{-\Delta G^{\circ\ddagger}/RT} \tag{8}$$

where k_0 is the pre-exponential factor and represents the maximum rate constant for a reaction in the absence of free energy barriers. Combination of Eqs (6) and (8) allows us to gain information on the free energy changes along the reaction coordinate (see Figure 12.1.1).

$$\Delta G^\circ = \Delta G_f^{\circ\ddagger} - \Delta G_u^{\circ\ddagger} \tag{9}$$

In the original Eyring theory the pre-exponential factor corresponds to the frequency of a single bond vibration [7, 8]:

$$k_0 = \kappa \frac{k_B T}{h} \approx 6 \cdot 10^{12} \text{ s}^{-1} \text{ at } 25 \text{ °C} \tag{10}$$

where κ represents the transmission coefficient, which is usually set to 1. Since protein folding involves formation and breakage of weak intramolecular interactions rather than changes in covalent bonds, the Eyring prefactor will not be useful for the analysis of protein folding reactions. The pre-exponential factor for protein folding reactions will be strongly influenced by intrachain diffusion processes but will also depend on the location of the transition state on the free energy landscape and on the nature of the rate-limiting step for a particular protein. It is likely around 10^7–10^8 s^{-1} [9]. This is discussed in detail in Chapter 22.

Experimental analysis of protein folding reactions usually makes use of the effect of chemical denaturants like urea and GdmCl on the rate and equilibrium constants (see Chapter 3). Empirically, a linear correlation between the denaturant concentration ([D]) and both the equilibrium free energy (ΔG°) [10, 11] and the activation free energies [12] for folding ($\Delta G_f^{\circ\ddagger}$) and unfolding ($\Delta G_u^{\circ\ddagger}$) is found (Figure 12.1.2). Accordingly, for two-state folding a plot of the logarithm of the single observable (apparent) rate constant, λ, vs. [D] yields a V-shaped curve, commonly called chevron plot [13, 14] (Figure 12.1.2C). This plot gives information on the refolding reaction at low denaturant concentrations (the refolding limb) and information on the unfolding reaction at high denaturant concentrations (the unfolding limb). Accordingly, proportionality constants, m_x, are defined as

$$m_{eq} = \frac{\partial \Delta G^\circ}{\partial [\text{Denaturant}]} \tag{11}$$

$$m_{f,u} = \frac{\partial \Delta G_{f,u}^{\circ\ddagger}}{\partial [\text{Denaturant}]} \tag{12}$$

Since the m_{eq}-values were shown to be proportional to the changes in accessible surface area (ASA) upon unfolding of the protein [15], the kinetic m-values are believed to reflect the changes in ASA between the unfolded state and the transition state (m_f) and the native state and the transition state (m_u). This can be used to characterize the solvent accessibility of a protein folding transition state (see Chapter 12.2).

The m-values allow an extrapolation of the data to any given denaturant concentration using

$$\Delta G^\circ(D) = \Delta G^\circ(H_2O) + m_{eq} \cdot [D] \qquad (13)$$

$$\Delta G_f^{\circ\ddagger}(D) = \Delta G_f^{\circ\ddagger}(H_2O) + m_f \cdot [D] \qquad (14)$$

$$\Delta G_u^{\circ\ddagger}(D) = \Delta G_u^{\circ\ddagger}(H_2O) + m_u \cdot [D] \qquad (15)$$

where (H_2O) denotes the values in the absence of denaturant and (D) those at a given denaturant concentration $[D]$. Combining Eq. (9) with Eqs (13)–(15) shows that the comparison of the kinetic and equilibrium free energies and m-values provides a tool to test for the validity of the two-state mechanism (see Figure 12.1.2):

$$\Delta G^\circ(H_2O) = \Delta G_f^{\circ\ddagger}(H_2O) - \Delta G_u^{\circ\ddagger}(H_2O) \qquad (16)$$

and

$$m_{eq} = m_f - m_u \qquad (17)$$

12.1.2.3
Complex Folding Kinetics

For simple two-state folding reactions the microscopic rate constants are readily obtained from the chevron plot (cf. Figure 12.1.2) and the properties of the transition barriers can be analyzed using rate equilibrium free energy relationships [1, 16–19] (REFERs; see Chapter 12.2). However, folding of most proteins is more complex with several observable rate constants. These complex folding kinetics can have different origins. The transient population of partially folded intermediates, which frequently occurs under native-like solvent conditions (i.e., at low denaturant concentrations), will lead to the appearance of additional kinetic phases (see Section 12.1.3.2) and to a downward curvature in the chevron plot (Figure 12.1.3A) [13, 20–23]. Kinetic coupling between two-state folding and slow equilibration processes in the unfolded state provides another source for multiphasic folding kinetics [2, 24, 25] (Section 12.1.3.1). Since these possible origins for complex folding kinetics have far-reaching effects on the molecular interpretation of the kinetic data it is crucial to discriminate between them and to elucidate the origin of complex kinetics. With the knowledge of the correct folding mechanism the microscopic rate constants (k_{ij}) connecting the individual states can be determined.

12.1.2.3.1 Heterogeneity in the Unfolded State
The unfolded state of a protein consists of a large ensemble of different conformations which are in rapid equilibrium, and can thus be treated as a single kinetic

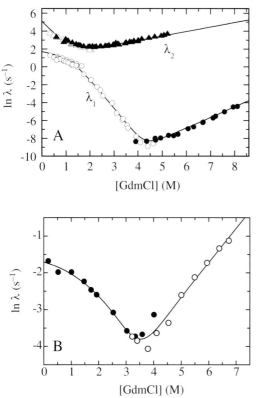

Fig. 12.1.3. Different origins for curvatures in chevron plots: A) Population of a transient intermediate during folding of hen egg white lysozyme. At low denaturant concentrations two rate constants are observed in refolding (λ_2 and λ_1; open triangles and open circles). For unfolding of native lysozyme a single kinetic phase is observed (filled circles). λ_2 at higher denaturant concentrations (filled triangles) corresponds to unfolding of the intermediate and can only be detected when the intermediate is transiently populated by a short refolding pulse and then unfolded at high denaturant concentrations in sequential mixing experiments [52] (interrupted refolding experiments). B) Chevron plot for ribonuclease T1 W59Y. The curvature in the refolding limb at low denaturant concentrations is caused by kinetic coupling between prolyl isomerization and protein folding [24, 25]. Only the slowest refolding phase is shown. The fraction of faster folding molecules with all prolyl peptide bonds in the native conformation could not be detected since the experiments were carried out by manual mixing. Data for lysozyme were taken from Ref. [52] and data for RNase T1 were taken from Ref. [25].

species as long as their interconversion is faster than the kinetic reactions leading to the native state. Studies on the time scale of intrachain diffusion in unfolded polypeptide chains in water showed that these processes occur on the 10–100 ns time scale [9, 26], which is significantly faster than protein folding reactions [27, 28]. On the other hand, slow interconversion reactions between the different unfolded conformations lead to kinetic heterogeneity. This was first observed by Garel and Baldwin [29] who showed that both fast and slow refolding molecules exist in unfolded ribonuclease A (RNase A). In 1975 Brandts and coworkers [30] showed with their classical double jump experiments that slow and fast folding forms of RNase A are caused by a slow equilibration process in the unfolded state and proposed *cis–trans* isomerization at Xaa-Pro peptide bonds as the molecular origin (see Chapter 1). This was confirmed later with proline mutations [31] and catalysis of the slow reactions by peptidyl-prolyl *cis-trans* isomerases [32]. The fast folding molecules have all Xaa-Pro peptide bonds in their native isomerization states, whereas slow-folding molecules contain nonnative prolyl isomers, which have to isomerize to their native isomerization state during folding. This isomerization process is slow with time constants ($\tau = 1/\lambda$) around 10–60 s at 25 °C, which usually limits the folding reaction of the unfolded molecules with nonnative prolyl isomers (see Chapter 25). Since the equilibrium population of the *cis* isomer in Xaa-Pro peptide bonds is between 7 and 36%, depending on the Xaa position, [33] all proteins with prolyl residues should show a significant population of slow folding molecules. Prolyl isomerization has indeed been observed in folding of many proteins [34]. Whenever folding and proline isomerization have similar rate constants or when folding is slower than isomerization, a pronounced curvature is observed in the refolding limb of the chevron plot, which looks similar to the effect of transiently populated intermediates (Figure 12.1.3B). The effect of prolyl isomerization on folding kinetics and its identification as a rate-limiting step is discussed in detail in Chapter 10 in Part II. The effect on chevron plots is treated quantitatively in Refs [2, 35, 36].

In addition to prolyl isomerization also nonprolyl peptide bond isomerization has recently been shown to cause slow parallel folding pathways [37, 38]. *Cis–trans* isomerization of nonprolyl peptide bonds has long been speculated to cause slow steps in protein folding, since it is an intrinsically slow process with a high activation energy [30, 33]. The large number of peptide bonds in a protein leads to a significant fraction of unfolded molecules with at least one nonnative *cis* isomer, although the *cis* isomer is only populated to about 0.15% in equilibrium in the unfolded state [39]. Studies on a slow folding reaction in a prolyl-free tendamistat variant revealed that folding of about 5% of unfolded molecules is limited by the *cis–trans* isomerization of nonprolyl peptide bonds [38] (Figure 12.1.4). The slow equilibrium between *cis* and *trans* nonprolyl peptide bonds in the unfolded state was revealed by double jump experiments [38] (see also Chapter 25). In these experiments (Figure 12.1.4C) the protein is rapidly unfolded at high denaturant concentrations often in combination with low pH. After various times, unfolding is stopped by dilution to refolding conditions. The resulting folding kinetics are monitored by spectroscopic probes. If a slow isomerization reaction in the unfolded state occurs, the slow folding molecules will be formed slowly after unfolding. Thus, the amplitude of the slow reaction will increase with unfolding time as observed for the proline free tendamistat variant shown in Figure 12.1.4C. Figure 12.1.4 further reveals that nonprolyl isomerization has a rate constant around 1 s^{-1}, which is significantly faster than prolyl peptide bond isomerization. This is in accordance with rate constants for nonprolyl isomerization measured in model

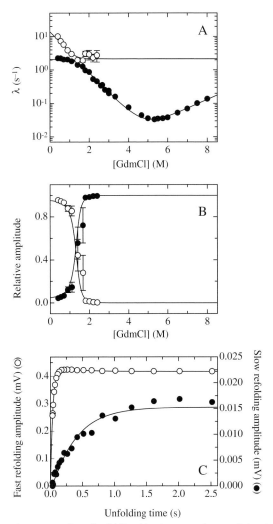

Fig. 12.1.4. Complex folding kinetics caused by nonprolyl *cis–trans* isomerization in the unfolded state of a proline-free variant of tendamistat. GdmCl dependence of the apparent rate constants, λ_1 (open circles), λ_2 (filled circles) for folding and unfolding (A) and the respective refolding amplitudes (A_1 (open circles), A_2 (filled circles) are shown. The lines in panels A and B represent the global fit of both apparent rate constants and amplitudes to the analytical solution of a linear three-state model. The fit reveals a GdmCl-independent isomerization reaction and a GdmCl-dependent folding reaction (see Ref. [38]). C) Double jump experiments to detect the slow isomerization process in the unfolded state. Appearance of the fast-refolding (open circles) and the slow-refolding (filled circles) molecules during unfolding. Native protein was unfolded at high denaturant concentrations for the indicated time. Under the applied conditions unfolding was complete after 150 ms. After various unfolding times refolding was initiated by a second mixing step and the amplitudes of the two refolding reactions (open circles, filled circles) in the presence of 0.8 M GdmCl (cf. panel A) were measured. The results show that the slow-refolding phase (filled circles) is formed slowly in the unfolded state whereas the fast-refolding phase is formed with the same time constant as that with which unfolding of tendamistat occurs. Data were taken from Ref. [38].

peptides [39]. The results on the effect of nonprolyl isomerization on protein folding imply that isomerization at non prolyl peptide bonds will dramatically effect folding of large proteins. For a protein with 500 amino acid residues more than 50% of the unfolded molecules have at least one nonnative peptide bond isomer. Since the isomerization rate constant is around $1\ \text{s}^{-1}$, it will mainly affect the early stages of folding and folding of fast folding proteins [38].

Another cause of kinetic heterogeneity in the unfolded state was identified in cytochrome c, where parallel pathways were shown to be due to the exchange of the heme ligands in the unfolded state [40–42] (see Chapter 15).

12.1.2.3.2 Folding through Intermediates

As discussed above, the number of kinetic species, n, is related to the number of observable rate constants, $n-1$, and thus, with unfolded and native state as the initial and final states, the number of intermediates is given by $n-2$. Thus, for a single observable rate constant there are no intermediates, corresponding to two-state folding. The observation of two apparent rate constants indicates a single intermediate, etc. It should be noted that these considerations only apply if there is a kinetically homogeneous unfolded state with rapidly interconverting conformations. In the case of heterogeneous populations of unfolded molecules separated by slow interconversion reactions like prolyl isomerization, the number of observable rate constants is additionally correlated with the number of unfolded species. In the following we will describe experiments that allow the elucidation of mechanisms of fast folding molecules (i.e., of molecules with all Xaa-Pro bonds in the native isomerization state).

The first step in elucidating a folding mechanism is the determination of the number of exponentials, i.e., the number of observable rate constants, λ_i, needed to describe the kinetics. In principle, the number of observable rate constants should not depend on the probe used to monitor the folding reaction, as observed for lysozyme refolding shown in Figure 12.1.5, which shows the same two observable rate constants measured by a number of different probes. However, some reactions may not cause a signal change in a specific probe, if, for example, an intermediate shows the same spectral properties as the native or unfolded state. Thus, kinetics should be monitored using different probes. In addition, it is crucial to test for burst-phase reactions (i.e., for processes occurring within the experimental dead-time) (Figure 12.1.5). These reactions are observed for many proteins during refolding at low denaturant concentrations and are usually caused by a considerable compaction of the polypeptide chain [43, 44] (Figure 12.1.5E) and concomitant formation of significant amounts of secondary structure [45] (see Chapter 23). Whether rapid collapse represents a distinct step on a folding pathway or whether it is the response of the unfolded state to the change in solvent conditions upon refolding is currently under discussion. Results on the folding of α-lactalbumin [46–48], apo-myoglobin [49], and lysozyme [44] suggest that burst-phase intermediates unfold cooperatively, indicating that they are separated by a barrier from the ensemble of random coil conformations, which are populated at high denaturant concentrations. To test for aggregation side reactions during refolding, which

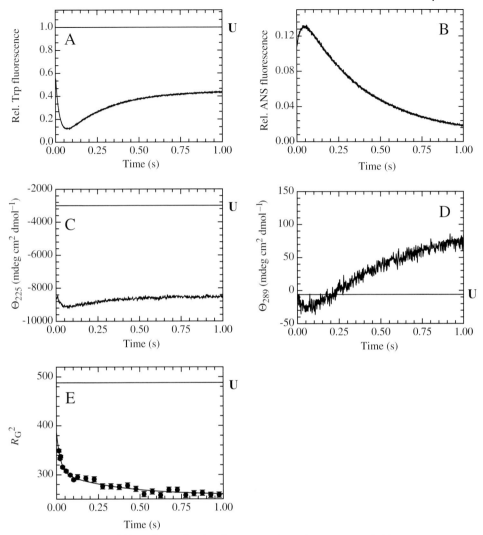

Fig. 12.1.5. Lysozyme refolding observed with different probes: A) Intrinsic tryptophan fluorescence above 320 nm after excitation at 280 nm; B) ANS fluorescence; C) far-UV circular dichroism at 225 nm; D) near-UV circular dichroism at 289 nm; E) changes in radius of gyration measured by small-angle X-ray scattering (SAXS). The lines of constant signal represent the signal of the unfolded state. The traces are recorded at 23 °C, pH 5.2. Data in panels A–D were recorded by stopped-flow mixing. SAXS data were recorded by continuous-flow and stopped-flow mixing. All data can be fitted globally with $\lambda_2 = 34$ s^{-1}; $\lambda_1 = 4.3$ s^{-1} demonstrating that all probes detect the same two kinetic phases. In addition, major signal changes in the dead time of mixing are detected by all probes except near-UV CD, indicating a fast reaction that cannot be resolved by the experiments (burst phase with $\lambda > 2000$ s^{-1}). Data were taken from Refs [43] and [44].

are occasionally also observed for small proteins, [50], the concentration dependence of the folding reaction should be tested.

The next step in determining a folding mechanism is to locate the intermediates on the folding pathway. If a single folding intermediate is observed two apparent rate constants will be observed for all possible three-state mechanisms shown in Scheme 12.1.1.

$$U \underset{k_{IU}}{\overset{k_{UI}}{\rightleftharpoons}} I \underset{k_{NI}}{\overset{k_{IN}}{\rightleftharpoons}} N \qquad U \underset{k_{NU}}{\overset{k_{UN}}{\rightleftharpoons}} N \qquad (A)$$

$$U \underset{k_{IU}}{\overset{k_{UI}}{\rightleftharpoons}} I \underset{k_{NI}}{\overset{k_{IN}}{\rightleftharpoons}} N \qquad (B)$$

$$I \underset{k_{UI}}{\overset{k_{IU}}{\rightleftharpoons}} U \underset{k_{NU}}{\overset{k_{UN}}{\rightleftharpoons}} N \qquad (C)$$

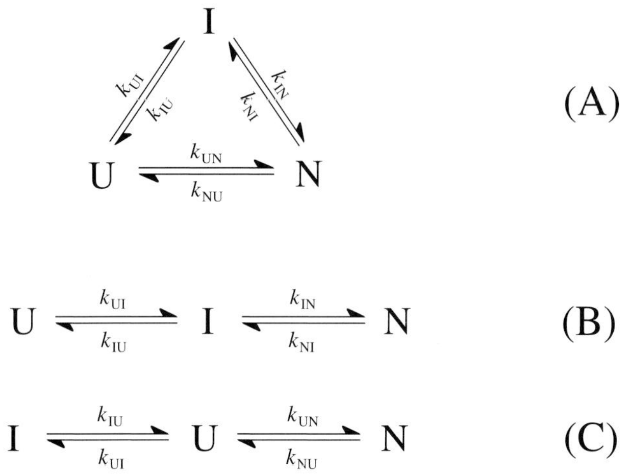

Scheme 12.1.1

The triangular mechanism shown in panel A represents the general three-state model whereas mechanisms B (on-pathway intermediate) and C (off-pathway intermediate) represent special cases of mechanism A with one of the three equilibria between the individual states being too slow to influence the kinetics. Thus, if only mechanisms B or C are considered for a particular protein, the assumption is made that one of the equilibration processes between the three states is significantly slower than the other two. A consequence from the mathematical analysis of the mechanisms shown in Scheme 12.1.1 is that the two observable rate constants for the triangular mechanism cannot be simply related to individual microscopic rate constants but are functions of several microscopic rate constants (see Protocols, Section 12.1.6). Thus it is essential to elucidate the correct folding mechanism and to determine all microscopic rate constants in order to draw conclusions on the individual transition states between the kinetic species.

Since all mechanisms shown in Scheme 12.1.1 give rise to two experimentally observable rate constants, it is impossible to exclude the triangular mechanism on the basis of direct spectroscopic measurements of the folding kinetics. However, in the triangular mechanism the native molecules are produced in both the faster and the slower kinetic phase whereas in the on-pathway mechanism the native state is produced with a lag phase. Thus, discrimination between these mechanisms requires to specifically monitor the time course of the native state. Experiments,

12.1.2 Analysis of Protein Folding Reactions using Simple Kinetic Models | 391

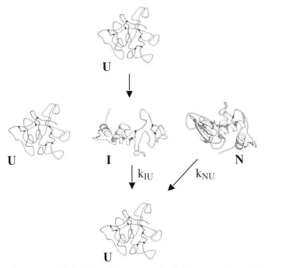

1) Unfolded protein

2) *Start refolding*

3) Mixture of U, I, N

4) *Interrupt refolding at t=t_i by transferring the solution to high denaturant concentrations and measure amplitudes of the unfolding reactions of N and I as a function of t_i*

Fig. 12.1.6. Principle of interrupted refolding experiments to measure the time course of the population of a folding intermediate (I) and of native proteins (N) starting from completely unfolded protein (U). The experiments are described in detail in the text.

which selectively measure the population of the individual kinetic species during folding, have been introduced by Schmid [51] and were initially used to detect slow and fast folding pathways during RNase A folding. They are commonly termed interrupted refolding experiments. These experiments make use of the fact that each state of the protein has its characteristic stability and unfolding rate constant. Interrupted refolding experiments consist of two consecutive mixing steps (Figure 12.1.6). In a first step refolding is initiated from completely unfolded protein. The folding reaction is allowed to proceed for a certain time (t_i) after which the solution is transferred to a high concentration of denaturant. This results in the unfolding of all native molecules and of partially folded states which have accumulated during the refolding step. Each state (N or I) is identified by its specific unfolding rate constant at high concentrations of denaturant, where unfolding is virtually irreversible, so that little kinetic coupling occurs. The amplitudes of the observed unfolding reactions reflect the amounts of the respective species present at time (t_i) when refolding was interrupted. Thus, varying the time (t_i) allowed for refolding gives the time course of the populations of native protein and of the intermediate during the folding process. With the use of highly reproducible sequential mixing stopped-flow instruments these experiments also became feasible on a faster time scale and allow the discrimination between sequential and triangular folding mechanism on fast folding pathways [22, 52].

The discrimination between the off-pathway mechanism and the triangular mechanism is usually difficult, since also in the off-pathway model some molecules may fold directly from U to N, depending on the ratio of the rate constants k_{UN} and k_{UI} [52]. The discrimination between these mechanisms usually requires analysis of the denaturant dependence of all folding and unfolding rate constants in combination with the time course of the different species [52]. The data are then fitted to the analytical solutions of the differential equations describing the different models. For folding reactions of monomeric proteins all reactions are of first order and the analytical solutions of the differential equations can be obtained for mechanisms with four species or less. These solutions are derived in many kinetic textbooks and rate constants and amplitudes for three-state and four-state mechanisms are given in Refs [24] and [52]. The solutions for the three-state models in Scheme 12.1.1 are given in the Protocols (Section 12.1.6.1–12.1.6.3). A particularly useful source for the analytical solution of a vast number of different kinetic mechanisms is the article by Szabo [4]. For more complex mechanisms numerical methods have to be applied to analyze the data. As an example for the discrimination between the different kinetic models shown in Scheme 12.1.1, the elucidation of the folding mechanism of lysozyme will be discussed in detail in Section 12.1.3.

12.1.2.3.3 Rapid Pre-equilibria

The analysis for mechanisms B and C in Scheme 12.1.1 simplifies when equilibration between U and I occurs on a much faster time scale than folding to the native state [53, 54]. In this case, formation of the intermediate can be treated as a rapid pre-equilibrium. The apparent rate constants for mechanism B (on-pathway intermediate) can thus be approximated by

$$\lambda_1 = k_{UI} + k_{IU}, \quad \lambda_2 = \frac{1}{1 + 1/K_{UI}} k_{IN} + k_{NI} \quad \text{with} \quad K_{UI} = \frac{k_{UI}}{k_{IU}} \tag{18}$$

and for mechanism C (off-pathway intermediate) as

$$\lambda_1 = k_{UI} + k_{IU}, \quad \lambda_2 = \left(1 - \frac{1}{1 + 1/K_{UI}}\right) k_{UN} + k_{NU} \tag{19}$$

where $1/(1 + (1/K_{UI}))$ represents the fraction of intermediate $(f(I))$ in the pre-equilibrium. Since I is productive in mechanism B and nonproductive in mechanism C, the rate of formation of N depends on $f(I)$ and $1 - f(I)$, respectively. Due to the commonly observed strong denaturant dependence of the microscopic rate constants, the simplifications made above might not be valid at some denaturant concentrations. Therefore, the simplified treatment of the data should only be performed if formation and unfolding of the intermediate is significantly faster than the k_{IN} and k_{NI} under all experimental conditions. Otherwise, the exact solutions of the three state model must be used to fit the data [52] (see Protocols, Section 12.1.6).

Under conditions where the simplifications given in Eqs (18) and (19) hold and the triangular mechanism has been ruled out, these equations can provide a simple tool for identifying off-pathway intermediates. Destabilization of the intermedi-

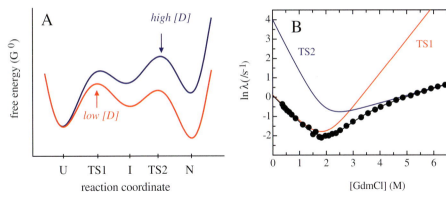

Fig. 12.1.7. Transition barrier for apparent two-state folding through a high-energy intermediate and the consequences on the shape of the chevron plot. A) The reaction coordinate of tendamistat comprises two discrete transition states and a high-energy intermediate. B) A single folding rate constant is observed which shows a kink in the denaturant dependence of the chevron-plot which indicate the switch between the two alternative transition states [57]. The black line represents a fit of the data to a three-state model as described in the Protocols (Eqs (35)–(42)). The lines labeled TS1 and TS2 correspond to constructed chevron plots for the two transition states. Data were taken from Ref. [57].

ate leads to a larger fraction of productive (unfolded) molecules in the off-pathway model. Thus, addition of denaturant will speed up folding, when the resulting destabilization of the intermediate exceeds the deceleration of the U → N reaction. As a consequence, an increase in the folding rate with increasing concentrations of denaturant points to the transient accumulation of an off-pathway intermediate. This behavior has been observed for intermediates trapped by nonnative disulfide bonds, [55] by nonnative proline isomers [56] and by nonspecific aggregation [50].

12.1.2.3.4 Folding through an On-pathway High-energy Intermediate

Many apparent two-state folders were shown to fold through on-pathway high-energy intermediates, which cannot be detected directly with spectroscopic methods, since they are less stable than N and U [57, 58] (Figure 12.1.7A). Their existence can be inferred from kinks in chevron plots, which are not accompanied by additional kinetic phases nor by burst-phase reactions [1, 18, 19, 57–61]. Figure 12.1.7B shows such a kink in the chevron plot for a tendamistat variant [57]. The changing slope in the unfolding limb of the chevron plot indicates the existence of two consecutive transitions states, TS1 and TS2, and a "hidden" high-energy intermediate (I*; see Chapter 12.2) and thus provides an effective way to identify on-pathway intermediates [1, 18, 19, 57–61]. The detection of a high-energy on-pathway intermediate results in a linear three-state model:

$$U \underset{k_{IU}}{\overset{k_{UI}}{\rightleftharpoons}} I^* \underset{k_{NI}}{\overset{k_{IN}}{\rightleftharpoons}} N \tag{20}$$

The kinetics at a single denaturant concentration cannot be distinguished from two-state folding (Eq. (3)). However, the presence of a kink in the chevron plot and the resulting altered slope at high denaturant concentrations allows a characterization of both transition states [57] (see Protocols, Section 12.1.6.4). Since I* does not become populated at any conditions, its stability can not be determined. However, the difference in free energy between both transition states ($\Delta G^\circ_{TS2/TS1}$) and the difference in their ASA ($m_{TS2} - m_{TS1}$) can be obtained by fitting the GdmCl dependence of $\ln \lambda$ to the analytical solutions of the three-state model shown in Eq. (20) (see Protocols, Section 12.1.6.4). Figure 12.1.7B shows the results of a three-state fit for the kinked chevron and the constructed chevron plots for the individual transition states for the tendamistat variant.

12.1.3
A Case Study: the Mechanism of Lysozyme Folding

In the following, we will use lysozyme as a case study to discuss how sequential mixing experiments in combination with the denaturant dependence of the apparent rate constants allow a quantitative analysis of complex folding kinetics and the determination of all microscopic rate constants.

12.1.3.1
Lysozyme Folding at pH 5.2 and Low Salt Concentrations

Lysozyme consists of two structural subdomains, the α-domain with exclusively α-helical structure and the β-domain with predominantly β-structure [62, 63]. Starting from GdmCl-unfolded disulfide intact protein, large changes in far-UV CD and fluorescence signals are observed within the first millisecond of refolding [43, 44, 64, 65] (Figure 12.1.5A–D). Time-resolved small angle X-ray scattering experiments show that this burst-phase reaction leads to a globular state, with a significantly smaller radius of gyration (R_G) than the unfolded protein [43, 44] (Figure 12.1.5E). In a subsequent reaction, with a time constant of 30 ms (at pH 5.2 and 20 °C), an intermediate state (I_N) is formed, as observed by strong quenching in tryptophan fluorescence to a level below that of the native state [43, 44, 66], by an increase in the far-UV CD signal [64, 65], by small changes in near-UV CD [44] and by a further compaction of the polypeptide chain [43, 44]. Pulsed hydrogen/deuterium exchange experiments showed that this intermediate has native-like helical structure in the α-domain whereas the β-sheet structures are not formed [67]. The intermediate converts to the native state with a relaxation time of about 400 ms. Double-jump experiments showed that the slow folding reaction of lysozyme is not caused by slow equilibration processes coupled to the folding reaction [52] (Figure 12.1.8A). There had been controversial reports as to the role of I_N in the folding process and it remained unclear whether it represents an obligatory intermediate on the folding pathway for all lysozyme molecules [66, 67]. In addition, a reaction on the 10 ms time scale was observed in pulsed hydrogen exchange experiments [67–69], which leads to formation of molecules with native hydrogen

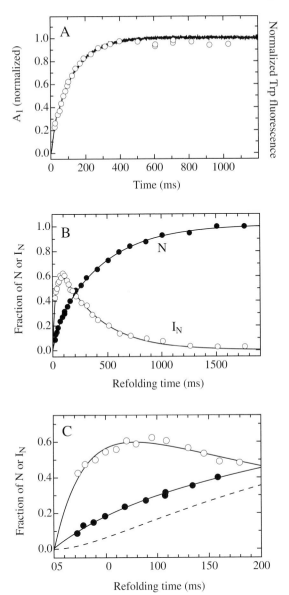

Fig. 12.1.8. Double jump (A) and interrupted refolding experiments (B, C) for lysozyme folding at pH 5.2. Double jump experiments show that amplitude of the slow folding reaction (A_1, open circles) is formed with the same time constant as that at which unfolding of native lysozyme occurs. For comparison the unfolding kinetics directly monitored by the change in intrinsic Trp fluorescence is shown (solid line). B) Time course of the population of native lysozyme, N (filled circles) and of the kinetic intermediate, I_N (open circles) measured in interrupted refolding experiments. C) The early time region of the kinetics shows that the faster process ($t = 30$ ms) produces both native molecules and the partially folded intermediate I_N. The lag phase for formation of N, expected for a linear on-pathway model shown in Scheme 12.1.1B (---), is not observed. Data were taken from Ref. [52].

bonding network and was thus proposed to produce an intermediate with native-like hydrogen bonding network [70]. However, this reaction was not observed with any other spectral probe or in SAXS experiments (see Figure 12.1.5), which obscured its interpretation.

The observation of two apparent rate constants (treating the burst-phase collapse as a pre-equilibrium uncoupled from the slower reactions) cannot rule out any of the mechanisms shown in Scheme 12.1.1. However, comparing mechanism A with mechanisms B and C offers a simple way to distinguish obligatory from nonobligatory folding intermediates. In the case of an obligatory intermediate, all molecules have to fold through this state, resulting in a lag phase in the formation of native molecules [5]. Direct spectroscopic measurements or measurements of changes in R_G are not able to monitor formation of native lysozyme directly, since changes in spectroscopic and geometric properties occur in all folding steps (see above). Further, the use of inhibitor binding to detect formation of active lysozyme did not give any clear-cut results since binding is too slow under the applied experimental conditions [22, 66]. Thus, interrupted refolding experiments were used to monitor the time course of native protein and of I_N. Two unfolding reactions were observed in interrupted refolding experiments [22, 52]. One of them corresponds to the well-characterized unfolding kinetics of native lysozyme, and a second much faster one corresponds to unfolding of the intermediate (see Figure 12.1.6). The faster reaction is only observed at short and intermediate refolding times but is not observed when native lysozyme is unfolded. The collapsed state unfolds too fast under all conditions to be measured by stopped-flow unfolding. The results show that the intermediate is not obligatory for lysozyme folding, since no lag phase in forming native lysozyme is observed [22] (Figure 12.1.8B,C). Rather, formation of I_N and formation of 20% of the native molecules occur with the faster kinetic phase ($\tau = 30$ ms). The slower process ($\tau = 400$ ms) reflects the interconversion of the intermediate to the native state, resulting in the remaining 80% of the molecules folding to the native state [22, 52]. As discussed above, identical apparent rate constants for forming the intermediate and native molecules on a fast pathway are expected, since any three-state model gives rise to only two observable rate constants.

The lack of an initial lag phase during formation of native lysozyme ruled out a linear on-pathway model, but it could not distinguish between the triangular model and the off-pathway model [22]. Performing a least-squares fit of the denaturant dependence of the two apparent rate constants to the analytical solutions of both models (see Protocols, Section 12.1.6) revealed that only the triangular model was able to describe the data [52] (Figure 12.1.9). In the case of the off-pathway model, unfolding of the intermediate should become the rate-limiting step for folding at very low denaturant concentrations. This would predict an increase in the folding rate with increasing denaturant concentration, which was, however, not observed (Figure 12.1.9). Fitting the data to the circular three-state model allowed the determination of all six microscopic rate constants (Figure 12.1.10). For this analysis it was crucial to know the unfolding rate constant of the intermediate at high denaturant concentrations (λ_2 in Figures 12.1.3A and 12.1.9). Since unfolding of

12.1.3 A Case Study: the Mechanism of Lysozyme Folding

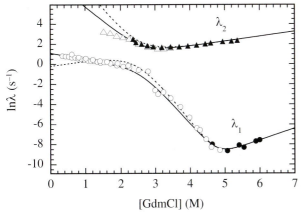

Fig. 12.1.9. GdmCl dependence of the apparent rate constants of lysozyme folding at pH 5.2. λ_2 (open and filled triangles) and λ_1 (open and filled circles) are fit to the analytical solutions of the triangular model (Scheme 12.1.1A, —) and to the linear off-pathway model (Scheme 12.1.1C, ----). The fits show that the off-pathway model is not able to describe the data. To increase the stability of the inter-mediate and thus allow a better distinction between the on- and the off-pathway model the experiments were carried out in the presence of 0.5 M Na$_2$SO$_4$. Data were taken from Ref. [52].

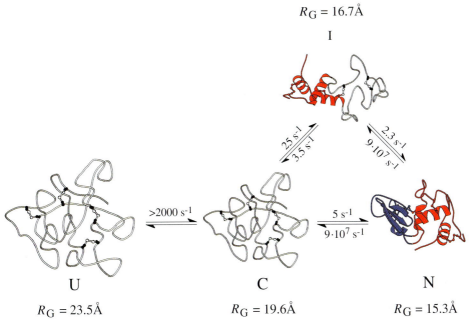

Fig. 12.1.10. Folding mechanism of lysozyme at pH 5.2 and low salt concentrations obtained by analyzing a large number of experimental data using various probes to detect refolding (see Figure 12.1.5), the GdmCl dependence of λ_1 and λ_2 (Figure 12.1.3A) and the results from interrupted refolding experiments (Figure 12.1.8B,C). Adapted from Refs [3], [43] and [52].

native lysozyme is single exponential at high denaturant concentrations and determined by λ_1, the faster rate constant (λ_2) had to be obtained in sequential mixing experiments [52]. In these experiments the intermediate was populated by a short (100 ms) refolding pulse and then unfolded at high denaturant concentrations to yield λ_2 (see Figure 12.1.6).

The triangular folding mechanism for lysozyme raises the question for the origin for the kinetic partitioning into a fast direct pathway and a slow pathway through I_N at the stage of the collapsed state. Fluorescence and NMR measurements revealed nonnative short- and long-range interactions around Trp residues in denaturant-unfolded lysozyme [44, 71] and in the collapsed state [43]. These interactions involve Trp62 and Trp63 which are located in the β-domain and may cause distinct subpopulations in the collapsed state and prevent structure formation in the β-domain of part of the collapsed molecules. This model is supported by the folding kinetics of a W62Y mutant, which has weakened hydrophobic interactions in the unfolded state and shows faster folding kinetics [72]. Also replacement of Trp108 by tyrosine accelerates folding of lysozyme [72] supporting a model of nonnative short-range and long-range interactions, which prevent fast folding of the β-domain.

The final conundrum in the folding mechanisms of lysozyme at pH 5.2 was the origin of the 10 ms reaction that was only observed in pulsed hydrogen exchange experiments [67–69]. This reaction was interpreted as formation of a native-like intermediate with secondary structure formed both an the α- and β-domains [70]. It is very difficult to imagine formation of a native-like state to occur without changes in any spectroscopic parameters. A quantitative analysis of all reactions occurring during the pulsed hydrogen exchange experiments revealed, however, that the 10 ms reaction is an artifact of incomplete hydrogen exchange, which was misinterpreted as fast protection of protons [73]. The reason for the incomplete exchange was an acceleration of all folding rate constants and a change in the folding mechanism under the conditions of the exchange pulse (pH 9–10; see below). The determination of the folding mechanism and all microscopic rate constants for folding and unfolding under the conditions of the exchange pulse allowed a quantitative calculation of the exchange kinetics expected on the basis of the rate constants determined from spectroscopic probes. All microscopic rate constants for all conditions applied during the exchange experiments and the rate constants for H/D exchange of the individual amide protons in lysozyme [74] were used to calculate exchange kinetics. The results quantitatively reproduced the experimentally measured time course of amide occupancy when the folding rate constants from spectroscopic experiments (Figures 12.1.3A and 12.1.5) were used [73]. This showed that no kinetic phases in addition to the ones observed by spectroscopic probes are required to quantitatively explain the hydrogen exchange kinetics.

12.1.3.2
Lysozyme Folding at pH 9.2 or at High Salt Concentrations

The procedure described above to elucidate the three-state mechanism for lysozyme folding at pH 5.2 can also be applied to elucidate the mechanism of more

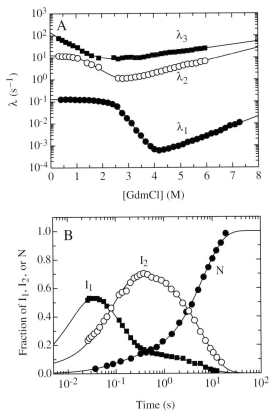

Fig. 12.1.11. Folding of lysozyme at pH 9.2. A) Effect of GdmCl on the three apparent rate constants λ_1, λ_2, and λ_3 measured by a combination of single and sequential mixing experiments [73]. B) Time course of the two kinetic intermediates and the native state during refolding at 0.6 M GdmCl. A global fit to the mechanism shown in Scheme 12.1.3 can describe the data (solid lines in all panels) and allowed all the microscopic rate constants to be determined (see Scheme 12.1.3).

complex folding reactions with more than one intermediate. For lysozyme folding, a third kinetic reaction is observed at high salt concentrations [75] and at high pH (>8.5) [73], indicating the stabilization of an additional intermediate under these conditions. Thus, a four-state model with an additional rapid pre-equilibrium is required to describe the folding kinetics under these conditions. Again interrupted refolding experiments were able to monitor the time course of the population of all intermediates and of native molecules (Figure 12.1.11B). Together with the denaturant dependence of all three rate constants obtained from single and sequential mixing stopped-flow refolding and unfolding experiments (Figure 12.1.11A) this allowed the identification of the minimal kinetic mechanism for lysozyme folding under these conditions. The results showed that the additional intermediate induced at high salt concentrations or at high pH is located on a third parallel folding pathway. The four-state mechanisms shown in Schemes 12.1.2 and 12.1.3

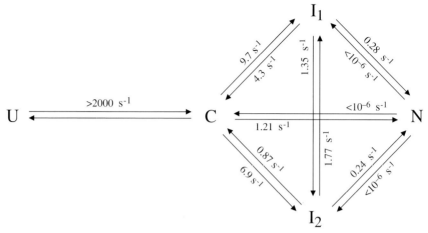

Scheme 12.1.2. Mechanism and rate constants for lysozyme folding at pH 5.2 in the presence of 0.85 M NaCl. Adapted from Ref. [75].

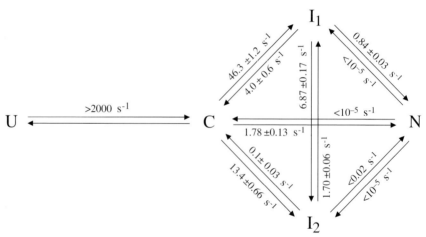

Scheme 12.1.3. Mechanism and rate constants for lysozyme folding at pH 9.2. Adapted from Ref. [73]

show the minimal models describing lysozyme folding at pH 9.2 and high salt concentrations, respectively. In addition the microscopic rate constants obtained from global fits of all data are shown.

This case study demonstrates the need for methods capable of detecting the time course of individual kinetic species during the folding process. Direct spectroscopic measurements are usually not able to provide this information. In addition, the folding studies on lysozyme show that for complex reactions it is usually impossible to assign experimentally observed rate constants directly to individual steps on folding pathways, since they are complex functions of several microscopic rate constants. In order to determine all microscopic rate constants for a mechanism it is essential to combine results from interrupted refolding experiments and from the denaturant-dependence of all observable folding and unfolding reactions in a global fit to the solutions of various possible models (see Protocols, Section 12.1.6).

12.1.4
Non-exponential Kinetics

Non-exponential kinetics were first reported by Kohlrausch [76] for the relaxation of glass fibers after stretching. He described a broad distribution of relaxation times covering time scales of several orders of magnitude. Such non-exponential kinetics can be treated with a stretched exponential term of the kind

$$A = A_0 e^{-(k \cdot t)^\beta} \tag{21}$$

The stretch factor (β) indicates a time-dependent change in the rate constant (k), with $\beta = 1$ corresponding to the special case of single exponential kinetics. The kinetics become increasingly stretched with decreasing β. Stretched exponential behavior, which has also been referred to as "strange" kinetics [77], has recently been observed both in theoretical [78, 79] and in experimental work on biological systems [80]. The best-studied experimental model is the structural relaxation of myoglobin at low temperature [80] or at high solvent viscosity [81]. A prerequisite for strange kinetics is a rough energy landscape with significant barriers between a large number of local minima [78]. Thus, it was argued that protein folding reactions might exhibit stretched behavior, under conditions where local minima on the folding landscape become stabilized (e.g., at low temperature). Up to date there are only a few experimental reports on strange kinetics in protein folding, which describe the refolding of ubiquitin [82], phosphoglycerate kinase [82], and a WW domain at low temperature [83] measured in laser temperature-jump experiments on the microsecond time scale. The difficulty of unambiguously assigning the observed kinetics to stretched behavior may be seen from the fact that the kinetics only cover about 2.5 orders of magnitudes of life-times and can also be described by the sum of two or three exponentials. Nevertheless, these first indications of non-exponential behavior in protein folding on the very fast time scale provide an interesting new aspect to the understanding of energy landscapes for protein folding (see Chapter 14 for a more detailed discussion of non-exponential kinetics in protein folding).

12.1.5
Conclusions and Outlook

A quantitative treatment of folding kinetics is a prerequisite for elucidating mechanisms of protein folding. We have seen that concepts from classical reaction kinetics can be applied to protein folding as long as different ensembles of states are separated by significant barriers ($>5kT$). In most cases a discrimination between different folding mechanisms is not possible without the determination of the time course of native molecules. Thus, interrupted refolding experiments in combination with the complete denaturant dependence of all observable rate constants should be routinely used to analyzed complex folding reactions.

12.1.6
Protocols – Analytical Solutions of Three-state Protein Folding Models

12.1.6.1
Triangular Mechanism

The most general three-state mechanism is the triangular model is given by

$$U \underset{k_{NU}}{\overset{k_{UN}}{\rightleftharpoons}} NH \quad \text{with } I \text{ via } k_{UI}, k_{IU}, k_{NI}, k_{IN} \tag{22}$$

The characteristic equation for this mechanism is

$$f(\lambda) = \lambda^2 - \lambda(k_{UI} + k_{IU} + k_{UN} + k_{NU} + k_{NI} + k_{IN}) + (\gamma_1 + \gamma_2 + \gamma_3) = 0 \tag{23}$$

with

$$\begin{aligned}\gamma_1 &= k_{NI}k_{IU} + k_{NU}k_{IU} + k_{IN}k_{NU} \\ \gamma_2 &= k_{NU}k_{UI} + k_{UI}k_{NI} + k_{UN}k_{NI} \\ \gamma_3 &= k_{UI}k_{IN} + k_{IU}k_{UN} + k_{UN}k_{IN}\end{aligned} \tag{24}$$

The apparent rate constants, λ_1 and λ_2 are the solution of Eq. (23) and are given by:

$$\lambda_{1,2} = \frac{-B \pm \sqrt{B^2 - 4C}}{2} \tag{25}$$

with $B = -(k_{UN} + k_{NU} + k_{UI} + k_{IU} + k_{IN} + k_{NI})$ and $C = \gamma_1 + \gamma_2 + \gamma_3$.

The time-dependent behavior of the different kinetic species for refolding starting from the unfolded state ($U_0 = 1; I_0 = 0; N_0 = 0$) is given by

$$\begin{aligned} U(t) &= U_0 \left(\frac{\lambda_1(k_{UN} + k_{UI}) - (\gamma_2 + \gamma_3)}{\lambda_1(\lambda_1 - \lambda_2)} e^{-\lambda_1 t} + \frac{\gamma_2 + \gamma_3 - \lambda_2(k_{UN} + k_{UI})}{\lambda_2(\lambda_1 - \lambda_2)} e^{-\lambda_2 t} + \frac{\gamma_1}{\lambda_1 \lambda_2} \right) \\ I(t) &= U_0 \left(\frac{\gamma_2 - k_{UI}\lambda_1}{\lambda_1(\lambda_1 - \lambda_2)} e^{-\lambda_1 t} + \frac{k_{UI}\lambda_2 - \gamma_2}{\lambda_2(\lambda_1 - \lambda_2)} e^{-\lambda_2 t} + \frac{\gamma_2}{\lambda_1 \lambda_2} \right) \\ N(t) &= U_0 \left(\frac{\gamma_3 - k_{UN}\lambda_1}{\lambda_1(\lambda_1 - \lambda_2)} e^{-\lambda_1 t} + \frac{k_{UN}\lambda_2 - \gamma_3}{\lambda_2(\lambda_1 - \lambda_2)} e^{-\lambda_2 t} + \frac{\gamma_3}{\lambda_1 \lambda_2} \right) \end{aligned} \tag{26}$$

where $U(t)$, $I(t)$, and $N(t)$ represent the concentrations of the respective species at a given time t. The pre-exponential factors give the amplitude of the observable rate constant (see Eq. (2)) for monitoring the time course of the respective species and the constant terms represent the equilibrium concentrations of each species. The principle of microscopic reversibility provides an additional constraint for fitting the experimental data to the triangular mechanism

$$\frac{k_{UI} + k_{IN} + k_{NU}}{k_{IU} + k_{NI} + k_{UN}} = 1 \tag{27}$$

The time-dependent behavior of the different species for starting from the native state ($N_0 = 1$; $U_0 = 0$; $I_0 = 0$) or from the intermediate ($I_0 = 1$; $U_0 = 0$; $N_0 = 0$) are obtained by substituting the respective microscopic rate constant in Eq. (26).

12.1.6.2
On-pathway Intermediate

$$U \underset{k_{IU}}{\overset{k_{UI}}{\rightleftharpoons}} I \underset{k_{NI}}{\overset{k_{IN}}{\rightleftharpoons}} N \tag{28}$$

The characteristic equation for the on-pathway mechanism is given by

$$f(\lambda) = \lambda^2 - \lambda \cdot (k_{UI} + k_{IU} + k_{IN} + k_{NI}) + (k_{UI}k_{IN} + k_{UI}k_{NI} + k_{IU}k_{NI}) = 0 \tag{29}$$

The apparent rate constants, λ_1 and λ_2 are the solution of Eq. (29) (see Section 12.1.6.1). The time-dependent behavior of the different kinetic species for refolding starting from the unfolded state ($U_0 = 1$; $I_0 = 0$; $N_0 = 0$) is given by

$$U(t) = U_0 \left(\frac{k_{UI}(\lambda_1 - k_{IN} - k_{NI})}{\lambda_1(\lambda_1 - \lambda_2)} e^{-\lambda_1 t} + \frac{k_{UI}(k_{NI} + k_{IN} - \lambda_2)}{\lambda_2(\lambda_1 - \lambda_2)} e^{-\lambda_2 t} + \frac{k_{NI}k_{IU}}{\lambda_1 \lambda_2} \right)$$

$$I(t) = U_0 \left(\frac{k_{UI}(k_{NI} - \lambda_1)}{\lambda_1(\lambda_1 - \lambda_2)} e^{-\lambda_1 t} + \frac{k_{UI}(\lambda_2 - k_{NI})}{\lambda_2(\lambda_1 - \lambda_2)} e^{-\lambda_2 t} + \frac{k_{UI}k_{NI}}{\lambda_1 \lambda_2} \right) \tag{30}$$

$$N(t) = U_0 \left(\frac{k_{UI}k_{IN}}{\lambda_1(\lambda_1 - \lambda_2)} e^{-\lambda_1 t} + \frac{-k_{UI}k_{IN}}{\lambda_2(\lambda_1 - \lambda_2)} e^{-\lambda_2 t} + \frac{k_{UI}k_{IN}}{\lambda_1 \lambda_2} \right)$$

For initial conditions $I_0 = 1$; $U_0 = 0$; $N_0 = 0$ we yield

$$U(t) = I_0 \left(\frac{k_{IU}(k_{NI} - \lambda_1)}{\lambda_1(\lambda_1 - \lambda_2)} e^{-\lambda_1 t} + \frac{k_{IU}(\lambda_2 - k_{NI})}{\lambda_2(\lambda_1 - \lambda_2)} e^{-\lambda_2 t} + \frac{k_{NI}k_{IU}}{\lambda_1 \lambda_2} \right)$$

$$I(t) = I_0 \left(\frac{(k_{UI} - \lambda_1)(k_{NI} - \lambda_1)}{\lambda_1(\lambda_1 - \lambda_2)} e^{-\lambda_1 t} + \frac{(\lambda_2 - k_{NI})(k_{UI} - \lambda_2)}{\lambda_2(\lambda_1 - \lambda_2)} e^{-\lambda_2 t} + \frac{k_{UI}k_{NI}}{\lambda_1 \lambda_2} \right) \tag{31}$$

$$N(t) = I_0 \left(\frac{k_{IN}(k_{UI} - \lambda_1)}{\lambda_1(\lambda_1 - \lambda_2)} e^{-\lambda_1 t} + \frac{k_{IN}(\lambda_2 - k_{UI})}{\lambda_2(\lambda_1 - \lambda_2)} e^{-\lambda_2 t} + \frac{k_{UI}k_{IN}}{\lambda_1 \lambda_2} \right)$$

12.1.6.3
Off-pathway Mechanism

$$I \underset{k_{UI}}{\overset{k_{IU}}{\rightleftharpoons}} U \underset{k_{NU}}{\overset{k_{UN}}{\rightleftharpoons}} N \tag{32}$$

The characteristic equation for the off-pathway mechanism is given by

$$f(\lambda) = \lambda^2 - \lambda(k_{UI} + k_{IU} + k_{UN} + k_{NU}) + (k_{IU}k_{UN} + k_{IU}k_{NU} + k_{UI}k_{NU}) = 0 \tag{33}$$

The apparent rate constants, λ_1 and λ_2 are the solution of Eq. (33) (see Section 12.1.6.1). The time-dependent behavior of the different kinetic species for refolding starting from the unfolded state ($U_0 = 1$; $I_0 = 0$; $N_0 = 0$) is given by

$$U(t) = U_0 \left(\frac{(k_{IU} - \lambda_1)(k_{NU} - \lambda_1)}{\lambda_1(\lambda_1 - \lambda_2)} e^{-\lambda_1 t} + \frac{(\lambda_2 - k_{NU})(k_{IU} - \lambda_2)}{\lambda_2(\lambda_1 - \lambda_2)} e^{-\lambda_2 t} + \frac{k_{IU}k_{NU}}{\lambda_1 \lambda_2} \right)$$

$$I(t) = U_0 \left(\frac{k_{UI}(k_{NU} - \lambda_1)}{\lambda_1(\lambda_1 - \lambda_2)} e^{-\lambda_1 t} + \frac{k_{UI}(\lambda_2 - k_{NU})}{\lambda_2(\lambda_1 - \lambda_2)} e^{-\lambda_2 t} + \frac{k_{NU}k_{UI}}{\lambda_1 \lambda_2} \right) \tag{34}$$

$$N(t) = U_0 \left(\frac{k_{UN}(k_{IU} - \lambda_1)}{\lambda_1(\lambda_1 - \lambda_2)} e^{-\lambda_1 t} + \frac{k_{UN}(\lambda_2 - k_{IU})}{\lambda_2(\lambda_1 - \lambda_2)} e^{-\lambda_2 t} + \frac{k_{IU}k_{UN}}{\lambda_1 \lambda_2} \right)$$

The solutions for starting from the native state ($N_0 = 1$; $U_0 = 0$; $I_0 = 0$) or from the intermediate ($I_0 = 1$; $U_0 = 0$; $N_0 = 0$) are obtained in the same way as discussed in Section 12.1.6.2.

12.1.6.4
Folding Through an On-pathway High-Energy Intermediate

Apparent two-state folding through a high-energy intermediate (see Figure 12.1.7A) can be described by a three-state on-pathway model

$$U \underset{k_{IU}}{\overset{k_{UI}}{\rightleftharpoons}} I^* \underset{k_{NI}}{\overset{k_{IN}}{\rightleftharpoons}} N \tag{35}$$

As discussed above, the kinetics of folding through a high-energy intermediate cannot be distinguished from two-state folding, when monitored at a single denaturant concentration. However, the presence of a kink in the chevron plot and the resulting altered slope at high denaturant concentrations (Figure 12.1.7B) allows a characterization of both transition states [57]. Comparing two-state folding (Eq. (3))

12.1.6 Protocols – Analytical Solutions of Three-state Protein Folding Models

with Eq. (35) allows the rate constants k_f and k_u to be expressed for apparent two-state folding as a function of the microscopic rate constants k_{ij} for the three-state model for the different regimes shown in Figure 12.1.7A. At low denaturant concentrations (TS1 limit) formation of the high-energy intermediate (k_{UI}) is rate-limiting for folding ($k_{UI} \ll k_{IN}$) and thus we can approximate k_f in the TS1 limit ($k_f(\text{TS1})$) as

$$k_f(\text{TS1}) = k_{UI} \tag{36}$$

The unfolding reaction in the TS1 limit requires going through I*. Since I* is higher in free energy than N ($k_{NI} \ll k_{IN}$) and TS1 represents the highest barrier, the crossing of transition state 2 can be treated as a pre-equilibrium. Thus, k_u in the TS1 limit ($k_u(\text{TS1})$) is given by

$$k_u(\text{TS1}) = \frac{k_{NI}}{k_{IN}} k_{I^*U} = k_{NI} \left(\frac{k_{IN}}{k_{IU}}\right)^{-1} \tag{37}$$

For folding and unfolding at high denaturant concentrations TS2 represents the highest barrier (TS2 limit). Thus, in analogy to Eqs (36) and (37), k_f and k_u can be expressed in the TS2 limit as:

$$k_f(\text{TS2}) = \frac{k_{UI}}{k_{IU}} k_{IN} = k_{UI} \frac{k_{IN}}{k_{IU}} \tag{38}$$

$$k_u(\text{TS2}) = k_{NI} \tag{39}$$

These considerations show that the presence of a kink in the chevron plot allows the determination of the ratio k_{IU}/k_{IN}, which is equivalent to the equilibrium constant between the two transition states ($K_{\text{TS1/TS2}}$) and can be converted into the difference in free energy between the two transition states ($\Delta G°_{\text{TS1/TS2}}$)

$$k_{IN}/k_{IU} = K_{\text{TS1/TS2}} = e^{-(1/RT)(\Delta G°^{\ddagger}(\text{TS2}) - \Delta G°^{\ddagger}(\text{TS1}))} = e^{-\Delta G°_{\text{TS2/TS1}}/RT} \tag{40}$$

Accordingly, the stability of I* cannot be determined, since it does not become populated, but the difference in free energy between both transition states ($\Delta G°_{\text{TS2/TS1}}$) can be obtained. It should be noted that this analysis does not make any assumptions on the stability of the hypothetical intermediate besides that it is always less stable than U and N. Fitting the GdmCl dependence of ln λ to the analytical solutions of the three-state model (Eq. (29)) with λ_1 and λ_2 given by Refs [4] and [24]

$$\lambda_{1,2} = \frac{-B \pm \sqrt{B^2 - 4C}}{2} \tag{41}$$

with

$$B = -(k_{UI} + k_{IU} + k_{IN} + k_{NI})$$
$$C = k_{UI}(k_{IN} + k_{NI}) + k_{IU}k_{NI}$$
(42)

therefore allows the determination of k_{UI} and k_{NI} and their denaturant dependencies m_{UI} and m_{NI}, respectively. It further yields the parameters k_{IN}/k_{IU} and $m_{IN} - m_{IU}$. Since only the ratios k_{IN}/k_{IU} and $m_{IN} - m_{IU}$ are defined for folding through a high-energy intermediate, these ratios have to be used for data fitting [57]. Folding through a high-energy intermediate shows apparent two-state behavior and thus only the smaller one of the two apparent rate constants (λ_1) should have the complete folding and unfolding amplitude ($A_1 = 1$) whereas $A_2 = 0$ (see Ref. [57]).

Acknowledgments

We thank Manuela Schätzle and Beat Fierz for comments on the manuscript and all members of the Kiefhaber lab for help and discussion.

References

1 SÁNCHEZ, I. E. & KIEFHABER, T. (2003). Hammond behavior versus ground state effects in protein folding: evidence for narrow free energy barriers and residual structure in unfolded states. *J. Mol. Biol.* 327, 867–884.

2 KIEFHABER, T. (1995). Protein folding kinetics. In *Methods in Molecular Biology, Vol. 40: Protein Stability and Folding Protocols* (SHIRLEY, B. A., ed.), pp. 313–341. Humana Press, Totowa, NJ.

3 BIERI, O. & KIEFHABER, T. (2000). Kinetic models in protein folding. In *Protein Folding: Frontiers in Molecular Biology* 2nd edn (PAIN, R., ed.), pp. 34–64. Oxford University Press, Oxford.

4 SZABO, Z. G. (1969). Kinetic characterization of complex reaction systems. In *Comprehensive Chemical Kinetics* (BAMFORD, C. H. & TIPPER, C. F. H., eds), Vol. 2, pp. 1–80. 7 vols. Elsevier Publishing Company, Amsterdam.

5 MOORE, J. W. & PEARSON, R. G. (1981). *Kinetics and Mechanisms.* John Wiley & Sons, New York.

6 JACKSON, S. E. (1998). How do small single-domain proteins fold? *Folding Design* 3, R81–R91.

7 EYRING, H. (1935). The activated complex in chemical reactions. *J. Chem. Phys.* 3, 107–115.

8 EVANS, M. G. & POLANYI, M. (1935). Some applications of the transition state method to the calculation of reaction velocities, especially in solution. *Trans. Faraday Soc.* 31, 875–885.

9 KRIEGER, F., FIERZ, B., BIERI, O., DREWELLO, M. & KIEFHABER, T. (2003). Dynamics of unfolded polypeptide chains as model for the earliest steps in protein folding. *J. Mol. Biol.* 332, 265–274.

10 GREENE, R. F. J. & PACE, C. N. (1974). Urea and guanidine-hydrochloride denaturation of ribonuclease, lysozyme, alpha-chyomtrypsinn and beta-lactoglobulin. *J. Biol. Chem.* 249, 5388–5393.

11 SANTORO, M. M. & BOLEN, D. W. (1988). Unfolding free energy changes determined by the linear extrapolation method. 1. Unfolding of

phenylmethanesulfonyl alpha-chymotrypsin using different denaturants. *Biochemistry* 27, 8063–8068.

12 TANFORD, C. (1970). Protein Denaturation. Part C. Theoretical models for the mechanism of denaturation. *Adv. Prot. Chem.* 24, 1–95.

13 IKAI, A., FISH, W. W. & TANFORD, C. (1973). Kinetics of unfolding and refolding of proteins. II. Results for cytochrome c. *J. Mol. Biol.* 73, 165–184.

14 MATTHEWS, C. R. (1987). Effect of point mutations on the folding of globular proteins. *Methods Enzymol.* 154, 498–511.

15 MYERS, J. K., PACE, C. N. & SCHOLTZ, J. M. (1995). Denaturant m values and heat capacity changes: relation to changes in accessible surface areas of protein unfolding. *Protein Sci.* 4, 2138–2148.

16 LEFFLER, J. E. (1953). Parameters for the description of transition states. *Science* 117, 340–341.

17 LEFFLER, J. E. & GRUNWALD, E. (1963). *Rates and Equilibria of Organic Reactions*. Dover, New York.

18 JENCKS, W. P. (1969). *Catalysis in Chemistry and Enzymology*. McGraw-Hill, New York.

19 SÁNCHEZ, I. E. & KIEFHABER, T. (2003). Non-linear rate-equilibrium free energy relationships and Hammond behavior in protein folding. *Biophys. Chem.* 100, 397–407.

20 IKAI, A. & TANFORD, C. (1973). Kinetics of unfolding and refolding of proteins. I. Mathematical Analysis. *J. Mol. Biol.* 73, 145–163.

21 TANFORD, C., AUNE, K. C. & IKAI, A. (1973). Kinetics of unfolding and refolding of proteins. III. Results for lysozyme. *J. Mol. Biol.* 73, 185–197.

22 KIEFHABER, T. (1995). Kinetic traps in lysozyme folding. *Proc. Natl Acad. Sci. USA* 92, 9029–9033.

23 KHORASANIZADEH, S., PETERS, I. D. & RODER, H. (1996). Evidence for a three-state model for protein folding from kinetic analysis of ubiquitin variants with altered core residues. *Nat. Struct. Biol.* 3, 193–205.

24 KIEFHABER, T., KOHLER, H. H. & SCHMID, F. X. (1992). Kinetic coupling between protein folding and prolyl isomerization. I. Theoretical models. *J. Mol. Biol.* 224, 217–229.

25 KIEFHABER, T. & SCHMID, F. X. (1992). Kinetic coupling between protein folding and prolyl isomerization. II. Folding of ribonuclease A and ribonuclease T1. *J. Mol. Biol.* 224, 231–240.

26 BIERI, O., WIRZ, J., HELLRUNG, B., SCHUTKOWSKI, M., DREWELLO, M. & KIEFHABER, T. (1999). The speed limit for protein folding measured by triplet-triplet energy transfer. *Proc. Natl Acad. Sci. USA* 96, 9597–9601.

27 BIERI, O. & KIEFHABER, T. (1999). Elementary steps in protein folding. *Biol. Chem.* 380, 923–929.

28 KUBELKA, J., HOFRICHTER, J. & EATON, W. A. (2004). The protein folding speed limit. *Curr. Opin. Struct. Biol.* 14, 76–88.

29 GAREL, J. R. & BALDWIN, R. L. (1973). Both the fast and slow folding reactions of ribonuclease A yield native enzyme. *Proc. Natl Acad. Sci. U.S.A.* 70, 3347–3351.

30 BRANDTS, J. F., HALVORSON, H. R. & BRENNAN, M. (1975). Consideration of the possibility that the slow step in protein denaturation reactions is due to *cis-trans* isomerism of proline residues. *Biochemistry* 14, 4953–4963.

31 KIEFHABER, T., GRUNERT, H. P., HAHN, U. & SCHMID, F. X. (1990). Replacement of a cis proline simplifies the mechanism of ribonuclease T1 folding. *Biochemistry* 29, 6475–6480.

32 LANG, K., SCHMID, F. X. & FISCHER, G. (1987). Catalysis of protein folding by prolyl isomerase. *Nature* 329, 268–270.

33 REIMER, U., SCHERER, G., DREWELLO, M., KRUBER, S., SCHUTKOWSKI, M. & FISCHER, G. (1998). Side-chain effects on peptidyl-prolyl cis/trans isomerization. *J. Mol. Biol.* 279, 449–460.

34 BALBACH, J. & SCHMID, F. X. (2000). Proline isomerization and its catalysis

in protein folding. In *Protein Folding: Frontiers in Molecular Biology* (PAIN, R., ed.). Oxford University Press, Oxford.

35 KIEFHABER, T., QUAAS, R., HAHN, U. & SCHMID, F. X. (1990). Folding of ribonuclease T1. 1. Existence of multiple unfolded states created by proline isomerization. *Biochemistry* 29, 3053–3061.

36 KIEFHABER, T., QUAAS, R., HAHN, U. & SCHMID, F. X. (1990). Folding of ribonuclease T1. 2. Kinetic models for the folding and unfolding reactions. *Biochemistry* 29, 3061–3070.

37 ODEFEY, C., MAYR, L. & SCHMID, F. X. (1995). Non-prolyl cis/trans peptide bond isomerization as a rate-determinig step in protein unfolding and refolding. *J. Mol. Biol.* 245, 69–78.

38 PAPPENBERGER, G., AYGÜN, H., ENGELS, J. W., REIMER, U., FISCHER, G. & KIEFHABER, T. (2001). Nonprolyl cis peptide bonds in unfolded proteins cause complex folding kinetics. *Nat. Struct. Biol.* 8, 452–458.

39 SCHERER, G., KRAMER, M. L., SCHUTKOWSKI, M., REIMER, U. & FISCHER, G. (1998). Barriers to rotation of secondary amide peptide bonds. *J. Am. Chem. Soc.* 120, 5568–5574.

40 COLON, W., WAKEM, L. P., SHERMAN, F. & RODER, H. (1997). Identification of the predominant non-native histidine ligand in unfolded cytochrome c. *Biochemistry* 36, 12535–12541.

41 YEH, S.-R., TAKAHASHI, S., FAN, B. & ROUSSEAU, D. L. (1998). Ligand exchange in unfolded cytochrome c. *Nat. Struct. Biol.* 4, 51–56.

42 YEH, S.-R. & ROUSSEAU, D. L. (1998). Folding intermediates in cytochrome c. *Nat. Struct. Biol.* 5, 222–228.

43 SEGEL, D., BACHMANN, A., HOFRICHTER, J., HODGSON, K., DONIACH, S. & KIEFHABER, T. (1999). Characterization of transient intermediates in lysozyme folding with time-resolved small angle X-ray scattering. *J. Mol. Biol.* 288, 489–500.

44 BACHMANN, A. & KIEFHABER, T. (2002). Test for cooperativity in the early kinetic intermediate in lysozyme folding. *Biophys. Chem.* 96, 141–151.

45 KUWAJIMA, K. (1989). The molten globule state as a clue for understanding the folding and cooperativity of globular-protein structure. *Proteins Struct. Funct. Genet.* 6, 87–103.

46 WU, L. C., PENG, Z.-Y. & KIM, P. S. (1995). Bipartite structure of the α-lactalbumin molten globule. *Nat. Struct. Biol.* 2, 281–286.

47 WU, L. C. & KIM, P. S. (1998). A specific hydrophobic core in the α-lactalbumin molten globule. *J. Mol. Biol.* 280, 175–182.

48 LUO, Y. & BALDWIN, R. L. (1999). The 28–111 disulfide bond contrains the a-lactalbumin molten globule and weakens its cooperativity of folding. *Proc. Natl Acad. Sci. USA* 96, 11283–11287.

49 LUO, Y., KAY, M. S. & BALDWIN, R. L. (1997). Cooperativity of folding of the apomyoglobin pH 4 intermediate studied by glycine and proline mutations. *Nat. Struct. Biol.* 4, 925–929.

50 SILOW, M. & OLIVEBERG, M. (1997). Transient aggregates in protein folding are easilty mistaken for folding intermediates. *Proc. Natl Acad. Sci. USA* 94, 6084–6086.

51 SCHMID, F. X. (1983). Mechanism of folding of ribonuclease A. Slow refolding is a sequential reaction via structural intermediates. *Biochemistry* 22, 4690–4696.

52 WILDEGGER, G. & KIEFHABER, T. (1997). Three-state model for lysozyme folding: triangular folding mechanism with an energetically trapped intermediate. *J. Mol. Biol.* 270, 294–304.

53 HILL, T. L. (1974). The sliding filament model of contraction of striated muscle. *Progr. Biophys. Mol. Biol.* 28, 267–340.

54 HILL, T. L. (1977). *Free Energy Transduction in Biology*. Academic Press, London.

55 WEISMANN, J. S. & KIM, P. S. (1991). Reexamination of the folding of BPTI: predominance of native intermediates. *Science* 253, 1386–1393.

56 KIEFHABER, T., GRUNERT, H. P., HAHN, U. & SCHMID, F. X. (1992).

Folding of RNase T1 is decelerated by a specific tertiary contact in a folding intermediate. *Proteins Struct. Funct. Genet.* 12, 171–179.

57 BACHMANN, A. & KIEFHABER, T. (2001). Apparent two-state tendamistat folding is a sequential process along a defined route. *J. Mol. Biol.* 306, 375–386.

58 SÁNCHEZ, I. E. & KIEFHABER, T. (2003). Evidence for sequential barriers and obligatory intermediates in apparent two-state protein folding. *J. Mol. Biol.* 325, 367–376.

59 JENCKS, W. P. (1980). When is an intermediate not an intermediate? Enforced mechanisms of general acid-base catalyzed, carbonation, carbanion, and ligand exchange reactions. *Acc. Chem. Res.* 13, 161–169.

60 KIEFHABER, T., BACHMANN, A., WILDEGGER, G. & WAGNER, C. (1997). Direct measurements of nucleation and growth rates in lysozyme folding. *Biochemistry* 36, 5108–5112.

61 WALKENHORST, W. F., GREEN, S. & RODER, H. (1997). Kinetic evidence for folding and unfolding intermediates in staphylococcal nuclease. *Biochemistry* 36, 5795–5805.

62 BLAKE, C. C. F., KOENIG, D. F., MAIR, G. A., NORTH, A. C. T., PHILLIPS, D. C. & SARMA, V. F. (1967). Structure of hen egg-white lysozyme. *Nature* 206, 757–761.

63 BLAKE, C. C. F., MAIR, G. A., NORTH, A. C. T., PHILLIPS, D. C. & SARMA, V. F. (1967). On the conformation of the hen egg-white lysozyme molecule. *Proc. R. Soc. B* 167, 365–377.

64 KUWAJIMA, K., HIRAOKA, Y., IKEGUCHUI, M. & SUGAI, S. (1985). Comparison of the transient folding intermediates in lysozyme and α-lactalbumin. *Biochemistry* 24, 874–881.

65 CHAFFOTTE, A. F., GUILLOU, Y. & GOLDBERG, M. E. (1992). Kinetic resolution of peptide bond and side-chain far-UV CD during folding of HEWL. *Biochemistry* 31, 9694–9702.

66 ITZHAKI, L. S., EVANS, P. A., DOBSON, C. M. & RADFORD, S. E. (1994). Tertiary interactions in the folding pathway of hen lysozyme: kinetic studies using fluorescent probes. *Biochemistry* 33, 5212–5220.

67 RADFORD, S. E., BUCK, M., TOPPING, K. D., DOBSON, C. M. & EVANS, P. A. (1992). Hydrogen exchange in native and denatured states of hen egg-white lysozyme. *Proteins* 14, 237–248.

68 MIRANKER, A. D., ROBINSON, C. V., RADFORD, S. E., APLIN, R. T. & DOBSON, C. M. (1993). Detection of transient protein folding populations by mass spectroscopy. *Science* 262, 896–900.

69 MATAGNE, A., CHUNG, E. W., BALL, L. J., RADFORD, S. E., ROBINSON, C. V. & DOBSON, C. M. (1998). The origin of the α-domain intermediate in the folding of hen lysozyme. *J. Mol. Biol.* 277, 997–1005.

70 KULKARNI, S. K., ASHCROFT, A. E., CAREY, M., MASSELOS, D., ROBINSON, C. V. & RADFORD, S. E. (1999). A near-native state on the slow refolding pathway of hen lysozyme. *Protein Sci.* 8, 35–44.

71 KLEIN-SEETHARAMAN, J., OIKAWA, M., GRIMSHAW, S. B., WIRMER, J., DUCHARDT, E., UEDA, T., et al. (2002). Long-range interactions within a nonnative protein. *Science* 295, 1719–1722.

72 ROTHWARF, D. M. & SCHERAGA, H. A. (1996). Role of non-native aromatic and hydrophobic interactions in the folding of hen egg white lysozyme. *Biochemistry* 35, 13797–13807.

73 BIERI, O. & KIEFHABER, T. (2001). Origin of apparent fast and non-exponential kinetics of lysozyme folding measured in pulse labeling experiments. *J. Mol. Biol.* 310, 919–935.

74 BAI, Y., MILNE, J. S., MAYNE, L. & ENGLANDER, W. (1993). Primary structure effects on peptide group hydrogen exchange. *Proteins Struct. Funct. Genet.* 17, 75–86.

75 BIERI, O., WILDEGGER, G., BACHMANN, A., WAGNER, C. & KIEFHABER, T. (1999). A salt-induced intermediate is on a new parallel pathway of lysozyme folding. *Biochemistry* 38, 12460–12470.

76 KOHLRAUSCH, R. (1847). Über das

Dellmann'sche Elektrometer. *Ann. Phys. Chem.* 11, 353–405.

77 SHLESINGER, M. F., ZASLAVSKY, G. M. & KLAFTER, J. (1993). Strange kinetics. *Nature* 363, 31–37.

78 SAVEN, J. G., WANG, J. & WOLYNES, P. G. (1994). Kinetics of protein folding: The folding dynamics of globally connected rough energy landscapes with biases. *J. Chem. Phys.* 101, 11037–11043.

79 SKOROBOGATIY, M., GUO, H. & ZUCKERMANN, M. (1998). Non-Arrhenius behavior in the realaxation of model proteins. *J. Chem. Phys.* 109, 2528–2535.

80 AUSTIN, R. H., BEESON, K. W., EISENSTEIN, L., FRAUENFELDER, H. & GUNSALUS, I. C. (1975). Dynamics of ligand binding to myoglobin. *Biochemistry* 14, 5355–5373.

81 HAGEN, S. J., HOFRICHTER, J. & EATON, W. A. (1995). Protein kinetics in a glass at room temperature. *Science* 269, 959–962.

82 SABELKO, J., ERVIN, J. & GRUEBELE, M. (1999). Observation of strange kinetics in protein folding. *Proc. Natl Acad. Sci. USA* 96, 6031–6036.

83 YANG, W. Y. & GRUEBELE, M. (2003). Folding at the speed limit. *Nature* 423, 193–197.

12.2
Characterization of Protein Folding Barriers with Rate Equilibrium Free Energy Relationships

Thomas Kiefhaber, Ignacio E. Sánchez, and Annett Bachmann

12.2.1
Introduction

The characterization of barriers and mechanisms of chemical reactions has a long tradition in physico-organic chemistry [1]. Several concepts have been developed to gain information on the structural and thermodynamic properties of transition states. Rate equilibrium free energy relationships (REFERs) have proved to be a powerful tool for the characterization of the properties and the shape of free energy barriers. REFERs have also been applied to the study of complex biochemical reactions like enzyme kinetics [2]. This chapter will show how REFERs can be used to gain information on free energy barriers for protein folding reactions and will discuss results in terms of a general picture of the properties of protein folding transition states. This analysis implies that protein folding encounters major free energy barriers which separate ensembles of different kinetic states on the free energy landscape. This assumption seems to be supported by experimental results (see Chapter 12.1).

12.2.2
Rate Equilibrium Free Energy Relationships

For many reactions in organic chemistry the changes in activation free energy ($\Delta G^{\circ\ddagger}$) induced by changes in solvent conditions or by modifications in structure are linearly related to the corresponding change in equilibrium free energy (ΔG°) between reactant and product [3]. A proportionality constant, α_x, was defined by Leffler to quantify the energetic sensitivity of the transition state relative to the ground states with respect to a perturbation, ∂x [3]:

$$\alpha_x = \frac{\partial \Delta G^{\circ\ddagger}/\partial x}{\partial \Delta G^\circ/\partial x} \qquad (1)$$

where α_x is a measure for the position of the transition state along the reaction co-

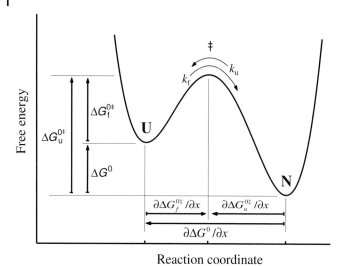

Fig. 12.2.1. Free energy profile for a two-state folding reaction. The sensitivity of the reaction to changes in a parameter ∂x as a test for the respective reaction coordinate according to the Leffler postulate (see Eq. (1)) are indicated.

ordinate probed by ∂x and it can be used to gain structural and thermodynamic information on the transition state. If $\alpha_x = 1$ the transition state has the same property as the product and for $\alpha_x = 0$ it has the same properties as the educt. Figure 12.2.1 shows the relationship between the equilibrium free energies and the activation free energies for a two-state reaction as well as the relationship between $\partial \Delta G°/\partial x$ and $\partial \Delta G°^{\ddagger}/\partial x$. Obviously, $\partial \Delta G°/\partial x$ is a measure for the length of the reaction coordinate, whereas $\partial \Delta G°^{\ddagger}/\partial x$ is a measure for the respective change between the educt and the transition state. Frequently used perturbations are changes in pressure, temperature, or structure, which will yield information on the position of the transition state in respect to its volume, its entropy, or its interactions at the site of structural change, respectively (see below). Since these relationships are empirical rather than rigorously derived from the laws of thermodynamics they are commonly termed "extrathermodynamic relationships" [1].

For many reactions the α_x-values are constant over a broad range of $\Delta G°$ indicating little effect of the perturbations on the structure of the transition state (linear REFERs) [1]. This can be used to infer structural and thermodynamic properties of a transition state (see Section 12.2.2.1). For other reactions, α_x is sensitive to changes in $\Delta G°$ which leads to nonlinear REFERs indicating changes in the position of the transition state relative to the ground states (cf. Figure 12.2.1). It was shown that the characterization of nonlinear REFERs can give valuable information on the mechanism of a reaction and on the shape of the transition barrier since nonlinearities can have different origins: (1) a movement of the position of

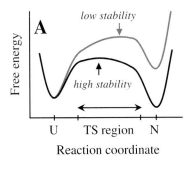
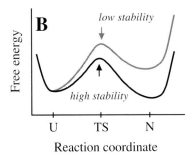

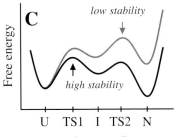

Fig. 12.2.2. Schematic representation of the response of different types of free energy barriers to the same perturbation. The position of the transition state along the reaction coordinate is more sensitive to the perturbation if the free energy shows a broader maximum (A) than if the maximum is narrow (B). An apparent movement of the position of the transition state can also be due to a switch between consecutive transition states on a linear pathway (C). The position of the highest point along the barrier region is indicated by an arrow.

the transition state along a broad barrier region (Hammond behavior; Figure 12.2.2A,B) [4, 5]; (2) a change in the rate-limiting step on a sequential pathway (Figure 12.2.2C) [1]; (3) a change in the mechanism of the reaction due to a switch between parallel pathways [1]; or (4) structural changes in the ground state(s) (ground state effects) [6]. If the rate-limiting step (Figure 12.2C) or the mechanism of a reaction changes, a discrete jump in the position of the transition state along the reaction coordinate will be observed. A gradual transition state movement (Figure 12.2.2A) will result in a rather smooth structural shift of the position of transition state upon perturbation. Detailed analysis of nonlinear REFERs have been widely used to characterize the shape of free energy barriers and to elucidate reaction mechanisms in organic chemistry and of complex biochemical reactions like the catalytic mechanisms of enzymes [2]. In the following, we will discuss how these concepts can be applied to characterize the barriers for protein folding reactions and results from linear and nonlinear REFERs will be presented.

12.2.2.1
Linear Rate Equilibrium Free Energy Relationships in Protein Folding

REFERs for protein folding reactions can be derived using various perturbations. Let us consider the Gibbs fundamental equation

$$d\Delta G^\circ = \Delta V^\circ \, dp - \Delta S^\circ \, dT + \sum \Delta \mu_i^\circ \, dn_i \tag{2}$$

where $\Delta G^\circ, \Delta V^\circ, \Delta S^\circ$, and $\sum \Delta \mu_i^\circ$ are the differences in Gibbs free energy, volume, entropy, and chemical potential, respectively, between the unfolded and the native states. In protein folding the most common perturbation of chemical potential is the addition of chemical denaturants (D) like urea and guanidinium chloride (GdmCl), which were shown to have a linear effect on ΔG° [7, 8] (see Chapters 3 and 12.1). Including the empirically observed effect of a chemical denaturant, Eq. (2) can be written as

$$d\Delta G^\circ = \Delta V^\circ \, dp - \Delta S^\circ \, dT + m \cdot [D] \tag{3}$$

Assuming a free energy barrier between the unfolded and the native protein, the Gibbs equation can be applied to the activation free energies of the folding ($\Delta G_f^{\circ\ddagger}$) and unfolding ($\Delta G_u^{\circ\ddagger}$) reaction:

$$d\Delta G_{f,u}^{\circ\ddagger} = \Delta V_{f,u}^{\circ\ddagger} \, dp - \Delta S_{f,u}^{\circ\ddagger} \, dT + m_{f,u} \cdot [D] \tag{4}$$

since $\Delta G_f^{\circ\ddagger}$ and $\Delta G_u^{\circ\ddagger}$ were also shown to be linearly dependent on [D] [9] (see also Chapter 12.1). Combining Eqs (1)–(3) shows that different properties of the transition state and thus different reaction coordinates can be probed by applying REFERs to protein folding. α_p and α_T-values can be obtained by changes in pressure at constant temperature or changes in temperature at constant pressure, respectively [10, 11].

$$\alpha_p = \frac{\partial \Delta G_f^{\circ\ddagger}/\partial p}{\partial \Delta G^\circ/\partial p} = \frac{\Delta V_f^{\circ\ddagger}}{\Delta V^\circ} \tag{5}$$

$$\alpha_T = \frac{\partial \Delta G_f^{\circ\ddagger}/\partial T}{\partial \Delta G^\circ/\partial T} = \frac{\Delta S_f^{\circ\ddagger}}{\Delta S^\circ} \tag{6}$$

This gives information on the volume and on the entropy of the transition state, respectively. From Eq. (6) we can further calculate the change in enthalpy for formation of the transition state using the Gibbs-Helmholtz equation. The value of α_T changes with temperature due to the change in molar heat capacity (ΔC_p°) typically associated with protein folding reactions [12]. This allows the definition of another reaction coordinate [13]

$$\alpha_C = \frac{\Delta C_{p(f)}^{o\ddagger}}{\Delta C_p^o} \tag{7}$$

α_C allows the characterization of a transition state in terms of its relative solvent exposure, since ΔC_p^o mainly arises from differences in interactions of the protein with the solvent [14].

For two-state folding the logarithm of the single observable apparent rate constant, λ ($\lambda = k_f + k_u$) vs. chemical denaturant concentration yields a V-shaped curve [15, 16], commonly called a chevron plot (Figure 12.2.3B, see also Chapter 12.1). This plot gives information on the refolding reaction at low denaturant concentrations (the refolding limb) and information on the unfolding reaction at high denaturant concentrations (the unfolding limb). The V-shaped form of the chevron plot is the result of linear changes in $\Delta G_f^{o\ddagger}$ and $\Delta G_u^{o\ddagger}$ with denaturant concentration. Since ΔG^o is also linearly dependent on denaturant concentration (Figure 12.2.3A), the proportionality constants, m_x, are defined as:

$$m_{eq} = \frac{\partial \Delta G^o}{\partial [\text{Denaturant}]} \tag{8}$$

$$m_{f,u} = \frac{\partial \Delta G_{f,u}^{o\ddagger}}{\partial [\text{Denaturant}]} \tag{9}$$

This allows us to use the m_f- and m_{eq}-values to calculate a denaturant-induced REFER to obtain α_D [9] according to

$$\alpha_D = \frac{\partial \Delta G_f^{o\ddagger}/\partial [\text{Denaturant}]}{\partial \Delta G^o/\partial [\text{Denaturant}]} = \frac{m_f}{m_{eq}} \tag{10}$$

α_D reflects the relative sensitivity of the transition state to changes in denaturant concentration. Since the m_{eq}-value was shown to be proportional to the changes in accessible surface area upon unfolding of the protein [17], α_D is interpreted as the relative amount of accessible surface area buried in the transition state and is expected to correlate with α_C [17].

A practical way to analyze REFERs and to test for linearity is to plot the rate constant for folding (k_f) vs. the equilibrium constant (K_{eq}) determined under the same conditions. The slope of this plot (Leffler plot) gives α_x. As an example, Figure 12.2.3C shows a Leffler plot of the effect of GdmCl concentration on k_f and K_{eq} for tendamistat folding. The refolding and unfolding limbs of the chevron plot for tendamistat are perfectly linear over the complete range of denaturant concentrations (Figure 12.2.3B) and thus also the slope of the Leffler plot is constant between 0 and 8 M GdmCl with an α_D-value of 0.67 ± 0.02. This indicates that 67% of the change in accessible surface area (ASA) between U and N has already occurred in the transition state. The same results will be obtained by plotting $\Delta G^{o\ddagger}$ vs. ΔG^o, as in the original Leffler formalism (cf. Figure 12.2.5). However, calculation of absolute values for $\Delta G_f^{o\ddagger}$ requires the knowledge of the pre-exponential fac-

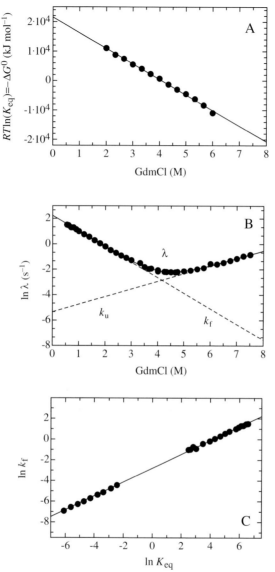

Fig. 12.2.3. Relationship between equilibrium ($\Delta G°$) and activation free energies ($\Delta G^{o\ddagger}$) for tendamistat folding at pH 2. A) Change in $-\Delta G°$ with GdmCl concentration calculated using the equilibrium constant determined from the data shown in Chapter 12.1, Fig. 12.1.2A. B) Effect of GdmCl on the apparent folding rate constant ($\lambda = k_f + k_u$). The V-shaped plot is commonly termed "chevron plot" and allows calculation of k_f and k_u as indicated (cf. Eqs (8) and (9) and Chapter 12.1. C) Leffler plot comparing the effect of GdmCl concentration on the equilibrium constant (K_{eq}) and folding rate constant (k_f) calculated from the data shown in panels A and B. The slope corresponds to the α_D-value and a linear fit (solid line) gives a value of 0.67. The folding rate constant (k_f) under unfolding conditions was calculated from the unfolding rate constant (k_u) using $k_f = K_{eq} \cdot k_u$. Only data from the linear regions of the chevron plot ($|\ln K_{eq}| > 2$) were used since near the minimum of the chevron plot both k_f and k_u significantly contribute to λ.

tor for folding. This will not influence the slope of the Leffler plot but it poses some uncertainty on the absolute value of $\Delta G_f^{o\ddagger}$.

Other solvent additives such as alcohols (2,2,2-trifluoroethanol), polyols (glycerol, sugars), salts (Na_2SO_4, NaCl), and D_2O as solvent can also lead to a change in protein stability and can therefore be used to define the corresponding α-values [10, 11] (see Table 12.2.1). Specific ligand binding is another source of changes in protein stability and thus allows characterization of the transition state [9]. Many proteins specifically bind ions, substrates, or cofactors, and all bind hydrogen ions at ionizable side chains, which can be used to compare the effect on kinetics and thermodynamics. The changes in free energy upon binding are commonly proportional to the logarithm of the ligand concentration and can be used to determine an α_L-value [18–23], which represents the ligand-binding ability of the transition state.

Equations (5)–(7) and (10) can be considered as medium- or solvent-induced REFERs. A detailed structural information on interactions formed in the transition state can be obtained by analyzing structure-induced REFERs [16, 24, 25], which compare the effect of amino acid replacements introduced by site-directed mutagenesis on $\Delta G_f^{o\ddagger}$ and ΔG°.

$$\alpha_S = \phi_f = \frac{\partial \Delta G_f^{o\ddagger}/\partial \text{Structure}}{\partial \Delta G^\circ/\partial \text{Structure}} \tag{11}$$

α_S, which is commonly called ϕ_f in protein folding, reports on the energetics of the interactions formed by a side chain with the rest of the protein in the transition state relative to the native state (see Chapter 13). The free energy of the unfolded state serves as a reference. Usually, only a single-point mutation is made at an individual position to calculate a ϕ_f-value, which is a major source for uncertainty in ϕ_f-value analysis and does not allow to test for linearity in the REFER [26]. To gain more detailed information on the interactions in the transition state of a certain amino acid side chain, multiple mutations at the same site should be analyzed in a Leffler plot. Figure 12.2.4 shows a Leffler plot of 14 mutations at position 24 of the fyn SH3 domain. The data reveal a linear REFER over a 400-fold change in K_{eq} (16 kJ mol^{-1} change in protein stability) and give a ϕ_f-value (slope of the plot) of 0.33 ± 0.01. Similar results were obtained for two other sites in SH3 domains [26–28]. In addition to analyzing ϕ_f-values for a single residue (ϕ_f^i), information on larger parts of the protein can be obtained by determining the ϕ_f-value using all residues in a protein [29] ($\langle \phi_f^{prot} \rangle$; Figure 12.2.5) or in substructures such as subdomains or individual secondary structure elements [30].

It should be noted that the type of REFERs described here are commonly termed "Brønsted plots" in protein folding literature, which is misleading. "Brønsted plots" were originally used to relate the effect of a change in the rate constant of an acid- or base-catalyzed reaction to the dissociation constant of the catalyst [31]. The rate equilibrium relationships considered here are rather Leffler-type relationships, since they directly relate the rate constants of a reaction to the equilibrium constants of the *same* reaction [3].

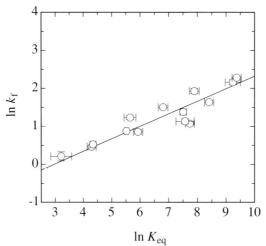

Fig. 12.2.4. Structure-induced Leffler plot for multiple mutations at position 24 of the fyn SH3 domain. The line corresponds to a linear fit of the data and gives a slope (ϕ_f-value) of 0.33 ± 0.01. The relationship between k_f and K_{eq} is linear over the complete stability range indicating negligible effect of the mutations on transition state structure even for highly destabilizing mutants. Data were taken from Ref. [28] and the plot was adapted from Ref. [26].

12.2.2.2
Properties of Protein Folding Transition States Derived from Linear REFERs

The most popular REFER in protein folding studies is the study of the effect of denaturants on rate and equilibrium constants (see Figure 12.2.3). Experimentally

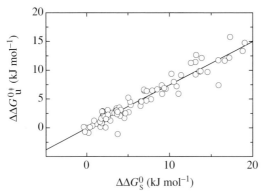

Fig. 12.2.5. Structure-induced Leffler plot for all residues in CI2 to determine the average ϕ_f-value ($\langle \phi_f \rangle$). The line represents a linear fit of the data. Since $\Delta G_u^{o\ddagger}$ is plotted vs. ΔG_S^o the slope corresponds to $1 - \langle \phi_f \rangle$. The plot was adapted from Ref. [26] and data were taken from Ref. [30].

determined α_D-values are usually native-like with values between 0.6 and 1, indicating that protein folding transition states have a rather native-like ASA. Temperature and pressure have less frequently been used to determine REFERs in protein folding and are more difficult to interpret since both have large and compensating effects on the solvent, on the protein and on solvent–protein interactions. The most straightforward parameter to interpret is the α_C-values derived from heat capacity changes (Eq. (7)), which reflects the interactions of the solvent (water) with the protein. All reported α_C-values are also native-like but generally slightly lower than α_D-values. Analysis of the effect of temperature further revealed both entropic and enthalpic contributions to protein folding barriers at room temperature. However, due to the large changes in $\Delta C^{\circ \ddagger}_{p(f)}$ these values are strongly temperature dependent. Even fewer data are available for the effect of pressure on protein folding transition states. High-pressure stopped-flow measurements on tendamistat revealed a rather native-like volume of the transition state with an α_p-value of 0.6, which increases with increasing denaturant concentration (see Section 12.2.3.1).

Structural information on the properties of transition states from ϕ_f ($= \alpha_S$)-value analysis has been obtained for several proteins (see Chapter 13). A major problem in these studies is the small data set at a single position with usually just a single mutation and the wild-type protein determining a REFER. It was recently shown that ϕ_f-values from such a two-point analysis are highly inaccurate if the stability change of a mutation ($\Delta \Delta G^{\circ}_S$) is smaller than 6–7 kJ mol^{-1} [26]. This uncertainty is in part due to intrinsic statistical errors, which should, however, give reliable ϕ_f-values for $|\Delta \Delta G^{\circ}_S| > 3$ kJ mol^{-1} for high-quality data sets [26]. The additional uncertainty in ϕ_f-values probably arises from changes in the structure of the unfolded state, which is observed in many mutants [11, 26] (see Section 12.2.4.1). This not only changes the structure of the ground state but very likely also influences the pre-exponential factor for folding by changing the dynamics of the unfolded state [32, 33]. The contributions from these effects will be more pronounced for mutations with small $|\Delta \Delta G^{\circ}_S|$ [26].

The results from reliable ϕ_f-values give a picture of transition states as distorted native states for the major part of a protein (diffuse transition states) or for large substructures (polarized transitions states). For diffuse transition states the ϕ_f-values are around 0.1–0.5 throughout the structured regions of the protein and average ϕ_f-values are between 0.2 and 0.4, indicating partial formation of the native set of interactions [26]. For these proteins a Leffler plot of all mutations throughout the protein usually gives a linear REFER supporting a structurally homogeneous transition state (Figure 12.2.5). In proteins with polarized transition states ϕ_f-values are significantly higher (up to 0.8) in a large substructure of a protein whereas they are 0 or close to 0 in other parts. These results suggest that the formation of the native topology for the whole protein or for a large substructure of the protein is a major part of the rate-limiting step in folding of small single domain proteins and might explain the correlation between the contact order and protein folding rate constants [34, 35]. The results contradict a nucleation-condensation model with a small structural nucleus formed by a few specific interactions in the transition state [36].

In ϕ_f-value analysis structural changes are typically introduced in amino acid side chains and the results consequently report on side-chain interactions although they are commonly interpreted in terms of secondary structure. Replacement of alanine by glycine residues, which leads to a more flexible backbone, was recently suggested to give information on the backbone interactions in the transition state [37]. However, side-chain interactions may also be changed upon replacing an amino acid by glycine and it thus questionable whether these ϕ_f-values are indicative of secondary structure formation. To gain more information on the secondary structure in protein folding transition states it will be necessary to introduce isosteric changes into the amino acid backbone. A backbone ϕ_f-value analysis using cytochrome c as well as monomeric and dimeric versions of GCN4 investigated amide deuterated proteins and gave ϕ_f-values around 0.5. This was interpreted as formation of 50% of the hydrogen bonds in the transition state [38]. However, this result is also compatible with formation of all native backbone hydrogen bonds at reduced strength, which would be compatible with the results on the diffuse nature of protein transition states from side chain ϕ_f-value analysis [26]. This discussion reveals a major problem in the quantitative interpretation of fractional ϕ_f-values, which may originate in formation of partial interactions in a single ensemble of transition states or in parallel pathways with different degrees of structure formation in the different transition states. However, there is no evidence for parallel pathways in the majority of proteins (see Section 12.2.4.2), which, for most proteins, favors the model of a native-like transition state topology throughout the protein with weakened interactions [26, 34, 35]. However, as discussed in Section 12.2.4.2, parallel pathways would be difficult to detect if they have similar α_x-values [11].

12.2.3
Nonlinear Rate Equilibrium Free Energy Relationships in Protein Folding

As pointed out above, the characterization of nonlinear REFERs can give information on various properties of the transition barriers. In the following we will first describe methods to detect and to analyze nonlinearities in REFERs and will then discuss the possible origins of nonlinear REFERs in more detail. Finally, results from nonlinear REFERs will be discussed in terms of general properties of protein folding barriers.

12.2.3.1
Self-Interaction and Cross-Interaction Parameters

Jencks and coworkers [5] proposed self-interaction and cross-interaction parameters as practical ways to detect and to analyze nonlinear REFERs. A self-interaction parameter (p_x) is the direct measure for a curvature in a REFER caused by a shift in the position of the transition state along the reaction coordinate with changing $\Delta G°$ upon a perturbation ∂x:

$$p_x = \frac{\partial \alpha_x}{\partial \Delta G_x^\circ} = \frac{\partial^2 \Delta G^{\circ\ddagger}}{(\partial \Delta G_x^\circ)^2} \tag{12}$$

By definition, a shift in the position of the transition state towards the destabilized state (e.g., as a result of Hammond behavior or due to sequential barriers) will give a positive p_x-value.

For a folding reaction perturbed by addition of denaturant or destabilized by mutations, the corresponding self-interaction parameters p_D and p_S are [10, 11]:

$$p_D = \frac{\partial \alpha_D}{\partial \Delta G_D^\circ} = \frac{\partial^2 \Delta G^{\circ\ddagger}}{(\partial \Delta G_D^\circ)^2} \tag{13}$$

$$p_S = \frac{\partial \phi_f}{\partial \Delta G_S^\circ} = \frac{\partial^2 \Delta G^{\circ\ddagger}}{(\partial \Delta G_S^\circ)^2} \tag{14}$$

Thus, in the denaturant-induced Leffler plot shown in Figure 12.2.3C, a positive p_D-value will result in a downward curvature (transition state becomes more native-like with decreasing protein stability) and a negative p_D-value would result in an upward curvature. The same curvatures would consequently be observed in the chevron plots.

The use of self-interaction parameters in the analysis of transition state movements is often not sensitive enough, because the energy range of the measurements is too narrow or the curvatures are too small [5]. A more sensitive test for transition state movements is provided by cross-interaction parameters [5], which measure changes in the position α_x of the transition state of a reaction (measured using the perturbation ∂x) caused by a second perturbation ∂y:

$$p_{xy} = \frac{\partial \alpha_x}{\partial \Delta G_y^\circ} = \frac{\partial^2 \Delta G^{\circ\ddagger}}{(\partial \Delta G_x^\circ)(\partial \Delta G_y^\circ)} = \frac{\partial \alpha_y}{\partial \Delta G_x^\circ} = p_{yx} \tag{15}$$

For nonlinear REFERs the value of α_x changes with the amount of a second perturbation ∂y, resulting in a nonzero p_{xy}-value. A shift in the position of the transition state towards the destabilized state will yield positive p_{xy}-values.

For tendamistat, the denaturant dependence of the folding reaction was measured at different pressures [39]. This allowed the calculation of the pressure/denaturant cross-interaction parameter (p_{Dp}). Figure 12.2.6 shows equilibrium transition curves and chevron plots at pressures between 20 and 1000 bar. Increasing pressure destabilizes tendamistat leading to a shift of the GdmCl-induced unfolding transition to lower GdmCl concentrations (Figure 12.2.6A) and indicating that the native state has a larger volume than the unfolded state. Interestingly, the volume of the transition state exceeds the volume of the native state at high denaturant concentrations as indicated by a decrease in the unfolding rate constant with increasing pressure above 5 M GdmCl (Figure 12.2.6B). This scenario is not treated in the original Leffler formalism, which assumes that the properties of the transition state are between those of the reactants and the products. The

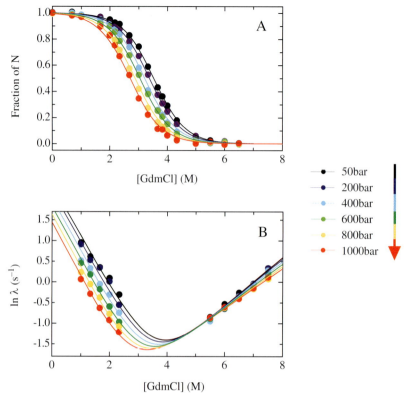

Fig. 12.2.6. Effect of pressure on A) the GdmCl-induced unfolding transition and B) the GdmCl dependence of the apparent rate constant for tendamistat folding at pH 2. The lines represent global fits of the data with $\partial m_f/\partial p = \partial \Delta V_f^{\circ\ddagger}/\partial [\text{GdmCl}] = 2.5 \pm 0.6 \text{ cm}^3 \text{ mol}^{-1} \text{ M}^{-1}$. Data and fits were taken from Ref. [39].

denaturant/pressure-induced nonlinear activation free energy relationships in tendamistat indicate a transition state movement towards a less solvent-exposed structure when the protein is destabilized [39]. The data were used to calculate the denaturant/pressure cross-interaction parameter

$$p_{Dp} = \frac{\partial \alpha_D}{\partial \Delta G_p^\circ} = \frac{\partial^2 \Delta G^{\circ\ddagger}}{(\partial \Delta G_p^\circ)(\partial \Delta G_D^\circ)} = \frac{\partial \alpha_p}{\partial \Delta G_D^\circ} = p_{pD} \quad (16)$$

Figure 12.2.7 shows the effect of $\partial \Delta G_p^\circ$ on the α_D-value for the denaturant/pressure-dependent folding data shown in Figure 12.2.5. The α_D-value increases significantly with decreasing stability indicating a positive p_{Dp}-value (Figure 12.2.7A). Figure 12.2.7B shows that the effect of $\partial \Delta G_p^\circ$ is only observed in the kinetic m-values while m_{eq} is unchanged due to compensating changes in m_f and m_u. This indicates a true denaturant/pressure-induced transition state movement.

12.2.3 Nonlinear Rate Equilibrium Free Energy Relationships in Protein Folding | 423

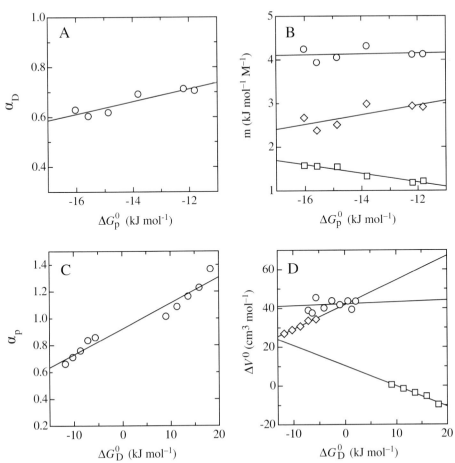

Fig. 12.2.7. Analysis of the pressure-denaturant cross-interaction parameter p_{pD} and p_{Dp} for tendamistat folding at pH 2 (see Figure 12.2.6). Panels A and B show the effect of $\partial \Delta G_p^\circ$ on α_D (A) and on m_{eq} (open circles), m_f (diamonds), and m_u (squares), respectively. Panels C and D show the effect of $\partial \Delta G_D^\circ$ on α_p (C) and on ΔV° (circles), $\Delta V_f^{\circ\ddagger}$ (diamonds), and $\Delta V_u^{\circ\ddagger}$ (squares), respectively. As expected for real Hammond behavior, the same effect is observed for $\partial \alpha_D / \partial \Delta G_p^\circ \partial \alpha_p / \partial \Delta G_D^\circ$ and for $\partial \alpha_p / \partial \Delta G_D^\circ$, i.e., $p_{pD} = p_{Dp}$; $V = 0.025 \pm 0.006$ mol/kJ. The original data used to construct the cross-interaction parameters are shown in Figure 12.2.6 and were taken from Ref. [39].

As postulated by Eq. (16) the same effect is observed when the effect of $\partial \Delta G_D^\circ$ on the α_p-value is analyzed (Figure 12.2.7C,D). Here a change in α_p is caused by compensating changes in $\Delta V_f^{\circ\ddagger}$ and $\Delta V_u^{\circ\ddagger}$ with constant ΔV°.

In mutational studies the changes in $\Delta G^{\circ\ddagger}$ and ΔG° are usually determined as a function of the denaturant concentration for each mutant. This directly allows the calculation of the denaturant/structure cross-interaction parameter (p_{DS}) [10, 11].

$$p_{DS} = \frac{\partial \alpha_D}{\partial \Delta G_S^\circ} = \frac{\partial^2 \Delta G^{\circ\ddagger}}{(\partial \Delta G_S^\circ)(\partial \Delta G_D^\circ)} = \frac{\partial \phi_f}{\partial \Delta G_D^\circ} = p_{SD} \quad (17)$$

p_{DS} tests whether changes in the stability of the native state caused by mutations ($\partial \Delta G_S^\circ$) have an effect on the relative solvent exposure of the transition state for folding (α_D).

If the effect of a change in structure by mutation is tested in the background of the wild type (∂Structure) and of another variant (∂Structure′), we can calculate a structure/structure cross-interaction parameter.

$$p_{SS'} = \frac{\partial \phi_f^S}{\partial \Delta G_{S'}^\circ} \quad (18)$$

$p_{SS'}$ indicates whether ϕ_f ($= \alpha_S$) changes when the protein stability is altered by an additional mutation [10, 11, 40].

These considerations show that various self-interaction and cross-interaction parameters can be used to detect and to analyze nonlinear REFERs along different reaction coordinates. In the following, several possible origins of nonlinear REFERs will be discussed.

12.2.3.2
Hammond and Anti-Hammond Behavior

According to the Hammond postulate the position of a transition state is shifted towards the ground state that is destabilized by the perturbation. Accordingly, α_x increases if the native state is destabilized relative to the unfolded state, which will lead to curvatures in REFERs. In principle, Hammond behavior should be observed for any transition state. If the free energy landscape in the vicinity of the transition state is a broad and smoothly curved maximum, α_x will be strongly influenced by changes in protein stability [5] (Figure 12.2.2A). If the transition region represents a rather narrow free energy maximum on the reaction coordinate, the changes in α_x will usually be too small to be detected experimentally [5] (Figure 12.2.2B). The sensitivity of α_x to changes in protein stability can thus be used to obtain information on the broadness of the free energy barriers.

It is worth discussing Thornton's theoretical explanation of transition state movements [41]. It assumes that the transition state is a saddle point in the energy landscape of the reaction. It represents an energy maximum in the direction of the reaction coordinate and an energy minimum in all other directions. A second assumption is that the potential for each vibrational mode of the molecule can be approximated by a quadratic function. The vibration of the reacting bond(s), the one(s) that are formed or broken in the reaction, is considered to be parallel to the reaction coordinate. Since the transition state is located at a maximum of the parabolic potential along the reaction coordinate, it will shift towards the destabilized state if a linear free energy perturbation is applied (Figure 12.2.2A). All other vibrational modes of the molecule are considered to be perpendicular to the reaction coordinate. For those, the transition state is at the minimum of the parabolic potential, which will shift towards the stabilized side of the coordinate if a linear perturbation is applied (Figure 12.2.2B). These are the two "reacting bond rules" in the IUPAC nomenclature [42], which are sometimes referred to as "Hammond effect" and "anti-Hammond effect." The effect of the perturbation on the structure of the transition state will be the sum of the individual effects on all coordinates. This theory is valid for any transition state along the reaction coordinate and does not require the product to have nearly the same energy content as the transition state. Significant movements of the transition state (i.e., both Hammond and anti-Hammond behavior) are thus expected if the energy landscape around it has a smooth curvature and different regions become limiting after a perturbation [5] (e.g., in reactions taking place with a continuum of mechanisms or when the free energy would consist of a continuum of consecutive states with different structure but similar free energy) (Figure 12.2.2A). A perturbation will not yield an observable change in the position of the maximum along the coordinate if the transition state of a reaction is a narrow free energy maximum in the reaction coordinate (Figure 12.2.2B). In organic reactions the transition state movements probed by different perturbations can be decoupled (i.e., a certain p_x or p_{xy} can have a positive value while others are zero). For these reactions the experimental reaction coordinates x and y describe the formation or breakage of different covalent bonds in a very direct way and thus x and y can change from totally synchronous to totally asynchronous [5]. In protein folding many noncovalent protein–protein, protein–solvent, and solvent–solvent interactions are formed and broken during the reaction and many "microscopic" reaction coordinates are averaged into the various "macroscopic" α_x-values, which therefore are likely to be highly correlated. Thus, the different p_x- and p_{xy}-values should generally be coupled in protein folding reactions if they are due to global structural changes in the transition state (i.e., transition state movements should be observable with independent perturbations) [10].

12.2.3.3
Sequential and Parallel Transition States

Nonlinear REFERs provide a unique tool to characterize complex transition state regions and especially to distinguish between sequential and parallel reaction pathways [1]. If destabilization of the native state leads to a shift in the rate-limiting step to a more native-like transition state (larger α_x; Figure 12.2.2C) a discrete change in α_x-value will be observed. This will lead to downward kink in the chevron plot and consequently also in the Leffler plot (positive p_D-value). Such a behavior implies that an on-pathway high-energy intermediate is "hidden" on the free energy landscape, which is not directly detectable due to its low stability relative to the native and unfolded state (see Figure 12.2.2C). Figure 12.2.9A shows chevron plots between 5 °C and 45 °C for a tendamistat variant that exhibits a nonlinear unfolding limb. The α_D-value changes from 0.4 at low denaturant concentration

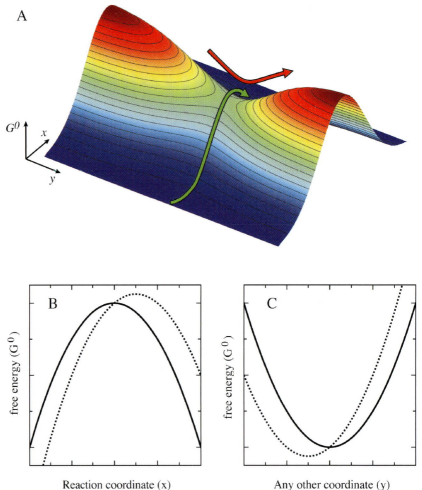

Fig. 12.2.8. Schematic representation of a simple transition region (A) and illustration of Thornton's first (B) and second reacting bond rule (C). The energy landscape in the vicinity of a transition state is approximated by a parabolic potential, both before (continuous line) and after (dotted line) a linear perturbation. B) In the direction of the reaction coordinate (green trajectory in panel A) the transition state is a maximum on the energy landscape, which will shift towards the state that is destabilized. C) In a direction perpendicular to the reaction coordinate (red trajectory in panel A) the transition state is a minimum on the energy landscape, which will shift towards the state that is stabilized. Panels B and C were adapted from Ref. [41].

(< 2 M GdmCl) to 0.9 at high denaturant concentrations (>5 M GdmCl). At high and low denaturant concentrations the chevron plots are linear, indicating a denaturant-induced switch between two distinct transition states [43]. This suggests a sequential folding model with consecutive transition states and a metastable high

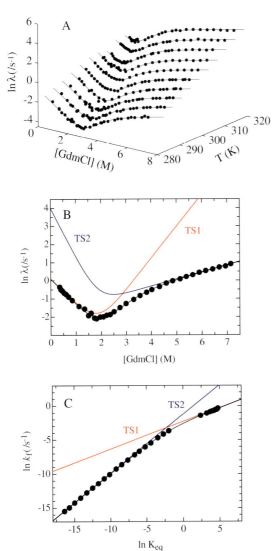

Fig. 12.2.9. Nonlinear rate equilibrium free energy relationship for folding of the tendamistat C45A/C73A variant caused by two consecutive transition states with a high-energy intermediate (cf. Figure 12.2.2C.). A) Temperature dependence of the chevron plots indicating a kink in the unfolding limb. B) Chevron plot at 25 °C. The black line represents a fit to a linear three-state mechanism with a high-energy intermediate (see Chapter 12.1). The thin lines represent hypothetical chevron plots for two-state folding limited either by the early transition state (TS1) of the late transition state (TS2). Plots were adapted from Ref. [43]. C) Leffler plot of the data shown in panel B. The kink in the chevron plots causes a downward kink in the Leffler plot. The red and the blue lines represent hypothetical Leffler plots for the early transition state (TS1 with $\alpha_D = 0.45$) and for the late transition state (TS2 with $\alpha_D = 0.88$), respectively. The folding rate constants (k_f) under unfolding conditions were calculated from the unfolding rate constant (k_u) using $k_f = K_{eq} \cdot k_u$. Data in the region $|\ln K_{eq}| < 2.5$ were not used since here both k_f and k_u significantly contribute to the apparent rate constant (λ).

energy intermediate (Figure 12.2.2B). Figure 12.2.9B shows the chevron plot at 25 °C with the fit of the data to a three-state model with a high-energy intermediate (see Chapter 12.1). This allows information to be gained on both transition states. Figure 12.2.9C shows the same data displayed as a Leffler plot, also showing a clear kink in the slope from 0.4 to 0.9 from going from low to high protein stability. Thus, a kink in the Leffler plot with a positive p_x value provides a simple way to show that an intermediate is on-pathway if the intermediate is higher in free energy than the unfolded and the native state. As discussed in Chapter 12.1, an on-pathway intermediate is more difficult to distinguish from an off-pathway intermediate when it is more stable than the unfolded state and thus becomes populated to detectable amounts. Depending on the α-values of the consecutive transition states and on the rate constants it might be difficult to distinguish a kink from a gradual shift in the position of the transition state caused by Hammond behavior.

If folding occurs through parallel pathways, the reaction with the most native-like transition state will be most sensitive to a reduced stability of the native state and an alternative parallel pathway with a more unfolded-like transition state may become rate limiting with increasing denaturant concentrations. This will lead to a decrease in α_x with decreasing protein stability resulting in a negative p_x-value. Thus parallel pathways will lead to an upward kink in the chevron plot (Figures 12.2.10A,B) and consequently also in the Leffler plot (Figure 12.2.10C,D) [11]. This phenomenon is sometimes referred to as anti-Hammond behavior in the protein folding literature [30, 44] but should be clearly distinguished from genuine anti-Hammond behavior discussed in Section 12.2.3.2, which is based on the property of a single barrier (see Figure 12.2.8).

12.2.3.4
Ground State Effects

Changes in the length of the reaction coordinate due to structural changes in the native or unfolded protein caused by the effects of a mutation or by a change in solvent conditions (ground state effects) can easily be mistaken for genuine transition state movement [6, 10, 11]. Both phenomena have similar effects on the experimentally observable REFERs in Leffler plots but they are based on completely different free energy barriers. Figure 12.2.11 illustrates the difference between Hammond behavior and ground state effects, showing as an example the effect of a change in protein stability caused by mutations ($\partial \Delta G_S^\circ$) on the location of the transition state in terms of its ASA measured in Chevron plots (α_D-value). This corresponds to the determination of a structure/denaturant cross-interaction parameter (p_{SD}) defined in Eq. (17). The reference conditions of the wild-type protein are given in Figure 12.2.11D. In the case Figure 12.2.11B the mutation leads to a transition state movement along the reaction coordinate relative to both ground states. The length of the reaction coordinate is unchanged (i.e., the structure of both ground states is not affected by the mutation, or both states are affected to the same extent). This scenario is in accordance with Hammond behavior if we as-

12.2.3 Nonlinear Rate Equilibrium Free Energy Relationships in Protein Folding

sume that the native state is destabilized by the mutation. Figure 12.2.11C and D show apparent transition state movements caused by ground state effects. In both cases the absolute position and the structure of the transition state remains unchanged but the structure of either the unfolded state (Figure 12.2.11C) or of the native state (Figure 12.2.11D) changes, which leads to a change in length of the reaction coordinate. In this case an apparent transition state movement will be observed if the position of the transition state is normalized against the length of the reaction coordinate. This demonstrates that the characterization of transition state movements requires the determination of the effects of a perturbation on the rate constants for the forward and backward reaction, on the equilibrium constant and on the length of the reaction coordinate. The effect on the length of the reaction coordinate can easily be tested by determining the effect of ∂x on the respective equilibrium properties. In the case discussed above, the length of the reaction coordinate is reflected by m_{eq}.

The same considerations apply for other self- and cross-interaction parameters. For pressure- and temperature-induced REFERs the length of the reaction coordinate can be tested by determining the effect of $\Delta V°$, and $\Delta S°/\Delta C_P°$, respectively. Changes in these parameters with ∂x indicate ground state effects induced by the change in conditions or by mutation. Structural changes in the unfolded state should also be detectable by NMR spectroscopy. However, it would be very time consuming to determine NMR structures of the unfolded state of each mutant and would not directly allow a correlation with the observed ground state effects, since they are defined by changes in free energy.

Figure 12.2.12 shows the results of a test for ground states effects and transition state movement in mutants of the Sso7d SH3 domain (Figure 12.2.12 upper panels) and for CI2 (Figure 12.2.12 lower panels). The effect of a change in $\Delta G°$ induced by mutations ($\partial \Delta G_S°$) on the α_D-values and on the m_{eq}-, m_f-, and m_u-values are shown. Both proteins show increasing α_D-values with decreasing protein stability in accordance with Hammond behavior. However, the origins for the changing α_D-values are different in the two proteins. In the SH3 domain the m_{eq}-value changes significantly upon mutation, indicating structural changes it at least one of the ground states. The value of m_f changes by the same amount as m_{eq} while m_u is essentially independent of protein stability. This is seen more clearly in a plot of m_f and m_u vs. m_{eq}, which shows that m_f correlates with m_{eq} whereas m_u is virtually independent of m_{eq} (Figure 12.2.13A). This indicates that the structure of the transition state does not move relative to the structure of the native state (cf. Figure 12.2.11). The increase in m_{eq}- and m_f-values point at the disruption of interactions in the unfolded state upon mutation which leads to a less compact unfolded state. These results show that the apparent shift in the position of the transition state along the reaction coordinate seen for the SH3 domain is a ground state effect on the unfolded state rather than Hammond behavior. Structural changes in the unfolded state can be rationalized in the light of recent results on residual structure of unfolded proteins (see Section 12.2.4.1). These interactions may be disrupted by mutations and thus lead to changes in the m_{eq}-values [11].

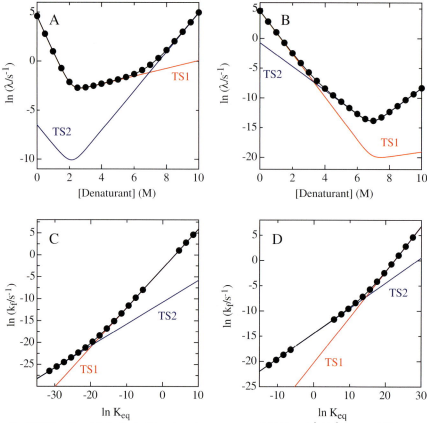

Fig. 12.2.10. Simulated effect of denaturant concentration on the apparent rate constant (filled circles and black line) for the folding of a protein through two parallel pathways with the transition states TS1 and TS2. The individual chevron plots corresponding to folding over TS1 and TS2 are shown. For two parallel pathways with the folding and unfolding rate constants of k_{f1}, k_{u1} (for TS1) and k_{f2}, k_{u2} (for TS2) the single observable rate constant (λ) is given by the sum of the two apparent rate constants ($\lambda_1 = k_{f1} + k_{u1}$ and $\lambda_2 = k_{f2} + k_{u2}$) for the parallel pathways: $\lambda = \lambda_1 + \lambda_2$. Data were simulated with the rate constants for folding and unfolding of A) $k_{f1} = 0.0015$ s^{-1}, $k_{u1} = 3 \cdot 10^{-7}$ s^{-1}; $k_{f2} = 100$ s^{-1}; $k_{u2} = 0.02$ s^{-1} and B) $k_{f1} = 5$ s^{-1}, $k_{u1} = 3 \cdot 10^{-7}$ s^{-1}; $k_{f2} = 100$ s^{-1}; $k_{u2} = 0.02$ s^{-1}. The m_i-values ($m_i = RT(\partial \ln k_i / \partial [\text{Denaturant}])$) were $m_{f1} = -4.95$ kJ mol^{-1} M^{-1}, $m_{u1} = 4.95$ kJ mol^{-1} M^{-1}; $m_{f2} = -8.92$ kJ mol^{-1} M^{-1}; $m_{u2} = 0.99$ kJ mol^{-1} M^{-1} in both cases. This corresponds to α_D-values of 0.5 and 0.9 for TS1 and TS2, respectively. The plots show that significantly different α_D-values for the two parallel pathways are required to obtain a clear nonlinearity. Panels C and D show Leffler plots of the data displayed in panels A and B, respectively. Both Leffler plots show a clear upward kink indicating the shift between the two parallel pathways. The red and blue lines represent hypothetical Leffler plots for the two parallel pathways with α_D-values of 0.5 (TS1) and 0.9 (TS2). Data in the region $|\ln K_{eq}| < 2.5$ were not used since here both k_f and k_u significantly contribute to the apparent rate constant (λ).

12.2.3 Nonlinear Rate Equilibrium Free Energy Relationships in Protein Folding

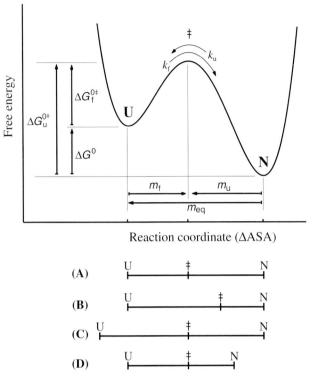

Fig. 12.2.11. Schematic representation of the possible effect of a perturbation on the position of the ground states and on the transition state along a reaction coordinate probed by the change in denaturant concentrations (∂[Denaturant]). The reference conditions (A) are compared with real Hammond behavior (B) and with apparent transition state movements caused by ground state effects due to changes in the structure of the unfolded (C) and native state (D). In all three cases (B–D) the position of the transition state will change by the same amount relative to the ground states (i.e., identical $p_{DS} = \partial \alpha_D / \partial \Delta G_S^\circ$ values). However, for genuine Hammond behavior (B) m_{eq} will not be affected by mutation and m_f and m_u will show compensating changes. Changes in the ASA of the ground states (C, D) will lead to changes in m_{eq}.

In CI2 a change in stability results in opposing effects on m_f and m_u accompanied by a significant change in the m_{eq}-values. The small change in m_u indicates a minor transition state movement in addition to a large ground state effect. As for the SH3 domain, m_f correlates with m_{eq} whereas m_u is independent of m_{eq}, indicating that mutations in CI2 also change the structure of the unfolded state with little effect on native state and transition state (Figure 12.2.13C). An example for genuine transition state movement without additional ground state effects is the nonzero pressure/denaturant cross-interaction parameter for tendamistat folding shown in Figures 12.2.6 and 12.2.7. Here the ΔV° and m_{eq} are independent of denaturant concentration and pressure, respectively.

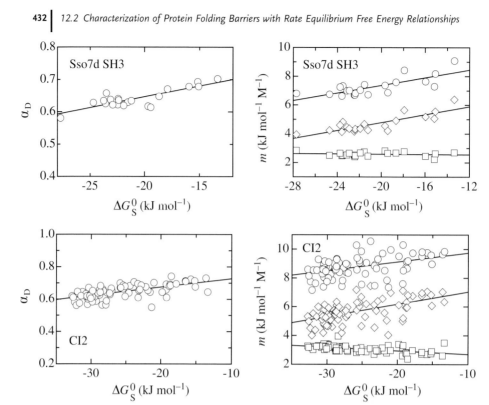

Fig. 12.2.12. Effect of changes in ΔG_S^0 on α_D (left panels) and on m_{eq} (open circles), m_f (diamonds) and $-m_u$ (squares; right panels) for the Sso7d SH3 domain and for CI2. Linear fits of the data show that in both proteins α_D, m_{eq}, and m_f are sensitive to changes in ΔG_S^0, whereas $-m_u$ is constant for Sso7d SH3 and slightly changing in the opposite direction as m_f for CI2. This indicates that the effect of mutations ΔG_S^0 on α_D is caused by ground state effects on the unfolded state in Sso7d SH3 and by major ground state effects in addition to small Hammond behavior in CI2. Plots were adapted from Ref. [11], and constructed based on data from Refs [29, 87].

12.2.4
Experimental Results on the Shape of Free Energy Barriers in Protein Folding

12.2.4.1
Broadness of Free Energy Barriers

From the observation of nonlinear chevron plots for apparent two-state folders [45, 46] and from changing α_D-values in mutants of several proteins [45–48] it was originally concluded that folding of many proteins shows Hammond behavior and thus a single broad barrier region with a continuum of states with similar free energy was proposed (Figure 12.2.1A). As discussed above for tendamistat (Figure 12.2.9), the frequently observed curvature in chevron plots may also be caused by a switch between a few consecutive narrow barriers on a sequential folding path-

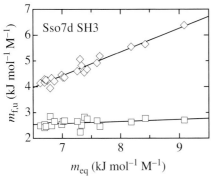 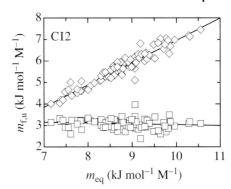

Fig. 12.2.13. Correlation of the changes in m_{eq} with changes in m_u (squares) and m_f (diamonds) for Sso7d SH3 and CI2. Linear fits of the data indicate that the variation in m_{eq} is correlated with the changes in m_f whereas m_u is only little affected by changes in m_{eq}. Correlation coefficients between m_f and m_{eq} are 0.97 for Sso7d SH3 and 0.95 for CI2. The correlations coefficients between m_u and m_{eq} are 0.34 Sso7d SH3 and −0.12 for CI2. Plots were adapted from Ref. [11].

way (Figure 12.2.2C) [43, 49–51]. Analysis of curved chevron plots reported for a large number of different proteins showed that a sequential folding model with consecutive transition states and at least one metastable high-energy intermediate is in better accordance with experimental data than a gradual transition state movement along a broad barrier [51]. This suggests that folding of many apparent two-state folders proceeds along a few defined consecutive barriers and through high-energy intermediates.

In ϕ_f-value analysis the folding and unfolding rate constants of mutants are typically obtained from chevron plots. This allowed a test for ground state effect and Hammond behavior in a large number of variants in several proteins by investigating the effect of the mutation on α_D, and the kinetic and equilibrium m-values as shown in Figure 12.2.12 [10, 11]. The analysis revealed that in about half of the proteins that have been extensively mutagenized the position of the transition state changes with mutation in accordance with apparent Hammond behavior (increased α_D-value with decreased protein stability; cf. Figure 12.2.12). However, the large majority of proteins with apparent transition state movement showed the same behavior as the SH3 domain displayed in Figure 12.2.12, indicating that the change in the position of the transition state is due to ground state effects on the structure of the unfolded state and not caused by genuine Hammond behavior [11]. In these proteins the structure of the transition state does not move relative to the structure of the native state (constant m_u-values), indicating that both the native state and the transition state are structurally well-defined and robust against mutations. In contrast, both the m_{eq}- and the m_f-values increase with decreasing in protein stability, which argues for a disruption of interactions in the unfolded state upon mutation leading to a less compact unfolded state.

Structural changes in the unfolded state can be rationalized in the light of the increasing evidence for residual structure of unfolded proteins. Recently, several

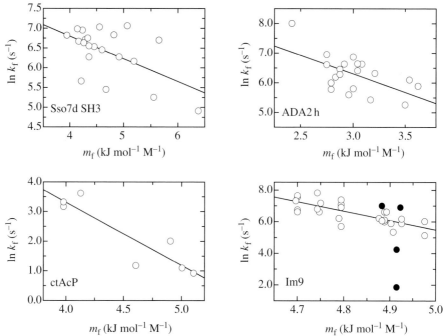

Fig. 12.2.14. Effect of changes in m_f on $\ln k_f$ for Sso7d SH3, ADA2h, ctAcP, and Im9 which exhibit ground state effects on the unfolded state. The filled symbols indicate mutants that have significantly different m_f-values and are thus likely to have a different rate-limiting transition stats. The correlation coefficients are -0.57, -0.59, -0.90, and -0.77, respectively. Plots were adapted from Ref. [11]. Original data used to calculate the correlations were taken from Refs [75, 87–89].

high-resolution NMR studies have shown that both native-like and nonnative-like interactions are present in unfolded states of several proteins [52–63] (see also Chapter 21). These interactions may be disrupted by mutations and thus lead to changes in the m_{eq}-values [32]. Only for a few proteins like CI2 weak Hammond behavior is observed, which is usually accompanied by significantly stronger ground state effects (Figures 12.2.12 and 12.2.13).

The presence of interactions in the unfolded state that are sensitive to mutation raises the question of which way these interactions influence the rate of protein folding. Figure 12.2.14 shows the correlation between the rate constant for folding (k_f) and the folding m-value (m_f) for four proteins (Ss07d SH3, ADAh2, ctAcP, and Im9) [11]. These proteins show significantly decreased folding rates with increasing m_{eq}- and m_f-values (significance of correlation larger than 90%). The same effect was observed for CI2, protein G, and protein L [11]. None of the proteins shows increased folding rates with increasing m_f and m_{eq} values.

Obviously, the disruption of residual structure in the unfolded state on average slows down folding, although the structure of the transition state is unchanged. This has three important consequences. First, it shows that the majority of the residual interactions in the unfolded state of these proteins are also present in the transition state. The presence of interactions in the unfolded state that are not present in the transition state would result in an energetic cost for the folding reaction, since the unfolded state would be stabilized relative to the transition state. This would lead to increased folding rates when these interactions are destabilized, which is not observed. The data do not allow, however, discrimination between native and nonnative interactions. Since ϕ-value analysis showed that transition states commonly contain a subset of the native interactions at reduced strength, our results indicate that the residual interactions in the unfolded state of these seven proteins are predominantly native-like. The second consequence of the results is that residual interactions in unfolded proteins can accelerate folding. It has been proposed that the accumulation of native-like structure in the unfolded state should stabilize the unfolded state relative to the transition state and therefore slow down folding [64]. This reasoning assumes, however, that the free energy of the transition state is independent of the strength of a specific interaction in the unfolded state, which can only be assumed for interactions that are not present in the transition state. Changes in stability of an interaction will affect the free energies of the unfolded state and of the transition state, if the interaction is present in both states. The destabilization and disruption of residual interactions obviously has on average a stronger effect on the free energy of the transition state compared with the unfolded state. This indicates that the interactions are stronger in the transition state than in the unfolded state and thus more sensitive to mutations. The third conclusion from these results is that a subset of the interactions present in the transition state for folding can be directly pointed out from structural studies on the unfolded state.

As discussed in Section 12.2.3.2, transition state movements in protein folding should be highly correlated since different reaction coordinates should be sensitive to protein–protein and protein–solvent interactions in the transition state. Transition state movements should thus show up in different p_x- and p_{xy}-values. Analysis of other self- and cross-interaction parameters in available folding data (Table 12.2.1) indicates the absence of genuine transition state movements for folding of most of the proteins [11]. Out of 21 investigated proteins only CI2 showed a small but coherent transition state movement. These results suggest that Hammond behavior is not common in protein folding reactions. Most protein folding transition states seem to be robust, conformationally restricted maxima in the free energy landscape and, similar to native states, they usually do not undergo extensive structural rearrangements upon mutation or changes in solvent conditions. These results argue for the generality of free energy landscapes with a limited number of sequential but structurally well-defined transition states in apparent two-state folding. The clearest examples for significant transition states movement is the effect of pressure and denaturant on the transition state for tendamistat folding (Figures 12.2.6 and 12.2.7). Due to the lack of other data on pressure-induced REFERs and the scarcity of data on temperature-induced REFERs it is difficult to judge whether transition state movements are frequently induced by perturbants other than denaturant or mutation.

Tab. 12.2.1. Proteins for which more than one self- or cross-interaction coefficient can be determined. A p_x or p_{xy}-value is considered to be larger (> 0) or smaller (< 0) than zero if it differs from zero in more than two standard deviations.

Protein	$p_D = \frac{\partial a_D}{\partial \Delta G_D^\circ}$	$p_S = \frac{\partial \phi_f^{prot}}{\partial \Delta G_S^\circ}$	$p_{DS} = \frac{\partial a_D}{\partial \Delta G_S^\circ}$	$p_{DT} = \frac{\partial a_D}{\partial \Delta G_T^\circ}$	$p_{DpH} = \frac{\partial a_D}{\partial \Delta G_{pH}^\circ}$	$p_{Dp} = \frac{\partial a_D}{\partial \Delta G_p^\circ}$	$p_{DGI} = \frac{\partial a_D}{\partial \Delta G_{GI}^\circ}$	$p_{ST} = \frac{\partial \phi_f}{\partial \Delta G_T^\circ}$	$p_{SpH} = \frac{\partial \phi_f}{\partial \Delta G_{pH}^\circ}$	$p_{SS'} = \frac{\partial \phi_f}{\partial \Delta G_{S'}^\circ}$
ACBP [74]	≅0	≅0								
ADA2h [75]	≅0	≅0	>0							
Barnase [30, 44, 47, 48, 76, 77]	>0	<0 for helix 1	>0 for core	>0				>0 for helix 1		<0 for helix 1
CI2 [29, 45, 48]	>0 for some mutants	≅0	>0	>0				>0		
CspB [78]	>0 for some mutants		≅0							
CTL9 [23]	≅0	≅0			≅0					
FKBP12 [79, 80]	>0 for some mutants	≅0	≅0	≅0						
NTL9 [21, 81]	≅0		≅0	≅0	≅0					
Protein L [82–84]	>0	≅0	≅0	≅0						
Spectrin SH3 [40, 85]	>0	≅0	≅0		≅0		≅0		≅0	≅0
Tendamistat [39, 43, 86]	>0 for some mutants			>0 for some mutants	≅0	>0				

In several proteins, some mutations lead to discrete jumps in the position of the transition state in a p_{DS} analysis. This indicates changes in the rate-limiting step of the folding process [11] and might be due to parallel folding pathways or to consecutive barriers on a sequential pathway [10, 11]. Sequential pathways with a few discrete transition states are in agreement with the results from the analysis of proteins with nonzero p_D-values as discussed above. The existence of multiple consecutive barriers may further explain transition state movements observed only with some p_x- or p_{xy}-values. If two consecutive transition states have a very similar α_x-value in that particular reaction coordinate, the range of experimentally accessible perturbations may lie in a region where the switch between the two transition states occurs. Therefore, conditions where only one of the transition states is limiting may be missed. This will show up as an apparent gradual movement of the transition state position [39, 43].

12.2.4.2
Parallel Pathways

Theoretical studies on protein-like model heteropolymers suggested a highly heterogeneous transition state ensemble with a manifold of parallel routes leading to the native state [65, 66]. In this case a perturbation should affect the substates of the transition state ensemble to different degrees. The free energy of more native-like transition states would be more strongly increased when the native state is destabilized, which would lead to an apparent movement to more unfolded-like transition states (see Figure 12.2.10). This would result in negative self- and cross-interaction parameters in our data analysis. The same reasoning applies to a scenario with defined parallel pathways of similar free energy leading to the native state (Figure 12.2.10). At least three parallel pathways were described for lysozyme folding. In this case one (low salt conditions) or two of the pathways (high salt conditions) transiently populate an intermediate, which facilitated the detection and characterization of the parallel routes [67] (see Chapter 12.1).

The clearest example for two parallel pathways in two-state folding came from a titin domain, where a clear upward curvature in the chevron plot was reported [68]. The existence of two parallel pathways was also deduced for folding of a variant of GCN4, a dimeric coiled-coil [22] with an engineered Zn^{2+}-binding site. In this case the presence of Zn^{2+} induced a parallel pathway detected by an upward curvature in the Leffler plot. Negative p_S- and $p_{SS'}$-values also indicated parallel pathways for the formation of the first helix of barnase [30, 44] and negative $p_{SS'}$-values were also found for protein G [69]. The analysis of other two-state folders did not detect any additional negative p_x- and p_{xy}-values. The lack of evidence for negative p_x- and p_{xy}-values in the majority of proteins could indicate that parallel folding pathways are rare or that they have transition states with similar α_x-values. This would lead to only minor shifts in the apparent position of the transition state which would escape detection in our data analysis.

Negative p_x-values observed for helix 1 in barnase (negative p_S-value) [44] and for the titin domain (negative p_D-value) [70] were originally interpreted as anti-

Hammond behavior. However, a more detailed analysis of the effects revealed that in both proteins the negative p_x-values are compatible with parallel pathways [11, 68]. Up to date there is no clear example for anti-Hammond behavior for protein folding reactions supporting the picture of narrow and defined transition state regions.

12.2.5
Folding in the Absence of Enthalpy Barriers

Theoretical models have proposed that folding is an energetically downhill process limited only by entropic barriers [71]. Zwanzig treated protein folding as a rapid barrier-less equilibration reaction between many unfolded conformations linked to a barrier-less escape process, which can only take place from a limited number of unfolded conformations or "escape states" [72], similar to the flow out of a bath tub. Simulation of the dynamic behavior of such a system shows that single exponential kinetics result when the transitions between the different unfolded states are fast and the number of escape states is small compared with the total number of unfolded conformations. This process can thus not be distinguished experimentally from two-state folding kinetics with an energy barrier. A more general treatment of the effect of entropy barriers on chemical dynamics shows that single exponential kinetics should always be observed (at least for reactions in solution) in the absence of energetic (enthalpic) barriers, as long as entropic barriers exist [73]. These considerations highlight the difficulties in discriminating between different folding models on the basis of single exponential behaviour, since the same experimentally observed kinetics can result from different folding landscapes.

12.2.6
Conclusions and Outlook

A quantitative treatment of folding kinetics is a prerequisite for elucidating mechanisms of protein folding. Even simple kinetics, like single exponential behavior can be caused by a variety of different folding landscapes. Therefore the interplay between theoretical models and experimental results is essential in the effort to understand protein folding. We have seen that concepts from classical reaction rate theory developed for simple chemical reactions can be applied to protein folding. Analysis of self-interaction and cross-interaction parameters prooved to be useful tools to detect nonlinear REFERs in protein folding. This gave valuable information on the shape of free energy barriers and thus on the mechanism of protein folding. The analysis of transition state movements in a large number of well-characterized proteins using self-interaction and cross-interaction parameters shows that Hammond behavior is rare in protein folding reactions, indicating that folding transition states are narrow regions on the free energy landscape which are robust against perturbations. Apparent transition state movements detected by p_{DS} cross-interaction parameters are in most proteins due to ground state effects and are most commonly caused by structural changes in the unfolded state. This supports the presence of residual structure in unfolded proteins. The analysis of the denaturant-induced REFERs for a large number of proteins further supports sequential folding pathways with obligatory high-energy intermediates as a simple and general explanation for curvatures in chevron plots. This suggests that apparent two-state folding and folding through transiently populated intermediates share similar free energy landscapes with consecutive barriers on sequential pathways. The major difference between two-state and multistate folding is the relative stability of partially folded intermediates.

Up to date mainly mutational analysis in combination with denaturant dependence of folding rate constants has been applied to characterize protein folding barriers. It will be interesting to see whether protein folding transition states are also robust along other reaction coordinates, which can be probed by changes in pressure, temperature and by ligand binding or protonation of specific side chains. In addition, more experimental results on backbone interactions in the transition state and on the existence of parallel transition states will be required to get a more complete picture of the structure of protein folding transition states and on the properties of the free energy barriers.

Acknowledgments

We thank Andi Möglich for preparing Figure 12.2.8A, Manuela Schätzle and Beat Fierz for comments on the manuscript and all members of the Kiefhaber lab for help and discussion.

References

1 LEFFLER, J. E. & GRUNWALD, E. (1963). Rates and Equilibria of Organic Reactions. Dover, New York.

2 JENCKS, W. P. (1969). Catalysis in Chemistry and Enzymology. McGraw-Hill, New York.

3 LEFFLER, J. E. (1953). Parameters for the description of transition states. Science 117, 340–341.

4 HAMMOND, G. S. (1955). A correlation of reaction rates. J. Am. Chem. Soc. 77, 334–338.

5 JENCKS, W. P. (1985). A primer for the Bema Hapothle. An empirical approach to the characterization of changing transition-state structures. Chem. Rev. 85, 511–527.

6 FARCASIU, D. (1975). The use and misuse of the Hammond postulate. J. Chem. Ed. 52, 76–79.

7 GREENE, R. F. J. & PACE, C. N. (1974). Urea and guanidine-hydrochloride denaturation of ribonuclease, lysozyme, alpha-chyomtrypsinn and beta-lactoglobulin. J. Biol. Chem. 249, 5388–5393.

8 SANTORO, M. M. & BOLEN, D. W. (1988). Unfolding free energy changes determined by the linear extrapolation method. 1. Unfolding of phenylmethanesulfonyl alpha-chymotrypsin using different denaturants. Biochemistry 27, 8063–8068.

9 TANFORD, C. (1970). Protein denaturation. Part C. Theoretical models for the mechanism of denaturation. Adv. Protein Chem. 24, 1–95.

10 SÁNCHEZ, I. E. & KIEFHABER, T.

(2003). Non-linear rate-equilibrium free energy relationships and Hammond behavior in protein folding. *Biophys. Chem.* 100, 397–407.

11 SÁNCHEZ, I. E. & KIEFHABER, T. (2003). Hammond behavior versus ground state effects in protein folding: evidence for narrow free energy barriers and residual structure in unfolded states. *J. Mol. Biol.* 327, 867–884.

12 POHL, F. M. (1976). Temperature-dependence of the kinetics of folding of chymotrypsinogen A. *FEBS Lett.* 65, 293–296.

13 CHEN, B. L., BAASE, W. A. & SCHELLMAN, J. A. (1989). Low-temperature unfolding of a mutant of phage T4 lysozyme. 2. Kinetic investigations. *Biochemistry* 28, 691–699.

14 PRIVALOV, P. L. & MAKHATADZE, G. I. (1992). Contribution of hydration and non-covalent interactions to the heat capacity effect on protein unfolding. *J. Mol. Biol.* 224, 715–723.

15 IKAI, A., FISH, W. W. & TANFORD, C. (1973). Kinetics of unfolding and refolding of proteins. II. Results for cytochrome c. *J. Mol. Biol.* 73, 165–184.

16 MATTHEWS, C. R. (1987). Effect of point mutations on the folding of globular proteins. *Methods Enzymol.* 154, 498–511.

17 MYERS, J. K., PACE, C. N. & SCHOLTZ, J. M. (1995). Denaturant m values and heat capacity changes: relation to changes in accessible surface areas of protein unfolding. *Protein Sci.* 4, 2138–2148.

18 KUWAJIMA, K. (1989). The molten globule state as a clue for understanding the folding and cooperativity of globular-protein structure. *Proteins Struct. Funct. Genet.* 6, 87–103.

19 SANCHO, J., MEIERING, E. M. & FERSHT, A. R. (1991). Mapping transition states of protein unfolding by protein engineering of ligand-binding sites. *J. Mol. Biol.* 221, 1007–1014.

20 TAN, Y. J., OLIVEBERG, M. & FERSHT, A. R. (1996). Titration properties and thermodynamics of the transition state for folding: comparison of two-state and multi-state folding pathways. *J. Mol. Biol.* 264, 377–389.

21 LUISI, D. L. & RALEIGH, D. P. (2000). pH-dependent interactions and the stability and folding kinetics of the N-terminal domain of L9. Electrostatic interactions are only weakly formed in the transition state for folding. *J. Mol. Biol.* 299, 1091–1100.

22 KRANTZ, B. A. & SOSNICK, T. R. (2001). Engineered metal binding sites map the heterogeneous folding landscape of a coiled coil. *Nat. Struct. Biol.* 8, 1042–1047.

23 SATO, S. & RALEIGH, D. P. (2002). pH-dependent stability and folding kinetics of a protein with an unusual alpha-beta topology: the C-terminal domain of the ribosomal protein L9. *J. Mol. Biol.* 318, 571–82.

24 GOLDENBERG, D. P., FRIEDEN, R. W., HAACK, J. A. & MORRISON, T. B. (1989). Mutational analysis of a protein-folding pathway. *Nature* 338, 127–32.

25 FERSHT, A. R., MATOUSCHEK, A. & SERRANO, L. (1992). The folding of an enzyme. I. Theory of protein engineering analysis of stability and pathway of protein folding. *J. Mol. Biol.* 224, 771–782.

26 SÁNCHEZ, I. E. & KIEFHABER, T. (2003). Origin of unusual phi-values in protein folding: Evidence against specific nucleation sites. *J. Mol. Biol.* 334, 1077–1085.

27 MOK, Y. K., ELISSEEVA, E. L., DAVIDSON, A. R. & FORMAN-KAY, J. D. (2001). Dramatic stabilization of an SH3 domain by a single substitution: roles of the folded and unfolded states. *J. Mol. Biol.* 307, 913–928.

28 NORTHEY, J. G., MAXWELL, K. L. & DAVIDSON, A. R. (2002). Protein folding kinetics beyond the phi value: using multiple amino acid substitutions to investigate the structure of the SH3 domain folding transition state. *J. Mol. Biol.* 320, 389–402.

29 ITZHAKI, L. S., OTZEN, D. E. & FERSHT, A. R. (1995). The structure of

the transition state for folding of chymotrypsin inhibitor 2 analyzed by protein engineering methods: evidence for a nucleation-condesation mechanism for protein folding. *J. Mol. Biol.* 254, 260–288.
30. FERSHT, A. R., ITZHAKI, L. S., elMASRY, N. F., MATTHEWS, J. M. & OTZEN, D. E. (1994). Single versus parallel pathways of protein folding and fractional formation of structure in the transition state. *Proc. Natl Acad. Sci. USA* 91, 10426–10429.
31. BRØNSTED, J. N. & PEDERSEN, K. J. (1924). *Z. physikal. Chem.* 108, 185–.
32. KLEIN-SEETHARAMAN, J., OIKAWA, M., GRIMSHAW, S. B., WIRMER, J., DUCHARDT, E., UEDA, T., et al. (2002). Long-range interactions within a non-native protein. *Science* 295, 1719–1722.
33. KRIEGER, F., FIERZ, B., BIERI, O., DREWELLO, M. & KIEFHABER, T. (2003). Dynamics of unfolded polypeptide chains as model for the earliest steps in protein folding. *J. Mol. Biol.* 332, 265–274.
34. MAKAROV, D. E., KELLER, C. A., PLAXCO, K. W. & METIU, H. (2002). How the folding rate constant of simple single-domain proteins depends on the number of native contacts. *Proc. Natl Acad. Sci. USA* 99, 3535–3539.
35. MAKAROV, D. E. & PLAXCO, K. W. (2003). The topomer search model: A simple quanitative theory of two-state protein folding kinetics. *Protein Sci.* 12, 17–26.
36. VENDRUSCOLO, M., PACI, E., DOBSON, C. M. & KARPLUS, M. (2001). Three key residues form a critical contact network in a protein folding transition state. *Nature* 409, 641–645.
37. SATO, S., RELIGA, T. L., DAGGETT, V. & FERSHT, A. R. (2004). Testing protein-folding simulations by experiment: B domain of protein A. *Proc. Natl Acad. Sci. USA* 101, 6952–6956.
38. KRANTZ, B. A., MORAN, L. B., KENTSIS, A. & SOSNICK, T. R. (2000). D/H amide kinetic isotope effects reveal when hydrogen bonds form during protein folding. *Nat. Struct. Biol.* 7, 62–71.
39. PAPPENBERGER, G., SAUDAN, C., BECKER, M., MERBACH, A. E. & KIEFHABER, T. (2000). Denaturant-induced movement of the transition state of protein folding revealed by high pressure stopped-flow measurements. *Proc. Natl Acad. Sci. USA* 97, 17–22.
40. MARTINEZ, J. C., PISABARRO, M. T. & SERRANO, L. (1998). Obligatory steps in protein folding and the conformational diversity of the transition state. *Nat. Struct. Biol.* 5, 721–9.
41. THORNTON, E. R. (1967). A simple theory for predicting the effects of substituent changes on transition-state geometry. *J. Am. Chem. Soc.* 89, 2915–2927.
42. MÜLLER, P. (1994). Glossary of terms used in physical organic chemistry. *Pure Appl. Chem.* 66, 1077–1184.
43. BACHMANN, A. & KIEFHABER, T. (2001). Apparent two-state tendamistat folding is a sequential process along a defined route. *J. Mol. Biol.* 306, 375–386.
44. MATTHEWS, J. M. & FERSHT, A. R. (1995). Exploring the energy surface of protein folding by structure-reactivity relationship and engineered proteins: Observation of Hammond behavior for the gross structure of the transitions state and anti-hammond behavior for structural elements for unfolding/folding of barnase. *Biochemistry* 34, 6805–6814.
45. OLIVEBERG, M., TAN, Y.-J., SILOW, M. & FERSHT, A. R. (1998). The changing nature of the protein folding transition state: implications for the free-energy profile for folding. *J. Mol. Biol.* 277, 933–943.
46. SILOW, M. & OLIVEBERG, M. (1997). High-energy channeling in protein folding. *Biochemistry* 36, 7633–7637.
47. MATOUSCHEK, A. & FERSHT, A. R. (1993). Application of physical organic chemistry to engineered mutants of proteins: Hammond postulate behavior in the transition state of protein folding. *Proc. Natl Acad. Sci. USA* 90, 7814–7818.
48. MATOUSCHEK, A., OTZEN, D. E.,

Itzhaki, L., Jackson, S. E. & Fersht, A. R. (1995). Movement of the transition state in protein folding. *Biochemistry* 34, 13656–13662.

49 Kiefhaber, T., Bachmann, A., Wildegger, G. & Wagner, C. (1997). Direct measurements of nucleation and growth rates in lysozyme folding. *Biochemistry* 36, 5108–5112.

50 Walkenhorst, W. F., Green, S. & Roder, H. (1997). Kinetic evidence for folding and unfolding intermediates in staphylococcal nuclease. *Biochemistry* 36, 5795–5805.

51 Sánchez, I. E. & Kiefhaber, T. (2003). Evidence for sequential barriers and obligatory intermediates in apparent two-state protein folding. *J. Mol. Biol.* 325, 367–376.

52 Evans, P. A., Topping, K. D., Woolfson, D. N. & Dobson, C. M. (1991). Hydrophobic clustering in nonnative states of a protein: interpretation of chemical shifts in NMR spectra of denatured states of lysozyme. *Proteins* 9, 248–266.

53 Neri, D., Billeter, M., Wider, G. & Wüthrich, K. (1992). NMR determination of residual structure in a urea-denatured protein, the 434-repressor. *Science* 257, 1559–1563.

54 Logan, T. M., Theriault, Y. & Fesik, S. W. (1994). Structural characterization of the FK506 binding protein unfolded in urea and guanidine hydrochloride. *J. Mol. Biol.* 236, 637–648.

55 Smith, C. K., Bu, Z. M., Anderson, K. S., Sturtevant, J. M., Engelman, D. M. & Regan, L. (1996). Surface point mutations that significantly alter the structure and stability of a protein's denatured state. *Protein Sci.* 5, 2009–2019.

56 Sari, N., Alexander, P., Bryan, P. N. & Orban, J. (2000). Structure and dynamics of an acid-denatured protein G mutant. *Biochemistry* 39, 965–977.

57 Wong, K. B., Clarke, J., Bond, C. J., Neira, J. L., Freund, S. M., Fersht, A. R. & Daggett, V. (2000). Towards a complete description of the structural and dynamic properties of the denatured state of barnase and the role of residual structure in folding. *J. Mol. Biol.* 296, 1257–1282.

58 Teilum, K., Kragelund, B. B., Knudsen, J. & Poulsen, F. M. (2000). Formation of hydrogen bonds precedes the rate-limiting formation of persistent structure in the folding of ACBP. *J. Mol. Biol.* 301, 1307–1314.

59 Kortemme, T., Kelly, M. J., Kay, L. E., Forman-Kay, J. D. & Serrano, L. (2000). Similarities between the spectrin SH3 domain denatured state and its folding transition state. *J. Mol. Biol.* 297, 1217–1229.

60 Garcia, P., Serrano, L., Durand, D., Rico, M. & Bruix, M. (2001). NMR and SAXS characterization of the denatured state of the chemotactic protein CheY: implications for protein folding initiation. *Protein Sci.* 10, 1100–1112.

61 Choy, W. Y. & Forman-Kay, J. D. (2001). Calculation of ensembles of structures representing the unfolded state of an SH3 domain. *J. Mol. Biol.* 308, 1011–1032.

62 Kazmirski, S. L., Wong, K. B., Freund, S. M., Tan, Y. J., Fersht, A. R. & Daggett, V. (2001). Protein folding from a highly disordered denatured state: the folding pathway of chymotrypsin inhibitor 2 at atomic resolution. *Proc. Natl Acad. Sci. USA* 98, 4349–4354.

63 Yi, Q., Scalley-Kim, M. L., Alm, E. J. & Baker, D. (2000). NMR characterization of residual structure in the denatured state of protein L. *J. Mol. Biol.* 299, 1341–1351.

64 Fersht, A. R. (1995). Optimization of rates of protein folding: The nucleation-condensation mechanism and its implications. *Proc. Natl Acad. Sci. USA* 92, 10869–10873.

65 Guo, Z. & Thirumalai, D. (1995). Kinetics of protein folding: nucleation mechanism, time scales, and pathways. *Biopolymers* 36, 83–102.

66 Wolynes, P. G., Onuchic, J. N. & Thirumalai, D. (1995). Navigating the folding routes. *Science* 267, 1619–1620.

67 Bieri, O., Wildegger, G., Bachmann, A., Wagner, C. &

KIEFHABER, T. (1999). A salt-induced intermediate is on a new parallel pathway of lysozyme folding. *Biochemistry* 38, 12460–12470.

68 WRIGHT, C. F., LINDORFF-LARSEN, K., RANDLES, L. G. & CLARKE, J. (2003). Parallel protein-unfolding pathways revealed and mapped. *Nat. Struct. Biol.* 10, 658–662.

69 NAULI, S., KUHLMAN, B. & BAKER, D. (2001). Computer-based redesign of a protein folding pathway. *Nat. Struct. Biol.* 8, 602–605.

70 FOWLER, S. B. & CLARKE, J. (2001). Mapping the folding pathway of an immunoglobulin domain: structural detail from Phi value analysis and movement of the transition state. *Structure (Camb.)* 9, 355–366.

71 WOLYNES, P. G. (1997). Folding funnels and energy landscapes of larger proteins within the capillarity approximation. *Proc. Natl Acad. Sci. USA* 94, 6170–6175.

72 ZWANZIG, R. (1997). Two-state models for protein folding. *Proc. Natl Acad. Sci. USA* 94, 148–150.

73 ZHOU, H.-X. & ZWANZIG, R. (1991). A rate process with an entropic barrier. *J. Chem. Phys.* 94, 6147–6152.

74 KRAGELUND, B. B., OSMARK, P., NEERGAARD, T. B., SCHIODT, J., KRISTIANSEN, K., KNUDSEN, J. & POULSEN, F. M. (1999). The formation of a native-like structure containing eight conserved hydrophobic residues is rate limiting in two-state protein folding of ACBP. *Nat. Struct. Biol.* 6, 594–601.

75 VILLEGAS, V., MARTINEZ, J. C., AVILÉS, F. X. & SERRANO, L. (1998). Structure of the transition state in the folding process of human procarboxypeptidase A2 activation domain. *J. Mol. Biol.* 283, 1027–1036.

76 MATOUSCHEK, A., MATTHEWS, J. M., JOHNSON, C. M. & FERSHT, A. R. (1994). Extrapolation to water of kinetic and equilibrium data for the unfolding of barnase in urea solutions. *Protein Eng.* 7, 1089–1095.

77 DALBY, P. A., OLIVEBERG, M. & FERSHT, A. R. (1998). Movement of the intermediate and rate determining transition state of barnase on the energy landscape with changing temperature. *Biochemistry* 37, 4674–4679.

78 PERL, D., HOLTERMANN, G. & SCHMID, F. X. (2001). Role of the chain termini for the folding transition state of the cold shock protein. *Biochemistry* 40, 15501–15511.

79 FULTON, K. F., MAIN, E. R., DAGGETT, V. & JACKSON, S. E. (1999). Mapping the interactions present in the transition state for unfolding/folding of FKBP12. *J. Mol. Biol.* 291, 445–461.

80 MAIN, E. R. G., FULTON, K. F. & JACKSON, S. E. (1999). Folding pathway of FKBP12 and characterization of the transition state. *J. Mol. Biol.* 291, 429–444.

81 KUHLMAN, B., LUISI, D. L., EVANS, P. A. & RALEIGH, D. P. (1998). Global analysis of the effects of temperature and denaturant on the folding and unfolding kinetics of the N-terminal domain of the protein L9. *J. Mol. Biol.* 284, 1661–1670.

82 SCALLEY, M. L. & BAKER, D. (1997). Protein folding kinetics exhibit an Arrhenius temperature dependence when corrected for the temperature dependence of protein stability. *Proc. Natl Acad. Sci. USA* 94, 10636–10640.

83 PLAXCO, K. W. & BAKER, D. (1998). Limited internal friction in the rate-limiting step of a two-state protein folding reaction. *Proc. Natl Acad. Sci. USA* 95, 13591–13596.

84 KIM, D. E., FISHER, C. & BAKER, D. (2000). A breakdown of symmetry in the folding transition state of protein L. *J. Mol. Biol.* 298, 971–984.

85 MARTINEZ, J. C. & SERRANO, L. (1999). The folding transition state between SH3 domains is conformationally restricted and evolutionarily conserved. *Nat. Struct. Biol.* 6, 1010–1016.

86 SCHÖNBRUNNER, N., PAPPENBERGER, G., SCHARF, M., ENGELS, J. & KIEFHABER, T. (1997). Effect of preformed correct tertiary interactions on rapid two-state tendamistat folding: evidence for hairpins as initiation sites

for β-sheet formation. *Biochemistry* 36, 9057–9065.

87 GUEROIS, R. & SERRANO, L. (2000). The SH3-fold family: experimental evidence and prediction of variations in the folding pathways. *J. Mol. Biol.* 304, 967–982.

88 TADDEI, N., CHITI, F., FIASCHI, T., BUCCIANTINI, M., CAPANNI, C., STEFANI, M., et al. (2000). Stabilisation of alpha-helices by site-directed mutagenesis reveals the importance of secondary structure in the transition state for acylphosphatase folding. *J. Mol. Biol.* 300, 633–647.

89 FRIEL, C. T., CAPALDI, A. P. & RADFORD, S. E. (2003). Structural analysis of the rate-limiting transition states in the folding of Im7 and Im9: similarities and differences in the folding of homologous proteins. *J. Mol. Biol.* 326, 293–305.

13
A Guide to Measuring and Interpreting ϕ-values

Nicholas R. Guydosh and Alan R. Fersht

13.1
Introduction

Arguably the most important state to characterize on a folding pathway, and indeed in any reaction, is the transition state. But, its ephemeral existence and composition of unstable interactions render conventional tools inadequate in determining an atomic level structure. But the structures of many transition states have been solved using protein engineering techniques developed over the last two decades: ϕ-value analysis [1, 2]. The folding pathways for numerous proteins have now been experimentally characterized with ϕ-values [1, 3–5]. These data have in turn been used for benchmarking molecular dynamics simulations of protein folding [6].

This chapter outlines the basic theory behind a ϕ-value analysis and discusses how structural information is derived from it. This includes a discussion on choosing mutations for minimizing error in calculations. The techniques for measuring the thermodynamic and kinetic parameters used to compute ϕ-values are also presented.

13.2
Basic Concept of ϕ-Value Analysis

The concept of ϕ-value analysis grew out of extensive study on structure–activity relationships used throughout organic chemistry and enzymology. These techniques include Brønsted analysis to analyze covalent bond formation and protein engineering methods to quantify noncovalent bond formation in enzymic transition states [7, 8]. Indeed, all the basic principles for analyzing protein folding were laid down in studies analysing the noncovalent interactions that stabilize the transition state for the formation of tyrosyl adenylate catalyzed by the tyrosyl-tRNA synthetase [2]. Small perturbations are made to the molecule that cause small changes in structure and hence thermodynamic and kinetic properties.

These perturbations take the form of systematic point mutations of amino acids

Protein Folding Handbook. Part I. Edited by J. Buchner and T. Kiefhaber
Copyright © 2005 WILEY-VCH Verlag GmbH & Co. KGaA, Weinheim
ISBN: 3-527-30784-2

in a ϕ-value analysis. Kinetic and thermodynamic data from these mutants are then used to determine the extent of native-like interactions formed in a protein folding transition state. For the wild-type protein and each mutant, two quantities are required: the free energy stability of the native state with respect to the denatured state (ΔG_{N-D}) and rate of folding from the denatured state to the native state (rate constant k_f).

The stability change for the folding reaction upon mutation is defined as

$$\Delta\Delta G_{N-D} = \Delta G_{N-D} - \Delta G_{N'-D'} \tag{1}$$

Here, ΔG_{N-D} typically ranges from -3 to -15 kcal mol^{-1} for small, single-domain proteins. The primed terms represent the quantities for the mutant.

Next, transition state theory can be applied to kinetic rate constants for folding to obtain the energetic difference between denatured and transition states, in terms of exponential prefactors. When taking the ratio of the rate constants for the mutant and wild-type proteins (k'_f and k_f, respectively), the prefactors cancel, with the assumption that they do not change upon mutation. This gives an expression for $\Delta\Delta G_{TS-D}$ as

$$\Delta\Delta G_{TS-D} = \Delta G_{TS-D} - \Delta G_{TS'-D'} = RT \ln \frac{k'_f}{k_f} \tag{2}$$

The notation is as before, but with TS symbolizing the transition state. The ratio of these two values is then defined as ϕ.

$$\phi = \frac{\Delta\Delta G_{TS-D}}{\Delta\Delta G_{N-D}} \tag{3}$$

The value of ϕ is simply a measure of the destabilization of the transition state relative to the denatured state, normalized to the destabilization between the ground states. A value of 1 indicates that the native and transition states have energies that are perturbed equally upon mutation. A value of 0 indicates that the perturbation to the ground and transition states is isoenergetic. Structurally, a ϕ-value of 1 corresponds to native-like interactions for the mutated residues in the transition state. A ϕ-value of 0 corresponds to denatured-like interactions of the mutated residues in the transition state.

For example, consider a deletion mutation of Ile to Val, where the deleted methyl group exists between two helices. If the helices have native-like packing at this position in the transition state, the expectation would be that the mutation has isoenergetic effects on the transition and native states. The rate of folding would be reduced, giving $\phi = 1$. If the helices are not packed in the transition state, the expectation would be that the rate of folding does not change, giving $\phi = 0$. This is illustrated in Figure 13.1.

Choice of mutation is crucial. For the value of ϕ to have any meaning in relation to the native folding pathway, mutations must not change the pathway for folding

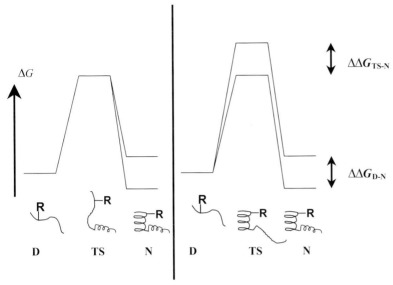

Fig. 13.1. Graphical representation of ϕ-value analysis. The panel on the left signifies a ϕ-value of 0. The residue R is not structured in the transition state. The panel on the right signifies a ϕ-value of 1. The residue is structured in the transition state.

or introduce new interactions. Ideally, they should not rearrange the structure of the protein around the site of mutation in either the denatured or native states, though information can still be obtained if these latter two conditions are not met (see below). Mutations found to have minimal effects on the pathway and structure are small, hydrophobic deletions without change in stereochemistry ("nondisruptive deletions" [2]). By making mutations such as Ile to Val, Thr to Ser, Ala to Gly, Leu to Ala, or Phe to Ala, cavities are created in the hydrophobic core, leaving the folding pathway largely the same. Correlation between native state stability and points of contact of the deleted atoms within a small, 6-Å radius supports the assumption that the structural region probed is defined locally around the deleted atoms [9].

More radical mutations, particularly those affecting electrostatic interactions, are more difficult to interpret because new, unpredictable interactions are introduced, which could be taking place more distant from the site of interest. In cases where these sorts of mutations are not possible, the residue of interest can be first truncated to Ala and then to Gly, to get some idea of the role of that particular position to the transition state.

Mutation of Ala to Gly is a special case because the Gly amino acid is much more conformationally flexible about its ϕ and ψ bond angles. Such a mutation not only probes the effect of the deleted steric bulk of the alanine methyl group, but also backbone conformational flexibility at the site of mutation. Estimates for the conformational entropy contribution of Gly are ~1 kcal mol^{-1}, which can be a significant percentage of a protein's typical stability [10, 11]. By carefully making Ala to Gly mutations, the role of backbone flexibility can be probed. Along with other single methyl group deletions (Ile to Val and Thr to Ser), Ala to Gly mutations can probe secondary structural interactions. The extent of formation of the hydrophobic core can be probed by larger deletions from fully buried residues.

The stability of mutated proteins is an indicator of possible error in computation of ϕ-values. Mutants that have a stability change (ΔG_{N-D}) of less than ~0.6 kcal mol^{-1} run the risk of producing ϕ-values with significant error since the ratio used to compute ϕ is of two small numbers. Similarly, mutations that substantially destabilise the protein are of concern because larger energetic changes are more likely to contain significant structural changes.

13.3
Further Interpretation of ϕ

To get a better understanding of the interpretation of ϕ, thermodynamic cycles for the folding and unfolding reactions can be written as shown in Figure 13.2. Here, D represents the denatured state, N the native state, and TS the transition state. The primed states are those corresponding to the mutant.

Using the cycle below to rearrange Eq. (3), we can then rewrite the definition of ϕ

$$\phi = \frac{\Delta\Delta G_{TS-D}}{\Delta\Delta G_{N-D}} = \frac{\Delta G_{TS'-TS} - \Delta G_{D'-D}}{\Delta G_{N'-N} - \Delta G_{D'-D}} \tag{4}$$

where the notation is as before. Here, the term $\Delta G_{D'-D}$ is simply the change in denatured state relative free energy upon mutation. Since the denatured state is unstructured for the most part, this term reflects the change in solvation and conformational entropy of the denatured state as a result of a mutation. This term is often negligible for small hydrophobic deletions. However, in cases where polar or

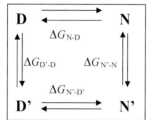
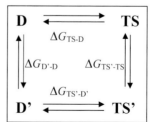

Fig. 13.2. Thermodynamic cycles between the denatured state (D) and transition state (TS), or native state (N) for a wild-type (unprimed) and mutant (primed) protein.

charged groups are exchanged for hydrophobic groups, solvation energies are no longer simply fractions of a kcal, as in the case of hydrophobic deletions [5]. Cases where glycine is deleted or introduced also introduce conformational entropy terms that are of similar magnitude [11].

There are two cases, as given earlier, of simple interpretation. When $\Delta G_{TS'-TS} = \Delta G_{N'-N}$, $\phi = 1$. Here, the native and transition states are equally perturbed. Whether or not the denatured state energy, $\Delta G_{D'-D}$, changes is not critical. Structurally, the residues mutated have the same structural interactions in the transition and native states.

In the second case, $\Delta G_{TS'-TS} = \Delta G_{D'-D}$ and $\phi = 0$. In this situation, changes in interactions in the native state are largely not important (so long as they do not change the folding pathway). Since the energy $\Delta G_{D'-D}$ is often very small, the transition and denatured state mutated atoms have the same unstructured form. However, it could also indicate equal breakdown of residual structure in the denatured and transition states. In such a case, however, the interpretation is the same: native interactions do not form until a position past the transition state on the folding pathway.

Fractional ϕ-values are more difficult to interpret. In one simplifying case, some speculation can be made. When the energy of solvation for a particular mutation is zero ($\Delta G_{D'-D} = 0$), the expression for ϕ simplifies to $\Delta G_{TS'-TS}/\Delta G_{N'-N}$. This is the case when the particular mutation is buried within the protein and makes no contact with solvent. This fractional ϕ-value is interpreted as an indication of a partially formed structure in the transition state. However, the energetics of the structural interactions is unlikely to be simply linear, so exact structural interpretation is difficult. Some ϕ-values are close to 0 or 1 in value, making interpretation easier. Fractional values are more likely to suggest partial structure formation when they appear in groups within a particular region of a protein structure, such as the hydrophobic buried core.

A large number of ϕ-values that are fractional over a specific region of a protein structure can help rule out the possibility of parallel folding pathways: one with ϕ-values around 1 and one with ϕ-values around 0 [12]. This is most readily accomplished by graphically plotting the ϕ-value data. Data are plotted with ΔG_{TS-D} on the vertical axis and ΔG_{N-D} on the horizontal axis (or $\Delta \ln k_{D-N}$ against $\Delta \ln K_{D-N}$). Points that fall onto a linear plot likely fold with the mutated atoms structured to a similar extent in the transition state. If more mutations are made in the same element of structure, the expectation is that points continue to fall on the same line. However, in the case of multiple pathways, additional mutations, which stabilize the native state of the protein to a greater or lesser extent, may favor one of the folding pathways and change the slope of the line to a new value.

Beyond multiple pathways, it is also important to rule out the possibility of the addition of new interactions upon mutation. In such a situation, a double mutant cycle can elucidate the problem. A known interaction between two residues is broken by mutating each residue in separate experiments and then both residues in a double mutant experiment [13]. A rough estimate of the energy of interaction between residues can be made, giving some indication of possible structural rearrangements upon mutation.

Occasionally, uncanonical ϕ-values (greater than 1 or less than 0) are reported. Such cases can be explained by an overpacked, and hence destabilized hydrophobic core, which may be preferentially stabilized in comparison to the transition state when a deletion mutant is made. Such problems are more common in engineered sequences [14, 15]. Nonnative interactions in the transition state can also lead to these sorts of unconventional ϕ-values.

It should also be noted that ϕ-values are not limited to characterization of the transition state. It is also possible to determine ϕ-values for intermediates along the folding pathway, substituting ΔG_{I-D} for ΔG_{TS-D} in Eq. (3), where I represents the intermediate state. In the case of the barnase, ϕ-value analysis of the intermediate showed a structure consistent with the complete folding mechanism assembled from NMR studies on the denatured state and ϕ-values determined of the transition state [16].

Thus far, discussion of ϕ-values has focused on the forward, folding reaction. It should be noted, however, that the reverse (unfolding) reaction also has an associated ϕ-value, which can be understood with a similar analysis. For a simple two-state reaction, this is simply $1 - \phi$ for folding. Often, this unfolding ϕ-value is simpler to compute, because intermediate states are most likely to occur between the denatured state and transition state, which affect the rate of folding and not unfolding. The unfolding pathway from the folded state to the transition state is more typically a first-order reaction that can be readily determined by fitting the unfolding arm of a chevron plot (see below).

13.4
Techniques

Protein stability is frequently measured using either chemical or thermal means to allow the protein to undergo denaturation under equilibrium conditions [17]. In general, the two methods for determining changes in stability upon mutation provide similar results for canonical two-state folding proteins.

Thermal denaturation is typically measured using differential scanning calorimetry (DSC). Fitting a DSC trace (or a trace created by spectroscopically following a thermal unfolding curve) will provide the enthalpy (ΔH_{N-D}) and entropy (ΔS_{N-D}), as well as a precise measure of the melting temperature, T_m. The entropy and enthalpy here are assumed to be at the melting temperature. Such a fit is best done with knowledge of the specific heat capacity change during folding (ΔC_P). If the change in stability upon mutation is approximated to be temperature independent, the following equation can be used to calculate it.

$$\Delta\Delta G_{D-N} \approx \Delta T_m \Delta S_{D-N} = \frac{\Delta T_m}{T_m} \Delta H_{D-N} \tag{5}$$

In a spectroscopically followed chemical denaturation using urea or guanidinum chloride as a denaturant, the free energy change with the concentration of denaturant is taken to be linear.

$$\Delta G_{D-N} = m_{D-N}([D]_{50\%} - [D]) \tag{6}$$

Similar to the thermal unfolding case, fits to denaturation curves will provide two numbers: the m-value (m_{D-N}) and midpoint of the denaturation ($[D]_{50\%}$). The m-value is usually taken to be a constant that scales with the change in solvent-accessible surface area of the protein upon unfolding. Often, the mutated and wild-type proteins have negligible differences in m-values, given that so few atoms are deleted. However, significant errors can be introduced when the stability is extrapolated to zero denaturant concentration. To minimize this error, the average m-value is computed. This is then multiplied against the change in the denaturation midpoint (a precisely determined value like the change in T_m) to give the change in stability upon mutation.

$$\Delta\Delta G_{D-N} \approx 0.5(m_{D-N} + m'_{D-N})\Delta[D]_{50\%} \tag{7}$$

The kinetics of protein folding can be followed using numerous techniques: stopped flow and continuous flow mixing, temperature jump, and NMR line-broadening analysis. As before, spectroscopic probes such as circular dichroism or tryptophan fluorescence can be used to indicate the population of folded and unfolded protein. The observed rate in the folding reaction is k_{obs} and is the sum of the folding and unfolding rate constants: $k_f + k_u$. It is typically taken as a function of denaturant concentration and plotted as a v-shaped chevron plot.

As in the equilibrium situation, a linear m-value is used to describe the dependence of the energetics of folding and unfolding on denaturant concentration.

$$\Delta G_{TS-D} \propto m_{TS-D}[D] \tag{8a}$$
$$\Delta G_{TS-N} \propto m_{TS-N}[D] \tag{8b}$$

where m_{TS-N} and m_{TS-D} are the kinetic m-values. Summation of the kinetic m-values should give the equilibrium m-value. A common explanation for any shortfall is the existence of intermediate states on the folding pathway. The ratio of the kinetic m-value to the equilibrium m-value gives a measure of the solvent exposure, or compactness, of the transition state. This information, in the form of Tanford β, defined below as β_T, is complementary to the ϕ-value, which is a structurally more specific measure of transition state structure.

$$\beta_T = \frac{m_{TS-N}}{m_{D-N}} \tag{9}$$

Application of transition state theory to Eq. (8) gives equations describing the dependence of rate on denaturant concentration, $[D]$

$$\ln k_{\mathrm{f}} = \ln k_{\mathrm{f}}^{\mathrm{H_2O}} - m_{\mathrm{TS-D}}[\mathrm{D}] \tag{10a}$$

$$\ln k_{\mathrm{u}} = \ln k_{\mathrm{u}}^{\mathrm{H_2O}} + m_{\mathrm{TS-N}}[\mathrm{D}] \tag{10b}$$

where $k_{\mathrm{f}}^{\mathrm{H_2O}}$ and $k_{\mathrm{u}}^{\mathrm{H_2O}}$ are the rate constants for folding and unfolding, respectively, in the absence of denaturant. These equations describe the characteristic chevron plot of k_{obs} as a function of denaturant concentration. Deviations from a v-shaped plot with two linear arms, representing the folding and unfolding parts of the reaction, are strong indicators that the folding reaction is more complicated than a simple two-state process. Further, it should be understood that a v-shaped chevron plot cannot alone be taken as evidence for a two-state transition without first fitting it to a standard equation and recovering reasonable values for kinetic constants. Fitting the entire plot also reduces error that may be present in single rate constant at a particular denaturant concentration.

The particular denaturant concentration used for computation of ϕ-values is not critical in the case of a two-state folding protein, where the value of the Tanford β does not change upon mutation and the folding pathway is unchanged. In general, it is wisest to compute values of $\Delta\Delta G_{\mathrm{TS-D}}$ at a variety of denaturant concentrations to observe whether the folding pathway changes with denaturant concentration [18]. In cases where intermediate states are believed to exist between the denatured and transition states, it may be best to use values exclusively at high denaturant concentrations or to simply fit the unfolding arm of the chevron plot with a linear fit, since such intermediates are contributing little to the observed rate constant, k_{obs}, at high denaturant concentration.

13.5
Conclusions

The method described here for carrying out a ϕ-value analysis is the only experimental method available for atomic level structural information of folding transition states. Extensive ϕ-value analyses have been carried out on the proteins barnase and chymotrypsin inhibitor 2 [1, 3–5]. The transition state structural information obtained in such studies, when combined with molecular dynamics simulation and NMR structural information, can then be used to determine the complete, atomistic picture of protein folding [19]. As computing power further improves to bring simulated folding temperatures into agreement with experimental temperatures, a combination of ϕ-value analysis and computational techniques will be essential for increasing the repertoire of proteins for which the complete, atomistic folding pathway is known.

References

1 FERSHT, A. R., MATOUSCHEK, A. & SERRANO, L. The folding of an enzyme I. Theory of protein engineering analysis of stability and pathway of protein folding. *J. Mol. Biol.* 224, 771–782 (1992).

2 FERSHT, A. R., LEATHERBARROW, R. J. & WELLS, T. N. Quantitative-analysis of structure activity relationships in engineered proteins by linear free-energy relationships. *Nature* 322, 284–286 (1986).

3 ITZHAKI, L. S., OTZEN, D. E. & FERSHT, A. R. The structure of the transition state for folding of chymotrypsin inhibitor 2 analysed by protein engineering methods: evidence for a nucleation-condensation mechanism for protein folding. *J. Mol. Biol.* 254, 260–288 (1995).

4 MATOUSCHEK, A., JR., KELLIS, J. T. K., SERRANO, L. & FERSHT, A. R. Mapping the transition state and pathway of protein folding by protein engineering. *Nature* 340, 122–126 (1989).

5 OTZEN, D. E., ITZHAKI, L. S., ELMASRY, N. F., JACKSON, S. E. & FERSHT, A. R. Structure of the transition state for the folding/unfolding of the barley chymotrypsin inhibitor 2 and its implications for mechanisms of protein folding. *Proc. Natl Acad. Sci. USA* 91, 10422–10425 (1994).

6 MAYOR, U., GUYDOSH, N. R., JOHNSON, C. M., GROSSMANN, J. G., SATO, S., JAS, G. S., FREUND, S. M. U., DAGGETT, U. & FERSHT, A. R. The complete folding pathway of a protein from nano-seconds to microseconds. *Nature* 421, 863–867 (2003).

7 FERSHT, A. R. Dissection of the structure and activity of the tyrosyl-tRNA synthetase by site-directed mutagenesis. *Biochemistry* 26, 8031–8037 (1987).

8 CARROLL, F. A. *Perspectives on Structure and Mechanism in Organic Chemistry*. Brooks/Cole Publishing, London (1998).

9 JACKSON, S. E., MORACCI, M., ELMASRY, N., JOHNSON, C. M. & FERSHT, A. R. Effects of cavity-creating mutations in the hydrophobic core of chymotrypsin inhibitor 2. *Biochemistry* 32, 11259–11269 (1993).

10 SERRANO, L., NEIRA, J.-L., SANCHO, J. & FERSHT, A. R. Effect of alanine versus glycine in α-helices on protein stability. *Nature* 356, 453–455 (1992).

11 BURTON, R. E., HUANG, G. S., DAUGHERTY, M. A., CALDERONE, T. L. & OAS, T. G. The energy landscape of a fast-folding protein mapped by Ala-Gly Substitutions. *Nature Struct. Biol.* 4, 305–310 (1997).

12 FERSHT, A. R., ITZHAKI, L. S., ELMASRY, N. F., MATTHEWS, J. M. & OTZEN, D. E. Single versus parallel pathways of protein folding and fractional formation of structure in the transition state. *Proc. Natl Acad. Sci. USA* 91, 10426–10429 (1994).

13 MATOUSCHEK, A. & FERSHT, A. R. Protein engineering in analysis of protein folding pathways and stability. *Methods Enzymol.* 202, 82–112 (1991).

14 VIGUERA, A. R., VEGA, C. & SERRANO, L. Unspecific hydrophobic stabilization of folding transition states. *Proc. Natl Acad. Sci. USA* 99, 5349–5354 (2001).

15 VENTURA, S., VEGA, M. C., LACROIX, E., ANGRAND, I., SPAGNOLO, L. & SERRANO, L. Conformational strain in the hydrophobic core and its implications for protein folding and design. *Nature Struct. Biol.* 9, 485–493 (2002).

16 BOND, C. J., WONG, K. B., CLARKE, J., FERSHT, A. R. & DAGGETT, V. Characterization of residual structure in the thermally denatured state of barnase by simulation and experiment: description of the folding pathway. *Proc. Natl Acad. Sci. USA* 9, 13409–13413 (1997).

17 FERSHT, A. *Structure and Mechanism in Protein Science*. W.H. Freeman and Company, New York (1998).

18 MATOUSCHEK, A., OTZEN, D. E., ITZHAKI, L. S., JACKSON, S. E. & FERSHT, A. R. Movement of the position of the transition state in protein folding. *Biochemistry* 17, 13656–13662 (1995).

19 DAGGETT, V. & FERSHT, A. R. The present view of the mechanism of protein folding. *Nature Rev. Mol. Cell Biol.* 4, 497–502 (2003).

14
Fast Relaxation Methods

Martin Gruebele

14.1
Introduction

In a chemical system at equilibrium, the forward and backward reactions occur at the same rate. For a simple two-state reaction D $\rightleftarrows$ F, the equilibrium concentrations and rate coefficients are thus related by the equilibrium "constant" $K_{eq} = [F]/[D] = k_f/k_d$. The quotation marks are warranted because K_{eq} is not really a constant, it depends on the environment of the chemical system. According to Le Châtelier, an endothermic reaction shifts towards higher concentrations of product when the temperature is raised, thus raising K_{eq}. Similarly, a reaction producing a reduced-volume product shifts towards the product when pressure is applied. As a final example, addition of a denaturant such as urea or guanidinium hydrochloride to a protein solution (see Chapter 3) favors exposure of hydrophobic surfaces, and would increase the concentration of denatured state D (the reactant in the equations shown above).

The sensitivity of K_{eq} to the environment can be exploited to study reaction rates. The idea behind relaxation methods is straightforward: If a parameter such as temperature, pressure, or denaturant concentration is switched to a new value so rapidly that the chemical system cannot respond during the switching process, the subsequent relaxation towards equilibrium can be monitored. This idea was first turned into a practical experimental technique by Manfred Eigen and coworkers [1]. Eigen received the 1967 Nobel prize in chemistry for this work.

Two ingredients are required to make a relaxation experiment work: a fast initiation step and a fast detection scheme. Protein folding has been initiated by pressure jumps, temperature jumps, electrochemical potential jumps, sudden binding/unbinding of substrates, and denaturant jumps, to name just the most prominent examples [2–4]. Optical probes have contributed the bulk of fast detection methods, including infrared and visible absorption, Raman scattering, X-ray scattering, fluorescence, and circular dichroism [4, 5].

What is "fast" in the context of protein folding? A somewhat arbitrary but commonly used definition is given here: Beginning in the 1980s, a number of protein-folding reactions were investigated by stopped-flow mixing with a time resolution

---

Protein Folding Handbook. Part I. Edited by J. Buchner and T. Kiefhaber
Copyright © 2005 WILEY-VCH Verlag GmbH & Co. KGaA, Weinheim
ISBN: 3-527-30784-2

of ca. 1 ms. It was found that many proteins have an unresolved burst phase, even though the complete folding process may require from milliseconds to seconds [6]. More recently, proteins and peptides have been discovered that fold completely on the submillisecond time scale, so none of the folding kinetics can be resolved by conventional stopped-flow techniques [7]. Relaxation techniques were developed to break the "ms-barrier," and so processes in the nanosecond to microsecond range will be considered "fast." Relaxation and other fast techniques have proved that none of the elementary events for folding take more than a millisecond, and far more stringent limits on the time scales have been set in the meantime.

Fast relaxation studies of biologically relevant systems have a long history. Eigen himself had studied imidazole protonation by 1960 [8]. Flynn and coworkers studied the association of proflavine in 1972 by laser temperature jump [9]. Holzwarth and coworkers studied lipid phase transitions [10] and dye–DNA binding in the 1980s, and Turner's group extensively investigated relaxation of nucleic acid oligomers in the 1980s [11]. Submillisecond relaxation was first applied to peptide and protein folding in the 1990s [12–16], including photolysis-induced, temperature, electrochemical [17], and later pressure jumps [3].

The relaxation experiments discussed here are not to be confused with dielectric relaxation spectroscopy [18], which does not involve a sudden initiation step, or with NMR relaxation experiments [7], which can study protein folding kinetics on a microsecond time scale via analysis of the lineshapes. These, as well as stopped-flow techniques and submillisecond continuous flow techniques are discussed in Chapter 23.

14.2
Techniques

14.2.1
Fast Pressure-Jump Experiments

Pressure (or solvent density) is a useful variable for controlling protein folding. Although folded proteins are more compact than unfolded proteins in terms of radius of gyration, the cores are not perfectly packed and the unfolded state generally has a smaller partial volume of solvation than the folded state. According to Le Châtelier's principle, proteins thus unfold at higher pressure, and this generally occurs at pressures of $1-5 \times 10^8$ Pa. Increased pressure also raises the cold-denaturation temperature of proteins and decreases the heat-denaturation temperature. To a first approximation, the folding–unfolding phase diagram can be derived from a quadratic expansion of the free energy in T and P. Supporting evidence shows that in the low-pressure regime, the stability of some proteins is actually enhanced by pressure [19, 20] (see Chapter 5).

A number of kinetic studies using pressure relaxation with millisecond or slower time resolution have been reported [21]. Submillisecond time resolution was achieved by Schmid and coworkers [3], who used a piezoelectric driver design first

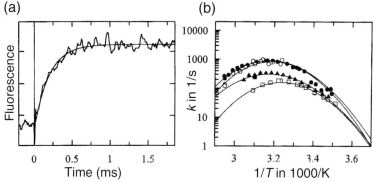

Fig. 14.1. a) Refolding of 6 μM *Bc*-Csp after a pressure jump from 100 to 3 bar at 75.3 °C. b) Arrhenius plot of the microscopic rate constant k of refolding of (open circle) wild-type *Bs*-CspB and the (filled circle) Phe27Ala, (triangle) Phe17Ala, and (square) Phe15Ala variants. From Ref. [3].

proposed by Clegg et al. [22] In this design, a sample contained in a 2 mm bore sapphire ring covered on each side by a flexible membrane is compressed in <100 μs by a piezoelectric element. Pressure changes up to 10^7 Pa can be achieved in both directions, and can be maintained essentially indefinitely. The sapphire ring provides optical access around the whole sample.

A prototypical result is shown in Figure 14.1 [3]. The cold shock protein from *Bacillus subtilis* relaxes in as little as 41 μs via an apparent two-state mechanism. (How to obtain folding rates from relaxation rates is briefly discussed in the protocols, and in Chapter 12.1.) The unfolding rate is Arrhenius-like (a straight line on a ln k versus $1/T$ plot), but the folding rate is highly curved with a maximum rate at the temperature of maximum hydrophobicity.

14.2.2
Fast Resistive Heating Experiments

The first successful fast temperature-jump relaxation experiments were based on resistive heating. Hoffman developed an instrument capable of 50 ns time resolution in high ionic strength solvents [23]. Commercial instruments (DIA-LOG) are available based on the work of Jovin and coworkers [24]. These instruments have a time resolution of <5 μs under physiologically relevant conditions (low ionic strength solvent).

In a resistive temperature jump, a 10–50 nF capacitor at about 500 V is discharged into a ≈1 ml sample cell. The risetime of the jump depends on solvent conductivity, and ranges down to 1 μs in solutions at a moderate ionic strength yielding 100 Ω cell resistance.

A prototypical result obtained by heating the cold-denatured protein barstar in about 5 μs is shown in Figure 14.2 [14, 25]. Two phases are resolved in the unfold-

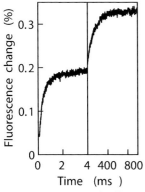

Fig. 14.2. Biphasic submillisecond and millisecond relaxation of barstar, showing accumulation of an intermediate ensemble. From Ref. [14].

ing reaction, so a minimal scheme involving a denatured state D, intermediate state I and folded state N could be constructed. A mutational phi-value analysis reveals the differing contributions of amino acid side chains to the stability of the intermediate state. More recent work on WW domain and homeodomain folds has revealed two-state refolding times as fast as 19 µs [26–30]. A study of *Drosophila* calmodulin, which folds in approximately 100 µs, confirms a behavior also seen in slower folding proteins [31, 32]: the unfolding rate is Arrhenius-like (linear plot of $\ln k$ vs. $1/T$ with negative slope), indicating that the transition state free energy shifts with the folded state, and implying that the transition state is not very different structurally from the folded state.

14.2.3
Fast Laser-induced Relaxation Experiments

Lasers can be used to induce relaxation via a variety of photochemical and photophysical processes. A substrate can be removed by photolysis, thereby causing rearrangements of the binding protein. Redox agents or proton donors can be light-activated, changing the oxidation or protonation state of a protein. Disulfide bridges or other linkages can be broken, releasing the strain of a nonnative conformation. Other photochemical processes include pH jumps and dissociation of unnatural amino acids. Among photophysical processes, chromophore relaxation has been used to study the folding speed limit, and the most commonly used is heating of the solvent (temperature jump). We discuss these approaches in turn.

14.2.3.1 Laser Photolysis
Laser photolysis was first used by Eaton and coworkers to induce protein refolding [12]. They used the increase of protein stability when the CO ligand is photodissociated to induce folding events. The same group also induced a "protein quake" by

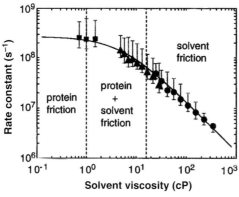

Fig. 14.3. Rate constant of myoglobin relaxation as a function of solvent viscosity. An inverse relationship holds except at the very lowest viscosities. From Ref. [33].

ligand dissociation [33]. By laser-dissociating CO from myoglobin in glycerol/water mixtures ranging in viscosity from 0.7 to 300 cP, they were able to show that protein self-friction contributes ≈ 1 cP (mPa s) to the effective viscosity in a Kramer's theory for the folding rate (Figure 14.3). (See Experimental Protocols and Chapter 22 for more details on Kramer's theory.)

14.2.3.2 Electrochemical Jumps

Electrochemical jumps were pioneered by Gray, Winkler and coworkers [17, 34–38]. In these experiments, redox active complexes (often based on Ru^{2+} ions) reduce a prosthetic group (such as the heme group in cytochrome c). Depending on the choice of complex, the electron transfer reaction can be complete in under 1 μs. The fastest redox agents are typically limited to <1 ms by charge recombination.

Barrier-limited refolding has been studied in cytochrome c by reducing unfolded ferricytochrome in ca. 3 M GuHCl buffer. Under these denaturant conditions, ferricytochrome c is more stable in the unfolded state, but ferrocytochrome c is more stable in the folded state; the protein refolds after the electron transfer occurs, with the fastest phase observed of 40 μs duration [17, 37]. Because electron transfer processes can be tuned to have large negative free energies, they can also be used to probe submicrosecond processes corresponding to diffusional motions of proteins (see also "Chromophore Excitation" below) [38]. Figure 14.4 shows the 250 ns contact time obtained under highly denaturing conditions between a ruthenium complex bound to histidine 33 of cytochrome c, and a zinc-modified porphyrin 15 residues away. After applying a steric correction, the authors estimate a ≈ 60 ns contact formation rate.

14.2.3.3 Laser-induced pH Jumps

Laser-induced pH jumps rely on changes of the pK_a induced by electronic excitation. Systems have been developed in which the pH is lowered through the reac-

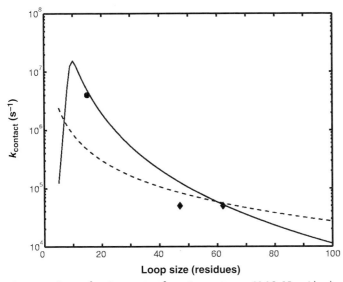

Fig. 14.4. Rates of tertiary contact formation in denatured cyt c have been extracted from measurements of ET rates in Ru(His33)-Zn-cyt c (filled circle [Gdm_HCl] = 5.4 M, temperature = 22 °C, 15-residue loop). The solid line is from the Camacho-Thirumalai theory, the dashed line from the Szabo model. From Ref. [38].

tion $AH \rightarrow A^- + H^+$, and where the pH is raised through the recombination $A^- + H^+ \rightarrow AH$ [39, 40]. The bimolecular recombination process is generally diffusion limited, so the original pH is reestablished upon deexcitation of the chromophore. This occurs within 10 s of microseconds at concentrations useful for protein-folding studies. Some reactions also yield stable photoproducts following a reaction of the photoproduced anion (e.g., *ortho*-nitrosobenzoic acid = *o*-NBA), so the pH is stable after the jump [41].

Recent pH-jump relaxation studies investigated the formation of the molten globule intermediate of apomyoglobin using *o*-NBA [42, 43]. The two observed kinetic processes were attributed to histine imidazole (fast) and carboxylate (slow) protonation by one set of authors, while the others found a significant denaturant-induced reduction in the slower rate process.

14.2.3.4 Covalent Bond Dissociation

Covalent bond dissociation overcomes the main limitation of the two previous approaches: the need for denaturant tuning. In order to poise the protein at the brink of (un)folding, so substrate removal or electron transfer can have a large effect, molar quantities of denaturant often need to be added. By photolyzing disulfide bridges used to constrain peptide or protein structure to a nonnative state, experiments can be carried out in denaturant-free buffer. This has been applied to the recombination dynamics of aminothiotyrosine "unnatural" amino acids in model peptides: the aqueous sample can be photolyzed at 270 nm with a femtosecond or

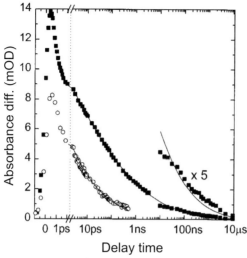

Fig. 14.5. Decay of thiyl radical absorbance in the 17-mer peptide discussed in the text. The dynamics can be fitted by powerlaws $t^{-\alpha}$ with α near 0.1. From Ref. [44].

picosecond pulse [44, 45]. Disulfide bridges cleave by a directly dissociative process of <1 ps duration, providing an enormous dynamic range over which two point correlations of the thiyl termini can be studied. Figure 14.5 shows the relaxation of an alanine-rich 17-mer initially cyclized by a disulfide bridge. The relaxation process is highly nonexponential and extends from picoseconds to microseconds.

14.2.3.5 Chromophore Excitation

Chromophore excitation has been used as a direct probe of the fastest diffusive events that can occur during protein folding [46–51]. A typical example would be diffusion of a peptide loop until end-to-end contacts are formed. In such experiments, a chromophore is excited at $t = 0$. The chromophore excitation then relaxes, but the relaxation rate is accelerated by quenching when the peptide loop forms contacts. Experiments quenching tryptophan triplet states with cysteine have yielded loop contact times ranging from 40 to 140 ns as the chain length is increased from 5 to 20 [49, 50]. Experiments utilizing triplet–triplet energy transfer between thioxanthone and naphtyl groups have yielded even shorter times of 20 ns for small loops [48, 51]. Such experiments are described in more detail in Chapter 22.

14.2.3.6 Laser Temperature Jumps

Laser temperature jumps are perhaps the most commonly used relaxation method for peptide and protein folding experiments [5, 13, 15, 16, 30, 52–66]. The first successful laser temperature-jump studies of chemical reactions were carried out in the 1970s, but the technique was not applied to protein folding until 20 years later.

Laser temperature-jumps can achieve dead times as short as 2 ps [13]. The lower limit is imposed by the need for thermal equilibration of the solvent after a specific solvent mode has been excited by the laser. (Backbone modes of the protein should of course not equilibrate, since one wishes to observe them after the temperature jump.) The achievable temperature jumps generally lie in the range of 5–30 K, sufficient to cause large changes in protein K_{eq} values. Most laser temperature-jump experiments are carried out with nanosecond heating pulses in the 1.5–1.9 µm range to pump directly H_2O or D_2O stretching vibrations, which thermalize within the nanosecond pump profile. These wavelengths can be obtained from Nd:YAG lasers Raman-shifted in hydrogen (1.9 µm) or methane (1.5 µm) [67], or by difference frequency generation in $LiNbO_3$ crystals [53]. Fast temperature jumps can potentially cause pressure and temperature fronts to propagate through the sample, but such effects can be mitigated by using the proper counterpropagating pumping geometry, initial temperature, pump spot size, and a noncollinear detection method [67].

Protein unfolding, peptide relaxation, and protein refolding were initially observed by laser temperature jumps in the mid-1990s [13, 15, 16]. In apomyoglobin, formation of the hydrophobic AGH helical core happened 5 orders of magnitude faster (<10 µs) than complete refolding of the protein, resolving the burst phase seen by stopped-flow kinetics [6, 16]. Small peptides of well-defined secondary structure exhibit even faster kinetics [15, 55, 68–72]. Figure 14.6 shows the relaxation kinetics of a β-hairpin and of a short helix, common motifs in protein secondary structure. Analysis of the hairpin data is compatible with formation of the turn as a rate-limiting step [70], although models with early interstrand contacts have similar time scales [73]. The helix data use ^{13}C isotopic labeling to dissect local peptide kinetics, and confirms nonexponential dynamics previously observed for nonnatural helical polymers and expected from helix formation models [71, 74, 75].

Two-state and more complex folding mechanisms of larger proteins have also been observed by laser temperature jumps. Phi value and other mutation analyses (see Chapter 13) have been reported [59], as well as solvent-tuning studies [76]. Like resistive temperature-jump experiments and temperature-dependent pressure jumps, laser temperature-jump studies have shown strong curvatures in Arrhenius plots for the folding rate of fast folders [59]. Figure 14.7 illustrates laser temperature jumps of proteins with the relaxation kinetics of two very fast folding λ-repressor fragment mutants [65]. All mutants of this molecule with >50 µs folding times exhibit single exponential kinetics, while all mutants with $\tau < 50$ µs, irrespective of the location of the mutations, exhibit an additional fast phase lasting for a few microseconds.

14.2.4
Multichannel Detection Techniques for Relaxation Studies

Relaxation experiments are easily combined with a large variety of single-channel probes such as integrated fluorescence [14], one-color absorption [77], photoacous-

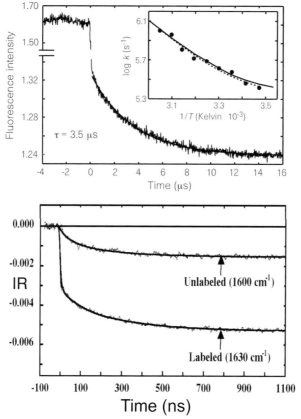

Fig. 14.6. Relaxation response of 16-mer beta hairpin (top) and of unlabeled and ^{13}C-labeled helix peptides (bottom), showing a significant change in the fast-phase amplitude. From Refs [69] and [71].

tic spectroscopy [42], or transient gratings [43]. Here multichannel techniques will be described in more detail. Multichannel techniques provide an array of data at each kinetic measurement, such as a full infrared spectrum or a time-resolved fluorescence decay. They can extract additional information not available when only rate constants from a single probe are used to characterize folding. These probes address different aspects of protein structure.

14.2.4.1 Small Angle X-ray Scattering or Light Scattering

This can be used to characterize protein compactness by plotting the intensity vs. scattering vector. Time-resolved Kratky plots can follow the evolution from highly denatured to globular states, and the radius of gyration can be used as a single measure of protein compactness. These detection techniques have achieved submillisecond resolution when combined with mixing techniques (see Chapter 15)

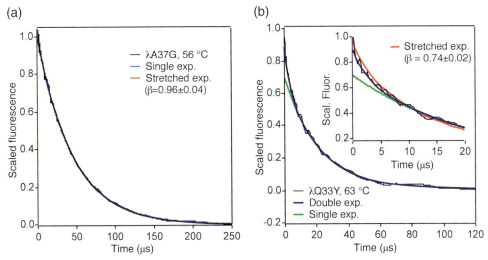

Fig. 14.7. a) Relaxation of a very fast folding mutant of lambda-repressor fragment (Y22W/Q33Y/G46A/G48A), showing deviation from single exponential relaxation below 5 μs. b) The slower Y22W/Q33Y/A37G mutant follows single-exponential relaxation with high precision. From Ref. [65].

[78], but they have not been combined with the relaxation experiments described here.

14.2.4.2 Direct Absorption Techniques

These have been applied to submillisecond protein folding in the infrared and visible [12, 31, 79] IR bands at ≈ 1650 cm^{-1} (amide I′) and ≈ 1300 cm^{-1} (amide III′) are sensitive to overall secondary structure, with specific signatures for native and solvated α-helices, β-sheets, and aggregates. Three approaches have been developed for obtaining mid-infrared spectra with submicrosecond time resolution from relaxation experiments. In the scanning method, a high-resolution laser is tuned across the spectral region in steps, and a relaxation experiment is performed at each step [15]. In step-scan FTIR, a Fourier-transform infrared spectrometer is stopped at regular intervals in the scan of its moveable mirror, and a kinetic transient is collected at n points in time. The resulting n interferograms can be Fourier-transformed to yield the infrared spectrum at each of the n kinetic time intervals [80]. In the third technique, a nanosecond-spaced sequence of broadband infrared pulses is generated by a femtosecond laser, sent through the sample after relaxation has initiated protein kinetics once, and finally dispersed onto an array detector [81]. Perhaps the most common use of IR techniques has been in time-resolved secondary structure melting via the amide I′ band of both peptides and proteins. Experiments have shown the existence of a new α-helical signature [53], ascribed to a solvated helix that occurs during unfolding. Particularly noteworthy is the fast secondary structure loss examined in detail for the folded and several partially folded forms of apomyoglobin [5].

In the near-IR and visible region, the local environment of prosthetic groups such as heme (Soret band at ≈ 420 nm) or side chains such as tyrosine [31] can be probed. Parallel data collection is possible with a broadband excitation source, such as a femtosecond Ti:Sapphire laser or fluorescence from a laser-pumped dye (typically 50–100 nm bandwidth) dispersed onto an array detector. This again has the advantage that the entire absorption spectrum is collected after a single relaxation event. The first submillisecond measurements of diffusional dynamics of a protein (cytochrome c) were carried out using absorption measurements and laser photolysis (see above) in 1993 [12]. Recently the method has been used to study the dissociation dynamics of light-harvesting complexes into dimer and monomer subunits (see Chapter 14.5 for precautions in the analysis of such relaxation) [64].

Some relaxation experiments (e.g. temperature jumps and pressure jumps) can induce significant variations in the solvent refractive index. In such cases, great care must be taken when applying direct absorption to detect protein kinetics, as the probe beam may wander or may be lensed as it passes through the solvent [64].

14.2.4.3 Circular Dichroism and Optical Rotatory Dispersion

These techniques measure the difference in absorbance or refractive index of left and right circularly polarized light due to chirality of the absorber. Two regions are of particular interest for studying protein backbones and side chains: at 250–300 nm, asymmetries in the environment of aromatic side chains cause CD or ORD signals; at 190–230 nm, absorption of the amide group probing the backbone itself leads to distinctive CD signals for random coils (small or positive from 210 to 220 nm), sheets (negative peak near 215 nm), and helices (two negative peaks at 208 and 222 nm). Nanosecond time-resolution methods have been developed for relaxation experiments: a highly elliptically polarized probe pulse (to avoid birefringence artifacts) is passed through the sample and modulated between left- and right-handed, then analyzed by a linear polarizer. This is repeated with different time delays after initiation of relaxation to yield a kinetic transient, and then at different wavelengths to yield a spectrum. The method has been used to directly measure the appearance of secondary structure on a nanosecond time scale in model peptides as well as microsecond time scale formation of some secondary structure in cytochrome c [82–85].

14.2.4.4 Raman and Resonance Raman Scattering

These have been used to determine both secondary structure content [86, 87] and changes in the local environment of prosthetic groups [88]. UV pulses near 210 nm resonantly excites amide electronic states of the backbone, creating scattered light about 5 nm away that can be selected by a monochromator (corresponding to the 1300 cm^{-1} amide III band). Pumping the Soret band of heme near 400 nm yields a combination of vibrational peaks very sensitive to ligation, and specific side chain vibrational modes can also be probed [88]. It is worth noting here that this detection technique has been applied extensively to fast folding coupled with continuous mixers (Chapter 15) [89]. Compared with absorption techniques, Raman and fluorescence emission have an important advantage: their relatively isotropic light collection makes them less sensitive to refractive index fluctuations and other variations in the sample induced by the initiation step. In addition, neither the probe beam nor the scattered Raman light are absorbed by the aqueous solvent, and Raman is an inherently parallel probe technique which does not require scanning to obtain an entire spectrum. The main disadvantage is that dispersed Raman spectra are quite weak. Intense pump lasers can help, but photodamage must then be taken into account [88].

14.2.4.5 Intrinsic Fluorescence

Intrinsic fluorescence can be used to probe solvent exposure, motional anisotropy, the formation of specific tertiary contacts via quenching, and pair distances via engineered quenchers. Intrinsic fluorescence of proteins without prosthetic groups is due to the aromatic residues [90]. Tryptophan fluorescence dominates at 290–300 nm excitation, but tyrosine fluorescence has also been detected in fast relaxation experiments [56, 63]. Tryptophan fluorescence can be quenched by short-range electron or proton energy transfer (1–5 Å) from/to protonated histidine, cysteine and a few other side chains [91], or by longer range Förster energy transfer [92] (10–50 Å) to heme, nitrotyrosine, and cysteines derivatized with acceptor dyes [93]. This results in a decrease of the fluorescence lifetime from as long as 10 ns to as short as 0.1 ns, making tryptophan a useful probe of tertiary contact formation [42, 94]. Tryptophan fluorescence shifts from $\lambda_{max} \approx 350$ nm to ≈ 330 nm upon burial in the hydrophobic core, and quenchers such as acrylamide can also be used to probe solvent accessibility in fast relaxation experiments [95]. A number of proteins hyperfluoresce during the initial stage of denaturation, indicating that fluorescence intensity is a very sensitive probe of the native-like packing [96, 97].

Instruments capable of detecting an entire fluorescence spectrum every 100 ns, and an entire fluorescence decay every 14 ns have been designed for use with relaxation experiments [67, 98]. The resulting series of fluorescence spectra or fluorescence decays at nanosecond intervals require parallel data analysis methods, such as singular value decomposition (see Chapter 14.5) [99]. The advantage compared with an analysis of the rate constants only is that general inferences about the structural changes of the protein can be drawn. For example, a change in fluorescence lifetime unaccompanied by a wavelength shift indicates that tryptophan is coming into contact with a quencher, but without being desolvated.

14.2.4.6 Extrinsic Fluorescence

Extrinsic fluorescence or phosphorescence can be used to extract structural information in the form of pair-distances. The protein is labeled with a fluorescent donor dye, and with an acceptor that may or may not be fluorescent. Distances between donor and acceptor can be extracted using the Förster resonant energy transfer (FRET) theory, although some caution is required when attempting to extract quantitative results because assumptions about orientational effects are often necessary. FRET techniques and single molecule techniques are described in detail in Chapter 17.

14.3
Protein Folding by Relaxation

14.3.1
Transition State Theory, Energy Landscapes, and Fast Folding

A general overview of the classical transition state theory and Kramer's theory as applied to protein folding is given in Chapters 12.1 and 22. Here certain aspects of energy landscape theory addressed by fast folding experiments are described in more detail.

Figure 14.8 illustrates the connection between folding funnels, folding free energy surfaces, and smoothed free energy surfaces for folding. Energy landscape theory, based on random heteropolymer theory with an energetic bias towards the native state [100, 101], posits that the energy of a protein gradually decreases as it becomes more compact. This gives a plot of energy versus configurational entropy per residue the funnel-like shape shown in Figure 14.8 (column 1).

Superimposed on the funnel shape are energy fluctuations caused by interactions of the backbone, side chain, and solvent if some coordinates are not averaged over, but kept as reaction coordinates 'x'. Nonnative interactions over a wide energy scale are the main cause of landscape roughness. Among the largest of these is the classic example of proline *cis–trans* barrier (other side chains can also create conformational traps; see Chapter 25) [102, 103]. Although proline isomerization is not a necessary feature of folding [104, 105], even proteins whose energy landscape has been smoothed by design as much as possible retain some roughness [65]. This is so because not all nonnative interactions can be removed – in the language of energy landscapes, a residual frustration persists [106]. A typical examples of such an unavoidable residual interaction would be the steric clashes of different side chains [107, 108].

Among the traps on the funnel (assumed to have a Gaussian depth-distribution in random-energy theory) some can be rather deep [109], thus qualifying for specific treatment as "intermediates" rather than a statistical treatment. The best example is the noncoincidence between chain collapse and core drying – the latter a transition where water molecules are squeezed from the protein core. A larger energy fluctuation is shown at the entrance of the top funnel in Figure 14.8. Rapidly formed folding intermediates are discussed further in a separate section below.

Free energies are related to energies and entropies by $F = E - TS$, and so the funnel can be transformed into a picture more amenable for experimental work, which is generally carried out at constant temperature, not at constant entropy. Of

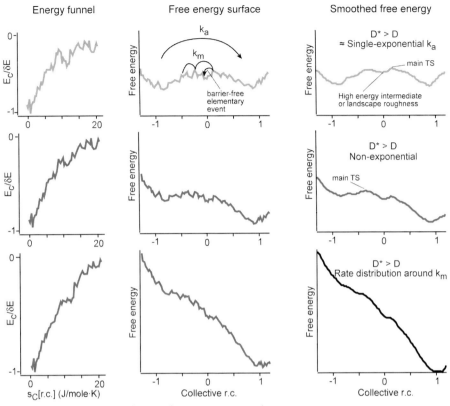

Fig. 14.8. Folding energy landscapes, free energy reaction plots, and smoothed free energy plots in one dimension. From top to bottom, the native bias increases. Important points and types of transitions are annotated, r.c. stands for reaction coordinate.

course the computation of the Gibbs free energy is somewhat more involved because it depends on enthalpy and total entropy (not energy and configurational entropy alone). The middle column of Figure 14.8 shows an example with a reaction coordinate linearly related to the configurational entropy.

The three free energy surfaces shown differ by the degree of compensation between energy and entropy. If entropy reduction during collapse is not fully compensated by contacts, a barrier results (top middle in Figure 14.8). In addition the free energy surface has a rough structure, and some of the local minima may be sufficiently prominent to deserve classification as "intermediates" or "high-energy intermediates" [110, 111]. The two are distinguished simply by whether the intermediate comes within 2–3 $k_B T$ of the denatured or native states (is populated), or not. The top middle case is very common in relaxation studies of small proteins [3, 14, 16, 26, 28, 31, 53, 57, 59].

It is not obvious that a free energy surface such as the one in the top middle of Figure 14.8 should give rise to single exponential kinetics. As surmised by Creighton and discussed by Zwanzig it does, if the interconversion among pairs of states is fast compared to traversing the bottleneck to the folded state [112, 113]. In that case, the activated folding rate coefficient k_a is much smaller than the diffusional time scale k_m for moving among the local minima, allowing the protein to equilibrate locally on the surface compared to the time scale required for barrier crossing. The local diffusion events characterized by k_m occur among the unavoidable minima left over when the protein is "minimally frustrated". Computations using Go models have shown that the free energy fluctuations caused by minimal frustration are generally $<3\, k_B T$" [114–116]. The three arrows in Figure 14.8 show part of a sequence of events occurring on the time scale k_m. Because of the residual roughness of the free energy surface, k_m is slower [65] than the fastest events the backbone can undergo (such as formation of a hairpin or the collision of two residues in a random peptide of length N in a θ-solvent tuned to eliminate all barriers) [15, 48, 49, 51, 68, 117]. A lower limit on k_m is set by such elementary-event peptide experiments, and ranges from $k_e = (10–100\text{ ns})^{-1}$, depending on peptide length and residue composition.

In the large barrier case (Figure 14.8 top) only k_a can be observed, not k_m. This leads to an ambiguity in using transition state-like theories, which contain two time scales: the prefactor, and the smaller rate coefficient, which differs from the prefactor by a Boltzmann factor (see Chapter 12.1 and equation 3 in Section 14.5.4.1). The choice of prefactor then depends on how we treat the free energy surface: on the rough free energy surface in the top middle of Figure 14.8, we would use $(10–100\text{ ns})^{-1}$, and treat the small barrier crossings caused by frustration explicitly as an additional free energy barrier. On the smoothed free energy surface to the right, we would use k_m as our slower prefactor, which has been shown to correspond to a renormalization of the diffusion constant to a smaller value D^* [118]. Experimentally, this renormalization factor can be derived by combining peptide relaxation studies [48, 68] and protein relaxation studies. For lambda repressor, it turns out to be about 40, equivalent to residual free energy fluctuations of $0.6\, k_B^2 T^2$ caused by frustration [65].

Can k_m be measured directly? The middle and lower rows of Figure 14.8 show what happens when the energy and entropy decrease are better compensated in the funnel: the free energy barrier disappears. In the completely downhill case on a rough surface, Zwanzig has shown with well-defined assumptions (see below) that the kinetics can be exponential, but with the renormalized slower diffusion constant D^* [118]. In the intermediate case (center row of Figure 14.8), the assumption that $k_m \gg k_a$ discussed earlier breaks down when the main barrier becomes comparable to the minimal roughness of the free energy surface in one dimension [56, 65, 119]. Both time scales can then be observed simultaneously by relaxation experiments with proteins engineered to fold downhill in a few microseconds, yielding a value of $k_m = (2\text{ μs})^{-1}$ for the case of lambda repressor [65]. Explicit calculations for lambda repressor [120], as well as general theoretical calculations also yield speed limits in the 0.5–2 μs range [121].

The speed limit k_m is not a universal quantity, but depends on the complexity of the protein's fold or "topology," as it is often referred to in the folding literature. The reason is that more complex proteins have to undergo more elementary events to fold and are also more likely to be subject to residual frustration [114, 119]. For example, the most optimized fast-folding sequence for a two-helix bundle should fold faster (<1 µs) than the most optimized 8-helix myoglobin fold (>7 µs). This has been expressed in a relationship between the contact order (average loop length between to native contacts) and logarithm of the rate constant [122, 123]; a rigorous analysis must also include chain-length effects [124]. ln k vs. contact order is usually plotted for any two-state folders at hand, resulting in a large amount of scatter (2–3 orders of magnitude) along the vertical axis. The analysis outlined above shows that one should use instead the largest value of k_m, obtained with the most optimized sequence for a given fold, i.e., the actual speed limit of a fold subject to minimal frustration. Measurements of the fastest folding rates will allow a more quantitative experimental analysis of the rate–contact order relationship by revealing the degree of energetic frustration a given sequence imparts on a fold.

Changing the bias of the free energy by optimizing folding conditions has an effect even when a significant barrier persists. Consider the top and middle cases in Figure 14.8, and imagine for the moment that the barriers in the middle case are still significant (by making the denatured minimum a bit deeper). Then the top transition state is "late," and the middle transition state is "early." Whether one describes the motion of the transition state as a smooth shift [125], or as a switch across a high-energy intermediate [111], depends on the degree of smoothing applied to the free energy surface. Thus it is necessary to define a rule for how much the free energy surface should be smoothed in kinetic descriptions. A useful rule is that the smoothing should leave intact minima deeper than 2–3 $k_B T$, since this is the crossover where fast relaxation experiments and Langevin simulations have shown Kramer's exponential rate theory to be applicable [65]. This degree of smoothing removes the roughness caused by minimal frustration, but leaves more pronounced intermediates intact. When smaller barriers have to be treated (middle right in Figure 14.8), Langevin simulations are required to obtain the correct dynamics on the one-dimensional potential [65, 126].

One can choose to not apply any coarse-graining at all, and pick the classical value $k_B T/h \approx 300$ fs for the prefactor [28], along with an atomic scale motion as the reaction coordinate. In such a description of the kinetics, the reaction coordinate can change dramatically from realization to realization (e.g., in a series of MD simulations, each successive folding trajectory has a different atomic-level reaction coordinate at the free energy maximum). Thus, a structural justification for smoothing of the free energy surface is to apply smoothing until a consistent set of collective coordinates emerges.

The question then arises: how many coordinates should be used for a good description – are the one-dimensional cartoons in Figure 14.8 sufficient? As discussed earlier, Zwanzig has shown that with certain assumptions, even diffusion on a rough surface can be treated as an exponential process [118]. A one-dimensional reaction coordinate (such as the plot in Figure 14.8 bottom) is an important assumption invoked to derive exponential diffusive behavior [118] $n > 1$ dimensions more easily accommodate the local minima needed so that a single bottleneck does not necessarily control the dynamics, and anomalous diffusion can result in stretched exponential or power-law dynamics [45, 127, 128]. In these nonexponential models, randomly distributed local minima of the free energy surface lead to "logarithmic oscillations" superimposed on the smooth nonexponential dynamics. These oscillations are analogous to the steps observed in multiexponential fits, and attributed to specific intermediates there. The conceptual difference between the two approaches hinges on either enumerating the fewest local minima needed to fit the data, or considering the entire hierarchy of minima constrained by scaling laws. The current computational and experimental evidence shows that at most a few coordinates should be required to accommodate the local minima necessary for robust nonexponential kinetics. For example, for a tetrapeptide investigated by Becker and coworkers [129, 130], two coordinates provided a rather complete description of the folding dynamics.

Experimental data, with few exceptions [44], are not highly nonexponential and can be fitted equally well by multiexponentials and logarithmic oscillations [56, 58]. This is partly due to the limited dynamic time range of experimental data. However, phosphoglycerate kinase data under a wide range of conditions always fits on average to stretched exponentials [63]. There is no a priori reason why a single trap should always yield rates that approximate a stretched exponential, whereas the hierarchical trap model naturally explains this because traps of many depths coexist. Both types of models agree that addition of denaturants will either smooth out the surface or raise the energy of traps in the barrier region, resulting in more exponential kinetics. This is observed experimentally [63]. It has also been shown experimentally that cold denaturation has a similar smoothing effect [56].

In a very recent experiment on λ-repressor, early time deviations from exponential kinetics are ascribed to diffusion events, but the deviation itself can be fitted by a stretched exponential with $\beta \geqslant 0.7$, again favoring a low-dimensional surface for the diffusion process [65]. No firm limits can be set at present, but it appears that at most a handful of collective reaction coordinates should be sufficient to describe folding processes rather completely.

14.3.2
Viscosity Dependence of Folding Motions

Viscosity effects on folding rates provide important clues to the mechanism. This is discussed in detail in Chapter 22, but a few observations from relaxation studies are worth reemphasizing here. A proper discussion requires at least Kramer's theory [131] because the same solvent–protein coupling both activates the protein for reaction and slows down the reaction when the coupling (and hence the viscosity) becomes large.

Viscogenic agents, such as sucrose or ethylene glycol, can affect both the prefactor ($\sim \eta^{-1}$ in the strongly damped Kramer's regime) and free energy barriers, as well as the compactness of the denatured state. Early observations of submillisecond refolding kinetics indicated a decrease of the rate with increasing viscosity, but did not compensate for barrier effects [16, 33, 132]. Recent work has carefully calibrated the free energy at different viscogen concentrations, finding that the viscosity dependence of the prefactor is very close to η^{-1} [133–135]. These authors also make the argument that barriers are required to protect the native state from continuous unfolding, supported by the recent observation that very fast folding proteins without a substantial free energy barrier are aggregation prone [65].

There has been an ongoing discussion of deviations from Kramer's strongly damped regime caused by self-friction of the protein molecule during folding [16, 33]. Such deviations have been modeled by phenomenological equations of the type $k \sim (\eta + \eta_0)^{-1}$ or $k \sim \eta^{-d}$ where $d < 1$. Small peptides indeed have a length-dependent viscosity dependence, with powers deviating from -1 to as low as -0.8 [136]. In larger proteins, the evidence generally favors the simple Kramer's result. Some of the insights from simulation are discussed further below.

14.3.3
Resolving Burst Phases

As discussed in the section on transition states, some local minima in the free energy surface are deep enough to be amenable to individual analysis by relaxation experiments. Even when they lie at high energy, they can have subtle effects on the folding kinetics [137]. It has been postulated that a discrepancy between the time scale for collapse and the time scale for desolvation of the protein core could give rise to compact but solvated globular states [138]. Fast relaxation experiments support this general idea. An intermediate with native-like quenching of the apomyoglobin AGH helices forms on a microsecond time scale, and amide exchange experiments show that parts of this intermediate are protected, while others are still undergoing rapid exchange [6, 16]. Several nonnative forms of apomyoglobin have been investigated by Dyer and coworkers, and some have been shown to form a compact core at the diffusion limit [76]. Resistive temperature-jump experiments have been used to study the biphasic folding of barstar. Phi-value analysis shows that two of the helices consolidate rapidly (≈ 300 μs), but a fluorescence shift towards a native-like structure requires nearly a second [25].

The question of the barrier height separating the unfolded state from the intermediate has been discussed extensively in the literature. Stretched- or multiexponential formation of a folding intermediate of the two-domain protein phosphoglycerate kinase indicates that for one of the two domains, the formation of the intermediate is limited by a hopping process on a rough free energy surface, rather than by a single barrier crossing [56, 63]. This is also supported by the observation that cold denaturation, which should smooth out the free energy surface, brings the kinetics back towards single-exponential folding. More work has been done using flow techniques (see Chapter 15 for details): Using a prefactor of 50 ns (to include free energy landscape roughness in the activation barrier) yields values below $8 k_B T$ for molecules such as cytochrome c or ubiquitin [138–141]. It has been argued that in many cases, no barrier beyond residual roughness exists [56, 142, 143], or that these barriers come after the main transition state [144, 145]. In one case, it has also been convincingly shown that the intermediate contains significant nonnative structure, but nonetheless has direct access to the native state [146].

14.3.4
Fast Folding and Unfolded Proteins

It has long been known that unfolded states contain sequence-dependent residual structure [147]. As discussed in Chapters 20 and 21, this has come to be much better understood in recent years. Residual unfolded structure is particularly important when interpreting fast relaxation experiments: the barriers are small, and changes in the unfolded structural ensembles can have a major effect on folding kinetics [31, 148]. For example, a variety of relaxation experiments have shown that unfolding rates tend to have "normal" linear Arrhenius plots, while folding rates have highly curved plots (see for example Figure 14.1). This has been taken as a sign that transition states remain relatively native like upon temperature tuning [31, 32], while the unfolded state becomes more native-like at low temperature. The effect of unfolded states on the rate constant can be reduced, by making them equally rigid at all temperatures. In that case motions of the transition state make a larger contribution to the rate turnover effect. Many natural proteins that are not highly optimized for folding probably fall in this category.

When the contributions of the unfolded state to the rate are not too large, a continuous analog of Φ-value analysis (see Chapter 13) can be used to analyze protein folding. This has the advantage that arbitrarily small perturbations can be applied, while mutations apply a defined-size perturbation [59]. Denaturant, pressure, and temperature have been tuned to this effect [57, 125, 137]. By itself, tuning thermodynamic variables provides no structural information, but when combined with mutations, structural information can be revealed. One useful application of continuous tuning is to test mutations for "conservativeness": a mutant with a very different temperature dependence of the free energy from the wild type is less likely to be a conservative mutation one wishes to use in Φ-value analysis [59].

The bottom line is this: whenever unfolded states are significantly altered by any perturbation, whether it is a mutation or the change of a thermodynamic variable, caution must be exercised when using free energy derivative analysis [32, 119]. An example in the case of temperature has already been given above, and is discussed in more detail in Ref. [148]. As an example of a mutation, truncation of a large side chain to a smaller side chain may reduce a nonnative interaction in the unfolded state, causing apparent shifts in Φ-values that are not related to changes in the transition state [32]. The characterization of unfolded proteins has only scratched the surface so far, but the number of experiments characterizing unfolded states of proteins is steadily increasing.

14.3.5
Experiment and Simulation

A hierarchy of simulation techniques has been used to describe fast relaxation data, some of which are described in more detail in Chapters 32 and 33. At the simplest level, master equations are used, such as the *n*-state kinetic schemes discussed throughout this volume [4, 149]. Two common ones, prototypical of fast folding and association reactions, are discussed in more detail in the Experimental Protocols (Chapter 14.5). Their breakdown, outlined qualitatively earlier in the discussion, is also discussed more mathematically in terms of linear response theory [150], which explains rigorously when the transition from activated kinetics to the molecular time scale $(k_m)^{-1}$ occurs. More sophisticated kinetic schemes, which provide a more microscopic treatment of the dynamics by including large numbers of states, have yielded insights into the formation of small segments of secondary structure, as well as the folding of proteins [70, 113, 123, 151]. These models trace their ancestry to classic treatments of secondary structure such as the Zimm-Bragg model [152]. An important concept in such models is how many groups of states can independently undergo a transition. In "single sequence approximations," native-like structure must spread out from a single locus. In "*n*-sequence approximation" models, more sites are allowed where native structure can form [123]. The latter models yield lower barriers, and are generally in close accord with full theoretical treatments and experimental data.

Motion on continuous folding free energy surfaces can be treated by Langevin dynamics, revealing some interesting behavior. For example, it has been shown that small minima in the barrier region can actually accelerate folding if they increase the local curvature at the barriers [126]. Simulations on smoothed downhill surfaces such as Figure 14.8 (middle right) have verified nonexponential decays with renormalized diffusion constants ranging as small as $D^* \approx 0.00016$ nm^2 ns^{-1}, considerably smaller than the free diffusion constant for a small segment of secondary structure, in accord with analytical models [65].

Lattice models [116, 153–155] have proved very fruitful in comparison with fast folding experiments. Originally two-dimensional and allowing only for two types of residues, they have been extended to include more residue types and different lattice geometries [156, 157], for example to describe nonexponential helix relaxation kinetics [74]. Lattice models were among the first to show that proteins could fold via a rapid non-cooperative collapse, followed by crossing a relatively small barrier [155].

Off-lattice models are useful because they still coarse-grain the folding problem, thus making larger systems accessible, yet they do so with fewer restrictions than lattice models [158]. The viscosity dependence of folding reactions has been investigated by Klimov and Thirumalai in a model coupling a coarse-grained model of the protein with Langevin dynamics to represent the solvent. In agreement with several experiments, the normal Kramer's regime ($k \sim \eta^{-1}$) provides a good fit for α-helical and β-sheet peptides [159]. There are however certain conditions where a fractional power smaller than -1 can occur [160]. Off-lattice models have also revealed a rough free energy surface which can give rise to nonexponential kinetics [114, 127].

Molecular dynamics (MD) simulation has become an increasingly important tool for protein folding, as fast relaxation time scales and the capabilities of MD simulation on large computers are beginning to merge on the microsecond time scale [161]. A detailed overview is given in Chapters 32 and 33; here a few examples of direct comparisons between fast relaxation experiments and MD simulation illustrate the synergy between them. The first detailed comparisons were made for unfolding rates extrapolated to higher temperature to match nanosecond time scale MD simulations. Excellent agreement between the extrapolated experimental and simulated unfolding time scales, as well as transition state structures was obtained [28, 162]. Folding rates are more challenging, but have become accessible with the development of very fast folding mini-proteins. A recent study compared 700 μs of trajectories of the molecule BBA5 with thermodynamic and kinetic data, and obtained good agreement for the folding rate constant and equilibrium constant [61]. The simulation shows that BBA5 does not fold via a single pathway, but by a funnel-like coalescence of secondary structures towards the native minimum. One caveat in this study is that folding was observed as a rare event in thousands of nanosecond-long trajectories, rather than in a few longer trajectories. MD simulations have also proved useful in providing a microscopic mechanism for experimental relaxation studies [66, 163]. They have also revealed that the free energy surface of denatured proteins is indeed complex, as also inferred from analytical models: [164]. Brooks and coworkers, and other groups, have found a rich structure of traps and nonnative local minima in two-dimensional free energy plots [165]. Figure 14.9 shows a one-dimensional plot for the 12-mer hairpin trpzip2 computed by Swope and coworkers. The free energy plot for this system at 250–300 K is reminiscent of the random free energy surface plot in Figure 14.8 (middle right): a plateau of denatured local minima, separated by a small barrier from the native state.

14.4
Summary

Fast relaxation experiments are testing predictions of the free energy landscape theory [166], such as roughness of the free energy surface and downhill folding.

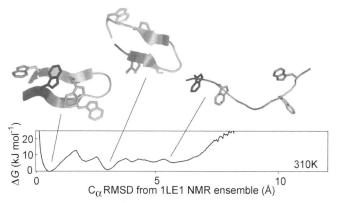

Fig. 14.9. Free energy surface for the hairpin trpzip2 with the C_α RMS deviation from the native state as one-dimensional reaction coordinate. The unfolded state is a plateau with multiple shallow minima, like the middle case in Figure 14.8 and the fitted surface for lambda repressor in Ref. [4].

They have directly measured the time scales of elementary folding processes, including secondary structure formation and loop formation. The low barriers of fast folding proteins allow their free energy surfaces to be explored beyond the native and unfolded basins. Experimental folding kinetics and molecular dynamics simulations are meeting in the nanosecond to microsecond regime.

14.5
Experimental Protocols

The following protocols provide some more technical information on laser temperature-jump relaxation experiments, and specific references where more information can be found.

14.5.1
Design Criteria for Laser Temperature Jumps

Water absorption in the 1.5 µm region is about 200 m^{-1}, limiting pathlengths to about 250 µm in single pass, or 500 µm in counterpropagating geometry, if a temperature uniformity of 10% is to be maintained in a 10 °C temperature jump. About 100 mJ of near infrared light is required to achieve sufficiently large temperature differentials. This output level can be obtained from a 1.5 m long Raman cell at 20 atm (2×10^6 Pa) of methane, with a collimated input beam diameter of 0.5 cm and input power levels of 500 mJ. Currently, an unamplified single-oscillator Nd:YAG laser is capable of meeting these specifications without requiring optical isolation to prevent feedback from reflections into the laser cavity. Care needs to be taken in the choice of optics: ≥ 3 mm sapphire or ≥ 6 mm fused silica windows with 1.06 µm antireflection coating are sufficient to withstand the pressure and laser intensity. In designing a Galilean (convex/concave lens) beam telescope, one must keep in mind that at 500 mJ, even second-surface reflections on optical curvatures can focus down to a damaging level. Upon output, a Pellin-Broca or equilateral glass prism can be used to separate the YAG fundamental from the Raman-shifted light. An IR image intensifier and Kodak Linagraph paper provide convenient means of tracing the Raman beam, which can be split in a 50:50 beam splitter for a counterpropagating pump geometry. For heating purposes, the beam needs to be brought to ca. 1–2 mm diameter when interacting with the cell. This can be achieved at the beamwaist of a long-focal length lens (ca. 1 m), or by placing a shorter lens sufficiently close to the sample so the focus lies beyond the sample. Because of the Fourier-transform property of focusing, the resulting beamprofile is generally of much higher quality than the original beamprofile, as verified by knife-edge measurements or a CCD camera.

The following two temperature-jump calibration methods are straightforward to implement: A ≈ 1.5 µm wavelength diode laser beam can be chopped, sent through the sample, and demodulated by a lock-in amplifier. The change in water absorbance can simply be calibrated by reading diode transmission changes over the desired temperature interval. Alternatively, tryptophan or NATA at pH 7 have a well-defined temperature dependence of the fluorescence lifetime, which can be measured if a fast lifetime detector is included in the system. A convenient way of reducing jump size is to use D_2O/H_2O mixtures (this can also be applied to D_2O jumps when using H_2 Raman-shifted 1.9 µm light), or to use thin films of water, so the laser system does not need to be adjusted. Additional details (superseded by the above information when different) can be found in Refs [67] and [55].

14.5.2
Design Criteria for Fast Single-Shot Detection Systems

In a single-shot detection system, the light source either provides sufficiently short pulses so that fluorescence decays can be directly resolved, enough bandwidth so a full absorption spectrum can be measured, and sufficiently high pulse repetition rates that kinetics can be fully resolved. Currently, the most convenient laser systems for excitation are mode-locked Ti:sapphire lasers pumped by intracavity-doubled diode-pumped YAG lasers. Such systems easily achieve 1 W average output power in <100 fs pulses at ca. 80 MHz pulse repetition rate. They are tunable in the 750–950 nm wavelength range, as well as at half, third, and quarter wavelengths by nonlinear sum-frequency generation (with maximum output levels typically 200 mW, 50 mW, and 10 mW, respectively). This provides at least 10^4 photon/pulse, or $1:10^2$ shot noise. Systems with sufficiently high bandwidth can also be difference-frequency generated in silverthiogallatre crystals to produce up to 200 cm^{-1} bandwidth mid-infrared pulses (1000–1800 cm^{-1}) [81]. Commercial lasers and sum-frequency generators are available.

The most important distinction between detection by transmission or by fluorescence lies in the allowed cell pathlength. Backround absorption in mid-IR absorption, even in D_2O, limits pathlengths to 50–100 µm. In addition, a thermal lens created by the variation of the refractive index about the ca. 1–2 mm diameter heating profile limits the pathlength in transmission experiments to <100 µm. The effect can be minimized by heating an at least 10× larger diameter of the sample than is probed, but even this precaution is not sufficient when cell pathlengths significantly >100 m are used.

Fluorescence detection is less sensitive. To avoid polarization artifacts, the pump beam should be depolarized and fluorescence should be detected at the magic angle. However, artifacts caused by cavitation (formation of vapor bubbles) during large temperature jumps are generally a more severe limiting factor. Light guides filled with a UV-transmitting fluid can be used in lieu of lens-based imaging optics to collect the fluorescent output. Models with f-ratios as large as 0.8 and diameters ≤4 mm are available, allowing efficient collection of the fluorescence and piping to remote detection systems.

Details of the electronics of a detector which can detect a fluorescence decay every 14 ns with a 500 ps risetime are given in Ref. [67], details of a detector which can collect an entire fluorescence spectrum from 300 to 400 nm every 100 ns are given in Ref. [98], and details of a detector that can detect a mid-IR spectrum of

200 cm^{-1} with 300 ns time resolution are given in Ref. [81]. Additional circuit diagrams and software are available from the author.

For experiments with cold-denatured proteins, the choice of cell is critical to avoid freezing supercooled samples because of nucleation on cell defects. In extruded fused silica cells, temperatures as low as -15 °C are routinely possible in water, and -20 °C have been observed. Such cells are available in pathlengths from 0.3 mm to 1 mm. The author has observed laser-induced temperature jumps between two temperatures below freezing without nucleation. Cell temperaures are most conveniently adjusted by thermoelectric coolers (for fast adjustment) in combination with a recirculating bath.

14.5.3
Designing Proteins for Fast Relaxation Experiments

Three design criteria have proved particularly useful in increasing the refolding rate of protein mutants:

1. Increase unfolded structure: this leads to a more structured unfolded state. For example, Oas has replaced glycine residues in helical positions 46 and 48 of lambda repressor fragment λ_{6-85} by alanines, thereby greatly increasing the folding rate [167].
2. Eliminate functional residues: a significant fraction of the energetic frustration in proteins probably comes from functional residues that are not optimal for folding. For example, replacing the distal histidine in apomyoglobin by a phenylalanine increases the folding significantly [168], and grafting the shorter FBP WW loop onto the Pin WW domain increases the folding rate by a factor of 3 [169].
3. Decrease the contact order: simplifying the fold reduces residual roughness of the free energy. As a rough guide, 2–3 helix bundles can be expected to reach a speed limit around 0.5 μs (W. F. DeGrado, private comminication, 2003), 5 helix bundles around 1–2 μs [65], and 8 helix bundles around 2–5 μs [52]. The time scales observed thus far for beta-rich proteins are somewhat longer (e.g., ≥ 1 μs for a simple double-stranded hairpin) [69].

14.5.4
Linear Kinetic, Nonlinear Kinetic, and Generalized Kinetic Analysis of Fast Relaxation

Many complicated kinetic schemes can arise in the discussion of relaxation experiments [170] (see Chapter 22). Here we discuss in detail three cases that illustrate the type of information that can be extracted, and the pitfalls that arise in the analysis. Other cases can be found in texts on reaction kinetics. Only kinetic schemes are discussed here; the actual analysis of raw experimental data is discussed in the next protocol.

14.5.4.1 The Reaction D $\rightleftarrows$ F in the Presence of a Barrier

This case corresponds to the reversible folding of a monomeric protein. Rate differential equations can be used with rate coefficients given by Kramer's theory:

$$\frac{d[D]}{dt} = -k_f[D] + k_d[F] \quad \text{with } [D] + [F] = P_{\text{tot}}. \tag{1}$$

The solution to this equation, using the equilibrium constant $K_{eq} = F_{eq}/D_{eq} = k_f/k_d$, is given by

$$[D] = D_{eq} + (D_0 - D_{eq})e^{-(k_f+k_d)t} \tag{2}$$

Kramer's version of transition state theory (which allows for recrossings of the barrier induced by the solvent) yields a forward rate constant [131]

$$k_f = v^\dagger[\eta(T)]e^{-\Delta G^\ddagger/kT} \tag{3}$$

where the barrier crossing prefactor $v^\dagger$ depends on the solvent viscosity η. ($v^\dagger$ generally decreases with increasing solvent viscosity. It may level off at small solvent viscosities because the protein self-friction becomes dominant. It may even decrease with decreasing solvent viscosity because the solvent–reactant collisions no longer are strong enough to activate the reactant when the solvent viscosity is very low; this "inverted Kramer's" regime has not been observed during protein folding.)

According to the rate law given above, the kinetics remain single exponential for any size relaxation; the forwards and backwards rate coefficients can be extracted independently if the equilibrium constant is known. However, under certain conditions nonsingle exponential kinetics can be observed: if the relaxation significantly changes the properties of the states D or F, and if their spectroscopic signatures change, then the populations before the jump are no longer local equilibrium populations. As a result, rapid relaxation within the states D and F may be observed before the barrier crossing kinetics [56].

14.5.4.2 The Reaction 2A ⇌ A₂ in the Presence of a Barrier

This is a prototype case for protein association reactions studied by relaxation experiments, such as the assembly/disassembly equilibrium of a light-harvesting complex containing multiple subunits. Assuming again that rate is limited by barrier crossing, one can write the differential equation

$$\frac{d[A]}{dt} = -k_a[A]^2 + 2k_d[A_2] \quad \text{with } [A] + 2[A_2] = A_{\text{tot}} \tag{4}$$

whose solution, with the initial condition $[A](t=0) = A_0$ and $[A_2](t=0) = 0$ can be written in compact form as

$$[A](t) = \frac{k'}{k_f}\tanh\left(k't + \tanh^{-1}\frac{k_b + A_0 k_f}{k'}\right) - \frac{k_b}{k_f}, \quad k' = \sqrt{k_b^2 + 2k_b k_f A_0} \tag{5}$$

(Note that this evaluates to a real quantity even in cases where $(k_b + A_0 k_f)/k' > 1$.) Clearly the functional form of the kinetics is not single exponential, and the shape of the relaxation actually depends on the initial concentration A_0. This differs from the case above, where the relaxation remains exponential for any initial condition, no matter how far from equilibrium.

If this system is near equilibrium however, then the relaxation becomes single exponential. This can be seen most easily by writing the rate equation in terms of $[A] = \delta A + A_{eq}$, where δA represents the small deviation of the concentration from the equilibrium value. Neglecting quadratic terms in δA, the rate equation is approximated by

$$\frac{d\delta A}{dt} \approx (-2k_f A_{eq} + k_b)\delta A \tag{6}$$

and by using mass conservation and the equilibrium constant $K_{eq} = A_{2,eq}/A_{eq} = k_f/2k_b$ the solution becomes

$$\delta A = \delta A_0 e^{-kt}, \quad \text{where } k = \sqrt{k_b^2 + 4A_{tot}k_b k_f} \tag{7}$$

Thus the rate constant depends on the concentration, but the kinetics are single exponential. By varying the concentration, the forwards and backwards rate coefficients can be extracted independently.

14.5.4.3 The Reaction D ⇌ F at Short Times or over Low Barriers

The phenomenological description of two-state kinetics above is satisfactory when applied to large barrier crossings at sufficiently long times. At short times, especially when the barrier is not very large compared with $k_B T$, simple rate theories no longer apply. This is true for the more sophisticated Kramer's model (which includes perturbation of the solute free energy by the solvent explicitly), as well as for the classical transition state theory. There are several related reasons for this failure.

At very short times, the rate coefficient k_f is no longer a constant [150]. At $t = 0$, its value $k_f(0)$ must be greater than the value at long times. The reason is that at $t = 0$, barrier recrossings and other diffusive motions that reduce the rate coefficient have not yet had a chance to occur. A more complete theory has been developed in the linear response limit (i.e., for small relaxation jumps). The rate coefficient then becomes

$$k_f(t) = \min\langle v(t=0)\delta(RC(t=0) - RC_0)n_F(t)\rangle/\chi_F \tag{8}$$

In this equation, RC is a reaction coordinate. RC_0 is the position along the reaction coordinate that separates the states D and F (the position of the transition state in conventional transition state theory). n_F is the folded population obtained by integrating over the population from reaction coordinate $RC_0 \approx 0$ towards the F side of the free energy surface. δ is the Dirac delta function so only trajectories at $RC = 0$ are counted. v is the velocity of the protein trajectory along the reaction coordinate.

min<> indicates that an average over many trajectories is to be performed and minimized by moving RC_0.

If the folding barrier is very low, a substantial fraction of the protein population occupies the barrier region. Also, transitions among denatured states are no longer fast compared to transitions over the barrier region. If the spectroscopic signature of the activated population differs from that of the states D and F, a sizable kinetic amplitude will be observed for motions of this population. The relaxation of such a population is faster than the barrier-limited rate, but not arbitrarily fast. At a minimum, it is limited by free diffusion of the backbone, a process known to require 10–100 ns per contact for peptides of 10–100 chain length [48]. As discussed by Zwanzig, it may be slowed down further by roughness of the free energy surface. In his one-dimensional model, the diffusion constant on a rough surface is slowed down by a factor $\exp[-(\varepsilon/k_B T)^2]$, where ε is the root-mean-squared barrier height [118].

The barrier-free relaxation time scale for intermediate populations (termed "molecular time scale" by kineticists and "speed limit" in the protein folding community) determines when the rate coefficient $k_f(t)$ becomes a rate "constant" [150].

14.5.5
Relaxation Data Analysis by Linear Decomposition

Relaxation data can contain a single channel of information, or multiple channels. Linear decomposition techniques are very convenient for analyzing multichannel data [67, 98, 99]. For example, when an infrared spectrum shifts at successive kinetic time points, there are generally far fewer independent spectra than time points – in the case of a rigorous two-state folder, only two spectra. Linear decomposition techniques such as singular value decomposition can extract such "basis spectra" while minimizing assumptions about the nature of the fluorescence spectrum (λ) or fluorescence decays (t').

14.5.5.1 Singular Value Decomposition (SVD)

Both fluorescence decay transients and fluorescence/infrared spectra may be thought of as $m \times n$ matrices whose n columns track the spectroscopic profile (as a function of λ or t'), and whose m rows track the kinetics (as a function of $t =$ kinetic time). SVD reduces this matrix to a set of n basis functions that describe the profile, n singular values which describe the importance of each profile to the total signal, and n vectors of length m that describe how each of the basis functions contributes to the kinetic signal as a function of time t.

Pronounced singular values up to $n' \leq n$ indicate that at least an n'-state kinetic model is required to account for the data quantitatively. n' sets a lower limit on the complexity of the kinetics because some intermediates may have spectral signatures similar to reactant or product. For example, a double-exponential decay guarantees that the kinetics are not two-state, but it does not guarantee that the kinetics are three-state.

14.5.5.2 χ-Analysis

When the thermodynamic data for a folding reaction as a function of temperature (or some other variable) can be represented by a cooperative two-state folding model, the following question becomes relevant: Are the folding kinetics those of an activated two-state system, and hence single exponential? In that case the reaction proceeds as

$$D \rightleftarrows N, \quad [D](t) = ([D]_0 - [D]_\infty)e^{-kt} + [D]_\infty \tag{9}$$

A similar question can be asked if a long-lived intermediate I is formed from D because the reaction $D \rightleftarrows I$ is then a quasi-two-state reaction. The χ-fit is a powerful tool to test these questions without making any assumptions about the functional form of the fluorescence decays or spectra of D, N (or I).

The practical implementation of the procedure for time-resolved fluorescence decays requires two steps. (In the following discussion, λ may replace t' in any formula, if spectra are analyzed instead of decays.) First, two fluorescence profiles f_1 and f_2 are constructed. f_1 represents the fluorescence profile as a function of λ or t' just after initiation of the kinetics. f_2 represents the fluorescence long after initiation of the kinetics.

The sampled data matrix $f(t, t')$ has m rows indexed by t and n columns indexed by t' or λ. Each row of f is fitted by linear least-squares to a combination of the two basis spectra f_1 and f_2, which themselves are functions of t' or λ:

$$f(t, t' \text{ or } \lambda) = a_1(t)f_1(t' \text{ or } \lambda) + a_2(t)f_2(t' \text{ or } \lambda) \tag{10}$$

The function χ_1 is then defined as:

$$\chi_1(t) = \frac{a_1(t)}{a_1(t) + a_2(t)} \tag{11}$$

It describes how the shape of the fluorescence profile evolves from a more unfolded signature towards a more folded signature, independent of the signal amplitude. This function has three important properties, which make it an excellent choice for representing folding kinetics. It approaches a signal-to-noise ratio limited only by the Poisson statistics of photon numbers in each laser shot (like ideal fluorescence intensity analysis); it is immune to laser intensity fluctuations (like fluorescence lifetime analysis); it allows one to extract species populations for two-state folders, and distinguishes two-state from multi-state folding (unlike amplitude or lifetime analysis). Further details are discussed in Ref. [98].

Acknowledgments

The author was supported by NSF grant MCB-0316925 while this work was prepared, and would like to thank the research groups who have generously allowed figures from their publications to be used in this work. Additional material for this review was taken from other primary literature publications by the author listed in the references.

References

1 EIGEN, M. & MAEYER, L. D. (1963). Relaxation methods. In *Technique of Organic Chemistry* (WEISSBERGER, A., ed.), pp. 895–1054. Interscience, New York.
2 VALENTINE, J. S. (1998). Special Issue on Protein Folding. In *Acc. Chem. Research* (VALENTINE, J. S., ed.), Vol. 31, pp. 697–780.
3 JACOB, M., HOLTERMANN, G., PERL, D. et al. (1999). Microsecond folding of the cold shock protein measured by a pressure-jump technique. *Biochemistry* 38, 2882–91.
4 GRUEBELE, M. (1999). The physical chemistry of protein folding. *Annu. Rev. Phys. Chem.* 50, 485–516.
5 CALLENDER, R. & DYER, D. B. (2002). Probing protein dynamics using temperature jump relaxation spectroscopy. *Curr. Opin. Struct. Biol.* 12, 628–33.
6 JENNINGS, P. & WRIGHT, P. (1993). Formation of a molten globule intermediate early in the kinetic folding pathway of apomyoglobin. *Science* 262, 892–5.
7 HUANG, G. S. & OAS, T. G. (1995). Submillisecond folding of monomeric λ repressor. *Proc. Natl Acad. Sci. USA* 92, 6878–82.
8 EIGEN, M., HAMMES, G. G. & KUSTIN, K. (1960). Fast reactions of imidazole studied with relaxation spectrometry. *J. Am. Chem. Soc.* 82, 3482–3.
9 TURNER, D. H., FLYNN, G. W., LUNDBERG, S. K., FALLER, L. D. & SUTIN, N. (1972). Dimerization of proflavin by the laser Raman temperature-jump method. *Nature* 239, 215–17.
10 HOLZWARTH, J. F., ECK, V. & GENZ, A. (1985). Iodine laser temperature-jump: relaxation processes in phospholipid bilayers on the picosecond to millisecond time-scale. In *Spectroscopy and the Dynamics of Molecular Biological Systems* (BAYLEY, P. M. & DALE, R. E., eds), pp. 351–377. Academic Press, London.
11 WILLIAMS, A. P., LONGFELLOW, C. E., FREIER, S. M., KIERZEK, R. & TURNER, D. H. (1989). Laser temperature-jump, spectroscopic, and thermodynamic study of salt effects on duplex formation by dGCATGC. *Biochemistry* 28, 4283–91.
12 JONES, C. M., HENRY, E. R., HU, Y. et al. (1993). Fast events in protein folding initiated by pulsed laser photolysis. *Proc. Natl Acad. Sci. USA* 90, 11860–64.
13 PHILLIPS, C. M., MIZUTANI, Y. & HOCHSTRASSER, R. M. (1995). Ultrafast thermally induced unfolding of RNase A. *Proc. Natl Acad. Sci. USA* 92, 7292–6.
14 NÖLTING, B., GOLBIK, R. & FERSHT, A. R. (1995). Submillisecond events in protein folding. *Proc. Natl Acad. Sci. USA* 92, 10668–72.
15 WILLIAMS, S., CAUSGROVE, T. P., GILMANSHIN, R. et al. (1996). Fast events in protein folding: helix melting and formation in a small peptide. *Biochemistry* 35, 691–7.
16 BALLEW, R. M., SABELKO, J. & GRUEBELE, M. (1996). Direct observation of fast protein folding: The initial collapse of apomyoglobin. *Proc. Natl Acad. Sci. USA* 93, 5759–64.
17 PASCHER, T., CHESICK, J. P., WINKLER, J. R. & GRAY, H. B. (1996). Protein folding triggered by electron transfer. *Science* 271, 1558–60.
18 SCHWARZ, G. & SEELIG, J. (1968). Kinetic properties and the electric field effect of the helix-coil transition of poly(γ-benzyl L-glutamate) determined from dielectric relaxation measurements. *Biopolymers* 6, 1263–77.

19 Hawley, S. A. (1971). Reversible pressure-temperature unfolding of chymotrypsinogen. *Biochemistry* 10, 2436–42.
20 Royer, C. A. (2002). Revisiting volume changes in pressure-induced protein unfolding. *Biochim. Biophys. Acta* 1595, 201–9.
21 Desai, G., Panick, G., Zein, M., Winter, R. & Royer, C. A. (1999). Pressure-jump Studies of the Folding/Unfolding of trp Repressor. *J. Mol. Biol.* 288, 461–7.
22 Clegg, R. M., Elson, E. L. & Maxfield, B. W. (1975). New technique for optical observation of the kinetics of chemical reactions perturbed by small pressure changes. *Biopolymers* 14, 883–7.
23 Hoffman, G. W. (1971). A nanosecond temperature-jump apparatus. *Rev. Sci. Instrum.* 42, 1643–7.
24 Rigler, R., Rabl, C. R. & Jovin, T. M. (1974). A temperature-jump apparatus for fluorescence measurements. *Rev. Sci. Instrum.* 45.
25 Nölting, B., Golbik, R., Neira, J. L., Soler-Gonzalez, A. S., Schreiber, G. & Fersht, A. R. (1997). The folding pathway of a protein at high resolution from microseconds to seconds. *Proc. Natl Acad. Sci. USA* 94, 826–30.
26 Ferguson, N., Johnson, C. M., Macias, M., Oschkinat, H. & Fersht, A. (2001). Ultrafast folding of WW domains without structured aromatic clusters in the denatured state. *Proc. Natl Acad. Sci. USA* 98, 13002–7.
27 Ferguson, N., Pires, J. R., Toepert, F. et al. (2001). Using flexible loop mimetics to extend Phi-value analysis to secondary structure interactions. *Proc. Natl Acad. Sci. USA* 98, 13008–13.
28 Mayor, U., Johnson, C. M., Daggett, V. & Fersht, A. R. (2000). Protein folding and unfolding in microseconds to nanoseconds by experiment and simulation. *Proc. Natl Acad. Sci. USA* 97, 13518–22.
29 Gillespie, B., Vu, D. M., Shah, P. S. et al. (2003). NMR and temperature-jump measurements of de novo designed proteins demonstrate rapid folding in the absence of explicit selection for kinetics. *J. Mol. Biol.* 4.
30 Mayor, U., Guydosh, N. R., Johnson, C. M. et al. (2003). The complete folding pathway of a protein from nanoseconds to microseconds. *Nature* 421, 863–7.
31 Rabl, C. R., Martin, S. R., Neumann, E. & Bayley, P. M. (2002). Temperature jump kinetic study of the stability of apo-calmodulin. *Biophys. Chem.* 101, 553–464.
32 Sanchez, I. E. & Kiefhaber, T. (2003). Hammond behavior versus ground state effects in protein folding: Evidence for narrow free energy barriers and residual structure in unfolded states. *J. Mol. Biol.* 327, 867–84.
33 Ansari, A., Jones, C. M., Henry, E. R., Hofrichter, J. & Eaton, W. A. (1992). The role of solvent viscosity in the dynamics of protein conformational changes. *Science* 256, 1796–8.
34 Mines, G. A., Pascher, T., Lee, S. C., Winkler, J. R. & Gray, H. B. (1996). Cytochrome c folding triggered by electron transfer. *Chem. Biol.* 3, 491–7.
35 Wittung-Stafshede, P., Gray, H. B. & Winkler, J. R. (1997). Rapid formation of a four-helix bundle, cytochrome b_{562} folding triggered by electron transfer. *J. Am. Chem. Soc.* 119, 9562–3.
36 Telford, J. R., Wittung-Stafshede, P., Gray, H. B. & Winkler, J. R. (1998). Protein folding triggered by electron transfer. *Acc. Chem. Res.* 31, 755–63.
37 Lee, J. C., Gray, H. B. & Winkler, J. R. (2001). Cytochrome c' folding triggered by electron transfer: fast and slow formation of four-helix bundles. *Proc. Natl Acad. Sci. USA* 98, 7760–4.
38 Chang, I. J., Lee, J. C., Winkler, J. R. & Gray, H. B. (2003). The protein-folding speed limit: Intrachain diffusion times set by electron-transfer rates in denatured Ru(NH3)(5)(His-33)-Zn-cytochrome c. *Proc. Natl Acad. Sci. USA* 100, 3838–40.
39 Gutman, M., Huppert, D. & Pines, E. (1981). The pH jump: a rapid

modulation of pH of aqueous solutions by a laser pulse. *J. Am. Chem. Soc.* 103, 3709–13.

40 PINES, E. & HUPPERT, D. (1983). pH jump: a relaxation approach. *J. Phys. Chem.* 87, 4471–8.

41 GEORGE, M. V. & SCAIANO, J. C. (1980). Photochemistry of o-nitrobenzaldehyde and related studies. *J. Phys. Chem.* 84, 492–5.

42 ABBRUZZETTI, S., CREMA, E., MASINO, L. et al. (2000). Fast events in protein folding: Structural volume changes accompanying the early events in the N → I transition of apomyoglobin induced by ultrafast pH jump. *Biophys. J.* 78, 405–15.

43 CHOI, J., HIROTA, N. & TERAZIMA, M. (2001). Enthalpy and volume changes on the pH jump process studied by the transient grating technique. *Anal. Sci.* 17, s13–s15.

44 VOLK, M., KHOLODENKO, Y., LU, H. S. M., GOODING, E. A., DEGRADO, W. F. & HOCHSTRASSER, R. M. (1997). Peptide conformational dynamics and vibrational stark effects following photoinitiated disulfide cleavage. *J. Phys. Chem.* 101, 8607–16.

45 METZLER, R., KLAFTER, J., JORTNER, J. & VOLK, M. (1998). Multiple time scales for dispersive kinetics in early events of peptide folding. *Chem. Phys. Lett.* 293, 477–84.

46 HAGEN, S. J., HOFRICHTER, J., SZABO, A. & EATON, W. A. (1996). Diffusion-limited contact formation in unfolded cytochrome c: Estimating the maximum rate of protein folding. *Proc. Natl Acad. Sci. USA* 93, 11615–17.

47 HAGEN, S. J., HOFRICHTER, J. & EATON, W. A. (1997). Rate of intrachain diffusion of unfolded cytochrome c. *J. Phys. Chem. B* 101, 2352–65.

48 BIERI, O. & KIEFHABER, T. (1999). Elementary steps in protein folding. *Biol. Chem.* 380, 923–9.

49 LAPIDUS, L. J., EATON, W. A. & HOFRICHTER, J. (2000). Measuring the rate of intramolecular contact formation in polypeptides. *Proc. Natl Acad. Sci. USA* 97, 7220–5.

50 LAPIDUS, L., EATON, W. & HOFRICHTER, J. (2001). Dynamics of intramolecular contact formation in polypeptides: distance dependence of quenching rates in room temperature glass. *Phys. Rev. Lett.* 87.

51 BIERI, O. & KIEFHABER, T. (2000). Kinetic models in protein folding. In *Mechanisms of Protein Folding*, 2nd edn (PAIN, R. H., ed.), Vol. 32, pp. 34–64. Oxford University Press, Oxford.

52 BALLEW, R. M., SABELKO, J. & GRUEBELE, M. (1996). Observation of distinct nanosecond and microsecond protein folding events. *Nat. Struct. Biol.* 3, 923–6.

53 GILMANSHIN, R., WILLIAMS, S., CALLENDER, R. H., WOODRUFF, W. H. & DYER, R. B. (1997). Fast events in protein folding: Relaxation dynamics of secondary and tertiary structure in native apomyoglobin. *Proc. Natl Acad. Sci. USA* 94, 3709–13.

54 THOMPSON, P. A., EATON, W. A. & HOFRICHTER, J. (1997). Laser temperature jump study of the helix reversible arrow coil kinetics of an alanine peptide interpreted with a 'kinetic zipper' model. *Biochemistry* 36, 9200–10.

55 DYER, R. B., GAI, F., WOODRUFF, W. H., GILMANSHIN, R. & CALLENDER, R. H. (1998). Infrared studies of fast events in protein folding. *Acc. Chem. Res.* 31, 709–16.

56 SABELKO, J., ERVIN, J. & GRUEBELE, M. (1999). Observation of strange kinetics in protein folding. *Proc. Nat. Acad. Sci. USA* 96, 6031–6.

57 CRANE, J. C., KOEPF, E. K., KELLY, J. W. & GRUEBELE, M. (2000). Mapping the transition state of the WW domain beta sheet. *J. Mol. Biol.* 298, 283–92.

58 LEESON, D. T., GAI, F., RODRIGUEZ, H. M., GREGORET, L. M. & DYER, R. B. (2000). Protein Folding on a Complex Energy Landscape. *Proc. Natl Acad. Sci. USA* 97, 2527–32.

59 JÄGER, M., NGUYEN, H., CRANE, J., KELLY, J. & GRUEBELE, M. (2001). The folding mechanism of a β-sheet: The WW domain. *J. Mol. Biol.* 311, 373–93.

60 GULOTTA, M., GILMANSHIN, R., BUSCHER, T. C., CALLENDER, R. H. & DYER, R. B. (2001). Core formation in

apomyoglobin: probing the upper reaches of the folding energy landscape. *Biochemistry* 40, 5137–43.

61 SNOW, C., NGUYEN, H., PANDE, V. & GRUEBELE, M. (2002). Absolute comparison of simulated and experimental protein folding dynamics. *Nature* 420, 102–6.

62 OSVÁTH, S. & GRUEBELE, M. (2003). Proline can have opposite effects on fast and slow protein folding phases. *Biophys. J.* 85, 1215–22.

63 OSVÁTH, S., SABELKO, J. & GRUEBELE, M. (2003). Tuning the heterogeneous early folding dynamics of phosphoglycerate kinase. *J. Mol. Biol.* 333, 187–99.

64 PANDIT, A., MA, H., STOKKUM, I. v., GRUEBELE, M. & GRONDELLE, R. v. (2003). The time resolved dissociation reaction of the light-harvesting 1 complex of Rhodospirillum rubrum, studied with an infrared laser-pulse temperature jump. *Biochemistry* 41, 15115–20.

65 YANG, W. & GRUEBELE, M. (2003). Folding at the speed limit. *Nature* 423, 193–7.

66 NGUYEN, H., JÄGER, M., GRUEBELE, M. & KELLY, J. (2003). Tuning the free-energy landscape of a WW domain by temperature, mutation and truncation. *Proc. Natl Acad. Sci. USA* 100, 3948–53.

67 BALLEW, R. M., SABELKO, J., REINER, C. & GRUEBELE, M. (1996). A single-sweep, nanosecond time resolution laser temperature-jump apparatus. *Rev. Sci. Instrum.* 67, 3694–9.

68 EATON, W. A., MUÑOZ, V., THOMPSON, P. A., HENRY, E. R. & HOFRICHTER, J. (1998). Kinetics and dynamics of loops, α-helices, β-hairpins, and fast-folding proteins. *Acc. Chem. Res.* 31, 745–53.

69 MUÑOZ, V., THOMPSON, P. A., HOFRICHTER, J. & EATON, W. A. (1997). Folding dynamics and mechanism of β-hairpin formation. *Nature* 390, 196–9.

70 MUÑOZ, V., HENRY, E. R., HOFRICHTER, J. & EATON, W. A. (1998). A Statistical Mechanical Model for β-Hairpin Kinetics. *Proc. Natl Acad. Sci. USA* 95, 5872–9.

71 HUANG, C. Y., GETAHUN, Z., WANG, T., DEGRADO, W. F. & GAI, F. (2001). Time-resolved infrared study of the helix-coil transition using C-13-labeled helical peptides. *J. Am. Chem. Soc.* 123, 12111–12.

72 GETAHUN, Z., HUANG, C. Y., WANG, T., LEON, B. D., DEGRADO, W. F. & GAI, F. (2003). Using nitrile-derivated amino acids as infrared probes of local environment. *J. Am. Chem. Soc.* 125.

73 KLIMOV, D. K. & THIRUMALAI, D. (2000). Mechanisms and kinetics of beta-hairpin formation. *Proc. Natl Acad. Sci. USA* 97, 2544–9.

74 YANG, W., PRINCE, R., SABELKO, J., MOORE, J. S. & GRUEBELE, M. (2000). Transition from exponential to nonexponential kinetics during formation of an artificial helix. *J. Am. Chem. Soc.* 122, 3248–9.

75 HUMMER, G., GARCIA, A. E. & GARDE, S. (2000). Conformational diffusion and helix formation kinetics. *Phys. Rev. Lett.* 85, 2637–40.

76 GILMANSHIN, R., CALLENDER, R. H. & DYER, R. B. (1998). The core of apomyoglobin E-form folds at the diffusion limit. *Nat. Struct. Biol.* 5, 363–5.

77 LOW, D. W., WINKLER, J. R. & GRAY, H. B. (1996). Photoinduced oxidation of microperoxidase-9: generation of ferryl and cation-radical prophyrins. *J. Am. Chem. Soc.* 118, 117–20.

78 POLLACK, L., TATE, M. W., DARNTON, N. C., KNIGHT, J. B., GRUNER, S. M., EATON, W. A. & AUSTIN, R. H. (1999). Compactness of the denatured state of a fast-folding protein measured by submillisecond small-angle X-ray scattering. *Proc. Natl Acad. Sci. USA* 96, 10115.

79 GILMANSHIN, R., CALLENDER, R. H., DYER, R. B. & WOODRUFF, W. H. (1998). Fast event in protein folding: the time evolution of primary processes. *Annu. Rev. Phys. Chem.* 49, 173–202.

80 WANG, J. & EL-SAYED, M. A. (1999). Temperature jump-induced secondary structural change of the membrane

protein bacteriorhodopsin in the premelting temperature region: a nanosecond time-resolved Fourier transform infrared study. *Biophys. J.* 76, 2777–83.

81 MA, H., ERVIN, J. & GRUEBELE, M. (2003). Multichannel, single-sweep infrared detection for nanosecond relaxation kinetics experiments. *Rev. Sci. Inst.* 75, 486–491.

82 CHEN, E. F. & KLIGER, D. S. (1996). Time-resolved near UV circular dichroism and absorption studies of carbonmonoxymyoglobin photolysis intermediates. *Inorg. Chim. Acta* 242, 149–58.

83 CHEN, E. F., GOLDBECK, R. A. & KLIGER, D. S. (1997). Nanosecond time-resolved spectroscopy of biomolecular processes. *Annu. Rev. Biophys. Biomol. Struct.* 26, 327–55.

84 CHEN, E., WOOD, M. J., FINK, A. L. & KLIGER, D. S. (1998). Time-resolved circular dichroism studies of protein folding intermediates of cytochrome c. *Biochemistry* 37, 5589–98.

85 CHEN, E. F., GENSCH, T., GROSS, A. B., HENDRIKS, J., HELLINGWERF, K. J. & KLIGER, D. S. (2003). Dynamics of protein and chromophore structural changes in the photocycle of photoactive yellow protein monitored by time-resolved optical rotatory dispersion. *Biochemistry* 42, 2062–71.

86 LEDNEV, I. K., KARNOUP, A. S., SPARROW, M. C. & ASHER, S. A. (1999). α-Helix peptide folding and unfolding activation barriers: a nanosecond UV resonance study. *J. Am. Chem. Soc.* 121, 8074–86.

87 LEDNEV, I. K., KARNOUP, A. S., SPARROW, M. C. & ASHER, S. A. (1999). Nanosecond UV resonance raman examination of initial steps in α-helix secondary structure evolution. *J. Am. Chem. Soc.* 121, 4076–7.

88 YAMAMOTO, K., MIZUTANI, Y. & KITAGAWA, T. (2000). Nanosecond temperature jump and time-resolved Raman study of thermal unfolding of ribonuclease A. *Biophys. J.* 79, 485–95.

89 YEH, S., HAN, S. & ROUSSEAU, D. L. (1998). Cytochrome c Folding and Unfolding: A Biphasic Mechanism. *Acc. Chem. Res.* 31, 727–36.

90 LAKOWICZ, J. R. (1986). Fluorescence studies of structural fluctuations in macromolecules as observed by fluorescence spectroscopy in the time, lifetime, and frequency domains. In *Methods in Enzymology*, Vol. 131, pp. 518–567. Academic Press, New York.

91 CHEN, Y. & BARKLEY, M. D. (1998). Toward understanding tryptophan fluorescence in proteins. *Biochemistry* 37, 9976–82.

92 FÖRSTER, T. (1948). Zwischenmolekülare Energiewanderung und Fluoreszenz. *Ann. Physik* 2, 55–75.

93 RISCHEL, C. & POULSEN, F. M. (1995). Modification of a specific tyrosine enables tracing of the end-to-end distance during apomyoglobin folding. *FEBS Lett.* 374, 105–9.

94 BALLEW, R. M. (1996). Direct observation of fast protein folding: distinct nanosecond and microsecond events in the folding of apomyoglobin. Thesis, University of Illinois.

95 GRUEBELE, M., SABELKO, J., BALLEW, R. & ERVIN, J. (1998). Laser temperature jump induced protein refolding. *Acc. Chem. Res.* 31, 699–707.

96 GARCIA, P., DESMADRIL, M., MINARD, P. & YON, J. M. (1995). Evidence for residual structures in an unfolded from of yeast phosphoglycerate kinase. *Biochemistry* 34, 397–404.

97 ERVIN, J., LARIOS, E., OSVATH, S., SCHULTEN, K. & GRUEBELE, M. (2002). What causes hyperfluorescence: Folding intermediates or conformationally flexible native states? *Biophys. J.* 83, 473–83.

98 ERVIN, J., SABELKO, J. & GRUEBELE, M. (2000). Submicrosecond real-time fluorescence detection: application to protein folding. *J. Photochem. Photobiol.* B54, 1–15.

99 HENRY, E. R. & HOFRICHTER, J. (1992). Singular value decomposition: application to analysis of experimental data. *Methods Enzymol.* 210, 129–92.

100 BRYNGELSON, J. D. & WOLYNES, P. G. (1987). Spin glasses and the statistical mechanics of protein folding. *Proc. Natl Acad. Sci. USA* 84, 7524–8.

101 ONUCHIC, J. N., WOLYNES, P. G., LUTHEY-SCHULTEN, Z. & SOCCI, N. D. (1995). Toward an outline of the topography of a realistic protein folding funnel. *Proc. Natl Acad. Sci. USA* 92, 3626–30.

102 KIEFHABER, T., SCHMID, F. X., WILLAERT, K., ENGELBORGHS, Y. & CHAFFOTTE, A. (1992). Structure of a rapidly formed intermediate in ribonuclease T1 folding. *Protein Sci.* 1, 1162–72.

103 PAPPENBERGER, G., AYGUN, H., ENGELS, J. W., REIMER, U., FISCHER, G. & KIEFHABER, T. (2001). Nonprolyl cis peptide bonds in unfolded proteins cause complex folding kinetics. *Nat. Struct. Biol.* 8, 452–8.

104 KELLEY, R. F. & RICHARDS, F. M. (1987). Replacement of proline-76 with alanine eliminates the slowest kinetic phase in thioredoxin folding. *Biochemsitry* 26, 6765–74.

105 PLAXCO, K. W., GUIJARRO, J. I., MORTON, C. J., PITKEATHLY, M., CAMPBELL, I. D. & DOBSON, C. M. (1998). The folding kinetics and thermodynamics of the Fyn-SH3 domain. *Biochemistry* 37, 2529–37.

106 ONUCHIC, J. N., LUTHEY-SCHULTEN, Z. & WOLYNES, P. G. (1997). Theory of protein folding: the energy landscape perspective. *Annu. Rev. Phys. Chem.* 48, 545–600.

107 PAPPU, R. V., SRINIVASAN, R. & ROSE, G. D. (2000). The Flory isolated-pair hypothesis is not valid for polypeptide chains: Implications for protein folding. *Proc. Natl Acad. Sci. USA* 97, 12565–70.

108 BROMBERG, S. & DILL, K. A. (1994). Side-chain entropy and packing in proteins. *Protein Sci.* 3, 997–1009.

109 BRYNGELSON, J. D. & WOLYNES, P. G. (1989). Intermediates and barrier crossing in random energy model (with applications to protein folding). *J. Phys. Chem.* 93, 6902–15.

110 KIM, P. S. & BALDWIN, R. L. (1990). Intermediates in the folding reactions of small proteins. *Annu. Rev. Biochem.* 59, 631–60.

111 BACHMANN, A. & KIEFHABER, T. (2001). Apparent two-state tendamistat folding is a sequential process along a defined route. *J. Mol. Biol.* 306, 375–86.

112 CREIGHTON, T. E. (1988). Toward a better understanding of protein folding pathways. *Proc. Natl Acad. Sci. USA* 85, 5082–6.

113 ZWANZIG, R. (1997). Two-state models of protein folding kinetics. *Proc. Natl Acad. Sci. USA* 94, 148–50.

114 NYMEYER, H., GARCÍA, A. E. & ONUCHIC, J. N. (1998). Folding funnels and frustration in off-lattice minimalist protein landscapes. *Proc. Natl Acad. Sci. USA* 95, 5921–8.

115 PLOTKIN, S. S., WANG, J. & WOLYNES, P. G. (1997). Statistical mechanics of a correlated energy landscape model for protein folding funnels. *J. Chem. Phys.* 106, 2932–48.

116 GO, N. (1983). Theoretical studies of protein folding. *Annu. Rev. Biophys. Bioeng.* 12, 183–210.

117 THOMPSON, P. A., EATON, W. A. & HOFRICHTER, J. (1997). Laser temperature jump study of the helix × coil kinetics of an alanine peptide interpreted with a 'kinetic zipper' model. *Biochemistry* 36, 9200–10.

118 ZWANZIG, R. (1988). Diffusion in a rough potential. *Proc. Natl Acad. Sci., USA* 85, 2029–30.

119 SHEA, J. E., ONUCHIC, J. N. & C L BROOKS, I. (2000). Energetic frustration and the nature of the transition state ensemble in protein folding. *J. Chem. Phys.* 113, 7663–71.

120 PORTMAN, J. J., TAKADA, S. & WOLYNES, P. G. (2001). Microscopic theory of protein folding rates. II. Local reaction coordinates and chain dynamics. *J. Chem. Phys.* 114, 5082–96.

121 THIRUMALAI, D., KLIMOV, D. K. & DIMA, R. I. (2002). Insights into specific problems in protein folding using simple concepts. *Adv. Chem. Phys.* 120, 35–76.

122 ALM, E. & BAKER, D. (1999). Prediction of protein-folding mechanisms from free energy landscapes derived from native structures. *Proc. Natl Acad. Sci. USA* 96, 11305–10.

123 MUÑOZ, V. & EATON, W. A. (1999).

A simple model for calculating the kinetics of protein folding from three-dimensional structures. *Proc. Natl Acad. Sci. USA* 96, 11311–16.

124 KOGA, N. & TAKADA, S. (2001). Roles of native topology and chain-length scaling in protein folding: a simulation study with a Go-like model. *J. Mol. Biol.* 313, 171–80.

125 SILOW, M. & OLIVEBERG, M. (1997). High-energy channeling in protein folding. *Biochemistry* 36, 7633–7.

126 WAGNER, C. & KIEFHABER, T. (1999). Intermediates can accelerate protein folding. *Proc. Natl Acad. Sci. USA* 96, 6716–21.

127 SKOROBOGATIY, M., GUO, H. & ZUCKERMAN, M. (1998). Non-Arrhenius modes in the relaxation of model proteins. *J. Chem. Phys.* 109, 2528–35.

128 METZLER, R., KLAFTER, J. & JORTNER, J. (1999). Hierarchies and logarithmic oscillations in the temporal relaxation patterns of proteins and other complex systems. *Proc. Natl Acad. Sci. USA* 96, 11085–9.

129 BECKER, O. M. & KARPLUS, M. (1997). The topology of multidimensional potential energy surfaces: theory and application to peptide structure and kinetics. *J. Chem. Phys.* 22, 1495–517.

130 BECKER, O. M. (1998). Principal coordinate maps of molecular potential energy surfaces. *J. Comput. Chem.* 19, 1255–67.

131 KRAMERS, H. A. (1940). Brownian motion in a field of force and the diffusion model of chemical reactions. *Physica* 7, 284.

132 PLAXCO, K. W. & BAKER, D. (1998). Limited internal friction in the rate-limiting step of a two-state protein folding reaction. *Proc. Natl Acad. Sci. USA* 95, 13591–6.

133 JACOB, M., SCHINDLER, T., BALBACH, J. & SCHMID, F. X. (1997). Diffusion control in an elementary protein folding reaction. *Proc. Natl Acad. Sci. USA* 94, 5622–7.

134 JACOB, M., GEEVES, M., HOLTERMANN, G. & SCHMID, F. X. (1999). Diffusional barrier crossing in a two-state protein folding reaction. *Proc. Natl Acad. Sci. USA* 94, 5622–7.

135 BHATTACHARYYA, R. P. & SOSNICK, T. R. (1999). Viscosity dependence of the folding kinetics of a dimeric and monomeric coiled coil. *Biochemistry* 38, 2601–9.

136 BIERI, O., WIRZ, J., HELLRUNG, B., SCHUTKOWSKI, M., DREWELLO, M. & KIEFHABER, T. (1999). The speed limit of protein folding measure by triplet-triplet energy transfer. *Proc. Natl Acad. Sci. USA* 96, 9597–601.

137 PAPPENBERGER, G., SAUDAN, C., BECKER, M., MERBACH, A. E. & KIEFHABER, T. (2000). Denaturant-induced movement of the transition state of protein folding revealed by high-pressure stopped-flow measurements. *Proc. Natl Acad. Sci. USA* 97, 17–22.

138 KHORASANIZADEH, S., PETERS, I. & RODER, H. (1996). Evidence for a three-state model of protein folding from kinetic analysis of ubiquitin variants with altered core residues. *Nat. Struct. Biol.* 3, 193–205.

139 SHASTRY, M. C. R. & RODER, H. (1998). Evidence for barrier-limited protein folding kinetics on the microsecond time scale. *Nat. Struct. Biol.* 5, 385–92.

140 HAGEN, S. J. & EATON, W. A. (2000). Two-state expansion and collapse of a polypeptide. *J. Mol. Biol.* 297, 781–9.

141 QIN, Z., ERVIN, J., LARIOS, E., GRUEBELE, M. & KIHARA, H. (2002). Formation of a compact structured ensemble without fluorescence signature early during ubiquitin folding. *J. Phys. Chem. B* 106, 13040–6.

142 FINKELSTEIN, A. V. & SHAKHNOVICH, E. I. (1989). Theory of cooperative transitions in protein molecules. ii. phase diagram for a protein molecule in solution. *Biopolymers* 28, 1681–94.

143 PARKER, M. J. & MARQUSEE, S. (1999). The cooperativity of burst phase reactions explored. *J. Mol. Biol.* 293, 1195–210.

144 ENGLANDER, S. W., SOSNICK, T. R., MAYNE, L. C., SHTILERMAN, M., QI, P. X. & BAI, Y. (1998). Fast and slow folding in cytochrome c. *Acc. Chem. Res.* 31, 767–74.

145 Krantz, B. A. & Sosnick, T. R. (2000). Distinguishing between two-state and three-state models for ubiquitin folding. *Biochemistry* 39, 11696–701.

146 Capaldi, A. P., Shastry, M. C. R., Kleanthous, C., Roder, H. & Radford, S. E. (2001). Ultrarapid mixing experiments reveal that Im7 folds via an on-pathway intermediate. *Nat. Struct. Biol.* 8, 68–72.

147 Miller, W. G. & Goebel, C. V. (1968). Dimensions of protein random coils. *Biochemistry* 7, 3925–34.

148 Ervin, J. & Gruebele, M. (2002). Quantifying protein folding transition states with Phi-T. *J. Biol. Phys.* 28, 115–28.

149 Pain, R. H., ed. (2000). *Mechanisms of Protein Folding*, Vol. 32. *Frontiers in Molecular Biology*. Oxford University Press, Oxford.

150 Berne, B. J. (1993). Theoretical and numerical methods in rate theory. In *Activated Barrier Crossing: Applications in Physics, Chemistry and Biology* (Hänggi, P. & Fleming, G. R., eds), pp. 82–119. World Scientific, Singapore.

151 Muñoz, V. (2002). Thermodynamics and kinetics of downhill protein folding investigated with a simple statistical mechanical model. *Int. J. Quantum Chem.* 90, 1522–8.

152 Poland, D. & Scheraga, H. A. (1970). *Theory of Helix-Coil Transitions in Biopolymers*, Academic Press, New York.

153 Dill, K. A. (1985). Theory for the folding and stability of globular proteins. *Biochemistry* 24, 1501–9.

154 Yue, K., Fiebig, K. M., Thomas, P. D., Chan, H. S., Shakhnovich, E. I. & Dill, K. A. (1995). A test of lattice protein folding algorithms. *Proc. Natl Acad. Sci. USA* 92, 325–9.

155 Shakhnovich, E., Farztdinov, G., Gutin, A. M. & Karplus, M. (1991). Protein folding bottlenecks: a lattice Monte Carlo simulation. *Phys. Rev. Lett.* 67, 1665–8.

156 Onuchic, J. N., Nymeyer, H., Garcia, A. E., Chahine, J. & Socci, N. D. (2000). The energy landscape theory of protein folding: Insights into folding mechanisms and scenarios. *Adv. Protein Chem.* 53, 87–152.

157 Kaya, H. & Chan, H. S. (2000). Energetic components of cooperative protein folding. *Phys. Rev. Lett.* 85, 4823–6.

158 Guo, Z. & Thirumalai, D. (1996). Kinetics and thermodynamics of folding of a de novo designed four-helix bundle protein. *J. Mol. Biol.* 263, 323–43.

159 Klimov, D. K. & Thirumalai, D. (1997). Viscosity dependence of the folding rates of proteins. *Phys. Rev. Lett.* 79, 317–20.

160 Srinivas, G., Yethiraj, A. & Bagchi, B. (2001). Nonexponentiality of time dependent survival probability and the fractional viscosity dependence of the rate in diffusion controlled reactions in a polymer chain. *J. Chem. Phys.* 114, 9170–8.

161 Duan, Y. & Kollman, P. A. (1998). Pathways to a protein folding intermediate observed in a 1-microsecond simulation in aqueous solution. *Science* 282, 740–4.

162 Ladurner, A. G., Itzhaki, L. S., Daggett, V. & Fersht, A. R. (1998). Synergy between simulation and experiment in describing the energy landscape of protein folding. *Proc. Natl Acad. Sci. USA* 95, 8473–8.

163 Karanicolas, J. & III, C. L. B. (2003). The structural basis for biphasic kinetics in the folding of the WW domain from a formin-binding protein: lessons for protein design. *Proc. Natl Acad. Sci. USA* 100, 3954–9.

164 Kaya, H. & Chan, H. S. (2000). Polymer principles of protein calorimetric two-state cooperativity. *Proteins Struct. Funct. Genet.* 40, 637–61.

165 Shea, J. & Brooks, C. L. (2001). From folding theories to folding proteins: a review and assessment of simulation studies of protein folding and unfolding. *Annu. Rev. Phys. Chem.* 52, 499–535.

166 Clementi, C., Nymeyer, H. & Onuchic, J. N. (2000). Topological and energetic factors: What deter-

mines the structural details of the transition state ensemble and "en-route" intermediates for protein folding? An investigation for small globular proteins. *J. Mol. Biol.* 298, 937–53.

167 BURTON, R. E., HUANG, G. S., DAUGHERTY, M. A., CALDERONE, T. L. & OAS, T. G. (1997). The energy landscape of a fast-folding protein mapped by Ala → Gly Substitutions. *Nat. Struct. Biol.* 4, 305–10.

168 GARCIA, C., NISHIMURA, C., CAVAGNERO, S., DYSON, H. J. & WRIGHT, P. E. (2000). Changes in the apomyoglobin folding pathway caused by mutation of the distal histidine residue. *Biochemistry* 39, 11227–37.

169 JÄGER, M., NGUYEN, H., KELLY, J. & GRUEBELE, M. (2003). Redesigning the turns of WW domain: the function-folding relationship. Submitted.

170 FERSHT, A. (1999). *Structure & Mechanism in Protein Science: A Guide to Enzyme Catalysis and Protein Folding.* WH Freeman, New York.

15
Early Events in Protein Folding Explored by Rapid Mixing Methods

Heinrich Roder, Kosuke Maki, Ramil F. Latypov, Hong Cheng, and M. C. Ramachandra Shastry

15.1
Importance of Kinetics for Understanding Protein Folding

As with any complex reaction, time-resolved data are essential for elucidating the mechanism of protein folding. Even in cases where the whole process of folding occurs in a single step, which is the case for many small proteins [1], the kinetics of folding and unfolding provide valuable information on the rate-limiting barrier. The effects of temperature and denaturant concentration give insight into activation energies and solvent-accessibility of the transition state ensemble, and by measuring the kinetic effects of mutations, one can gain more detailed structural insight [2–4]. If the protein folding process occurs in stages, i.e., if partially structured intermediate states accumulate, kinetic studies can potentially offer much additional insight into the structural and thermodynamic properties of intermediate states and intervening barriers [5–9]. Rapid mixing techniques have played a prominent role in kinetic studies of protein folding [5–7, 10–12]. The combination of quenched-flow techniques with hydrogen exchange labeling and NMR has proven to be particularly fruitful for the structural characterization of transient folding intermediates [13–15].

Theoretical models and computer simulations describe the process of protein folding in terms of a diffusive motion of a particle on a high-dimensional free energy surface [16–18]. This "landscape" description of protein folding predicts that a protein can choose among a large number of alternative pathways, which eventually converge toward a common free energy minimum corresponding to the native structure. In contrast, the time course of protein folding monitored by optical and other experimental probes generally shows relaxation kinetics with one or a few exponential phases, which are adequately described in terms of a simple kinetic scheme with a limited number of populated states (the chemical kinetics description). These apparently conflicting models can be consolidated if the free energy surface is divided into several regions (basins) separated by substantial free energy barriers due to unfavorable enthalpic interaction or entropic factors (conformational bottlenecks). The protein can rapidly explore conformational space within each basin corresponding to a broad ensemble of unfolded or partially folded states,

Protein Folding Handbook. Part I. Edited by J. Buchner and T. Kiefhaber
Copyright © 2005 WILEY-VCH Verlag GmbH & Co. KGaA, Weinheim
ISBN: 3-527-30784-2

but has to traverse substantial kinetic barriers before entering another basin. This type of free energy surface can thus give rise to multi-exponential folding kinetics.

Dissecting the sequence of structural events associated with the folding of a protein poses formidable technical challenges. Important structural events occur on the microsecond time scale, which cannot be accessed by conventional kinetic techniques, such as stopped-flow mixing. Temperature-jump and other rapid perturbation methods have shown that isolated helices and β-hairpins form and decay over a time window ranging from about 50 ns to several microseconds [19–23], which is short compared with the time it takes to complete the process of folding, even for the most rapidly folding proteins (reviewed in Ref. [24]). While these results demonstrate that secondary structure formation is not the rate-determining step in folding, they do not rule out the possibility that these elementary structural events affect the overall rate of folding if they represent energetically unfavorable, but obligatory, early steps. Recent advances in rapid mixing techniques combined with structurally informative spectroscopic probes made it possible to resolve conformational events on the submillisecond time scale preceding the rate-limiting step in the folding of globular proteins [25–27]. Historically, some of the earliest rapid kinetic measurements with millisecond time resolution used a continuous-flow arrangement combined with absorbance measurements of the reaction progress at different points downstream [28]. However, continuous-flow experiments were later replaced by the more versatile and economic stopped-flow experiment, which can be coupled with a wide range of spectroscopic probes to monitor reactions with millisecond time resolution [29, 30]. Continuous-flow techniques have experienced a renaissance in recent years due to advances in mixer design and detection methods, which made it possible to push the time resolution into the microsecond time range [25, 26, 31, 32]. By coupling an efficient capillary mixer with a digital camera, we can routinely extend fluorescence- or absorbance-based kinetic measurements to times as short as 50 μs, which has yielded a wealth of new insight into early stages of protein folding [33–40]. Other techniques make use of two or more consecutive mixing steps in order to prepare the system in a particular initial state (double-jump stopped-flow), or to execute multiple reaction steps in sequence (quenched-flow). If a reaction can be quenched by manipulating solution conditions (e.g., pH) or lowering temperature, quenched-flow or freeze-quench protocols can be used in combination with slower analytical techniques, such as NMR, electron paramagnetic resonance (EPR), or mass spectrometry [41]. To achieve efficient turbulent mixing conditions requires high flow rates and relatively large channel dimensions, which can consume substantial amounts of material. A promising alternative to turbulent mixing with improved sample economy uses hydrodynamic focusing to mix solutions under laminar flow conditions [42, 43].

15.2
Burst-phase Signals in Stopped-flow Experiments

In many cases, stopped-flow and quenched-flow measurements of protein folding reactions show evidence for unresolved rapid processes occurring within the dead

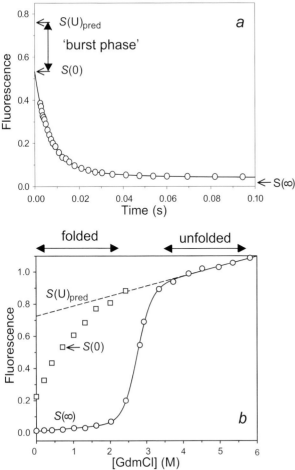

Fig. 15.1. Stopped-flow fluorescence evidence for an unresolved rapid process (burst phase) during folding of cyt c (pH 5, 10 °C). a) Tryptophan fluorescence changes during refolding of acid-unfolded cytochrome c (pH 2, ~15 mM HCl) at a final GdmCl concentration of 0.7 M. The initial signal S(0) at t = 0 (determined on the basis of a separate dead-time measurement) falls short of the signal for the unfolded state under refolding conditions, $S_{pred}(U)$, obtained by linear extrapolation of the unfolded-state baseline (see dashed line in b). b) Effect of the denaturant concentration on the initial (squares) and final (circles) fluorescence signal, S(0) and S(∞), measured in a series of stopped-flow refolding experiments at different final GdmCl concentration.

time (e.g., Refs [44, 45–49]). This is illustrated by Figure 15.1, which shows the kinetics of refolding of horse cytochrome c (cyt c) measured by stopped-flow fluorescence along with equilibrium fluorescence data vs. denaturant concentration [50]. The protein was unfolded by addition of 4.5 M guanidine hydrochloride (GdmCl), which lies in the baseline region above the cooperative unfolding transition (Figure 15.1b). The refolding reaction was triggered by sixfold dilution with buffer (0.1 M sodium acetate, pH 5), resulting in a final GdmCl concentration of 0.75 M, well within the folded baseline region. The data points in Figure 15.1a were recorded by sampling the fluorescence emission above 325 nm (using a glass cutoff filter) at logarithmically spaced time intervals. The first time point corresponds to the instrumental dead time of 2.5 ms, which was calibrated using a standard test reaction [51] (see Appendix). The observed time course describes a double-exponential decays (solid line) consisting of a major phase with a time constant of about 8 ms and a minor one with a time constant in the 100 ms range.

Extrapolation of the observed kinetics back to $t = 0$ yields the initial signal, $S(0)$, which is compared in Figure 15.1b (arrow) with the equilibrium unfolding transition plotted on the same fluorescence scale (relative to unfolded cyt c at 4.5 M GdmCl). The initial signal observed in this and a series of additional stopped-flow experiments at different final GdmCl concentrations is consistently below the relative fluorescence of the unfolded state, $S_{pred}(U)$, predicted by linear extrapolation from the unfolded baseline region to lower GdmCl concentrations (dashed line in panel b). The difference between the predicted and observed initial amplitude, $S_{pred}(U) - S(0)$, often called the burst phase, reflects conformational events occurring within the dead time of the stopped-flow experiment. Similar observations on many different proteins using various spectroscopic parameters gave clear evidence for the existence of rapid conformational events that cannot be resolved with conventional mixing techniques, and provided a strong incentive for the development of faster methods for triggering and observing structural changes during the first millisecond of refolding.

15.3
Turbulent Mixing

Most rapid mixing schemes rely on turbulent mixing to achieve complete mixing of two (or more) solutions. Mixers of various design are in use ranging from a simple T-arrangements to more elaborate geometries, such as the Berger ball mixer [52]. The goal is to achieve highly turbulent flow conditions in a small volume. The turbulent eddies thus generated can intersperse the two components down to the micrometer distance scale. However, the ultimate step in any mixing process relies on diffusion in order to achieve a homogeneous mixture at the molecular level. Given that the diffusion time t varies as the square of the distance r over which molecules have to diffuse, it takes a molecule with a diffusion constant $D = 10^{-5}$ cm^2 s^{-1} about 1 ms to diffuse over a distance of 1 μm ($t = r^2/D$). Thus, the mechanical mixing step has to intersperse the two components on a length scale of less than 1 μm in order to achieve submillisecond mixing times. The onset of turbulence is governed by the Reynolds number, Re, defined as

$$Re = \rho v d/\eta \qquad (1)$$

where ρ is the density (g cm^{-3}), v is the flow velocity (cm s^{-1}), d describes the characteristic dimensions of the channel (cm), and η is the viscosity of the fluid (e.g.,

0.01 poise for water at 20 °C). To maintain turbulent flow conditions in a cylindrical tube, Re has to exceed values of about 2000.

Turbulence is important not only for achieving efficient mixing, but also for maintaining favorable flow conditions during observation. In stopped-flow and quenched-flow experiments, turbulent flow insures efficient purging of the flow lines. In continuous-flow measurements, turbulent flow conditions in the observation channel lead to an approximate "plug flow" profile, which greatly simplifies data analysis compared to the parabolic profile obtained under laminar flow conditions. The time resolution of a rapid mixing experiment is governed not only by the mixing time, which in practice is difficult to quantify, but also the delay between mixing and observation. The effective delay between initiation of the reaction and the first reliably measurable data point is defined as the dead time, Δt_d. In both stopped- and continuous-flow experiments, any unobservable volume (dead volume), ΔV, between the point where mixing is complete and the point of observation contributes an increment $\Delta t = \Delta V/(dV/dt)$ to the dead time (dV/dt is the flow rate). Additional contributions to the effective dead time include the time delay to stop the flow and any artifacts that can obscure early parts of the kinetic trace (see Appendix).

Increasing the flow rate promotes more efficient mixing by generating smaller turbulent eddies, and yields shorter time delays Δt, and thus should lead to shorter dead times. However this trend does not continue indefinitely. Aside from practical problems due to back pressure and, in the case of stopped-flow measurements, various stopping artifacts, the time resolution of a rapid mixing experiment is ultimately limited by cavitation phenomena [53]. Under extreme conditions, the pressure gradients across turbulent eddies can become so large that the solvent begins to evaporate, forming small vapor bubbles that can take a long time to dissolve. The result is an intensely scattering plume that makes meaningful detection of the kinetic signal virtually impossible.

15.4
Detection Methods

The design principles and performance tests of common rapid mixing instruments, including a typical commercially available stopped-flow apparatus and the continuous-flow instrument developed in our laboratory [26], are described in the Appendix. A major strength of rapid mixing methods is that they can be combined with a wide range of detection methods. Table 15.1 lists common detection methods used in rapid mixing studies of the kinetics of folding, which are illustrated here with selected examples.

15.4.1
Tryptophan Fluorescence

The fluorescence emission properties of tryptophan and tyrosine side chains provide information on the local environment of these intrinsic chromophores.

Tab. 15.1. Common detection methods used in kinetic studies of protein folding.

Method	Probe	Properties probed	Sensitivity
Fluorescence	Trp, Tyr	Solvent shielding, tertiary contacts (quenching)	+++
	ANS	Hydrophobic clusters, collapse	+++
	FRET	Donor–acceptor distance	++
Absorbance	Trp, Tyr, cofactors	Polarity, solvent effects	++
Far-UV CD	Peptide bond	2° structure	–
Near-UV CD	Tyr, Trp, cofactor	Side-chain packing, moblity	–
Vibrational IR res. Raman	Peptide bond, cofactor	2° structure metal coordination	+
SAXS	Heavy atoms	Size (R_G), shape	–

For example, a fully solvent-exposed tryptophan in the denatured state of a protein typically shows a broad emission spectrum with a maximum near 350 nm and quantum yield of ∼0.14, similar to that of free tryptophan or its derivative, NATA. Burial of the tryptophan side chain in an apolar environment within the native state or a compact folding intermediate can result in a substantial blue-shift of the emission maximum (by as much as 25 nm) and enhanced fluorescence yield. These changes are a consequence of the decrease in local dielectric constant and shielding from quenchers, such as water and polar side chains. In other cases, close contact with a polar side chain or backbone moiety gives rise to a decrease in fluorescence yield upon folding. Most polar amino acid side chains (as well as main chain carbonyl and the terminal amino and carboxyl groups) are known to quench tryptophan fluorescence, probably via excited-state electron or proton transfer [54, 55]. Thus, the straightforward measurement of fluorescence intensity vs. folding or unfolding time can provide useful information on solvent accessibility and proximity to quenchers of an individual fluorescence probe. Complications due to the presence of multiple fluorophores can be avoided by using mutagenesis to replace any additional tryptophans (see, for example, Refs [56, 57]). Because Trp is a relatively rare amino acid, proteins with only one tryptophan are not uncommon; in the case of Trp-free proteins, a unique fluorophore can be introduced by using site-directed mutagenesis (see, for example, Refs [58, 59]).

The use of tryptophan fluorescence to explore early stages of protein folding is illustrated by our recent results on staphylococcal nuclease (SNase) [60] shown in Figure 15.2. A variant with a unique tryptophan fluorophore in the N-terminal β-barrel domain (Trp76 SNase) was obtained by replacing the single typtophan in wild-type SNase, Trp140, with His in combination with Trp substitution of Phe76. The fluorescence of Trp76 is strongly enhanced and blue-shifted under native conditions relative to the denatured state in the presence of urea (Figure 15.2a), indicating that upon folding the indole ring of Trp76 moves from a solvent-exposed location to an apolar environment within the native structure. An intermediate state with a fluorescence emission spectrum similar to, but clearly distinct from the

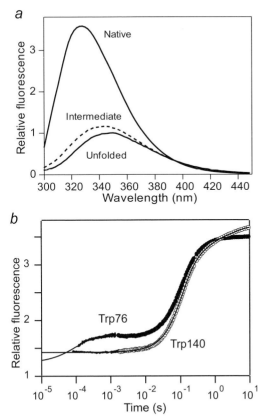

Fig. 15.2. Folding mechanism of SNase probed by tryptophan fluorescence. a) Fluorescence emission spectra of the Trp76 variant of SNase under native and denaturing conditions (solid) and a folding intermediate populated at equilibrium (dashed). The spectrum of the intermediate was determined by global analysis of the fluorescence spectra as a function of urea concentration (pH 5.2, 15 °C). b) Time course of folding (triggered by a pH jump from 2 to 5.2) for WT* SNase (Trp140) and a single-tryptophan variant (Trp76) measured by continuous-flow ($< 10^{-3}$ s) and stopped-flow ($> 10^{-3}$ s) fluorescence.

native state was detected in equilibrium unfolding experiments (dashed line in Figure 15.2a). In contrast to WT* SNase (P47G, P117G and H124L background), which shows no changes in tryptophan fluorescence prior to the rate-limiting folding step (~100 ms), the F76W/W140H variant shows additional changes (enhancement) during an early folding phase with a time constant of about 80 μs (Figure 15.2b). The fact that both variants exhibit the same number of kinetic phases with very similar rates confirms that the folding mechanism is not perturbed by the F76W/W140H mutations. However, the Trp at position 76 reports on the rapid formation of a hydrophobic cluster in the N-terminal β-sheet region while the wild-type Trp140 is silent during this early stage of folding.

15.4.2
ANS Fluorescence

Valuable complementary information on the formation of hydrophobic clusters at early stages of folding can be obtained by using 1-anilino-8-naphthalene-sulfonic acid (ANS) as extrinsic fluorescence probe [61–63]. Figure 15.3 illustrates this with recent results on the Trp76 variant of SNase introduced above [60]. Panel a shows continuous-flow measurements of the enhancement in ANS fluorescence that accompanies early stages of SNase folding under native conditions (U → N),

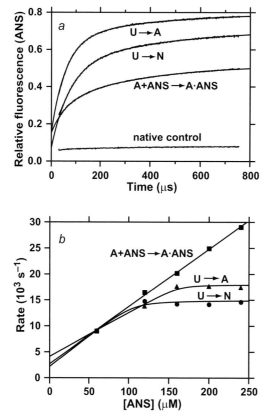

Fig. 15.3. An early folding event in Trp76 SNase detected by ANS fluorescence. a) ANS fluorescence changes during ANS binding/folding of Trp76 SNase measured by continuous-flow experiments at 15 °C in the presence of 160 μM ANS. U → A: Salt concentration jump from 0 M to 1 M KCl at pH 2.0; U → N: refolding induced by a pH-jump from 2.0 to 5.2; A + ANS → A · ANS: ANS binding kinetics in the presence of 1 M KCl at pH 2.0; native control: ANS binding kinetics under the native condition (pH 5.2). b) ANS concentration dependence of the rates for the major (fast) kinetic phases observed during the U → A (triangles), U → N (filled circles), and A + ANS → A · ANS (filled squares) reactions.

and during formation of the A-state, the compact acid-denatured state of SNase (U → A). Also shown is the kinetics of ANS binding to the preformed A-state. While the rate of ANS binding to the A-state shows the linear dependence on ANS concentration characteristic of a second-order binding process (panel b), the rates observed under refolding conditions (both to the native state and compact A-state) level off at ~120 μM ANS. The limiting ANS-independent rate at higher concentrations thus is due to an intramolecular conformational event that precedes ANS binding. The rate of this process closely matches that of the earliest phase detected by intrinsic fluorescence of Trp76 (Figure 15.3b), confirming that both processes reflect a common early folding step. The results are consistent with the rapid accumulation of an ensemble of states containing a loosely packed hydrophobic core involving primarily the β-barrel domain. In contrast, the specific interactions in the α-helical domain involving Trp140 are formed only during the final stages of folding.

15.4.3
FRET

Fluorescence resonant energy transfer (FRET) can potentially give more specific information on the changes in average distance between fluorescence donors and acceptors. For example, cyt c contains an intrinsic fluorescence donor–acceptor pair, Trp59 and the covalently attached heme group, which quenches tryptophan fluorescence via excited-state energy transfer [64]. We have made extensive use of this property to characterize the folding mechanism of cyt c [14, 50, 65], including the initial collapse of the chain on the microsecond time scale [33, 34].

We recently combined ultrafast mixing experiments with FRET in order to monitor large-scale structure changes during early stages of folding of acyl-CoA-binding protein (ACBP), a small (86-residue) four-helix bundle protein [38]. ACBP contains two tryptophan residues on adjacent turns of helix 3, which served as fluorescence donors, and an AEDANS fluorophore covalently attached to a C-terminal cystein residue was used as an acceptor (Figure 15.4a). Earlier equilibrium and kinetic studies, using intrinsic tryptophan fluorescence, showed a cooperative unfolding transition and single-exponential (un)folding kinetics consistent with an apparent two-state transition. Even when using continuous-flow mixing to measure intrinsic tryptophan fluorescence changes on the submillisecond time scale (Figure 15.4b), we found only minor deviations from two-state folding behavior. However, when we monitored the fluorescence of the C-terminal AEDANS group while exciting the tryptophans, we observed a large increase in fluorescence during a fast kinetic phase with a time constant of 80 μs, followed by a decaying phase with a time constant ranging from about 10 ms to 500 ms, depending on denaturant concentration (Figure 15.4c). The large enhancement in FRET efficiency is attributed to a major decrease in the average distance between helix 3 and C-terminus of ACBP. The fact that the early changes are exponential in character suggests that the initial compaction of the polypeptide is limited by an energy barrier rather than chain diffusion. The subsequent decrease in AEDANS fluorescence during the final stages

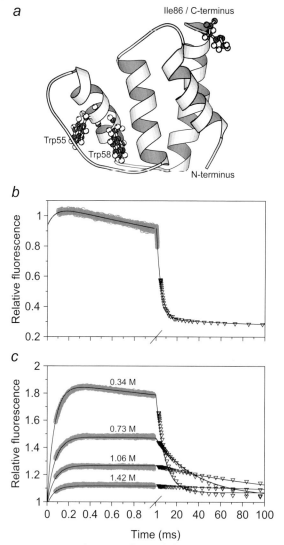

Fig. 15.4. FRET-detection of an early folding intermediate in a helix-bundle protein, ACBP. a) Ribbon diagram of ACBP, based on an NMR structure. The two tryptophan residues and the mutated C-terminal isoleucine are shown in ball and stick. The two lower panels show refolding kinetics of unmodified ACBP (b) and AEDANS-labeled ACBP,I86C (c) in pH 5.3 buffer containing 0.34 M GdmCl at 26 °C. In both panels data from continuous-flow (open circles) and stopped-flow (open triangles) experiments were matched and combined.

of folding is attributed to a sharp decrease in the intrinsic fluorescence yield of the two tryptophans due to intramolecular quenching.

The specific side-chain interactions responsible for quenching are established only in the close-packed native structure and are not present during the initial folding event. These observations indicate that the early (80 μs) folding phase marks the formation of a collapsed, but loosely packed and highly dynamic ensemble of states with overall dimensions (in terms of fluorescence donor–acceptor distance) similar to that of the native state. Accumulation of partially structured states with some native-like features may facilitate the search for the native conformation. Because of their short lifetime and low stability, such intermediates can easily be missed by conventional kinetic techniques, whereas the continuous-flow FRET technique offers the temporal resolution and structural sensitivity to detect even marginally stable intermediates populated during early stages of folding.

15.4.4
Continuous-flow Absorbance

Although fluorescence is inherently more sensitive, our capillary mixing instrument can also be adapted for continuous-flow absorbance measurements on the microsecond time scale. The fully transparent flow cell used for fluorescence measurements is replaced with a custom-made partially opaque absorbance flow cell of the same dimensions (0.25 mm pathlength). Relatively uniform illumination with minimal changes to the optical arrangement (see Figure 15.16) was achieved by using a 2-mm fluorescence cuvette filled with a highly turbid suspension (non-dairy creamer works well) as scattering cell. As in fluorescence measurements, a complete reaction profile can be recorded in a single 2–3 s continuous-flow run by imaging the flow channel onto the CCD chip. Using the reduction of 2,6-dichlorophenolindophenol (DCIP) by ascorbic acid as a test reaction ([66]; see Appendix), we measured dead times as short as 40 μs at the highest flow rate tested (1.1 mL s^{-1}).

To validate the technique, we measured the changes in heme absorbance in the Soret region (∼360–430 nm) associated with the folding of oxidized horse cyt *c*. The reaction was initiated by a rapid jump from pH 2, where the protein is fully unfolded, to pH 4.7, where folding occurs rapidly with minimal complications due to non-native histidine-heme ligands. A series of kinetic traces covering the time window from 40 μs to ∼1.5 ms were measured at different wavelengths spanning the Soret region (Figure 15.5). A parallel series of stopped-flow experiments (data not shown) was performed under matching conditions to extend the data to longer times (2 ms to 10 s). Global fitting of the family of kinetic traces to sums of exponential terms yielded three major kinetic phases with time constants of 65 μs, 500 μs, and 2 ms, respectively, consistent with accumulation of two intermediate species, I_1 and I_2, with absorbance properties distinct from both the initial (U) and final (N) states. In previous continuous-flow fluorescence measurements on horse cyt *c* [33, 34], we also observed three kinetic phases with very similar time constants, indicating that a basic four-state mechanism is sufficient to describe the folding process of cyt *c* in the absence of complications due to nonnative heme ligation and other slow events, such as *cis–trans* isomerization of peptide bonds.

15.4.5
Other Detection Methods used in Ultrafast Folding Studies

Continuous-flow measurements have been coupled with several other biophysical techniques, including resonance Raman spectroscopy [25], CD [67, 68], EPR [69, 70], and SAXS [27, 68, 71, 72]. In their pioneering work, Takahashi et al. [25] used resonance Raman spectroscopy to monitor changes in heme coordination during folding of cyt c on the submillisecond time scale. Their findings confirmed and extended prior results on the involvement of heme ligation in folding of cyt c, based on stopped-flow absorbance and fluorescence measurements [73–75].

CD spectroscopy in the far-UV (peptide) region provides an overall measure of secondary structure content, and is thus an especially valuable technique for protein folding studies (see, for example, Refs [44, 45]). However, the low inherent sensitivity of the technique, together with various flow artifacts, such as strain-induced birefringence, has limited the resolution of stopped-flow CD measurements to the 10-ms time range [76]. Akiyama et al. [67] were able to extend the time resolution down to the 400 μs range by coupling an efficient turbulent mixer (T-design) with a commercial CD spectrometer. Their continuous-flow measurements of CD spectral changes in the far-UV region revealed the formation of (helical) secondary structure during the second and third (final) stage of cyt c folding. The same group recently designed a mixer/flow-cell assembly with a dead time as short as 160 μs for continuous-flow SAXS measurements on a synchrotron [68, 72]. They were thus able to follow the changes in size (radius of gyration, R_g) and shape (pair distribution derived from scattering profiles) associated with refolding of acid-denatured cyt c under conditions similar to those used in our absorbance measurements (Figure 15.5). The intermediate formed within their dead time, which corresponds to the product of the 65 μs process in Figure 15.5, is substantially more compact ($R_g \sim 20$ Å) than the acid-denatured state ($R_g = 24$ Å). This finding clearly shows that cyt c undergoes a partial chain collapse during the initial folding phase, confirming earlier fluorescence data [33, 77].

15.5
A Quenched-Flow Method for H-D Exchange Labeling Studies on the Microsecond Time Scale

H-D exchange labeling experiments coupled with NMR detection [13–15, 78, 79] are important sources of structural information on protein folding intermediates. These experiments generally rely on commercial quenched-flow equipment to carry out two or three sequential mixing steps, which limits the time resolution to a few milliseconds or longer. In order to push the dead time of quenched-flow measurements into the microsecond time range, we made use of our highly efficient capillary mixers [26]. The device (illustrated in Figure 15.6a) uses a quartz capillary mixer similar to that used for optical measurements (see Appendix), but without observation cell, in order to generate a homogeneous mixture of solutions

15.5 A Quenched-Flow Method for H-D Exchange Labeling Studies on the Microsecond Time Scale

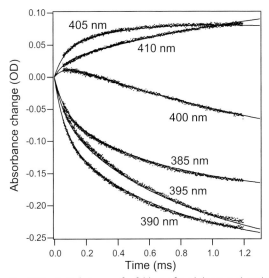

Fig. 15.5. Initial stages of refolding of acid-denatured oxidized cyt c at pH 5 monitored by continuous-flow absorbance measurements at different wavelengths spanning the Soret heme absorbance band. The lines represent a global fit of a four-state folding mechanism to the family of kinetic traces.

A and B. The mixture emerges from the capillary as a fine (200 μm diameter) jet with a linear velocity of up to 40 m s^{-1} at the highest flow rate used (1.25 mL s^{-1}). A second mixing event can be achieved simply by injecting the jet into a test tube containing a third solution C; the extremely high flow velocity ensures very efficient mixing.

To determine the dead time (i.e., the shortest delay between the two mixing events), we carried out a series of H-D exchange experiments on a pentapeptide (YGGFL). Rapid exchange of the backbone amide protons with solvent deuterons was achieved by mixing an H$_2$O solution of the peptide with fivefold excess of D$_2$O buffered at pH* 9.7 (uncorrected pH meter reading). The exchange reaction was quenched by injecting the mixture into ice-cooled acetate buffer at pH* 3. Under these quench conditions, the rate of exchange for some of the peptide NH groups (Gly3, Phe4, and Leu5) is sufficiently slow (10, 45, and 70 min, respectively) to determine their residual NH intensity by recording one-dimensional ^{1}H NMR spectra. Figure 15.6b shows the results of a series of experiments in which the capillary was raised from direct contact with the quench solutions to a distance of about 40 mm corresponding to a "time-of-flight" of ~1 ms. For Gly3 and Phe4, exponential fits of the decay in residual NH intensity with the incremented time delay yields exchange rates of 5600 and 4400 s^{-1}, respectively, in agreement with published intrinsic exchange rates [80] (the rate of the C-terminal amide group is too slow to be measurable over a 1-ms time window, and the Gly2 NH continues to

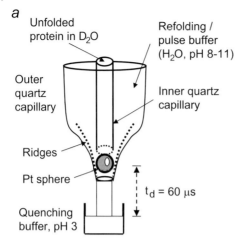

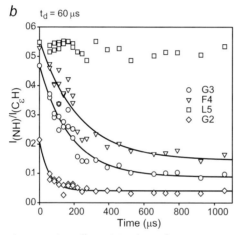

Fig. 15.6. A capillary mixing device for quenched-flow measurements on the microsecond time scale. a) A capillary mixer similar to that described in the Appendix (Figure 15.16), but without flow cell, is used to generate a fast free-flow jet. A second mixing event occurs upon impact of the jet with a quench buffer solution. b) Quenched-flow NMR measurement of the H-D exchange reaction of backbone NH groups of a model peptide (YGLFG). Extrapolation of the exponential decay in normalized NH resonance intensity (solid curves) yields an estimated dead time of 60 µs.

exchange under quench conditions). Extrapolation of the fits up to the NH intensity expected at $t = 0$ (measured in a separate control) indicates that the first measurement corresponds to an effective exchange time of 60 ± 10 µs, thus defining the dead time of the measurement. We used a similar set-up to measure the protection of amide protons at early times of refolding of β-lactoglobulin [37].

15.6
Evidence for Accumulation of Early Folding Intermediates in Small Proteins

15.6.1
B1 Domain of Protein G

Observation of a burst-phase signal, such as that shown in Figure 15.1, suggested deviations from two-state behavior for many proteins, including some small ones [7, 9, 81]. An illustrative example is the 57-residues B1 domain of protein G (GB1), which is among the smallest globular protein domains that do not depend on disulfide bonds or metals for stabilization (Figure 15.7a). Like many other small proteins [1], GB1 was initially thought to fold according to a two-state mechanism [82]. However, continuous-flow fluorescence measurements of the GB1 folding kinetics showed clear deviations from the first-order (single-exponential) kinetics expected for a simple two-state reaction [35]. The time course of refolding from the GdmCl-denatured state revealed a prominent exponential phase with a time constant of 600–700 μs followed by a second, rate-limiting process with a time constant of 2 ms or longer, depending on denaturant concentration (Figure 15.7b). The fast phase dominates the kinetics at low denaturant concentrations and accounts for the total fluorescence change associated with the burial of Trp43 upon folding, including the previously unresolved burst-phase signal [83]. In Figure 15.8, the rates of the two observable folding phases and the corresponding amplitudes are plotted vs. denaturant concentration. The biphasic kinetics of folding observed over a range of GdmCl concentrations can be modeled quantitatively on the

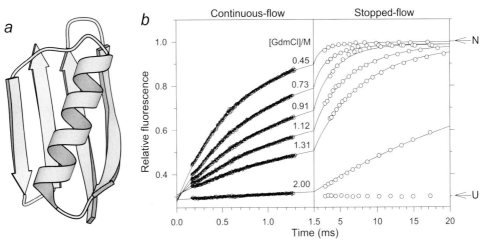

Fig. 15.7. a) Ribbon diagram of GB1 [145]. b) Folding kinetics of GB1 at different final GdmCl concentrations in the presence of 0.4 M sodium sulfate (pH 5.0, 20 °C) monitored by continuous-flow (left) and stopped-flow (right) fluorescence. The lines represent double-exponential fits to the the combined traces, which are normalized with respect to the native state.

basis of a three-state folding mechanism (Scheme 15.1),

$$U \underset{k_{iu}}{\overset{k_{ui}}{\rightleftharpoons}} I \underset{k_{ni}}{\overset{k_{in}}{\rightleftharpoons}} N$$

Scheme 15.1

where I represents an ensemble of intermediate states with native-like fluorescence properties (i.e., Trp43 is buried). Alternative mechanisms with nonproductive or nonobligatory intermediates lead to somewhat poorer fits of the data at low denaturant concentration, but cannot be ruled out definitively on the basis of available data (see Section 15.7.4 and Chapter 12.1).

The dependence of elementary rate constants, k_{ij}, on denaturant concentration c (dashed lines in Figure 15.8) is governed by the following relationship

$$\ln k_{ij} = \ln k_{ij}^0 + (m_{ij}^{\ddagger}/RT) \times c \qquad (2)$$

where k_{ij}^0 represents the elementary rate constant in the absence of denaturant, and $m_{ij}^{\ddagger}$ describes its dependence on denaturant concentration (kinetic m-value). The system of linear differential equations describing Scheme 15.1 was solved by determining the eigenvalues and eigenvectors of the corresponding rate matrix, using standard numeric methods [84, 85]. Although a three-state kinetic mechanism can be solved analytically (see, for example, Ref. [86]), the rate-matrix approach has the advantage that it can be readily expanded to more complex first-order kinetic mechanisms. After optimizing the four elementary rate constants and corresponding m-values, the two observable rates (eigenvalues) and associated amplitudes predicted by the model (solid lines in Figure 15.8) simultaneously fit both the observed rate profile (log(rate) vs. GdmCl concentration) and kinetic amplitudes at each denaturant concentration, as well as the midpoint and slope of the equilibrium unfolding transition (diamonds in panel b).

The three-state mechanism explains the kinetic behavior at low ($<$ 1 M) and intermediate (1–3 M) GdmCl concentrations in terms of two distinct kinetic limits reminiscent of the EX1 and EX2 limits of hydrogen exchange [87]. At low denaturant concentration, the intermediate is stable and transiently populated (the free energy of I in Figure 15.8c is lower than that of U), and the rate of the slower folding phase approaches the elementary rate constant of the final folding step, k_{IN}. At the same time, the fast (submillisecond) folding phase gains amplitude at the expense of the slow (rate-limiting) phase. At intermediate denaturant concentrations, the intermediate is destabilized (Figure 15.8c) and no longer accumulates. However, if I is an obligatory intermediate, the observed folding rate approaches the limiting value $k_f = K_{UI} \times k_{IN}$, where $K_{UI} = k_{UI}/k_{IU}$ is the equilibrium constant of the $U \leftrightarrow I$ transition. Thus, the sharp decrease in the net folding rate as the denaturant concentration approaches the midpoint of the unfolding transition is explained in terms of the unfavorable $U \leftrightarrow I$ pre-equilibrium involving a high-energy intermediate.

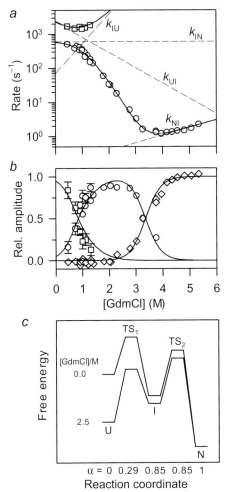

Fig. 15.8. GdmCl-dependence of the rate constants (a) and kinetic amplitudes (b) of the fast (squares) and slow (circles) kinetic phases observed during folding of GB1. (c) Free energy diagrams for folding of GB1 under conditions where the intermediate, I is well populated (0 M) and unstable (2.5 M GdmCl). α represents the change in solvent-accessible surface area relative to the unfolded state U.

Three-state analysis of a complete set of kinetic data, such as that of GB1 reported by Park et al. [35], provides a comprehensive description of the folding mechanism in terms of the size of the free-energy barriers separating the intermediate from the unfolded and native states (labeled TS$_1$ and TS$_2$ in Figure 15.8c) and the stability of the intermediate ($\Delta G = -RT \ln(k_{UI}/k_{IU})$). In addition, the denaturant dependence of the four elementary rate constants provides valuable insight into the changes in solvent-accessible surface area associated with each transition. In Figure 15.8c, each state and intervening transition state is labeled with the corresponding α-value (sometimes called β_T in reference to Tanford [88]) obtained by the cumulative kinetic m-values with respect to the total equilibrium m-value. According to this analysis, the initial barrier, TS_1, represents a well solvated ensemble of states ($\alpha = 0.29$) while both I and TS_2 are nearly as solvent-shielded as the native state ($\alpha = 0.85$). The relative changes in fluorescence associated with the two folding phases of GB1 provide additional insight into the structural properties of the intermediate. The large increase in fluorescence during the fast phase (Figure 15.7b), which accounts for nearly all of the fluorescence change at equilibrium, indicates that Trp43 becomes largely buried already during the initial phase of folding. Given the central location of Trp43 at the interface between the C-terminal β-hairpin and the α-helix, this indicates that the intermediate contains a well-developed hydrophobic core.

Krantz et al. recently questioned the validity of our analysis and argued that the folding kinetics of GB1 should be modeled as a two-state process [89]. However, this requires that the complete time course of folding can be fitted by a single exponential. In contrast, we find that a satisfactory fit of our combined continuous- and stopped-flow data requires two exponential phases, while a single exponential fit leads to nonrandom residuals of the order of 10%, which is unacceptable, given the quality of the data (Figure 15.9). It should be noted that the two phases differ sufficiently in rate (> 3.5-fold) to make their separation unambiguous. This conclusion is strengthened by the absence of additional, slower phases in GB1, which contains no proline residues. Reanalyzing a manually digitized version of our data, Krantz et al. [89] were able to fit single exponentials to the separate continuous-flow and stopped-flow traces, but obtained different rates for the two experiments, which confirms our conclusion that the overall kinetics is biphasic. Below 1 M GdmCl, where the slower phase of our double-exponential fit levels off (Figure 15.8), the apparent rate obtained by single-exponential fitting continues to increase, approximating the linear chevron behavior of a two-state system.

This phenomenon is explained by the fact that the approximate rate obtained by fitting a single exponential represents a weighted average of the rates obtained by double-exponential fitting (see next section). Because the fast phase dominates at low and the slow phase at higher denaturant concentrations, the result is a relatively linear rate profile. We further note that the population of the intermediate becomes negligible at GdmCl concentrations approaching the transition region, resulting in an apparent two-state unfolding equilibrium. Thus, contrary to Krantz et al. [89], the predicted equilibrium behavior cannot be used to discriminate between the two models. Finally, the folding kinetics of GB1 was found to be independent of protein concentration [83], indicating that intermolecular interactions are not involved in stabilizing the intermediate.

15.6.2
Ubiquitin

The 76-residue α/β protein ubiquitin is another well-studied small protein for which three-state folding behavior has been reported under some conditions. Khor-

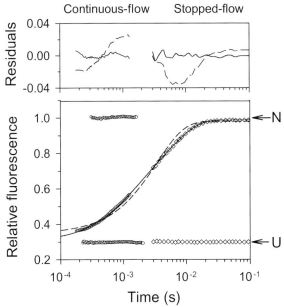

Fig. 15.9. Folding kinetics of GB1 at 1.12 M GdmCl (pH 5.0, 20 °C, 0.4 M sodium sulfate) monitored by continuous-flow (circles) and stopped-flow (diamonds) fluorescence. Single- and double-exponential fits and residuals are shown with solid and dashed lines, respectively.

asanizadeh et al. [48, 58] found deviations from two-state behavior in the folding kinetics of a tryptophan-containing ubiquitin variant (F45W mutant), including a downward curvature in the rate profile (log(rate) vs. [denaturant] plot) and a concomitant drop in the relative amplitude of the main folding phase at low denaturant concentration. They were able to account for both phenomena (rollover and burst phase) in terms of a three-state mechanism (U ↔ I ↔ N) with an obligatory on-path intermediate. As detailed above for GB1, this simple kinetic mechanism explains the leveling-off of the rate constant and diminishing amplitude of the principal (rate-limiting) folding phase at GdmCl concentrations below 1 M. However, without direct observation of the inferred fast folding phase, it is not possible to rule out alternative mechanisms with off-path or non-obligatory intermediates (further discussed in Section 15.7.3). Moreover, the uncertainty in the burst-phase amplitude is substantial if the main observable folding phase approaches the dead time of the kinetic experiment, which is the case for ubiquitin under stabilizing conditions. Krantz and Sosnick [90] remeasured the folding kinetics of F45W ubiquitin, using a stopped-flow instrument with a dead time of ~1 ms. They found a linear rate profile for the main folding phase and no indications of a burst phase, and concluded that our earlier evidence for a folding intermediate was based on fitting artifacts.

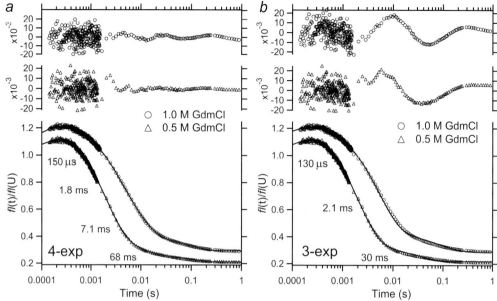

Fig. 15.10. Comparison of quadruple (a) and triple (b) exponential fitting of the kinetics of refolding of F45W ubiquitin at final GdmCl concentrations of 0.5 and 1.0 M (pH 5, 25 °C). Fluorescence traces measured in continuous- and stopped-flow experiments were normalized with respect to the unfolded protein in 6 M GdmCl. The residuals (top two traces in each panel) indicate that four exponentials are required to obtain a satisfactory fit of the data over the time window shown.

In an effort to settle this debate, we recently revisited the folding kinetics of F45W ubiquitin, combining continuous-flow fluorescence measurements with stopped-flow experiments with an improved dead time, which allowed us to continuously monitor the time course of folding from about 100 μs to 100 s (see Figure 15.10a for two representative kinetic traces). The initial increase in fluorescence relative to the GdmCl-unfolded state with a time constant of 150 μs becomes more pronounced in the presence of sodium sulfate (data not shown) and is consistent with a decrease in the solvent accessibility of the fluorophore, Trp45, at an early stage of folding. However, further studies are required to determine whether this process reflects formation of a folding intermediate or a less specific collapse event. The subsequent fluorescence decay is attributed to intramolecular quenching of Trp45 fluorescence upon folding [91].

A thorough kinetic analysis, using multi-exponential fitting functions, indicates that a minimum of four distinct phases are required to obtain a satisfactory fit of the data over the 0.1–1 s time window (Figure 15.10a); a minor additional fluorescence decay at longer times has been attributed to proline isomerization [58]. In contrast, if we use only three exponentials (including one increasing and two decaying phases) to fit the data over the same time window, we obtain non-random

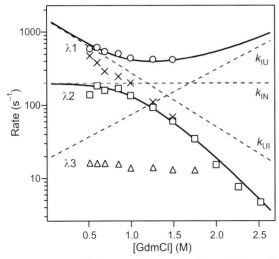

Fig. 15.11. Expanded region of the rate profile (log rate versus [GdmCl]) for the two main phases (circles and squares) and a minor slower phase (triangles) observed during folding of F45W ubiquitin (pH 5, 25 °C). The solid lines show the rates predicted by a three-state model, and the dashed lines indicate the elementary rate constants used. The X-symbols show the apparent rates obtained by triple-exponential fitting (Figure 15.10b).

residuals with amplitudes much larger than the scatter of the data points (Figure 15.10b). Under the most stabilizing conditions studied (0.5 M GdmCl), a satisfactory fit of the main fluorescence decay between 300 µs to 10 ms requires two distinct phases, λ_1 and λ_2, with time constants $\tau_1 = 1.8$ and $\tau_2 = 7.1$ ms, respectively. A fourth phase ($\tau_3 \sim 70$ ms) is necessary to account for a minor decay observed between 20 and 200 ms. The fact that both the rate and amplitude of λ_3 are essentially independent of denaturant concentration points toward a heterogeneous process, perhaps involving a minor population of molecules containing nonnative (cis) proline isomers.

A plot of the rate constants for the three decaying phases vs. GdmCl concentration (Figure 15.11) shows a pronounced rollover for λ_2 (squares) with a denaturant-independent regime below 0.75 M followed by a linear decrease above 1 M GdmCl, whereas λ_1 shows a shallow upward curvature. As in the case of GB1 (Figure 15.8), this behavior is fully consistent with a three-state mechanism (Scheme 15.1). Figure 15.11 also shows the apparent rates for the main decaying phase obtained by fitting only three exponentials (symbol x). Since they represent an average of λ_1 and λ_2 weighted according to the relative amplitudes, they continue to increase with decreasing GdmCl concentration below 1 M. Thus, the controversy whether an intermediate accumulates during folding of ubiquitin boils down to a rather subtle curve-fitting problem.

In our initial study, we detected only the slower process, λ_2, which exhibits the rollover and missing-amplitude effects indicative of a burst-phase intermediate

[48]. By extending their stopped-flow measurements to shorter times (~1 ms), Krantz and Sosnick [90] obtained a faster rate without apparent rollover corresponding to a weighted average of the two underlying processes. In a recent stopped-flow study, Went et al. [92] found that F45W ubiquitin exhibits three-state folding kinetics under some conditions (in the presence of GdmCl or urea plus salt) and two-state behavior under other conditions (in the presence of urea at low ionic strength). They further report that the folding kinetics of F45W ubiquitin (unmodified N-terminus) varies with protein concentration, suggesting that transient association may stabilize the intermediate. All of these observations, including our earlier findings on the effects of core mutations [48], can be explained by the presence of a marginally stable intermediate, which affects the kinetics of folding only under sufficiently stabilizing conditions.

15.6.3
Cytochrome c

Cyt c has played a central role in the development of new kinetic approaches for exploring early events in protein folding. The presence of a covalently attached heme group in this 104-residue protein (Figure 15.12) serves as a useful optical marker, and its redox and ligand binding properties provide unique experimental opportunities for rapid initiation and observation of folding [19, 25, 93]. Its sole tryptophan residue, Trp59, is located within 10 Å of the heme iron (Figure 15.12), resulting in efficient fluorescence quenching through a Förster-type energy transfer mechanism. Strongly denaturing conditions (e.g., 4.5 M GdmCl, or acidic pH

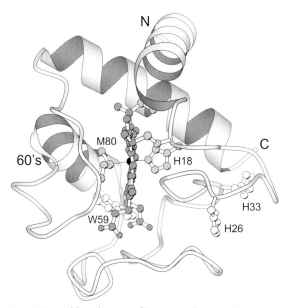

Fig. 15.12. Ribbon diagram of horse cytochrome c, based on the crystal structure [146].

at low ionic strength) result in a large increase in Trp59 fluorescence (up to ~60% of that of free tryptophan in water) indicative of an expanded chain conformation with an average tryptophan–heme distance greater than 35 Å [32, 64]. While numerous studies have shown that folding of oxidized cyt c is accompanied by changes in coordination of the heme iron [14, 25, 45, 50, 73, 74, 94–96], these complications can be largely avoided by working at mildly acidic pH (4.5–5) where the protein is still stable, but histidine residues are protonated and no longer can bind to the heme iron [73, 97].

Our capillary mixing apparatus [26] enabled us to resolve the entire fluorescence-detected folding kinetics of cyt c, including the elusive initial collapse of the chain [33]. Figure 15.13a shows the decay in Trp59 fluorescence observed during refolding of acid-unfolded cyt c (pH 2, 10 mM HCl) induced by a pH jump to native conditions (pH 4.5, 22 °C). The continuous-flow data covering the time range from 45 µs to ~1 ms are accurately described by a biexponential decay with a major rapid phase (time constant 59 ± 6 µs) and a minor process in the 500 µs range. This fit extrapolates to an initial fluorescence of 1.0 ± 0.05 (relative to the acid-unfolded protein) at $t = 0$, indicating the absence of additional, more rapid fluorescence changes that remain unresolved in the 45 µs dead time. The same data are plotted in Figure 15.13b on a logarithmic time scale, along with a stopped-flow trace measured under matching conditions, which accounts for the final 10% of the total fluorescence change.

The combined kinetic trace covers six orders of magnitude in time and accounts for the complete change in Trp59 fluorescence associated with refolding of cyt c. An Arrhenius plot of the rate of the initial phase (Figure 15.13a, inset) yields an apparent activation enthalpy of 30 kJ mol^{-1}, which is significantly larger than that expected for a diffusion-limited process [98]. In subsequent laser T-jump studies, Hagen, Eaton and colleagues [77, 99] detected a relaxation process with similar rates and activation energy, confirming the presence of a free energy barrier between unfolded and collapsed conformations of cyt c.

In other continuous-flow experiments, we measured the kinetics of folding of cyt c starting from either the acid-unfolded (pH 2, 10 mM HCl) or the GdmCl-unfolded state (4.5 M GdmCl, pH 4.5 or pH 7), and ending under various final conditions (pH 4.5 or pH 7 and GdmCl concentrations from 0.4 to 2.2 M) [33]. In each case, we observed a prominent initial decay in fluorescence with a time constant ranging from 25 to 65 µs. In particular, the rate of the initial phase, measured under the same final conditions (pH 4.5, 0.4 M GdmCl), was found to be independent of the initial state (acid- or GdmCl-unfolded). These observations clearly indicate that a common rate-limiting step is encountered during the initial stages of cyt c folding. The large amplitude of the initial phase (as much as 70% of the total change in fluorescence under strongly native conditions) is consistent with the formation of an ensemble of compact states, which was confirmed in subsequent small-angle X-ray (SAXS) studies [71, 72].

Using their continuous-flow CD instrument to follow refolding of acid-denatured cyt c after a dead time of ~400 µs, Akiyama et al. [67] observed significant changes in the far-UV region during the 500 µs and 2 ms folding phases of cyt

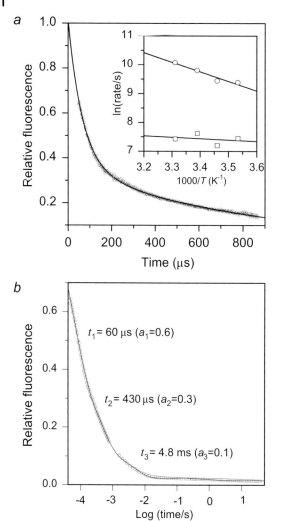

Fig. 15.13. Refolding kinetics of acid-unfolded cyt c at pH 4.5, 22 °C [33]. a) Submillisecond kinetics measured by continuous-flow mixing, indicating heme-induced quenching of Trp59 fluorescence associated with chain collapse. Inset: Arrhenius plot for the rates of the major (circles) and the minor (squares) submillisecond phases. b) Combined plot of continuous-flow and stopped-flow kinetic traces measured under identical conditions. The time constants and relative amplitudes (in parenthesis) are indicated for the three major phases.

c, indicating that most of the native helical secondary structure is acquired during the second and third (final) stages of folding. Extrapolation of the CD signal at 222 nm to shorter times indicated that the product of the 60 μs collapse phase is about 20% more helical than the GdmCl-unfolded state, but has a similar helix content as the acid-unfolded state. In their subsequent continuous-flow SAXS measurements, Akiyama et al. [72] found that a partially collapsed cyt c intermediate ($R_g = 20.5$ Å) accumulates within the 160 µs dead time of their experiment, which corresponds to a significant compaction relative to the acid-denatured state ($R_g = 24$ Å). In a subsequent phase on the millisecond time regime, the protein passes through a second, more compact intermediate ($R_g = 18$ Å) before reaching the native state ($R_g = 13.9$ Å). These findings confirm that cyt c undergoes a major chain collapse during the initial folding phase on the 10–100 µs time scale detected in previous continuous-flow and T-jump measurements of the heme-induced quenching of Trp59 fluorescence [33, 77, 99].

15.7
Significance of Early Folding Events

15.7.1
Barrier-limited Folding vs. Chain Diffusion

In Sections 15.4 and 15.6, we presented continuous-flow results on the folding kinetics of five proteins (GB1, cyt c, ubiquitin, ACBP, and SNase) all of which show significant fluorescence changes in one or more rapid phases on the submillisecond time scale that precede the rate-limiting folding step [33, 35, 36, 38, 60]. Other examples include the bacterial immunity protein Im7 [36], which will be discussed further in Section 15.7.3, cytochrome c_{551} from *Pseudomonas aeruginosa* [39], β-lactoglobulin [37] and the Y92W mutant of ribonuclease A (Welker, Maki, Shastry, Juminaga, Bhat, Roder & Scheraga, submitted). For most proteins (GB1, ubiquitin, Im7, β-lactoglobulin, SNase, and RNase), we observed an increase in Trp fluorescence during the fast phase, which we attribute to the burial of the tryptophan fluorophores within an apolar environment. For three proteins (cyt c, cytochrome c_{551}, and ACBP), we measured a large increase in energy transfer efficiency during the fast phase, confirming that these early conformational events include a large-scale collapse of the polypeptide chain. Thus, we conclude that for all of these structurally diverse proteins, major conformational changes, including a large-scale collapse of the chain, precede the rate-limiting step in folding. In every case the initial phase follows an exponential time course, indicating that a discrete free energy barrier separates the ensemble of compact states from the more expanded ensemble of unfolded states. In contrast, if the collapse of the polypeptide chain were a continuous process governed by diffusive dynamic, as predicted for homopolymers [100], it would be characterized by a broad distribution of time constants likely to give rise to nonexponential ('distributed') kinetics [16, 101].

In fact, Sabelko et al. [102] observed such distributed folding kinetics in temperature-jump experiments under conditions where folding can proceed downhill from a saddle point in the free energy surface. While the observation of exponential kinetics may not always be a sufficient condition for the presence of an activation barrier [103], the case for a barrier-limited process is strengthened considerably if the process also exhibits a significant activation enthalpy, as observed for cyt c [33]. Moreover, comparison of the pH-induced folding traces of cyt c monitored by Trp59 fluorescence (Figure 15.13) and heme absorbance (Figure 15.5) indicates that the time constant of the initial phase is independent of the probe used (59 ± 6 μs by fluorescence compared to 65 ± 4 μs by absorbance), which strongly supports the presence of a barrier (a diffusive collapse process is likely to lead to probe-dependent kinetics).

The existence of a prominent free energy barrier between expanded and compact states justifies the description of the initial folding events in terms of a model involving an early folding intermediate. The first large-scale conformational change during folding can, thus, be described as a reversible two-state transition. The finding that the rate of collapse in cyt c is insensitive to initial conditions [33] suggests that the barrier may represent a general entropic bottleneck encountered during the compaction of the polypeptide chain, which is consistent with the moderate activation enthalpy associated with the initial transition (Figure 15.13). Bryngelson et al. [16] classified this type of folding behavior as a "type I" scenario where a free energy barrier arises because of the unfavorable reduction in conformational entropy, which can be compensated by favorable enthalpic and solvent interactions only during the later stages of collapse.

15.7.2
Chain Compaction: Random Collapse vs. Specific Folding

Although denatured proteins often retain far more structure than expected for a random coil polymer [104], they are substantially more expanded than the native state [105]. The folding process therefore must be accompanied by a net decrease in chain dimensions. However, there are persistent controversies surrounding the question whether chain contraction occurs prior to or concurrent with the rate-limiting folding step, and whether compact states are the result of random hydrophobic collapse or a more specific structural events. Our earlier conclusion that the optical changes on the submillisecond time scale (burst-phase events) seen for cyt c and many other proteins [7, 9] reflect the formation of productive folding intermediates has been challenged by Englander and coworkers [106, 107] on the basis of fluorescence and CD measurements on two heme-containing peptide fragments of cyt c (residues 1–65 and 1–80, respectively). Although the fragments are unable to assume a folded structure, they nevertheless show rapid (<ms) denaturant-dependent fluorescence and far-UV CD changes resembling the burst-phase behavior of the intact polypeptide chain (cf. Figure 15.1).

Assuming that the fragments remain fully unfolded under all conditions, Sosnick et al. [106, 107] concluded that both the optical properties of the fragments and the burst-phase changes seen during refolding of intact cyt c reflect a rapid solvent-dependent readjustment of the unfolded polypeptide chain rather than formation of partially folded states. However, this conclusion is inconsistent with our observation that refolding of both GdmCl- and acid-denatured cyt c is accompanied by an exponential fluorescence decay [33], which indicates that the ensemble of states formed on the submillisecond time scale are separated by a free energy barrier from the unfolded states found immediately after adjustment of the solvent conditions. That these states are not just part of a broad distribution of more or less expanded denatured conformations is further supported by the observation of nonrandom amide protection patterns [79]. Apparently, both the full-length and truncated forms of cyt c can assume a compact ensemble of states upon lowering of the denaturant concentration. This leads to the prediction that the fragments will also exhibit an exponential fluorescence change on a time scale similar to the intact protein (40–60 μs at 22 °C and GdmCl concentrations of 0–0.4 M [33]). This prediction was recently confirmed by Qiu et al. [99], who observed an exponential relaxation process with a time constant of 20–30 μs at room temperature and a substantial activation energy (60–70 kJ mol^{-1}) for the 1–65 and 1–80 fragments of cyt c, using a laser T-jump apparatus.

In a subsequent study on ribonuclease A (RNase A), Englander and colleagues [108] reported that the GdmCl-dependence of the CD signal for the fully disulfide-reduced form closely matches the initial kinetic amplitude observed in stopped-flow refolding experiments on the protein with all four disulfide bonds intact. As in the case of the cyt c fragments, Qi et al. [108] assumed that reduced RNase remains in a random denatured state even under nondenaturing solvent conditions, which needs to be explored further. As an alternative interpretation, we again suggest that both in the presence and absence of disulfide bonds, RNase A rapidly forms a nonrandom ensemble of states, which is consistent with the shallow, but distinctly sigmoidal GdmCl dependence of the far-UV CD signal [108]. Since the native RNase A structure relies on disulfide bonds for its stability, folding of the reduced form cannot proceed beyond this early intermediate while the oxidized protein continues to fold to the native state.

15.7.3
Kinetic Role of Early Folding Intermediates

Several small proteins exhibit two-state behavior under certain conditions (e.g., elevated denaturant concentration, destabilizing mutations), but show evidence for populated intermediates (burst phase and/or nonlinear rate profiles) under other, more stabilizing, conditions [32, 38, 48, 58, 93, 109–111]. Thus, apparent two-state behavior can be considered a limiting cases of a multistate folding mechanism with unstable (high-energy) intermediates [48]. On the other hand, there are many well-documented examples of small proteins that show two-state folding behavior even under stabilizing conditions (reviewed in Refs [1, 112]). In several cases, the rate of folding at low denaturant concentrations was found to extend into the submillisecond time range and could be resolved only by methods such as NMR lineshape analysis [113, 114], electron transfer triggering [115], pressure-jump [116] or temperature-jump techniques [24].

While these findings clearly indicate that many small proteins can fold rapidly without going through intermediates, there is little support for the notion that early intermediates slow down the search for the native conformation [74, 117, 118], except in cases where they contain features inconsistent with the native fold, such as nonnative proline isomers, metal ligands [73, 74, 119], or intermolecular interactions [120]. In fact, solution conditions that favor accumulation of intermediates, such as low denaturant concentrations or stabilizing salts, generally accelerate the overall rate of folding [7]. Moreover, mutations that destabilize or abolish early intermediates often result in much slower rates of folding (see, for example, Refs [48, 121, 122]), which is a strong indication that they are productive states. Although continuous-flow results are fully consistent with Scheme 15.1 [33, 35, 38], it has been difficult to rigorously demonstrate that early intermediates are obligatory states on a direct path to the native state.

Alternative mechanisms involving formation of nonproductive states (Scheme 15.2) or mechanisms with parallel pathways and nonobligatory intermediates (Scheme 15.3) can be ruled out only if (i) the two transitions are kinetically coupled (i.e., they have similar rates), (ii) both phases are directly observable and kinetically resolved, and (iii) the experimental probe used to monitor folding can discriminate native from intermediate and unfolded populations. Under these circumstances, the transient accumulation of an obligatory intermediate leads to a detectable lag in the appearance of the native population whereas an off-path intermediate gives rise to a rapid increase in the population of N during the initial phase. Clear evidence for such a lag phase was obtained in a stopped-flow fluorescence study on a proline-free variant of staphylococcal nuclease [123].

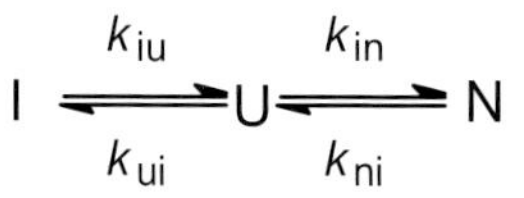

$$I \underset{k_{ui}}{\overset{k_{iu}}{\rightleftharpoons}} U \underset{k_{ni}}{\overset{k_{in}}{\rightleftharpoons}} N$$

Scheme 15.2

$$U \underset{k_{nu}}{\overset{k_{un}}{\rightleftharpoons}} N$$ with I connected via $k_{ui}, k_{iu}, k_{in}, k_{ni}$

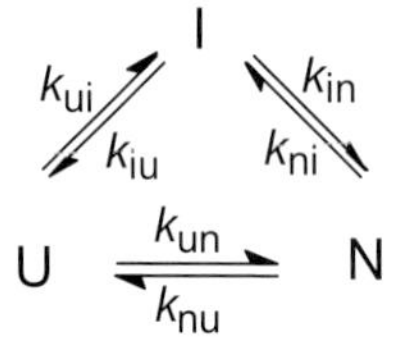

Scheme 15.3

On the other hand, a thorough kinetic analysis, using a double-jump protocol, indicated that a late intermediate during folding of lysozyme is nonobligatory [124]. Other efforts to detect a lag phase during folding of interlukin-1β [125, 126] and apomyoglobin [127] were inconclusive because of the disparate time scales of early and rate-limiting folding events. In order to determine the kinetic role of early folding intermediates, it is necessary to directly measure the population of native molecules on the submillisecond time scale, either by double-jump experiments or via a specific spectroscopic probe for the native state.

A particularly favorable protein for investigating the kinetic importance of early folding intermediates is the bacterial immunity protein Im7, an 86-residue protein with a simple four-helix bundle structure (Figure 15.14a). Radford and colleagues have previously shown that Im7 populates a hyperfluorescent intermediate during

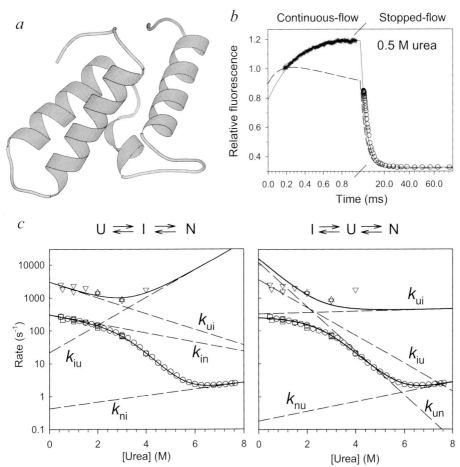

Fig. 15.14. Kinetic mechanism of Im7 folding. a) Ribbon diagram of Im7. b) Representative kinetic trace measured by continuous-flow (●) and stopped-flow (○) fluorescence. The kinetics at this and all other urea concentrations measured is accurately predicted by on on-pathway mechanism (solid line) while schemes with off-pathway intermediates fail to reproduce the data (dashed line). c) Observed (symbols) and predicted (solid lines) rates of folding and unfolding, based on mechanisms with on-pathway (left) and off-pathway (right) intermediates. Dashed lines indicate the corresponding elementary rate constants.

the 2.5-ms dead time of their stopped-flow instrument. However, the data could be fitted equally well to models involving either on- or off-pathway intermediates. Capaldi et al. extended these measurements into the microsecond time range [36], using our continuous-flow mixing instrument to detect the changes in fluorescence of a single tryptophan (Trp75) associated with folding under various conditions (Figure 15.14b). The initial increase in fluorescence above the level of the denatured state in 6 M urea over the 100 μs to 1 ms time range is consistent with the rapid formation of a compact state resulting in a solvent-shielded environment for

Trp75. The native structure is formed within a few milliseconds or longer (depending on the final concentration of urea), which is accompanied by partial quenching of Trp75 fluorescence due to close contact with a histidine. The tight kinetic coupling of the two folding events, together with the distinct fluorescence properties of the three conformational states involved, allowed us to discriminate among various possible kinetic mechanisms. The time course of folding predicted by solving the kinetic equations corresponding to the on-path mechanism (Scheme 15.1) accurately describes the observed behavior (Figure 15.14b, solid line). By contrast, the off-path mechanism (Scheme 15.2) failed to reproduce the data, especially at low urea concentrations where the intermediate is well populated (Figure 15.14b, dashed line). Likewise, the quality of the fits does not improve when a second, parallel pathway is introduced (Scheme 15.3) and deteriorates when more than 25% of the flux bypassed the intermediate.

These findings were further confirmed by a more detailed analysis of the dependence of the two observable rate constants on urea concentration (Figure 15.14c). While Scheme 15.1 reproduces all of the kinetic data, there are major discrepancies in the rate of the fast phase predicted by Schemes 15.2 and 15.3, which can thus be ruled out.

By eliminating alternative mechanisms, we have thus been able to show that accumulation of a compact intermediate on the submillisecond time scale is a productive and obligatory event in folding of Im7 [36]. This supports the notion that rapid formation of compact states can facilitate the search for the native conformation. Interestingly, a recent mutation study revealed that the Im7 folding intermediate contains non-native tertiary interactions among the three major α-helices, A, B and D, while the short helix C forms only during the final stages of folding [128]. Apparently, intermediates can be kinetically productive even if they contain some structural features not found in the final native state.

15.7.4
Broader Implications

Table 15.2 lists the time constants of the earliest folding phase measured by continuous-flow fluorescence for some of the proteins discussed in this chapter along with their size and structural type. The initial folding times vary by an order of magnitude, but show no apparent correlation with protein size. This argues against the notion that the early stages of folding are dominated by a nonspecific hydrophobic polymer collapse, in which case the rate of the initial phase would depend primarily on the size of the hydrophobic core and would be insensitive to structural details.

There appears to be no simple relationship between the time scale of early folding events and secondary structure content. For example, the β-sheet containing proteins, GB1 and SNase, are near opposite ends of the time range, and Im7 has a ~fivefold slower initial rate compared to the structurally similar ACBP. Thus, the individual structural and topological features have a profound effect already during early stages of folding.

The finding that the earliest detectable folding event in GB1 domain is an order

Tab. 15.2. Comparison of the time constants of the initial folding phase with protein size and secondary structure type for proteins with multi-state folding kinetics.

Protein	Residues	Structure type	τ (initial)/μs	Ref.
GB1	57	α/β	600	35
ACBP	86	α	80	38
Im7	87	α	450	36
cyt c	104	α	60	33
SNase (F76W)	149	α/β	75	60

of magnitude slower than that of cyt c may well be related to the fact that GB1 contains β-sheet structure (Figure 15.7a) while cyt c is primarily α-helical (Figure 15.12). With its parallel pairing of β-strands from opposite ends of the chain, GB1 has a relatively high contact order, a metric of fold complexity that was found to correlate with the logarithm of the folding time of proteins with two-state mechanisms [129]. Such a trend is not expected for proteins with multistate folding mechanisms, since the rate-limiting step in structurally and kinetically more complex proteins is likely to involve structural events unrelated to the overall fold, such as side chain packing and specific tertiary interactions. However, if compact early intermediates have a native-like chain topology, one might expect a correlation between the time constant of the initial phase and contact order. Some of the trends seen in Table 15.2 are consistent with this idea. For instance, cyt c and ACBP have low contact order and collapse much faster than GB1, which has the highest contact order among the proteins studied. The surprisingly fast initial phase of SNase (75 μs) in comparison to the much smaller GB1 (600 μs) may be related to the fact that the folding step in SNase detected by Trp76 and ANS fluorescence reflects a relatively local structural event within its antiparallel β-barrel domain, whereas the fluorescence of Trp43 in protein G may report on a more global conformational change involving the parallel pairing of N- and C-terminal β-strands.

Finally, we note that several small proteins (or isolated domains of larger proteins) can complete the process of folding on a comparable time scale as the early stages of folding for some of the multistate proteins discussed here (see, for example, Refs [130–133]; see Ref. [24] for a review). The underlying conformational events may be similar, but the fast-folding two-state proteins can directly reach the native state in a concerted collapse/folding event while most proteins encounter additional steps after crossing the initial barrier.

Appendix

A1 Design and Calibration of Rapid Mixing Instruments

A1.1 Stopped-flow Equipment

In a typical stopped-flow experiment, a few hundred microliters of solution are delivered to the mixer via two syringes driven by a pneumatic actuator or stepper-motors. Total flow rates in the range of 5–10 mL s^{-1} with channel diameters of the order of 1 mm insure turbulent flow conditions ($Re > 5000$). After delivering a volume sufficient to purge and fill the observation cell with freshly mixed solution, the flow is stopped abruptly when a third syringe hits a stopping block, or by closing a valve. Commercial instruments can routinely reach dead times of a few milliseconds. Recent improvements in mixer and flow-cell design by several manufacturers of stopped-flow instruments resulted in dead times well under 1 ms.

The upper end of the time scale that can be reliably measured in a stopped-flow experiment is determined by the stability of the mixture in the flow cell, which is limited by convective flow or diffusion of reagents in and out of the observation volume. For slow reactions with time constants longer than a few minutes, manual mixing experiments are generally more reliable. Stopped-flow mixing is usually coupled with real-time optical observation using absorbance (UV through IR), fluorescence emission or circular dichroism spectroscopy, but other biophysical techniques, including fluorescence lifetime measurements [134, 135], NMR [136, 137], and small-angle X-ray scattering (SAXS; [138]), have also been implemented.

The interpretation of stopped-flow data requires a careful calibration of the instrumental dead time by measuring a pseudo-first-order reaction tuned to the time scale of interest (i.e., a single-exponential process with a rate-constant approaching the expected dead time) and an optical signal matching the application. Common test reactions for absorbance measurements include the reduction of 2,6-dichlorophenolindophenol (DCIP) or ferricyanide by ascorbic acid [66].

A convenient test reaction for tryptophan fluorescence measurements is the irreversible quenching of N-acetyltryptophanamide (NATA) by N-bromosuccinimide (NBS). For fluorescence studies in or near the visible range, one can follow the pH-dependent association of the Mg^{2+} ion with 8-hydroxyquinoline, which results in a fluorescent chelate [139], or the binding of the hydrophobic dye 1-anilino-8-naphthalene-sulfonic acid (ANS) to bovine serum albumin (BSA), which is associated with a large increase in fluorescence yield [140].

As a practical example, Figure 15.15 shows a series of DCIP absorbance measurements at several ascorbic acid concentrations used to estimate the dead time of our Bio-Logic SFM-4 stopped-flow instrument equipped with a FC-08 microcuvette accessory (Molecular Kinetics, Indianapolis, IN, USA). The reaction was started by mixing equal parts of DCIP (0.75 mM in water at neutral pH, where the dye is stable) with L-ascorbic acid at pH 2 at final concentrations ranging from 6 to 25 mM. Absorbance changes measured at 525 nm, near the isosbestic point between the acidic and basic forms of DCIP, are plotted relative to the absorbance of the reactant in water, measured in a separate control (Figure 15.15, inset). After the flow comes to a full stop, the absorbance decays exponentially with a rate constants of about 290, 780, and 1400 s^{-1} at ascorbate concentrations of 6, 14, and 25 mM, respectively, which is consistent with a pseudo-first-order reduction process with a second-order rate constant $k'' \approx 5.7 \times 10^5$ s^{-1} M^{-1}.

At shorter times, the exponential fits intersect at an absorbance $\Delta A = 0$, which corresponds to the level of the oxidized DCIP measured in the control. The delay between this intercept and the first data point that joins the fitted exponential pro-

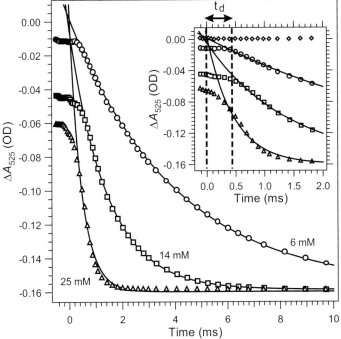

Fig. 15.15. Estimation of the dead time of stopped-flow absorbance measurement on a Biologic SFM-4 instrument with FC-08 microcuvette accessory, using the reduction of DCIP by ascorbic acid as a test reaction. Equal parts of DCIP in water (pH 7) and sodium ascorbate (pH 2) at final concentrations of 6 (circles), 14 (squares) and 25 mM (triangles) were mixed at a total flow rate of 12 mL s^{-1}. The inset shows an expanded plot at early times with dashed lines indicating the dead time (~0.5 ms). Absorbance changes at 525 nm are plotted relative to the absorbance of a control (diamonds) measured by mixing DCIP with 10 mM HCl.

vides an estimate of the instrumental dead time, $t_d = 0.55 \pm 0.1$ ms. Alternatively, the dead time can be estimated from the signal level of the continuous-flow regime prior to closure of the stop valve, I_{cf}, using the following equation:

$$t_{cf} = -\ln[(I_{cf} - I_\infty)/(I_0 - I_\infty)]/k \tag{3}$$

where I_∞ is the baseline at long times, I_0 is the signal of the reactant (in this case, DCIP mixed with an equal volume of water), and k is the first-order rate constant obtained by exponential fitting. Note that t_{cf} does not account for the finite stop time and other stop-related artifacts, and is thus always shorter than t_d, which explains why some manufacturers of stopped-flow instruments prefer to cite t_{cf} over the more realistic operational dead time, t_d. In the present example, Eq. (3) yields t_{cf} in the range of 0.25 and 0.4 ms, compared with $t_d \approx 0.55$ ms obtained by the extrapolation (Figure 15.15).

A1.2 Continuous-flow Instrumentation

In a continuous-flow experiment, the reaction is again triggered by turbulent mixing, but, in contrast to stopped-flow, the progress of the reaction is sampled under steady-state flow conditions as a function of the distance downstream from the mixer [141, 142]. This avoids artifacts related to arresting the flow and makes it possible to use relatively insensitive detection methods. Thus, continuous-flow measurements can achieve shorter dead times compared to stopped-flow, but this comes at the expense of sample economy. Most earlier versions of this experiment involved point-by-point sampling of the reaction profile while maintaining constant flow at high rates (several ml/s for a conventional mixer). The prohibitive amounts of sample consumed limited the impact of continuous-flow techniques until advances in mixer design made it possible to achieve highly efficient mixing at lower flow rates [26, 31, 32, 143], and an improved detection scheme allowed simultaneous recoding of a complete reaction profile in a few seconds [26]. These developments lowered both the dead time and sample consumption by at least an order of magnitude, and made routine measurements on precious samples with dead times as short as 50 µs possible.

In 1985, Regenfuss et al. [31] described a capillary jet mixer consisting of two coaxial glass capillaries with a platinum sphere placed at their junction. The reaction progress was monitored in a free-flowing jet, using conventional photography to measure fluorescence vs. distance from the mixer. Measurements of the binding kinetics of ANS to BSA indicated that dead times less than 100 µs can be achieved with this mixer design. More recently, several laboratories reported continuous-flow resonance Raman and fluorescence studies of enzyme and protein folding reactions on the submillisecond time scale, using machined mixers with dead-times of about 100 µs [25, 32, 143]. More widespread use of these methods has been hampered by a number of technical and experimental difficulties. Continuous-flow experiments involving a free-flowing jet [31, 32, 143, 144] are fraught with difficulties due to instability and scattering artifacts. The use of a conventional camera with high-speed monochrome film for fluorescence detection is inadequate due to the low sensitivity of the film in the UV region, limited dynamic range and the nonlinearity of the film response. Finally, prohibitive sample consumption makes continuous-flow experiments that record a kinetic trace one point at a time feasible only for highly abundant proteins [25, 67, 72, 97].

We were able to overcome many of these limitations by combining a highly efficient quartz capillary mixer, based on the design of Regenfuss et al. [31], with a flow cell and an improved detection system involving a digital camera system with a UV-sensitized CCD detector [26]. A diagram of the experimental arrangement is shown in Figure 15.16. Two Hamilton syringes driven by an Update (Madison, WI, USA) quenched-flow apparatus deliver the reagents to be studied at moderate pressure (< 10 atm) into each of the two coaxial capillaries. The outer capillary consists of a thick-walled (6 mm o.d., 2 mm i.d.) quartz tube, which is pulled to a fine tip ($\sim$200 µm i.d. at the end), using a glassblowing lathe or a simple gravity method. The inner capillary (360 µm o.d., 150–180 µm i.d., purchased from Polymicro Technologies, Phoenix, AZ, USA) with a $\sim$250 µm platinum sphere suspended at the end is positioned inside the tapered end of the outer capillary.

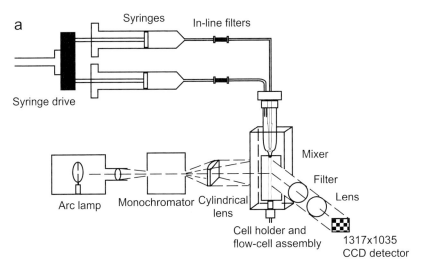

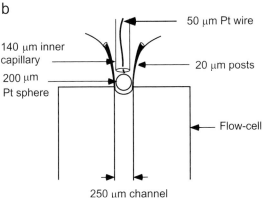

Fig. 15.16. Continuous-flow capillary mixing apparatus in fluorescence mode. a) Schematic diagram of the solution delivery system, mixer, observation cell and optical arrangement. b) Expanded view of the mixer/flow cell assembly.

The sphere is formed by melting the end of 50 µm diameter platinum wire. Thin glass rods fused to the inner wall of the outer capillary (tapering down to diameter of ~20 µm) prevent the sphere from plugging the outlet. The reagents are forced through the narrow gap between the sphere and the outer wall where mixing occurs under highly turbulent flow conditions. The fully homogeneous mixture emerging from the mixer is injected into the 250 µm × 250 µm flow channel of a fused-silica observation cell (Hellma) joined to the outer capillary by means of a hemispherical ground-glass joint. Typical flow rates are 0.6–1.0 mL s^{-1} resulting in a linear flow velocities of 10–16 m s^{-1} through the 0.25 × 0.25 mm^2 channel of the observation cell.

The reaction progress in a continuous-flow mixing experiment is measured by recording the fluorescence profile vs. distance downstream from the mixer (Figure

15.16). A conventional light source consisting of an arc lamp (we currently use a 350 W Hg arc lamp in a lamp housing from Oriel, Stratford, CT, USA), collimating optics and monochromator, is used for fluorescence excitation. Relatively uniform illumination of the flow channel over a length of 10–15 mm is achieved by means of a cylindrical lens. A complete fluorescence vs. distance profile is obtained by imaging the fluorescent light emitted at a 90° angle onto the CCD detector of a digital camera system (Micromax, Roper Scientific, Princeton, NJ, USA) containing a UV-coated Kodak CCD chip with an array of 1317 × 1035 pixels. The camera is equipped with a fused silica magnifying lens and a high-pass glass filter or a band-pass interference filter to suppress scattered incident light.

Figure 15.17 illustrates a typical continuous-flow experiment, using the quenching of NATA by NBS as a test reaction. The upper panels show raw data obtained by averaging the intensity across the flow channel vs. the distance d downstream from the mixer. Panel a shows the intensity profile for the quenching reaction, $I_e(d)$, at final NATA and NBS concentrations of 40 μM and 4 mM, respectively (mixing 1 part of 440 μM NATA in water with 10 parts of 4.4 mM NBS in water). Panel b shows the distribution of incident light intensity, $I_c(d)$, measured by mixing the same NATA solution with water. Panel c shows the scattering background, $I_b(d)$, measured by passing water through both capillaries. Panel d shows the corrected kinetic trace, $fl_{rel}(t)$, obtained according to

$$fl_{rel} = (I_e - I_b)/(I_c - I_b) \qquad (4)$$

Distance was converted into time on the basis of the known flow rate (0.8 mL s^{-1} in this example), the cross-sectional area of the flow channel (0.0625 mm^2) and the length of the channel being imaged (12.5 mm, or 9.5 μm per pixel). The signal measured at points below the entrance to the flow channel are well fitted by a single-exponential decay. The trace at $fl_{rel} = 1$ in panel d of Figure 15.17 represents a control for the mixing efficiency measured as follows. The 440 μM NATA stock solution used in the experiment above was diluted 11-fold with water and filled into both syringes. The continuous-flow trace recorded at the same flow rate was then background-corrected and normalized according to Eq. (3), using the I_c trace recorded previously in which the same NATA solution was diluted 11-fold in the capillary mixer. Immediately after entering the flow channel, the ratio approaches unity, indicating that NATA is completely mixed with water. Alternatively, mixing efficiency could be assessed using a very fast reaction completed within the dead time, such as the quenching of tryptophan (or NATA) by sodium iodide.

To estimate the dead time of the continuous-flow experiment, we measured the pseudo-first-order NATA-NBS quenching reaction at a final NATA concentration of 40 μM and several NBS concentrations in the range 2–32 mM. Figure 15.18a shows a semilogarithmic plot of the relative fluorescence, fl_{rel}, calculated according to Eq. (3), along with exponential fits (solid lines). Since $fl_{rel} = 1$ corresponds to the unquenched NATA signal expected at $t = 0$, the intercept of the fits with $fl_{rel} = 1$ indicates the time point $t = 0$ where the mixing reaction begins. The delay between this point and the first data point that falls onto the exponential fit corresponds to

Appendix | 527

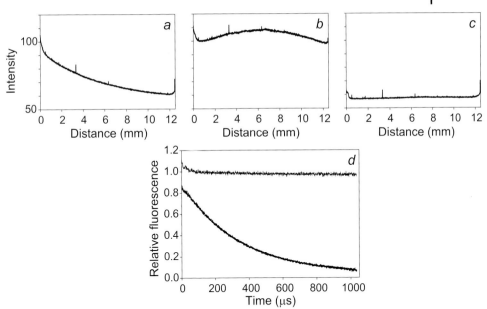

Fig. 15.17. A typical continuous-flow mixing experiment. The upper panels show raw intensity profiles vs. distance from the mixing region. a) Raw kinetic trace (NATA mixed with NBS). b) Fluorescence control (NATA mixed with water). c) Scattering background (water deliverd from both syringes). Panel d shows the corrected kinetic trace calculated according to Eq. (3).

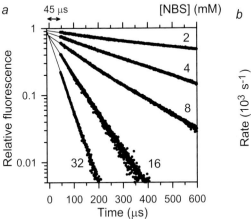 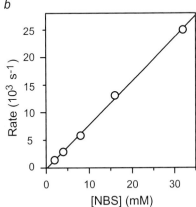

Fig. 15.18. Continuous-flow measurements of the quenching of NATA fluorescence by NBS used to determine the experimental dead-time. a) Semi-logarithmic plot of NATA fluorescence (> 324 nm) vs. time at several NBS concentrations. b) NATA-NBS reaction rates from exponential fitting of the data in panel a vs. NBS concentration. Linear regression (line) yields a second-order rate constant of 7.9 10^5 M^{-1} s^{-1}.

the dead time of the experiment, Δt_d, which in this example is 45 ± 5 μs. In Figure 15.18b, the rate constants obtained by exponential fitting are plotted as a function of NBS concentration. The slope of a linear fit (solid line) yields a second-order rate constant for the NBS-induced chemical quenching of NATA of 7.9×10^5 M^{-1} s^{-1}, which is consistent with data obtained by stopped-flow measurements at lower NBS concentration [51]. This agreement, together with the linearity of the second-order rate plot (Figure 15.18b), demonstrates the accuracy of the kinetic data.

Acknowledgments

This work was supported by grants GM56250 and CA06927 from the National Institutes of Health, grant MCB-079148 from the National Science Foundation, and an Appropriation from the Commonwealth of Pennsylvania.

References

1 JACKSON, S. E. (1998). How do small single-domain proteins fold? *Fold. Des.* 3, R81–R91.
2 MATTHEWS, C. R. (1987). Effect of point mutations on the folding of globular proteins. *Meth. Enzymol.* 154, 498–511.
3 MATOUSCHEK, A., KELLIS, J. T., JR., SERRANO, L. & FERSHT, A. R. (1989). Mapping the transition state and pathway of protein folding by protein engineering. *Nature* 340, 122–126.
4 JACKSON, S. E., ELMASRY, N. & FERSHT, A. R. (1993). Structure of the hydrophobic core in the transition state for folding of the chymotrypsin inhibitor-2: A critical test of the protein engineering method of analysis. *Biochemistry* 32, 11270–11278.
5 MATTHEWS, C. R. (1993). Pathways of protein folding. *Annu. Rev. Biochem.* 62, 653–683.
6 EVANS, P. A. & RADFORD, S. E. (1994). Probing the structure of folding intermediates. *Curr. Opin. Struct. Biol.* 4, 100–106.
7 RODER, H. & COLÓN, W. (1997). Kinetic role of early intermediates in protein folding. *Curr. Opin. Struct. Biol.* 7, 15–28.
8 BALDWIN, R. L. & ROSE, G. D. (1999). Is protein folding hierarchic? II. Folding intermediates and transition states. *Trends Biochem. Sci.* 24, 77–83.
9 RODER, H., ELÖVE, G. A. & SHASTRY, R. M. C. (2000). Early stages of protein folding. In *Mechanisms of protein folding* (PAIN, R. H., ed.), pp. 65–104. Oxford University Press, New York.
10 TANFORD, C., AUNE, K. C. & IKAI, A. (1973). Kinetics of unfolding and refolding of proteins III. Results for lysozyme. *J. Mol. Biol.* 73, 185–197.
11 KIM, P. S. & BALDWIN, R. L. (1982). Specific intermediates in the folding reactions of small proteins and the mechanism of protein folding. *Annu. Rev. Biochem.* 51, 459–489.
12 RODER, H. & SHASTRY, M. C. R. (1999). Methods for exploring early events in protein folding. *Curr. Opin. Struct. Biol.* 9, 620–626.
13 RODER, H. & WÜTHRICH, K. (1986). Protein folding kinetics by combined use of rapid mixing techniques and NMR observation of individual amide protons. *Proteins* 1, 34–42.
14 RODER, H., ELÖVE, G. A. & ENGLANDER, S. W. (1988). Structural characterization of folding intermediates in cytochrome c by H-exchange labelling and proton NMR. *Nature* 335, 700–704.
15 UDGAONKAR, J. B. & BALDWIN, R. L. (1988). NMR evidence for an early

framework intermediate on the folding pathway of ribonuclease A. *Nature* 335, 694–699.
16. BRYNGELSON, J. D., ONUCHIC, J. N., SOCCI, N. D. & WOLYNES, P. G. (1995). Funnels, pathways, and the energy landscape of protein folding: A synthesis. *Proteins* 21, 167–195.
17. DOBSON, C. M. & KARPLUS, M. (1999). The fundamentals of protein folding: bringing together theory and experiment. *Curr. Opin. Struct. Biol.* 9, 92–101.
18. ONUCHIC, J. N. & WOLYNES, P. G. (2004). Theory of protein folding. *Curr. Opin. Struct. Biol.* 14, 70–75.
19. JONES, C. M., HENRY, E. R., HU, Y. et al. (1993). Fast events in protein folding initiated by nanosecond laser photolysis. *Proc. Natl Acad. Sci. USA* 90, 11860–11864.
20. CALLENDER, R. H., DYER, R. B., GILMANSHIN, R. & WOODRUFF, W. H. (1998). Fast events in protein folding: the time evolution of primary processes. *Annu. Rev. Phys. Chem.* 49, 173–202.
21. GRUEBELE, M. (1999). The fast protein folding problem. *Annu. Rev. Phys. Chem.* 50, 485–516.
22. EATON, W. A., MUNOZ, V., HAGEN, S. J. et al. (2000). Fast kinetics and mechanisms in protein folding. *Annu. Rev. Biophys. Biomol. Struct.* 29, 327–359.
23. HUANG, C. Y., HE, S., DEGRADO, W. F., MCCAFFERTY, D. G. & GAI, F. (2002). Light-induced helix formation. *J. Am. Chem. Soc.* 124, 12674–12675.
24. KUBELKA, J., HOFRICHTER, J. & EATON, W. A. (2004). The protein folding 'speed limit'. *Curr Opin Struct Biol* 14, 76–88.
25. TAKAHASHI, S., YEH, S.-R., DAS, T. K., CHAN, C.-K., GOTTFRIED, D. S. & ROUSSEAU, D. L. (1997). Folding of cytochrome *c* initiated by submillisecond mixing. *Nature Struct. Biol.* 4, 44–50.
26. SHASTRY, M. C. R., LUCK, S. D. & RODER, H. (1998). A continuous-flow capillary mixer to monitor reactions on the microsecond time scale. *Biophys. J.* 74, 2714–2721.
27. POLLACK, L., TATE, M. W., DARNTON, N. C. et al. (1999). Compactness of the denatured state of a fast-folding protein measured by submillisecond small-angle x-ray scattering. *Proc. Natl Acad. Sci. USA* 96, 10115–10117.
28. HARTRIDGE, H. & ROUGHTON, F. J. W. (1923). The velocity with which carbon monoxide displaces oxygen from combination with heamoglobin. I. *Proc. R. Soc. (Lond.) B Biol. Sci.* 94, 336–367.
29. GIBSON, Q. H. & MILNES, L. (1964). *Biochem. J.* 91, 161.
30. CHANCE, B. (1964). *Rapid Mixing and Sampling Techniques in Biochemistry.* (CHANCE, B., EISENHARDT, R. H., GIBSON, Q. H. & LONBERG-HOLM, K. K., Eds), Academic Press, New York.
31. REGENFUSS, P., CLEGG, R. M., FULWYLER, M. J., BARRANTES, F. J. & JOVIN, T. M. (1985). Mixing liquids in microseconds. *Rev. Sci. Instrum.* 56, 283–290.
32. CHAN, C.-K., HU, Y., TAKAHASHI, S., ROUSSEAU, D. L., EATON, W. A. & HOFRICHTER, J. (1997). Submillisecond protein folding kinetics studied by ultrarapid mixing. *Proc. Natl Acad. Sci. USA* 94, 1779–1784.
33. SHASTRY, M. C. R. & RODER, H. (1998). Evidence for barrier-limited protein folding kinetics on the microsecond time scale. *Nature Struct. Biol.* 5, 385–392.
34. SHASTRY, M. C. R., SAUDER, J. M. & RODER, H. (1998). Kinetic and structural analysis of submillisecond folding events in cytochrome *c*. *Acc. Chem. Res.* 31, 717–725.
35. PARK, S.-H., SHASTRY, M. C. R. & RODER, H. (1999). Folding dynamics of the B1 domain of protein G explored by ultrarapid mixing. *Nature Struct. Biol.* 6, 943–947.
36. CAPALDI, A. P., SHASTRY, R. M. C., KLEANTHOUS, C., RODER, H. & RADFORD, S. E. (2001). Ultra-rapid mxing experiments reveal that Im7 folds *via* an on-pathway intermediate. *Nature Struct. Biol.* 8, 68–72.
37. KUWATA, K., SHASTRY, R., CHENG, H. et al. (2001). Structural and kinetic

characterization of early folding events in beta-lactoglobulin. *Nature Struct. Biol.* 8, 151–155.

38 TEILUM, K., MAKI, K., KRAGELUND, B. B., POULSEN, F. M. & RODER, H. (2002). Early kinetic intermediate in the folding of acyl-CoA binding protein detected by fluorescence labeling and ultrarapid mixing. *Proc. Natl Acad. Sci. USA* 99, 9807–9812.

39 GIANNI, S., TRAVAGLINI-ALLOCATELLI, C., CUTRUZZOLA, F., BRUNORI, M., SHASTRY, M. C. & RODER, H. (2003). Parallel pathways in cytochrome c(551) folding. *J. Mol. Biol.* 330, 1145–1152.

40 KHAN, F., CHUANG, J. I., GIANNI, S. & FERSHT, A. R. (2003). The kinetic pathway of folding of barnase. *J. Mol. Biol.* 333, 169–186.

41 JOHNSON, K. A. (1995). Rapid quench kinetic analysis of polymerases, adenosinetriphosphatases, and enzyme intermediates. *Meth. Enzymol.* 249, 38–61.

42 KNIGHT, J. B., VISHWANATH, A., BRODY, J. P. & AUSTIN, R. H. (1998). Hydrodynamic focusing on a silicon chip: Mixing nanoliters in microseconds. *Phys. Rev. Lett.* 80, 3863–3866.

43 PABIT, S. A. & HAGEN, S. J. (2002). Laminar-flow fluid mixer for fast fluorescence kinetics studies. *Biophys. J.* 83, 2872–2878.

44 KUWAJIMA, K., YAMAYA, H., MIWA, S., SUGAI, S. & NAGAMURA, T. (1987). Rapid formation of secondary structure framework in protein folding studied by stopped-flow circular dichroism. *FEBS Lett.* 221, 115–118.

45 ELÖVE, G. A., CHAFFOTTE, A. F., RODER, H. & GOLDBERG, M. E. (1992). Early steps in cytochrome c folding probed by time-resolved circular dichroism and fluorescence spectroscopy. *Biochemistry* 31, 6876–6883.

46 JENNINGS, P. A. & WRIGHT, P. E. (1993). Formation of a molten globule intermediate early in the kinetic folding pathway of apomyoglobin. *Science* 262, 892–895.

47 JONES, B. E. & MATTHEWS, C. R. (1995). Early intermediates in the folding of dihydrofolate reductase from *Escherichia coli* detected by hydrogen exchange and NMR. *Protein Sci.* 4, 167–177.

48 KHORASANIZADEH, S., PETERS, I. D. & RODER, H. (1996). Evidence for a three-state model of protein folding from kinetic analysis of ubiquitin variants with altered core residues. *Nature Struct. Biol.* 3, 193–205.

49 WALKENHORST, W. F., EDWARDS, J. A., MARKLEY, J. L. & RODER, H. (2002). Early formation of a beta hairpin during folding of staphylococcal nuclease H124L as detected by pulsed hydrogen exchange. *Protein Sci* 11, 82–91.

50 COLÓN, W., ELÖVE, G. A., WAKEM, L. P., SHERMAN, F. & RODER, H. (1996). Side chain packing of the N- and C-terminal helices plays a critical role in the kinetics of cytochrome c folding. *Biochemistry* 35, 5538–5549.

51 PETERMAN, B. F. (1979). Measurement of the dead time of a fluorescence stopped-flow instrument. *Anal. Biochem.* 93, 442–444.

52 BERGER, R. L. & CHAPMAN, H. F. (1968). High resolution mixer for the study of the kinetics of rapid reactions in solution. *Rev. Sci. Instrum.* 39, 493–498.

53 ZEFF, B. W., LANTERMAN, D. D., MCALLISTER, R., ROY, R., KOSTELICH, E. J. & LATHROP, D. P. (2003). Measuring intense rotation and dissipation in turbulent flows. *Nature* 421, 146–149.

54 CHEN, Y. & BARKLEY, M. D. (1998). Toward understanding tryptophan fluorescence in proteins. *Biochemistry* 37, 9976–9982.

55 VAN GILST, M., TANG, C., ROTH, A. & HUDSON, B. (1994). Quenching interactions and nonexponential decay: Tryptophan 138 of bacteriophage T4 lysozyme. *Fluorescence* 4, 203–207.

56 CLARK, P. L., WESTON, B. F. & GIERASCH, L. M. (1998). Probing the folding pathway of a beta-clam protein with single-tryptophan constructs. *Fold. Des.* 3, 401–412.

57. Shao, X. & Matthews, C. R. (1998). Single-tryptophan mutants of monomeric tryptophan repressor: optical spectroscopy reveals nonnative structure in a model for an early folding intermediate. *Biochemistry* 37, 7850–7858.
58. Khorasanizadeh, S., Peters, I. D., Butt, T. R. & Roder, H. (1993). Folding and stability of a tryptophan-containing mutant of ubiquitin. *Biochemistry* 32, 7054–7063.
59. Eftink, M. R. & Shastry, M. C. R. (1997). Fluorescence methods for studying kinetics of protein-folding reactions. *Meth. Enzymol.* 278, 258–286.
60. Maki, K., Cheng, H., Dolgikh, D. A., Shastry, M. C. & Roder, H. (2004). Early events during folding of wild-type staphylococcal nuclease and a single-tryptophan variant studied by ultrarapid mixing. *J. Mol. Biol.* 338, 383–400.
61. Stryer, L. (1968). Fluorescence spectroscopy of proteins. *Science* 162, 526–533.
62. Semisotnov, G. V., Rodionova, N. A., Kutyshenko, V. P., Ebert, B., Blanck, J. & Ptitsyn, O. B. (1987). Sequential mechanism of refolding of carbonic anhydrase B. *FEBS Lett.* 224, 9.
63. Engelhard, M. & Evans, P. A. (1995). Kinetics of interaction of partially folded proteins with a hydrophobic dye: evidence that molten globule character is maximal in early folding intermediates. *Protein Sci.* 4, 1553–1562.
64. Tsong, T. Y. (1974). The Trp-59 fluorescence of ferricytochrome c as a sensitive measure of the over-all protein conformation. *J. Biol. Chem.* 249, 1988–1990.
65. Colón, W. & Roder, H. (1996). Kinetic intermediates in the formation of the cytochrome *c* molten globule. *Nature Struct. Biol.* 3, 1019–1025.
66. Tonomura, B., Nakatani, H., Ohnishi, M., Yamaguchi-Ito, J. & Hiromi, K. (1978). Test reactions for a stopped-flow apparatus. Reduction of 2,6-dichlorophenolindophenol and potassium ferricyanide by L-ascorbic acid. *Anal. Biochem.* 84, 370–383.
67. Akiyama, S., Takahashi, S., Ishimori, K. & Morishima, I. (2000). Stepwise formation of alpha-helices during cytochrome c folding. *Nature Struct. Biol.* 7, 514–520.
68. Uzawa, T., Akiyama, S., Kimura, T. et al. (2004). Collapse and search dynamics of apomyoglobin folding revealed by submillisecond observations of α-helical content and compactness. *Proc. Natl Acad. Sci. USA* 101, 1171–1176.
69. Grigoryants, V. M., Veselov, A. V. & Scholes, C. P. (2000). Variable velocity liquid flow EPR applied to submillisecond protein folding. *Biophys. J.* 78, 2702–2708.
70. DeWeerd, K., Grigoryants, V., Sun, Y., Fetrow, J. S. & Scholes, C. P. (2001). EPR-detected folding kinetics of externally located cysteine-directed spin-labeled mutants of iso-1-cytochrome c. *Biochemistry* 40, 15846–15855.
71. Pollack, L., Tate, M. W., Finnefrock, A. C. et al. (2001). Time resolved collapse of a folding protein observed with small angle x-ray scattering. *Phys. Rev. Lett.* 86, 4962–4965.
72. Akiyama, S., Takahashi, S., Kimura, T. et al. (2002). Conformational landscape of cytochrome c folding studied by microsecond-resolved small-angle x-ray scattering. *Proc. Natl Acad. Sci. USA* 99, 1329–1334.
73. Elöve, G. A., Bhuyan, A. K. & Roder, H. (1994). Kinetic mechanism of cytochrome c folding: involvement of the heme and its ligands. *Biochemistry* 33, 6925–6935.
74. Sosnick, T. R., Mayne, L., Hiller, R. & Englander, S. W. (1994). The barriers in protein folding. *Nature Struct. Biol.* 1, 149–156.
75. Colón, W., Wakem, L. P., Sherman, F. & Roder, H. (1997). Identification of the predominant non-native histidine ligand in unfolded cytochrome *c*. *Biochemistry* 36, 12535–12541.
76. Kuwajima, K. (1996). Stopped-flow

circular dichroism. In *Circular Dichroism and the Conformational Analysis of Biomolecules* (Fasman, G. D., Ed.). Plenum Press, New York.
77 Hagen, S. J. & Eaton, W. A. (2000). Two-state expansion and collapse of a polypeptide. *J. Mol. Biol.* 297, 781–789.
78 Gladwin, S. T. & Evans, P. A. (1996). Structure of very early protein folding intermediates: new insights through a variant of hydrogen exchange labelling. *Fold. Des.* 1, 407–417.
79 Sauder, J. M. & Roder, H. (1998). Amide protection in an early folding intermediate of cytochrome c. *Fold. Des.* 3, 293–301.
80 Bai, Y., Milne, J. S. & Englander, S. W. (1993). Primary structure effects on peptide group exchange. *Proteins* 17, 75–86.
81 Clarke, A. R. & Waltho, J. P. (1997). Protein folding pathways and intermediates. *Curr. Opin. Biotechnol.* 8, 400–410.
82 Alexander, P., Orban, J. & Bryan, P. (1992). Kinetic analysis of folding and unfolding the 56 amino acid IgG-binding domain of streptococcal protein G. *Biochemistry* 31, 7243–7248.
83 Park, S.-H., O'Neil, K. T. & Roder, H. (1997). An early intermediate in the folding reaction of the B1 domain of protein G contains a native-like core. *Biochemistry* 36, 14277–14283.
84 Benson, S. W. (1960). The foundations of chemical kinetics. In *Advanced Topics in Chemistry*, 1st edn, p. 725. McGraw-Hill, New York.
85 Pogliani, L. & Terenzi, M. (1992). Matrix formulation of chemical reaction rates. *J. Chem. Educ.* 69, 278–280.
86 Sanchez, I. E. & Kiefhaber, T. (2003). Evidence for sequential barriers and obligatory intermediates in apparent two-state protein folding. *J. Mol. Biol.* 325, 367–376.
87 Hvidt, A. & Nielsen, S. O. (1966). Hydrogen exchange in proteins. *Adv. Protein Sci.* 21, 287–386.
88 Tanford, C. (1968). Protein denaturation. *Adv. Protein Chem.* 23, 121–282.
89 Krantz, B. A., Mayne, L., Rumbley, J., Englander, S. W. & Sosnick, T. R. (2002). Fast and slow intermediate accumulation and the initial barrier mechanism in protein folding. *J. Mol. Biol.* 324, 359–371.
90 Krantz, B. A. & Sosnick, T. R. (2000). Distinguishing between two-state and three-state models for ubiquitin folding. *Biochemistry* 39, 11696–11701.
91 Laub, P. B., Khorasanizadeh, S. & Roder, H. (1995). Localized solution structure refinement of an F45W variant of ubiquitin using stochastic boundary molecular dynamics and NMR distance restraints. *Protein Sci.* 4, 973–982.
92 Went, H. M., Benitez-Cardoza, C. G. & Jackson, S. E. (2004). Is an intermediate state populated on the folding pathway of ubiquitin? *FEBS Lett.* 567, 333–338.
93 Pascher, T., Chesick, J. P., Winkler, J. R. & Gray, H. B. (1996). Protein folding triggered by electron transfer. *Science* 271, 1558–1560.
94 Brems, D. N. & Stellwagen, E. (1983). Manipulation of the observed kinetic phases in the refolding of denatured ferricytochromes c. *J. Biol. Chem.* 258, 3655–3660.
95 Muthukrishnan, K. & Nall, B. T. (1991). Effective concentrations of amino acid side chains in an unfolded protein. *Biochemistry* 30, 4706–4710.
96 Hammack, B., Godbole, S. & Bowler, B. E. (1998). Cytochrome c folding traps are not due solely to histidine-heme ligation: Direct demonstration of a role for N-terminal amino group-heme ligation. *J. Mol. Biol.* 275, 719–724.
97 Yeh, S.-R., Takahashi, S., Fan, B. & Rousseau, D. L. (1997). Ligand exchange during cytochrome c folding. *Nature Struct. Biol.* 4, 51–56.
98 Hagen, S. J., Hofrichter, J., Szabo, A. & Eaton, W. A. (1996). Diffusion-limited contact formation in unfolded cytochrome c: Estimating the maximum rate of protein folding. *Proc. Natl Acad. Sci. USA* 93, 11615–11617.
99 Qiu, L., Zachariah, C. & Hagen, S. J. (2003). Fast chain contraction during protein folding: "foldability"

and collapse dynamics. *Phys Rev Lett* 90, 168103.
100 DILL, K. A., BROMBERG, S., YUE, K. et al. (1995). Principles of protein folding – a perspective from simple exact models. *Protein Sci.* 4, 561–602.
101 NYMEYER, H., GARCIA, A. E. & ONUCHIC, J. N. (1998). Folding funnels and frustration in off-lattice minimalist protein landscapes. *Proc. Natl Acad. Sci. USA* 95, 5921–5928.
102 SABELKO, J., ERVIN, J. & GRUEBELE, M. (1999). Observation of strange kinetics in protein folding. *Proc. Natl Acad. Sci. USA* 96, 6031–6036.
103 HAGEN, S. J. (2003). Exponential decay kinetics in "downhill" protein folding. *Proteins* 50, 1–4.
104 SHORTLE, D. (2002). The expanded denatured state: an ensemble of conformations trapped in a locally encoded topological space. *Adv. Protein Chem.* 62, 1–23.
105 TANFORD, C., KAWAHARA, K. & LAPANJE, S. (1967). Proteins as random coils. I. Intrinsic viscosities and sedimentation coefficients in concentrated guanidine hydrochloride. *J. Am. Chem. Soc.* 89, 729–749.
106 SOSNICK, T. R., MAYNE, L. & ENGLANDER, S. W. (1996). Molecular collapse: the rate-limiting step in two-state cytochrome c folding. *Proteins* 24, 413–426.
107 SOSNICK, T. R., SHTILERMAN, M. D., MAYNE, L. & ENGLANDER, S. W. (1997). Ultrafast signals in protein folding and the polypeptide contracted state. *Proc. Natl Acad. Sci. USA* 94, 8545–8550.
108 QI, P. X., SOSNICK, T. R. & ENGLANDER, S. W. (1998). The burst phase in ribonuclease A folding and solvent dependence of the unfolded state. *Nature Struct. Biol.* 5, 882–884.
109 CHOE, S. E., MATSUDAIRA, P. T., OSTERHOUT, J., WAGNER, G. & SHAKHNOVICH, E. I. (1998). Folding kinetics of villin 14T, a protein domain with a central beta-sheet and two hydrophobic cores. *Biochemistry* 37, 14508–14518.
110 FERGUSON, N., CAPALDI, A. P., JAMES, R., KLEANTHOUS, C. & RADFORD, S. E. (1999). Rapid folding with and without populated intermediates in the homologous four-helix proteins Im7 and Im9. *J. Mol. Biol.* 286, 1597–1608.
111 GORSKI, S. A., CAPALDI, A. P., KLEANTHOUS, C. & RADFORD, S. E. (2001). Acidic conditions stabilise intermediates populated during the folding of Im7 and Im9. *J. Mol. Biol.* 312, 849–863.
112 IVANKOV, D. N., GARBUZYNSKIY, S. O., ALM, E., PLAXCO, K. W., BAKER, D. & FINKELSTEIN, A. V. (2003). Contact order revisited: influence of protein size on the folding rate. *Protein Sci* 12, 2057–2062.
113 HUANG, G. S. & OAS, T. G. (1995). Submillisecond folding of monomeric λ repressor. *Proc. Natl Acad. Sci. USA* 92, 6878–6882.
114 BURTON, R. E., HUANG, G. S., DAUGHERTY, M. A., FULLBRIGHT, P. W. & OAS, T. G. (1996). Micro-second protein folding through a compact transition state. *J. Mol. Biol.* 263, 311–322.
115 WITTUNG-STAFSHEDE, P., LEE, J. C., WINKLER, J. R. & GRAY, H. B. (1999). Cytochrome b562 folding triggered by electron transfer: approaching the speed limit for formation of a four-helix-bundle protein. *Proc. Natl Acad. Sci. USA* 96, 6587–6590.
116 JACOB, M., HOLTERMANN, G., PERL, D. et al. (1999). Microsecond folding of the cold shock protein measured by a pressure-jump technique. *Biochemistry* 38, 2882–2891.
117 CREIGHTON, T. E. (1994). The energetic ups and downs of protein folding. *Nature Struct. Biol.* 1, 135–138.
118 FERSHT, A. R. (1995). Optimization of rates of protein folding: the nucleation-condensation mechanism and its implications. *Proc. Natl Acad. Sci. USA* 92, 10869–10873.
119 RODER, H. & ELÖVE, G. A. (1994). Early stages of protein folding. In *Mechanisms of Protein Folding: Frontiers in Molecular Biology* (PAIN, R. H., Ed.), pp. 26–55. Oxford University Press, New York.
120 SILOW, M. & OLIVEBERG, M. (1997). Transient aggregates in protein folding are easily mistaken for folding

intermediates. *Proc. Natl Acad. Sci. USA* 94, 6084–6086.
121 MATOUSCHEK, A., SERRANO, L. & FERSHT, A. R. (1994). Analysis of protein folding by protein engineering. In *Mechanisms of Protein Folding: Frontiers in Molecular Biology* (PAIN, R. H., Ed.), pp. 137–159. Oxford University Press, Oxford.
122 RASCHKE, T. M., KHO, J. & MARQUSEE, S. (1999). Confirmation of the hierarchical folding of RNase H: a protein engineering study. *Nature Struct. Biol.* 6, 825–831.
123 WALKENHORST, W. F., GREEN, S. M. & RODER, H. (1997). Kinetic evidence for folding and unfolding intermediates in Staphylococcal nuclease. *Biochemistry* 63, 5795–5805.
124 KIEFHABER, T. (1995). Kinetic traps in lysozyme folding. *Proc. Natl Acad. Sci. USA* 92, 9029–9033.
125 HEIDARY, D. K., GROSS, L. A., ROY, M. & JENNINGS, P. A. (1997). Evidence for an obligatory intermediate in the folding of interleukin-1β. *Nature Struct. Biol.* 4, 725–731.
126 JENNINGS, P., ROY, M., HEIDARY, D. & GROSS, L. (1998). Folding pathway of interleukin-1 beta. *Nature Struct. Biol.* 5, 11.
127 TSUI, V., GARCIA, C., CAVAGNERO, S., SIUZDAK, G., DYSON, H. J. & WRIGHT, P. E. (1999). Quench-flow experiments combined with mass spectrometry show apomyoglobin folds through and obligatory intermediate. *Protein Sci.* 8, 45–49.
128 CAPALDI, A. P., KLEANTHOUS, C. & RADFORD, S. E. (2002). Im7 folding mechanism: misfolding on a path to the native state. *Nature Struct. Biol.* 9, 209–216.
129 PLAXCO, K. W., SIMONS, K. T. & BAKER, D. (1998). Contact order, transition state placement and the refolding rates of single domain proteins. *J. Mol. Biol.* 277, 985–994.
130 MYERS, J. K. & OAS, T. G. (2001). Preorganized secondary structure as an important determinant of fast protein folding. *Nature Struct. Biol.* 8, 552–558.
131 QIU, L., PABIT, S. A., ROITBERG, A. E. & HAGEN, S. J. (2002). Smaller and faster: the 20-residue Trp-cage protein folds in 4 micros. *J. Am. Chem. Soc.* 124, 12952–12953.
132 MAYOR, U., GUYDOSH, N. R., JOHNSON, C. M. et al. (2003). The complete folding pathway of a protein from nanoseconds to microseconds. *Nature* 421, 863–867.
133 ZHU, Y., ALONSO, D. O., MAKI, K. et al. (2003). Ultrafast folding of alpha3D: a de novo designed three-helix bundle protein. *Proc. Natl Acad. Sci. USA* 100, 15486–15491.
134 BEECHEM, J. M., JAMES, L. & BRAND, L. (1990). Time-resolved fluorescence studies of the protein folding process: new instrumentation, analysis and experimental approaches in time-resolved laser spectroscopy in biochemistry. *SPIE Proc.* 1204, 686–698.
135 JONES, B. E., BEECHEM, J. M. & MATTHEWS, C. R. (1995). Local and global dynamics during the folding of *Escherichia coli* dihydrofolate reductase by time-resolved fluorescence spectroscopy. *Biochemistry* 34, 1867–1877.
136 BALBACH, J., FORGE, V., VAN NULAND, N. A. J., WINDER, S. L., HORE, P. J. & DOBSON, C. M. (1995). Following protein folding in real time using NMR spectroscopy. *Nature Struct. Biol.* 2, 865–870.
137 FRIEDEN, C. (2003). The kinetics of side chain stabilization during protein folding. *Biochemistry* 42, 12439–12446.
138 SEGEL, D. J., BACHMANN, A., HOFRICHTER, J., HODGSON, K. O., DONIACH, S. & KIEFHABER, T. (1999). Characterization of transient intermediates in Lysozyme folding with time resolved small-angle X-ray scattering. *J. Mol. Biol.* 288, 489–499.
139 BRISSETTE, P., BALLOU, D. P. & MASSEY, V. (1989). Determination of the dead time of a stopped-flow fluorometer. *Anal. Biochem.* 181, 234–238.
140 GIBSON, Q. H. & ANTONINI, E. (1966). Kinetics of heme–protein intractions. In *Hemes and Hemeproteins* (CHANCE, B., ESTABROOK, R. W. & YONETANI, T.,

Eds), pp. 67–78. Academic Press, New York.
141 HARTRIDGE, H. & ROUGHTON, F. J. W. (1923). Method of measuring the velocity of very rapid chemical reactions. *Proc. R. Soc. (Lond.): A Math. Phys. Sci.* 104, 376–394.
142 GUTFREUND, H. (1969). *Meth. Enzymol.* 16, 229–249.
143 TAKAHASHI, S., CHING, Y.-c., WANG, J. & ROUSSEAU, D. L. (1995). Microsecond generation of oxygen-bound cytochrome c oxidase by rapid solution mixing. *J. Biol. Chem.* 270, 8405–8407.
144 PAENG, K., PAENG, I. & KINCAID, J. (1994). Time-resolved resonance raman spectroscopy using a fast mixing device. *Anal. Sci.* 10, 157–159.
145 GALLAGHER, T., ALEXANDER, P., BRYAN, P. & GILLILAND, G. L. (1994). Two crystal structures of the B1 immunoglobulin-binding domain of streptococcal protein G and comparison with NMR. *Biochemistry* 33, 4721–4729.
146 BUSHNELL, G. W., LOUIE, G. V. & BRAYER, G. D. (1990). High-resolution three-dimensional structure of horse heart cytochrome c. *J. Mol. Biol.* 214, 585–595.

16
Kinetic Protein Folding Studies using NMR Spectroscopy

Markus Zeeb and Jochen Balbach

16.1
Introduction

The mechanism by which polypeptides fold to their native conformation remains an area of active research in structural biology. Our understanding of the protein folding reaction and the deduction of protein folding models has been strongly influenced by the data accessible at the time when they were formulated. During all stages of progression in our understanding of protein folding, nuclear magnetic resonance (NMR) spectroscopy played an important role. NMR combines high spatial resolution with dynamic and kinetic analyses on an enormous time scale ranging from picoseconds to days (Figure 16.1). Therefore, NMR has emerged as an especially fruitful technique to study the protein folding reaction on molecular grounds.

The first milestones were certainly the use of NMR to define the distributions of deuterons at labile sites of proteins. Site-specific H/D exchange experiments [1, 2] and the calculation of protection factors revealed the thermodynamics of local and global unfolding of the native state as well as protein folding intermediates [3]. These states show distinct protection factors at different denaturant concentrations [4]. The determination of proton occupancies on a residue level by NMR following quenched-flow pulse labeling experiments provided important information about the structure formation during the first steps in protein folding [5–7].

Early direct NMR investigations of protein folding included studies of the equilibrium conversion between the native and the unfolded state at elevated temperatures, high denaturant concentrations, or extreme pH values [8, 9]. These experiments provided insights into the cooperativity and thermodynamics of protein folding. Additionally, NMR spectroscopy of nonnative states emerged. Although the system is at equilibrium, the rates of conversion between different states of folding can be determined by NMR spectroscopy [10]. Low rates in the range between 0.1 s^{-1} and 10 s^{-1} can be determined directly by NMR exchange spectroscopy. Fast processes on a microsecond to millisecond time scale affect the relaxation of the NMR active nuclei by chemical exchange contributions to the transversal relaxation rates (Figure 16.1). Therefore, the microscopic un- and re-

Protein Folding Handbook. Part I. Edited by J. Buchner and T. Kiefhaber
Copyright © 2005 WILEY-VCH Verlag GmbH & Co. KGaA, Weinheim
ISBN: 3-527-30784-2

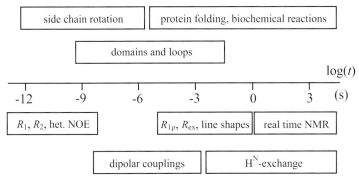

Fig. 16.1. Logarithmic NMR time scale. Illustration of accessible protein motions and reactions by the respective NMR technique: R_1, spin-lattice relaxation rate; R_2, spin-spin relaxation rate; het. NOE, heteronuclear Overhauser enhancement; $R_{1\rho}$, relaxation in the rotating frame; R_{ex}, chemical exchange contribution to R_2; line shapes, line shape analysis of proton resonances; dipolar couplings, residual dipolar couplings in different alignment media; H^N-exchange, proton–deuterium exchange of amide protons.

folding rates are accessible by line shape analyses of NMR resonances [11, 12], from ^{15}N $R_{1\rho}$ relaxation rates [13], from ^{15}N transverse cross-correlation relaxation rates [14], or from ^{15}N R_2 relaxation dispersion experiments [15]. These fast folding rates can be well predicted by the diffusion-collision model for several small proteins [16] or derived experimentally by extrapolating the denaturant dependence of the un- and refolding rates from stopped-flow experiments to 0 M denaturant [12, 14]. Recently, the analysis of residual dipolar couplings measured in different alignment media have allowed the determination of the dynamics between nanoseconds and milliseconds and filled the gap between very fast and slower motions of proteins [17, 18]. All these methods have in common that the populations of unfolded and folded polypeptide chains are at equilibrium but differ in their chemical shifts of the NMR active nuclei. During the detection or delay times in the pulse sequence (between about 1 ms and 1 s), fluctuations between the populations cause additional contributions to the natural relaxation of the different states from Brownian motions. These contributions can be converted into un- and refolding rates.

Nonequilibrium reactions can be directly followed in the NMR probe, where rates above 5 min^{-1} require a stopped-flow device to be observable. Since very early NMR studies of protein reactions in "real time" several of these devices have been designed [19–21]. Unfolding and refolding of proteins can be initiated by rapid mixing of solutions inside the NMR spectrometer with short experimental dead times [22–25]. One very powerful approach has been to use ^{19}F NMR to study proteins with ^{19}F-labeled tryptophan residues and a data collection started within 100 ms of the initiation of the mixing. Distinct steps of the folding reaction of dihydrofolate reductase could be resolved by these pioneering studies by Frieden and coworkers [23, 26, 27]. Recently, this method was successfully applied to resolve the consecutive steps of folding and assembly of the two-domain chaperone PapD using a 5-mm ^{19}F cryoprobe [28].

Very slow protein folding reactions, such as those limited by isomerization of peptidyl-prolyl bonds [29] can be initiated by manual mixing [30–33] and followed by the sequential accumulation of one-dimensional spectra. These studies revealed important information about the properties of intermediates during un- and refolding. An extension to two-dimensional techniques has been possible in a few cases, to increase the number of specific spin labels in well-resolved NMR spectra [34]. Most recently, Mizuguchi et al. succeeded in recording a series of 3D NMR spectra during the refolding reaction of apoplastocyanin [35] and were able to resolve changes of distances between protons during refolding.

For several proteins denaturant-induced unfolding transitions monitored by NMR uncovered equilibrium intermediates, which were silent in circular dichroism (CD)- or fluorescence-detected transitions [36–38]. This demonstrates the unique sensitivity of the chemical shifts of NMR resonances towards the local environment even in poorly structured protein states. Many features of the molten globule state, the archetypal protein folding intermediate, have been characterized by NMR spectroscopy [39] and especially the close correspondence between equilibrium and kinetic molten globules [40].

The use of NMR spectroscopy to study protein folding has been extensively reviewed [9, 23, 41–49]. The first part of this chapter concentrates on the developments of real-time NMR spectroscopy during the last 5 years in following very slow protein folding reactions. Earlier achievements have been already reviewed [24, 50, 51]. The very rapid development of the interpretation of NMR relaxation data in terms of micro- and millisecond motions during the last few years has been excellently reviewed [10]. Based on this theoretical background the second part of this contribution will focus on the derivation of millisecond protein folding rates from NMR relaxation experiments.

16.2
Following Slow Protein Folding Reactions in Real Time

The development of 1D and 2D real-time NMR experiments focuses on generally suitable methods of studying the structural and dynamic changes occurring during the protein folding reaction. A relatively simple mixing device has turned out to be very useful and generally applicable to initiate folding reactions, because no modifications of the spectrometer or the probe head are required [24]. This device has recently been optimized to reduce the dead time of mixing from 800 ms [40] down to tens of milliseconds (Figure 16.2) [25]. Typically, 50 µL of unfolded protein at high denaturant concentrations are injected into 250–450 µL refolding buffer already present in the NMR tube inside the magnet. To avoid premature mixing of the denatured protein with the refolding buffer and with the solvent in the tube that connects to the syringe, the solutions are separated by two air bubbles. Pushing the syringe located outside the magnet by 10 bar results in a complete mixing

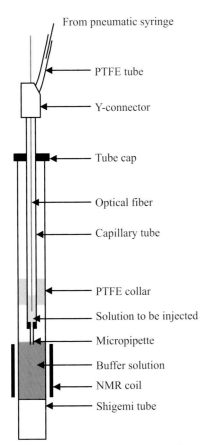

Fig. 16.2. Schematic representation of the rapid mixing device used for real-time NMR experiments. The NMR tube in the probe of the spectrometer contains the refolding or unfolding buffer, which is separated from the denatured or native protein solution, respectively, in the micropipette and capillary tube by an air bubble to avoid mixing of the solutions before injection. The transfer line to the syringe, which is outside the magnet, has to be filled with solvent and the piston is pneumatically triggered by 10 bar for a rapid transfer of the protein solution into the NMR tube.

within the active volume of the detection coil of the probe head during the dead time of the device (for further details see Section 16.6.1). Most presented examples in this report are multiscan NMR experiments with proteins, which require a relaxation period of at least 300 ms between scans for a sufficient build-up of polarization. Therefore, the present dead time of this simple stopped-flow mixing device is satisfactory.

The first proteins studied by real-time NMR employing such a device were bovine α-lactalbumin (α-LA) and hen lysozyme [24]. α-LA was particularly suitable for setting up the device [40], because it is a well-defined system for many aspects of protein folding and its folding kinetics can be varied over several orders of magnitude by control of the Ca^{2+} concentration of the solution [52, 53]. The folding reaction is initiated by changing the solvent conditions such as the pH [54, 55] or the concentration of denaturants [40, 56] or stabilizing agents [57, 58]. Subsequently, various NMR techniques can be applied to monitor the change of different properties of the polypeptide chain during the folding reaction. In the simplest case, a series of 1D spectra follows the rapid mixing step, allowing kinetic studies for processes with rates below 0.1 s^{-1}. These 1D real-time NMR experiments have been successfully applied to more than a dozen proteins: bovine α-LA [40, 59, 60], ribonuclease T1 of *Aspergillus oryzae* [61], human CDK inhibitor P19^{INK4d} [38], phosphocarrier HPr of *Escherichia coli* [62], SH3 domain of phosphatidylinositol 3′-kinase [63], human muscle acylphosphatase [64], French bean apoplastocyanin [32], thioredoxin of *E. coli* [33], rat intestinal fatty acid binding protein [23], dihydrofolate reductase of *E. coli* [27], bovine pancreatic ribonuclease A [31], hen lysozyme [25, 65, 66], chymotrypsin inhibitor 2 [67], barnase [67], barstar [68, 69], staphylococcal nuclease [70], yeast prion Sup35 [71], HIV I protease [72, 73] as well as peptides [74]. All studies make use of distinct chemical shifts for the different states changing with time, to determine the kinetics of the respective protein folding reaction. The dispersion of the resonances and the line width can also be used to distinguish possible folding intermediates from the unfolded and folded state. The appearance of 1D spectra of folding intermediates range between mainly unfolded, molten-globule-like, and native-like. Intermediate populations below 5% of the protein molecules are difficult to detect.

Figures 16.3 and 16.4 illustrate one example for the use of 1D real-time NMR spectroscopy to study the refolding of P19^{INK4d}, human cyclin D-dependent kinase inhibitor [38]. After dilution from 6 M to 2 M urea two fast refolding phases produce 83% of native molecules during the dead time of the NMR experiment. These two fast phases have been revealed by stopped-flow far-UV circular dichroism spectroscopy during the formation of the five native ankyrin repeats comprising ten α-helices. Refolding of the remaining 17% of unfolded molecules is retarded by a slow *cis/trans* isomerization of prolyl residues resulting in a refolding rate of 0.017 s^{-1}. This slow phase can be accelerated by the prolyl isomerase cyclophilin 18 from the cytosol of *E. coli*. The native population of P19^{INK4d} can be directly detected by the high-field shifted resonance at 0.7 ppm. To analyze the decay of the unfolded population at 0.9 ppm or 1.45 ppm, the native contribution to the 1D spectra has to be subtracted [38, 61]. First, this subtraction revealed identical un- and refolding rates. Secondly, each 1D spectrum can be represented by a linear combination of the spectrum of the unfolded and the folded population. This suggests that no further intermediate gets significantly populated during this final folding event of P19^{INK4d} and that the polypeptide chain cannot form a native-like ordered structure prior to the prolyl isomerization.

A similar analysis was possible for the S54G/P55N variant of ribonuclease T1 from *Aspergillus oryzae* [61, 75]. About 15% of the unfolded S54G/P55N-RNase T1 molecules contain the native *cis* prolyl peptide bond Tyr38-Pro39 and fold to the native state within the dead time of the NMR experiment. They give rise to a burst

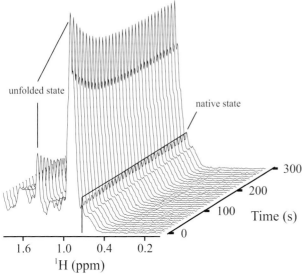

Fig. 16.3. Stacked plot of the high field region of the 1D ^{1}H NMR spectra at different times of refolding. Human CDK inhibitor P19^{INK4d} in 50 mM Na-phosphate pH 7.4 (10% ^{2}H$_2$O) at 15 °C was refolded by a tenfold dilution from 6 M urea to yield a final urea concentration of 2 M. For this representation two 1D experiments with four transients were averaged to yield a resolution of 11 seconds per spectrum. Resonances of the unfolded and native state are indicated and the integral of the latter is plotted in Figure 16.4a.

of all resonances of the native state in the very first spectrum recorded after initiation of the refolding reaction. The refolding of the remaining 85% unfolded molecules is very slow and limited in rate by the *trans* → *cis* isomerization of Pro39. In contrast to the above-mentioned example of P19^{INK4d}, where the molecules were largely unfolded before the very slow rate-limiting step, S54G/P55N-RNase T1 rapidly forms a folding intermediate arrested at this stage of folding because of the nonnative prolyl peptide bond. The major fraction of their intensities decays with a rate of 1.4×10^{-4} s^{-1} at 10 °C, which is very close to the formation of the native state with a rate constant of 1.25×10^{-4} s^{-1}. A minor fraction (about 7%) reveals 50 times higher rates and belongs to a second shorter-lived intermediate. The entire dataset could be decomposed according to the time dependence of a basis set of 1D spectra of the native state, the major intermediate I^{39t} and the minor intermediate (for a detailed description see Ref. [61]). The dispersion and line widths of the resonances indicated that the long-lived intermediate I^{39t} has a defined tertiary structure already close to the native state, but several resonances of I^{39t} revealed not-yet-native chemical shifts. The shorter lived transient species is mainly unstructured.

These examples of S54G/P55N-RNase T1 and P19^{INK4d} substantiate that nonnative prolyl peptide bonds can either prevent the formation of native protein or can arrest the protein in an already well-structured but not-yet-native conformation.

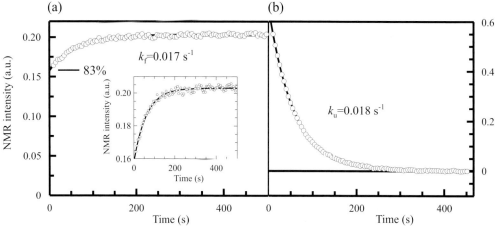

Fig. 16.4. Folding kinetics of P19[INK4d] extracted from the experiment depicted in Figure 16.3. a) The build-up of the native state is represented by the integral of the NMR intensities between 0.65 ppm and 0.70 ppm. The continuous line corresponds to a single exponential fit with a rate constant of 0.017 s^{-1} and a burst phase in the dead time of the experiment of 83%. The inset shows the same data with enlarged ordinate. b) Decay of the contribution of the unfolded state to the NMR spectra. A single exponential fit was applied to the integral of the NMR signals between 0.78 ppm and 0.90 ppm and reveals a rate constant of 0.018 s^{-1}, which is shown as a continuous line. Prior to the integration, the native component was subtracted from every 1D spectrum (see Section 16.6.1).

Therefore, for a protein function in the cell several levels of regulation based on prolyl peptide bonds are possible. Ubiquitous peptidyl prolyl isomerases with and without substrate specificity have been identified and proposed as key for this kind of regulation [29, 76]. Most recently that kind of a proline-driven conformational switch has been structurally characterized [77, 78].

A particularly important approach to the characterization of transient protein folding intermediates is to monitor nuclear Overhauser effects (NOEs) between protons of these states and how they evolve during the folding process. NOEs can, in principle, be converted into interproton distances, which is the basis for structure determination of proteins by NMR spectroscopy. The molten globule state formed rapidly after initiation of refolding of α-LA can be studied by real-time 1D NOE experiments [57]. Although the resolution of the spectrum is limited, the overall NOE on the aliphatic proton resonances as a consequence of saturating all aromatic resonances can be measured. It had been shown previously that this approach works for the A-state of α-LA, which is populated at equilibrium at pH 2 [55]. The intensity of the steady-state NOE of the kinetic molten globule observed transiently during refolding at pH 7.0 is already 80% of the steady-state NOE intensity of the native state. This is even larger than the value of 70% measured previously for the equivalent NOE effect in the A-state at equilibrium. The average radius of the species formed after initiation of folding can be estimated from the steady-state NOE to be within 4% of that of the native state [57]. This experiment shows that the folding of α-LA involves the rapid formation of a highly collapsed but partially disorganized molten globule, which subsequently reorganizes slowly to form the fully native state mainly by packing of the side chains.

Additional evidence for a native-like compactness of the molten globule of α-LA came from 1D real-time NMR diffusion experiments [56]. This method allows the determination of the hydrodynamic radius of kinetic folding intermediates. The experiment is based on a pulsed field gradient spin echo pulse sequence to determine the diffusion coefficient of molecules in liquids [79, 80]. The latter can subsequently be converted into the effective hydrodynamic radius for nonspherical molecules (R_h) such as proteins (for further details see Section 16.6.1 and Refs [81, 82]). The application of these spin echoes during refolding of α-LA from 6 M GdmHCl revealed R_h of the kinetic molten globule, which is only 8% larger than the R_h of apo α-LA [56]. The equivalent experiment was performed during refolding of S54G/P55N-RNase T1 and showed that R_h of the I^{39t} intermediate is only about 5% larger than the radius of the native state [75]. Consequently, the major compaction of the extended unfolded polypeptide chain occurs before the rate-limiting folding step regardless whether the side chains are already in a well-defined conformation (S54G/P55N-RNase T1) or not (α-LA).

In parallel to the successful formation of the native state, polypeptide chains can follow misfolding pathways. This failure of proteins to fold correctly or to remain folded is implicated in the onset of a range of diseases [83, 84]. To characterize these misfolding reactions, it is necessary to use a variety of complementary spectroscopic methods. Real-time NMR spectroscopy can be used to study the association and aggregation of proteins in solution by monitoring the aggregation-competent but still soluble states during these processes. Carver and coworkers used α-LA as a model system to study protein aggregation, which occurs during unfolding of the protein by reducing its four disulfide bonds. They initiated the aggregation reaction by rapid mixing of apo α-LA or holo α-LA with a large excess of the reducing agent dithiothreitol (DTT) [85–87]. Reduction of the disulfide bonds promotes unfolding of α-LA through several well-characterized intermediate states, which expose hydrophobic surfaces prone to aggregation. The well-dispersed NMR spectrum of the native apo protein gets lost during the dead time of the experiment (1.2 s) after the addition of DTT. The first obtained spectrum shows broad resonances with significantly reduced signal intensity and resembles the molten globule or A-state of α-LA. After several minutes the resonance intensity disappears since aggregation occurs leading to substantial precipitation out of the solution with an apparent rate of 1.5×10^{-3} s^{-1} at 37 °C. Aggregation can be significantly decelerated and precipitation totally prevented by adding chaperones such as α-crystallin [85, 87] or clusterin [86] to the refolding buffer of the real-time NMR experiment. Both chaperones bind to the aggregation competent monomeric molten globule state of apo α-LA and therefore stabilize the protein. Reduction and concomitant aggregation of holo α-LA is less efficiently prevented by α-crystallin because it is only partially reduced and more structured compared with the fully reduced and disordered form of apo α-LA. Additionally, holo α-LA precipitates more rapidly, which is important for the chaperone action of α-crystallin, since it is much more efficient with slowly precipitating species [87].

Solution aggregation of acid-denatured cold shock protein A (CspA) from *E. coli* is also a very slow process, which results in formation of insoluble fibrils [88]. Several 2D ^{1}H–^{15}N heteronuclear single-quantum coherence (HSQC) spectra, ^{15}N relaxation experiments and even four 3D HSQC-NOESY spectra can be recorded during this self-assembly process. A combined analysis of these data reveal that mainly residues from the β-strands of CspA monitor a fast exchange on the NMR chemical shift time scale between acid-denatured monomers and soluble aggregates during the lag phase of aggregation. During the exponential phase of aggregation the first three β-strands are predominantly involved in the association process.

The aggregation of the prion-determining region of the NM domain of Sup35 in yeast, Sup35$^{5-26}$, has been followed recently by 1D real-time NMR spectroscopy [71]. As expected, only signals originating from soluble peptide molecules are visible in the spectra and after 80 min 50% of the peptide became insoluble. Adding catalytic amounts of Hsp104 prevents aggregation for about 80% of the peptide molecules by binding to the hexameric/tetrameric species of Sup35$^{5-26}$ and releasing intermediate and monomeric species. These conclusions could be drawn because binding of Hsp104 causes significant changes of chemical shifts of several tyrosine resonances. Based on these resonances, the molecular weight of the Sup35$^{5-26}$ oligomers could be estimated by NMR diffusion experiments.

In photo-CIDNP (chemically induced dynamic nuclear polarization) NMR experiments, the polarization of solvent-exposed protons is induced more rapidly than the natural spin-lattice relaxation via a dye excited by a laser flash [89]. This method reduces the time required to polarize spins transferred into the NMR probe from a lower field region of the magnet [66]. Together with the very small portion of the sample illuminated by the laser beam, the dead time of the device shown in Figure 16.2 can be further reduced [25]. The strength of the photo-CIDNP approach, however, is not only in its time resolution but also in its spectral resolution. The latter is significantly improved because the induced polarization for example by flavin mononucleotide as dye is only generated in a limited number of residues (tyrosine, tryptophan, and, under some conditions, histidine). Moreover, polarization occurs only when these residues are accessible to the photoexcited dye molecules in the solution, enabling the environment of individual residues to be probed directly in real time during folding. Hen lysozyme for example forms initially (after 30 ms experimental dead time) a relatively disordered collapsed state with largely buried tryptophan residues [25, 66]. The structurally closely related bovine α-LA also forms very early a collapsed state during refolding, which very closely resembles the photo-CIDNP spectrum of the pH 2 molten globule state of this protein [59, 60]. Subsequently, a reorganization of these aromatic residues occurs during the course towards the fully native proteins in both cases.

Histidine residues only get excited by the photo-CIDNP process when the protein contains no tryptophan residues or when all tryptophan residues are buried and the histidine residues are exposed. These very rare conditions are met by the histidine-containing phosphocarrier protein HPr from *E. coli*. The wild-type protein contains no tryptophan and tyrosine residues, giving rise to several well-resolved histidine resonances in the photo-CIDNP spectrum [90]. In several tryptophan-containing variants of HPr no histidine resonances of the native and unfolded state were detectable because of the exposed Trp. Directly after refolding has been initiated, in most variants the Trp side chain was buried and some His residues could be transiently excited by the dye until the native conformation was formed.

16.3
Two-dimensional Real-time NMR Spectroscopy

A major limitation of 1D NMR experiments with proteins is, of course, the low resolution due to severe signal overlap. Therefore, only a few well-resolved resonances allow a detailed analysis with molecular resolution. In many cases the general appearance of the 1D spectrum rather than single resonances is analyzed. Thus, current work focuses on the extension of real-time NMR experiments to utilize the power of multidimensional NMR in kinetic experiments. Very slow folding reactions can be directly followed by sequentially recorded 2D spectra such as 2D ^{1}H–^{15}N HSQC spectra [34]. An equivalent proton–nitrogen correlation spectrum can be recorded in much less time (in about 200 s per 2D spectrum) if a heteronuclear multiple-quantum coherence (HMQC) pulse sequence is used in combination with Ernst angle pulses [56, 91]. During the refolding of S54G/P55N-RNase T1 from 6 M GdmHCl at 1 °C, for example, a series of 128 of these fast-HMQC spectra could be recorded [75]. After correction for the dead time events, the complete 2D spectrum of the kinetic folding intermediate I^{39t} could be obtained. The spectrum exhibits the same dispersion of the resonances and the same line widths as the spectrum of the native state, indicating a well-defined tertiary structure of I^{39t}. The backbone resonances of 22 (out of 104) residues of I^{39t} show native chemical shifts and identify therefore regions in which a native environment is already present in the intermediate state. For 66 backbone amide probes, which differ in chemical shifts between the native state and I^{39t}, single exponential refolding kinetics could be extracted from the 2D real-time NMR experiment. All showed the same time constant within experimental error indicating that the *trans* → *cis* isomerization of Pro39 fully synchronizes the rate-limiting step of refolding, which is the transition of not-yet-native regions in I^{39t} towards the native conformation. Interestingly, amide protons in not-yet-native environments are not only located close to the nonnative *trans* Tyr38-Pro39 peptide bond, but are spread throughout the entire protein.

The slowest step during the refolding of barstar, the intracellular inhibitor of barnase, is also dominated by the formation of a *cis* peptidyl prolyl bond of Tyr47-Pro48 in the native state [92, 93]. A set of 16 2D ^{1}H–^{15}N HSQC spectra recorded in real time revealed that the intermediate with the *trans* Tyr47-Pro48 bond has a predominantly native-like conformation [68]. Only the *trans* prolyl peptide bond and three residues in close proximity to Pro48 rearrange upon complete refolding.

Therefore, this intermediate is silent in conventional CD and fluorescence spectra. Addition of cyclophilin from the human cytosol leads to a rapid interconversion of the *trans* intermediate to the native form within the dead time of the NMR experiment. This helped to identify not-yet-native backbone amides in the intermediate by calculating double difference spectra. In contrast to the intermediate of S54G/P55N-RNase T1, where a nonnative prolyl peptide bond prevented the formation of a native conformation at various close and remote sites, the intermediate of barstar tolerates such a peptide bond and therefore affects the conformation only locally.

The collagen triple helix folds slowly, on a time scale of minutes to hours, making this system amenable to 2D real-time NMR monitoring of the kinetics of peptide fragments (see reviews [49, 50]). Triple helix formation of these peptides can be initiated by rapid cooling of the sample from 50 °C to 10 °C. Residues located at the termini exhibit rates and folding kinetics that are distinct from residues in the central region of the peptide. The NMR data are consistent with an association/nucleation mechanism at the peptide termini and subsequent propagation [34]. The slow propagation results most likely from the conversion of a mixed *cis/trans* ensemble of numerous imino acids of the monomers to the final all-*trans* form in the triple helix [94]. A recent 2D real-time NMR study of peptides comprising 18 residues from human collagen type I proved that both the nucleation at the C-terminus and the propagation towards the N-terminus are limited by *cis/trans* isomerizations of imide bonds [95]. These peptides contained ^{15}N spin labels at various positions along the sequence, allowing a detailed analysis of consecutive steps during folding of the collagen triple helix. Point mutations known from collagen diseases such as osteogenesis imperfecta, which can be easily included in these model peptides, revealed during which steps misfolding events occur [49, 96].

The collagen triple helix formation was further investigated by residue-specific real-time NMR diffusion experiments [97]. This very powerful 2D development of the 1D real-time diffusion experiment described above also allows determination of the diffusion coefficient and therefore the association states of unfolded, folded, and kinetic intermediates with transient lifetimes simultaneously. The authors could confirm their previous results from the kinetic studies that early nucleated trimers are well ordered only at the C-terminus and that the zipper-like propagation follows in the expected direction.

As mentioned above, a major limitation of 1D experiments is of course the limited resolution of the spectra. Series of 2D experiments, on the other hand, are time demanding, because the acquisition time of each 2D dataset requires at least several minutes. Therefore, we developed methods to reconstruct the kinetic history of folding reactions from a single 2D NMR spectrum recorded during the entire time course of the reaction [54, 61]. This information arises because different time courses of the respective protein states modulate the line widths and intensities of their NMR resonances differently. In general, 2D NMR spectra are recorded by incrementing a time delay between the preparation and observation pulses while a series of free induction decays (FIDs) is recorded. If a chemical re-

(a) Kinetic spectrum

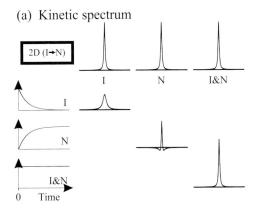

(b) Reference spectrum

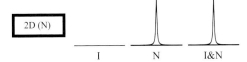

(c) Difference spectrum

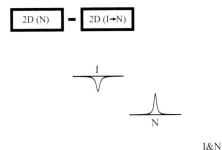

Fig. 16.5. Schematic line shapes and intensities in the indirect dimension F1 for different cross-peaks in a) the kinetic spectrum recorded during the refolding reaction, b) the reference spectrum of the native state after refolding, and c) the result of subtracting the kinetic spectrum from the reference spectrum.

action occurs during the accumulation of data in the experiment, it influences the amplitude of the different signals recorded at various incrementation times. This modulation determines both the line shape and intensity of the cross-peaks in the resulting 2D spectrum, namely in the indirect dimension (Figure 16.5). The principle and effectiveness of this strategy have been demonstrated for α-LA using a ^{1}H–^{15}N HSQC experiment [54]. The resonances of the native state of α-LA showed in the kinetic spectrum the expected lower intensities and negative components in

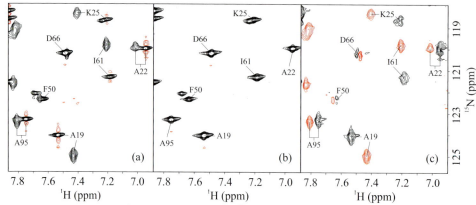

Fig. 16.6. Two-dimensional real-time ^{1}H-^{15}N HSQC spectra of S54G/P55N-RNase T1 recorded during and after the refolding reaction from 6 M GdmHCl at 15 °C. Section of a) the kinetic HSQC, b) the reference HSQC of the native protein, and c) the difference spectrum between the kinetic HSQC and the reference HSQC. Positive NMR intensities in (a) and (c) are indicated with black lines, negative signals by red lines. D66 and F50 are in a native region of the intermediate. The cross-peaks of the amide protons depict equivalent chemical shifts, sharp lines and equivalent intensities in the intermediate and the native state. Therefore, the cross-peaks are present in (a) and (b) and cancel out in the difference spectrum (c). A19, A22, K25, I61, and A95 are in a nonnative environment in the intermediate and depict different chemical shifts as well as line broadening. These cross-peaks of the intermediate state are absent in (b) and show negative intensities in (c). Correspondingly, the emerging cross-peaks of the native state can be identified by the two flanking negative components along the ^{15}N dimension in (a) and reveal positive cross-peaks in (c).

the wings of the peaks compared to the reference spectrum (see the line shapes and intensities of N in panel (a) in the schematic illustration in Figure 16.5). In the special case of α-LA, only very few signals of the molten globule (represented by I in Figure 16.5) are observable, because most resonances are widened.

All three line shapes shown in panel (a) of Figure 16.5 could be experimentally observed in a single 2D real-time ^{1}H-^{15}N-HSQC NMR experiment during the refolding of S54G/P55N-RNase T1 from 6 M GdmHCl at 15 °C [98]. The resonances of F50 and D66 show native chemical shifts in the folding intermediate I^{39t} in both, the ^{1}H and ^{15}N dimension of the spectrum. Thus, sharp lines have been observed in the 2D real-time NMR experiment (Figure 16.6a) along the ^{15}N dimension. For A19, A22, K25, I61, and A95, the resonances of I^{39t} differ in their chemical shifts from the final native state and therefore depict line broadening along the indirect F1 dimension. The corresponding resonances of the native state can be identified by the two flanking negative components along the ^{15}N dimension (negative intensities are shown in red in Figure 16.6a).

Beside identification and assignment of the resonances of transient species, real-time 2D NMR can be used to determine folding kinetics from these described line shapes (see Section 16.6.3) [54, 61]. Here, the key aspect is that folding kinetics can be monitored residue-by-residue to identify regions of the peptide chain that fold cooperatively according to kinetic arguments. Towards this end, the rate constants from each secondary structure element of α-LA have been determined and it was concluded that the closely packed native environment of the main-chain amide groups of different residues is achieved cooperatively, and with the same kinetics as observed for the side chains [54].

This approach was further developed for homonuclear 2D techniques, where we mainly use the line shape analysis to assign the resonances and NOEs to the respective state of folding. Cross-peaks in a conventional 2D NOESY spectrum reveal protons that are closer than about 5 Å apart. This distance information is used as constraints in structure calculations. Therefore, the obvious goal is the development of NMR experiments to determine NOEs between protons in transient protein folding intermediates to gain high-resolution structural data. The time requirements for well-resolved 2D and 3D NOESY spectra are about 10-fold above the recording time for respective heteronuclear correlation spectra. Rates of amide proton exchange could be determined from a set of 2D NOESY spectra (with a recording time of 5.2 min per spectrum), but both the resolution and the signal-to-noise ratio was very poor in these experiments [99].

Very detailed distance information, however, can be extracted from a single 2D NOESY spectrum recorded during very slow folding reactions with time constants of at least 30 min. Such an experiment was measured during the first 10 hours of refolding of S54G/P55N-RNase T1 at 10 °C to further characterize the major folding intermediate I^{39t} [61]. Three kinds of NOEs can be differentiated in this kinetic 2D NOESY according to the observed F1 line shape of respective cross-peaks. As illustrated in Figure 16.5: (i) NOEs between protons, which are close to each other only in the intermediate state but not in the native state, are specified as nonnative NOEs. They show decaying intensities during the refolding reaction because the concentration of the intermediate decreases. This results in an apparent line broadening of the respective cross-peaks. (ii) NOEs between protons that are close in space only in the native state exhibit increasing intensities during refolding as the native protein is formed. These cross-peaks show distinct line shapes as described above with two flanking negative components [54]. (iii) Native NOEs already present in the intermediate state also exist in the native state, of course, and are therefore specified as native-like NOEs. Their intensity does not change during the entire recording of the spectra so that the intensity and the line shape remain unchanged.

To classify the NOEs to one of the three groups, every single NOE cross-peak has to be compared with the respective signal in the reference NOESY experiment by means of intensity and line shape, which is very time consuming. Therefore, a method was developed to convert the line shape information of each cross-peak into positive and negative intensities [61]. A reference spectrum has to be recorded immediately after the refolding reaction and the kinetic spectrum has to be subtracted afterwards. The result is shown schematically in Figure 16.5c and experimentally in Figure 16.6c for the 2D real-time ^{1}H–^{15}N HSQC. Along these lines, the above-mentioned kinetic NOESY was subtracted from a reference NOESY of the native protein to facilitate the classification of NOE effects. This approach, in principle, allows the 3D structure of a protein folding intermediate to be determined. About 200 NOEs have been analyzed for a detailed characterization of the I^{39t} intermediate of S54G/P55N-RNase T1 [61]. They revealed, for example, that the β-strands involved in disulfide bonds already have the native conformation in I^{39t} as well as the N-terminal half of the α-helix. The C-terminal half of the α-helix has a not-yet-native conformation together with many other regions interspersed among the native-like sections of this protein folding intermediate. This very detailed structural information about I^{39t} considerably extended the qualitative picture we got by just comparing its chemical shifts with the native state as discussed above.

16.4
Dynamic and Spin Relaxation NMR for Quantifying Microsecond-to-Millisecond Folding Rates

Most folding reactions are too fast to be monitored directly by real-time NMR spectroscopy, because the time required to record one 1D ^1H spectrum of a protein is at least 100 ms. For signal-to-noise reasons, usually more than one scan is needed with an interscan delay for sufficient relaxation of excitable magnetization of at least 300 ms. Therefore, not more than three 1D scans are possible within one second after a sufficient fast mixing to initiate refolding. Reactions with time constants below 500 ms cause line broadening of the proton resonances in 1D spectra, but even with a fast stopped-flow mixing device only single-scan experiments are possible, which is not very practicable for protein NMR [21]. Interleaved methods with different delay times after mixing to record the first 1D spectrum as well as continuous-flow experiments require very large amounts of protein, because NMR with biological samples works at millimolar concentrations, and are therefore rare [55, 66, 100].

Because of these technical problems of nonequilibrium experiments, fast protein folding reactions have been measured under equilibrium conditions. Fluctuations on a 10 ms to 10 μs time scale between two or more states, where the NMR active nuclei experience different chemical environments, have profound effects on the shape and position of their resonances [101]. For proteins, the analysis of line shapes has been used to study ligand binding, local fluctuations as well as global fluctuations such as complete un- and refolding of the polypeptide chain (a very good overview and references are given in Ref. [10]). One major drawback during the quantification of NMR line shapes in terms of intrinsic unfolding and refolding rates is that a folding model has to be assumed. In most cases simple two-state protein folding systems have been studied or an apparent two-state model such as free and complexed protein has been applied to analyze the line shapes. The first protein folding system that could be quantitatively studied by dynamic NMR spectroscopy on the above-mentioned time scale was the N-terminal domain of phage λ repressor, λ_{6-85} [11]. The urea dependence of the intrinsic un- and refolding rate constants could be determined between 1.3 M and 3.2 M urea with a maximum value for $k_f = 1200$ s^{-1} and a minimum value for $k_u = 100$ s^{-1} at 1.3 M urea. From a linear extrapolation of the chevron plot to 0 M urea, $k_f^0 = 3600 \pm 400$ s^{-1} and $k_u^0 = 27 \pm 6$ s^{-1} have been obtained. Later, Oas et al. verified these extrapolated rates by ^{1}H transverse relaxation experiments (see below) [12]. Typically, the resonances of aromatic protons are analyzed in ^{2}H$_2$O samples with completely exchanged amide protons, because only then can the former resonances between 6 and 11 ppm be assigned for both folded and unfolded state. These chemical shifts are needed at any urea concentration for an accurate line shape analysis (see Section 16.6.2). According to a two-state folding mechanism, only aromatic resonances of the folded and unfolded state are present within this region. In the case of λ_{6-85}, which contains only two tyrosine and two phenylalanine residues, the entire aromatic region of the 1D spectra could be simulated at any urea concentration (for example by using the program ALASKA [102]), which significantly increases the reliability of the extracted folding rates [12].

The cold shock protein CspB from Bacillus subtilis also follows a two-state folding mechanism with $k_f^0 = 1070 \pm 20$ s^{-1} and $k_u^0 = 12 \pm 7$ s^{-1} at 25 °C and 0 M urea [103]. From a urea transition monitored by 1D NMR in ^{2}H$_2$O, the resonance of His29$^{\varepsilon 1}$ could be used to determine unfolding and refolding rates between 2.9 M and 5.6 M urea (Figures 16.7 and 16.8). The simulation of the line shape and the position of this resonance according to the extracted parameters closely resembles the experimental data and is depicted in the right panel of Figure 16.7. The extrapolated folding rates from the dynamic NMR experiment ($k_f^0 = 1490 \pm 370$ s^{-1} and $k_u^0 = 16 \pm 3$ s^{-1}) correspond nicely to the rates derived from the stopped-flow fluorescence experiments. The Tanford factors $\beta_T = m_f/(m_f + m_u)$ obtained from the linear slopes of log k_f and log k_u versus urea concentration (Fig-

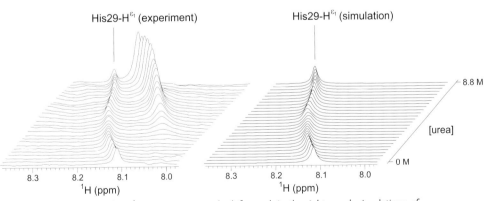

Fig. 16.7. Stacked plot of 1D ^{1}H-NMR spectra of a complete urea-induced unfolding transition of CspB from B. subtilis in ^{2}H$_2$O at 25 °C. Experimental 1D spectra of the His29$^{\varepsilon 1}$ resonance in the low field region is shown in the left panel. In the right panel, simulations of the 1D spectra using the extracted and extrapolated folding rates from the line shape analysis described in Section 16.6.2 at the respective urea concentration are depicted.

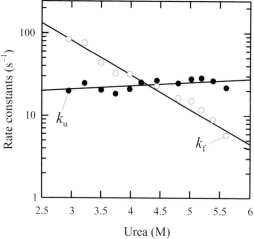

Fig. 16.8. Urea dependence of the folding rates of CspB determined by the line shape analysis approach applied to the 1D ^{1}H-NMR spectra of His29^{ε1} in ^{2}H$_2$O (Figure 16.7). The logarithm of k_f and k_u in the transition region (between 2.9 ppm and 5.6 ppm) is plotted versus the urea concentration. The linear slope of the continuous lines represents the *m*-value of refolding and unfolding (m_f, m_u) from which the Tanford value $β_T$ is derived with 0.9. Extrapolation to the absence of denaturant reveals folding rates of $k_f{}^0 = 1490 ± 370$ s^{-1} and $k_u{}^0 = 16 ± 3$ s^{-1}.

ure 16.8) are 0.9 from both the NMR and fluorescence experiment and indicate that the activated state of unfolding of CspB still resembles the native state in the accessibility to the solvent [103].

The folding and unfolding kinetics of the N-terminal domain of the ribosomal protein L9 have been measured at temperatures between 7 °C and 85 °C and between 0 M and 6 M guanidine deuterium chloride. The joint analysis of stopped-flow fluorescence (between 7 °C and 55 °C) and dynamic NMR data (between 55 °C and 85 °C) revealed thermodynamic parameters of the activated state of folding [104, 105]. The dynamic NMR data were required because the apparent folding rate constant for the latter temperature range was 700–3000 s^{-1} too high for traditional stopped-flow experiments. Following the same lines a small all-helix protein psbd41 could be analyzed by the line shape approach revealing maximum folding rates of 21 600 s^{-1} and maximum unfolding rates of 24 600 s^{-1} [106].

The upper time limit for dynamic NMR to study protein folding reactions are tens of microseconds if the difference in chemical shift of an NMR nuclei between the unfolded and native state is sufficient (at least 500 Hz). Raleigh and coworkers found refolding rates between $3 × 10^4$ s^{-1} and $2 × 10^5$ s^{-1} in a thermal unfolding transition of the villin headpiece, which confirmed experimentally recent all-atoms molecular dynamics calculations to simulate protein folding reactions (see Ref. [107] and references therein). A similar link between theory and experiment was possible for the three-helix bundle-forming B-domain of staphylococcal protein A.

16.4 Dynamic and Spin Relaxation NMR for Quantifying Microsecond-to-Millisecond Folding Rates

Dynamic NMR analyses revealed folding rate constants of 120 000 s^{-1}, which are in good agreement with predictions from diffusion–collision theory [108].

One limitation of the use of dynamic NMR to study fast protein folding rates is that reliable rates can only be obtained from the central region of denaturant- or temperature-induced unfolding transitions, where at least 15% of both states are populated. The wide linear extrapolation of log k_f to conditions with a maximal population of the native state might miss nonlinear effects such as a "roll-over" under strong native conditions. One way to bypass this problem is the use of NMR relaxation data to determine protein folding rates. If the rates of interconversion are on the millisecond-to-microsecond time scale and large chemical shift differences between the states are present, transverse relaxation rates can be sensitive to the presence of the minor conformation with populations as low as 1% [10].

One early application was the direct determination of folding rates of the λ repressor head piece λ_{6-85} under strong native conditions by ^{1}H transverse relaxation using a 1D Carr-Purcell-Meiboom-Gill (CPMG)-based spin echo pulse sequence (see Section 16.6.4) [12]. The chemical exchange contribution (R_{ex}) to the transverse relaxation rate (R_2) leads to an apparent increase of the latter from which the folding rates can be extracted. At 0 M urea the authors found for λ_{6-85} $k_f = 4000 \pm 340$ s^{-1} and $k_u = 32 \pm 3.2$ s^{-1}, which is in good agreement with the values obtained by extrapolation from the line shape analysis ($k_f = 4900 \pm 600$ s^{-1} and $k_u = 30 \pm 4.6$ s^{-1}).

Most current applications facilitate CPMG-based ^{15}N-relaxation measurements to determine the chemical exchange contribution R_{ex} to R_2. The advantage of this approach is that 2D ^{1}H–^{15}N correlation spectra are used to extract the relaxation rates. Therefore, R_{ex} can be determined on a residue-by-residue basis to discriminate for example local and global unfolding of the peptide chain. The calculation of k_f and k_u from R_{ex} requires some further parameters to be determined beforehand (see Section 16.6.4). Among several approaches to determine the R_{ex} contributions to R_2 (such as an extended Lipari-Szabo approach, $R_{1\rho}$ measurements, or the interference of dipolar ^{1}H–^{15}N and ^{15}N chemical shift anisotropy relaxation), R_2 dispersion experiments are the most popular ones (for an overview and further literature see Ref. [10]). The basic idea is that the R_{ex} contributions to R_2 can be modulated in a CPMG-based sequence by varying the delay time τ_{cp} between consecutive 180° pulses on the ^{15}N nuclei (see Section 16.6.4). The so-called dispersion curve is a plot of R_2 over $1/\tau_{cp}$ (Figure 16.9). At very small $1/\tau_{cp}$ values, the transverse ^{15}N relaxation rate R_2 contains the full contribution of R_{ex} and at high $1/\tau_{cp}$ values no R_{ex} contributions are present. Therefore, the difference between R_2 at the lowest and highest $1/\tau_{cp}$ value provides an estimate for R_{ex}. Fitting of the entire dispersion curve reveals the apparent folding rate k_{ex}, which is the sum of k_f and k_u. If the two-state model is valid and the equilibrium constant under the respective conditions is known explicit values for k_f and k_u can be calculated [10, 12, 13]. R_2 dispersion curves depend on the magnetic field strength [109] and therefore a joint fit of data obtained at various field strengths decreases the errors of the derived dynamic parameters.

Figure 16.9 shows the ^{15}N R_2 dispersion curves for A32 of the cold shock protein

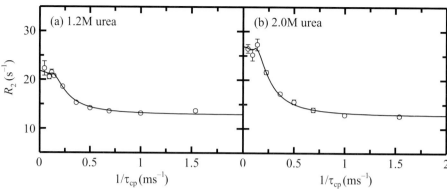

Fig. 16.9. Relaxation dispersion curves of A32 of CspB from B. subtilis in 90% H_2O/10% ^2H_2O at 25 °C. Values for $R_2(1/\tau_{cp})$ in the presence of a) 1.2 M and b) 2 M urea at $B_0 = 14.1$ T (60.8 MHz ^{15}N frequency) are depicted. The solid lines represent fits of the data to Eq. (10) with the known difference in ^{15}N chemical shifts between the cross-peaks of A32 in native and unfolded CspB (119 Hz and 122 Hz for 1.2 M and 2 M urea, respectively) as well as the population of the native state (0.96 and 0.89, respectively). It revealed k_f (217 s^{-1} at 1.2 M urea and 124 s^{-1} at 2.0 M urea) and k_u (9.8 s^{-1} at 1.2 M urea and 15 s^{-1} at 2.0 M urea).

CspB at 1.2 M urea and 2.0 M urea with native populations of 0.96 and 0.89, respectively. From a urea transition under NMR detection and from ZZ-exchange spectra [110] the difference in chemical shifts of the A32 backbone amide resonances between the native and unfolded state at the two urea concentrations are known (119 Hz and 122 Hz, respectively). The analysis of the dispersion curves according to the protocol described below (see Section 16.6.4) reveals k_{ex} (227 s^{-1} at 1.2 M urea and 139 s^{-1} at 2.0 M urea) and subsequently k_f (217 s^{-1} at 1.2 M urea and 124 s^{-1} at 2.0 M urea) and k_u (9.8 s^{-1} at 1.2 M urea and 15 s^{-1} at 2.0 M urea) in good agreement with the rates from conventional stopped-flow folding experiments under fluorescence detection [103]. In principle, this kind of NMR relaxation experiment can be used to measure chevron plots of proteins on a residue-by-residue basis, which should be very useful to map the transition state of folding in terms of homogeneity or heterogeneity and to determine the cooperativity within the folding peptide chain.

Several applications of ^{15}N and ^{13}C R_2 dispersion curves have been reported to determine rates of local motions under strong native conditions [109, 111–115]. In this case, two conformations have to be assumed in this region of the protein to gain apparent rates of interconversion. Additionally, the populations of the two conformations and the difference in chemical shifts of the respective nuclei have to be extracted entirely from the NMR relaxation data, which makes this kind of analysis very difficult and might limit the interpretation of the data.

Studies of global two-state folding–unfolding reactions allow an accurate determination of folding rates from R_2 dispersion experiments, because the populations of the native and unfolded state under the respective conditions can be determined experimentally [13, 15, 116]. Additionally, the differences of their chemical shifts are accessible from independent NMR experiments or can be at least extrapolated (see also Section 16.6.4). For global unfolding reactions all resonances of the protein should show R_2 dispersion curves as long as these differences are above 2 Hz, which makes joint fittings possible. In the case of the N-terminal SH3 domain of the *Drosophila* protein drk, on top of the global unfolding reaction, a local conformational exchange process of the unfolded state could be revealed by a systematic analysis of ^{15}N R_2 dispersion data [15]. For the de novo designed dimeric four-helix bundle protein α_2D, refolding rates of $(4.7 \pm 0.9) \times 10^6$ M^{-1} s^{-1} and unfolding rates of 15 ± 3 s^{-1} could be determined for this bimolecular reaction from ^{13}C R_2 dispersion curves [116]. The ^{13}C$^{\alpha}$ chemical shifts for unfolded and folded forms of α_2D indicate that the ensemble of unfolded states includes transiently structured helical conformations.

16.5
Conclusions and Future Directions

NMR spectroscopy is an important, well-established technique in structural studies of protein folding. We have highlighted here in Sections 16.2 and 16.3 the use of NMR to follow protein folding in real-time mode, where the reaction occurs within the NMR spectrometer. The step from 1D real-time NMR to 2D real-time NMR dramatically increased the accessible structural information about protein folding and misfolding reactions. Recently, the first 3D real-time NMR experiment following the refolding of apoplastocyanin, as mentioned above, has been described [35]. The authors were able to follow structure-specific NOEs in a series of 12 3D ^{1}H–^{15}N HSQC-NOESY-HSQC spectra. Local and long-range interactions in the native apoplastocyanin are formed simultaneously, consistent with highly cooperative formation of the native structure. One future direction to increase the dimensionality without increasing the recording time per spectrum might be the use of real-time G-matrix Fourier transform (GFT) NMR spectroscopy [117]. Another recent experimental development that might become fruitful in the near future is the induction of un- and refolding reactions inside the NMR tube by pressure-jump experiments [118]. Here several transitions can be accumulated, because the un- and refolding under high and low pressure, respectively, is reversible and does not require mixing of solutions. All the presented real-time NMR experiments are powerful methods for the characterization of the aggregation-competent conformation of proteins on misfolding pathways [87, 88]. Thus, we expect key information about malfunctioning protein folding reactions and pathological fibril formation to be revealed in the near future. These novel methodologies will further increase the versatility of real-time NMR.

Our studies of the I^{39t} intermediate of S54G/P55N-RNase T1 allow us to explore experimentally parts of the energy landscape for the refolding of this protein. These free energy landscapes calculated from molecular simulations can be put up by simple progress coordinates such as the number of native contacts (Q) and the radius of gyration [119]. From real-time NOESY experiments we can count the number of native and nonnative NOEs for I^{39t} resulting in a Q value of 0.6. The hydrodynamic radius of I^{39t} was calculated from real-time NMR diffusion experiments to be only 5% above R_h of the native state. The relative Gibbs free energy of I^{39t} was determined from the protection factors of 45 amide protons derived from a competition between refolding and H/D exchange under NMR detection [75]. The intermediate has already gained 40% of the total Gibbs free energy change during refolding of 36.6 kJ mol^{-1}, which is the global minimum of the native state with $Q = 1$. I^{39t} has a well-defined tertiary structure, implying that only a very narrow ensemble of states is present at this very deep local minimum of the funnel. As predicted [45, 120], NMR can provide very detailed contributions combining experiment and theory within the new view of protein folding.

Recent developments have gone one step further in this direction. On the one hand, extremely long molecular dynamics (MD) simulations became possible by using clusters of thousands of computers [121, 122]. On the other hand, dynamic NMR spectroscopy can reveal experimentally folding rates of proteins up to 200 000 s^{-1} [107]. At such high rates, several folding–unfolding interconversions occur during the now accessible length of one MD simulation. Therefore the gap between experiments and simulations has been closed and many exciting new insights into the elementary steps of protein folding can be expected in the near future.

16.6
Experimental Protocols

16.6.1
How to Record and Analyze 1D Real-time NMR Spectra

Refolding experiments are typically started with a dilution of a protein solution in high concentrations of denaturants. Here, the protein solution has to be injected into the refolding buffer, which is already present in the NMR tube (see Figure 16.2). For a rapid change of the pH for starting the folding reaction, the protein is already in the NMR tube and the respective puffer has to be injected. The concentrations and pH values of buffer of the protein sample and of the injected buffer have to be properly adjusted beforehand. Low pH values ($< $ pH 2) can be achieved by titrating the protein sample with HCl or DCl without buffer. Typically, final protein concentrations between 0.5 mM and 1 mM are used.

16.6.1.1 Acquisition
One-dimensional real-time NMR experiments start with the initiation of the protein folding reaction inside the NMR spectrometer. If reactions are followed with time constants of seconds, fast mixing devices are required with dead times below 100 ms [25, 123]. A dead time around 1 s can be achieved with a much simpler device, which needs a pneumatically driven gas-tight syringe (Hamilton, Bonaduz,

Switzerland) working between 8 bar and 10 bar [24]. Dead times between 2 s and 5 s are possible by the same device but manual pulling of the syringe. In the latter case, after the first shot, 50 μL should be sucked back in the syringe and pulled again to gain a homogeneous mixture in the NMR tube. For 2D real-time NMR, this has to be performed during the dummy scans before recording the spectrum. For 1D applications, the shots can occur during the first 1D spectra, which allows the first spectrum to be selected without mixing artifacts. For a fast recovery of the shim system, one shot should be performed in advance with the same buffers and volumes but with a test protein such as commercially available bovine α-lactalbumin or hen lysozyme. After this shot with the NMR tube in the magnet, a proper matching and tuning of the probe head is possible, as well as shimming of the field. For the actual experiment a tube with exactly the same geometry has to be used. Due to the induced field inhomogeneity during the mixing process the lock signal should be electronically increased by using a very high "lock gain" to avoid unlocking at the beginning of the experiment. For denaturant-induced refolding experiments, urea should be used in preference to GdmCl if the stability of the protein allows a choice, because the salt properties of GdmCl reduce the sensitivity of NMR coils significantly. For experiments in ^2H_2O, no suppression of the residual resonances of the denaturant (between 6 ppm und 6.7 ppm) is needed if the protons of urea and GdmCl have been properly exchanged against deuterons by several cycles of solving in ^2H_2O and freeze drying. A suppression of these signals can even be avoided for experiments in H_2O, if temperatures above 20 °C are used. The fast chemical exchange between the protons of water and of the denaturant causes an effective saturation transfer and suppression of water is sufficient. At lower temperatures, the water resonance can be suppressed for example by a WATERGATE sequence [124] and the resonance of the denaturant by an off-resonance presaturation during the relaxation delay between scans [38].

16.6.1.2 Processing

Conventional 1D NMR data processing can be applied to each of the series of 1D experiments. It is most convenient to run a pseudo 2D experiment by storing the 1D spectra in one serial file and to apply the Fourier transformation of this file only in the F2 dimension. The phase correction should be the same for all 1D spectra and the values for zero order and first order should be determined for the last 1D dataset of the real-time NMR experiment. If possible, only a zero order polynomial function should be used for baseline correction and pivot baseline points should be selected manually in regions where no signals occur during the entire NMR experiment.

16.6.1.3 Analysis

After the mixing dead time constant buffer and temperature conditions are reached and the chemical shift position of the resonances of the different protein states are fixed because we are usually working in the slow exchanging regime. Thus, well-resolved resonances can be integrated within identical limits in every 1D spectrum to obtain input data for the kinetic analyses. Each 1D spectrum is a sum of spectra of the respective protein populations present at the time point of recording. The kinetic information obtained from well-resolved resonances of these states allows a reconstruction of the full 1D spectrum of every state [38, 61]. Usually the final native state has the largest deviations from random coil chemical shifts and well-resolved resonances of aromatic residues or high field shifted CH_2 and CH_3 groups are unique for this state. Secondly, a 1D spectrum with a very good signal-to-noise ratio can be recorded after the reaction has finished. This spectrum has to be scaled according to the refolding kinetics of the native state, which has been determined from the integrals of its well-resolved resonances. This scaled 1D spectrum has to be subtracted from every 1D spectrum of the real-time NMR experiment. This step eliminates all contributions of the native state to the 1D spectra. If a two-state reaction is analyzed, the remaining NMR intensity represents entirely the 1D spectrum of the population from which the reaction has started. Now, these 1D spectra can be averaged to get a decent 1D spectrum of this state. Secondly, the decay kinetics should reveal the same rate constant as the build-up of the native state. Thirdly, this approach allows determination of the result of fast folding events during the dead time of the NMR experiment, which cause a burst of one of these two states [38, 61]. Finally, the 1D contributions of the decaying population can be eliminated in a second subtraction step by scaling the above-mentioned averaged 1D spectrum according to the decay kinetics. If the two-state approximation is valid, no residual NMR intensity should be obtained [38]. It is a sensitive test to add all 1D spectra after the second subtraction step. In the case of RNase T1, this test showed a third protein species populated only by 7% at the beginning of the refolding reaction [61]. Its full 1D spectrum and time course could be determined by analyzing the remaining NMR intensities after the second subtraction step.

16.6.1.4 Analysis of 1D Real-time Diffusion Experiments

The diffusion coefficient of proteins in protein folding reactions can be determined by PFG-SLED NMR experiments [79–82]. If the diffusion delay in this stimulated echo experiments remains constant, Eq. (1) can be fitted to measured NMR intensities $I(g)$ at various gradient strength g.

$$I(g) = A \cdot \exp(-dg^2) \tag{1}$$

The diffusion delay (for example 30 ms) and the variation of the gradient strength (a constant gradient pulse length between 5 ms and 10 ms are common) have to be adjusted before the real-time NMR experiment to make sure that the intensities of the protein signals decay by more than 50% at the highest gradient strength. If dioxan is used as internal standard at 3.6 ppm, a modified version of Eq. (1) has to be used [14] and a dioxan signal should remain for at least half of the varied gradient strengths. A fit of Eq. (1) to the Gaussian decay curves (gradient strength plotted against NMR intensity) reveals d, which is proportional to the diffusion coefficient of the protein. d can be converted into the hydrodynamic radius R_h of the protein by assuming R_h of dioxan of 2.12 Å [14, 82]. From the empiric correlations $R_h =$

$4.75 \times N^{0.29}$ and $R_h = 2.21 \times N^{0.57}$ the hydrodynamic radius determined by NMR diffusion experiments can be estimated from the number of residues N of a globular folded protein and an unfolded peptide chain, respectively [82].

For the determination of the hydrodynamic radius of a transient protein folding intermediate in 1D real-time NMR diffusion experiments, a set of diffusion experiments have to be recorded during the refolding reaction [56, 75]. The gradient strength has to be varied continuously between low and high values typically in 10 steps. For a two-state model, where the intermediate forms in the dead time of the time-resolved NMR experiment, the total intensity of all protein resonances $I(g,t)$ in Eq. (2) depend on the known gradient strength g at every time point t. The relative populations of the intermediate and the native state can be calculated from the known reaction rate constant k. The differences in the relaxation rates between the intermediate and the native state are reflected in the respective amplitudes A_I and A_N.

$$I(g,t) = A_I \cdot \exp(-d_I g^2) \cdot \exp(tk) + A_N \cdot \exp(-d_N g^2) \cdot (1 - \exp(tk)) \qquad (2)$$

From d_I the hydrodynamic radius of the folding intermediate can be determined via the internal standard dioxan or from the ratio d_I/d_N the extend of compaction of the intermediate compared to the native state.

16.6.2
How to Extract Folding Rates from 1D Spectra by Line Shape Analysis

For an accurate line shape analysis the system under investigation has to fulfill several requirements. Most important is the existence of a two-state folding process where only folded (N) and unfolded (U) protein molecules are present at equilibrium and in the kinetics at all denaturant concentrations (Eq. (3)). Additionally it is important that the analyzed resonance signals are well resolved throughout the entire equilibrium transition and do not overlap with other resonances. In rare cases, where only few resonances are in a narrow region of the 1D spectra under all conditions, the entire 1D spectrum can be simulated to overcome signal overlap problems [12, 102]. Denaturation of the protein can be achieved either by successively adding chaotropic agents (for example urea, guanidinium chloride) or by increasing the temperature or by a combination of both. A small difference between the resonance frequency of the native and unfolded state (below 100 Hz) allows a determination of folding time constants of tens of milliseconds, whereas a difference above 1500 Hz can cover time constants in a submillisecond range. This difference can be varied for the same system by using different field strengths [107]. Since the exchange between the two states is in the fast exchange regime in respect to the NMR chemical shift time scale $((k_f + k_u) > \Delta\nu)$ only one resonance is detectable over the entire transition with a line shape and resonance frequency that represents the relative populations.

$$N \underset{k_f}{\overset{k_u}{\rightleftharpoons}} U \qquad (3)$$

16.6.2.1 Acquisition

NMR spectra of high quality at reasonably high protein concentrations (ensure the absence of aggregation) with a very good signal-to-noise ratio should be obtained with at least 256 transients and 4096 complex points. Small receiver gains should be used to avoid disadvantageous baseline effects. To prevent saturation of ^{1}H with long spin-lattice relaxation times (t_1) the recycle delay between individual transients should be at least 3 s. This ensures that the magnetization reaches the Boltzmann equilibrium. Using ^{2}H$_2$O as solvent for the protein and deuterated buffers should be preferred because a selective suppression of individual resonances (for example from water or denaturants) reduces the quality of the fit by producing resonance intensities, which are not proportional to the protein populations as well as baseline problems. For example the WATERGATE sequence can effectively suppress the solvent signal but has a nonlinear excitation profile at the edges of the spectral window. Therefore, resonances that undergo a change in chemical shift during the unfolding transition are not excited equally throughout the entire transition. If chemically induced denaturation of the protein is achieved by adding urea or guanidinium chloride the exchangeable ^{1}H are substituted by NMR nonactive ^{2}H, which reduces the solvent suppression scheme to residual water. Therefore, a long presaturation pulse with a duration of 1–1.5 s on the water signal is sufficient. To minimize the residual water one should also perform several deuteration steps of the denaturation agent and the protein before mixing the sample. Another advantage of ^{2}H$_2$O is that the exchangeable ^{1}H of the protein, namely the amide protons, are not detectable, which greatly reduces spectral overlap in the low field region of the ^{1}H spectrum (6–11 ppm). Thus, well-resolved resonances of ^{1}H of the aromatic amino acids Phe, Tyr, Trp, and His can be used to perform the line shape analysis approach. High-field shifted resonances in the native state have the disadvantage that these protons cluster at their random coil positions in the unfolded state.

16.6.2.2 Processing

The NMR data are converted to Felix format or to equivalent processing software such as NMR pipe, XEASY, etc. The data must be dc-offset corrected and zero-filled once in respect to the acquired complex points. No window function should be applied to prevent distortion of the original line shape of the resonance. If a window function is required, only a single-exponential function can be used, which increases the adjustable T_{2N}^* and T_{2U}^*. Individual spectra are separately phase corrected because a correct phasing is essential for the extraction of reliable rates in the further analysis. Baseline correction of each spectrum in the region around the resonance of interest is also performed manually. The order of the polynomial function depends on the quality of the baseline but should be invariant for all spectra. The referencing of the ^{1}H spectrum to an internal standard like dioxane, DSS, or TSP is recommended. The choice of the standard should depend on the denaturation method (chaotropes, pH, temperature) due to different properties of the respective compound and on the distance between the resonance frequency of the standard and the resonance of interest to avoid unwanted disturbance. To reduce the amount of data the spectra are truncated to the region of the analyzed signal and the real data points are saved in an ASCII format.

16.6.2.3 Analysis

The line shape analysis can be performed with any standard processing software like Sigmaplot (SPSS Inc.), Kaleidagraph (Synergy software), or Grafit (Erithacus software), which is capable of nonlinear least square fitting using the Levenberg-Marquardt algorithm. Especially for this purpose Oas and coworkers designed the ALASKA package, which is based on Mathematica [102]. This package is also very useful for simulating complete aromatic 1D ^{1}H spectra. The formalism to determining folding rates from line shape analysis (Eq. (4)) was introduced by Oas and coworkers [11], and is only applicable when the two-state approximation (Eq. (3)) is fulfilled. In Eq. (4), $I(v)$ represents the intensity as a function of the frequency v and C_0 is a normalization constant, which is proportional to the protein concentration. $I(v)$ also depends on the population of the native and unfolded state (p_N, p_U), the apparent spin–spin relaxation times of the resonance in the respective state under nonexchanging conditions (T_{2N}^*, T_{2U}^*), the resonance frequency of both states (v_N, v_U) and the refolding and unfolding rate constant (k_f, k_u).

$$I(v) = \frac{C_0}{P^2 + R^2} \cdot \left(P \cdot \left(1 + \tau \cdot \left(\frac{p_N}{T_{2U}^*} + \frac{p_U}{T_{2N}^*}\right)\right) + Q \cdot R \right)$$

$$P = \tau \cdot \left(\left(\frac{1}{T_{2N}^* T_{2U}^*}\right) - 4\pi^2 \cdot \Delta v^2 + \pi^2 \cdot (\delta v)^2 \right) + \frac{p_N}{T_{2N}^*} + \frac{p_U}{T_{2U}^*}$$

$$Q = \tau \cdot (2\pi \cdot \Delta v - \pi \cdot \delta v \cdot (p_N - p_U))$$

$$R = 2\pi \cdot \Delta v \cdot \left(1 + \tau \cdot \left(\frac{1}{T_{2N}^*} + \frac{1}{T_{2U}^*}\right) + \pi \cdot \delta v \cdot \tau \cdot \left(\frac{1}{T_{2U}^*} - \frac{1}{T_{2N}^*}\right)\right)$$

$$+ \pi \cdot \delta v \cdot (p_N - p_U)$$

$$\Delta v = (v_N + v_U)/2 \quad \delta v = (v_N - v_U) \quad \tau = (k_f - k_u)^{-1}$$

$$p_N = k_f \cdot \tau \quad p_U = k_u \cdot \tau \tag{4}$$

In the following, the line shape analysis procedure is described step by step.

1. The apparent transverse relaxation rates of the resonance in the native and unfolded state (T_{2N}^*, T_{2U}^*) have to be determined. Therefore, the signal of the fully native or unfolded protein, respectively, is simulated with a function corresponding to an absorptive Lorentzian signal ($I(v) = C_0 R_2 / \{R_2^2 + (v_i - v)\}$) where $R_2 = 1/T_2^*$ and v_i represents the chemical shift of the signal.
2. The dependence of the resonance frequency of the signal of the native and the unfolded state from the denaturant concentration or the temperature is also required. Thus, the resonance frequency of the respective signal in the spectra under nonexchanging conditions (that is in the beginning of the baselines of the transition under strong native and strong nonnative conditions) is measured.

Assuming a linear dependence of the resonance frequency on the denaturant concentration or the temperature the actual resonance frequency of the signal of the native and the unfolded state (v_N, v_U) under all conditions can be calculated by a linear extrapolation method.

3. With the determined parameters ($T_{2N}^*, T_{2U}^*, v_N, v_U$) the simulation of the resonance signals in the transition region with Eq. (4) reveals the normalization constant C_0 and the folding rates (k_f, k_u) for every individual 1D spectrum (Figures 16.7 and 16.8). Most reliable results are obtained if the population of both states is above 15%.

4. The error estimated by the Levenberg-Marquardt algorithm is not very meaningful. Therefore usually a systematic variation of k_f and k_u around their fitted values gives a much more realistic error. A comparison of the residuals between the calculated and experimental line shapes reveals errors typically around 10% [11, 104, 107].

16.6.3
How to Extract Folding Rates from 2D Real-time NMR Spectra

In principle, there are two ways to determine folding rates by 2D real-time NMR. First, a series of individual 2D NMR experiments can be recorded during the folding reaction, which reveals cross-peaks with increasing or decreasing intensities depending on the respective state. This method is useful if the folding reaction is slow enough to record a set of reasonably good 2D spectra. The processing of these spectra follows standard protocols. Individual cross-peaks can be integrated in every 2D spectrum or its maximum can be used to determine the folding rates [75]. The second method to obtain folding rates from 2D real-time NMR spectra is, in terms of acquisition, even more straightforward but processing and analysis are more complex. Technically, only a single 2D NMR spectrum is recorded during the entire refolding reaction, which has been initiated by a rapid mixing device (see Section 16.6.1). This has the advantage that significantly more transients and t_1-increments can be recorded compared with the first method and therefore the signal-to-noise ratio and spectral resolution in the indirect dimension is much higher. As discussed in Section 16.3, the folding reaction leads to different line shapes and intensities of cross-peaks assigned to the respective species (Figure 16.5 and Figure 16.6). Comparing the kinetic experiment with a reference experiment under identical conditions after the reaction is finished reveals three different read-outs. Cross-peaks of the starting conformation (for example the intermediate or unfolded state) lose intensity during the reaction and exhibit line broadening. Cross-peaks of the emerging final state (for example the native state) show increasing intensities accompanied by characteristic line shape in the indirect dimension. Finally, cross-peaks with invariant chemical shifts in the starting and the final conformation depict constant intensity and narrow line widths. The assignment of individual cross-peaks to the respective species can be achieved by using the different line shape, which works nicely for 2D ^{1}H–^{15}N HSQC spectra (Figure 16.6). In the case of 2D NOESY or 2D TOCSY spectra this approach becomes very time consuming because hundreds of cross-peaks have to be analyzed and signal overlap can alter the expected theoretical line shapes. A more straightforward procedure is to subtract the kinetic experiment from the reference experiment, revealing positive or negative intensities for the final or starting state, respectively. Cross-peaks with invariant chemical shifts show the same intensity in both experiments and cancel out. A mathematical description is provided below. The power of this method was shown for 2D ^{1}H-^{1}H NOESY [61] and 2D ^{1}H-^{15}N HSQC experiments [54] and can theoretically be extended to any other two- or three-dimensional NMR experiment.

16.6.3.1 Acquisition

Basically, one single 2D or 3D NMR experiment is recorded after initiation of the refolding reaction depending on the time constant of the folding reaction. Since the folding kinetics modulates the FIDs of the indirect dimension a reasonable number of t_1-increments should be recorded to achieve high spectral resolution. The dead time of the experiment can be greatly reduced by using a mixing device (see Section 16.6.1) because tuning and matching of the probe and shimming as well as temperature calibration can be performed before starting the reaction. No modifications of standard pulse programs are needed.

16.6.3.2 Processing

The NMR data can be converted and analyzed by any standard processing software. The FIDs in both dimensions should be zero-filled once and an apodization in the time domain applied. Only single-exponential window functions can be used if a quantitative line shape analysis is to be performed to extract protein folding rates.

16.6.3.3 Analysis

The assignment of each cross-peak to the respective state can be achieved by analyzing the different line shapes. For a more straightforward assignment of the cross-peaks a simple subtraction method can be used. The complex FID for a signal with a frequency offset v' and an apparent transversal relaxation rate constant R_2^* is given by Eq. (5).

$$F(t) = A \cdot \exp\{2\pi i v' t\} \cdot \exp\{-R_2^* t\} \tag{5}$$

As discussed before, three cases of modulation of the FID in the indirect dimension are possible and are defined by Eq. (6) with the assumption that k is the rate constant of a two-state transition.

$$G^1(t) = \exp(-kt)$$
$$G^2(t) = 1 - \exp(-kt)$$
$$G^3(t) = 1 \tag{6}$$

Fourier transformation of the respective products $S^x = G^x(t)F(t)$ yield the following real parts in the frequency domain (Eq. (7)).

$$\text{Re}\{FT[S^1(t)]\} = A \cdot \left[\frac{R_2^* + k}{(R_2^* + k)^2 + 4\pi^2(v - v')^2}\right]$$

$$\text{Re}\{FT[S^2(t)]\} = A \cdot \left[\frac{\frac{1}{R_2^*}}{1 + \left(\frac{1}{R_2^*}\right)^2 \cdot 4\pi^2(v - v')^2} - \frac{R_2^* + k}{(R_2^* + k)^2 + 4\pi^2(v - v')^2}\right]$$

$$\text{Re}\{FT[S^3(t)]\} = A \cdot \left[\frac{\frac{1}{R_2^*}}{1 + \left(\frac{1}{R_2^*}\right)^2 \cdot 4\pi^2(v - v')^2}\right] \quad (7)$$

Subtracting the three equations (Eq. (7)) from the reference experiment of the fully native protein reveals Eq. (8), which corresponds to the three different read-outs discussed above.

$$\text{Re}\{FT[S^1(t)]\} = -A \cdot \left[\frac{R_2^* + k}{(R_2^* + k)^2 + 4\pi^2(v - v')^2}\right]$$

$$\text{Re}\{FT[S^2(t)]\} = A \cdot \left[\frac{R_2^* + k}{(R_2^* + k)^2 + 4\pi^2(v - v')^2}\right]$$

$$\text{Re}\{FT[S^3(t)]\} = 0 \quad (8)$$

The refolding rate is obtained from the 2D NMR spectra (for example ^{1}H-^{1}H NOESY, ^{1}H-^{15}N HSQC) by a two-step line shape analysis procedure. For this approach the 1D spectrum in the indirect dimension has to be extracted from the two-dimensional matrix of the kinetic and reference experiment, respectively. In the first step, the resonance signal of the reference spectrum is fitted with the real part of the Fourier transformed function $S^3(t)$ of Eq. (7), which yields the amplitude, R_2^* and v' of the native signal under equilibrium conditions (Figure 16.10a). In the second step, the resonance signal of the native state in the kinetic experiment is fitted by $\text{Re}\{FT[S^2(t)]\}$ of Eq. (7) using the determined R_2^* and v' as fixed parameters (Figure 16.10b). The thus derived apparent folding rate k has to be further scaled because the time scale of the acquisition of the indirect dimension ($t_{\text{dwell}} = 1/2SW_{\text{indirect}}$ with SW_{indirect} as the spectral width in the incremented dimension) and the "real" time resolution between two points in the FID of the indirect dimension (that is the explicit time between two sequential t_1-increments, t_{1D}, which is the total length of the refolding experiment divided by the number t_1-increments) differ. The actual refolding rate k_f can therefore be calculated by Eq. (9):

$$k_f = k \cdot \frac{k_{\text{dwell}}}{k_{1D}} \quad (9)$$

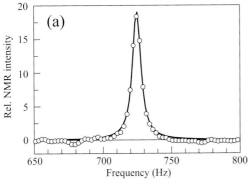 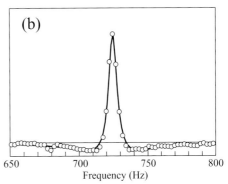

Fig. 16.10. Line shape analysis of A19 from S54G/P55N-RNase T1. The folding reaction at 15 °C was followed by 2D ^{1}H-^{15}N NMR correlation spectroscopy. a) The ^{15}N 1D NMR spectrum of A19 in the native state (trace along in the indirect F1 dimension from the reference spectrum (Figure 16.6b)). The continuous line represents the fit of the data to Re{FT[$S^3(t)$]} of Eq. (7) (see Section 16.6.3), which reveals R_2^* and v' with 23.4 s^{-1} and 725 Hz, respectively. b) The equivalent ^{15}N 1D NMR spectrum of A19 from the kinetic ^{1}H-^{15}N HSQC experiment (Figure 16.6a). The resonance shows the characteristic line shape with the two flanking negative components. Fitting the data to Re{FT[$S^2(t)$]} of Eq. (7) with constant R_2^* and v' from (a) reveals the continuous line and a rate k of 130.6 s^{-1}. The actual folding rate k_f is then calculated by Eq. (9) and yields 4.3×10^{-4} s^{-1}.

An alternative way to determine the rate constant k of folding would be a simulation of cross-sections of the resonances of individual residues along the indirect F1 dimension [54, 125]. For this purpose the cross-sections from the kinetic 2D spectrum can be simulated by multiplying the respective cross-section from the reference spectrum with a single-exponential function prior to the second Fourier transformation of the 2D data set.

16.6.4
How to Analyze Heteronuclear NMR Relaxation and Exchange Data

As described in the preceding sections, the determination of explicit un- and refolding rates requires a protein that follows a two-state folding model. One of the first applications of R_2 relaxation dispersion curves to study fast folding reactions was performed by homonuclear 1D ^{1}H NMR spectroscopy [12]. To overcome the limitations of 1D NMR like severe spectral overlap, more recently heteronuclear 2D relaxation methods have been established [10], which provide many site-specific probes. Therefore, a further requirement is to enrich the protein under investigation with the NMR active heteronuclei ^{15}N and/or ^{13}C.

One common approach is the measurement of so-called ^{15}N R_2 dispersion curves. Despite the high spectral resolution of the 2D ^{1}H-^{15}N correlation spectra used to determine the R_2 relaxation rates there are some further advantages of this method in respect to the above-mentioned 1D experiments (for example line shape analysis, see Section 16.6.2). First, the determination of the R_2 rates is

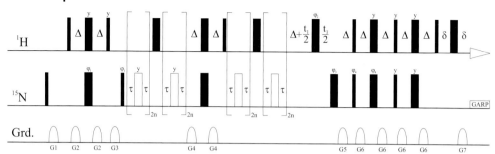

Fig. 16.11. Pulse sequence of a relaxation-compensated ^{15}N CPMG experiment for the determination of conformational and chemical exchange. Variation of τ ($\tau = \tau_{cp}/2$) and a successive determination of R_2 for each τ_{cp} reveals the so called R_2 relaxation dispersion curves (Figure 16.9). Narrow and wide black bars depict 90° and 180° pulses, respectively. The wide white bars represent 180° ^{15}N pulses in the spin echo sequence. All pulses are x-phase unless indicated otherwise and gradients are shown as half-ellipsoids. Spectra for initial intensities are obtained by omitting the sequences in brackets. For more experimental details and theoretical description see Ref. [132].

straightforward and robust, because instrument effects such as magnetic field inhomogeneities might introduce significant errors in 1D line shape analyses. More important, it is possible to extract folding rates down to a population of one state of about 1%. The major drawback of this method is the time-consuming acquisition. R_2 rates are usually measured with pulse sequences based on the Carr-Purcell-Meiboom-Gill (CPMG) spin echo element ($[\tau - 180° - \tau]_n$) with $\tau_{cp} = 2 \times \tau$ [126, 127]. An example of a useful pulse sequence and further description is provided in Figure 16.11. Principally, the spin echo element contains a single 180° pulse on the nuclei of interest, which is separated by two equivalent delays $\tau = \tau_{cp}/2$. Contributions from chemical exchange (R_{ex}) to R_2 can be effectively eliminated by the CPMG sequence if the exchange process is on a slower time scale in respect to the delay time τ_{cp}. For very short τ_{cp} values (that is the fast pulsing limit) only very fast processes on a picosecond to nanosecond time scale contribute to the apparent R_2 value, which is R_2^0. However, if τ_{cp} is made long enough that multiple folding/unfolding events occur during τ_{cp}, chemical exchange R_{ex} contributes to R_2: $R_2 = R_2^0 + R_{ex}$. Therefore, a dispersion of R_2 regarding R_{ex} can be generated by a variation of τ_{cp}.

16.6.4.1 Acquisition

High-quality heteronuclear 2D NMR spectra can be recorded with pulse sequences as shown in Figure 16.11. The recycle delay between individual transients should be as long as possible and at least 1 s. For each 2D ^{1}H-^{15}N spectrum at least 128 t_1-increments in the indirect dimension should be acquired with 16–32 transients for each t_1-increment depending on the protein concentration. The relaxation delay using a fixed spin echo delay τ_{cp} has to be varied by multiple repetition of the spin echo sequence. This relaxation delay (calculated by $4 \times (2n \times (2\tau + \tau_{180(N)})) + 2 \times \Delta + \tau_{180(N)} + \tau_{90(H)}$ according to Figure 16.11, where $\tau_{180(N)}$ and $\tau_{90(H)}$ denote the length of the 180° pulse on ^{15}N and the 90° pulse on ^{1}H, respectively) varies typically between 10 ms and 250 ms. The decay of the intensity for an accurate determination of R_2 should be defined by at least six different relaxation delays with a minimal loss of 80% of the initial intensity.

16.6.4.2 Processing

The 2D NMR data can be processed with any available processing software such as Felix, NMR pipe, XEASY, etc. and follows closely standard protocols. The FID is zero-filled once in both dimensions and any window function can be applied before Fourier transformation. Normally phase correction in the ^{1}H dimension remains constant over sequentially acquired data sets if the pulse program is not changed. Phase correction in the indirect dimension should be avoided by properly adjusting the first t_1-increment. The spectral resolution in the indirect dimension can be enhanced by linear prediction methods especially if only 128 t_1-increments are recorded. A baseline correction can also be performed. However, most important is that all spectra are processed identically to minimize uncertainties of the extracted peak intensities.

16.6.4.3 Analysis

For the determination of the R_2 rates for every τ_{cp} value the fitting of the intensity decay curve is applicable with a single exponential function without offset ($I = A \exp(-R_2 t)$) and can be performed using any standard software. However, the programs of Palmer and coworkers such as CURVEFIT and for further analysis CPMGfit (http://cpmcnet.columbia.edu/dept/gsas/biochem/labs/palmer/software/modelfree.html) are strongly recommended because they are specially designed for this purposes and due to the extensive error analysis by various methods such as Monte-Carlo or jack-knife simulations [128, 129]. The derived R_2 values are plotted on the ordinate of the R_2 dispersion curve (see Figure 16.9). The exchange rate k_{ex} and the chemical exchange contribution R_{ex} to R_2 can be derived by fitting Eq. (10) to the R_2 dispersion curves [10, 130]. This formalism represents the general phenomenological description of R_2 of site N in Eq. (3), which is independent of the exchange regime (slow, intermediate or fast in respect of the NMR chemical shift time scale) as well as the present population ratio.

$$R_2(1/\tau_{cp}) = \frac{1}{2}\left(R_{2N} + R_{2U} + k_{ex} - \frac{1}{\tau_{cp}} \cosh^{-1}[D_+ \cosh(\eta_+) - D_- \cosh(\eta_-)]\right) \tag{10}$$

with

$$D_\pm = \frac{1}{2}\left[\pm 1 + \frac{\psi + 2\Delta\omega^2}{\sqrt{\psi^2 + \zeta^2}}\right]; \quad \eta_\pm = \frac{\tau_{cp}}{\sqrt{2}}\sqrt{\pm\psi + \sqrt{\psi^2 + \zeta^2}}$$

$$\psi = (R_{2N} - R_{2U} - p_N k_{ex} + p_U k_{ex})^2 - \Delta\omega^2 + 4p_N p_U k_{ex}^2$$

$$\zeta = 2\Delta\omega \cdot (R_{2N} - R_{2U} - p_N k_{ex} + p_U k_{ex})$$

$$R_{ex} = \Delta R_2(0, \infty) = R_2(1/\tau_{cp} \to 0) - R_2(1/\tau_{cp} \to \infty)$$

$$= \frac{1}{2}\left\{\sqrt{\psi + \Delta\omega^2} - \frac{1}{\sqrt{2}}\sqrt{\psi + \sqrt{\psi^2 + \zeta^2}}\right\}$$

where $\Delta\omega$ is the difference of the ^{15}N chemical shift of the cross-peak between the native and the unfolded state ($\Delta\omega = 2\pi\Delta\nu$ with $\Delta\nu$ given in Hz), p_N and p_U the population of the respective state and $\tau_{cp} = 2 \cdot \tau$ as well as $k_{ex} = k_f + k_u$. $\Delta\omega$ defines the exchange regime for a given k_{ex} ($k_{ex} < \Delta\omega$: slow exchange; $k_{ex} \approx \Delta\omega$: intermediate exchange; $k_{ex} > \Delta\omega$: fast exchange). A less complex approximation was proposed by Ishima and Torchia, which is valid for all time scales if $p_N \gg p_U$ (Eq. (11)).

$$R_2(1/\tau_{cp}) = R_2(1/\tau_{cp} \to \infty) + p_N p_U \Delta\omega^2 k_{ex}/[k_{ex}^2 + (p_N^2 \Delta\omega^4 + 144/\tau_{ex}^4)^{1/2}] \quad (11)$$

This expression is especially helpful to estimate the expected contributions of R_{ex} to R_2 for a protein folding system, where kinetic rates and populations are already available. If R_{ex} is smaller than 2 Hz, dispersion curves become very difficult to analyze. If R_{ex} exceeds significantly 2 Hz, errors for the folding rates can drop below 10%.

The precise determination of $\Delta\omega$ is essential for the reliability of the extracted parameters, which could be achieved by recording for example a denaturant-induced unfolding transition followed by a series of ^{1}H-^{15}N HSQC spectra. Therein, nuclei in the slow exchange regime depict a cross-peak for each state. Unfolding transition data have the advantage that the dependence of the chemical shifts from the concentration of the denaturant can be determined. The same holds for temperature-induced transitions. A straightforward assignment of the unfolded state can be performed by correlating the cross-peaks of the native and the unfolded state via the respective exchange cross-peaks in a 2D ZZ-exchange spectrum [110]. However, only a fraction of amide protons can be assigned by this method due to severe spectral overlap of the cross-peaks of the unfolded state with the exchange cross-peaks or by exhibiting fast or intermediate exchange phenomena (such as averaged chemical shifts or extreme line broadening). If the chemical exchange rates are below 0.1 s^{-1}, no exchange cross-peaks in a 2D ZZ-exchange spectrum are observed, which requires a time-consuming assignment of the unfolded state by other methods [131].

In many cases, $\Delta\omega$ cannot be determined directly. In this case, $\Delta\omega$ has to become an adjustable parameter during the fitting procedure. If dispersion curves at two magnetic fields are available, reliable interconversion rates can be extracted, if the populations are known or if the populations can be estimated within certain limits. Here, typical errors of the rates are around 30%. A global fit of all dispersion curves might help to estimate the populations [115]. Extremely good dispersion curves are required if the populations have to be adjusted as well during the fit and an iterative scheme has to be developed to successively reduce the lower and upper limits for the respective parameters.

Acknowledgments

We thank F. X. Schmid and C. M. Dobson for many stimulating discussions about protein folding and real-time NMR. This research was supported by grants from the Deutsche Forschungsgemeinschaft (Ba 1821/2-1; Ba 1821/3-1), the ARC program of the Deutscher Akademischer Austauschdienst, the INTAS-2001 program, and the SON UMR in Utrecht (HPRI-CT-2001-00172).

References

1 WAGNER, G., WÜTHRICH, K. *J. Mol. Biol.* 1982, *160*, 343–361.
2 WAND, A. J., RODER, H., ENGLANDER, S. W. *Biochemistry* 1986, *25*, 1107–1114.
3 ENGLANDER, S. W., SOSNICK, T. R., ENGLANDER, J. J., MAYNE, L. *Curr. Opin. Struct. Biol.* 1996, *6*, 18–23.
4 BAI, Y. W., SOSNICK, T. R., MAYNE, L., ENGLANDER, S. W. *Science* 1995, *269*, 192–197.
5 RADFORD, S. E., DOBSON, C. M., EVANS, P. A. *Nature* 1992, *358*, 302–307.
6 UDGAONKAR, J. B., BALDWIN, R. L. *Nature* 1988, *335*, 694–699.
7 RODER, H., ELÖVE, G. A., ENGLANDER, S. W. *Nature* 1988, *335*, 700–704.
8 DOBSON, C. M., EVANS, P. A. *Biochemistry* 1984, *23*, 4267–4270.
9 RODER, H. *Methods Enzymol.* 1989, *176*, 446–473.
10 PALMER, A. G., III, KROENKE, C. D., LORIA, J. P. *Methods Enzymol.* 2001, *339*, 204–238.
11 HUANG, G. S., OAS, T. G. *Proc. Natl Acad. Sci. U.S.A.* 1995, *92*, 6878–6882.
12 BURTON, R. E., HUANG, G. S., DAUGHERTY, M. A., FULLBRIGHT, P. W., OAS, T. G. *J. Mol. Biol.* 1996, *263*, 311–322.
13 VUGMEYSTER, L., KROENKE, C. D., PICART, F., PALMER, A. G., III, RALEIGH, D. P. *J. Am. Chem. Soc.* 2000, *122*, 5387–5388.
14 ZEEB, M., JACOB, M. H., SCHINDLER, T., BALBACH, J. *J. Biomol. NMR* 2003, *27*, 221–234.
15 TOLLINGER, M., SKRYNNIKOV, N. R., MULDER, F. A., FORMAN-KAY, J. D., KAY, L. E. *J. Am. Chem. Soc.* 2001, *123*, 11341–11352.
16 MYERS, J. K., OAS, T. G. *Annu. Rev. Biochem.* 2002, *71*, 783–815.
17 MEILER, J., PETI, W., GRIESINGER, C. *J. Am. Chem. Soc.* 2003, *125*, 8072–8073.
18 MEILER, J., PROMPERS, J. J., PETI, W., GRIESINGER, C., BRÜSCHWEILER, R. *J. Am. Chem. Soc.* 2001, *123*, 6098–6107.
19 GRIMALDI, J., BALDO, J., MCMURRAY, C., SYKES, B. D. *J. Am. Chem. Soc.* 1972, *94*, 7641–7645.
20 GRIMALDI, J. J., SYKES, B. D. *J. Am. Chem. Soc.* 1975, *97*, 273–276.
21 KÜHNE, R. O., SCHAFFHAUSER, T., WOKAUN, A., ERNST, R. R. *J. Magn. Reson.* 1979, *35*, 39–67.
22 MCGEE, W. A., PARKHURST, L. J. *Anal. Biochem.* 1990, *189*, 267–273.
23 FRIEDEN, C., HOELTZLI, S. D., ROPSON, I. J. *Protein Sci.* 1993, *2*, 2007–2014.
24 VAN NULAND, N. A. J., FORGE, V., BALBACH, J., DOBSON, C. M. *Acc. Chem. Res.* 1998, *31*, 773–780.
25 MOK, K. H., NAGASHIMA, T., DAY, I. J., JONES, J. A., JONES, C. J., DOBSON, C. M., HORE, P. J. *J. Am. Chem. Soc.* 2003, *125*, 12484–12492.
26 HOELTZLI, S. D., FRIEDEN, C. *Biochemistry* 1996, *35*, 16843–16851.
27 HOELTZLI, S. D., FRIEDEN, C. *Biochemistry* 1998, *37*, 387–398.
28 BANN, J. G., PINKNER, J., HULTGREN, S. J., FRIEDEN, C. *Proc. Natl Acad. Sci. USA* 2002, *99*, 709–714.
29 BALBACH, J., SCHMID, F. X. In *Mechanisms of Protein Folding*; 2nd edn; PAIN, R. H., Ed.; Oxford

University Press: Oxford, 2000; pp 212–237.
30 BLUM, A. D., SMALLCOMBE, S. H., BALDWIN, R. L. *J. Mol. Biol.* 1978, *118*, 305–316.
31 KIEFHABER, T., LABHARDT, A. M., BALDWIN, R. L. *Nature* 1995, *375*, 513–515.
32 KOIDE, S., DYSON, H. J., WRIGHT, P. E. *Biochemistry* 1993, *32*, 12299–12310.
33 WISHART, D. S., SYKES, B. D., RICHARDS, F. M. *Biochim. Biophys. Acta* 1993, *1164*, 36–46.
34 LIU, X. Y., SIEGEL, D. L., FAN, P., BRODSKY, B., BAUM, J. *Biochemistry* 1996, *35*, 4306–4313.
35 MIZUGUCHI, M., KROON, G. J., WRIGHT, P. E., DYSON, H. J. *J. Mol. Biol.* 2003, *328*, 1161–1171.
36 ROPSON, I. J., FRIEDEN, C. *Proc. Natl Acad. Sci. USA* 1992, *89*, 7222–7226.
37 RUSSELL, B. S., MELENKIVITZ, R., BREN, K. L. *Proc. Natl Acad. Sci. USA* 2000, *97*, 8312–8317.
38 ZEEB, M., RÖSNER, H., ZESLAWSKI, W., CANET, D., HOLAK, T. A., BALBACH, J. *J. Mol. Biol.* 2002, *315*, 447–457.
39 KUWAJIMA, K., ARAI, M. In *Mechanisms of Protein Folding*; 2nd edn; PAIN, R. H., Ed.; University Press: Oxford, 2000; pp 138–174.
40 BALBACH, J., FORGE, V., VAN NULAND, N. A. J., WINDER, S. L., HORE, P. J., DOBSON, C. M. *Nat. Struct. Biol.* 1995, *2*, 865–870.
41 ENGLANDER, S. W., MAYNE, L. *Annu. Rev. Biophys. Biomol. Struc.* 1992, *21*, 243–265.
42 BALDWIN, R. L. *Curr. Opin. Struct. Biol.* 1993, *3*, 84–91.
43 SHORTLE, D. R. *Curr. Opin. Struct. Biol.* 1996, *6*, 24–30.
44 DYSON, H. J., WRIGHT, P. E. *Annu. Rev. Phys. Chem.* 1996, *47*, 369–395.
45 ONUCHIC, J. N. *Proc. Natl Acad. Sci. USA* 1997, *94*, 7129–7131.
46 CLARKE, A. R., WALTHO, J. P. *Curr. Opin. Biotechnol.* 1997, *8*, 400–410.
47 DYSON, H. J., WRIGHT, P. E. *Nat. Struct. Biol.* 1998, *5* Suppl., 499–503.
48 BARBAR, E. *Biopolymers* 1999, *51*, 191–207.
49 BAUM, J., BRODSKY, B. *Curr. Opin. Struct. Biol.* 1999, *9*, 122–128.
50 BAUM, J., BRODSKY, B. *Fold Des.* 1997, *2*, R53–60.
51 DOBSON, C. M., HORE, P. J. *Nat. Struct. Biol.* 1998, *5* Suppl., 504–507.
52 IKEGUCHI, M., KUWAJIMA, K., MITANI, M., SUGAI, S. *Biochemistry* 1986, *25*, 6965–6972.
53 KUWAJIMA, K. *FASEB J.* 1996, *10*, 102–109.
54 BALBACH, J., FORGE, V., LAU, W. S., VAN NULAND, N. A. J., BREW, K., DOBSON, C. M. *Science* 1996, *274*, 1161–1163.
55 BALBACH, J., FORGE, V., LAU, W. S., JONES, J. A., VAN NULAND, N. A., DOBSON, C. M. *Proc. Natl Acad. Sci. USA* 1997, *94*, 7182–7185.
56 BALBACH, J. *J. Am. Chem. Soc.* 2000, *122*, 5887–5888.
57 FORGE, V., WIJESINGHA, R. T., BALBACH, J., BREW, K., ROBINSON, C. V., REDFIELD, C., DOBSON, C. M. *J. Mol. Biol.* 1999, *288*, 673–688.
58 KÜHN, T., SCHWALBE, H. *J. Am. Chem. Soc.* 2000, *122*, 6169–6174.
59 MAEDA, K., LYON, C. E., LOPEZ, J. J., CEMAZAR, M., DOBSON, C. M., HORE, P. J. *J. Biomol. NMR* 2000, *16*, 235–244.
60 WIRMER, S. L., KÜHNE, T., SCHWALBE, H. *Angew. Chem* 2001, *113*, 4378–4381.
61 BALBACH, J., STEEGBORN, C., SCHINDLER, T., SCHMID, F. X. *J. Mol. Biol.* 1999, *285*, 829–842.
62 VAN NULAND, N. A. J., MEIJBERG, W., WARNER, J., FORGE, V., SCHEEK, R. M., ROBILLARD, G. T., DOBSON, C. M. *Biochemistry* 1998, *37*, 622–637.
63 GUIJARRO, J. I., MORTON, C. J., PLAXCO, K. W., CAMPBELL, I. D., DOBSON, C. M. *J. Mol. Biol.* 1998, *276*, 657–667.
64 VAN NULAND, N. A. J., CHITI, F., TADDEI, N., RAUGEI, G., RAMPONI, G., DOBSON, C. M. *J. Mol. Biol.* 1998, *283*, 883–891.
65 LAURENTS, D. V., BALDWIN, R. L. *Biochemistry* 1997, *36*, 1496–1504.
66 HORE, P. J., WINDER, S. L., ROBERTS, C. H., DOBSON, C. M. *J. Am. Chem. Soc.* 1997, *119*, 5049–5050.

67 KILLICK, T. R., FREUND, S. M., FERSHT, A. R. *FEBS Lett.* 1998, *423*, 110–112.
68 KILLICK, T. R., FREUND, S. M., FERSHT, A. R. *Prot. Sci.* 1999, *8*, 1286–1291.
69 BHUYAN, A. K., UDGAONKAR, J. B. *Biochemistry* 1999, *38*, 9158–9168.
70 KAUTZ, R. A., FOX, R. O. *Protein Sci* 1993, *2*, 851–858.
71 NARAYANAN, S., BOSL, B., WALTER, S., REIF, B. *Proc. Natl Acad. Sci. USA* 2003, *100*, 9286–9291.
72 PANCHAL, S. C., BHAVESH, N. S., HOSUR, R. V. *FEBS Lett.* 2001, *497*, 59–64.
73 PANCHAL, S. C., HOSUR, R. V. *Biochem. Biophys. Res. Commun.* 2000, *269*, 387–392.
74 REIMER, U., SCHERER, G., DREWELLO, M., KRUBER, S., SCHUTKOWSKI, M., FISCHER, G. *J. Mol. Biol.* 1998, *279*, 449–460.
75 STEEGBORN, C., SCHNEIDER-HASSLOFF, H., ZEEB, M., BALBACH, J. *Biochemistry* 2000, *39*, 7910–7919.
76 YAFFE, M. B., SCHUTKOWSKI, M., SHEN, M., ZHOU, X. Z., STUKENBERG, P. T., RAHFELD, J. U., XU, J., KUANG, J., KIRSCHNER, M. W., FISCHER, G., CANTLEY, L. C., LU, K. P. *Science* 1997, *278*, 1957–1960.
77 BRAZIN, K. N., MALLIS, R. J., FULTON, D. B., ANDREOTTI, A. H. *Proc. Natl Acad. Sci. USA* 2002, *99*, 1899–1904.
78 MALLIS, R. J., BRAZIN, K. N., FULTON, D. B., ANDREOTTI, A. H. *Nat. Struct. Biol.* 2002, *9*, 900–905.
79 STEJSKAL, E. O., TANNER, J. E. *J. Chem. Phys.* 1965, *42*, 288–292.
80 GIBBS, S. J., JOHNSON, C. S., JR. *J. Magn. Reson.* 1991, *93*, 395–402.
81 JONES, J. A., WILKINS, D. K., SMITH, L. J., DOBSON, C. M. *J. Biomol. NMR* 1997, *10*, 199–203.
82 WILKINS, D. K., GRIMSHAW, S. B., RECEVEUR, V., DOBSON, C. M., JONES, J. A., SMITH, L. J. *Biochemistry* 1999, *38*, 16424–16431.
83 DOBSON, C. M. *Trends Biochem. Sci.* 1999, *24*, 329–332.
84 RASO, S. W., KING, J. In *Mechanisms of Protein Folding*; 2nd edn; PAIN, R. H., Ed.; Oxford University Press: Oxford, 2000; pp 406–428.

85 LINDNER, R. A., KAPUR, A., CARVER, J. A. *J. Biol. Chem.* 1997, *272*, 27722–27729.
86 POON, S., TREWEEK, T. M., WILSON, M. R., EASTERBROOK-SMITH, S. B., CARVER, J. A. *FEBS Lett.* 2002, *513*, 259–266.
87 CARVER, J. A., LINDNER, R. A., LYON, C., CANET, D., HERNANDEZ, H., DOBSON, C. M., REDFIELD, C. *J. Mol. Biol* 2002, *318*, 815–827.
88 ALEXANDRESCU, A. T., RATHGEB-SZABO, K. *J. Mol. Biol.* 1999, *291*, 1191–1206.
89 HORE, P. J., BROADHURST, R. W. *Prog. NMR Spectrosc.* 1993, *25*, 345–402.
90 CANET, D., LYON, C. E., SCHEEK, R. M., ROBILLARD, G. T., DOBSON, C. M., HORE, P. J., VAN NULAND, N. A. *J. Mol. Biol.* 2003, *330*, 397–407.
91 ROSS, A., SALZMAN, M., SENN, H. *J. Biomol. NMR* 1997, *10*, 389–396.
92 SCHREIBER, G., FERSHT, A. R. *Biochemistry* 1993, *32*, 11195–11203.
93 GOLBIK, R., FISCHER, G., FERSHT, A. R. *Protein Sci.* 1999, *8*, 1505–1514.
94 MAYO, K. H., PARRA-DIAZ, D., MCCARTHY, J. B., CHELBERG, M. *Biochemistry* 1991, *30*, 8251–8267.
95 BUEVICH, A. V., DAI, Q. H., LIU, X., BRODSKY, B., BAUM, J. *Biochemistry* 2000, *39*, 4299–4308.
96 BHATE, M., WANG, X., BAUM, J., BRODSKY, B. *Bichemistry* 2002, *41*, 6539–6547.
97 BUEVICH, A. V., BAUM, J. *J. Am. Chem. Soc.* 2002, *124*, 7156–7162.
98 HAGN, F. Diplom thesis, Universität Bayreuth, 2003.
99 MORIKIS, D., WRIGHT, P. E. *Eur. J. Biochem.* 1996, *237*, 212–220.
100 WINDER, S. L. PhD thesis, University of Oxford, 1997.
101 SANDSTRÖM, J. *Dynamic NMR Spectroscopy*; Academic Press: New York, 1982.
102 BURTON, R. E., BUSBY, R. S., OAS, T. G. *J. Biomol. NMR* 1998, *11*, 355–360.
103 SCHINDLER, T., HERRLER, M., MARAHIEL, M. A., SCHMID, F. X. *Nat. Struct. Biol.* 1995, *2*, 663–673.
104 KUHLMAN, B., BOICE, J. A., FAIRMAN, R., RALEIGH, D. P. *Biochemistry* 1998, *37*, 1025–32.

105 Kuhlman, B., Luisi, D. L., Evans, P. A., Raleigh, D. P. *J. Mol. Biol.* 1998, *284*, 1661–1670.
106 Spector, S., Raleigh, D. P. *J. Mol. Biol.* 1999, *293*, 763–768.
107 Wang, M., Tang, Y., Sato, S., Vugmeyster, L., McKnight, C. J., Raleigh, D. P. *J. Am. Chem. Soc.* 2003, *125*, 6032–6033.
108 Myers, J. K., Oas, T. G. *Nat. Struct. Biol.* 2001, *8*, 552–558.
109 Millet, O., Loria, J. P., Kroenke, C. D., Pons, M., Palmer, A. G. *J. Am. Chem. Soc.* 2000, *122*, 2867–2877.
110 Farrow, N. A., Zhang, O., Forman-Kay, J. D., Kay, L. E. *J. Biomol. NMR* 1994, *4*, 727–734.
111 Akke, M., Liu, J., Cavanagh, J., Erickson, H. P., Palmer, A. G., III. *Nat. Struct. Biol.* 1998, *5*, 55–59.
112 Ishima, R., Wingfield, P. T., Stahl, S. J., Kaufman, J. D., Torchia, D. A. *J. Am. Chem. Soc.* 1998, *120*, 10534–10542.
113 Mulder, F. A., van Tilborg, P. J., Kaptein, R., Boelens, R. *J. Biomol. NMR* 1999, *13*, 275–288.
114 Mulder, F. A., Mittermaier, A., Hon, B., Dahlquist, F. W., Kay, L. E. *Nat. Struct. Biol.* 2001, *8*, 932–935.
115 Mulder, F. A., Hon, B., Mittermaier, A., Dahlquist, F. W., Kay, L. E. *J. Am. Chem. Soc.* 2002, *124*, 1443–1451.
116 Hill, R. B., Bracken, C., DeGrado, W. F., Palmer, A. G. *J. Am. Chem. Soc.* 2000, *122*, 11610–11619.
117 Kim, S., Szyperski, T. *J. Am. Chem. Soc.* 2003, *125*, 1385–1393.
118 Kitahara, R., Royer, C., Yamada, H., Boyer, M., Saldana, J. L., Akasaka, K., Roumestand, C. *J. Mol. Biol.* 2002, *320*, 609–628.
119 Brooks, C. L. *Acc. Chem. Res.* 2002, *35*, 447–454.
120 Dobson, C. M., Sali, A., Karplus, M. *Angew. Chem. Int. Ed. Engl.* 1998, *37*, 868–893.
121 Snow, C. D., Nguyen, H., Pande, V. S., Gruebele, M. *Nature* 2002, *420*, 102–106.
122 Zagrovic, B., Snow, C. D., Shirts, M. R., Pande, V. S. *J. Mol. Biol.* 2002, *323*, 927–937.
123 Hoeltzli, S. D., Ropson, I. J., Frieden, C. *Tech. Prot. Chem.* 1994, *V*, 455–465.
124 Piotto, M., Saudek, V., Sklenar, V. *J. Biomol. NMR* 1992, *2*, 661–665.
125 Helgstrand, M., Härd, T., Allard, P. *J. Biomol. NMR* 2000, *18*, 49–63.
126 Carr, H. Y., Purcell, E. M. *Phys. Rev.* 1954, *94*, 630–638.
127 Meiboom, S., Gill, D. *Rev. Sci. Instrum.* 1958, *29*, 688–691.
128 Mandel, A. M., Akke, M., Palmer, A. G., III *J. Mol. Biol.* 1995, *246*, 144–163.
129 Palmer, A. G., III, Rance, M., Wright, P. E. *J. Am. Chem. Soc.* 1991, *113*, 4371–4380.
130 Carver, J. P., Richards, R. E. *J. Magn. Reson.* 1972, *6*, 89–105.
131 Yao, J., Chung, J., Eliezer, D., Wright, P. E., Dyson, H. J. *Biochemistry* 2001, *40*, 3561–3571.
132 Loria, J. P., Rance, M., Palmer, A. G. *J. Am. Chem. Soc.* 1999, *121*, 2331–2332.